“Topics in Mining, Metallurgy and Materials Engineering” welcomes manuscripts in these three main focus areas: Extractive Metallurgy/Mineral Technology; Manufacturing Processes, and Materials Science and Technology. Manuscripts should present scientific solutions for technological problems. The three focus areas have a vertically lined multidisciplinarity, starting from mineral assets, their extraction and processing, their transformation into materials useful for the society, and their interaction with the environment.

More information about this series at http://www.springer.com/series/11054

Stefano Gialanella • Alessio Malandruccolo

Aerospace Alloys

Springer

Stefano Gialanella
Industrial Engineering Department
University of Trento
Trento, Italy

Alessio Malandruccolo
Metallurgy Industrial Consultant
Bolzano, Italy

ISSN 2364-3293 ISSN 2364-3307 (electronic)
Topics in Mining, Metallurgy and Materials Engineering
ISBN 978-3-030-24439-2 ISBN 978-3-030-24440-8 (eBook)
https://doi.org/10.1007/978-3-030-24440-8

This Springer imprint is published by the registered company Springer Nature Switzerland AG.
The registered company address is: Gewerbestrasse 11, 6330 Cham, Switzerland

To my son Giulio Carlo
(SG)
To my friends and to all those who supported me
(AM)

Preface

The aerospace field is traditionally strategic for the development of novel materials needed for the specific requirements of aircraft structures and propulsion systems. A large number of advanced materials and relevant processing routes have been implemented first for aerospace needs. On many occasions, aerospace materials and technologies have subsequently been transferred successfully to a wider range of applications, like chemical industry, energy production, nuclear plants, terrestrial and marine transportation, and many more.

The aim of this book is to present an overview of the main classes of metallic materials, like light alloys (Al-, Mg-, and Ti-based, including Ti-aluminides), steels, superalloys, refractory alloys, oxide dispersion strengthened alloys, ordered intermetallics, shape memory alloys, and related systems, e.g., laminate composites and coatings.

The basic concepts concerning the structure of aircrafts and the working principles of gas turbine engines are recalled to provide the reader with indications on the specific needs and possible issues in this field, so that the adoption of given materials systems and relevant processing routes are better understood.

The content of the book is organized according to the main classes of materials that are of actual and potential interest for aerospace applications. In each chapter, materials properties and relevant technological aspects, particularly as concerns processing, are presented.

The motivations that have guided the research efforts for the development of new materials systems have changed over the comparatively short life of aeronautics. Nowadays, energy-saving and environmental issues are paramount: these stand points and their role in materials development are also highlighted.

A deliberate attempt, pursued in writing *Aerospace Alloys*, has been to include an updated bibliography and indication of further readings, both important in a continuously evolving field. A list of general references concerning materials science and engineering and physical metallurgy is included among the "Further Readings" lists at the end of each chapter. In this way, the fundamental aspects of these disciplines, which may contribute to a better understanding of the book content, can easily be recovered.

We believe the book can be used in advanced undergraduate courses in materials and mechanical engineering, considering that the starting point for writing this book has been a collection of lecture notes of courses that SG ran over several years at the University of Trento and the University of Florence in Italy. However, we would expect that the subject and relevant materials information may also be of interest to researchers working in R&D industrial laboratories, considering that in this perspective, we have included compositional tables, some in the appendices, so that the textbook turns out be reasonably self-consistent. We believe that the mixed academic (SG's) and industrial (AM's) background may provide original and complementary stand points on the different subjects, attracting the interest of researchers and professionals working in the different but still strongly interacting development laboratories in the academia and industry.

We conclude this *preface* with the acknowledgments. In the first place, we wish to thank all the students (AM has been one of them at some stage!), who during the lecture classes placed questions and contributed to a stimulating discussion, promoting in this way the improvement and correction of the lecture notes. We also thank our colleagues for stimulating discussions and suggestions on several subject dealt in the book.

Trento, Italy — Stefano Gialanella
Bolzano, Italy — Alessio Malandruccolo

Contents

Abbreviations

ACARE	Advisory Council for Aviation Research and Innovation in Europe
AISI	American Iron and Steel Institute
ALPS	Advanced low-pressure systems
AOD	Argon oxygen decarburization
APB	Antiphase boundary
APS	Air plasma spraying
ASM	American Society of Materials
ASTM	American Society for Testing and Materials
ATC	Atmosphere and temperature controlled plasma spraying
AUW	All-up weight
BC	Before Christ
bcc	Body centered cubic
bct	Body-centered tetragonal
BVI	Blade-vortex interaction
CBN	Cubic boron nitride
CCT	Continuous cooling transformation
CEVAM	Consumable electrode arc melting
CEVAR	Consumable electrode vacuum arc remelting
CF	Casting factor
CFRP	Carbon fiber reinforced polymer
CGDS	Columnar-grained directionally solidified
CHM	Cold heart melting
CMAS	Calcium-magnesium-aluminosilicate
CP	Commercially pure
CROR	Counter-rotating open rotor
CTE	Coefficient of thermal expansion
CTOL	Conventional take-off and landing
CV	Carrier variant
CVD	Chemical vapor deposition
DAS	Dendrite arm spacing
DB	Diffusion bonding

DBT	Ductile to brittle transition
DBTT	Ductile to brittle transition temperature
DC	Direct current
D-Gun	Detonation gun
DIN	Deutsche Institut für Normung
DMHT	Dual microstructure heat treatment
DOS	Density of states
DRX	Direct recrystallization
DS	Directionally solidified or directional solidification
EAF	Electric arc furnace
EB	Electron beam
EBCHR	Electron beam cold heart remelting
EBM	Electron beam melting
EB-PVD	Electron beam physical vapor deposition
EDXS	Energy-dispersive X-ray spectroscopy
EFTA	European Free Trade Association
ELI	Extra low interstitial
EN	European Norms (Euronorm)
ESR	Electroslag remelting
ESRR	Electroslag rapid remelting
ETEM	Environmental transmission electron microscopy
F/M	Ferritic martensitic
F1	Formula 1
fcc	Face-centered cubic
FML	Fiber-metal laminate
GAR	Grain aspect ratio
GB	Grain boundary
GE	General Electric
GFRP	Glass fiber reinforced polymer
GPZ	Guinier-Preston zone
GT	Gas turbine
GTE	Gas turbine engine
GTEs	Gas turbine engines
HAZ	Heat-altered zone or heat-affected zone
hBN	Hexagonal boron nitride
hcp	Hexagonal close packed
HCP	High-pressure compressor
HDI	High density inclusion
HE	Hydrogen embrittlement
HEAC	Hydrogen environmental assisted cracking
HEDE	Hydrogen-enhanced decohesion
HELP	Hydrogen-enhanced localized plasticity
HID	Hard interstitial defects
HIP	Hot isostatic pressing

HSLA	High-strength low-alloy steels
HTHC	High-temperature hot corrosion
HTLA	High-temperature low activity
HVOF	High-velocity oxy-fuel flame spraying
IADS	International Alloy Designation System
IHAC	Internal hydrogen-assisted cracking
IUPAC	International Union of Pure and Applied Chemistry
LBW	Laser beam welding
LCF	Low-cycle fatigue
LDI	Low-density inclusions
LPPS	Low-pressure plasma spraying
LPSO	Long-period stacking ordered
LPT	Low-pressure turbine
LTHA	Low-temperature high activity
LTHC	Low-temperature hot corrosion
MA	Mechanical alloying
MISS-NC	Mechanically induced surface self-nanocrystallization
M_s	Martensite start
NDT	Nondestructive testing
NEO	Near Earth objects
NEP	Nuclear electric propulsion
NS	Nanostructured
NTP	Nuclear thermal propulsion
ODS	Oxide dispersion strengthened
OEW	Operating empty weight
OPR	Overall pressure ratio
OTB	Oxygen top and bottom blowing
OWSME	One-way shape memory effect
PAM	Plasma arc melting
PCA	Process control agent
PFZ	Precipitation-free zone or precipitate-free zone
PGM	Platinum group metal
PH	Precipitation hardening or precipitation hardenable
PHSS	Precipitation hardening stainless steels or precipitation hardenable stainless steels
PM	Powder metallurgy
PS	Plasma spraying
PSZ	Partially stabilized zirconia
PTC	Plasma transferred arc
PVD	Physical vapor deposition
R&D	Research and development
RA	Reduced activation
RRB	Reconfigurable rotor blade

RS	Rapid solidification
SAE	Society of Automotive Engineers
SAP	Sinter-aluminum pulver
SC	Single crystal
SCC	Stress corrosion cracking
SE	Superelasticity
SEM	Scanning electron microscopy
SFC	Specific fuel consumption
SFE	Stacking fault energy
SFH	Simulated flying hours
SHE	Standard hydrogen electrode
SIM	Stress-induced martensite
SMA	Shape memory alloy
SMAT	Surface mechanical attrition treatment
SME	Shape memory effect
SNC	Surface nanocrystallization
SPF	Superplastic forming
SRZ	Secondary reaction zone
SSCP	Stress-strain calculation program
STOVL	Short take-off and vertical landing
TBC	Thermal barrier coating
TBCs	Thermal barrier coatings
TBO	Time between overhaul
TCP	Tetragonally close packed
TD	Thoriated
TEM	Transmission electron microscopy
TET	Turbine entry temperature
TGO	Thermally grown oxide
TISS-NC	Thermally induced surface self-nanocrystallization
TPS	Thermal protection system
TTT	Time-temperature-transformation
TWSME	Two-way shape memory effect
UNFCCC	United Nations Framework Convention on Climate Change
UNS	Unified Numbering System
USD	US dollar
UT	Ultrasonic testing
VAFN	Variable area fan nozzle
VAR	Vacuum arc remelting
VARTM	Vacuum assisted resin transfer molding
VARTMFML	Vacuum-assisted resin transfer molding fiber-metal laminate
VARTMPCL	Vacuum-assisted resin transfer molding-infused plasma coated laminates
VGC	Variable geometry chevron

VIM	Vacuum induction melting
VOCD	Vacuum oxygen converter decarburization
VOD	Vacuum oxygen decarburization
VPS	Vacuum plasma spraying
VSTOL	Vertical and/or short take-off and landing
XRD	X-ray diffraction
YSZ	Yttria-stabilized zirconia

Chapter 1
A Brief Introduction to Aerospace Applications

1.1 Historical Background

Aerospace applications represent an extremely challenging field, for the development of new materials, and for the improvement of the existing ones, with the important additional requirement of complying with highest safety standards, particularly as concerns civil and commercial aviation. A successful design and manufacturing of components, relevant performances, certification, maintenance, and safe operation are all directly influenced by the properties of materials used for engines and airframe structures. Therefore, mechanical strength, creep, corrosion, and oxidation resistance must be optimized for any specific application in this field (Mouritz 2012). Process technologies for materials and production of components are also subjected to a continuous evolution, with important reciprocal interaction.

Aerospace materials technology proceed very rapidly, in order to keep up with the momentum of changes in the air transportation that over the years have been driven by different forces. Nowadays, compositional refining and novel processing routes attain better properties in already existing material systems so that more demanding servicing conditions and higher safety standards can be met. The development of new materials and processing technologies, needed for the new engines, airframes, and control systems, not only resulted in improved aircraft performances but was also transferred and applied in other technological fields (King et al. 2009), as recalled in the next chapters.

The first engine-operated aircraft was the Flyer I (1903, Fig. 1.1), built by Wilbur and Orville Wright. In order to achieve an adequate rigidity, the wing structure was realized using a triangulated frame (see Fig. 1.2).

The triangulated structure of the wing provides several advantages, in addition to rigidity, especially interesting for the aircraft:

- High maneuverability.
- Low weight as compared to bulk structures.

S. Gialanella, A. Malandruccolo, *Aerospace Alloys*, Topics in Mining, Metallurgy and Materials Engineering, https://doi.org/10.1007/978-3-030-24440-8_1

Fig. 1.1 The Flyer I airplane of the Wright brothers (NASA 2003)

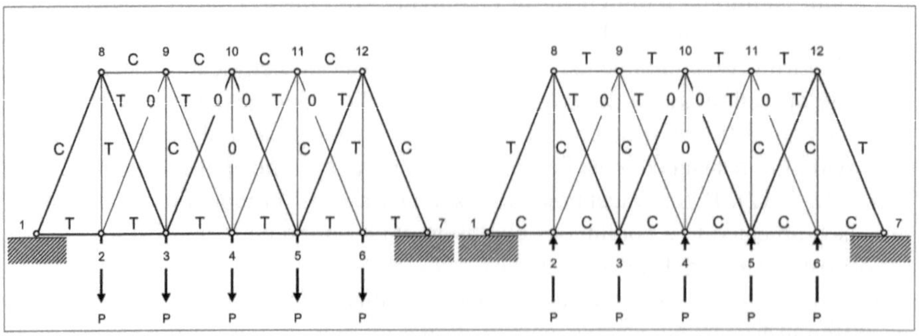

Fig. 1.2 Schematic of triangulated structures under two different loading regimes. The letter C indicates compressive load, the letter T indicates tensile load, the number 0 indicates the load-free condition, and P is the overall load acting on the structure. (Redrawn from Curtis 1997)

- Possibility to swap from tension to compression of the bracing wires to sustain the spars in all operational conditions.

At the time the Flyer I was built, no specific materials were available for the construction of an aircraft, mostly due to the absence of an industrial field dedicated to these applications. The materials used for the fabrication of the Flyer I were:

- Aluminum. An aspect not always mentioned is that the engine of the Flyer I airplane, to meet the weight reduction requirements, was made of cast aluminum.
- Cotton fabric used as coating for the wing surface.
- Steel used for the critical structural parts, like the bracing wires (i.e., tension-compression wires) and the wing struts.

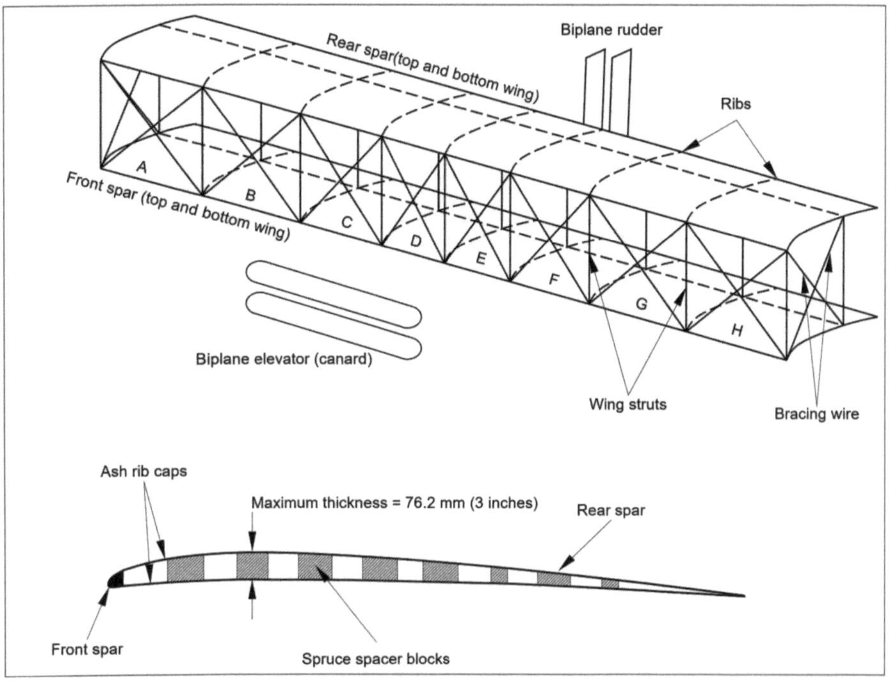

Fig. 1.3 Schematic of the overall structure of the Flyer I wing and detail of the wing rib. (Redrawn from Curtis 1997)

- Wood for the fabrication of the ash (inner part the wing), the ribs, the spars, and the propeller.

The maximum speed of the Flyer I was 56 km/h, due to the limitations caused by the aerodynamic drag associated with the shape of the aircraft, in particular with the structure of the double-wing with bracing wires (see Fig. 1.3). To overcome the drag effect of the double-wing surface and relevant wiring, needed to achieve a suitable rigidity of the structure, as a possible approach to reduce this effect, a cantilever wing was developed.

A sort of intermediate step toward this innovative structure can be found in the Blériot monoplane (1909). The Blériot monoplane had a single wing, although not fully self-sustaining, so that steel wires were needed to ensure the rigidity of the structure (see Fig. 1.4). This airplane was the first to feature a particularly rigid fuselage structure, a choice made to improve further its stability and maneuverability. The airframe was made of tension and compression wires (see Fig. 1.5) with a similar geometry to the wings of the Flyer I (see Fig. 1.3). The maximum speed of this aircraft was still below 100 km/h, due to the poor aerodynamics of the whole structure.

The situation was improved with the introduction of real cantilever wing structures. The main difference with respect to past geometries was that this new design

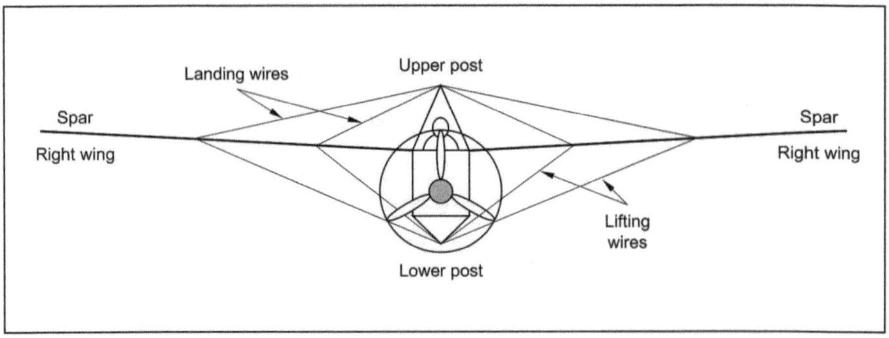

Fig. 1.4 1909 Blériot monoplane and overall schematic of the structure. (Redrawn from Curtis 1997)

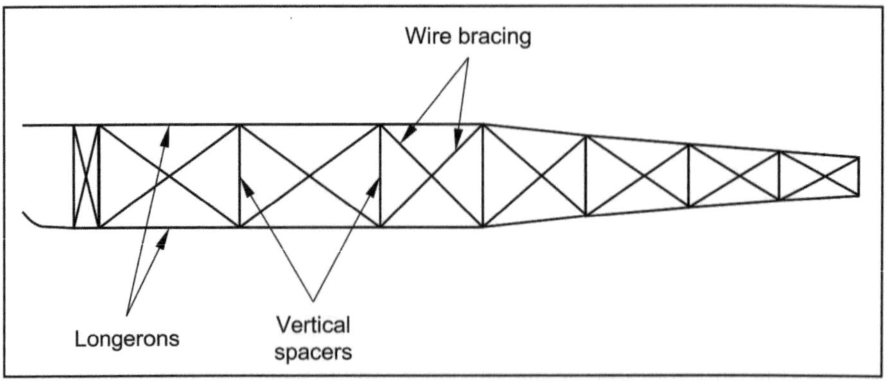

Fig. 1.5 Schematic of the Blériot's fuselage. (Redrawn from Curtis 1997)

was based on a self-standing structure, not requiring bracing wires and parts other than the wing itself, to guarantee an adequate rigidity. An example, probably the first, certainly the most popular, of the application of cantilever wings was the Fokker DR 1 (1917) (Fig. 1.6). The main feature of this kind of wing is its thickness that is the key factor as concerns its strength and rigidity. The internal structure of the wing, characterized by composite ribs, is sufficient to avoid any additional external structure (e.g., bracing wires). Thanks to the cantilever wing, the maximum speed increased up to 160 km/h, still maintaining substantially the same materials as the Flyer I. The situation evolved with the availability of more performing piston engines and, subsequently, toward the end of World War II, with the development of gas turbine engines, affording even higher speed and in principle capable to overcome the speed of sound.

This evolution required, in the first place, new wing geometries and, consequently, the selection of new materials.

The higher velocity had another important consequence: the overheating of the aircraft skin, particularly at the wing leading edge and other frontal parts of the

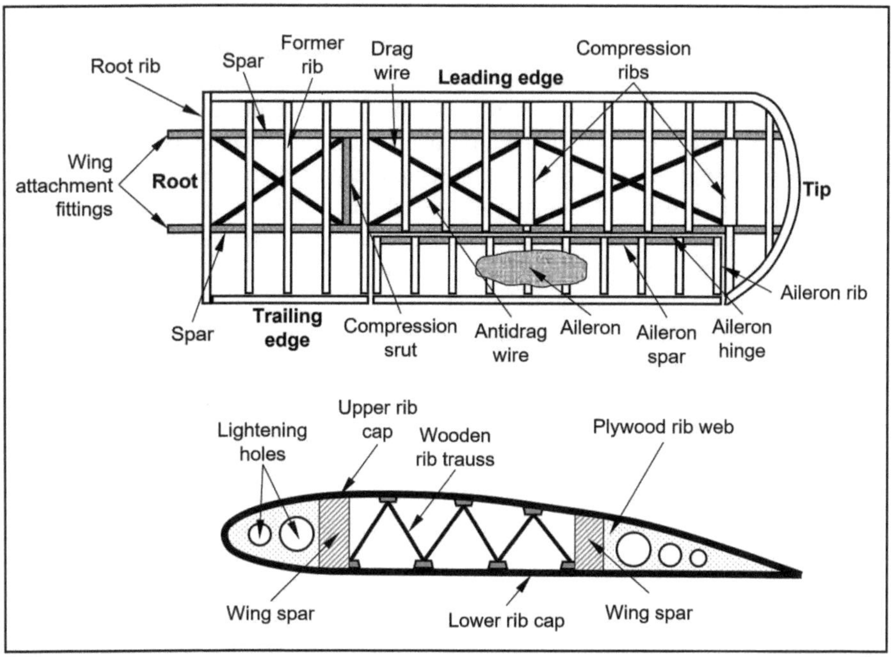

Fig. 1.6 The cantilever wing of the Fokker DR 1 (Redrawn from Curtis 1997)

aircraft in direct contact with the static air film that forms along the splitting line of the air flux. Therefore, the materials traditionally used for aircraft skin, especially fabric and wood, turned out to be absolutely inadequate to face the temperature reached under the new operational conditions.

The search for new materials was also encouraged by the contemporary shortage of wood, a factor that turned to be critical, particularly in the late years of World War II, so that alternative structural materials were increasingly needed. The solution to these two problems was the production of metallic aircrafts, like the Junkers J-1 (1915). This airplane was fully built using steel: quite an important demonstration of what might have been achieved with metal in aircraft structures. On the other hand, the exceedingly high density of the ferrous alloy, notwithstanding the adopted design solutions, rendered the airplane not very practical and flexible in operational maneuvering. Therefore, few years later (1917), the availability of a novel aluminum alloy was very much welcome, and a full-aluminum airplane was produced: the Junkers J-4. No pure aluminum but an Al-Cu alloy was used (see Sect. 3.2). The alloy had been developed a few years earlier by the German metallurgist Alfred Wilm, who could only confirm empirically the hardening effect of thermal treatments, without having the possibility of a direct observation of the precipitates responsible for this improvement in the mechanical properties (see Sect. 3.2.3.3). It can be noted that this was quite a fast technological transfer, from materials research to product development, of particular specific interest, considering the importance that Al-alloys were bound to have in the aerospace field, still lasting until the present times. These two cases are particularly important, since they have demonstrated for the first time the possibilities of metallic materials in the fabrication of aircrafts: airframe, fuselage, and engines. This was certainly an enabling factor for the development of commercial aviation that in fact, in between World War I and World War II, started its history. In this regard, a year that is worth mentioning is 1933. In February the first Boeing 247, a full-metal airliner, capable to carry ten passengers, took off for the first time, being regarded as the first modern transport airliner. Although on board comfort was very much limited by structural restrictions, the orders started to arrive soon, so numerous that the company was not able to keep up with the pace. The situation was better exploited by Douglas Aircraft, which launched in the second half of the same year 1933 the 12-passenger DC-1 airliner, more spacious and comfortable than the Boeing 247. Other DCs (DC stays for Douglas Commercial) were delivered in the following years: DC-2 (1934, 14 passengers) and DC-3 (1935, 21 passengers). The DC-3, thanks to its flight performances and payload capabilities, gained enormous success among the starting up airline companies that also thanks to this airplane, realized the possibility of making profit out of air transportation.

Although, particularly in the war years, the technological advancements in the military field were transferred to civilian air transportation, in some cases, the needs were rather divergent. This is the case of the maximum operational speed that with the advent of gas turbine engines disclosed the possibility to overcome the intrinsic limit for a propeller aircraft, i.e., the speed of sound. The supersonic flight implies a change both in the profile and in the dimension of the wings, in order to reduce air drag, being the lift anyway guaranteed by the higher accessible speed. A good example of this new philosophy was the Lockheed F-104 “Starfighter” (see Fig. 1.7) capable to reach a maximum speed of Mach 2.2. Shorter and thinner wings require a change not only in their design but also in the selected materials. Titanium alloys for the fabrication of structural components became an obvious choice. The development of aircraft structures, considering the interplay with the development of new materials, is actually quite a complex process, since a great number of variables must be considered. When increasing the maximum speed, the skin temperature becomes an issue. In this regard, the application of titanium alloys reached a particularly impressive level in the Lockheed YF 12 A and its successor, the Lockheed SR-71 “Blackbird” (see Fig. 1.8), in which not only structural parts but also the skin and control surfaces were made of newly developed β-Ti-alloys (see Chap. 4 and Sect. 4.3.1.4).

The adoption of titanium alloys also for the fabrication of the skin of the SR-71 was a necessary choice to stand the peak temperatures at its highest speed of Mach 3.35. It is worthwhile to note that conventional aluminum alloys would not have had a sufficient strength to resist above Mach 2.0 and steel would be too heavy. On this occasion too, the experience and knowledge developed for a military aircraft were important also for the design and its implementation in supersonic civil aircrafts. The Concorde structure and skin were made of aluminum alloys (see Chap. 3), suitable for a peak speed not exceeding Mach 2.

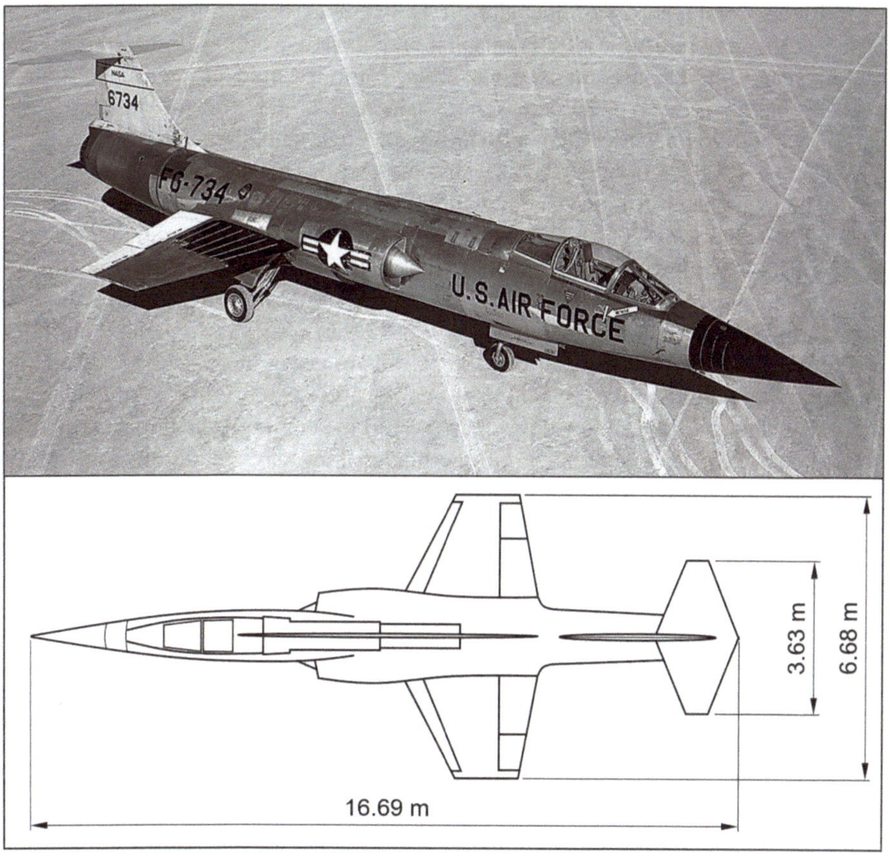

Fig. 1.7 Lockheed F-104 Starfighter. (Friend and Sefic 1972, NASA 2017, reproduced with permission of the Publisher)

Fig. 1.8 Lockheed SR-71 "Blackbird." (NASA 2008, reproduced with permission of the Publisher)

If higher temperatures of the aircraft skin are to be stood, considering, for instance, the conditions faced in atmosphere reentry after orbital flight, like in case of the Space Shuttle, a composite, reusable, ceramic-based protection system is needed, made of tiles of zirconia-glassy silica felt and, for the hottest parts, carbon-carbon composites. Thanks to this protective shield, surface temperatures in excess of 1300 °C do not provoke any harm to the underlying Al-alloy frame, unless faults in the protection system would intervene, as occurred in case of a tragic accident (Savino et al. 2005). The engine temperature is another parameter that has been sensitively influenced by the selected materials properties. The trend in the operational temperatures and capabilities of aircraft engines has always been increasing, starting from the in-line piston engines of the early propeller-operated aircrafts, reaching material temperature in the 200 °C range, going through the radial high-efficiency engines, with peak temperatures of 500 °C, arriving to the extreme thermal challenge of the newly developed gas turbine engines. The hottest structural parts, i.e., the high-pressure stage turbine blades, nowadays made of nickel-based superalloys (see Sect. 6.4), are operated at gas temperatures above the material capability, i.e., its incipient melting point. It is only thanks to the specific micro-structure of the components, i.e., single crystal, and optimized cooling design, i.e., hollow blades, that these conditions can be safely sustained.

1.2 Novel Design Criteria

In recent years, the development of aircraft structures has become increasingly complex in terms of metrics and design drives, with respect to the early times, when technical performances were by far the dominant aspects. Nonetheless, as in the past, contemporary materials are still a decisive enabling factor, since only suitable materials properties will render new design requirements feasible. A remarkable evolution has occurred in the selection criteria that, depending on the specific aircraft structure, will anyway be a trade-off choice.

The improvements achieved in aircraft structures have been continuously accompanied by a growing concern for safety. This aspect becomes increasingly important with the development of civil aviation, where the protection of passengers' lives is not only an ethical issue but also an economic parameter, considering the impact of insurance costs on the business plan budget of any company, not to mention the indirect costs that air accidents may provoke. This aspect appeared soon dramatically evident, for the accidents occurred at two de Havilland Comet aircrafts during take-off at the airport of Ciampino in Rome (Italy) in 1954. The de Havilland Comet (1952, see Fig. 1.9) was a commercial plane derived from a military cargo that originally had no windows in the fuselage that were present instead in the case of the commercial version.

The reason for the failure, occurring under the effect of pressurization-depressurization cycles, was an inappropriate design of the window frames, whose corners acted as initiation spots for the fracture, propagating to the rest of the

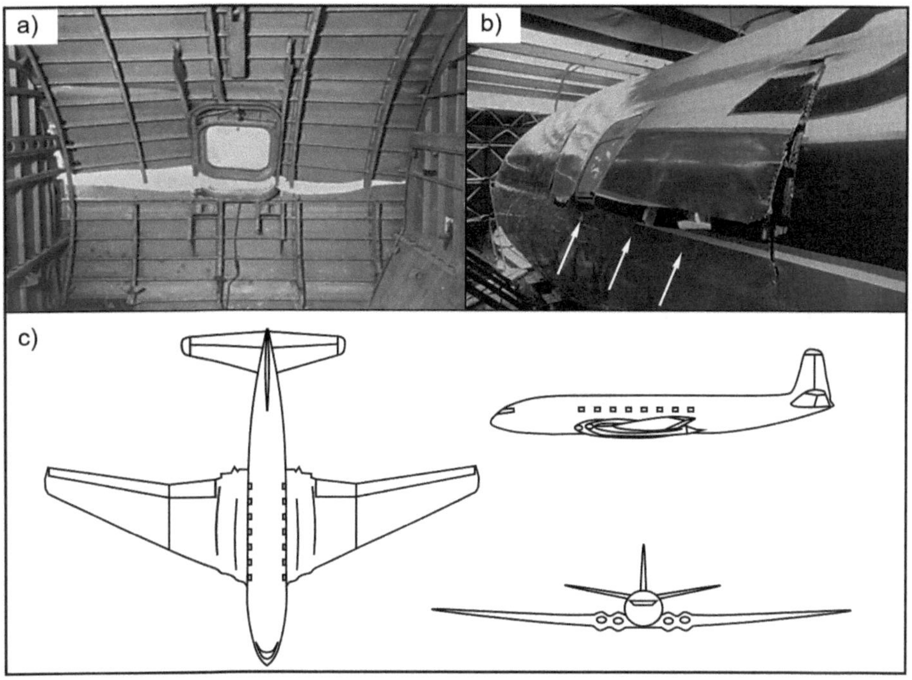

Fig. 1.9 RCAF de Havilland Comet. (**a**) Internal view of the aircraft showing the rupture around the window; (**b**) external view showing the same rupture in the skin of the fuselage (see arrows for the indication of the position of the window and the crack nearby); (**c**) schematic of the airplane (Darling 2001; The National Archives 2019)

fuselage. This failure could not occur in the original military model since no windows were there in the fuselage. The disasters blocked the fast expansion of British Airways that at the time was gaining a leading position in civil aero-transportation. Important consequence of this sort of structural failures was the adoption of two general design criteria for aircraft structures: *safe life* and *fail safe*, which can be regarded as the main guidelines for the development of passenger airliners and in general of all aircrafts for commercial transportation. *Safe life* criterion means the possibility to predict the useful lifetime of structures, components, and systems of an aircraft, in order to schedule maintenance, to proceed with limited repairs, and to replace heavily damaged parts. As to the *fail safe* criterion, it is paramount in aircraft structures since it should allow the airplane to retain sufficient operational capabilities, notwithstanding any possible failure, allowing, for instance, emergency landing procedures.

Looking at the present and past use of aluminum alloys in aircraft structures, it sounds strange that after the Comet disasters, Al-Zn alloys (see Fig. 1.10 and also Chap. 3) were banned, since they were considered the main responsible for the structural failure. Once the real reasons for the accidents were made clear, the use and consequent development of these alloys was resumed and is continuing until

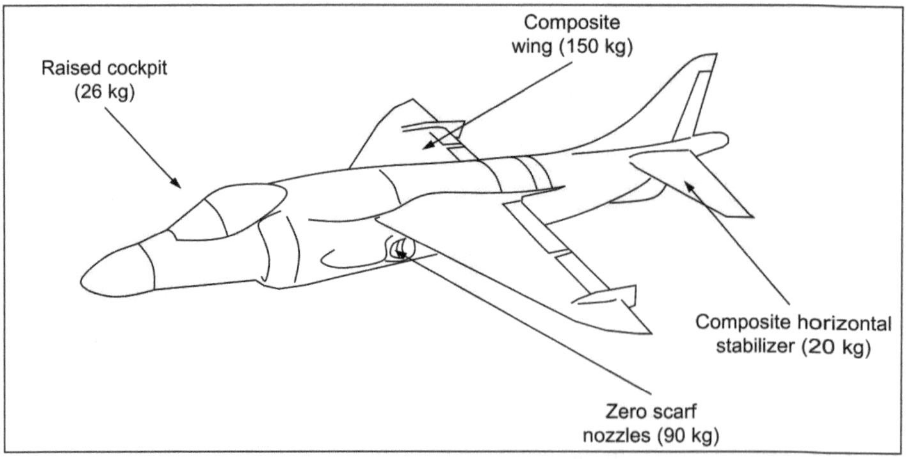

Fig. 1.12 Sea Harrier and some parts replaced by structural composites to achieve an adequate weight reduction. Values among brackets refer to the weight saving obtained by means of composite materials. (Redrawn from Calvert 1990)

Table 1.2 Weight percentages of structural materials used in the airframes of civilian and military aircrafts (Mouritz 2012)

Aircraft	Aluminum alloys	Composites	Steel	Titanium	Other materials
Boeing 737	81%	3%	6%	3%	7%
Airbus A330-340	68%	17%	9%	4%	2%
Airbus A380	61%	25%	10%		4%
McDonnell Douglas F-18 Hornet	51%	9%	16%	13%	11%
Lockheed F-22 Raptor	11%	35%	5%	33%	16%

As it is for other transportation fields, also aviation companies have to consider their own economical sustainability and competitiveness. A successful aerospace product should be valuable for the customers, with a low environmental impact, and still retain elevated safety standards. The cost for the end user, the passenger in case of civil aviation, results from a number of parameters, including initial cost and cost of ownership. The total cost of ownership includes several terms, according to the scheme in Fig. 1.13. It is worth saying that a more efficient management affords not only money savings to the companies but also a reduction of the impact of aviation on the environment and, thereby, on human health. Better designed and lighter aircraft structures will have better aerodynamic performances, lower exhaust, and noise emissions. Of course, for these latter aspects, an improved efficiency of the engines is paramount. This is the reason why also for the engines, a continuous search for new materials and part design have been pursued, as reflected by the evolution over the years of the turbine inlet temperature, even named turbine entry temperature (TET).

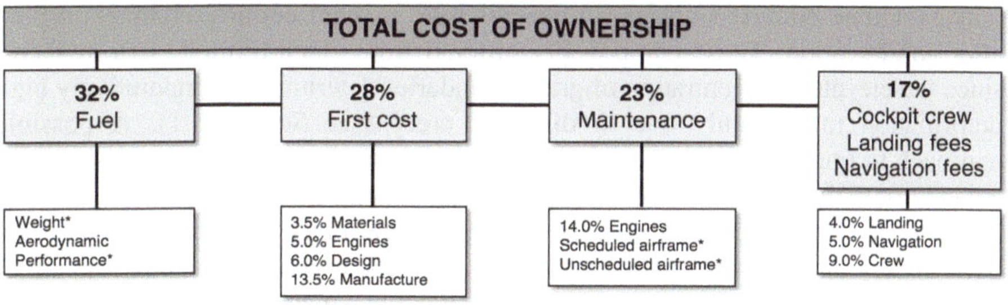

Fig. 1.13 Total cost of ownership and its split in the main components. * refers to parameters that are influenced by custom features such as aircraft type, age of the aircraft, operating conditions, and owner (Cantor et al. 2001)

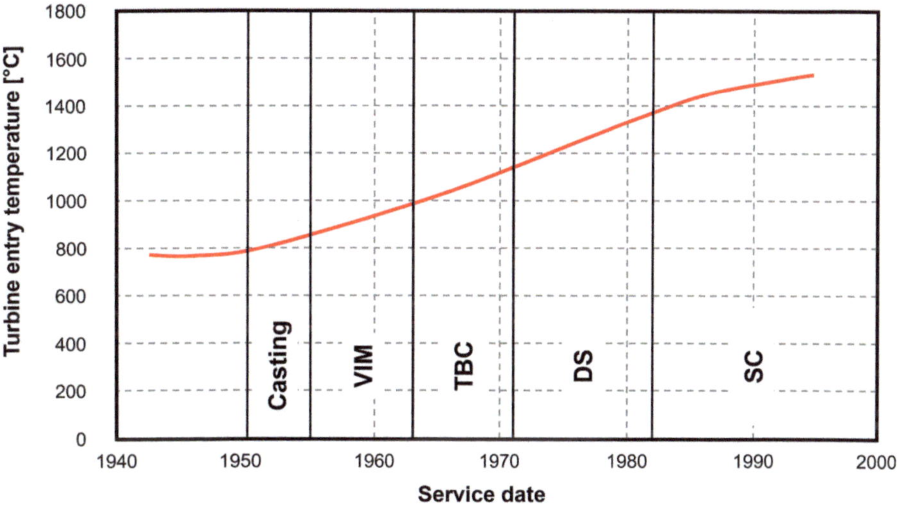

Fig. 1.14 Evolution of the turbine inlet temperature as a function of the in-service date. See main text for the legend of the processing techniques recalled in the graph. (Redrawn from King et al. 2009)

As it will be illustrated in Chap. 2, the turbine inlet temperature is also a very important parameter to assess the efficiency of a GT engine. An increase by 50 °C of the turbine inlet temperature can increase by 1% the engine efficiency, with a consequently lower fuel consumption and reduced exhaust emission. Considering that fuel costs are the most important expenses for airline companies (see Fig. 1.13), the interest in better performing engines becomes evident and may explain the remarkable development of the processing routes for the manufacturing of turbine blades. This is shown in Fig. 1.14, in which the main enabling process technologies used for the production of high-pressure blades, regarded as the most critical parts of a gas turbine engine, are associated with the relevant achieved TETs. Casting and vacuum induction melting (VIM, see Sect. 5.3.2) were used to produce polycrystalline equiaxed blades, in general solid or with cooling channels drilled after solidification. These components suffered from creep-limited performances, since the elevated concentration of grain boundaries determined comparatively high deformation rate, mainly due to diffusive creep (see Sect. 6.5.1). A possible approach to improve creep life was through the reduction of the metal temperature, attained with the deposition of a thermal barrier coating (TBC, see Sect. 7.3.2).

A lower blade temperature would display a lower creep deformation rate. This was actually the case, although the durability of TBC systems and the prediction of their useful lifetime were soon regarded as critical parameters for key components, as high-pressure turbine blades are. The creep problem was therefore tackled acting directly on the structural reason for its occurrence in turbine blades, i.e., the presence of grain boundaries, particularly those oriented normally to the maximum stress, applied along the blade main axis. Therefore, using directional solidification (DS) and, subsequently, single-crystal (SC) solidification (see Sect. 6.5), blades with grain boundaries aligned along the main axis of the blade, or missing at all, were obtained. These components, since the 1980s, are the standard for commercial aircrafts and are being adopted also for land-based gas turbines, again for enhancing their performances at the acceptable costs afforded by the now mature technology of the directional solidification and single-crystal processes.

1.3 Aerospace-Related Fields

The mentioned land-based gas turbine engines are just one of the many technologies and materials developed for the aerospace field and then transferred to terrestrial applications. The situation is actually not always so clear-cut: in fact, gas turbine engines exploit some of the working principles of turbo-charged piston engines, originally developed for automotive applications. It is therefore advisable thinking in terms of two-way exchange between aerospace and other technological fields.

It is certainly true that the aerospace field has always had a particular focus on the development of materials and systems complying with strict limitations, as concerns safety and reliability. These standards require important systemic research efforts to study problems in depth and to find optimal solutions. Therefore, materials and systems developed and/or optimized for the aerospace fields are most interesting candidates for applications also in other fields. In the present book, only some of those more directly involved with the use of aerospace alloys will be considered.

The field of energy production is certainly benefitting from the processing technologies for the production of components made of superalloys (see Chap. 6) and special steels (see Chap. 5), used in land-based gas turbines, vapor turbines, and other systems working at comparatively high temperatures, although generally lower than the aerospace counterpart, but still demanding for the alloy materials. This is so even because the environmental conditions that are faced by components of terrestrial plants may be more demanding in terms of corrosion resistance, considering that the quality of the fuels used in this field is definitely worse than the extremely refined aerospace ones.

Light alloys, in particular aluminum and magnesium (see Chap. 3), are being employed to increasing extent in the automotive field, for the same reason just discussed for aerospace applications, to reduce the weight of the vehicle in order to reduce fuel consumption and consequently the environmental pollution associated with the production of particulate matter and gaseous emissions.

Titanium alloys (see Chap. 4) are very important for biomedical applications, as concerns particularly their biocompatibility. Shape-memory alloys (see Sect. 9.5), which have been applied for the first time as cryo-fit joints in a military aircraft, are now having an important set of applications in the aerospace field, in some case really strategic, like the deployment system Frangibolt (see Sect. 9.5.2). At the same time, still remaining in the field of products with high added value, numerous biomedical applications are now available, like stents, orthodontic wires, surgery tools, and many more.

Structural polymer matrix composites are now finding their way in aerospace, already widely employed for terrestrial application, and are gaining more interest.

The aircraft brake systems, based on carbon-carbon composites, have provided fundamental knowledge and performance data for the application of similar principles and technical design to brakes of top and high-performance cars, like the F1 racing.

References

Boeing (2013) 787 Aircraft Rescue & Firefighting Composite Structure. Available via DIALOG. http://www.boeing.com/assets/pdf/commercial/airports/faqs/787_composite_arff_data.pdf. Accessed 18 Aug 2017

Calvert D J (1990) Harrier. Ian Allan Publishing Ltd

Cantor B, Assender H, Grant P (2001) Aerospace Materials. Institute of Physics Publishing

Curtis H D (1997) Fundamentals of aircraft structural analysis. Irwin Publ., Chicago

Darling K (2001) Airliner Tech Volume 7 – De Havilland Comet. Specialty Press Publishers and Wholesalers

Friend E L, Sefic W J (1972) Flight Measurements of Buffet Characteristics of the F-104 Airplane for Selected Wing-Flap Deflections. Nasa Technical Note D-6943. Available via DIALOG. https://www.nasa.gov/centers/dryden/pdf/87804main_H-666.pdf. Accessed 05 May 2019

King D et al (2009) Advanced Aerospace Materials: Past, Present and Future. Aviation and the Environment 3(9): 22–27

Krishnan K S (2006) Materials and the Aerospace Industry. In: Enhancing Innovation and Competitiveness Through Investments in Fundamental Research, Westin Hotel, Arlington, VA, 3–5 December 2006

Mouritz A P (2012) Introduction to Aerospace Materials. Woodhead Publishing Ltd

NASA (2003) Learning to Fly: The Wright Brothers' Adventure – A Guide for Educators and Students With Activities in Aeronautics. NASA Educational Division. Available via DIALOG. https://www.nasa.gov/audience/foreducators/topnav/materials/listbytype/Learning.to.FlyThe.Wright.Brothers.Adventure.html. Accessed 09 Jun 2017

NASA (2008) SR-71 Blackbird. NASA Facts. Available via DIALOG. https://www.nasa.gov/centers/dryden/pdf/495839main_FS-030_SR-71.pdf. Accessed 10 Jun 2017
NASA (2018) F-104 Starfighter Image Gallery. https://www.nasa.gov/centers/armstrong/multimedia/imagegallery/F-104/index.html. Accessed 05 May 2019
Savino R et al (2005) Aerothermodynamic Study of UHTC-based Thermal Protection Systems. Aerospace Science and Technology 9 (2): 151–160
Starke E A, Staley J T (1996) Application of Modern Aluminum Alloys to Aircraft. Progr Aerospace Sci 32:131–172
The National Archives (2019) Public Information Films | 1945 to 1951. Available via DIALOG. http://www.nationalarchives.gov.uk/search/results?_q=de+havilland+comet. Accessed 05 May 2019
Torenbeek E (2013) Advanced Aircraft Design. John Wiley & Sons Ltd

Further Reading

Bowman M W, Vogelsang M (2000) Lockheed F-104 Starfighter. The Crowood Press Ltd, Ramsbury
Cutler J (2005) Understanding Aircraft Structures. Blackwell Science
Federal Aviation Administration (2012) Aviation Maintenance Technician Handbook – Airframe Vol. 1. Federal Aviation Administration
Federal Aviation Administration (2012) Aviation Maintenance Technician Handbook – Airframe Vol. 2. Federal Aviation Administration
Federal Aviation Administration (2012) Aviation Maintenance Technician Handbook – Powerplant Vol. 1. Federal Aviation Administration
Federal Aviation Administration (2012) Aviation Maintenance Technician Handbook – Powerplant Vol. 2. Federal Aviation Administration
Gates D (2004) Sky Wars: A History of Military Aerospace Power. Reaktion Books Ltd
Megson T (2013) Aircraft Structures for Engineering Students. Elsevier
Withey P (2019) The Real Story of the Comet Disaster: De Havilland Comet – Structural Fatigue. In: Hamburg Aerospace Lecture Series. Available via DIALOG. https://zenodo.org/record/2551089#.XNHwSI5LiUl. Accessed 05 May 2019
Wixey K E (1987) Lockheed Constellation. Ian Allan Ltd

Chapter 2
Gas Turbine Aero-Engines

2.1 Introduction

Gas turbine engines are mechanical systems able to provide an extremely efficient conversion of the energy, produced by internal combustion, into mechanical work, to power an electric generator or to provide thrust for an aircraft to fly. Several features of gas turbine engines have been especially developed for aircraft propulsion, since the pioneering studies by René Lorin in 1913 (Rolls-Royce 1996), followed by Frank Whittle in the UK and Hans von Ohain in Germany, in the years preceding the burst of the World War II (Gunston 1995). A jet propulsion engine is a particular type of reaction engine, in which a fast-moving airflow (jet) generates a thrust, according to Newton's third law of motion. Several different types of jet engines do exist, named according to the relevant working principles. Strictly speaking, a pure jet propulsion occurs in turbojets, ramjets, pulse jets, and rockets too. In turbo-propeller and turbofan engines, the thrust is completely or largely produced by the compression waves generated by the propeller or ducted fan, respectively. Gas turbines have moving parts, which pose important challenges to materials owing to the combined effects of mechanical stresses, corrosive atmosphere, and high temperatures. In these air-breathing systems, a rotary compressor, powered by the turbine, provides high-pressure air to the combustor. Here, air is mixed up with the fuel and spark ignited to produce the flame and enthalpy output that through the gas is transferred to the turbine rotor. Although, working under continuous gas flux, the efficiency of these thermal machines can be evaluated referring to a closed Brayton cycle, thermal efficiency is not the only parameter that qualifies different classes of jet aero-engines. In fact, specific fuel consumption; specific thrust, particularly during takeoff; noise; and other issues pertaining to the environmental impact are all other important parameters to be considered.

S. Gialanella, A. Malandruccolo, *Aerospace Alloys*, Topics in Mining, Metallurgy and Materials Engineering, https://doi.org/10.1007/978-3-030-24440-8_2

2.2 Types of Aircraft Engines

The development of gas turbine engines stems from the need to overcome the intrinsic limits of piston engines, used since the early flight of the Wright brothers (see Chap. 1) to operate a standard wooden propeller. The two main drawbacks that have emerged in applying these engines to aircraft propulsion are efficiency reduction for speeds in excess of about 600 km/h and reduction in the propulsion efficiency determined by the so-called air starvation at high altitude flights. The first limit is due to the fact that the propeller cannot reach, and overtake, the speed of the pressure waves it produces. Therefore, the propulsion efficiency of the engine would progressively decrease as the speed of sound is approached. The second phenomenon is caused by the reduction of air density in the upper layers of the Earth's atmosphere, with a consequent effect on the compression ratio that can be reached inside the piston chamber.

The first patent for a jet propulsion system was issued by René Lorin (1913, see Fig. 2.1), although the early application of this new technology appeared more than 20 years later, due to the limitations in technological processes and materials properties. The main idea behind the development of jet propulsion engines can be retrieved in the working principle of a turbocharger engine (see Fig. 2.2).

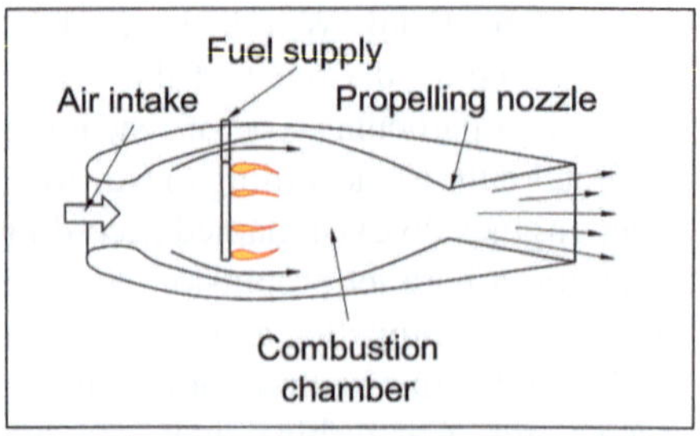

Fig. 2.1 Scheme of the first jet propulsion system developed by René Lorin, 1913. (Redrawn from Rolls-Royce 1996)

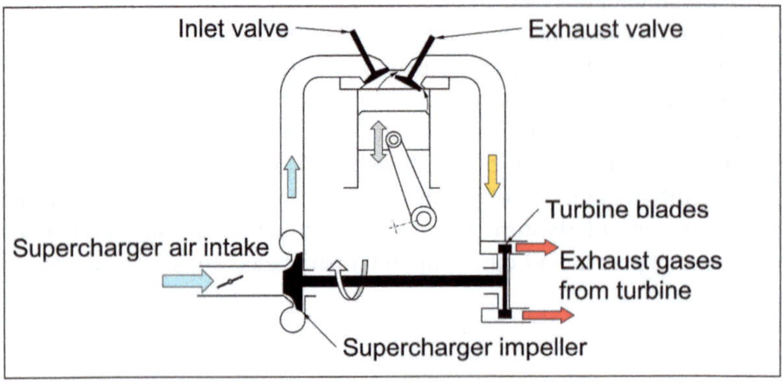

Fig. 2.2 Turbocharger engine with centrifugal compressor. (Redrawn from Meetham 1981)

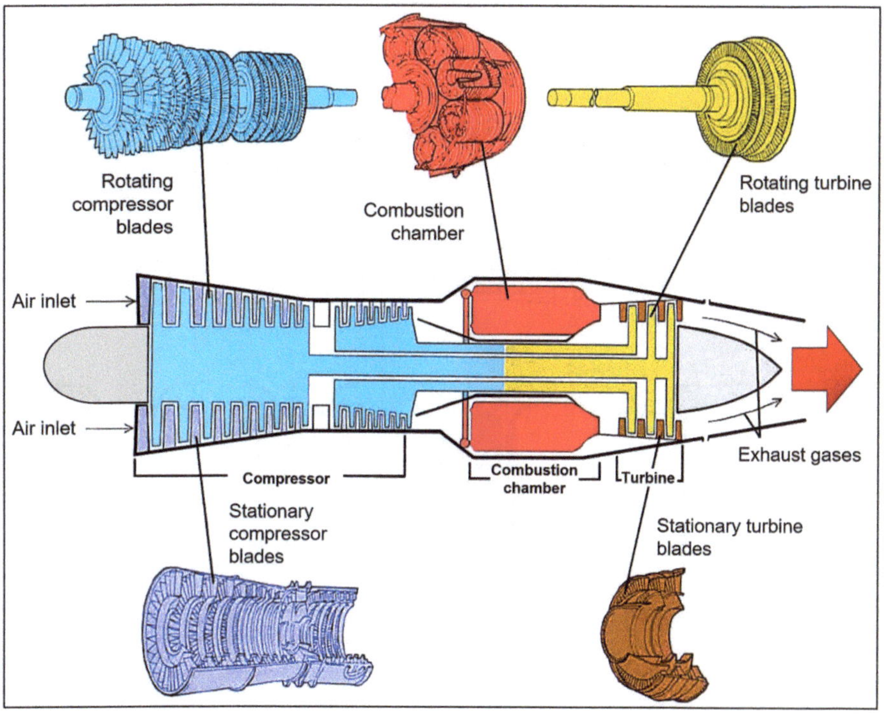

Fig. 2.3 Structure of a two-stage gas turbine jet engine

The pressure of the air entering the piston chamber is raised by a centrifugal compressor connected to the turbine. The turbine is operated by the exhaust emission ejected from the combustion chamber, so that further energy can be recovered from the gas before it is emitted from the engine. Similarly, in gas turbine engines, a compressor is meant to increase the pressure of the air entering the combustors to optimize the combustion under all operational conditions. Part of the energy carried by the gas flux is transferred back to the compressor by the turbine stages through coaxial multiple shafts. The rest of the combustion energy output is used to generate thrust or other forms of energy in case of non-aero applications, e.g., marine and terrestrial engines. The main operation stages of a gas turbine (GT, Fig. 2.3) can be summarized as follows:

- Air inlet into the compressor
- Air compression
- Compressed air inlet into the combustion chamber
- Fuel addition
- Ignition
- Entrance of the exhaust gas into the turbine
- Generation of thrust work

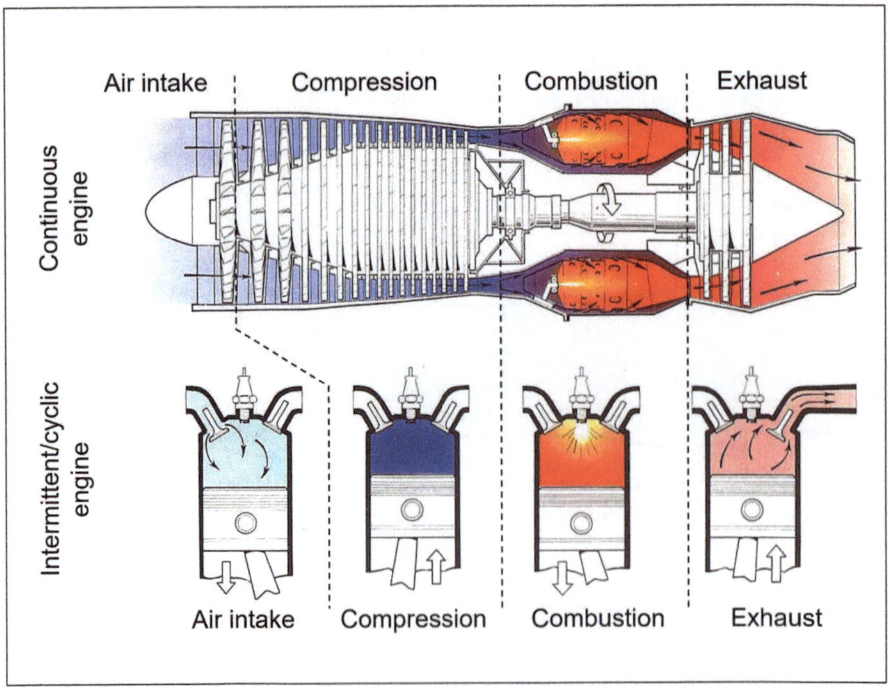

Fig. 2.4 Comparison between a gas turbine jet engine and a four-stroke piston engine. (Rolls-Royce 1996, images courtesy of Rolls-Royce plc)

GT engines may reach extremely high efficiencies, close to 100%. They operate according to a continuous process with a free expansion in the combustion area.

Piston engines are characterized by a lower efficiency, if compared with GTs, since just one out of the four strokes is active (see Fig. 2.4). Furthermore, they must be more massive in order to withstand an instantaneous increase in pressure associated with the constrained, fixed combustion volume.

These two features contribute to make these engines a non-ideal choice for aerospace applications. Notwithstanding the "open" character of a GT engine, it is common practice to analyze its working principles referring to a basic thermodynamic approach, using an equivalent thermodynamic cycle to estimate its efficiency and to evaluate the associated work output from the relevant cycle area, as depicted by the general Carnot cycle in Fig. 2.5.

As anticipated in the introductory paragraph of this chapter, the GT engine working principles can be described according to the Brayton cycle. It is evident that the thermodynamic cycle is different if compared to the piston engine, considering that this latter is better represented by the Otto cycle (Fig. 2.6).

One of the most important parameters to take into account while analyzing GT engine thermodynamic cycles is the turbine entry, or inlet, temperature (TET). This

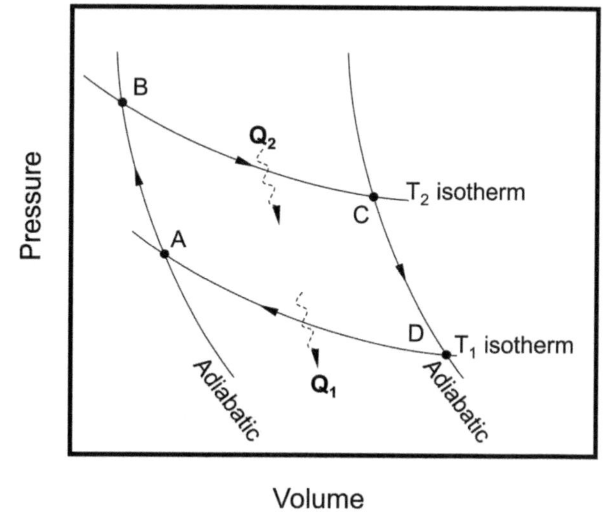

Fig. 2.5 P-V diagram of the Carnot cycle

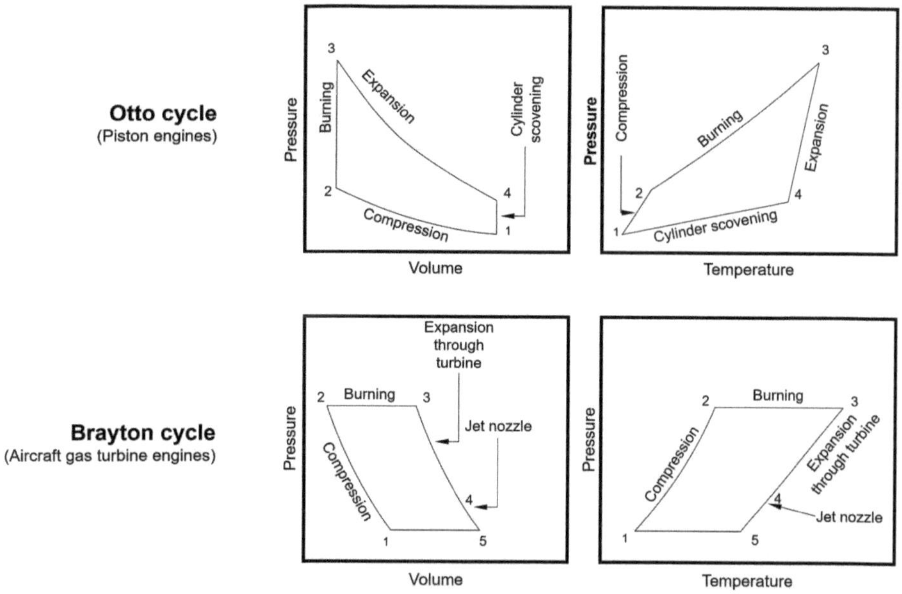

Fig. 2.6 Comparison between Otto and Brayton cycles using P-V and P-T diagrams

is a very important parameter for evaluating the efficiency of the propulsion system. The most severe operating conditions inside the jet engine are those corresponding to point 3 in Figs. 2.6 and 2.7. In fact, this point of the cycle features the highest temperature of the hot gases, impinging directly on the vanes and blades of the turbine.

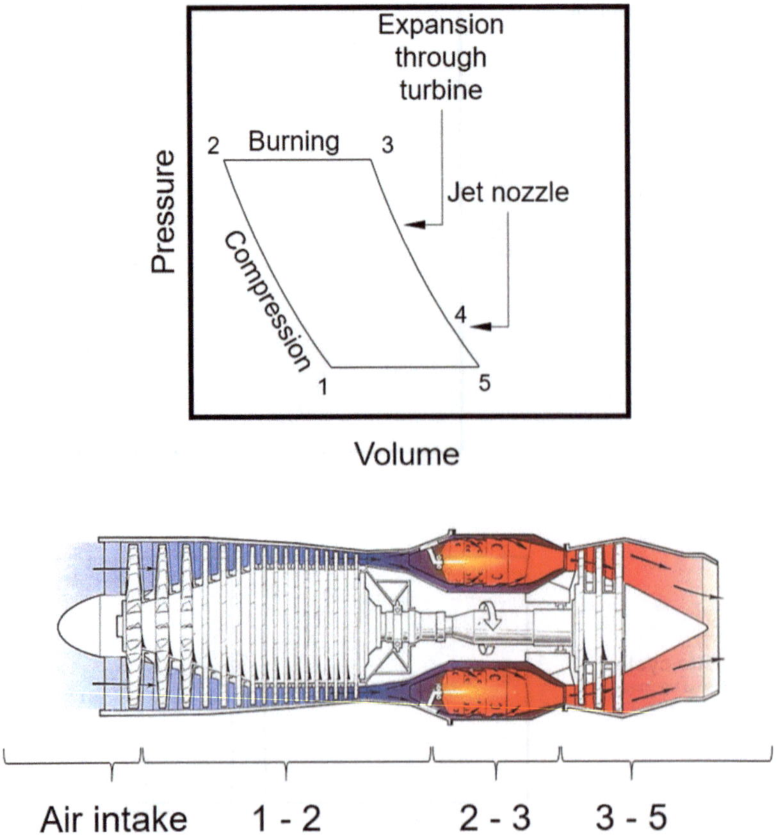

Fig. 2.7 Brayton cycle and correspondence with different GT operational stages. (Rolls-Royce 1996, images courtesy of Rolls-Royce plc)

In addition to the combustion gases and relevant thermomechanical stresses, blades are also under the effect of the centrifugal force generated by the disc and blade high-speed rotation.

The efficiency of the propulsion system is improved by increasing the difference in temperature between points 1 and 3 (see Fig. 2.6, Brayton cycle). Inlet air temperature is a parameter that cannot be changed straightaway, depending on the atmospheric conditions. Conversely, T_3, i.e., the turbine entry temperature, can be modified and, in particular, increased with an appropriate design of the engine and selection of suitable materials. However, the improvement in thermodynamic efficiency is accompanied by an unwanted increase in the specific fuel consumption (SFC). SFC can be regarded as a fuel-related efficiency parameter of the engine, and it is given by the amount of fuel consumed per unit thrust. This increase in SFC can be compensated for by a corresponding increase in the overall pressure ratio (OPR), i.e., the ultimate air pressure value that can be achieved in the latest stage of the compressor, before air injection into the combustion chamber. The situation is

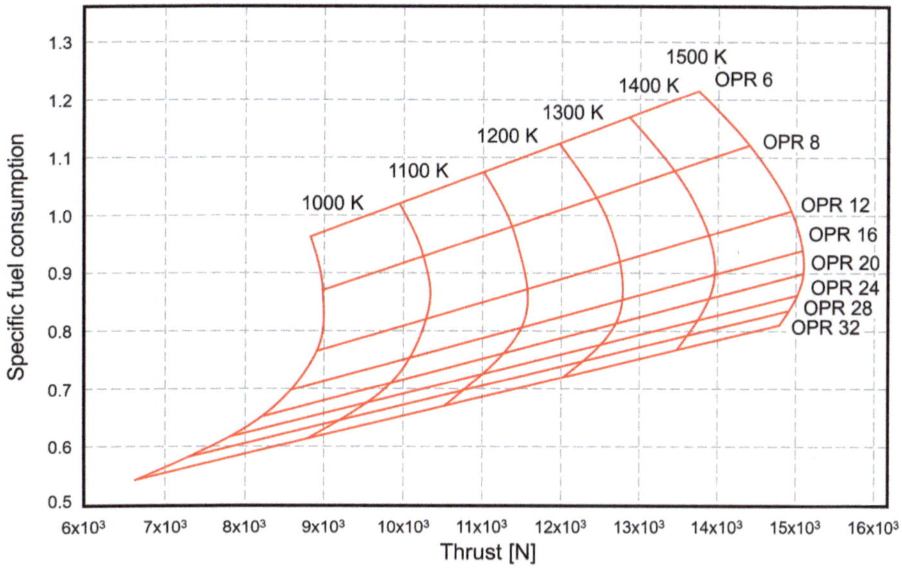

Fig. 2.8 Dependence of the specific fuel consumption and engine thrust as a function of the turbine entry temperature (TET) and overall pressure ratio (OPR). (Redrawn from Reed 2006)

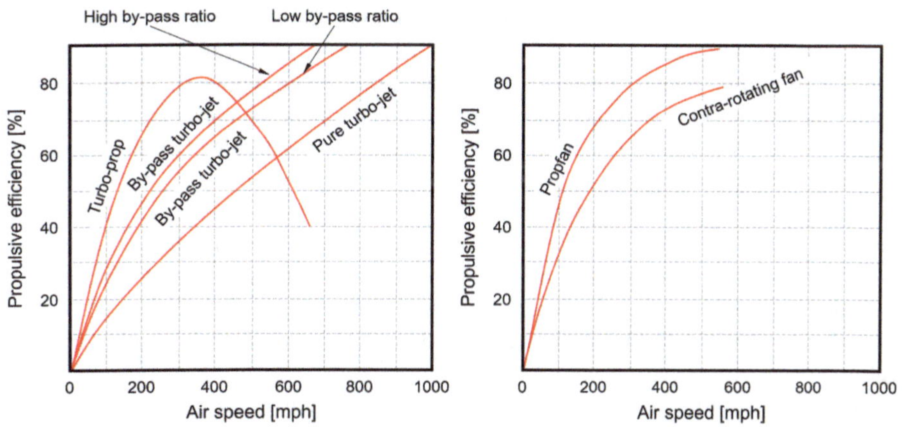

Fig. 2.9 Propulsive efficiency vs aircraft speed for different engines (1 mph = 1.609 km/h). (Redrawn from Rolls-Royce 1996)

depicted by the graph in Fig. 2.8, displaying the combined effect of TET and OPR on the engine performances. In state-of-the-art gas turbine engines, e.g., Rolls-Royce Trent XWB, the OPR has reached values of 50 or so. Concerning the efficiency of the propulsion system, turbojets show an increase in their efficiency with the aircraft speed, since this reduces the dissipation of the energy of the outlet gas flux against the surrounding static air (see Fig. 2.9). The peak followed by a decreasing trend at

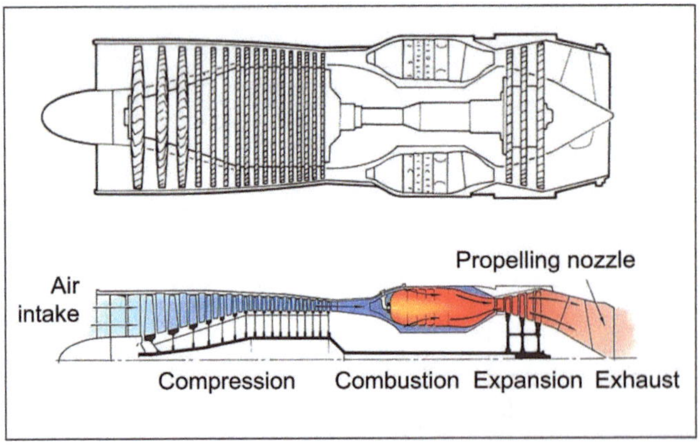

Fig. 2.10 Rolls-Royce Avon single-shaft engine. (Rolls-Royce 1996, images courtesy of Rolls-Royce plc)

higher speeds observed in the efficiency curve of the propeller-operated aircrafts is dependent instead on the reduction in efficiency that occurs once the advancement speed of the propeller in air is approaching the speed of sound.

Differences in gas turbine engines for aircraft applications can be found in the architecture and in the range of performances. Turbojet engines use the combustion gases directly for the propulsive thrust. In turbofans, the propulsion is provided by the fan and to minor extent only by the jet of the exhaust gas. This is the basis for the by-pass principle, which implies a split in the airflow. The whole air intake is given an initial low compression, and a large fraction is then ducted to the bypass, the rest being delivered to the combustion chamber (Fig. 2.11). The idea behind the development of by-pass engines is related to the following specific aspects: better fuel consumption, higher thrust (even at low-speed regimes), noise reduction, and reduced relative weight of the engine. These aspects have been achieved thanks to the evolution of the engine design, starting from a comparatively simple single-shaft turbojet, as shown by the schematic of the Avon engine (see Fig. 2.10), to reach the complexity of the multiple-shaft engines in Fig. 2.11, featuring also higher by-pass ratios.

The by-pass ratio, i.e., the ratio between cool air by-passed through the outer duct and the air flowing through the high-pressure inner part of the engine, is an important feature of these engines. With low bypass ratios (e.g., 1:1), the two air streams are generally mixed before being emitted through the exhaust duct of the engine. In turbofan engines requirements for higher bypass ratios, up to 5:1, are met by large front fans and multiple-stage compressor, using a twin- or triple-shaft configuration (see Fig. 2.11). In this case the front fan acts also as a low-pressure compressor stage. Bypass ratios in the order of 15:1 are achieved using prop-fans (Rolls-Royce 1996). Custom design has been developed for complying with the specific requirements of the different aircrafts and relevant operational conditions. Correspondingly, different pressure and temperature profiles are present along the engine like those shown in

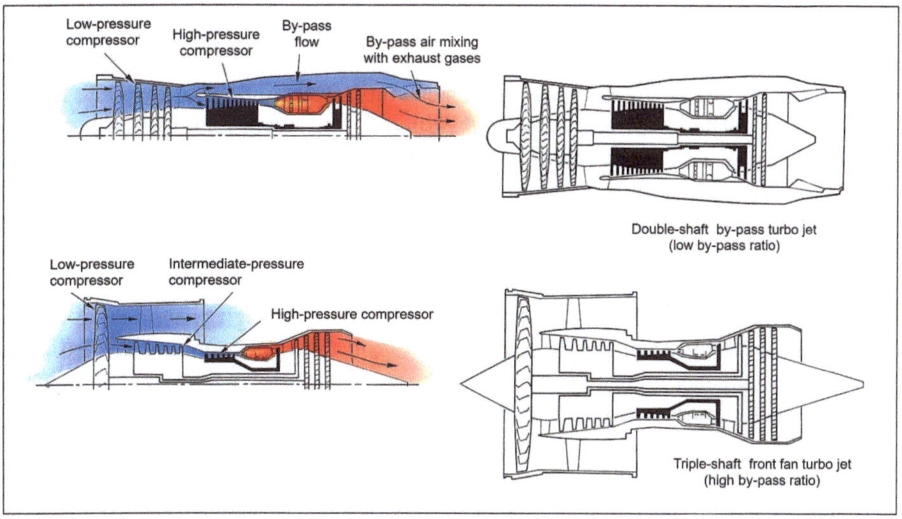

Fig. 2.11 Rolls-Royce engines with a multiple-shaft design. (Rolls-Royce 1996, images courtesy of Rolls-Royce plc)

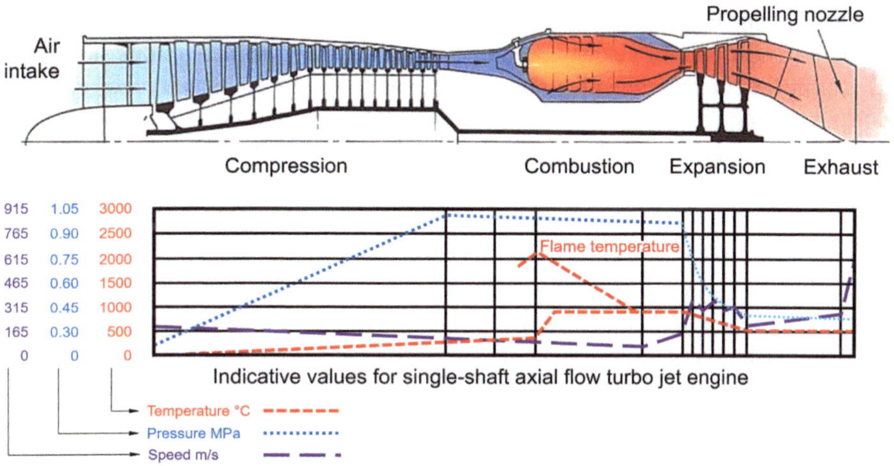

Fig. 2.12 Air speed, pressure, and temperature profiles in a turbojet engine. (Rolls-Royce 1996, image courtesy of Rolls-Royce plc)

Fig. 2.12 for a turbojet engine. These fundamental changes in the structure of the aircraft engines have been possible thanks to the availability not only of better performing materials but also of new technical choices in the design of the engine components. These aspects are considered in the following sections.

2.3 Compressors

In a gas turbine engine, the compressor converts the power from the turbine into kinetic energy of the airflow, by increasing its pressure. An upstream rotating compressor is coupled to a downstream turbine. The combustion chamber is placed in between these two components. Energy is added to the gas flow through the ignition of the fuel-air mixture that forms inside the combustor by spraying fuel into the compressed air. The resulting enthalpy output raises the gas temperature, the *flame temperature*, which in the most severe operating conditions can even reach values of 3000 °C.

The work output of the combustion is used to drive, through coaxial components of the shaft, the different stages of the turbine and the corresponding stages in the compressor. All energy that is not used for the shaft operation leaves the engine as an exhaust gas flux and contributes to the thrust of the aircraft.

Actually, a multi-stage compressor is rather common in case of axial geometry, but not so much in centrifugal compressors, the earliest to be used in aerospace gas turbines. In this regard, compare typical schemes for centrifugal (Fig. 2.13) and axial (Fig. 2.14) compressors, respectively. The complex trajectories of the air flux inside centrifugal compressors (see Figs. 2.13 and 2.15) lead to instabilities that can be regarded among the main reasons for their decline. In jet engines equipped with axial compressors, the pressure raise is due to the progressive reduction of the volume available to the fluid as it proceeds inside the engine (see Fig. 2.15). Axial compression produces a nearly linear velocity profile of the gas, and it is definitely more stable than in centrifugal compression. One important parameter is the compressor exit temperature, increasing over the years (see Fig. 2.16), following the trend of the OPR, also quoted in the same graph for different engines. The related increasing trend of the pressure ratio for both aerospace and industrial applications is also displayed (Fig. 2.16).

The two main components of the compressor rotor are the discs and the blades. Disc materials should exhibit adequate values of the following properties:

- Resistance to high-frequency fatigue
- Resistance to low-frequency fatigue (engine start and shutdown phases)
- Tensile strength, toughness, and creep resistance

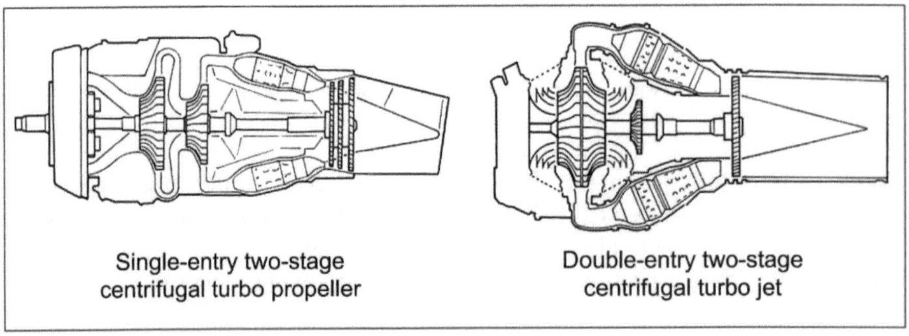

Fig. 2.13 Two examples of centrifugal compressors used in early aero gas turbine engines. (Rolls-Royce 1996, images courtesy of Rolls-Royce plc)

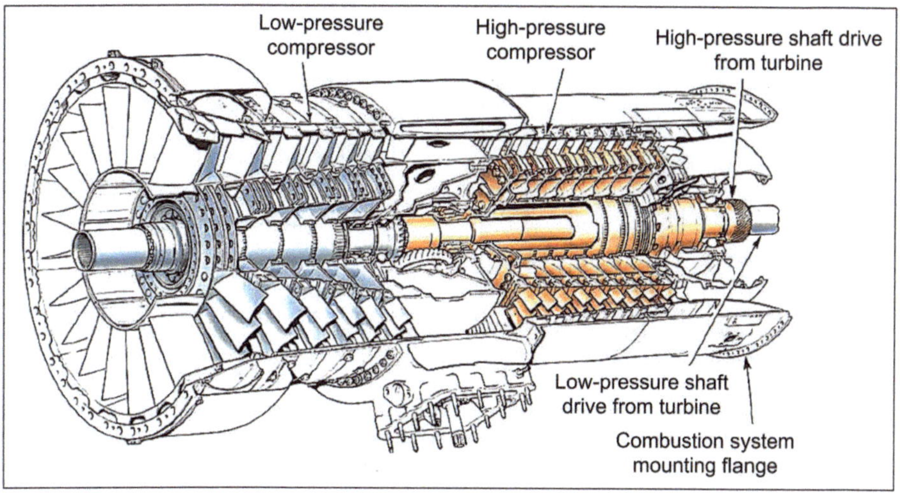

Fig. 2.14 Twin-shaft compressor. (Rolls-Royce 1996, image courtesy of Rolls-Royce plc)

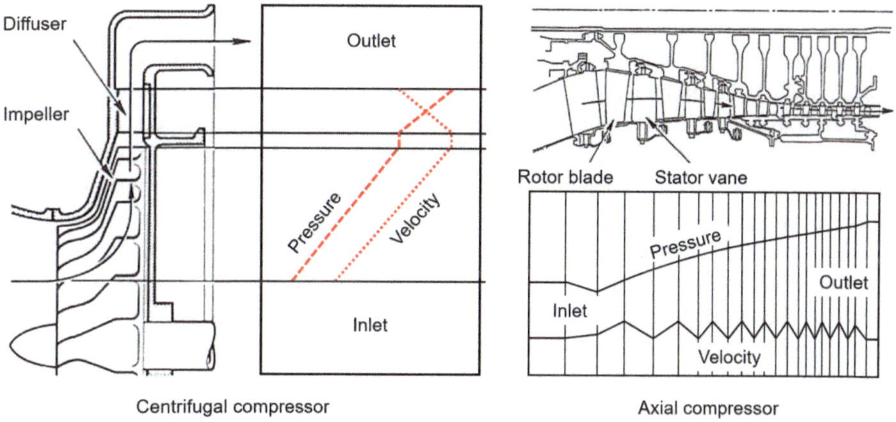

Fig. 2.15 Air velocity and pressure for centrifugal and axial compression systems. (Rolls-Royce 1996, images courtesy of Rolls-Royce plc)

Compressor blades have some specific requirement too. In particular, they must be characterized by good values of:

- Erosion resistance
- Resistance to high-frequency fatigue
- Tensile strength
- Fracture toughness

Erosion resistance can be improved using protective composite coatings; Co-WC-based coatings show excellent erosion resistance and tribological properties

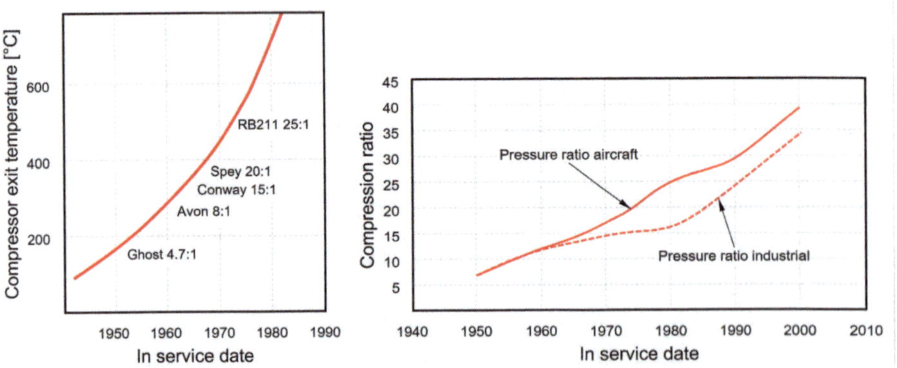

Fig. 2.16 Compressor exit temperature vs in-service date for several GT engine models (for each one of them, the OPR value is also indicated) and compression ratio vs in-service date for aircraft and industrial applications. (Redrawn from Boyce 2012; Meetham 1981)

(see Sect. 7.3.5). In Table 2.1 a list of materials used for the fabrication of compressor rotor parts of a commercial GT engine is presented.

Titanium alloys (see Chap. 4) are used for the first outer stages, both discs and blades, since they provide a good compromise of high specific mechanical strength, creep, and oxidation resistance. Where higher temperatures are achieved and render diffusion-assisted phenomena more important, nickel- and iron-based superalloys, like Inconel 718 (see Sect. 6.3), and precipitation hardened austenitic stainless steels (PHSS), like A286 (see Sect. 5.4.3.3), are preferably used. These materials indeed provide adequate mechanical stability even against creep deformation, and the definitely higher densities, as compared to Ti-alloys, are not resulting in too large inertial momenta, due to high-speed rotation, for the comparatively smaller dimensions of these parts. The intrinsically good oxidation resistance can be improved further with suitable coating systems, like Ni-aluminides diffusion coatings or MCrAlY overlay coatings (see Sect. 7.3.1).

2.4 Combustors

The combustor is the part of the GT engine in which the conversion of the chemical energy content of the fuel is turned into the thermal energy of the gas flow. This is achieved through the combustion of the air-fuel mixture with the as higher as possible efficiency, also to keep the emission of undesired pollutants at the lowest levels. To meet this target, a highly turbulent gas flow is very much welcome (Cumpsty 2003). In early GT engines, the combustor was made of separated chambers, with several inlet hole access for the compressed air, in order to improve the homogeneity of the composition of the combustion mixture (see Fig. 2.17).

Table 2.1 Compressor rotor materials, coating systems, and part finish

Part name	Material	Finishing	Part name	Material	Finishing
Blades, stage 1	Ti-6Al-4V	WC	Blades, stages 4–5	Ti-6Al-4V	Colloidal graphite
Adapter retainer	AMS 5754	None	Blades, stages 6–9	Ti-6Al-2Sn-4Zr-2Mo	Colloidal graphite
Air duct	A286	Cu-Ni-In	Bumper bearing	AMS 5612	Ni graphite
		Ag plate	Disc, stage 1	Ti-6Al-4V	Ni-aluminide
Assembly, air seal	Inconel 718	Ni-aluminide			Molydisulfide
		Aluminum oxide	Disc, stage 2	Ti-6Al-4V	Molydisulfide
Blades, stage 1	Ti-6Al-4V	Cu-Ni-In	Discs, stages 10 and 14	Inconel 718	None
		Molydisulfide	Hub tie clamps	Inconel 718	Ni-aluminide
Blades, stage 2	Ti-6Al-4V	Cu-Ni-In			Cu-Ni-In
	Ti-6Al-2Sn-4Zr-2Mo	Molydisulfide	Hub tie supports	Inconel 718	None
Blades, stage 3	Ti-6Al-2Sn-4Zr-2Mo	Colloidal graphite	Rear shaft	Inconel 718	Cr plate
		Cu-Ni-In			Ni-aluminide
		Molydisulfide	Spline adapter	AMS 6381	None
Blades, stages 10–12	A286	Colloidal graphite	Spools, stages 3–9	Ti-6Al-2Sn-4Zr-2Mo	Ni-aluminide
Blades, stages 13–14	A286	Colloidal graphite	Spools, stages 11–13	Inconel 718	Ni-aluminide
	Inconel 718				

Nowadays, annular combustors have been developed (see Fig. 2.18), although a multiple-chamber design afforded higher safety standards, since the failure or the malfunctioning of one component element is not necessarily jeopardizing the whole combustor. Air enters into the chamber at different places along the combustor, thanks to the presence of calibrated gaps left among the metal sheets of the annular chamber walls. A double-wall architecture is mandatory for this part of the engine where the peak temperatures of the whole thermodynamic cycle are achieved and an effective cooling is thus paramount, and in fact a fraction of the air flux from the compressor is used to cool down the combustor walls. Different operating regimes such as take-off, cruise, and landing would feature specific ranges for the actual air mass flow and of the air/fuel ratio. Requirements for combustor materials are:

- Oxidation/corrosion resistance. In the steady cruise regime, gas phase corrosion, i.e., oxidation and vanadation, may occur. During the engine shutdown and start-up phases, deposition of sodium sulfate may occur, introducing an additional surface instability process, i.e., hot corrosion (see Sect. 8.2.5).
- Thermal fatigue resistance.
- Thermal stability.

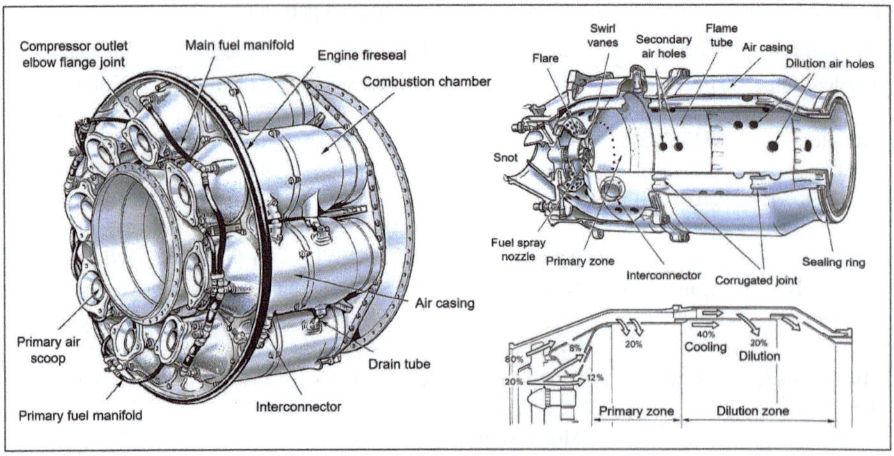

Fig. 2.17 Scheme of an early multiple-chamber system with detail regarding the single combustion chamber component and indication of the main functions of the inlet airflow coming from the compressor. (Rolls-Royce 1996, images courtesy of Rolls-Royce plc)

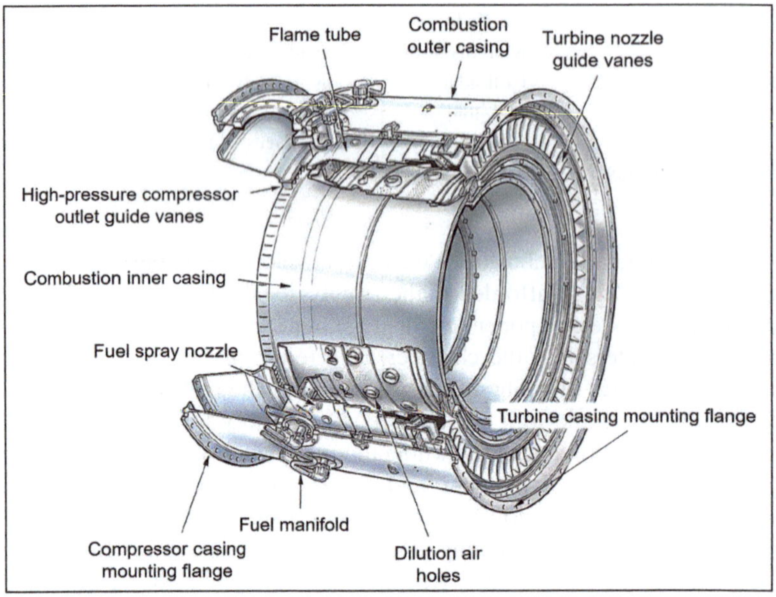

Fig. 2.18 Scheme of an annular combustion chamber. (Rolls-Royce 1996, image courtesy of Rolls-Royce plc)

According to these requirements, it is possible to identify some solutions that are highly reliable and, thereby, recommended for the fabrication of combustion chambers.

- Co-based superalloys (see Sect. 6.2). They usually display a lower strength than the companion materials, Ni-based superalloys (see Sect. 6.4). However, their strengthening mechanisms, mostly based on carbide precipitation, together with a comparatively lower coefficient of thermal expansion, render these materials preferable for the combustion chambers.
- Ceramic-based thermal barrier coatings (see Sect. 7.3.2). Coatings made of ZrO_2 with suitable stabilizer additions, like Y_2O_3 and ScO_2, are a standard choice in this respect; thanks to the thermomechanical compatibility with the underlying superalloy, they guarantee for these critical parts.

2.5 Turbines

The turbine converts the kinetic energy of the gases coming from the combustor into mechanical energy used to drive possible appliances and, most importantly, the compressor rotor stages and the fan, if present. In the turbine, a slight reduction, with respect to the combustion chamber, in the gas pressure and temperature takes place. The hot gases are at temperatures above 1500 °C, and the flow rate may exceed 750 m/s. The energy conversion process involves high stresses and high rotational speeds. Altogether, this renders the working conditions in the high-pressure stage of the GT extremely harsh. Vanes (static blades) and turbine blades of the first high-pressure stage are the most critical components of the whole engine, as concerns materials requirements (see Fig. 2.19). In fact, they are directly exposed to the jet of hot gases leaving the combustor, whose temperature (TET, see Fig. 2.20) has continuously increased over the years.

As depicted in Figs. 2.10 and 2.11, different architectures for jet engines are possible, with a different number of shafts and therefore of turbine stages. High bypass ratio fan engines may be equipped with a third intermediate-pressure turbine stage to operate the fan. This further stage is located between the high- and low-pressure turbines (*vide infra*, Fig. 2.22), according to a triple-spoon arrangement (Rolls-Royce 1996).

High peak temperatures and high-temperature gradients are not the only critical conditions for these turbine components, which have to withstand the high level of stresses, related to both the rotational motion and differential heating. Nickel-based superalloys (see Sect. 6.4) are the obvious choice, not only for the blades but also for other critical components of the turbine, requiring excellent strength associated with creep and fatigue resistance. Optimal performances have been achieved thanks to the use of newly developed casting techniques, leading to the manufacturing of directionally solidified and single-crystal components (see Sect. 6.5). It is important to highlight also the role of inner cooling channels. In cooled blades, the energy transferred from the flame to the material is removed so fast and efficiently that the alloy can even stand working temperatures that are important fractions of its incipient melting temperature.

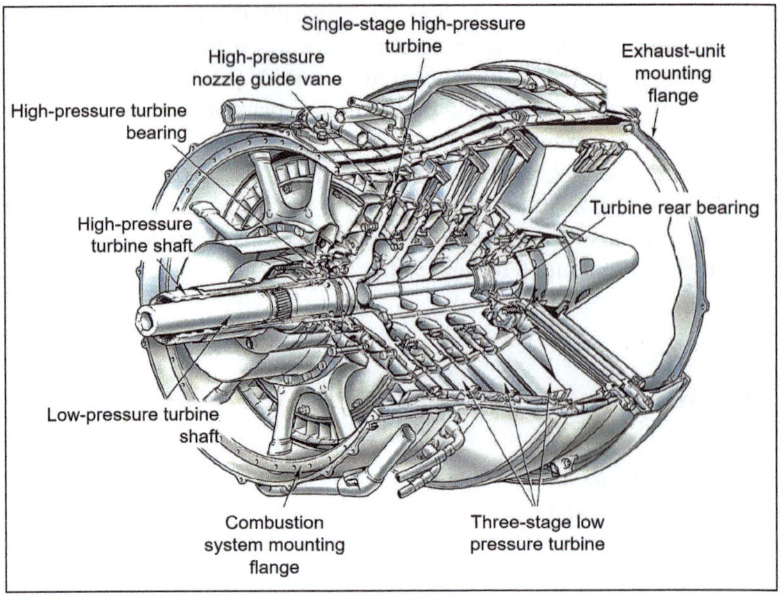

Fig. 2.19 Twin turbine and relevant double-shaft arrangement. (Rolls-Royce 1996, images courtesy of Rolls-Royce plc)

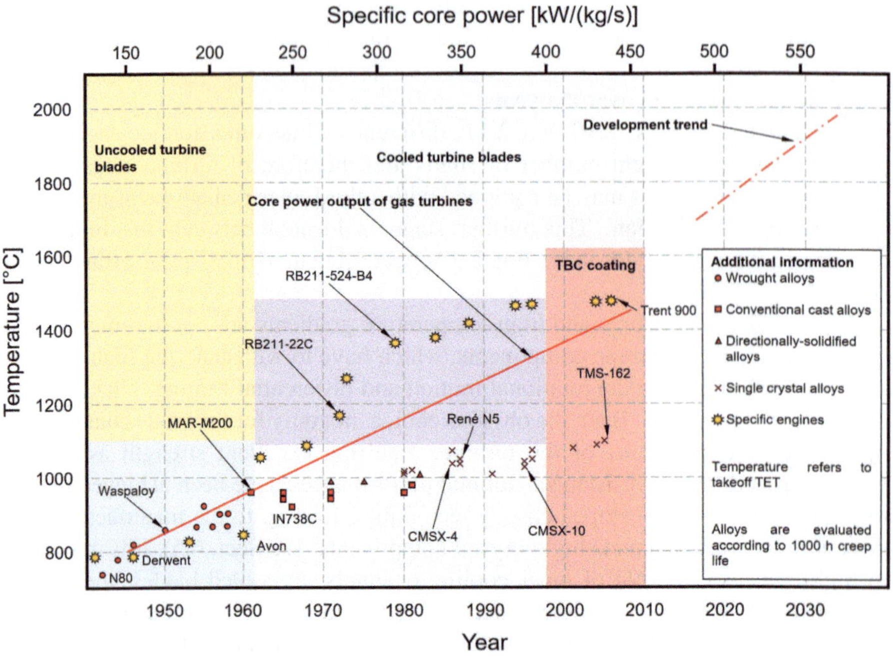

Fig. 2.20 Temperature capability of the blade superalloys and dependence also on the blade design: cooled or solid blades. In the graph the evolution of the take-off TET of civil aero-engines is considered. (Redrawn from Mukherji et al. 2011)

Eventually, surface reactions, induced by the interactions with the aggressive atmosphere, require an intrinsic resistance of the alloy against high-temperature corrosion (see Sect. 8.2). This feature can be improved with suitable coating systems. The already mentioned diffusion coatings, based on nickel aluminides, or the MCrAlY overlay coatings (M being Ni and/or Co and/or Fe) are excellent choices (see Sect. 7.3.1). As concerns coatings, turbine high-pressure blades have been in some engines also protected with thermal barrier coatings, using specific deposition techniques based on electron beam physical vapor deposition (see Sects. 7.2.2 and 7.3.2).

2.6 State of the Art and Future Trends

The aerospace industry has a great impact in terms of employees and output value. As an example, 14.9 billion USD in 2004 were invested in aviation gas turbine engines, and 3.7 billion USD have been used for military aircraft applications, while 11.2 billion USD have been invested in the field of commercial aviation (Giampaolo 2006). Aerospace industry permeates several other industries and is used as a fundamental benchmark for experimenting and testing new materials and technologies. Some of the most important driving forces pushing the efforts for innovation are associated with the improvement of aircraft performances, concerning safety and, in more recent times, pollution reduction. The reduction in emissions has become a major issue, considering that commercial aviation alone is responsible for the release in the atmosphere of 705 million of metric tons of CO_2 per year, corresponding to about 2% of the yearly global carbon emissions. Moreover, there is evidence that the greenhouse gases in jet fuel have greater effects on the atmosphere, since they are released at elevated altitudes (Martin 2016). The typical emission profile of a two-engine aircraft during 1 h flight with 150 passengers is given in Fig. 2.21. Pollutant substances and greenhouse gases in engine exhaust are present: CO_2, CO, H_2O, NO_x, SO_2, various hydrocarbons, and particulate matter, just to mention the main constituents. Some of these substances had attracted already investigation interests and research efforts, mainly in connection with their influence on corrosion (see Sect. 8.2) and erosion (see Sect. 7.3.5) aspects in gas turbine engines.

The European market is one of the largest for commercial aviation. According to the data reported by international agencies, like EU28 and European Free Trade Association (EFTA), and by the United Nations Framework Convention on Climate Change (UNFCCC), CO_2 emissions in the atmosphere have experienced a 77% growth between 1990 and 2005. Starting from the present aviation traffic forecast, notwithstanding the improvement rate of innovative technologies, an unavoidable 44% increase in CO_2 emissions from 2005 to 2035 period (European Commission 2016) is expected. For this reason, several international programs aiming at reducing atmospheric pollution are being conducted. ACARE (Advisory Council for Aviation and Innovation in Europe) Vision 2020 and Flightpath 2050 are two examples of European collaborative programs that set the targets for the future of the air

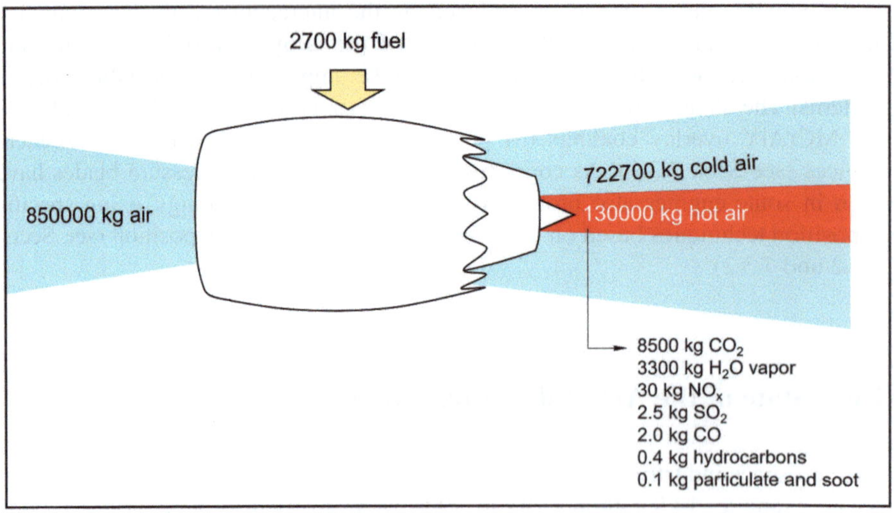

Fig. 2.21 Typical emission profile of a two-engine aircraft during 1 h flight with 150 passengers. (Redrawn from European Commission 2016)

transportation, also as concerns climate change and atmospheric pollution reduction. The two programs will set ambitious goals in terms of emissions cut:

- 90% reduction in NO_x emissions (15% achieved through operational efficiency and 75% from technology improvements)
- 75% reduction in CO_2 emissions per passenger per kilometer
- 25–30% overall CO_2 emissions reduction

For meeting these targets, multidisciplinary approaches are necessary. The main pillars on which research efforts are based can be summarized as follows:

- Average improvement in fuel efficiency, by at least 1.5% per year from 2009 to 2020
- Development of a more efficient air traffic management system
- Improvement in aircraft operational efficiency
- Technology improvement including sustainable low-carbon fuels and biofuels

As concerns low-carbon fuels, in 2009 the incidence of biofuels in the total fuel consumption was 0.05% only. A widespread use of biofuels for aircraft engines is not that easy to achieve, first of all due to the competitive demand by other transportation sectors. Furthermore, the hesitation of the industry concerning investments in dedicated production facilities for biomass to liquid fuels transformation can be justified by the limited foreseen supply of 0.05 million tons by 2020 (European Commission 2016). Therefore, biofuels introduction is still far from its potential target, and the main approach behind the development of more environmental-friendly jet engines is rather through the improvement of their efficiency. This has been highlighted already in the previous sections of this chapter, with reference to the improvement in the engine performances demonstrated by the evolution of parameters like TET, OPR, SFC, and bypass ratio.

However, increasing TET will result in higher emissions that can be limited by the adoption of catalytic combustors, capable to reduce the NO_x formation in the combustion chamber directly. Furthermore, higher TET would require the development of more resistant thermal barrier coatings and more efficient cooling design for the components, in order to withstand the more demanding operating temperatures. There are several examples of newly developed engines with improved performances. One of these is the Pratt & Whitney F135 engine for the F35 military aircraft. This engine is characterized by an OPR of 28 and is capable of 190 kN of takeoff thrust and, in general, maximum efficiency under different operational conditions: conventional take-off/landing (CTOL), carrier variant (CV), and short take-off/vertical landing (STOVL) (Pratt and Whitney 2012). In commercial aviation the latest Airbus A380 is powered by the General Electric-Pratt & Whitney GP 7000 or by the Rolls-Royce Trent 900 engines (Giampaolo 2006). The GP 7000 has a maximum thrust of 340 kN with an OPR of 43.9 and a bypass ratio of 8.7:1 (MTU Aero Engines 2016). The Trent 900 is a three-shaft turbofan engine capable of a maximum thrust of about 342 kN. Its architecture consists of an eight-stage intermediate-pressure compressor, a six-stage high-pressure compressor, a single-stage high-pressure turbine, a single-stage intermediate-pressure turbine, and a five-stage low-pressure turbine. Its OPR is 39, and the bypass ratio reaches a value of 8.5:1 (Rolls-Royce 2015). A further evolution of this engine, the Rolls-Royce Trent XWB (see Fig. 2.22) is a three-shaft turbofan engine capable of a thrust value between

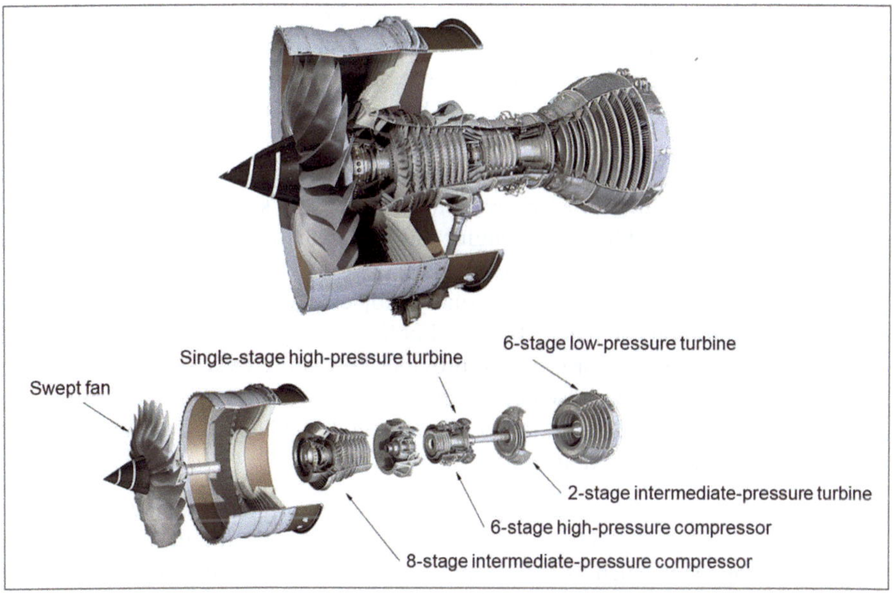

Fig. 2.22 Modern Rolls-Royce Trent XWB. (Rolls-Royce 2016, images courtesy of Rolls-Royce plc)

374 and 432 kN (84,000–97,000 lbf). A large diameter swept fan and an eight-stage intermediate-pressure compressor followed by a six-stage high-pressure compressor characterize the front part of the engine. The combustor has an innovative design, incorporating advanced ceramic coatings that can withstand combustion temperatures exceeding 2700 °C. A six-stage low-pressure turbine follows a single-stage high-pressure turbine. The power generated by the turbine system can reach values as high as 50,000 hp. Its overall pressure ratio (OPR) can climb up to 50, well above those given in Fig. 2.8. The bypass ratio of the engine is 9.6:1. These features allow the Trent XWB to improve several performances, including a fuel consumption reduction by 10% (Rolls-Royce 2016).

Future developments are already on their way, and the engine of the future should be lighter, less noisy, and more energy-efficient than conventional engines used today. Pratt & Whitney developed a new engine that might save up to 20% of fuel, with respect to the present average engine. One of the features that enable to achieve this result is a gearbox that slows down the fan. Meanwhile, CFM International is working on a conventional turbofan architecture engine that will reach similar results but without adding weight and drag through a gearbox. The increase in efficiency is obtained thanks to lightweight composite materials used in several components, such as carbon fiber-based fan blades (Martin 2016). Eventually, Rolls-Royce is developing an innovation plan, based on Advance® and UltraFan® engines (see Fig. 2.23).

Advance® will be ready for the market by 2020 and can be defined as a hybrid engine that will be manufactured with novel materials, like the light ceramic matrix composites (CMCs), 3D-printed components, and the innovative low-pressure system with carbon-titanium (C-Ti) composite fan system, including fan casing and blades. The engine will provide a 20% better efficiency as compared to the early Trent engine, fewer parts, and lower weight. UltraFan® engine will be available starting from 2025. One of the most impressive features of this engine will be its OPR, reaching a value of 70:1. It will be equipped with an innovative gearing system, i.e., *geared* architecture, located between the fan and the intermediate-pressure compressor. This system will be able to ensure that the fan runs at the optimum speed in any condition. Further improvements in the C-Ti composites will allow the removal of the thrust reverser and, thereby, a slim-line nacelle system. When compared with the first Trent engine, the UltraFan® engine will provide a 25% increase in fuel efficiency (Rolls-Royce 2017).

Aircrafts and their components are usually subjected to research and development cycles, whose duration can reach 20 years. For this reason, research efforts are based on a long-term planning approach that provides continuity over several years. The European Union Clean Sky research program comprises several projects aiming to meet both performance increase and environmental challenges. Two examples of R&D projects in this context are:

- Advanced low-pressure systems (ALPS). These engines are intended to demonstrate the feasibility of innovative lightweight materials for the production of engine components such as C-Ti composite fan blades and coatings. The use of these materials will be expected to lead to a weight saving of around 700 kg on a twin-shaft engine.

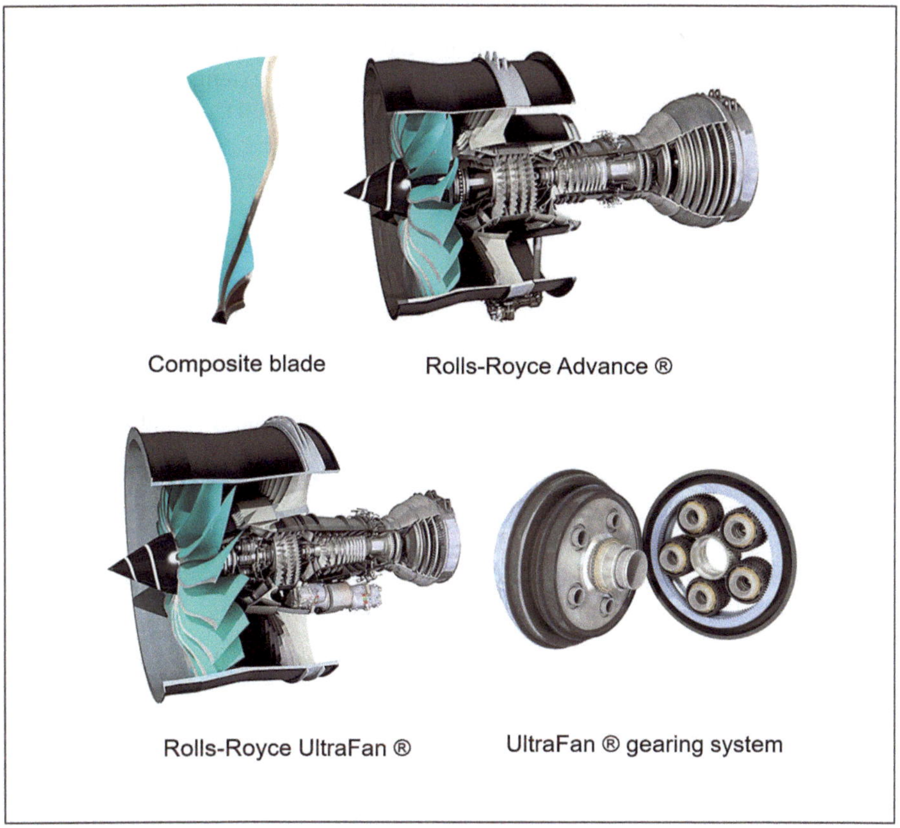

Fig. 2.23 Future Rolls-Royce engines. (Rolls-Royce 2017, images courtesy of Rolls-Royce plc)

- Counter-rotating open rotor engines (CROR). This aero-engine family should reach a 30% reduction in emissions compared to a year 2000 benchmark engine. The relevant noise level should be comparable to that of a modern turbofan engine. Airbus and Rolls-Royce are currently involved in research programs on this propulsion system (European Commission 2016).

CROR and ALPS are just two of the many research programs currently under development. Engine producers are now facing many other challenging tasks in the search for the "best engine:"

- Advanced cycles.
- Airframe optimization.
- Hybrid systems with separate power and thrust generation. They are turboelectrical hybrid propulsion system with a gas turbine for electricity generation and electrically driven fans coupled to batteries for energy storage.
- Pressure gain combustion.

Considering the above scenario, it is possible to conclude that gas turbine engines will still have a lot of potential in the future of aircraft propulsion. Furthermore, they will generate the power for aircraft propulsion to a great extent, possibly through hybrid systems and electrically driven engines (Wenger 2014).

References

Boyce M P (2012) Gas Turbine Engineering Handbook 4th edn. Elsevier, Oxford

Cumpsty N (2003) Jet Propulsion: A Simple Guide to Aerodynamic and Thermodynamic Design and Performance of Jet Engine 2nd edn. Cambridge University Press

European Commission (2011) Flightpath 2050 Europe's Vision for Aviation. Available via DIALOG. http://ec.europa.eu/transport/modes/air/doc/flightpath2050.pdf. Accessed 17 Aug 2017

European Commission (2016) European Aviation Environmental Report 2016. Available via DIALOG. https://ec.europa.eu/transport/sites/transport/files/european-aviation-environmental-report-2016-72dpi.pdf. Accessed 17 Aug 2017

Giampaolo A (2006) Gas Turbine Handbook Principles and Practices. The Fairmont Press Inc., Lilburn

Gunston B (1995) The Development of Jet and Turbine Aero Engines. Patrick Stephens Ltd Publ

Kear B H (1986) Advanced Metals. Scientific American 220: 130–139

Martin R (2016) The Race for the Ultra-Efficient Jet Engine of the Future. In: Sustainable Energy. MIT Technology Review. https://www.technologyreview.com/s/601008/the-race-for-the-ultra-efficient-jet-engine-of-the-future/. Accessed 15 Aug 2017

Meetham G W (1981) The Development of Gas Turbine Materials. Applied Science Publishers Ltd, London

MTU Aero Engines (2016) GP7000 turbofan engine. Available via Dialog. http://www.mtu.de/fileadmin/EN/7_News_Media/2_Media/Brochures/Engines/GP7000.pdf. Accessed 07 Sept 2017

Mukherji D et al (2011) Beyond Ni-based Superalloys: Development of CoRe-based Alloys for Gas Turbine Applications at Very High Temperatures. International Journal of Materials Research 102(9): 1125–1132

Pratt & Whitney (2012) F135 Specs Charts. Available via DIALOG. http://www.pw.utc.com/F135_Engine. Accessed 06 Sept 2017

Reed R C (2006) The Superalloys-Fundamentals and Applications. Cambridge University Press, New York

Rolls-Royce (1996) The Jet Engine 5th edn. Rolls-Royce plc, Derby

Rolls-Royce (2015) Trent 900. Available via DIALOG. https://www.rolls-royce.com/products-and-services/civil-aerospace/airlines/trent-900.aspx#training. Accessed 07 Sept 2017

Rolls-Royce (2016) Trent XWB. Available via DIALOG. http://www.rolls-royce.com/site-services/images/trent-xwb-infographic.aspx. Accessed 17 Aug 2017

Rolls-Royce (2017) Pioneering intelligent innovation for our customers. https://www.rolls-royce.com/products-and-services/civil-aerospace/future-products.aspx#/. Accessed 23 Aug 2017

The Washington Post (2017) The next generation of eco-friendly airplanes has arrived. . .quietly. http://www.washingtonpost.com/sf/brand-connect/wp/enterprise/the-next-generation-of-eco-friendly-airplanes-has-arrivedquietly/. Accessed 17 Aug 2017

Wenger U (2014) Rolls-Royce Technology for the Future Aircraft Engines. Lecture for RAeS, DGLR, VDI at the Hamburg University of Applied Sciences, Hamburg, 20 March 2014

Further Reading

Connors J (2010) The Engines of Pratt & Whitney. American Institute of Aeronautics and Astronautics, Inc.
Hill P G, Peterson C R (1992) Mechanics and Thermodynamics of Propulsion 2nd edn. Pearson
Kerrebrock J L (1992) Aircraft Engines and Gas Turbines 2nd edn. The MIT Press
Soares C (2015) Gas Turbines: A Handbook of Air, Land and Sea Applications. Elsevier

Chapter 3
Alloys for Aircraft Structures

3.1 Introduction

As seen in Chap. 1, the Flyer I of the Wright brothers was mostly made of wood and fabric. However, in addition to steel, used in the wing struts and the bracing wires, an aluminum alloy was used for the crankcase of the piston engine, in order to reduce its weight. So, aluminum started to fly since the very early beginning of modern aviation. After this start, aluminum attracted further interest and research efforts, since the need for new materials capable to resist the mechanical and thermal stresses associated with the increasing flight performances emerged soon (see Chap. 1). Moreover, the shortage of wood, particularly spruce and willow tree, caused by the massive aircraft production during World War I, created the conditions for the development and use of *light alloys* in aircraft structures. *Light* refers to the comparatively low densities of alloys based on elements like aluminum, magnesium, and titanium. Although definitely higher than the average density of wood, which was the main structural material of the early airplanes, the densities of aluminum alloys are still absolutely acceptable, even considering their effect on the specific structural properties, i.e., properties averaged by the material density. Aluminum alloys were among the first metallic alloys used in aircraft structures, and they still are the most used alloys in this field. Magnesium alloys are other interesting systems, particularly now that compositions displaying acceptable corrosion resistance and interesting mechanical properties have been developed. These alloys are nowadays gaining an increasing importance in the aeronautic field, where actually had already been used in the past. For them, some issues are still standing, like the search for mechanically workable compositions, which are introducing some obstacle to a more rapid widespread use of these materials. Titanium alloys, owing to their specific properties, are used not only for aircraft structures but also for engine parts. Therefore, they will be treated separately in Chap. 4. Laminate composite materials instead are included in this chapter. Many of them are based on aluminum alloys and represent a very important class of aerospace structural materials.

S. Gialanella, A. Malandruccolo, *Aerospace Alloys*, Topics in Mining, Metallurgy and Materials Engineering, https://doi.org/10.1007/978-3-030-24440-8_3

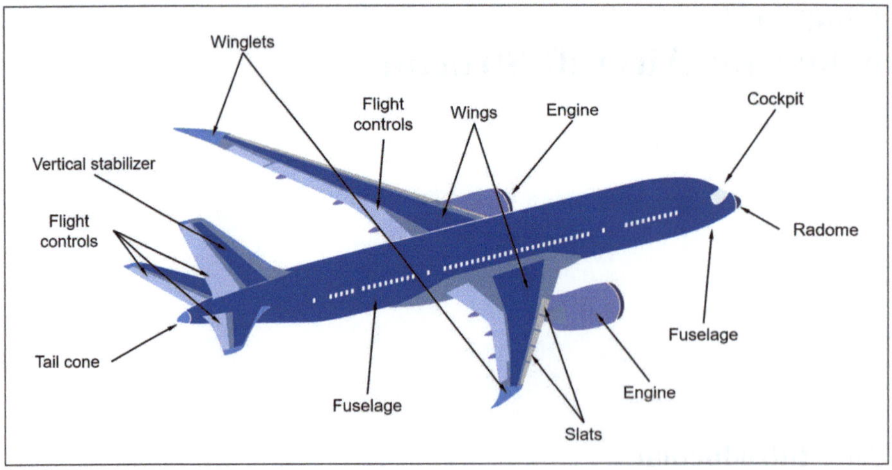

Fig. 3.1 Structure and main components of a commercial aircraft. (Redrawn from Boeing 2008)

The structure of modern aircrafts, although recalling to some extent early airplanes, exhibits obvious aerodynamic optimization. Commercial and military aircrafts share the airframe shape and the denomination of the main components (Fig. 3.1), although major differences in size, geometry, and structural properties are necessary to comply with the specific operational requirements.

The main structural parts of an airframe are the fuselage, the wings, the landing gear, and the flight control surfaces.

The wings are meant to provide the necessary lift but are also subjected to other mechanical stresses, like, for instance, the weight of fuel and engines. Their design and internal structure are therefore conceived to fit the specific needs of any aircraft (Fig. 3.2). Similar principles apply, obviously with different parameters, to fuselage, vertical stabilizers, and all the other parts of an aircraft (Fig. 3.3). In general terms, the components whose failure would endanger the safety of the whole aircraft are called *primary structures*, being *secondary structures* all the other. Fuselage, wings, empennage, and landing gears are typical *primary structures*, and for some of them, aluminum alloys are an excellent choice.

3.2 Aluminum Alloys

3.2.1 Introduction

Aluminum is a post-transition metal with atomic number 13 (13th group, 3rd period) of the periodic table. It is characterized by a face-centered cubic (*fcc*) crystal lattice, a density of 2.70 g/cm^3, and a Young's modulus of 70 GPa. The name aluminum derives from the term *alum*, used by Greeks and, afterward, by the Romans with

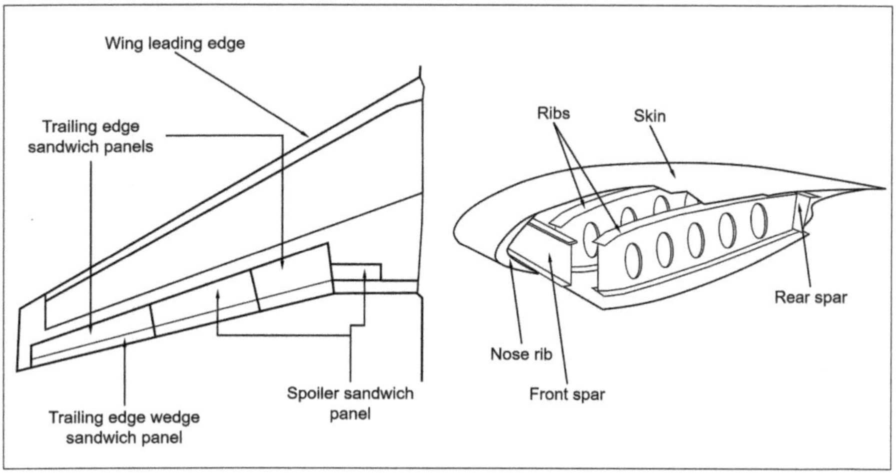

Fig. 3.2 Wing components and structure. (Redrawn from Federal Aviation Administration 2012)

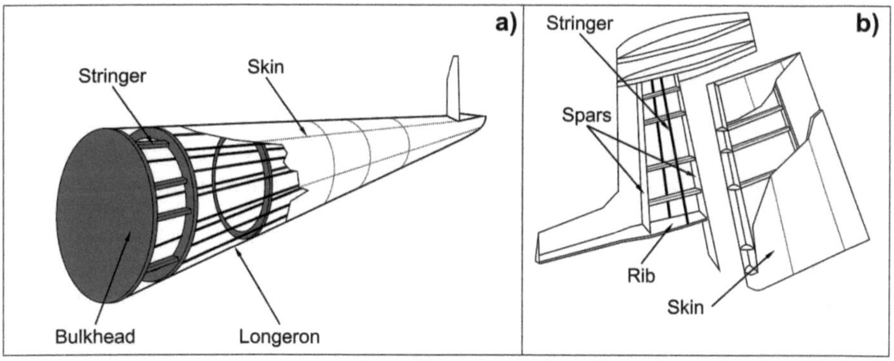

Fig. 3.3 (**a**) Semimonocoque tail part of the fuselage; (**b**) structure of the vertical stabilizer. (Redrawn from Federal Aviation Administration 2012)

reference to a bitter salt, based on aluminum double sulfate hydrated (e.g., KAl $(SO_4)_2$ H_2O), used in medicine as an astringent but also for skin tanning and glass production. The first who actually identified the presence of aluminum in alum salt was Sir Humphry Davy at the beginning of 1800, although he was not able to isolate it. In 1807, he proposed the name *aluminium* for the metal. Shortly after, the name aluminium was adopted by the IUPAC. However, in 1925 the American Chemical Society introduced the name *aluminum*, still used in the American English literature (Totten et al. 2003). The driving force for the development of Al-alloys for aerospace structures has been the improvement in static strength, as well as fracture toughness and resistance to crack growth, particularly in fatigue regimes, damage tolerance, and stress corrosion. As shown next, an incremental progress and an alloy-targeted design have been particularly successful so that Al-alloys are still strategic materials

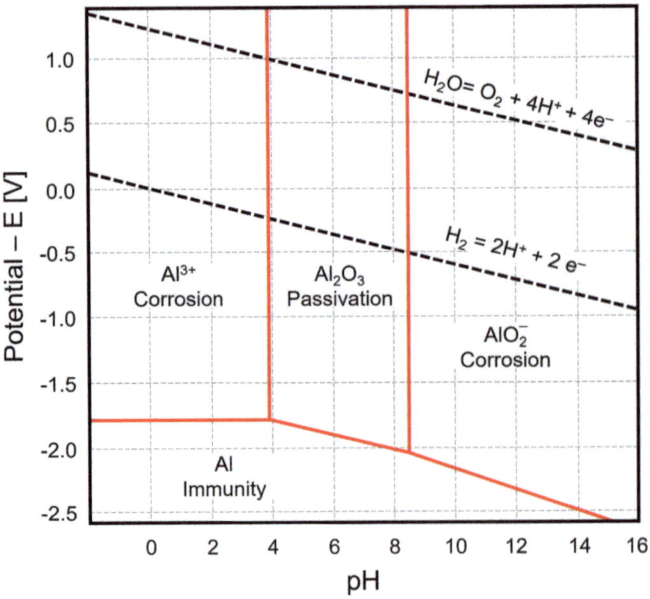

Fig. 3.4 Pourbaix diagram of pure aluminum. (Redrawn from McCafferty 2010)

for aerospace industry. This is also due to common features of these alloys, which can be largely ascribed, and to some extent retrieved, in pure aluminum, by far their majority component:

- Acceptable corrosion resistance.
- Good fatigue resistance, due to the high intrinsic ductility of the *fcc* structure given by the high number of slip systems.
- Good oxidation resistance.
- Good specific mechanical resistance.
- High specific stiffness ($E = 70$ GPa, average density of the alloys = 2.768 g/cm^3).

In standard conditions, the presence of a self-passivating and self-healing Al_2O_3 surface layer is the main reason for the fairly good surface stability of aluminum alloys, resulting in their good resistance to dry oxidation.

For similar reasons, it is possible to see from the Pourbaix diagram in Fig. 3.4 that Al-alloys still provide a positive response in not too aggressive corrosive environments, like aqueous solutions with a pH in the 3.9–8.6 range that do not determine localized corrosion phenomena, like pitting and crevice corrosion, discussed in Chap. 8 (ASM International 2003). As concerns fatigue resistance, Al-alloys are extremely sensitive to the presence of hard inclusions, which may act as stress concentrators. An intrinsic beneficial factor, as concerns fatigue resistance, is the face-centered cubic structure of aluminum and relevant large number of slip systems that help in accommodating local stresses by plastically straining, thus keeping their level relatively low.

The availability of aluminum and its alloys has remarkably influenced the aerospace industry since the beginning. Aluminum alloys were used during the World War I for the production of structural parts of the Zeppelin dirigibles and during the World War II for military and civil aircrafts. In these early applications, the hardenable Al-3.5Cu-0.5Mg-0.5 Mn alloy, named Duralumin, developed in the 1906 by the German chemist Alfred Wilm, was used. Hardening of Duralumin was attained thanks to a precipitation treatment, which induced in the alloy properties interesting not only for aerospace but also automotive applications (Lumely 2011). Notwithstanding the intrinsic corrosion resistance, a critical aspect of this alloy, that emerged soon in outdoor applications, was its tendency to exfoliation corrosion (see Sect. 8.1.4), which of course would not be acceptable for any application involving a long-term exposure to external atmosphere and meteorological events.

The problem was successfully solved with two alternative approaches, developed in the late 1920s: cladding the aluminum alloy with pure aluminum sheets and alloy protection through the formation of a surface film made of reaction products, using an anodizing process. Concerning this latter process, by means of appropriate bath composition and deposition parameters, it is possible to produce protective surface layers thicker than those forming spontaneously by the self-passivation tendency of the alloy. Moreover, by tuning the composition of the bath, it is possible to obtain deposits containing chemical agents, actively enhancing the corrosion resistance (Thompson et al. 1999). Thanks to these fundamental improvements, aluminum alloys rapidly became, and still are, the most used materials for aircraft structures, as outlined in the following.

Concerning the production of Al-alloys, raw materials usually are oxide-based ores with various grades of purity and levels of hydration. Bauxite, from the French village Les Baux, where alumina was first discovered, is the most important example. It is a sedimentary formation containing different aluminum hydroxides, like $AlO(OH)$ and $Al(OH)_3$. Chemical and mechanical enrichment of the ores is a common practice to increase the concentration of Al_2O_3 available to extraction (Totten et al. 2003). The main steps of aluminum extraction, still based on the Bayer process, developed and patented by Karl Bayer in 1888, are illustrated in Fig. 3.5. The output of the first step is hydrated alumina: $Al_2O_3 \cdot H_2O$. Pure aluminum requires the electrolytic reduction of Al_2O_3 dissolved in molten cryolite (Na_3AlF_6, Hall-Héroult process, see Fig. 3.5). Charles Hall and Paul Héroult developed independently this process in 1886, in Ohio (USA) and France, respectively (ASM International 1991). The overall electrochemical reaction involved in the extraction of aluminum is:

$$\frac{1}{2}Al_2O_3(\text{dissolved}) + \frac{3}{4}C(s) \rightarrow Al(l) + \frac{3}{4}CO_2(g) \quad (3.1)$$

The electrolytic dissolution process is conducted at about 960 °C in the Hall-Héroult cells. Typically, for each kilogram of primary aluminum, 1.93 kg of Al_2O_3 is needed, while 0.40–0.45 kg of C and 1.5 kg of CO_2 are eliminated as secondary products (Lumely 2011).

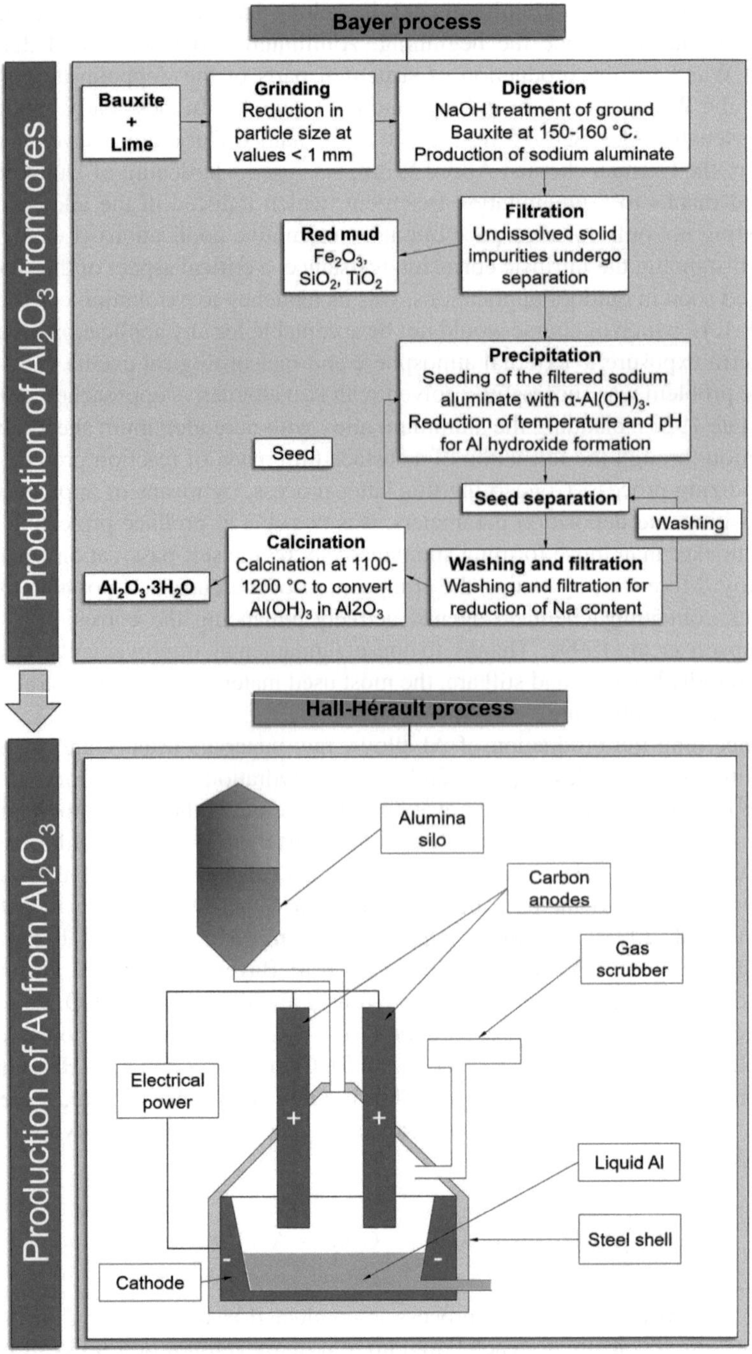

Fig. 3.5 Aluminum production route from ores to primary aluminum (Carter and Norton 2007; Lumely 2011)

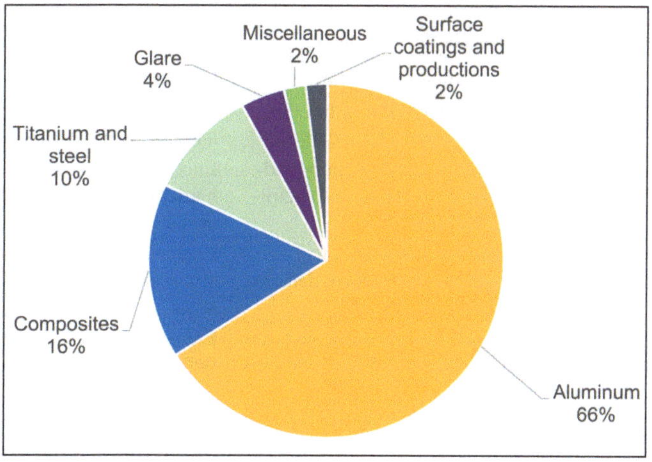

Fig. 3.6 Percentages of different materials used for the A380 Airbus, involving and promoting the development of the new materials and relevant processing routes. (Redrawn from Williams and Starke 2003)

Components made of aluminum alloys can sum up to more than a half of the total weight of a commercial airliner (see Fig. 3.6). Moreover, the possibility of varying and tailoring alloy's properties by suitable compositional changes, processing routes and post-processing thermomechanical treatments, is regarded as an attractive aspect of these materials, in addition to low density, a comparatively low cost, good availability, ease of manufacturing, reliability, and predictability of the in-service performances. Indeed, the leading role of Al-alloys in aerospace applications is mainly due to very well established alloy design principles. It is possible to obtain materials with specific properties for the different part requirements (see Table 3.1).

The density of Al-alloys, a primary selection criterion for these and other aerospace materials, varies from 2.643 g/cm^3 for commercially pure aluminum up to 2.803 g/cm^3 for highly alloyed compositions, like Al-Zn alloys. Metal working technologies, like milling, extrusion, and forging, lead to work-hardened products with higher specific strength (Starke and Staley 1996). The development occurring over the years that followed the end of World War II required a change of strategy after the accidents involving the Comet airliners at the Ciampino airport in Rome (1954, see Chap. 1). A premature fatigue failure was the main reason for the two accidents. However, an insufficient damage tolerance of the Al-Zn alloy, namely, the 7075 material used for the airframe of the Comet, was regarded as an additional factor behind the disasters. Therefore, these 7XXX alloys (see Sect. 3.2.3) were banned for several years from aerospace applications, and lower strength, but more damage-tolerant, Al-Cu alloys (precisely the 2024-T3 and 2014-T6) were used instead, with the exception for those components requiring highest tensile strength. The progress of commercial aviation has been accompanied by the request for larger payloads, larger number of passengers, and, thereby, larger airliners. Thicker sections of the structural parts started to be produced. At the same time,

Table 3.1 Summary table of the main properties of Al-alloys that can be attained by changing the composition and/or the microstructure of the alloy

Property	Desired microstructural feature	Function
Creep resistance	Thermally stable particles within the matrix and on the grain boundary	Inhibition of grain boundary sliding and coarse microstructure formation
Ductility and toughness	Fine structure with clean grain boundaries and absence of shearable precipitates or large particle	Encourage plasticity and work-hardening. Inhibition of void formation and growth
Fatigue crack initiation resistance	Fine grain size with absence of shearable particles. No surface defects	Prevent strain stress concentration, strain localization, and surface slip steps
Fatigue crack propagation resistance	Large grain size with shearable particles and no anodic phases or hydrogen traps	Encourage crack closure, branching, deflection, and slip reversibility
Pitting	Absence of anodic phases	Prevent preferential dissolution of second phase particles
Strength	Fine grain size with uniform dispersion of small, hard particles. Fine structure with clean grain boundaries and absence of shearable precipitates or large particles	Inhibition of dislocation motion
Stress corrosion cracking and HE	Hard particles and no anodic phases or interconnected hydrogen traps	Homogenization of slip and prevention of crack propagation due to anodic dissolution or HE

HE hydrogen embrittlement (Williams et al. 2003)

the incidence of damages due to stress corrosion started to be observed, with micro-cracks developing particularly when the load was applied along the short transverse direction. The most used alloys were 2014-T3, 7075-T6, and 7079-T6, where T3 and T6 represent two specific tempers, used to refine microstructure and relevant properties of aluminum alloys, as described in upcoming sections (Tables 3.9 and Table 3.10). Subsequently, thanks to alternative tempers for the 7075 alloy, codenamed T73 and T76, its stress and exfoliation corrosion problems were largely solved. These two tempers involve combinations of artificial aging at low and, as a subsequent step, at high temperature. The high-temperature treatment would generally result in a slight overaging of the alloy, with a consequent softening. However, the alternative beneficial effect of the treatment is the strengthening of the grain boundary region that enhances the fracture toughness of the alloy. This effect is determined by the enrichment of the grain boundary with Cu-rich precipitates. Nobler precipitates are less prone to microstructural changes and provide more effective obstacles to crack propagation. As an indirect toughening mechanism, there is also the lower affinity of the grain boundary precipitates and of the metallic matrix itself to hydrogen, whose embrittling action is thus neutralized or largely limited. In the A380, for assembling parts of the fuselage, laser beam welding (LBW) is used, with consequent requirements for a good weldability of the alloys. For this reason, Al-Mg-Si (6XXX alloys, see Sect. 3.2.3) materials have been considered. They exhibit excellent welding properties, and, in addition to that, they have a comparatively lower density and price, as compared to Al-Cu materials. The reference alloy 6056-T78, featuring a relatively low density, i.e., 2.702 g/cm^3, in comparison with 2.781 g/cm^3 of the 2024 alloy, is also a laser-weldable product, developed at a later stage to comply with the elevated damage tolerance requirements. The outcome of the evolution of this material was the 6156 alloy, extensively used in lower shell fuselage applications. It is worth saying that also the temper (T78) developed for the 6XXX alloys played an important role in overcoming the problem of intergranular corrosion, which was observed in T6-treated parts made of the same materials. In the A380, notwithstanding the availability and actual use of other materials, like composites used already in the Boeing 787 Dreamliner (see Fig. 1.11), Al-alloys still retain a leading role, as shown in Fig. 3.6.

An indirect proof of the technological reliability of standard aluminum alloys, like the 2024 and 7050 materials, is provided by the improvements attained thanks to new processing and manufacturing steps, still using the same alloy compositions. This choice had to be made on several occasions to comply with requests from aircraft producers in terms of better properties but also for a faster and, therefore, cheaper production. Larger monoliths, machined from bulk thick plates and beams, have replaced assemblies and built-up structures. Machining may involve the removal by up to 95% of the original volume of the part. On the other hand, stress-free components can be obtained in this way, and the scraps can be recycled straightaway by remelting. Through the optimization of the alloy composition, in association with controlled conditions during temper treatments, important enhancement in alloy response to rolling can be achieved. Larger and thinner Al-alloy sheets are very much interesting for the process and manufacture optimization that they can bring about (Heinz et al. 2000).

Another issue that promoted important research efforts for developing special aluminum alloys was the expected heating associated with air drag in supersonic flight, which started to be considered also for commercial airliners. The Concorde project, started in 1962, required the development of quite stable alloys, not affected by any kind of long term, in-service aging, due to peak temperatures close to 100 °C, for exposure times of 20,000 h, out of the 50,000 h expected total service lifetime, at the highest Mach 2 speed. As seen in Chap. 1, based on the experience gained with military aircraft, Mach 2 is still a speed affordable to Al-alloys. The starting point of this material development was an alloy, codenamed RR58, originally designed for components of the compressor of the early gas turbine jet engines. The alloy, that in the end was employed for the fuselage panels and other structural parts of the Concorde, has a modified composition with respect to the parent material, i.e., the 2618 alloy, in which copper was the main alloying element. In particular, the ratios of copper to iron and nickel were fixed in the deliberate attempt to limit the formation of equilibrium intermetallics, typical of the overaging condition of these hardenable alloys. Moreover, an accurate microstructural refinement, inducing large grain size, was pursued to optimize creep resistance and to reduce exfoliation and stress corrosion. Subsequently, the alloy development regarded other critical mechanical properties, like strength and fracture toughness, leading to the design of several alloys not only of the 2XXX (Al-Cu, see Sect. 3.2.3) but also of the 7XXX (Al-Zn,

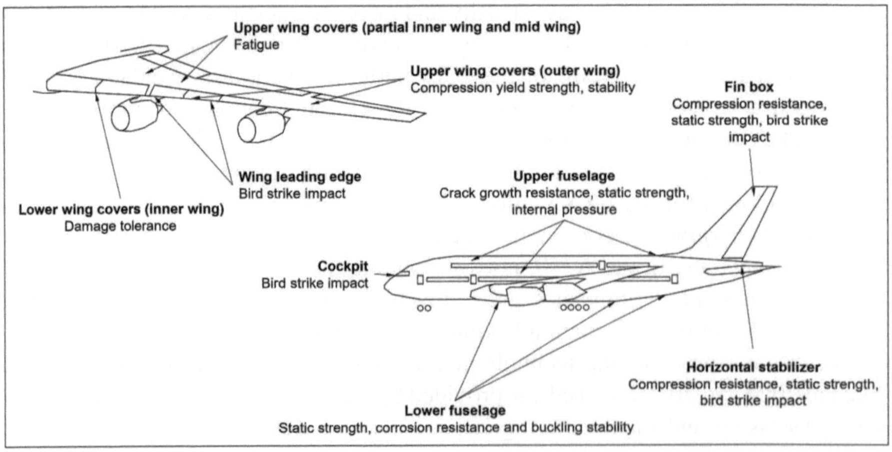

Fig. 3.7 General design aspects for the A380 Airbus aircraft, involving and promoting the development of the new materials and relevant processing routes. (Redrawn from Lequeu et al. 2007)

see Sect. 3.2.3) series, like the 7050-T74, 7050-T76, and 7150-T6, particularly recommended for the upper wing skin. For the Al-Cu series, it is worth mentioning the 2324-T39 and 2224-T351X materials, developed for the lower wing structures. The more recent developments of these alloys have been triggered by the design requirements for large size airliners, like the Airbus A380. Larger size and loads involved important improvements in static mechanical strength and damage tolerance, to be developed according to the leading selection criteria summarized in Fig. 3.7 for each specific component (Lequeu et al. 2007).

Moreover, thanks to a progressive implementation approach in alloy development, novel material products came up that could be processed relying on the knowledge background accumulated over the years. This is a fundamental prerequisite for a rapid technological transfer from research to applications. In this regard, to complete the picture of the main milestones in the development of aluminum-based materials, Al-Li alloys will be mentioned as well, together with the most recent class of aluminum alloys, containing scandium among alloying elements, and laminate composites, using aluminum as main metallic constituent, as illustrated in Sect. 3.4.

Al-alloys are generally divided into two large families: cast (see Sect. 3.2.2) and wrought (see Sect. 3.2.3) alloys. A great number of both cast and wrought alloys respond to thermal treatments that are essentially based on phase solubility, i.e., solutioning, quenching, and precipitation, and age hardening. These alloys are said to be *heat treatable* (or *hardenable*), since different phases and, thereby, different properties can be induced into the alloy through the control of the solid-state precipitation. Duralumin, the Al-3.5Cu-0.5Mg alloy (alloy 2017, according to the code that will be introduced in Sect. 3.2.3), already mentioned with reference to the early use of an aluminum alloys in aircraft structures (see Sect. 1.1), can be regarded as the prototype of the heat treatable alloys. If the composition of the alloy does not offer this chance and no precipitate phases can be formed in the alloy matrix, the alloy is said to be *non-heat treatable*. Both wrought and cast alloys can be classified according to an alloy nomenclature system. This coding system was developed by the Aluminum Association (ASM International 1991). The nomenclature of both wrought and cast alloys, used already in some part of this section dealing with specific alloys, will be discussed in details in the relevant sections of this chapter. The aluminum parts and components of aircrafts are supplied either as wrought or cast products, these latter typically requiring some machining and surface finishing before installation. As already mentioned, the fuselage, the wings, the empennage, and the supporting structure of an aircraft are regarded as *primary structures*, since their failure would endanger the whole aircraft. Usually, wrought alloys are used for these parts, although for military airplanes highly controlled casting products are on some occasion employed. For *secondary structures* both cast and wrought products are used, depending on the shape and required part finishing.

Before entering this broad subject, a few bibliographic notes, which might interest the reader to widen the knowledge on this most important class of alloys, concern some of the properties of aluminum alloys that render these materials excellent candidates for aerospace applications and to complete the picture emerging from the former sections. Metal handbooks, such as that of ASM International (2015), offer a wide description of Al-alloy properties, relevant compositions, and tempers developed to attain alloy properties suitable for specific applications. Another essential reading is the handbook edited by Davis (2001). The textbook of Polmear (2006) is particularly recommended for a comprehensive treatment of the main themes concerning aluminum but also other *light alloys*, like magnesium, titanium, and derived alloys. A particular focus on the aerospace applications of all these *light alloys* can be found in the monographic chapter by Prasad and Wanhill (2017).

3.2.2 Cast Alloys

Aluminum and its alloys can virtually be subjected to any casting and forming process. However, early casting of aluminum has been characterized by several problems. Because of gas pick-up during melting, castings were often characterized by the presence of porosity and other defects, easily visible also on eye inspection. Porosity in aluminum castings is still an issue, and several techniques have been developed over the years to control its impact on the final products, if not to eliminate it at all. Some examples include degassing, filtration, improved handling of the molten alloy, even using low pressures, and solidification simulation, for predicting the potential effects of technological changes. Thanks to these contributions, aluminum is now regarded as one of the more versatile among the foundry metals, even considering the continuous progress in alloy design (ASM International 1991; Lumely 2011; Polmear 2006). Since casting techniques allow for the

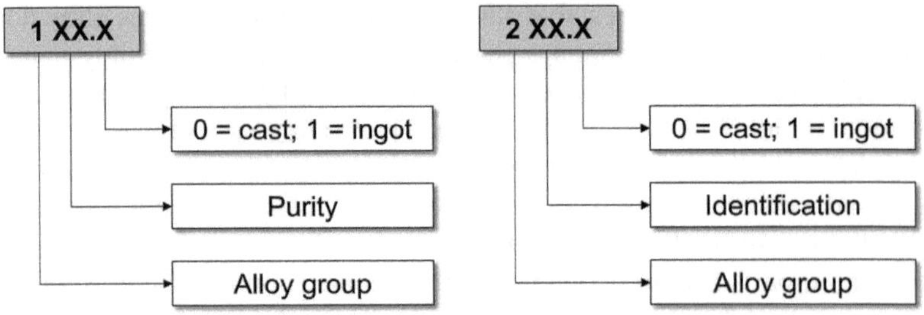

Fig. 3.8 Aluminum cast alloys nomenclature. Designation on the left is valid for 1XX.X alloys. Designation on the right holds for all the other cast aluminum alloys (ASM International 1991; Kaufman 2000). See also specifications in Table 3.2

Table 3.2 Aluminum cast alloy classification (ASM International 1991)

Series	Composition and brief description
1XX.X	Commercially pure aluminum with different grades
2XX.X	Copper is the main alloying element. Eventually, other alloying elements may be specified
3XX.X	Silicon is the main alloying element. Other alloying elements are present as specified, e.g., Cu and Mg
4XX.X	Silicon is the main alloying element
5XX.X	Magnesium is the main alloying element
6XX.X	The name of this series is not in use
7XX.X	Zinc is the main alloying element. Other elements, such as Mg, may be specified
8XX.X	Tin is the main alloying element
9XX.X	The name of this series is not in use

production of complex shapes and geometries, the choice of these materials is quite often driven by net shape considerations. Concerning the coding system for aluminum cast alloys, nine main groups do exist, although not all of them have been used so far: nXX.X, with n ranging from 1 to 9. The 1XX.X series is specific for the commercially pure (CP) aluminum alloys, whereas different codes have been developed for the other groups of alloys. The rationale of this coding system is provided in Fig. 3.8, together with the specification for the other alloy groups, listed in Table 3.2.

The selection of the best casting technique depends on the shape, the size, and the number of parts that have to be produced. High-pressure die-casting is used in more than 50% of the applications, whereas sand casting and permanent mold casting techniques are more often employed for the production of thick wall parts or for those characterized by internal hollow cavities for which high-pressure die-casting is not a suitable choice. High-pressure die-casting implies that the molten aluminum alloy is forced into a steel die at high speeds, ranging from 20 to 100 m/s. This value can be achieved thanks to a piston-cylinder system, based on a hydraulic ram that is able to reach pressures up to 100 MPa on the metal (Polmear 2006). Aluminum castings are widely spread in the automotive industry, like for the fabrication of wheels, engine blocks, and pistons. About 2% of the products obtained via high-pressure die-casting is used in aerospace applications. The percentage raises up to 10% of the total when considering aerospace applications of castings obtained via permanent mold casting and sand casting techniques. Permanent mold casting is the preferred process for the production of wheels and suspension components (Lumely 2011). The chemical composition of some cast alloys is given in Table 3.3.

Usually, castings have mechanical properties that are lower if compared with the wrought counterparts, except for the creep resistance. Furthermore, properties tend to have a certain variability throughout the component (Polmear 2006). However, the ratio of cast to wrought aluminum alloys is increasing thanks to the push of the automotive industry (ASM International 1991).

3.2.3 Wrought Alloys

Wrought products include extrusions, foils, forgings, plates, sheets, stampings, and wires. The most common form is that of the non-heat treatable sheet, largely used in the construction, packaging, and transportation industry (Lumely 2011). The feed-stock for metal working is typically an alloy ingot produced with a direct chill casting. Wrought aluminum sheet products have usually a thickness in between 0.15 and 6.3 mm. Plates feature a thickness above 6.3 mm. Concerning specific aerospace applications of the wrought alloys, the fuselage skin can be made of sheets and plates with thickness ranging from 1 up to 10 mm. Otherwise, wing covers, that are load-bearing structures, are generally 25–50 mm thick. The largest thicknesses can be found in bulkheads, wing spars, and other structural parts. The classification of extruded products is still made referring to the part thickness or to the diameter of the smallest circle in which the cross section of the piece is contained. Open- or close-die forgings are the two main options for these products. Open-die forged parts are obtained by repeatedly deforming the alloy using flat dies or with very simple geometries. The alternative approach (close-die) is used for parts with more complex geometries, for which impression dies are needed. Wrought alloys can be classified in nine different families according to their composition and in accordance with the International Alloy Designation System (IADS), introduced in 1970 and now shared worldwide. The classification of these Al-alloys is presented in Table 3.4.

Apart from the 1XXX family, the international coding is essentially referred to the main alloying elements (see Fig. 3.9), according to the specifications in Table 3.4. The second figure indicates the degree of development of a specific alloy and, indirectly, the "age" of the alloy. For instance, the 2124 alloy has been

Table 3.3 Cast alloys chemical composition. Compositions are intended as maximum weight percentage unless shown as a range of values. Except for 150.1 alloy, Al content is balance (Polmear 2006)

International designation	Si	Fe	Cu	Mn	Mg	Cr	Zn	Ti	Others
150.1	[a]	[a]	0.10	–	–	–	0.05	–	Al = 99.50 min
201.0	0.10	0.15	4.0–5.2	0.20–0.50	0.15–0.55	–	–	0.15–0.35	Ag = 0.40–1.0
208.0	2.5–3.5	1.2	3.5–0.45	0.50	0.10	–	1.0	0.25	Ni = 0.35
213.0	1.0–3.0	1.2	6.0–8.0	0.6	0.10	–	2.5	0.25	Ni = 0.35
238.0	3.5–4.5	1.5	9.0–11.0	0.6	0.15–3.5	–	1.5	0.25	Ni = 1.0
242.0	0.7	1.0	3.5–4.5	0.35	0.15–0.35	0.25	0.35	0.35	Ni = 1.7–2.3
295.0	0.7–1.5	1.0	4.0–5.0	0.35	0.03	–	0.35	0.25	–
308.0	5.0–6.0	1.0	4.0–5.0	0.50	0.10	–	1.0	0.25	–
319.0	5.5–6.5	1.0	3.0–4.0	0.50	0.10	–	1.0	0.25	Ni = 0.35
328.0	7.5–8.5	1.0	1.0–2.0	0.20–0.60	0.20–0.60	0.35	1.5	0.25	Ni = 0.25
355.0	4.5–5.5	0.60[b]	1.0-1.5	0.50[b]	0.40-0.60	0.25	0.35	0.25	–
356.0	6.5–7.5	0.6	0.25	0.35	0.20–0.40	–	0.35	0.25	–
357.0	6.5–7.5	0.15	0.05	0.03	0.45–0.60	–	0.05	0.20	–
360.0	9.0–10.0	2.0	0.60	0.35	0.40–0.60	–	0.50	–	Ni = 0.50
380.0	7.5–9.5	2.0	3.0–4.0		0.50	0.10	3.0	–	Ni = 0.50
390.0	16.0–18.0	1.3	4.0–5.0	0.10	0.45–0.65	–	0.10	–	–
413.0	11.0–13.0	2.0	1.0	0.35	0.10	–	0.50	–	Ni = 0.50
443.0	4.5–6.5	0.8	0.6	0.50	0.05	0.25	0.50	0.25	–
514.0	0.35	0.50	0.15	0.35	3.5–4.5	–	0.15	0.25	–
518.0	0.35	1.8	0.25	0.35	7.5–8.5	–	0.15	–	Ni = 0.15
520.0	0.25	0.30	0.25	0.15	9.5–10.6	–	0.15	0.25	–
535.0	0.15	0.15	0.05	0.10–0.25	6.2–7.5	–	–	0.10–0.35	–

705.0	0.20	0.80	0.20	0.40–0.60	1.4–1.8	0.20–0.40	2.7–3.3	0.25	–
707.0	0.20	0.60	0.20	0.40–0.60	1.4–1.8	0.20–0.40	4.0–4.5	0.25	–
712.0	0.15	0.50	0.25	0.10	0.50–0.65	0.40–0.60	5.0–6.5	0.10–0.25	–
713.0	0.25	1.18	0.40–1.0	0.60	0.20–0.50	0.35	7.0–8.0	0.25	Ni = 0.15
850.0	0.70	0.70	0.70–1.3	0.10	0.10	–	–	0.20	Ni = 070–1.3 Sn = 5.5–7.0

[a]Ratio Fe/Si minimum 2:1

[b]In case Fe content exceeds 0.45%, Mn content must be less than 0.5 times the Fe content

Table 3.4 Wrought alloys classification (ASM International 1991; Lumely 2011)

Series	Information about the composition	Heat treatment
1XXX	Aluminum. CP – commercially pure with a purity degree higher than 99%	Non-treatable
2XXX	Copper is the main alloying element. However, other elements may be present (Mg in particular)	Treatable
3XXX	Manganese is the main alloying element	Non-treatable
4XXX	Silicon is the main alloying element	Generally non-treatable. Some exceptions do exist (e.g., 4032)
5XXX	Magnesium is the main alloying element	Non-treatable
6XXX	Magnesium and silicon are the principal alloying elements	Treatable
7XXX	Zinc is the main alloying element. However, other spices can be present. It is the case of Cr, Cu, Mg, and Zr	Treatable
8XXX	Alloys in which elements such as lithium and tin are present	Both treatable (e.g., 8090) and non-treatable
9XXX	The name of this series is reserved for future use	–

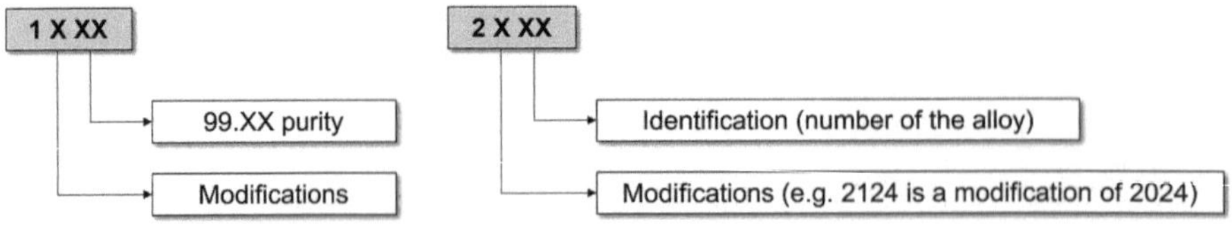

Fig. 3.9 Wrought alloy codename. Designation on the left is only valid for 1XXX alloys. Designation on the right holds for all the other wrought alloys (ASM International 1991; Kaufman 2000)

developed later, i.e., it is more recent, than 2024 alloy. However, both alloys have similar composition.

The latest two figures on the right indicate a serial number providing information on the specific set of alloying elements and relevant concentrations. For instance, alloy 7049 and 7050 are slightly different in composition. In case of 1XXX alloys, the last two figures of the code refer to the level of impurities of the alloy: e.g., 1155 alloy has a purity of at least 99.55% (Fig. 3.9). As seen in Table 3.4, concerning their composition, except for 1XXX series, wrought aluminum alloys can contain a wide range of alloying elements. The composition of selected wrought alloys of 1XXX series is given in Table 3.5.

A wider set of alloy compositions is given in Appendix 1; the indication of the main effect of any element on the alloy properties is given in Appendix 2. Aluminum wrought alloys can be split into two main subgroups: heat treatable and non-treatable alloys.

Heat treatments of aluminum alloys are used to improve two main properties: hardness and mechanical strength. This is attained through precipitation hardening, although other effects of thermomechanical treatments, e.g., strain hardening,

Table 3.5 1XXX alloys chemical composition. Compositions are intended as maximum weight percentage unless shown as a range of values

International designation	Si	Fe	Cu	Mn	Mg	Cr	Zn	Ti	Others
1050	0.25	0.40	0.05	0.05	0.05	–	0.05	0.03	Al = 99.50 V = 0.05
1070	0.20	0.25	0.04	0.03	0.03	–	0.04	0.03	Al = 99.70 V = 0.05
1098	0.01	0.006	0.003	–	–	–	0.015	0.003	Al = 99.98
1100	–	–	0.05–0.20	0.05	–	–	0.10	–	Al = 99.00
1275	0.08	0.12	0.05–0.10	0.02	0.02	–	0.03	0.02	Al = 99.75 Ga = 0.03 V = 0.03
1420	–	–	–	–	5.2	–	–	–	Li = 2.1 Zr = 0.11
1421	–	–	–	–	5.2	–	–	–	Li = 2.1 Sc = 0.17 Zr = 0.11
1440	–	–	1.5	–	0.80	–	–	–	Li = 2.4 Zr = 0.11
1441	–	–	1.65	–	0.90	–	–	–	Li = 2.4 Zr = 0.11
1450	0.25	0.40	0.05	0.05	0.05	–	0.07	0.10–0.20	–
1460	–	–	2.9	–	–	–	–	–	Li = 2.25 Sc = 0.09 Zr = 0.11

If not specified, Al content is balance (Polmear et al. 2017; Prasad et al. 2014; The International Aluminum Association 2015)

Table 3.6 Main families of wrought alloys and relevant strengthening mechanisms (ASM International 2015)

International designation	Alloys system	Strengthening mechanism
1XXX	Pure Al	Work-hardenable alloys
3XXX	Al-Mn	
4XXX	Al-Si	
5XXX	Al-Mg	
8XXX	Al-Fe	
8XXX	Al-Fe-Ni	
2XXX	Al-Cu	Precipitation-hardenable alloys
2XXX	Al-Cu-Mg	
2XXX	Al-Cu-Li	
6XXX	Al-Mg-Si	
7XXX	Al-Zn	
7XXX	Al-Zn-Mg	
7XXX	Al-Zn-Mg-Cu	
8XXX	Al-Li-Cu-Mg	

contribute to different extent to the overall characteristics of the alloys. Work-hardening is the main strengthening approach for non-heat treatable alloys. In Table 3.6, a further classification is proposed for the wrought alloys according to the main strengthening mechanisms and with the indication of the subgroups of the alloys depending on the reference compositional system.

3.2.3.1 Wrought Non-heat Treatable Alloys

Alloys of the series 1XXX, 3XXX, 4XXX, and 5XXX are non-treatable. This means that no alloy structural changes can be induced by thermal treatments so that the relevant properties are substantially insensitive to the thermal history of the alloy. In some respect, this is a positive feature, for instance, in applications involving uncontrolled thermal conditions, possibly inducing fast melting and solidification, like welding. On the other hand, cold working processes, even conducted on the component surface, like shot pinning and similar procedures, can drastically change the mechanical properties. Although not particularly used in aircraft structures, it is worth remembering some of their main application fields that are quite numerous and diverse (Davis 2001; Polmear 2006):

- **1XXX**: this family comprises un-alloyed aluminum alloys, with different grades of purity: commercially pure and super-pure aluminum (Belov et al. 2002). This series is used for the fabrication of several components, some interesting examples being architectural and decorations for structural parts in buildings, components of chemical plants, and electric wires and cables, considering the good electrical conductivity of pure aluminum.

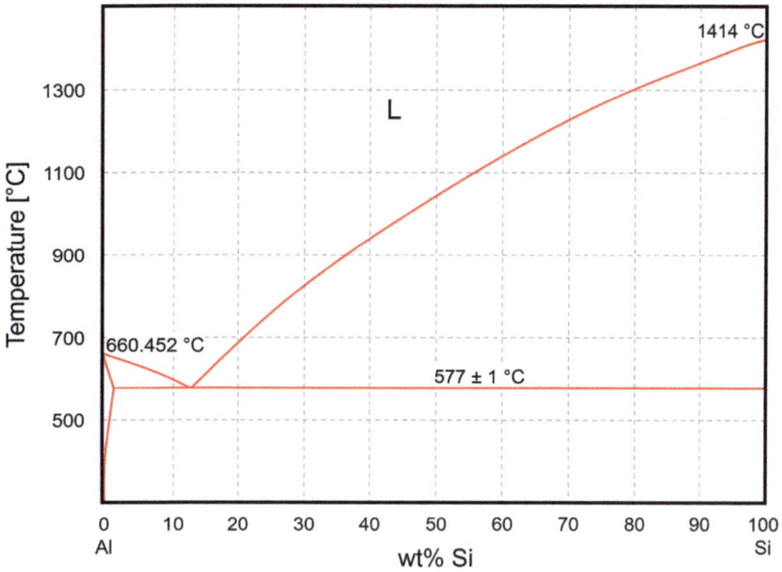

Fig. 3.10 Al-Si phase diagram. (Redrawn from ASM International 1992)

- **3XXX**: interesting characteristics of this family of alloys, having manganese as main addition, are a low mechanical resistance and a high ductility, in association with a good corrosion resistance. The most popular product is the 3003 alloy, whose strengthening mechanism relies on the formation of intermetallic compounds, e.g., Al_6Mn, dispersed within the alloy matrix. This intermetallic phase forms directly from the melt, and the structure of the phase diagram does not allow solutioning and thermal treatments inducing solid-state precipitation, necessary to refine the alloy microstructure and optimize its mechanical properties. Some common applications of the 3XXX alloys are production of multipurpose sheets, including building structures; furniture; cooking utensils, in particular pottery; cans for food and drinks; and thin foils.
- **4XXX**: the main alloying element is silicon, and these are two-phase alloys used for welding processes thanks to the presence of a low-temperature eutectic transformation (see Fig. 3.10). Silicon is not soluble in aluminum in the solid state, and this leads to the solidification of a two-phase mixture, consisting of mostly pure silicon grains dispersed into the aluminum matrix. The grain size and morphologies will depend upon the kinetic aspects of the transformation, although eutectic microstructures are very likely.
- **5XXX**: this series of alloys is characterized by the presence of magnesium as main alloying element. A particular feature of these alloys, interesting for some applications, is the fact that the *fcc* austenitic phase exhibits a very low ductile-to-brittle transition temperature. Therefore, these alloys are very much suitable for cryogenic use, like inner pressurized vessels for liquid nitrogen dewar containers, cryogenic sample holders, etc. They show a good weldability, so that complex

Table 3.7 Designation of the main temper condition used for all Al-alloys

Name	Main features
F	As fabricated. Corresponds to the strain-hardened condition due the production process. The letter F stands alone
O	Annealed condition to obtain the lowest strength temper, aiming at increasing the alloy workability. O can be applied to cast products that are annealed to improve dimensional stability and ductility. The letter O may be followed by a digit to indicate an annealed condition with special characteristics
H	Strain-hardened wrought products. Additional thermal treatment can be employed if a reduction of the alloy strength is needed. H is followed by two or more digits (see also Table 3.8)
W	Solution heat-treated alloys. This temper is applied only to alloys that undergo spontaneous aging after solution heat treatment. It induces a sort of metastable condition, followed by natural aging at room temperature. Digits to indicate the natural aging period, e.g., W1/2 h, are commonly used
T	Thermally heat-treated to produce stable tempers other than F, H, O. Supplementary strain hardening may be employed as well. The letter T is always followed by one or more digits (see Tables 3.9 and 3.10)

This scheme is adopted by the main international standards referring to these alloys (Benedyk 2010; Kaufman 2000)

structures and/or comparatively large pieces can be assembled: large tanks for milk but also petrol, oil, and other organic liquids. Small boat hulls have been also made using Al-Mg alloy sheets.

More information on wrought non-heat treatable alloys can be found in the references quoted at the end of the chapter.

3.2.3.2 Heat-Treatable Alloys and Designation of Tempers

Heat-treatable aluminum alloys are particularly interesting in aerospace applications. Different combinations of mechanical and corrosion resistance properties can be imparted to the same alloy thanks to suitable heat treatments. An international designation code, registered by the Aluminum Association, does exist for these tempers, and it is usually added to the alloy designation number. All changes induced into heat-treatable alloys are based on the solid-state formation of stable or metastable precipitates, as described in Sect. 3.2.3.3. The heat treatments are typically a combination of a solution stage and quenching, followed by natural and artificial aging. Considering the role that relevant defects may have in solid-state precipitation, cold-working is another aspect involved in the temper design. As shown by Table 3.7, the T temper is only one of the possible tempers used for Al-alloys. The designation, used since 1948 and updated over the years, is meant to assign to any heat-treatable alloy product the information on the thermal treatment used to set its properties.

Table 3.8 gives information about the codes used for strain-hardened, temper H, alloys.

Table 3.8 Description of the different "H" tempers. X indicates the residual strain hardening percentage, remaining in the alloy after each treatment, except that for H1X, where X is the percentage of imparted strain hardening

Type	Main features
H1X	Strain hardened only
H2X	Strain hardened and partially annealed. The strength acquired during strain hardening is reduced at the desired level by means of a partial annealing process. In this case X corresponds to the percentage of residual strain hardening after the partial annealing
H3X	Strain hardened and stabilized. Stabilization is used for improving ductility and can be carried out either thanks to heat introduced during the fabrication or thanks to a low temperature thermal treatment. Again, X corresponds to the percentage of residual strain hardening after the stabilization treatment
H4X	Strain hardened and lacquered or painted. It indicates the products that are strain hardened and subsequently subjected to heat during painting or lacquering. X corresponds to the percentage of residual strain hardening after lacquering or painting

The maximum value for $X = 8$ (Kaufman 2000)

As concerns the T tempers, a brief description is given in Table 3.9, with full details in Benedyk (2010). All T-treatments start with alloy solutioning. The subsequent quenching aims at freezing down to room temperature a super-saturated solid solution. The two main temper parameters are solution temperature and cooling rate. Precipitation occurs next, induced by either artificial or natural aging. Natural aging occurs spontaneously at room temperature, an aspect to be considered not only during the intermediate steps of the heat treatment but also as concerns the microstructural evolution occurring under operating conditions. Natural aging determines the useful lifetime of the alloy, corresponding to the time required to reach the overaging conditions, when material properties are no longer suitable for the relevant application.

The T tempers described in Table 3.9 can be rendered more complex by introducing further steps, mainly designed for relieving the stress accumulated into the alloy piece. Therefore, other digits can be added to the temper identification code, as specified in Table 3.10. In particular, stress relief by stretching is codenamed T51, by compressing T52, a combination of stretching and compressing T54. Stress relief can be performed either after solution heat treatment or after quenching, following a shaping process conducted at elevated temperatures. More digits can be used to provide additional information on other relevant parameters of the temper, like percentage of nominal permanent set, level of stress relieved, etc. (Table 3.10).

The selection of the right temper conditions is often the result of a compromise of diverging requirements. For example, tempers interesting for aircraft alloys of the 7XXX family are the T6 and T7. As shown by Fig. 3.11, T6 would induce excellent mechanical strength but a correspondingly lower corrosion resistance of the alloy. Intermediate conditions can be attained using suitable variants of the T7 temper.

Table 3.9 Description of the T temper treatments (Kaufman 2000)

Type	Main features
T1	Cooling from elevated temperature shaping process and natural aging. Natural aging leads to a substantially stable condition. Applications: products that do not undergo cold working after cooling from an elevated temperature shaping process
T2	Cooling from elevated temperature shaping process, cold working, and final natural aging process. Natural aging leads to a substantially stable condition. Applications: products that undergo cold working after cooling from an elevated temperature shaping process
T3	Solution treatment, cold working, and final natural aging process. Natural aging leads to a substantially stable condition. Applications: products that undergo cold working to improve the strength after being solution treated
T4	Solution treatment, natural aging. Natural aging leads to a substantially stable condition. Applications: products that do not undergo cold working after solution treatment
T5	Cooling from elevated temperature shaping process and artificial aging. Applications: products that do not undergo cold working after being cooled from elevated temperature shaping process
T6	Solution treatment and artificial aging. Applications: products that do not undergo cold working after solution treatment
T7	Solution treatment and overaging/stabilization. Applications: cast products that undergo artificial aging after solution treatment in order to give them stability in strength and dimension, wrought products that undergo artificial aging after solution treatment in order to give them an increase in strength beyond the maximum value achievable to control some specific characteristics or properties
T8	Solution treatment, cold working, and artificial aging. Applications: products that undergo cold working for increasing their strength
T9	Solution treatment, artificial aging, and cold working. Applications: products that undergo cold working operations for increasing their strength
T10	Cooling from elevated temperature shaping process, cold working, and artificial aging. Applications: products that undergo cold working for increasing their strength

For all T tempers, the solution heat treatment involves heating the alloy up to the solution temperature and holding at that temperature for a long enough time, so to obtain a complete solutioning of the alloying elements. The subsequent cooling (quenching) should be sufficiently fast to retain a supersaturated solid solution down to the quenching temperature, in order to have precipitation to take place during the subsequent artificial or natural aging

3.2.3.3 Precipitation Hardening

Solid-state precipitation is the phase transformation on which the hardening of *heat-treatable* Al-alloys is based. The same transformation and resulting strengthening mechanisms hold also for other classes of alloys that will be considered next: magnesium and titanium alloys (Sect. 3.3 and Chap. 4), PH stainless steels (Sect. 5.4.3.3), and iron- and nickel-based superalloys (Sects. 6.3 and 6.4). The fundamental aspects of these treatments, which will be illustrated herewith with reference to an aluminum alloy, are meant to provide indications, from the physical metallurgy point of view, for the heat treatments of the other alloy systems too, with obvious important differences as concerns both the process parameters and the precipitates that actually form in each specific alloy system.

Table 3.10 Additional digits for stress relief treatments in T-tempered aluminum alloy products (Kaufman 2000)

Additional digits for T tempers	Wrought product type	Stress relief process	Permanent set
TX51[a]	Plate	Stretching	1.5–3.0%
TX51[a]	Rolled or cold finished bars and rods	Stretching	1.0–3.0%
TX51[a]	Die or ring forgings and rolled rings	Stretching	1.0–5.0%
TX510[a]	Extruded bars, profiles, rods, and tubes	Stretching	1.2–3.0%
TX510[a]	Drawn tubes	Stretching	0.5–3.0%
TX511[b]	Extruded bars, profiles, rods, and tubes	Stretching	1.0–3.0%
TX511[b]	Drawn tubes	Stretching	0.5–3.0%
TX52	All products	Compressing	1.0–5.0%
TX54	Die forgings	Restriking cold in finish die	–

[a]No further straightening after stretching
[b]Minor straightening after stretching. This may be required by tolerance limitations

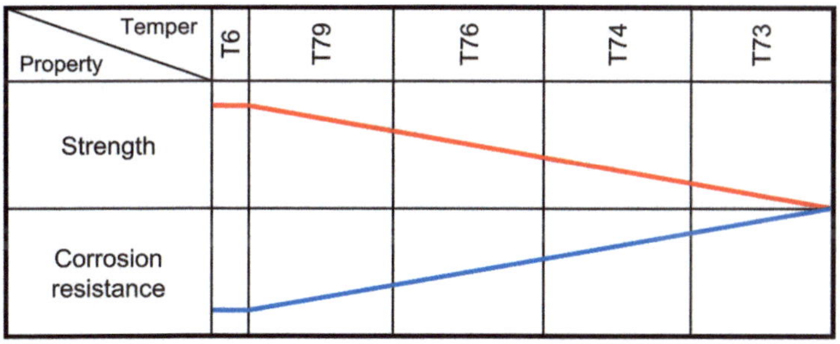

Fig. 3.11 Dependence of mechanical strength and corrosion resistance of the T6 and different T7 tempers. These tempers are typically used for alloys of the 7XXX group, including those for aircraft parts (redrawn from Benedyk 2010)

The main aspects of solid-state precipitation in hardenable aluminum alloys can be discussed considering the binary Al-Cu system, being an important reference also for alloys with more complex compositions. For technological purposes, the alloy composition never goes above the solubility limit of copper in aluminum (see Fig. 3.12); therefore the aluminum-rich corner is the interesting part of the phase diagram, as shown by Fig. 3.13. Solutioning of copper into the aluminum crystalline lattice results in a single-phase alloy, the α-phase in the phase diagram in Fig. 3.13. Quenching freezes the supersaturated alloy down to room temperature, and the subsequent aging, conducted either at room (natural aging) or higher (artificial aging) temperatures, is designed to reach the target conditions, in agreement with the requirements of the specific applications. The intermetallic compound θ-Al_2Cu, present in the Al-Cu equilibrium phase diagram, is not really interesting for the

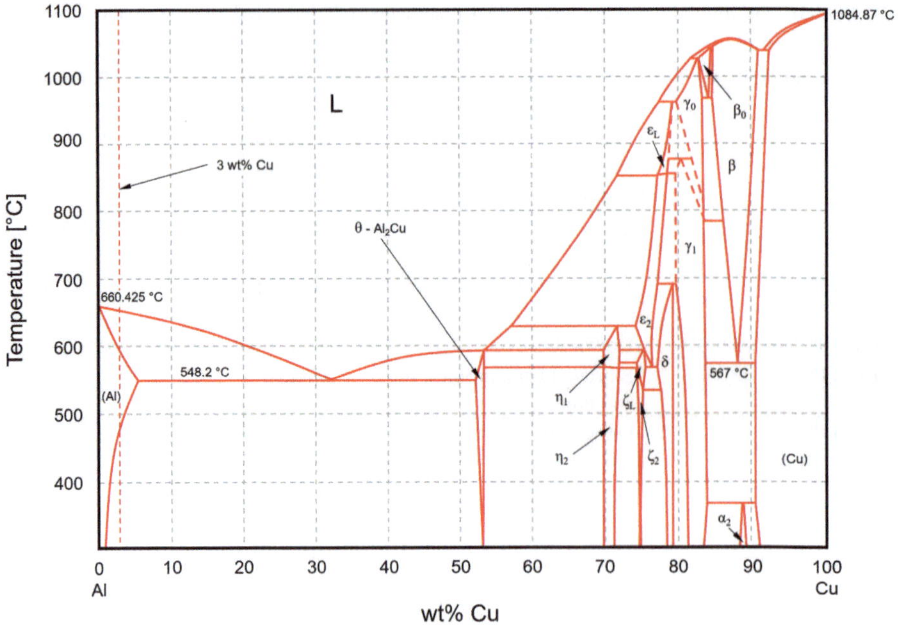

Fig. 3.12 Al-Cu phase diagram. (Redrawn from ASM International 1992)

improvement of the mechanical properties of the alloy and it is actually a typical phase observed in overaged conditions. Similar equilibrium phases, with different compositions, will form in other heat-treatable aluminum alloys (see Table 3.11). Several important metastable structures can be obtained indeed in real alloys, thanks to the kinetic constrains introduced by the heat treatment parameters. These structures, not present in the equilibrium phase diagram, are very important as concerns the alloy properties. In the Al-Cu system, these structures are θ', θ'', and GPZ (Guinier-Preston zone), described in the following.

The heat-treated condition of the alloy can be predicted from TTT curves (Fig. 3.14), once the main parameters of the treatment, like quenching rate, aging temperature, and time, have been selected. As illustrated in Fig. 3.14, there are two possibilities for obtaining GPZ. The first consists in quenching the alloy directly to the relevant GPZ field as shown by the cooling trajectory 1. A second approach is to cool down the alloy at a sufficiently high rate not to cut the GPZ directly from the solutioning temperature and then start an aging treatment, at constant temperature until the transformation is fully accomplished, according to the trajectory 2 in Fig. 3.14.

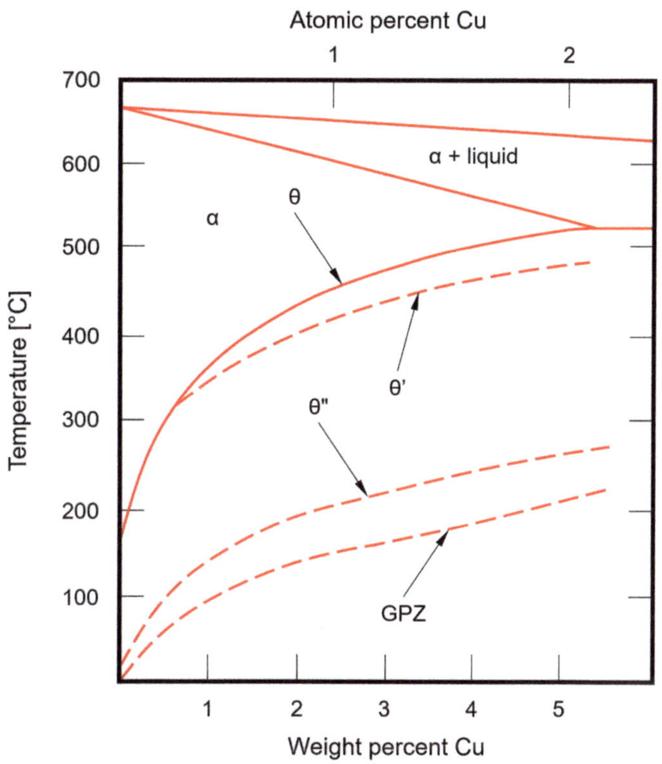

Fig. 3.13 Part of the Al-Cu phase diagram showing the stable θ-Al_2Cu precipitate and the metastable θ', θ'', and GPZ. (Redrawn from Porter et al. 2009)

Table 3.11 Solid-state precipitation in some alloy systems (Porter et al. 2009)

Base metal	Alloy	Precipitation sequence
Al	Al-Ag	GPZ (spheres) → γ' (plates) → γ (Ag_2Al)
	Al-Cu	GPZ (discs) → θ'' (discs) → θ' (plates) → θ (Cu Al_2)
	Al-Cu-Mg	GPZ (rods) → S' (laths) → S ($CuMgAl_2$, laths)
	Al-Zn-Mg	GPZ (spheres) → η' (plates) → η ($MhZn_2$, plates or rods)
	Al-Mg-Si	GPZ (rods) → β' (rods) → β (Mg_2Si, plates)
Cu	Cu-Be	GPZ (discs) → γ' → γ (CuBe)
	Cu-Co	GPZ (spheres) → β (Co, plates)
Fe	Fe-C	ε carbide (discs) → Fe_3C (plates)
	Fe-N	α'' (discs) → Fe_4N
Ni	Ni-Cr-Ti-Al	γ' (cubes or spheres)

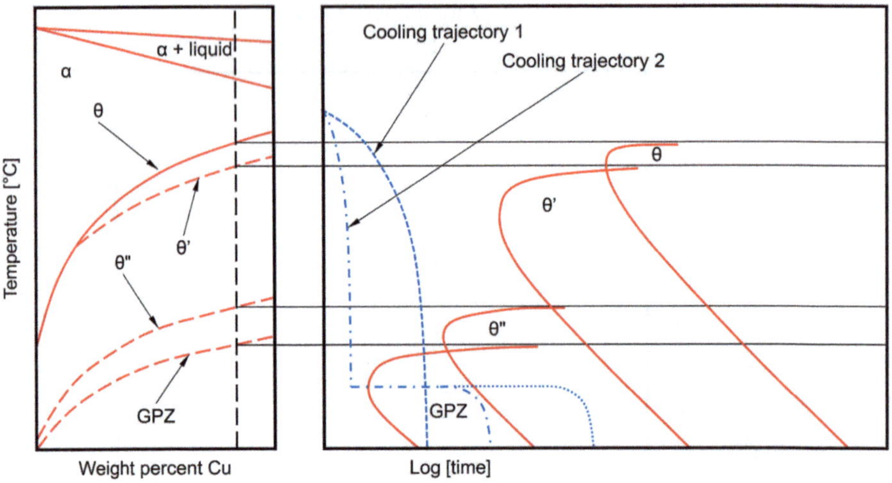

Fig. 3.14 Al-Cu phase diagram showing the stable θ precipitate and the metastable θ', θ'', and GPZ and related TTT diagram with different cooling trajectories. See main text for details. (Redrawn and adapted from Porter et al. 2009)

Similarly, the precipitation of other phases can be induced. Different precipitates feature different crystallographic structures that induce different kinds of interface with respect to the hosting aluminum matrix: coherent, semicoherent, or fully incoherent (Porter et al. 2009 – see also Sect. 6.4.1.1).

Figure 3.15 summarizes the situation for the precipitates' features in Al-Cu alloys. An important parameter that characterizes the matrix-precipitate interface is the lattice misfit δ (Geddes et al. 2010; Reed 2006):

$$\delta = \frac{a_{\theta''} - a_\alpha}{a_{\theta''} + a_\alpha} \times 2 \tag{3.2}$$

where

- δ is the lattice misfit.
- a_α is the lattice parameter of the matrix.
- $a_{\theta''}$ is the lattice parameter of the θ'' precipitate.

The lattice misfit depends on the difference of the lattice parameter, or interatomic distance, along specific crystallographic planes that are facing each other across the precipitate-matrix interface (see also Sect. 6.4.1.1). In the present example (Eq. 3.2), the lateral plane for the θ'' and cube plane for the α Al-*fcc* structure are considered. Other important features of the precipitates, like size, shape, composition, etc., are dependent on the overall alloy composition and on the elastic anisotropy along crystallographic directions. These intrinsic alloy parameters and the thermal treatment conditions act as kinetic constraints, so that not only stable but also metastable phases can form.

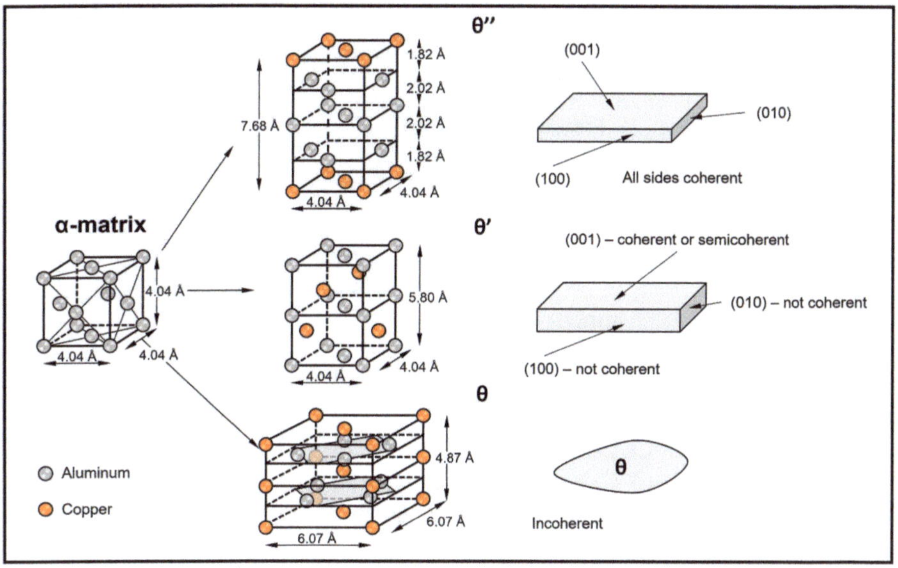

Fig. 3.15 α matrix, stable θ precipitate and metastable θ', θ'' structure and morphology. (Redrawn from Porter et al. 2009)

In Table 3.11, the precipitation transformations for different alloy systems are summarized. In case of *heat-treatable* aluminum alloys, the formation of metastable structures is quite a general feature. This is the case of the β' precipitates in the 6063 alloy (Al-Mg-Si group, Fig. 3.16). β' (Mg_2Si composition) is a coherent precipitate with preferential orientation along the <100> direction of the aluminum matrix. The precipitates are aligned along this particular crystallographic direction since it turns out to be the more energetically favored, as compared to the others. In fact, along this direction there is the lowest atomic linear density, the interatomic distances are larger, and so the elastic constant is lower. The shape of the precipitate is sensitive to the elastic anisotropy as well. In fact, the interfaces with the aluminum matrix having a lower elastic strain energy term will be wider than those with a higher elastic energy term. The hardening induced by the alloy heat treatments will depend on the resulting microstructure that influences the way in which precipitates and dislocations interact.

In agreement with the TTT curves (Fig. 3.14), at an aging temperature of 190 °C, the GPZs cannot form, since the relevant aging trajectory is not cutting the stability domain of these structures, meaning that the aging temperature is larger than the GPZ *solvus*. Direct precipitation of θ'' can take place instead, and these precipitates for longer aging times may transform into θ' precipitates. The peak hardness value scales up with the concentration of copper and, thereby, with the fraction of precipitates that may form. The hardening curves corresponding to a lower aging temperature (Fig. 3.17, $T = 130$ °C) may reach higher hardness peaks, for each one of the investigated alloys. This can be ascribed to the formation of finely

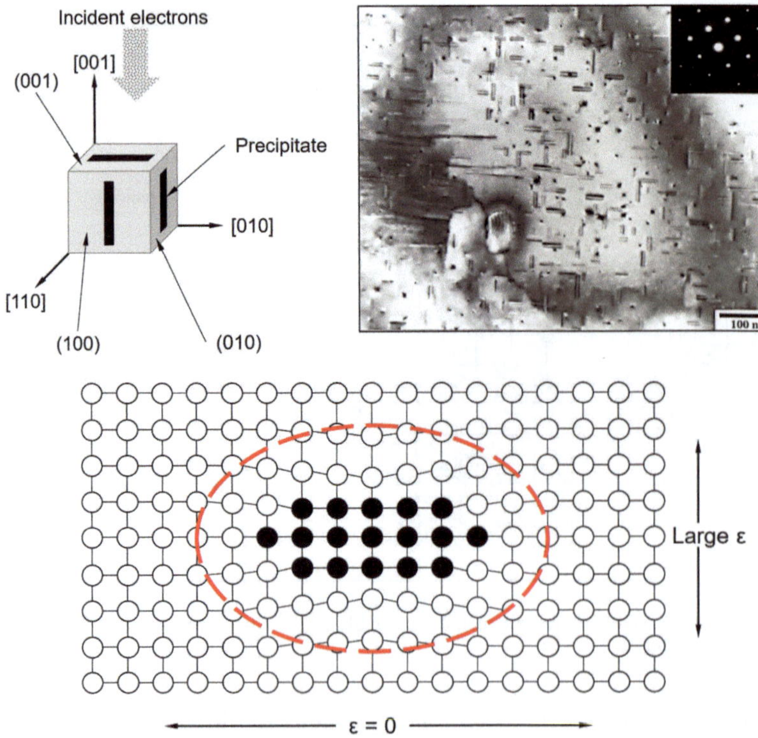

Fig. 3.16 Transmission electron microscopy micrograph of β' precipitate in 6063 alloy. The image contrast is due to the strain ε at the precipitate-alloy matrix coherent interface, as represented by the bottom scheme. (Photo property of the Authors, bottom scheme redrawn from Porter et al. 2009)

dispersed GPZs, not forming at the higher aging temperature. The evolution of hardness on aging is determined by the change of the dominant interaction mechanisms between dislocations and precipitates. In the early stages of the lower temperature aging, GPZs are the main strengthening agents. Dislocations are hindered in their movement by the elastically strained aluminum matrix. The same mechanism is active when the θ'' coherent precipitates start to form. Moreover, the coherency of the θ'' particles and their larger size, as compared to GPZs, allows also for the dislocation *cutting* mechanism to happen (Fig. 3.18). The resulting strengthening is due to the drag that the dislocations are subjected to when moving through the precipitates and to the resistance determined by the strain field at the precipitate-matrix interface.

The mechanisms described so far are responsible for the rising branch of the hardening curves in Fig. 3.17. Each one of these curves reaches a maximum and then starts to decrease. This is due to a change in the dislocation-precipitate interaction mechanism. The ripening of the precipitates with aging time results in a progressive alloying elements depletion from aluminum alloy matrix.

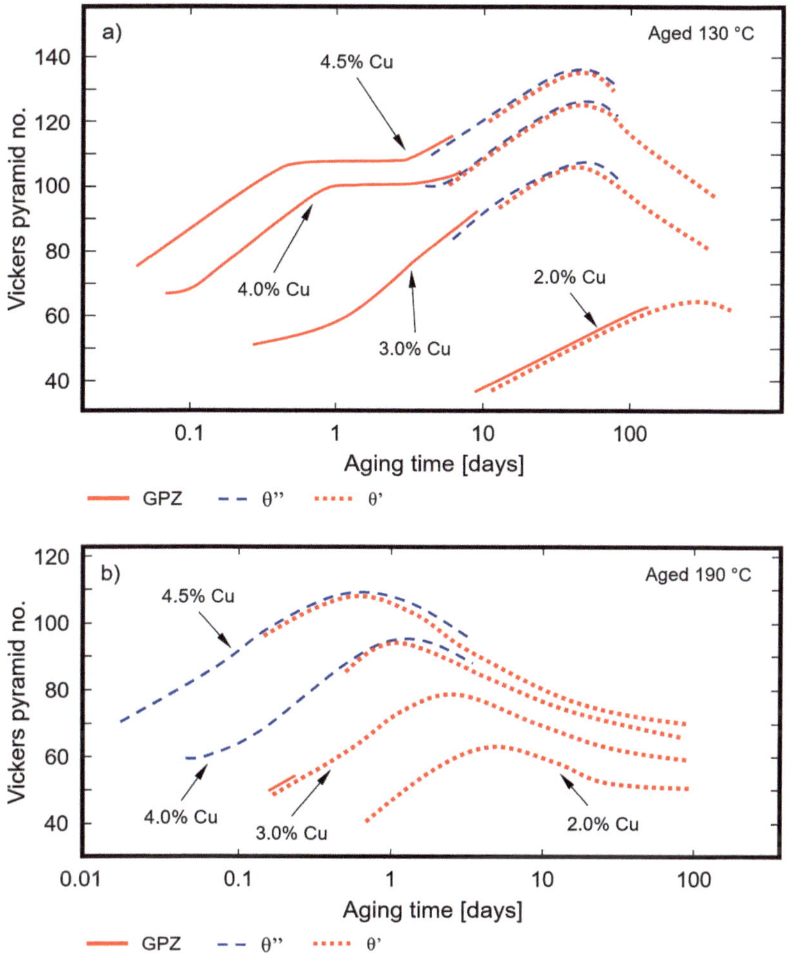

Fig. 3.17 Effect of the aging temperature on the hardness of binary Al-Cu alloys. (**a**) 130 °C aging; (**b**) 190 °C aging. Note that for the higher temperature, the precipitation of GPZ is nearly fully suppressed. (Redrawn from Porter et al. 2009)

This reduces the solid solution strengthening, and dislocations can now move more easily. Actually, they get pinned by the precipitates but can still continue to move if the applied stress is sufficient to increase their length. This is the strengthening mechanism active in the so-called dislocation *bowing* regime (also known as Orowan mechanism, Fig. 3.19; see Sect. 6.4.3 for further details).

Actually, additional strengthening may also come from the piling up of dislocation loops around each pinning precipitate. The Orowan mechanism is active in the Al-Cu alloys when incoherent θ' and equilibrium θ precipitates form. The occurrence of equilibrium precipitates in all hardenable aluminum alloys, i.e., those belonging to the 2XXX, 6XXX, and 7XXX groups (see Table 3.4), is often

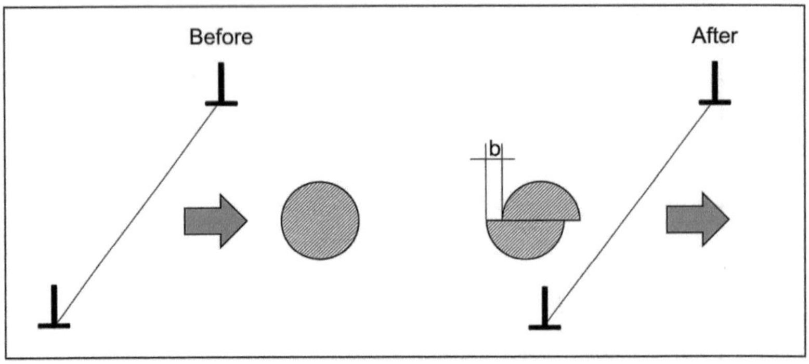

Fig. 3.18 Dislocation cutting through a spherical precipitate

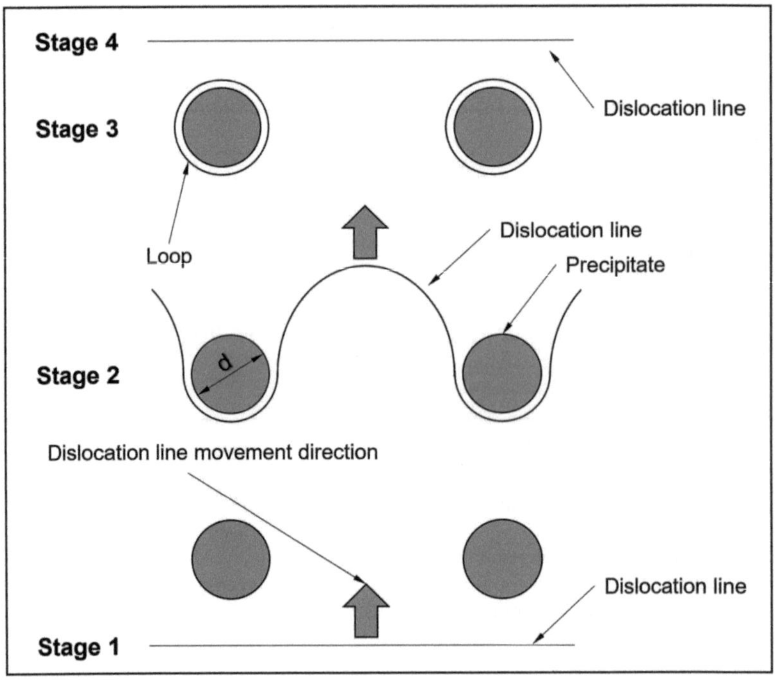

Fig. 3.19 Dislocation bowing. (Redrawn from Hull and Bacon 2011)

associated with the condition of overaging (Fig. 3.20 referring to a 6063 alloy with equilibrium Mg_2Si precipitates).

At this stage, which can be reached either for an incorrect heat treatment or for the prolonged exposure of the alloy to demanding servicing conditions, an important decay of the alloy hardness is observed. The transition from cutting to bowing as concerns the dominant dislocation-precipitate interaction mechanism is observed in other precipitation-hardened alloys, like Mg-alloys (see Sect. 3.3) and Ni-based

Fig. 3.20 Mg_2Si precipitate in an overaged 6063 alloy. (Photo property of the Authors)

superalloys (Chap. 6). For these latter, an important role as concerns strengthening is played by the atomic crystallographic order of the γ' precipitates (see Chap. 6 for further details). The heat treatments of aluminum alloys are important for the effects they have on the structure, composition, spatial distribution, and size of the strengthening precipitates. Also the microstructure of the alloy in the grain boundary regions is influenced by the heat treatment parameters and relevant precipitation kinetics. Figure 3.21 shows some examples of the so-called precipitation-free zone (PFZ) and relevant scheme.

Heterogeneous nucleation, driven by the grain boundary surface energy term, determines an anticipated precipitation along grain boundaries, which depletes the surrounding regions of alloying elements. When nucleation, mostly homogeneous, and subsequent growth of precipitates inside the grains start, they can no longer occur in the alloy-depleted regions that give place to the mentioned PFZs. As shown by the scheme in Fig. 3.21, PFZs become a preferential path for the movement of dislocations, resulting in a general weakening of the alloy. To get rid or at least to reduce the incidence of the PFZs, several approaches are possible. A higher undercooling, with respect to the solution temperature, enhances the homogeneous nucleation rate inside the grains. Alloying with insoluble atoms, like silicon, may introduce preferential nucleation sites in the alloy, which again favor a uniformly distributed precipitation within the alloy grains, by reducing the incidence of the preferential precipitation along grain boundaries. Another defect structure that can

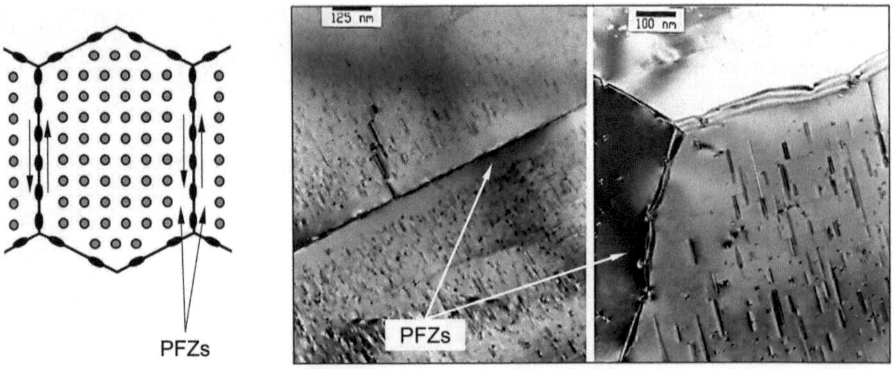

Fig. 3.21 Examples of precipitate-free zones. (Left picture redrawn from Polmear 2006; images on the right are photo property of the Authors)

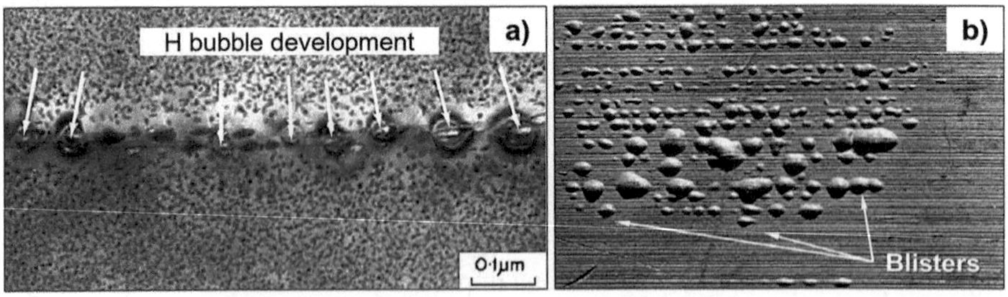

Fig. 3.22 (**a**) Transmission electron micrograph showing hydrogen inclusions development at the grain boundary of a thin foil of an artificially aged Al-Zn-Mg alloy; (**b**) blisters on the surface of an Al-alloy component after solution treatment in a humid atmosphere (Polmear 2006, reproduced with permission of the Publisher)

be observed along grain boundaries and subsurface regions is hydrogen micro-inclusions (Fig. 3.22). This phenomenon, called *blistering* for the particular morphology that induces on the alloy surface, occurs during water quenching as a consequence of the hydrolysis of H_2O catalyzed by the oxidation of aluminum, according to the reaction: $2Al + 3H_2O \rightarrow Al_2O_3 + 3H_2$.

The grade of the alloy, depending on the concentration of impurity elements, is very important because it determines the formation of intermetallic particles within the alloy itself (see Fig. 3.23). The size and intrinsic brittleness of these phases render their presence potentially detrimental for the properties of the alloy, in which they act as stress concentrators.

Fig. 3.23 Fe-rich inclusion in a 6063 alloy, isomorphous to the Al_3Fe intermetallic. (Photo property of the Authors)

3.2.4 Aluminum-Lithium Alloys

Lithium is present in several Al-alloys, but since it is not necessarily the majority alloying element, wrought Li-containing Al-alloys can be found in different alloy groups: 1XXX, 2XXX, and, finally, 8XXX, being the alloys of this latter group, those featuring larger Li concentrations (Table 3.12).

Although regarded as the most recently developed of the Al-alloys, actually, Al-Li alloys started their metallurgical history back in the 1920s, and lithium additions were used to enhance the strengthening of age-hardenable alloys (Starke et al. 1981). The very first Li-containing Al-alloy was developed in Germany in 1924 with the commercial name Scleron and nominal composition Al-12Zn-3Cu-0.6Mn-0.1Li (Prasad et al. 2014). The first results were very encouraging indeed, since the expected improvement in strength was accompanied also by a higher stiffness and, obviously, lower density (see Fig. 3.24). The production of Al-Li alloys requires a careful working condition control, in order to avoid the losses of lithium during melting, for which vacuum induction melting is regarded as a safe route (for additional information see Sect. 5.3.2) (ASM International 2008). Lithium content in the main Al-Li alloys is given in Table 3.12, while the complete chemical composition of the main alloys of the family can be found in Table 3.5 and Appendix 1. Three main different development stages can be identified. The *first-generation*

Table 3.12 Designation, Li content, density, and introduction of some relevant Al-Li alloys

Alloy	Generation	wt% Li	Density [g/cm^3]	Producer (introduction year)
2020	1	1.00	2.71	Alcoa (1958)
1420		2.10	2.47	Soviet (1965)
1421		2.10	2.47	Soviet (1965)
2090	2 (Li ≥ 2%)	2.10	2.59	Alcoa (1984)
2091		2.00	2.58	Pechiney (1985)
8090		2.40	2.54	EAA (1984)
1440		2.40	2.55	Soviet (1980s)
1441		1.95	2.59	Soviet (1980s)
1450		2.10	2.60	Soviet (1980s)
1460		2.25	2.60	Soviet (1980s)
2195	3 (Li ≤ 2%)	1.00	2.71	Lockheed Martin Corporation/Reynolds (1992)
2196		1.75	2.63	Lockheed Martin Corporation/Reynolds/McCook Metals (2000)
2297		1.40	2.65	Lockheed Martin Corporation/Reynolds (1997)
2397		1.40	2.65	Alcoa (1993)
2098		1.05	2.70	McCook Metals (2000)
2198		1.00	2.69	Reynolds/McCook Metals/Alcan (2005)
2099		1.80	2.63	Alcoa (2003)
2199		1.60	2.64	Alcoa (2005)
2050		1.00	2.70	Pechiney/Alcan (2004)
2296		1.60	2.63	Constellium Alcan (2010)
2060		0.75	2.72	Alcoa (2011)
2055		1.15	2.70	Alcoa (2011)
2065		1.20	2.70	Constellium (2012)
2079		1.50	2.64	Constellium (2012)

The complete chemical composition is given in Table 3.5 and Appendix 1 (Prasad and Wanhill 2017)

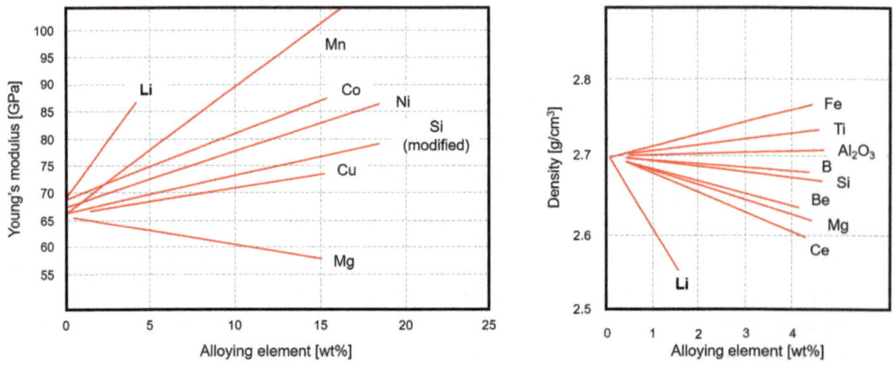

Fig. 3.24 Dependence of the Young's modulus and density of aluminum as a function of the concentration of different alloying elements. (Redrawn from Polmear et al. 2017)

Table 3.13 Main precipitate phases encountered in commercial Al-Li alloys (Prasad et al. 2014)

Phase	Chemistry	Crystal structure	Lattice parameters [nm]	Main features
δ	AlLi	Cubic (NaCl)	$a = 0.638$	Equilibrium phase with a plate morphology
δ'	Al_3Li	Cubic ($L1_2$)	$a = 0.401$	Non-equilibrium ordered phase, usually with spherical shape and coherent interface with the aluminum matrix
T_1	Al_2CuLi	Hexagonal	$a = 0.4965$	Equilibrium phase. Partially coherent interfaces
			$c = 0.9345$	
T_2	Al_6CuLi_3	Cubic	$a = 1.3914$	Please note: the cubic structure displays icosahedral symmetry
S'	Al_2CuMg	Orthorhombic	$a = 0.401$	Semicoherent phase
			$b = 0.925$	
			$c = 0.715$	
θ'	Al_2Cu	Tetragonal	$a = 0.404$	–
			$c = 0.580$	
–	Al_3Zr	Cubic ($L1_2$)	$a = 0.405$	Coherent, ordered, spherical dispersoids

alloys were developed over a decade across 1960. The potential of these materials was demonstrated and the critical aspects of their processing emerged. The *second-generation* Al-Li alloys featured a Li content in excess, or equal to, 2 wt% and were produced either via ingot metallurgy, mainly, or powder metallurgy. The development of the *third-generation* was initiated in the late 1980s, when the Martin Marietta Corporation was engaged in the design of a weldable, low-density aluminum alloy for cryogenic tanks and other components of aerospace vehicles (Prasad et al. 2014).

Lithium alloying reduces aluminum density and increases its stiffness (Fig. 3.24). The higher stiffness and, most importantly, the enhancement of the alloy strength are due to the formation of the ordered δ'-Al_3Li precipitates. These precipitates have the same $L1_2$ structure as the ordered γ' precipitates in Ni-based superalloys (see Chap. 6). The partial covalent character of the chemical bonds in the intermetallic phase enhances the Young's modulus of the alloy and, as counterpart, reduces the alloy ductility and its fracture toughness. Moreover, fracture toughness of Al-Li alloys is extremely sensitive to the content of alkali metal impurities. A few tens of ppm of Na and K, introduced during the ingot metallurgy casting for the production of the primary alloy, are sufficient to reduce the alloy fracture toughness below the safety values compatible with structural applications (Polmear 2006).

The main strengthening mechanism active in the early compositions (first generation, Table 3.12) of the Al-Li alloys is based on the dislocations cutting through the δ'-Al_3Li precipitates (Table 3.13). This intermetallic, not present in the relevant equilibrium phase diagram (Fig. 3.25), can be stabilized though using suitable thermal treatments.

The ordered structure of the δ'-Al_3Li precipitates introduces a further strengthening factor, associated with their interaction with dislocations and with the formation

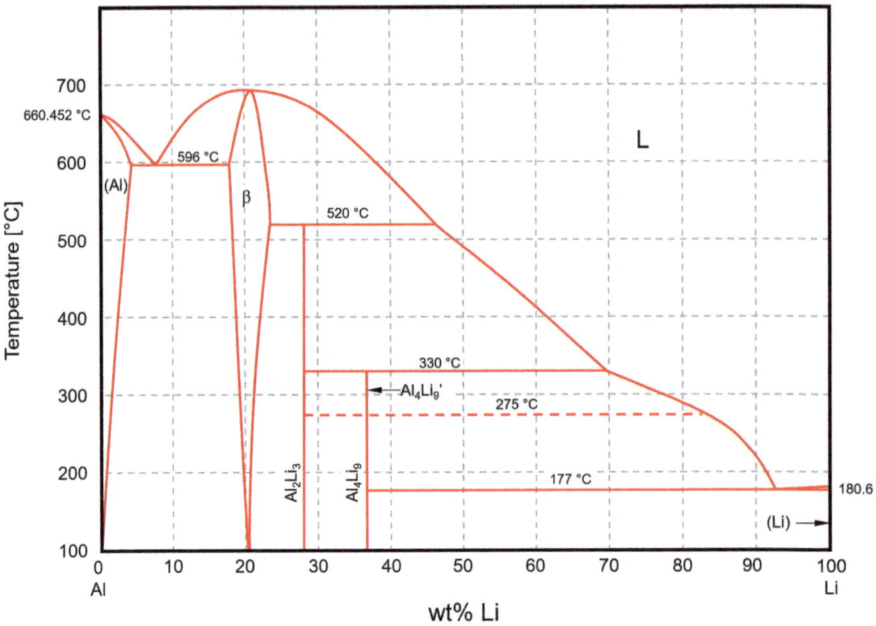

Fig. 3.25 Al-Li phase diagram. (Redrawn from ASM International 1992)

of complex defect structures, involving antiphase boundaries and super-dislocations, as in Ni-based superalloys (see Sect. 6.4.3). However, the localized shearing across the ordered precipitates may produce excessive strain hardening concentrated close to grain boundary, leading in the long run to diffuse cracking. This was one of the main reasons for the withdrawal of the *first-generation* alloys, e.g., the A2020 (see Table 3.12), used for the wing skin of the Northrop RA-5C Vigilante aircraft. The alloy compositions and relevant annealing treatments were developed further, aiming at the formation of other precipitates and dispersoids, capable to homogenize the dislocation structures. The reinforcing phases that formed in the novel alloy formulations were either finely distributed inside the alloy matrix, like the θ'-Al_2Cu, introducing lattice strain as obstacle to dislocation movement, or were semicoherent grains with a limited number of slip systems. This is the case of the T_1-Al_2CuLi and S'-Al_2CuMg intermetallics, not easily cut by dislocations, that thereby remain stuck at the precipitate-alloy interface. The formation of these phases is favored by the presence of alloying elements, like silver and zirconium, introduced in several of the *third-generation* Al-Li alloys. Another feature of these alloys is the optimum concentration of lithium, typically ranging from 1.0 to 1.3 wt%, anyway lower than 2.0%, or more, typical of the *second-generation* Al-Li alloys, for this reason featuring extremely low-density values and regarded as a promising metallurgical response to the early encouraging results obtained with the extra-light polymer matrix structural composites. Notwithstanding the interest in these alloys, their elevated brittleness and low toughness lead to a rapid abandon of these formulations and stopped any further development. A definitely better outcome was faced by the *third-generation* Al-Li alloys. Although relatively heavier, for the smaller concentrations of lithium, these alloys exhibited the right balance between workability and in-service performances. The development of these alloys received an important boost also from the results of the research conducted by metallurgists in the Soviet Union in the 1970s. The most recent Al-Li products, featuring the constant presence of copper, have been qualified for high strength and high damage-tolerant applications in aircraft structures, exhibiting even interesting corrosion resistance properties (Lequeu et al. 2010). Altogether the Al-Li alloys represent an important step in the development of light alloys. In fact, several Al-Li products have reached similar or better properties than the 2XXX and 7XXX traditional alloys, of which they represent excellent candidates for substitution (Table 3.14).

3.2.5 Aluminum Alloys in Aircraft Structures: Present State and Perspectives

The capability to meet quite complex and diverse requirements is an important aspect of the aluminum alloys and probably one of the main reasons for their success in aircraft structures. The situation is well depicted by the scheme in Fig. 3.26, summarizing the main properties for the principal parts and components of an aircraft structure, for which aluminum alloy represents an excellent choice.

Possible candidates like magnesium alloys (see Sect. 3.3) and structural polymer matrix composites that in the long run might take over the role of aluminum alloys are emerging. The B787-Dreamliner has demonstrated the possibility of using structural composites as reliable replacements for aluminum alloys (see Fig. 1.11 in Chap. 1). This result has been possible thanks to an integrated production approach, involving materials development and the implementation of novel manufacturing technologies. In some respect, this is what has also been made for the Airbus A380, this time with a priority for structural aluminum alloys. In this aircraft, an intelligent mix of improved new composition and workhorse alloys have been largely employed, as indicated by the list of aluminum parts given in Table 3.15 (Lequeu et al. 2007).

The large dimensions of the A380 (Fig. 3.27) and the optimization of the production costs, have required the development of new equipment for casting the primary alloys, for machining, for hot-rolling, and for handling the super-size slabs and plates (Lequeu et al. 2007). The use of several Al-Li alloys is coherent with the Airbus strategy that over the years has progressively reduced the percentage of conventional aluminum alloys used in their airplane structures and at the same time has increased the fraction of the lighter Al-Li alloys.

In other commercial airliners, like the B777 and A460-600, high strength and high damage-tolerant aluminum alloys belonging to the 2XXX and 7XXX groups, as those listed in Table 3.16, have been selected for the airframes, using products

Table 3.14 Products, requirements, applications of selected Al-Li alloys, with reference also to the substitution of alloys from other groups (Prasad et al. 2014)

Generation	Alloy	Temper	Product	Alloys to substitute	Requirements	Applications
2	8090	T3	Sheet	–	–	Floor installations, brackets, stiffeners, frames, spars, stringers, longerons, ribs, bulkheads
	8090C	T8		2014-T6	Medium strength	Flying control systems, avionics bay structures, cabin roof frames
	8090C	T81		2024-T3	Damage tolerance	Lower fuselage skin panels, flat roof panels, flying control structures
				2024-T42		
	8090	T852	Forgings	7010-T7451	Medium/high strength	Cabin roof and side frames
	8090	T8511	Extrusions	7075	Medium strength	Frames, brackets, stringers, bulkheads, door rails
3	2098	T8511	Sheet	2024-T3	Damage tolerance, medium strength	Fuselage/pressure cabin skin
				2524-T3		
	2198	T8		2524-T351		
	2199	T8E74				
	2060	T8E30				
	2199	T86	Plate	2024-T351 2324-T39	Damage tolerance	Lower wing covers
	2050	T84				
	2060	T8E86		2624-T351		
				2624-T39		
	2098	T82P		2024-T62	Medium strength	F-16 fuselage panels
	2297	T87		2124-T851		F-16 fuselage bulkheads
	2397	T87				
	2050	T84		7150-T7751 7055-T7751 7055-T7951 7255-T7951	High strength	Upper wing covers
	2055	T8X				
	2195	T82P				

	2050	T852	Forgings	7175-T7351	High strength	Wing/fuselage attachments, windows frames, crown frames
				7050-T7452		
	2099	T81	Extrusions	2024-T3511 2026-T3511 2024-T4312 6110-T6511	Damage tolerance	Lower wing stringers, fuselage/pressure cabin stringers
	2099	T83		7075-T73511 7075-T79511 7150-T6511 7175-T79511 7055-T77511 7055-T79511	Medium/high strength	Fuselage/pressure cabin stringers and frames, upper wing stringers, Airbus A380 floor beams and seat rails
	2055	T8R83				
	2065	T8511				

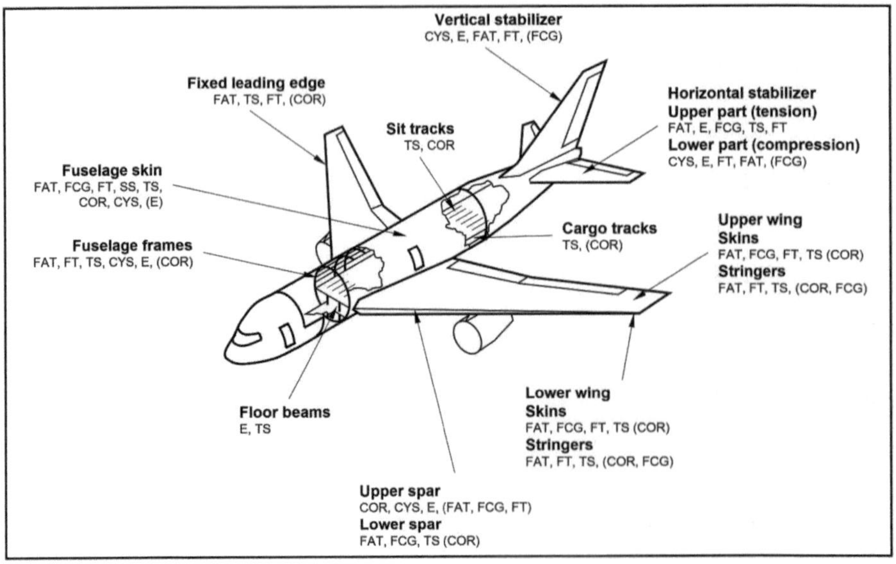

Fig. 3.26 Materials property requirements for a commercial airliner. Material properties abbreviations: *COR* corrosion, *CYS* compressive yield strength, *E* modulus, *FAT* fatigue, *FCG* fatigue crack growth, *FT* fracture toughness, *SS* shear strength, *TS* tensile strength. Material properties in brackets ( ) are regarded as important, but not critical, design requirements. (Redrawn from Polmear et al. 2006)

obtained by rolling, extrusion, or forging. In most cases, the starting batch is an ingot semi-continually direct-chill cast.

The extreme reliability of the conventional Al-alloys and the knowledge that has been accumulated over the years in more than one century represent also an important benchmarking for the development of new wrought alloys, like the recent formulations and tempers of Al-Li alloys discussed in Sect. 3.2.4 (Lequeu et al. 2010), particularly as concerns those mechanical properties important both for processing and in-service performances (Table 3.17).

Some structural aircraft components made of aluminum alloys are produced also by casting: sand casting, investment casting, and permanent mold casting. In this way, single-part components have been obtained, thus reducing or eliminating at all joining steps. The cast components are particularly well suited for replacing parts with complex geometries, requiring time-consuming and expensive machining. Another advantage of cast components is the elimination of all those defects that are typically associated with joints, like structural flaws and corrosion-initiation regions. However, since the resulting mechanical properties of cast products are on average lower than those of the corresponding wrought counterparts (see Sect. 3.2.2), the so-called *casting factor* (CF) has been introduced. This is a sort of safety factor, ranging from 1.0 to 2.0 used to scale down the reliable strength of the cast components, any time the material has an assigned CF value in excess of 1.0. A

Table 3.15 Aluminum alloy products used for different parts and components of the A380 (Lequeu et al. 2007)

Alloy	Temper	Form	A380 application
2024A	T351	Plates	Lower wing reinforcement
2027	T351		Lower outer wing panel (A380-800F)
2050	T84		Lower wing reinforcement
7010	T7651		Upper outer wing panel, heavier gauge wing ribs
7040	T7451		Fuselage main frames, cockpit, window frames, beams, fittings
7040	T7651		Wing spars (inner front and inner center part)
7056	T7951		Upper wing panels (A380-800F)
7449	T7951		Upper wing
7449	T7651		Lower gauge wing rib
2027	T3511	Heavy sections	Lower wing stringers
2196	T8511		Floor beams
7449	T79511		Upper wing stringers
2024HS	T432	Small sections	Fuselage frames
2196	T8511		Floor structure, fuselage stiffeners
6056	T78		Fuselage stiffeners
6056	T6		Fuselage stiffeners
7349	T6511		Seat rails, stiffeners of center wing box
7349	T76511		Fuselage stiffeners
6056	T78	Sheet	Pressure bulkhead below cockpit floor
6156Cl	T6		Fuselage panels

consolidated experience in using castings has been developed for military aircrafts. A few examples of quite an extensive use of Al-alloy castings in military aircrafts and helicopters are:

- The inlet duct of the General Dynamics F-16 Fighting Falcon.
- The pave tack pod of the General Dynamics F-111 and of the McDonnel Douglas F-4.
- The main support of the sensor for night vision in the Apache helicopter (Starke and Staley 1996).

The experience with military aircrafts has been exploited in the first place to reduce the impact of CF. Therefore, the use of castings has become a valuable and reliable choice also for primary structures. This has indeed happened for several Airbus aircrafts, starting from the A320 onward with the following main components:

- Baggage compartment door.
- Flap tracks.
- The inner structure of the passenger door.

The same choice has been adopted by Boeing too and has now become a standard (Rendigs 1994).

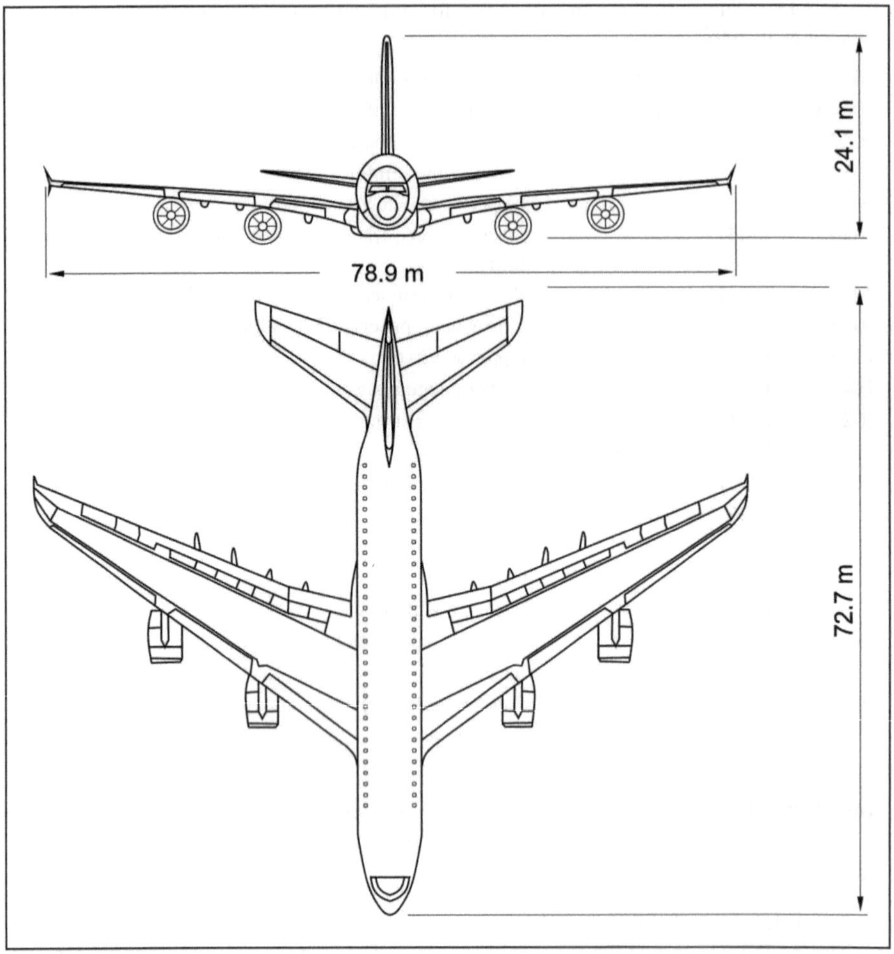

Fig. 3.27 Schematic of the Airbus A380

To complete the picture concerning the perspectives of aluminum alloys, a further alloying element is being considered for improving their structural properties: this is scandium. Scandium ($Z = 21$) is a light metal, having a density of 2.985 g/cm^3, and it is often associated with the group of RE elements, although it is not really part of this group. The early reported results on aluminum-scandium alloys date back to the early 1970s, both in the USA and in the Soviet Union. In 1971, the early Al-Sc alloys were patented. Meanwhile, several studies started at the VILS (All-Russian Institute of Light Alloys), demonstrating that it was possible to produce rather thick ingots of Sc-bearing Al-Mg alloys with non-dendritic structures. Subsequently, the investigations proceeded to assess the potential of these alloys in the aerospace field (Djukanovic 2017; Filatov et al. 2000; Röyset and Ryum 2005). The interest for Sc-containing alloys relies on the improved mechanical properties and processing

Table 3.16 Conventional wrought Al-alloys used for commercial aircraft parts (Polmear 2006; Staley and Lege 1993)

Alloy	Temper	Application	Alloy	Temper	Application
7090	T7E71	Aircraft parts	2024	T3	Fuselage skin
7091	T7E69		7075	T6	
2618	T61	Aircraft parts and structures for high-temperature applications	7475	T6	
2219	T62		7075	T6	Fuselage stringers
2219	T87		7075	T73	
2025	T6	Aircraft propellers, forgings	7475	T76	
6013	T4	Aircraft sheet	7150	T77	
2014	T6	Aircraft structures	6005	T5	General purpose extrusions
2124	T8		6060	T5	General purpose extrusions
2048	T85		6151	T6	Medium-strength forgings
7001	T6		2024	T3	Ribs and spars
7009	T6		7010	T76	
7010	T6		7150	T77	
7049	T73		2017	T4	Screw machine fittings
7050	T736		2011	T6	Screw machine parts
7075	T6		6061	T6	Welded structures
7075	T73		2024	T3	Wing lower panels
7075	T76		7075	T6	
7475	T7651		7175	T73	
7178	T6		2024	T3	Wing lower skin
7055	T7751		7475	T73	
7085	T7651		2024	T3	Wing lower stringers
2024	T4	Aircraft structures and sheets	7075	T6	
2024	T6		2224	T39	
2024	T73	Empennage (tail)	7075	T6	Wing upper skin
7075	T6		7150	T6	
7050	T76		7055	T77	
6063	T6	Extrusions	7075	T6	Wing upper stringers
2024	T6	Fuselage frames/bulkheads	7150	T6	
7075	T6		7055	T77	
7050	T6		7150	T77	

possibilities. In fact, scandium additions determine the highest specific increment of tensile strength in aluminum alloys than any other alloying element. Scandium has also attracted interest because its presence can lead to:

Table 3.17 Tensile properties of selected wrought Al-alloys and relevant tempers, widely used in aircraft structures (Polmear 2006)

Alloy	Temper	Tensile strength [MPa]	Yield strength [MPa]	%Elongation in 50 mm
2011	T6	390	295	17
2014	T6	480	410	13
2017	T4	425	275	22
2618	T61	435	330	10
2219	T62	415	290	10
2219	T87	475	395	10
2024	T4	470	325	20
2024	T6	475	395	10
2124	T8	490	440	8
2025	T6	400	255	19
2048	T85	480	440	10
6005	T5	305	270	12
6060	T5	215	195	17
6063	T6	240	215	12
6061	T6	310	275	12
6151	T6	330	295	17
6013	T4	315	185	25
7001	T6	675	625	9
7009	T6	535	470	12
7010	T6	545	485	12
7049	T73	530	470	11
7050	T736	550	510	11
7075	T6	570	500	11
7075	T73	500	430	13
7075	T76	540	470	12
7475	T7651	590	560	12
7178	T6	610	540	10
7055	T7751	630	610	12
7085	T7651	510	475	7
7090[a]	T7E71	620	580	9
7091[a]	T7E69	590	545	11

[a]Refers to alloys produced via powder metallurgy

- Better quality of welded joints due to the reduction in the incidence of cracks, with a consequent increase, up to 200%, in the fatigue life of the joints.
- Grain refinement during both casting and welding processes.
- Precipitation hardening.
- Recrystallization inhibition.

As concerns recrystallization, its temperature can be as high as 600 °C when scandium is added. This value is definitely above the typical heat treatment temperatures of aluminum alloys. A better understanding of grain refinement and hardening can be inferred from the Al-Sc phase diagram (see Fig. 3.28), featuring an eutectic on the Al-rich side, at 659 °C and 0.56 wt% scandium. On cooling from the melt a hypereutectic composition, the Al_3Sc intermetallic forms. This phase contributes to the strengthening of the alloy and also to the grain refinement during the solidification process. In fact, the intermetallic grains are preferred nucleation sites for the solidifying alloy, owing to the crystallographic affinity to aluminum. This aspect can be improved by the addition of titanium or zirconium, resulting in the formation of Al_3(Sc, Ti) or Al_3(Sc, Zr) particles, that, as shown by high resolution atom probe investigations, improves the control of the intermetallic formation kinetics, in agreement with the relevant transformation curves in Fig. 3.29 (Röyset and Ryum 2005).

Scandium has been added to non-heat treatable alloys of the 1XXX, 3XXX, 4XXX, and 5XXX series (Röyset and Ryum 2005). The percentage of added scandium goes from 0.1% to 0.45% (Scandium International Mining Corp 2017). The best combination of strength and other structural properties has been obtained with the Al-Mg alloys (5XXX series – Filatov et al. 2000). Thanks to the homogenous, fine-grained microstructure of the alloys, tensile strengths in the 300–450 MPa range are obtained, still retaining a 12–15% elongation, depending on the actual scandium concentration. The fine grain size has also positive effects on the extrusion flow rate (Scandium International Mining Corp 2017). Sc-containing Al-alloys are interesting for aerospace applications, with particular regard to non-treatable alloys, like welded gas tanks, structures for dashboard panels and compartments, fuselage stringers, and large stamped and welded structures (Djukanovic 2017; Röyset and Ryum 2005). The Russian military aircraft MIG 29 has the largest use of Sc-containing Al-alloys so far (Djukanovic 2017; Röyset and Ryum 2005). Airbus Group APWORKS Gmbh, in association with Airbus Group R&D, has developed recently the Scalmalloy®, a high-performance Al-Mg-Sc alloy specifically designed for the production of high-strength aerospace structures via laser powder bed additive manufacturing (see Table 3.18). The result is a material with exceptional high fatigue strength and specific strength approaching that of titanium alloys (APWORKS 2017; Djukanovic 2017).

It has been estimated that the use of high-strength Sc-containing alloys in aircrafts would afford a 10–15% weight reduction (Djukanovic 2017). Unfortunately, the extremely high cost of scandium (now higher than 2000 USD/kg) and its relatively low abundance and the relevant extraction difficulties (Djukanovic 2017) seem to be an important obstacle to a widespread application of this element, unless novel extraction technologies will maintain their improvement promises (Röyset and Ryum 2005). One approach is based on using, as raw material, ScO_3 from the “red mud” coming from the Bayer process (see Sect. 3.2.1). Alternatively, ScO_3 may be reduced directly in the liquid aluminum alloy. In this way the production cost is cut by one order of magnitude approximately (Röyset and Ryum 2005). Sc-containing alloys already on the market are the 1460, 2023, 5024, and 5025 (for full composition see Table 3.5 and Appendix 1). Other alloys, whose designation is not present in the IADS, are 1515 and 1570 (5XXX series; 0.30–0.50% and 0.15–0.35% Sc, respectively) and 1970 (7XXX series; 0.25% Sc).

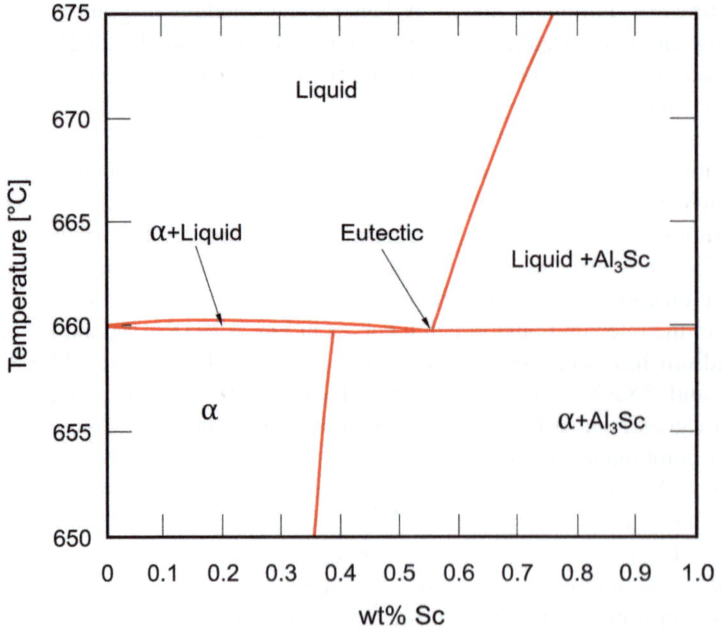

Fig. 3.28 Al-rich part of the Al-Sc phase diagram. (Redrawn from Röyset and Ryum 2005)

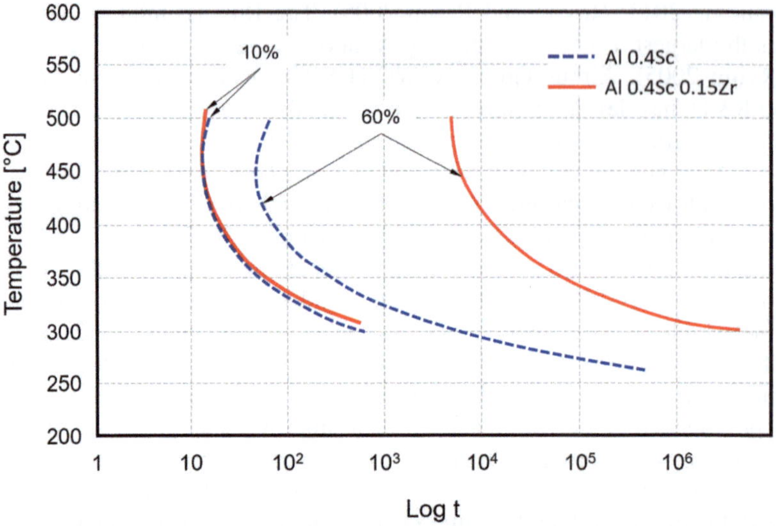

Fig. 3.29 Transformation curves for the formation of the intermetallic phases in Al-0.4Sc and Al-0.4Sc-0.15Zr (see main text for details). In the plot, the completion percentage of the transformation is also indicated. (Redrawn from Röyset and Ryum 2005)

Table 3.18 Typical mechanical properties of Scalmalloy®

Density [g/cm^3]	Young's modulus [GPa]	Ultimate tensile strength [MPa]	Yield strength [MPa]	% Elongation	Hardness vickers (HV 0.3)
2.67	70	520	470	13.0	180

Values are obtained according to the DIN EN ISO 2002-001 standard and measured from specimens that have been heat treated and machined. Values are typical of products obtained via ALM (Additive Layer Manufacturing) and are measured in the least strong direction (APWORKS 2017)

3.3 Magnesium Alloys

3.3.1 Introduction

Magnesium (atomic number $Z = 12$) is in the second group, third period of the periodic table. It is characterized by a density of 1.738 g/cm^3 and a Young's modulus of 45 GPa. Magnesium has several applications in the modern industry (e.g., steelmaking processes, Kroll's process for titanium extraction and alloying element in Al-alloys), and it is also an interesting metal for aerospace applications. Although aluminum and its alloys are holding a leading role in aerospace structures, this has not always been the case. Indeed, magnesium can be regarded as the less dense of the metals interesting for structural applications, leaving aside lithium and beryllium, not suited for bulk structural uses. The name magnesium derives from the Greek name *Magnesia*, a district of Thessaly, a traditional geographic and modern region of Greece (Davy 1808; Perkguleryuz et al. 2013). The existence of elemental magnesium was first claimed in 1755 by Joseph Black, who pointed out that magnesia mineral (MgO) contained a new element that, unfortunately, he was not able to isolate. In the past, magnesia was known as *white earth* or *white stone*. Magnesium was discovered by Sir Humphry Davy in 1775 (Davy also discovered aluminum, see Sect. 3.2.1). Sir Davy also suggested the name *magnium* for the metal, but magnesium was adopted instead. First extraction of the metal by Davy occurred in 1808, thanks to the electrolysis of moistened magnesium sulfate. A mercury cathode was employed in the process. The resulting product was a magnesium amalgam, from which magnesium was distilled by removing mercury. In 1828, Antoine Alexandre Brutus Bussy was able to isolate magnesium thanks to the fusion of dehydrated $MgCl_2$ with potassium at high temperature. Some years later, in 1833, Michael Faraday obtained metallic magnesium from the electrolysis of impure magnesium chloride in a molten state. However, it took two more decades to achieve production capabilities of commercially relevant quantities of metallic magnesium. The first who achieved this result was Robert Bunsen, who obtained pure magnesium using a small laboratory cell starting from anhydrous $MgCl_2$. At the beginning of its history, magnesium found considerable interest in Germany only, that in 1868 was the only producer in the world. The application of magnesium was not focused on structural components but on the production of powder for flash-lights and fireworks. Industrial production of magnesium started around the late 1920s in several countries.

Magnesium Elektron, a still existing company, was founded in 1934 and started its production 2 years later, using magnesia coming from Greece as raw material. However, when World War II was about to start, magnesia was directly extracted from seawater. In Italy, the SAMIS company built a production plant in Sardinia in 1935. The first owner was very much related to the aerospace field. His name was Giovanni Caproni regarded as one of the originators of the newborn Italian aircraft industry. In 1939 the production of magnesium also started in Bolzano, a town located in the Dolomites in the northern part of Italy. The plant was owned by the company Società Magnesio e Leghe di Magnesio. One interesting peculiarity of the Bolzano plant was its silicothermic production process, named Bolzano process (Horst and Mordike 2006; Perkguleryuz et al. 2013).

Magnesium is one of the most common elements on the Earth: according to the estimated total abundance, it is in the fourth place, after iron, oxygen, and silicon. Magnesium occupies the eighth position in the list of the most abundant elements in the Earth's crust (lithosphere), considering its actual availability to mining and extraction processes, using the so-called Clarke number. This is the concentration of each element referred to a surface crustal layer, and although magnesium is preceded by metals like iron and aluminum, it still has a larger Clarke number than other strategic raw materials, like nickel, chromium, and so on (Parker 1967). Furthermore, magnesium is the second most abundant metal in seawater, the first being sodium (Perkguleryuz et al. 2013). Primary magnesium can be extracted from very common carbonate ores, like magnesite ($MgCO_3$) and dolomite ($CaCO_3 \cdot MgCO_3$). In other words, the availability of magnesium as raw material seems not to be an issue, even in view of future developments and increase in the production. The extraction metallurgy of metallic magnesium is based on two main processes: electrolysis and thermal reduction of the ores.

The extraction of magnesium requires a relatively high energy consumption for the strong chemical bonds it has with other elements in the ores. From the 1970s to the 1990s, the production of Mg was primarily performed according to the electrolytic route (Table 3.19). Since the chemical reaction that takes place during the electrolytic process is a redox process, it can be considered as made of two semi-reactions, the cathodic and the anodic one (see Chap. 8). The electrolytic route for the production of magnesium is well established for both $MgCl_2$ coming from traditional raw materials (see Table 3.19) and for the recycled molten anhydrous magnesium chloride deriving from the Kroll's process used for the extraction of titanium (see Sect. 4.1). A few details on the thermal reduction of magnesium ores and major process parameters are given in Table 3.20.

As concerns the silicothermic extraction process, it is based on two main approaches: the Bolzano process and the Pidgeon process. In both cases, silicon coming from a ferroalloy, named ferrosilicon (FeSi), is used as reducing agent. This same agent is used for steel deoxidation and in the reduction stage during the steel conversion process (see Sect. 5.3) (Gasik 2013). The Bolzano process, using dolomite as raw material, was initiated by Edoardo Ravelli, who industrialized it on the basis of the silicothermic reduction process developed by Amati in 1938 at the University of Padua (Italy). The plant, based in Bolzano (Italy, from which the name

Table 3.19 Electrolytic processes for the production of magnesium

Process	Raw material	Raw material preparation	Chemical reactions involved	Temperature [°C]	Pressure [kPa]
AM	Magnesite	Mining, leaching with HCl, dehydration	Cathode semi-reaction $2Cl^- \rightarrow Cl_2 + 2e^-$ Anode semi-reaction $Mg^{2+} + 2e^- \rightarrow Mg$ Global redox reaction $MgCl_{2(l)} \rightarrow Mg_{(l)} + Cl_{2(g)}$	700–800	101.325
Dow	Seawater/ desalination brine	Neutralization, purification, dehydration			
IG Farben		Neutralization, prilling, dehydration chlorination			

AM indicates a process based on the chlorination of MgO obtained from thermal decomposition of magnesite ($MgCO_3$). In the IG Farben process, the dehydration step is optimized by prilling (Wulandari et al. 2010)

of the process comes), used ferrosilicon to reduce calcined dolomite. The production reached a peak in 1981 with 10,700 metric tons and continued until the closing down of the plant in 1992, when the costs of magnesium production became too high, thus not profitable any longer. In the Bolzano process, still used in Brazil by a company named Bramag, dolomite undergoes a calcination first, and it is then mixed and compacted with the ferrosilicon reduction agent. Bauxite may be employed as an additive in the process. The resulting material is prepared in the form of large blocks that are connected to the electric-heating conductors. Thanks to this particular design, the blocks only are heated up, thus avoiding heat dispersion over the entire furnace. The process is carried out at a pressure of about 0.4 kPa and at a temperature of 1200 °C, a relatively low value if compared to the typical values for non-silicothermic-based processes (Table 3.20), so that also the energy consumption is comparatively lower with respect to other thermal reduction processes. The alternative silicothermic extraction process for magnesium, i.e., the Pidgeon process, was developed by Lloyd Montgomery Pidgeon in the late 1940s. The reduction of dolomite, like in the Bolzano process, is carried out in a reactor using ferrosilicon. The process uses a furnace (retort) in which a briquetted mixture of ferrosilicon and calcined dolomite powder are heated in order to obtain the reduction of magnesium oxide coming from the carbonate decomposition. High-purity magnesium is produced and then remelted and cast in form of ingots. The process is carried out at a temperature of about 1500 °C, thus higher than that of the Bolzano process, and over a pressure range of 0.013–0.4 kPa. The retort can be coal-fired, gas-fired, or electrically heated (Horst and Mordike 2006). Notwithstanding specific differences, all magnesium extraction processes can be traced back to the same main functional stages, according to the schematic in Fig. 3.30.

Magnesium has a hexagonal close-packed (*hcp*) structure, whose limited number of slip systems (three slip systems) renders this metal and its alloys intrinsically difficult to be deformed, definitely more difficult than austenitic *fcc* alloys.

Table 3.20 Processes involving thermal reduction of Mg-containing ores (Horst and Mordike 2006; Wulandari et al. 2010)

Process	Raw materials	Raw material preparation	Main chemical reactions involved	Temperature [°C]	Pressure [kPa]
Aluminothermic	Dolomite, Al scrap	Calcination	$4MgO + 2Al \rightarrow 3Mg + MgAl_2O_4$	1700	86.126–101.325
Carbothermic	Magnesite, carbon	Calcination and pelleting	$MgO + C \rightarrow Mg + CO$	1700	101.325
Magnetherm	Dolomite, bauxite, FeSi	Calcination and FeSi formation	$2CaO \cdot MgO + (xFe)$ $Si + nAl_2O_3 \rightarrow 2CaO \cdot SiO_2 \cdot nAl_2O_3 + 2Mg + xFe$	1550	5.066
Mintek	Dolomite, bauxite, FeSi, Al scrap	Calcination	$2CaO \cdot MgO + (xFe)$ $Si + nAl_2O_3 \rightarrow 2CaO \cdot SiO_2 \cdot nAl_2O_3 + 2Mg + xFe$ $4MgO + 2Al \rightarrow 3Mg + MgAl_2O_4$	1700	86.126
Silicothermic	Dolomite, FeSi	Calcination, FeSi formation, pelleting	$MgO + CaO + FeSi \rightarrow Mg + Ca_2SiO + Fe$	1160–1500	0.013–0.4

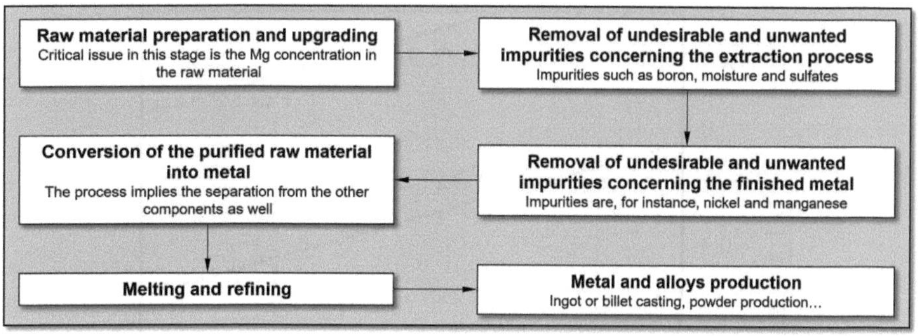

Fig. 3.30 Flowchart for a general magnesium production process (Perkguleryuz et al. 2013)

Table 3.21 A comparison among properties of key elements for structural alloys, including magnesium

Property	Mg	Al	Ti	Fe
Density [g/cm^3]	1.74	2.70	4.51	7.87
Melting *T* [°C]	650	660	1670	1538
Boiling *T* [°C]	1105	2470	3287	2862
Thermal conductivity (RT) [W/ °C · m]	160	235	22	80
Specific heat (RT) [J/kg °C]	1025	904	520	449
Young's modulus [GPa]	45	70	120	210
UTS [MPa]	90	98	150	275

UTS indicates the ultimate tensile strength (Donachie 2000; Pepperhoff and Acet 2001; Perkguleryuz et al. 2013)

Table 3.21 gives some properties of this metal, as compared with the same properties of other key metals: aluminum, titanium, and iron. The low density of magnesium stimulated the interest for this metal in the early days of the aerospace industry.

Starting from the military field, the first important example of the use of Mg-alloys in aeronautic structures was probably the prototype concept of all-magnesium aircraft, the Northrop XP-56, the so-called Black Bullet. It was tested in experimental flights in the latest years of World War II. The Black Bullet was important, since it demonstrated the real potentials of Mg-alloys in aircraft structures. Another full magnesium structure was there in the Lockheed F-80C. Magnesium and its alloys started to be used in large military transportation airplanes, like the Convair B-36, with a total amount of 8600 kg of magnesium, of which about 5600 kg of magnesium sheet and 700 kg of forged parts and 300 kg of castings (Ostrovsky and Henn 2007). Another notable example of application of Mg-alloys to aircraft structures is the Convair XC 99, which serviced from 1949 until 1957, and is considered as the first double-deck airliner. This airplane had several components of the structure and fuselage skin made of Mg-alloys. The exceptional payload and passenger capacity (i.e., 400 fully equipped troupes), allowed by the double-deck design, rendered this aircraft particularly attractive also for civil and commercial

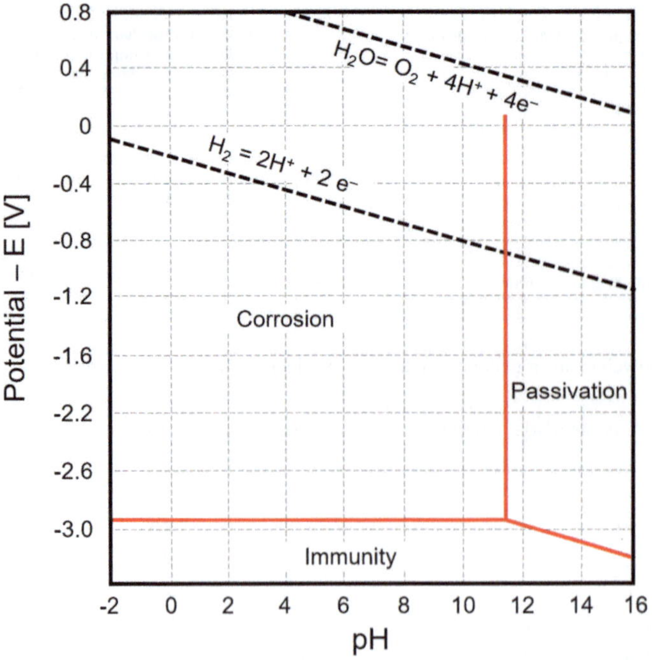

Fig. 3.31 Pourbaix diagram of pure magnesium. (Redrawn from ASM International 2003)

purposes. In fact, Pan Am ordered 15 of these aircrafts for its fleet, although no one of them was delivered and never started its service, because the program was first suspended and then cancelled. Among the reasons for this change of plans, there were the emerging limitations and critical aspects exhibited by magnesium alloys available at the time, particularly as concerns their surface durability.

The poor corrosion resistance is particularly evident when analyzing the Pourbaix diagram of pure Mg in aqueous environment (Fig. 3.31), where corrosion takes place over a wide range of pH and potential. Since it was not an issue in the military field, this feature was initially overlooked, although it emerged soon as an important aspect for commercial fleets in terms of reliability and costs associated with maintenance and repair. These aspects and the contemporary rapid, solid evolution of new aluminum alloys determined the progressive abandon of magnesium in air transportation. Quite different was the situation in the Soviet Union, where military aircraft structures were systematically used for prototyping solutions to be used for civil transportation purposes, with the possibility of faster development and validation loops for new materials. The Boeing 737 (produced since 1967) features an extremely limited number of magnesium components, summing up to a total weight of a few tens of kilograms, much lower if compared to the Tupolev TU-134,

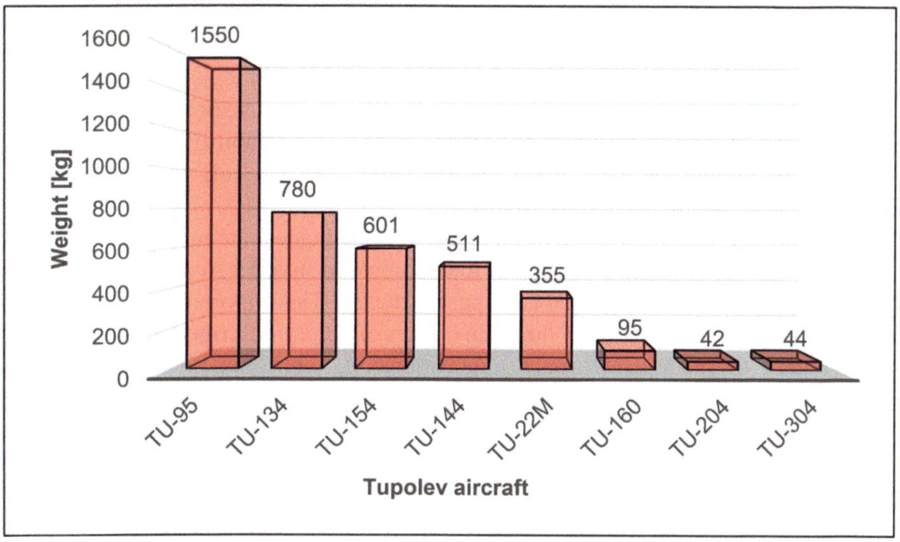

Fig. 3.32 Evolution of the weight of magnesium used for the different models of Tupolev planes, developed over the years (Gupta and Gupta 2017; Ostrovsky and Henn 2007)

developed over the same years (early production date 1963). In this case, the components made of magnesium alloys had a total weight of 780 kg. The Tupolev planes can be taken as important witnesses of the use of magnesium alloys for civilian air transportation. After the golden years for magnesium, 1960s and 1970s, a progressive reduction has been observed also in the Soviet Union, as can be seen in Fig. 3.32, referring to the use of magnesium in Tupolev airplanes. The situation in Russia in the 1990s was substantially similar to that of the Western aerospace industry, whose main players never used magnesium in their commercial and passenger airplanes intensively. The same is not true for helicopters, for which magnesium is a standard choice for gearboxes, transmission between engine and rotor, and other components, mainly nonstructural. In this case, weight saving is a decisive selection parameter. The comparatively limited applications of magnesium in aeronautic structures, if compared to aluminum alloys and notwithstanding the interesting properties and potential of this class of materials, can be ascribed to a set of concurrent reasons, in addition to the already mentioned poor corrosion resistance. The elevated oxidation kinetics, resembling combustion under particular conditions, has been obviously regarded as a negative issue in applications for which safety and reliability are paramount, although the magnesium combustion issue is to be regarded as an example of misconception. In real terms, combustion may become an actual risk for magnesium powder samples only and requiring some precautions in handling, for instance, scraps from machining of magnesium parts.

All commercial magnesium alloys are compliant and fully certified as concerns the resistance to ignition, although limitations to their use in specific components have been imposed, owing to the strict safety standards and legislation adopted for aerospace products. However, in recent times, a ban on the use of magnesium alloys for aircraft seats has been lifted, once the presumed flammability risk of candidate magnesium alloys has been ruled out with lab size and full-scale experiments (Marker 2013; Nature Materials 2016).

In addition to the limited corrosion resistance, the inadequate mechanical strength was identified as another key property to be improved in order to enable a wider use of magnesium alloys also in the aerospace industry. The important research efforts resulted in the developments of the main classes of alloys, described in the following with regard to specific aerospace applications. As concerns processing, casting appeared from the beginning as the obvious choice, considering the comparatively low-melting temperature of magnesium and its low latent heat of fusion, which renders solidification and cooling rather fast. The role of casting techniques was strengthened further by the intrinsically low deformability of magnesium for its limited number of slip systems. The development of more ductile alloys with novel compositions has incremented the number of wrought magnesium alloys available on the market. In this regard, the mechanical properties achieved by the state-of-the-art magnesium alloys, like the high-strength Elektron 21 and Elektron 675 materials, make them potential valid replacements for aluminum alloys, laminate, and polymer matrix composites in aeronautic structures (Lyon et al. 2007). This trend is encouraged by the always standing issue of the weight reduction, that is, nowadays considered also for the important beneficial effects, concerning fuel saving and consequent reduction of environmental pollution (see Chap. 2). Since the 1990s, there has been renewed interest in magnesium alloys for weight-sensitive applications, including the aerospace field. At the beginning of that period, there were a few commercial magnesium alloys systems: Mg-Al, Mg-Al-Zn, Mg-Zn-RE, and Mg-REs. Most magnesium alloys usage for structural applications was in die-casting, and a single alloy (AZ91D) accounted for around the 90% of the total market. A clear indication of the increasing, renewed interest for magnesium can be also inferred by the growing trend of magnesium production, started in the early 1990s, under the incisive drive by China, that is holding nowadays 80% of the global supply of this element (see the historical development in Fig. 3.33). When China entered the global economy in the early 1990s, it made magnesium one of the fundamental items of the raw materials and metallurgy industry portfolio. The base approach is the silicothermic Pidgeon production process, facilitated by the large availability of low-cost ferrosilicon and using coal as main heat source. This evolution has probably fostered the interest in magnesium also in the rest of the world, particularly as concerns its use in civilian transportation, as reported, for instance, in the research documents by the European Community AEROMAG program for the aerospace application of magnesium alloys (Perkguleryuz et al. 2013).

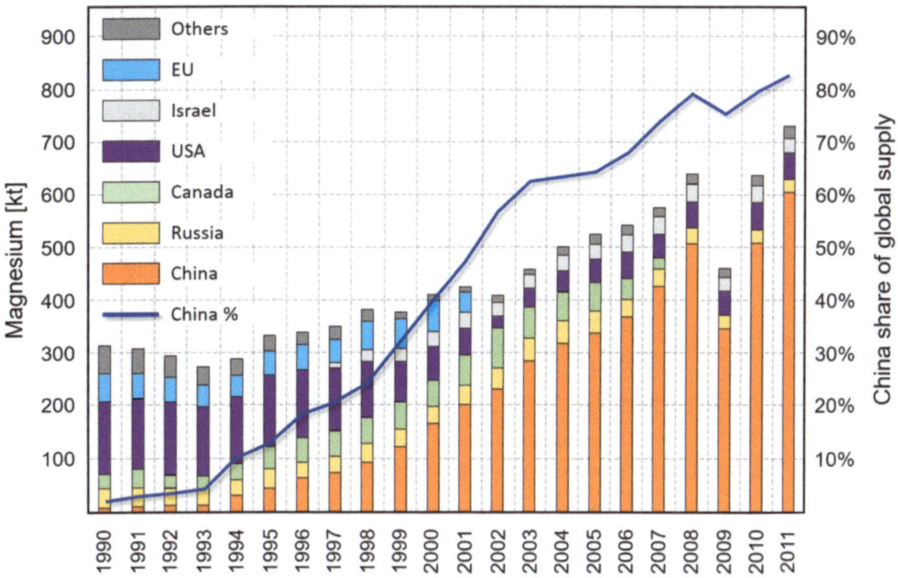

Fig. 3.33 The global magnesium industry in 2011. Quantities are expressed in thousands of metric tons. (Redrawn from The CM Group 2012)

3.3.2 Aerospace Magnesium Alloys: Metallurgy, Main Designation System, and Tempers

The ASTM (American Society for Testing and Materials) international designation system is the most widely accepted for both cast and wrought magnesium alloys, although it is not the only one, being a possible alternative of the so-called British system. The alphabetic part of the alloy name refers to the main alloying elements, according to the symbols listed in Table 3.22. The first digit of the code represents the rounded off concentration (in wt%) of the majority alloying element, identified by the first letter. The second figure corresponds to the concentration of the second element and so on. Additional letters, following the concentration reference numbers, represent possible modifications to the same alloy material. Some discrepancies are there, but it is not difficult to sort them out and identify the actual alloy product, also thanks to the conspicuous available literature on this matter (Cahn et al. 2005; Cardarelli 2008).

The AZ91 code indicates the Mg-9Al-1Zn cast alloy, with the relevant composition range and minority elements indicated in Table 3.23. The ZMC711 indicates the wrought alloy with composition: for Zn-6.5-Mn 0.75-Cu-1.25 (Polmear 2006). Therefore, there is no specific differentiation between the cast and wrought products, as seen instead for aluminum alloys. Other magnesium alloys are simply identified by the relevant trademark, like the Elektron alloys produced by Magnesium Elektron, and codenamed using a reference number, like the Elektron 21 alloy. A

Table 3.22 ASTM designation code for the main classes of magnesium alloys (ASM International 1991)

Code letter	Alloying element	Code letter	Alloying element
A	Aluminum	M	Manganese
B	Bismuth	N	Nickel
C	Copper	P	Lead
D	Cadmium	Q	Silver
E	Rare earths	R	Chromium
F	Iron	S	Silicon
G	Magnesium	T	Tin
H	Thorium	W	Yttrium
K	Zirconium	Y	Antimony
L	Lithium	Z	Zinc

Table 3.23 Chemical composition of some magnesium alloys

Alloy	Al	Mn	Th	Zn	Zr	Other elements
AM50	5.0	0.30	–	–	–	–
AZ31	3.0	0.20	–	1.0	–	–
AZ61	6.5	0.15	–	1.0	–	–
AZ81	7.0–8.0	0.13–0.35	–	0.40–1.0	–	Be = 0.0008 max
AZ91	8.1–9.2	0.17–0.35	–	0.40–1.0	–	–
AZ91E	9.0	0.10	–	0.50	–	–
AZ92	8.3–9.7	0.15–0.50	–	1.7–2.3	–	Be = 0.0008 max
AZ92A	9.0	0.10	–	2.0	–	–
ELECTRON 675	General chemical composition is property of Magnesium Elektron					
ELEKTRON 21	–	–	–	0.50	0.20–0.50	Gd = 1.0–1.7 Nd = 2.6–3.1
EZ33	–	–	–	2.0–3.0	0.50–1.0	RE = 2.5–4.0
EZ33A	–	–	–	2.0–3.0	0.40–1.0	RE = 2.5–4.0
HK31	3.0	–	3.25	0.30	0.70	–
HM21	–	0.45	2.0	–	–	–
LA141	1.0	–	–	–	–	Li = 14.0
LS141A	0.50–0.80	–	–	–	–	Li = 13–15
QE22	–	–	–	–	0.40–1.0	Ag = 2.0–2.7 RE = 2.0–3.0
WE43A	–	–	–	–	0.70	RE = 3.4; Y = 4.0
WE54	–	–	–	–	0.50	Ag = 5.25 Nd = 3.5
ZE41	–	–	–	4.2	–	RE = 1.2
ZE63	–	–	–	5.5–6.0	0.40–1.0	Cu = 0.10 max Nd = 2.1–3.0
ZK60	–	–	–	5.5	0.45	–
ZMC711	–	0.75	–	6.5	–	Cu = 1.25

Chemical compositions can slightly change according to the type of product (i.e., sheets, forgings, extrusions) (ASM International 1991; Czerwinski 2011; Magnesium Elektron 2012; Magnesium Elektron 2014, 2015; Perkguleryuz et al. 2013; Polmear et al. 2017; Prasad and Wanhill 2017)

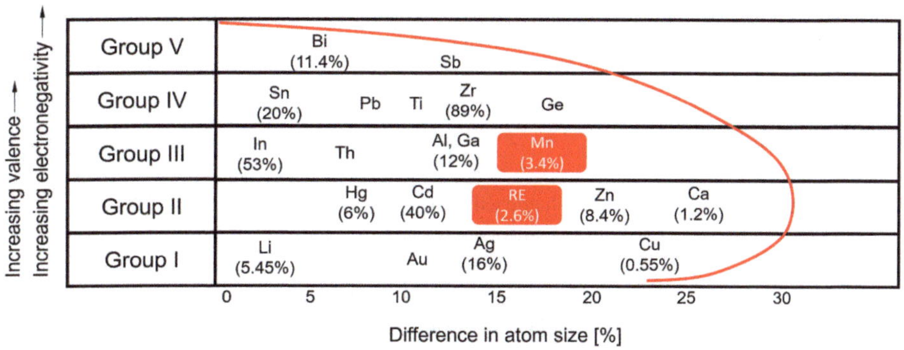

Fig. 3.34 Alloying behavior of magnesium. Values among brackets are the solid solubility limits of the element in Mg. (Redrawn from Pekguleryuz et al. 2013)

list of some of the most common magnesium alloys is given in Table 3.23, including their compositions.

The compositions of the magnesium alloys are designed according to the standard physical metallurgy principles for the formation of solid solutions and intermetallic phases, aiming at the optimization of the structural properties and corrosion resistance under different operating conditions. The graph in Fig. 3.34 shows how atomic size, electronegativity, and valence influence the solid solubility of several elements in magnesium, providing different effects, as specified in Appendix 3. Elements to the left of the curve have at least 0.5% maximum solid solubility in magnesium. The highest solubility belongs to those elements showing favorable size factor and valence, meaning a low difference in atom size and a valence equal to 2 or 3. Considerable solid solution strengthening occurs at around 12–15% size factor values, for instance, with Al, REs, and Zn. When Hume-Rothery rules are not satisfied, second phase formation occurs (Perkguleryuz et al. 2013). An important difference with respect to the tempers of Al-alloys is that the equilibrium phases foreseen by the relevant phase diagrams are not to be regarded as detrimental, as in the case of overaged aluminum alloy, but absolutely beneficial, being important strengthening agents. The equilibrium phases that are interesting as concerns precipitation hardening of magnesium alloys belong to three main intermetallic types. The first is the simple cubic CsCl structure such as that of MgTl, MgAg, CeMg, and SnMg. The second is that of the Laves phases, with an AB_2 stoichiometry like $MgCu_2$, $MgZn_2$, and $MgNi_2$, each featuring a different stacking sequence of the closest packed planes. Finally, the third type is that of the compounds with the antifluorite structure, like Mg_2Si, and with the fluorite structure, like Mg_2Sn. The equilibrium phases used as strengthening agents demonstrate the intrinsic stability of the alloy conditions induced by treatments in these alloys.

The equilibrium intermetallic phases are important in magnesium alloys also for their beneficial effect on creep resistance and for hindering alloy recrystallization (Pekguleryuz et al. 2013). Concerning the heat treatments of Mg-alloys, the temper condition is indicated in the alloy designation code, according to the specifications

Table 3.24 Designation letters of the main tempers for magnesium alloys (ASM International 1991; Pekguleryuz et al. 2013)

Code	Temper description	Details
F	As-fabricated	–
O	Annealed	–
H	Strain hardened	H1: strain-hardened only
		H2: strain-hardened and partially annealed
		H3: strain-hardened and stabilized
		The number following the letter H indicates the degree of strain hardening remaining in the alloy. Additional digits (H2X) indicate a subdivision of the H tempers giving information about the final degree of strain hardening between 0 (annealed) and 8 (totally hard)
T	Tempered	T4: solution heat treatment
		T5: artificial age hardening
		T6: solution heat treatment followed by artificial aging
		T7: solution heat treatment and subsequent stabilization
		T8: solution heat treatment, cold working, and artificial aging
W	Solution heat treated	–

contained in Table 3.24 with the main temper treatments for magnesium alloys. Those more commonly used for aerospace alloys are the T4 (solution-treated alloy), T5 (direct cooling and artificially aged), and T6 (solution-treated and artificially aged) (Eswara Prasad and Wanhill 2017).

3.3.3 Magnesium Alloy Processing

The alloy design and development have essentially concerned cast products, which still cover the majority of magnesium alloy production, notwithstanding the continuous progress in the development of deformable alloys, as discussed in the next section. Casting, using different techniques, with some restrictions and recommendations determined by the target grade and by other properties required by specific applications, is still the most flexible processing route for magnesium alloys. The main casting methods that are used for processing magnesium alloy parts are:

- Investment and die-casting.
- Permanent and semi-permanent mold casting.
- Sand casting.

A full description of these technologies, specifically applied to magnesium and its alloys, can be found in Polmear (2006), in the *ASM Specialty Handbook: Magnesium and Magnesium Alloys* (1999) and in the *ASM Handbook: Volume 15 – Casting* (2008). Wrought magnesium alloys, featuring in general better mechanical properties than cast alloys, although more challenging to be produced for the objective difficulties concerning their formability and therefore production costs, are gaining increasing importance. In fact, the possibility of using magnesium alloys as wrought products is paramount to sustain the general tendency to increase further the overall volume of magnesium products. As concerns aerospace applications, with particular regard to civil and commercial aircrafts, it is clear that the enhancement in the workability of magnesium alloys is a necessary step forward to widen the possibilities of these alloys in this field. Considering the standard parts of a commercial airliner, like the B757, it turns out that only 1% in weight are castings, whereas worked parts are the majority: 28% extruded components, 20% sheets, 18% forgings, 16% plates, and 13% bars (Kettner et al. 2007). The interesting aspect is that the main materials of which magnesium alloys are potential replacements, aluminum alloys and structural composites, sum up in this case to 78% and 3%, respectively, which is an extremely large volume. Continuous research efforts, backed by industrial and scientific projects, are actively working to find reliable solutions to this strategic issue. The EU project AEROMAG (6th Framework Program of the European Commission 2005) can be regarded as an interesting example of this challenge. The program was aimed at developing new magnesium wrought products (sheets and extrusions), particularly for aerospace structures and components, with excellent static strength and fatigue resistance. The benchmarks for these novel materials were two aluminum alloys: 5038 for nonstructural applications and 2024 for secondary structures. The three independent slip systems of the *hcp* Mg crystallographic structure are not sufficient to guarantee the homogeneous polycrystalline plastic deformation, since, according to the von Mises criterion, at least five of these slip systems are necessary (Hosford 2010). To overcome this intrinsic obstacle, higher temperatures are required to achieve adequate workability. Accordingly, temperatures close to 300 °C are absolutely recommended, and, indeed, magnesium alloys used in the aircraft industry are mechanically formed over a temperature range between 290 and 450 °C (Davis et al. 2007). Similarly, to make the process more effective, accurate, and fast, a preliminary heating up of the working tools at temperatures ranging from 250 to 350 °C is adopted, if necessary. In general, these temperatures are kept constant during the entire process, which is thus conducted under nearly isothermal conditions.

3.3.4 Development of Aerospace Magnesium Alloys

As seen, magnesium alloys experienced an alternating interest in the field of aerospace applications since the beginning of the aerospace era. Early compositions showed, in fact, several features that, in the long run, limited their application in aircraft structures. The main development lines of these alloys have been focused on the improvement of the following properties:

- Corrosion resistance, that in the early formulations was relatively poor.

- High-temperature creep resistance, a property that is particularly interesting for the relatively high-temperature applications of magnesium alloys, ranging from 150 up to 350 °C.
- Strength and elongation (Pekguleryuz and Celikin 2010).

A galvanic attack (see Sect. 8.1) is often observed at the contact region of a magnesium alloy part with dissimilar alloys. Heavy metals, reaching the magnesium surface from atmospheric deposition, may produce a generalized pitting attack, due to a localized galvanic coupling (Esmaily et al. 2017). The protection of the alloy surface by suitable coating systems, namely, synthetic and electrical insulating ones, is probably the most effective and practical strategy. Corrosion failures are usually associated with internal or surface contamination. In the first case, copper, iron, and nickel contamination can be introduced during casting operations. Another internal contamination phenomenon can be due to the inclusions coming from salt fluxing employed during casting and recycling operations. On the other side, concerning surface contamination, blast residues can have detrimental effects on the corrosion resistance of the material. These residues would typically come from blasting operations, used for improving the surface uniformity and appearance (Horst and Mordike 2006). In this case, a twofold corrosion protection strategy can be implemented: firstly, through the removal of the surface contaminants by surface treatments (e.g., chemical etching and sand blasting), possibly in association with the deposition of protective coatings. Secondly, a better control of the processing conditions, particularly during casting operations, will reduce the occurrence of contaminants and impurities automatically. The improvement in the alloy grade has generally resulted in better performing materials, also as concerns corrosion resistance. Of course, the search for better magnesium alloys, in terms of corrosion resistance, has been also the opportunity for developing novel coating systems and surface treatments. The surface protection strategies developed for these alloys are based on:

- Anodizing.
- Cold-spraying deposition.
- Conversion coatings.
- Electrochemical plating.
- Gas-phase deposition.
- Laser surface alloying and/or cladding.
- Organic coatings and paints.
- Plasma gel coatings.

Organic and composite coatings have also been specifically designed for aerospace industry and then transferred to other fields of applications. These coatings, being directly exposed to external conditions and undergoing structural and chemical evolution under environmental and operating conditions, are prone to release degradation products. Of course, also for organic coatings, the weight issue is important. Therefore, single rather than multiple layers are preferable. Possible alternatives are the wet methods, like conversion coatings, electrochemical plating, anodizing treatments, etc. (Blawert et al. 2006).

They are in general less expensive than advanced organic coatings, although still requiring particular precautions for the presence in their formulations of chromates and cyanides, both toxic to the environment and human health. In this regard, dry methods, using thermal spraying, laser melting, physical and chemical vapor deposition, and solid diffusion, are certainly more appealing. The best choice, as usual, is a trade-off of several factors, including the costs of the infrastructures and consumables. Considering the complexity of the requirements and conditions, composite coatings, like the Gardobond® coating, and surface treatments, like PGA ALGAN 2M, are attracting specific research interests in this field (Ostrovsky and Henn 2007). These coatings would not only make magnesium alloys as stable as aluminum but provide also an excellent surface finish for subsequent application of paints for an improvement of the surface barrier effect and also for decorative purposes.

In the aerospace industry, the first magnesium alloys that were developed contained Al, Mn, and Zn. AZ81, AZ91, and AZ92 are important early representatives of this group (Table 3.23). Referring the reader to Appendix 3 for a complete description of role of alloying elements, it is herewith reminded, as concerns these important aerospace alloys, that aluminum has positive effects on the mechanical properties of the alloy, enhancing its ultimate strength. Furthermore, aluminum improves creep resistance and the mechanical properties at elevated temperatures. Zinc has a strengthening effect too, although it is also important for the grain refinement, in particular, by preventing excessive grain growth during solidification and cooling. Zinc-containing alloys also show improved workability, quite a critical issue for these alloys. Zinc is generally associated to manganese. This latter element, in addition to the enhancement of the creep resistance of the alloy, has a central role in improving its corrosion resistance. It is now clear that even very little concentrations of iron (a tolerance level of 170 ppm of Fe is reported), generally deriving from contamination during primary casting, have an essential role in the corrosion behavior of magnesium alloys (Atrens et al. 2015). By controlling the alloy grade and the formation of stable Mn-Fe intermetallic phases, that neutralize the galvanic action of iron, a significant improvement in the corrosion resistance of the alloy can be attained. Very much the same effect of iron has been observed also with other impurities, particularly transition metals, determining detrimental galvanic couplings with the alloy. The intermetallic precipitates and secondary phases, like those obtained with temper treatments (Table 3.24) and having in general a higher electrochemical potential than the magnesium alloy matrix, affect the corrosion behavior of the alloy. Also in this case, the prevailing mechanism is a localized corrosion attack. Actually, the effect of impurity and secondary phases can be inhibited if their average grain size is kept below a critical value, estimated in 4 nm for standard aluminum alloys. The same can be assumed for magnesium alloys, considering that on this length scale, the effects of corrosion can be mended by the growth of surface reaction layers. The beneficial effects of a controlled refinement of the alloy microstructure will be recalled next with reference to Mg-Li alloys, being both these elements particularly reactive, and with a low

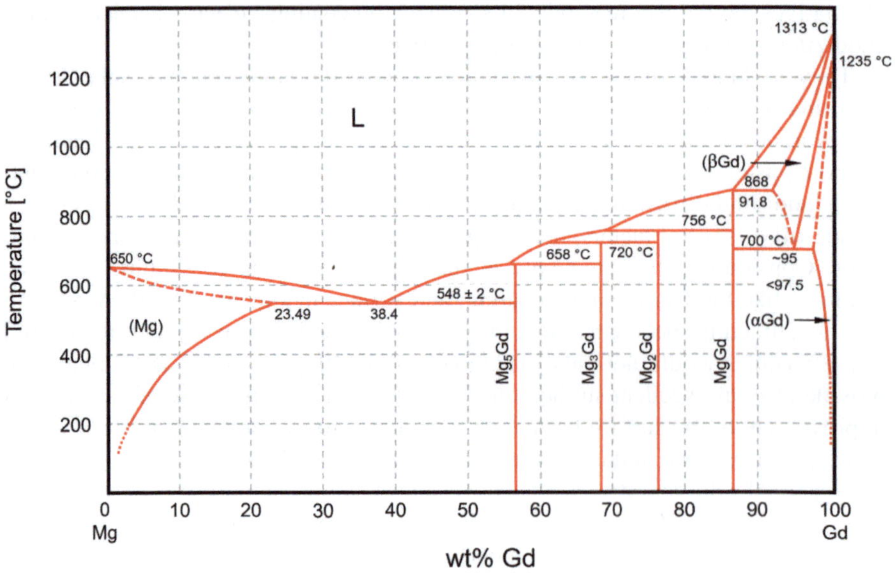

Fig. 3.35 Mg-Gd phase diagram. (Redrawn from ASM International 1992)

position in the galvanic series. Another important class of Mg-alloys, developed next, featured as main alloying elements rare earth (REs) elements and zirconium, in addition to zinc that is still present.

RE elements are interesting for their good solubility in magnesium (see Fig. 3.34). This aspect can be seen in the Mg-Gd phase diagram in Fig. 3.35, exhibiting also a low-temperature eutectic. A technological consequence of this behavior is an excellent castability of RE-containing Mg-alloys. To reduce the costs of the raw materials, RE elements are used in these alloys in the form of the so-called *misch* metals. These are mixtures of several RE elements based either on cerium (Ce55-La20-Nd15-Pr5) or neodymium and praseodymium. This latter mixture is often designated as *didymium misch* metal. This name derives from that of didymium, a mixture of neodymium and praseodymium that, in a first attempt, was considered as a new metal when Gustav Carl Mosander discovered it in 1840. Few years later, it was demonstrated that didymium was not really a new element in the periodic table but a mixture of oxides of the two elements, i.e., neodymium and praseodymium (Meyer 2001). The optimization of the mechanical properties, strength and creep resistance in particular, required alloying with zinc and zirconium, as anticipated. The already attested beneficial effect of zinc is enhanced by zirconium, which is added both as a grain refiner and a precipitation hardener. A refined and homogeneous microstructure, attained thanks to the right selection of the composition and subsequent temper, can stabilize the alloy microstructure and prevent fast recrystallization, an important aspect when the component is bound to be serviced in creep conditions. To this group of alloys, whose development started quite early (1940s) and continued until recently, belong several materials used in

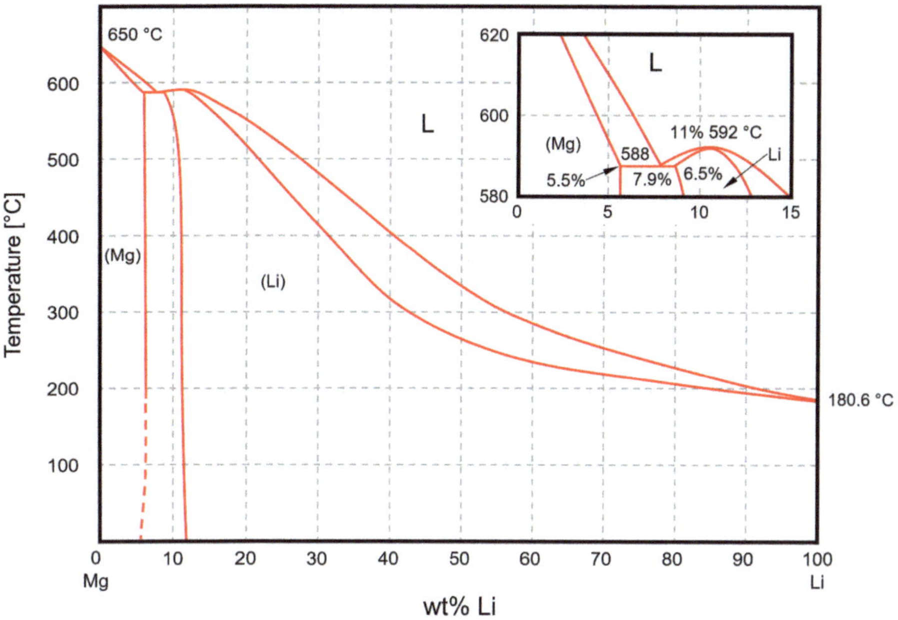

Fig. 3.36 Mg-Li phase diagram. (Redrawn from ASM International 1992; Perkguleryuz et al. 2013)

aircraft structures, like the EZ33, ZE41, ZE63, and the more recent ELEKTRON 21 (see Table 3.23). Strictly associated with REs-Mg-alloys are the alloys of the Mg-Ag system. The central feature of these alloys is the remarkable improvement of the mechanical properties, in particular tensile strength, observed, particularly in Mg-REs-Zr materials, with silver addition in concentrations close to 2.0%, which seems to be a critical value to differentiate Mg-Ag-REs-Zr alloys from standard Mg-REs-Zr (Polmear 2006). In fact, in the QE22 alloy (Table 3.23), one of the most used of this family, a concentration of silver equal to 2.5% is present. For silver-containing alloys, the use of the *didymium misch* metal is particularly recommended. Alloys combining excellent mechanical properties, also in creep regime up to 300 °C, and corrosion resistance have been obtained using yttrium as majority alloying element. In the WE54 alloy, for instance, a concentration of yttrium equal to 5.1% is present, attained using a particular *misch* metal, dominated by the presence of yttrium (75% approx.), in association with gadolinium and erbium. In analogy with Al-alloys, for which Al-Li compositions represent the latest development of their design; lithium has also been selected as alloying element for magnesium alloys. The addition of lithium reduces further the already low density of these systems, which represent at the present time the lightest metallic structural materials (Wu et al. 2015).

In the binary Mg-Li phase diagram (Fig. 3.36), the α-phase, i.e., the hexagonal solid solution of lithium in magnesium, is stable up to a concentration of 5.5 wt% Li in the phase diagram. Above this limit and up to 10.3 wt% Li, a mixture of α + β phases is present, being β a *bcc* solid solution of magnesium in lithium, (Li) in the phase diagram. The β-phase is the only one present at room temperature above 10.3 wt% up to pure lithium axis. Any raise in the concentration of β-phase improves the ductility of the alloy, an aspect that renders the Mg-Li alloys ideal candidates for metal forming, although strength and creep resistance tend to be reduced by the presence of the β-phase. Moreover, the high chemical reactivity of lithium, particularly in association with magnesium, requires a particular attention in managing the oxidation and corrosion resistance of these alloys. Initially, the alloy design regarded compositions in which standard alloying elements, already encountered in magnesium alloys, are present. Several families of alloys were developed: LA (Mg-Li-Al), LZ (Mg-Li-Zn), and LAZ (Mg-Li-Al-Zn), strengthened by the formation of intermetallics, like $MgLi_2Al$, $MgLi_2Zn$, AlLi, and MgLiZn (see Sect. 3.3.2). Specific additions have been explored in the attempt to improve the alloy properties and compensate for the partial instability exhibited by some of the above strengthening intermetallic phases. Starting from the reference compositions of the LA and LAZ alloys, the formation of stable intermetallic phases was attained thanks to the addition of Ag, Ca, Cu, and REs (Zhang et al. 2013). These alloy developments indeed achieved the expected results, although some limitations, pertaining surface stability, are still there and hinder the widespread, safe use of these alloys. However, strengthening mechanisms, which already proved successful in other Mg-alloys, have been applied to improve the properties of Mg-Li alloys too. Mg-RE-Zn alloys, in particular those with RE concentration in excess of the zinc concentration, exhibit an improved room- and high-temperature mechanical strength for the presence of the so-called long-period stacking ordered (LPSO) structures (Zhu et al. 2010). LPSO structures are based on intermetallic compositions, like $Mg_{96}Y_2Zn_2$, featuring a peculiar crystalline superlattice, as results from electron diffraction investigations. TEM observations have also demonstrated that the LPSO determine a fine lamellar structure within the relevant domains. The formation of the LPSO structures has been reported to occur in as-cast alloys, although not with optimized properties. For a complete exploitation of the benefits introduced by these ordered phases, solution heat treatments, possibly followed by extrusion, are recommended. The first reported application of this approach to Mg-Li alloys is the study on the cast Mg-8Li-6Y-2Zn (wt%) alloy, solution treated at 500 °C for 6 h in an inert atmosphere and then hot-extruded with an extrusion ratio of 14:1 (Zhang et al. 2013). The resulting mechanical properties turned out to be particularly good, as concerns the ultimate tensile strength, the tensile yield strength, and the elongation to fracture. If extrusion is preceded by the above mentioned thermal treatment, the refined microstructure shows mechanical properties up to 243 MPa, 187 MPa, and 30%, starting from 202 MPa, 142 MPa, to 19%, respectively, in alloy samples that were just extruded but not heat-treated. The microstructural refinement, resulting from the thermomechanical treatments imparted to the alloy samples, has also an indirect beneficial influence on its corrosion behavior, since it reduces the detrimental micro-galvanic effect associated with presence in the alloy of intermetallic particles and unwanted impurities. A similar approach has been applied also to fully β Mg-Li alloys, as reported in (Xu et al. 2015). Again, using specific thermomechanical treatments, remarkable improvements in the alloy mechanical strength and corrosion resistance have been achieved retaining at the same time an excellent ductility. The main processing steps adopted for Mg-Li alloy materials comprise hot-extrusion and solution heat treatments, followed by water quenching. Furthermore, aging treatments, conducted at low temperatures (<100 °C), and cold-rolling are imparted to the alloy in order to stabilize the microstructure and relevant properties. The newly developed alloy exhibits also a very attractive corrosion behavior that overcomes the intrinsic weakness of α- and α + β alloys, which is determined by combined effect of the alloy composition, in particular the elevated concentration of lithium, and the microstructural refinement induced by the processing treatments. The investigated Li-rich alloy is characterized by a single-phase *bcc* microstructure containing a fine Li-rich solute nanostructure. The solute nanostructure is likely to be formed thanks to the high diffusivity of Li in the Mg lattice at low temperatures. Furthermore, the most interesting feature of this alloy is probably its corrosion behavior, being the Li-rich solute nanostructure substantially electrochemically homogeneous, thus increasing the intrinsic corrosion resistance of the alloy, in which inhomogeneities acting as preferential corrosion nucleation sites have been eliminated (Xu et al. 2015). A comparative picture of the way Mg-Li alloys perform in air is shown in Fig. 3.37. The scheme illustrates how the surface region of different Mg-alloys reacts to standard atmospheric conditions. The poor oxidation resistance of Mg-rich α-alloy is due to the incomplete coverage by the oxide layer that spontaneously forms on the alloy surface (Fig. 3.37, *incomplete coverage*). Therefore, the alloy is not insulated from the surrounding environment, and oxidation can proceed, possibly accompanied by the hydration of the outer oxide. Higher concentrations of lithium, like in two-phase α + β alloys, result in the oxidation of this element too, to form Li_2O (Fig. 3.37, *preferential attack*). The α-grains, which are still present, behave similarly to what is observed in the case of a fully α-alloy. From the scheme, it is also evident that the finer is the microstructure of the alloy, the more homogeneous is the coverage of the alloy itself produced by the outer reaction layer. The interesting novelty introduced by these advanced Mg-Li alloys is the active role of lithium, which, thanks to the refined microstructure of the alloy, better if with a nanometric grain size, forms together with Mg a compact mixed oxide layer. Moreover, lithium oxide (Li_2O), reacting with carbon dioxide to form Li_2CO_3, improves the protection performances of the outer reaction layer. Thanks to the combined oxidation and carbonation reactions, a complete coverage of the metallic surface can be achieved in case of the fully *bcc* nanostructured Mg-Li alloy (Fig. 3.37, *complete coverage*). This latter alloy is definitely better performing as concerns corrosion resistance, and this is due both to the complete surface coverage and to the underlying matrix with the nanostructured features avoiding corrosion to initiate at inhomogeneities contained in the material (Xu et al. 2015).

Moreover, it turns out that this reaction layer has excellent self-healing features, so that it can also stand possible damages, thus behaving very much like the self-passivating film of native oxides, which form on the surface of aluminum, stainless steels, and titanium alloys. These novel Mg-Li alloys exhibit structural properties

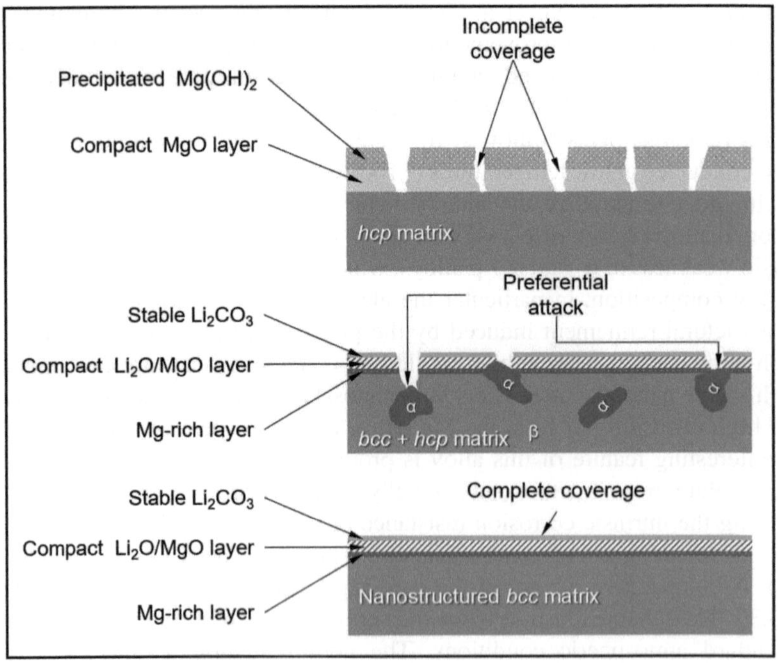

Fig. 3.37 Schemes of the surface reactions of different Mg-Li alloys when exposed to air with reference to three matrix conditions: conventional *hcp* Mg-rich α-alloy, *bcc* + *hcp* Mg-Li extruded alloy, and *bcc* nanostructured Mg-Li alloy. (Redrawn from Xu et al. 2015)

and surface stability that render them potentially interesting for real applications, including aerospace ones.

The research efforts that have been spent for this task indicate anyway the growing, in some respect, renovated interest for Mg-alloys, as proved also by the ever-growing trend in Mg production shares (Fig. 3.33) and by the variety of available mechanical properties, and related tempers, provided by state-of-the-art magnesium alloys (Table 3.25).

3.3.5 *Current Aerospace Applications*

The aircraft components in Fig. 3.38 provide an interesting overview of the potential afforded by metal working technologies of magnesium alloys. These encouraging results are promoting new investigations for the development of fully enabled production routes, suitable also for the aerospace field. The explorative manufacturing of aircraft brackets, with different geometries, has attained excellent results in terms of part control and resulting properties, absolutely comparable, if not better than those of the benchmark cast counterparts, that still represent the larger part of

Table 3.25 Selected Mg-alloys and tempers with resulting mechanical properties

Alloy	Thermal treatment	Tensile strength [MPa]	Tensile yield strength [MPa]	Compressive yield strength [MPa]	Elongation [%]	Hardness	Product form
AM50	F	268	168	–	18.0	–	Extrusions
AZ31	F	245	165	–	12.0	–	Extrusions
AZ31	H24	325	180	–	15.0	–	Sheets
AZ61	F	280	165	–	14.0	–	Extrusions
AZ91A	F	230	150	165	3.0	63 HRB	Castings
AZ92A	T6	275	150	150	3.0	84 HRB	Castings
ELEKTRON 21	T6	248–280	145–170	168	2.0–5.0	–	Castings
ELEKTRON 675	T5A	300–350[a]	200–230[a]	–	1.0–5.0[a]	–	Extrusions
ELEKTRON 675	T5B	340–400[a]	230–260[a]	–	1.0–3.0[a]	–	Extrusions
HK31	H24	285	160	–	9.0	–	Sheets
HM21	T8	270	130	–	11.0	–	Sheets
LA141	T7	132–165	105–160	–	11.0–24.0	–	Sheets
WE43A	T6	250	165	–	2.0	75–95 HRB	Castings
WE54A	T6	250	172	172	2.0	75–95 HRB	Castings
ZE41A	T5	205	140	140	3.5	62 HRB	Castings
ZE63A	T6	300	190	195	10.0	60–85 HRB	Castings

(continued)

Table 3.25 (continued)

Alloy	Thermal treatment	Tensile strength [MPa]	Tensile yield strength [MPa]	Compressive yield strength [MPa]	Elongation [%]	Hardness	Product form
ZK60	F	325	240	–	13.0	–	Extrusions
ZK60	T5	330	268	–	12.0	–	Extrusions
ZK60A	T5	365	305	250	11.0	88 HRB	Extrusions
ZMC711	T6	325	300	–	3.0	–	Extrusions

References and details about temper treatments are given in Table 3.24 (ASM International 1991; Magnesium Elektron 2014, 2016; Pekguleryuz et al. 2013)

[a]Properties of samples up to 100 mm in diameter. Lower value is measured in transverse direction, while higher value is measured in longitudinal direction

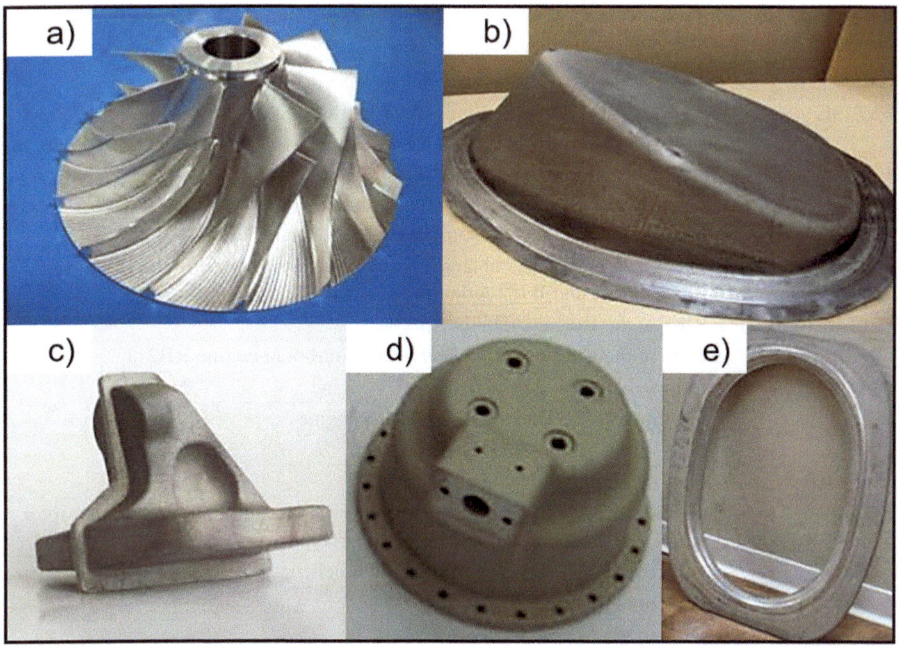

Fig. 3.38 Some aircraft components obtained by different forming techniques and made of different Mg-alloys, as specified in the following. (**a**) WE43 impeller with twisted blades; (**b**) AZ31B support of the Eurocopter antenna – deep drawing; (**c**) AZ80 door stop fitting – forging; (**d**) WE43 compressor upper case for air conditioning plant – closed-die forging; (**e**) AZ80 window frame for Airbus plane – forging (Dziubinska and Gontarz 2015, reproduced with permission of the Publisher)

magnesium alloy components (Dziubinska and Gontarz 2015; Dziubinska and Gontarz 2016). The weight saving introduced by magnesium alloys encouraged their widespread and successful use for helicopters. Some examples of applications of magnesium alloys for helicopter and airplane components are given in Table 3.26.

The future applications of magnesium alloys in the aerospace field seem to be liable of important expansion. In fact, as suggested by the examples in Fig. 3.38 and Table 3.26, the consolidation of metal working technologies for the manufacture of magnesium alloy components is widening the application possibilities of these materials. This process route is certainly benefitting from the established use of these alloys in the automotive industry and in other large-scale production industrial fields, leading to the creation of knowledge and solutions for high-quality components.

Table 3.26 List of some magnesium alloys used in the fabrication of components of helicopters and aircrafts (Gupta and Gupta 2017)

Alloy	Applications
AZ19E	Sikorsky CH33D-Seastallion with a part weight of 238 kg
	After the T6 temper, it is used for the sump of the six-cylinder 540 Lycoming engine used to power the Piper Chieftain
	Fabrication of some components for the Piper Comanche
	Fabrication of some components for the Sikorsky CH53D
AZ92A	Thrust reverser panels obtained by cascade casting used in the Boeing B737, B747, B757, and B767 airliners
ELEKTRON 21	Parts of the Boeing AH64D Apache
EZ33A	Parts of the gearbox of the Rolls-Royce turbofan engine RB211
QE22	Compressor casing of the ALF 507s engine
	Gearbox of the General Electric F110 engine
WE43	Parts of the MD 600 and Sikorsky S92
WE43	Pratt & Whitney F119 auxiliary casing
WE43	Castings used in the gearbox of the Bell Augusta 609 twin engine tilt rotor VTOL (vertical take-off and landing) aircraft
ZE41	Parts of the Sud-Aviation Super Frelon helicopter
	Parts of the Boeing AH 64 Apache
	Parts of the Boeing CH47 (Chinook)
	Parts of the Sikorsky UH60 (Black Hawk)
	Parts of the Pratt & Whitney PW 305 turbofan
	Castings used in the Rolls-Royce Allison AE-3007 engine

3.4 Composite Materials

3.4.1 Introduction

Composites are materials made of at least two phases, which are physically distinct. In general terms, any two or multiphase material can be classified as a composite, including the precipitation-hardened aluminum and magnesium alloys, discussed in Sects. 3.2 and 3.3, and other multiphase alloy systems that will be discussed next. A particular group of composites will be considered in this section, belonging to the broader family of structural composites. Structural composites are materials designed to respond better to specific requirements regarding mechanical properties and realized by mixing together two or more phases in order to achieve these properties, which each single component would not be able to reach by itself (Hull 1981). This aspect together with the low-density values of the main composite systems are the main reasons for the strong interest for structural composites in aircraft applications. The main classes of composites used in aircraft structures are based on epoxy polymer matrix, reinforced with carbon, glass, or aramid fibers. Since the early use of composites in aircraft structures, dating back to the end of the

1990s, their shares have continuously increased, starting, as usual, from military fighter airplanes and spreading to any aircraft and spacecraft. Actually, the path was not straight, and some diverting accident occurred. This is the case of the low toughness and impact resistance exhibited by the carbon fiber fans and blades of the early stages of the compressor of the RB211 Rolls-Royce engine, which did not resist bird strikes and for this reason had to be temporarily abandoned (see *fail safe* concept, Sect. 1.2). Moreover, the quest for a reliable assessment of the useful lifetime (*life safe* criterion, Sect. 1.2), especially, of primary structures, like wings and empennages, was not always met, particularly in case of early composites based on carbon fibers, not displaying any fatigue limit. The situation evolved positively, following two parallel paths: materials based on alternative reinforcements and better grade and microstructural control on carbon fibers. One interesting example of novel design component is the single-shell monocoque that provides higher strength and lower total weight, as compared to the same part assembled by joining together several parts. In addition to the improvement in the specific mechanical properties obtained both from the adoption of new materials and of the new structures that they allow to realize, the rapid spread of epoxy matrix structural composites has been sustained by other important advantages. With reference to the B787-Dreamliner (Sect. 1.2, Fig. 1.11), the use of composites requires also novel processing and manufacturing technologies that result in important cost reduction, thanks to the optimization of the material use. In fact, once the target properties (strength, stiffness, toughness, etc.) have been set, the lay-up design indicates precisely the necessary production steps and relevant materials quantities. The composite parts produced in this way can easily be assembled using lay-up machinery and rotational molding process. In general, composites exhibit a larger mechanical stability as a consequence of their relatively low coefficient of thermal expansion. Lighter structures obviously result in a reduction in fuel consumption and an increase in the payload and/or passenger number. The latest generation of composite materials, featuring high damage tolerance, has also improved the accident survivability of the components. Eventually, the electrically insulating character of the polymer matrix structural composites eliminates all drawbacks pertaining galvanic corrosion, typically initiated at the contact regions between dissimilar metal alloys, particularly in marine environments, featuring high concentrations of water vapor and salty aerosols. For the same reasons, the incidence of fatigue/corrosion problems is virtually eliminated. These latter aspects, influencing materials and structural durability, are of course very positive for the maintenance programs that may potentially result in important reduction in the fleet management costs.

For the military field, other benefits can be considered, related not only, or not necessarily, to the increase the payload, but also operational range, flight performances, and maneuverability under extreme conditions. Polymer matrix structural composites definitely have a number of interesting properties, and this can explain their continuously increasing usage in aerospace structures. However, a specific aspect descending from the electrically insulating character of these materials has to be underlined. Whereas an aircraft with fuselage and skin made of aluminum alloys is quite conductive so that it can dissipate the electrical currents produced by lightning strikes, the same would not be possible for a polymer matrix composite, unless some alternative solutions are implemented. In fact, all structural composites used for the outer surfaces of an airplane, those directly exposed to possible lightning strike, have in their lay-up a protection system, consisting of a ply of metallic mesh or other kinds of conductive layers. These metallic components have essentially a functional role, i.e., improving electrical conductivity, and their relative concentrations are comparatively small. A class of composites that has been specifically developed for aircraft structures and that features major fractions of metallic components are the *fiber-metal laminates* (FMLs). These materials are composed of stacks of metal sheets, made of aluminum or titanium alloys, and prepreg layers, made of continuous felts of aramid, glass, or carbon fibers. The potential of FMLs in aerospace structures appeared evident since the beginning, even considering the low density of their component phases: metallic alloys, reinforcing fibers (see next), and epoxy resin. The possibility to produce anisotropic parts through a suitable fiber orientation and to optimize the stress distribution within the load-bearing section is another positive feature of these important structural composites (Botelho et al. 2006; Vermeeren 2003).

3.4.2 Fiber-Metal Laminates: Development and Designation

The early developed FML was the aramid-fiber-reinforced Al-laminate, ARALL, for which the Dutch University of Delft and the AKZO company filed two patents in 1981. One of the patents covered the general concept of the FML, based on the combination of metal sheets and prepreg fiber felts. The other patent regarded the so-called *stretch forming* (stretching) process. Although flying components of this prototype FML were actually produced, any further development of ARALL was interrupted at the time, due to an insufficient fatigue resistance, particularly under compressive and tensile load conditions. A new class of laminates was developed next, whose name was GLARE, i.e., glass-fiber-reinforced Al-laminate. Also for this material, there is a patent by AKZO, now AKZO Nobel (1987). The general layout of an FML is shown in Fig. 3.39.

Further developments of FML were composites reinforced by carbon fibers:

- CARALL – carbon fiber reinforced aluminum laminate.
- Ti-Gr (titanium-graphite composites).

In this latter, developed by the Boeing company, aluminum is replaced by titanium. Other materials systems have been considered, but their actual use in aircraft structures is still far to come. This is the case of the composites like ALLIC (Al-Li reinforced by carbon fibers), VIRALL (Al-alloy reinforced by Vinylon), BRALL (Al-alloy reinforced by bamboo), and RAFAL (Al-alloy reinforced by ramie) (Laliberte et al. 2000).

For the identification of the different products, an international designation code has been established, according to the scheme illustrated in Fig. 3.40.

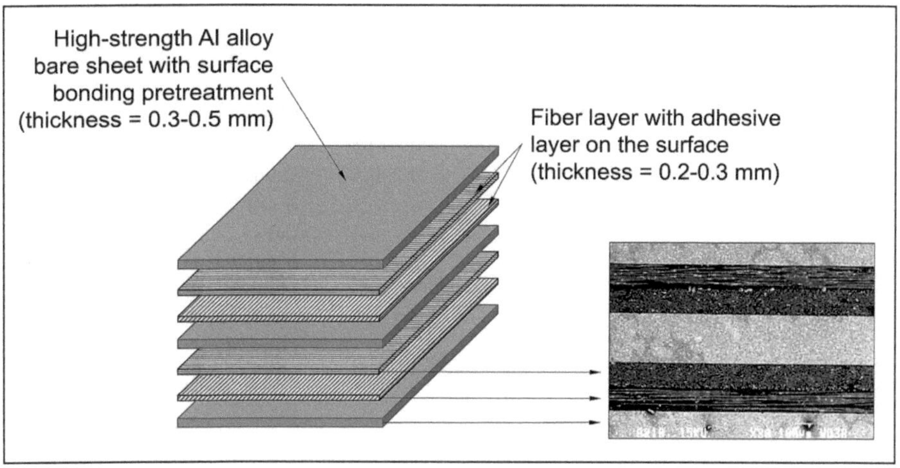

Fig. 3.39 General layout of a FML composite. On the right the cross-sectional view of a real FML. Please note the different orientation of each layer of fibers (Vermeeren 2003, reproduced with permission of the Publisher)

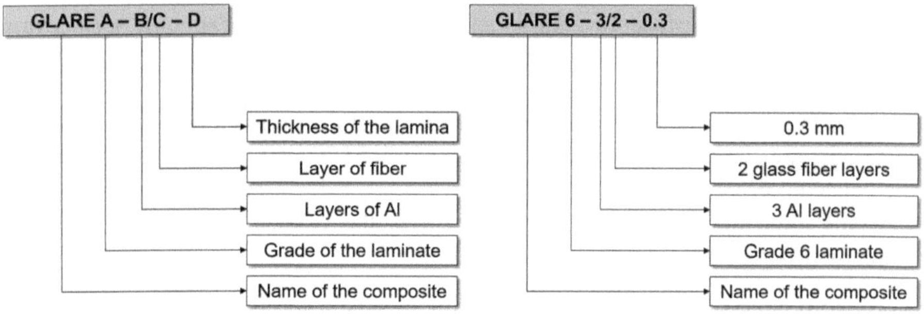

Fig. 3.40 Scheme for the FML international designation code, with an example

For a full qualification of FMLs, other features are important, in particular the orientation of the fibers (Table 3.27).

3.4.3 Main Production Steps and Resulting Characteristics

The processing of FML components typically involves the following steps: surface preparation, lay-up preparation, preparation of the lay-up for the curing treatment, curing treatment, inspection, and post-processing operations (Botelho et al. 2006). During the first step (surface preparation), the metallic surfaces are pre-treated in order to improve the bonding efficiency with the prepreg layer. Different approaches

Table 3.27 Specifications of the GLARE composites (Wu and Yang 2005)

Type	Typical density (g/cm^3)	Al layers characteristics			Fiber layer characteristics	
		Alloy	Temper	Thickness [mm]	Orientation	Thickness [mm]
GLARE 1	2.52	7475	T76	0.3–0.4	Unidirectional	0.25
GLARE 2	2.52	2024	T3	0.2–0.5	Unidirectional	0.25
GLARE 3	2.52	2024	T3	0.2–0.5	0°/90° cross-ply (50%)	0.25
GLARE 4	2.45	2024	T3	0.2–0.5	0°/90°/0° cross-ply (67%–33%)	0.375
GLARE 5	2.38	2024	T3	0.2–0.5	0°/90°/90°/0° cross-ply (50%)	0.50
GLARE 6	2.52	2024	T3	0.2–0.5	+45°/−45° cross-ply (50%)	0.25

have been developed in this regard, which can be grouped into the following main categories: mechanical, chemical, and electrochemical (Sinmazçelik et al. 2011).

The effects of a chemical etching, carried out on an aluminum alloy (2024-T3) foil, using a sulfuric acid solution, are visible in the SEM micrographs in Fig. 3.41. The increase in the surface irregularity and, consequently, in the specific surface area is an important parameter used to improve the adhesion between the aluminum foil and the prepreg fiber felt.

After treating the metallic surfaces, the lay-up of the composite is assembled. At this stage, the choice of the right orientation of the fiber layers is paramount to obtain the expected properties.

Some indications have been developed over the years as concerns this aspect, depending on the different loading conditions:

- 0°/0° to enhance fatigue resistance, ultimate tensile strength and yield strength;
- 0°/90° to obtain fatigue resistance, impact resistance;
- +45°/−45° to obtain shear resistance.

Concerning the metallic alloy layers, they are integrated into the composite starting from already formed foils. The alternative involves the direct deposition of Al-foil directly from the melt using a method that will be fully discussed next. Once the composite lay-up has been prepared, it is transferred to a vacuum bag for the curing treatment (Fig. 3.42). The vacuum bag is typically made of nylon with rubber seals. A release film, e.g., fluorinated ethylene propylene, wraps the composite stacks and helps its detachment from the vacuum bag and other parts of the set up at the end of the curing. A bleeder, that might be woven fabric or felts, adsorbs the excess of resin, possibly ejected during the treatment (Gutowski 1997). The autoclave curing occurs at pressures and temperatures suitable for a full consolidation of the composite. Temperature and pressure profiles suitable for ARALL (2024 alloy, with a T3 temper) composites are shown by the graph in Fig. 3.42, as a function of

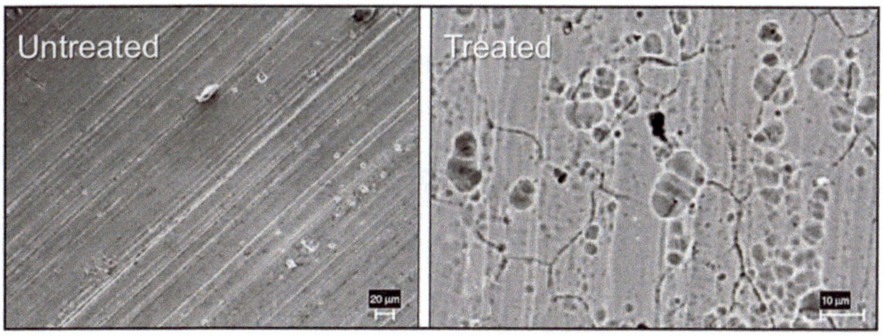

Fig. 3.41 SEM micrograph showing the surface of a 2024 T3 alloy before (untreated) and after a H_2SO_4 chemical etching used to improve the adhesion with the prepreg fibers. (Bothelo et al. 2009, reproduced with permission of the Publisher)

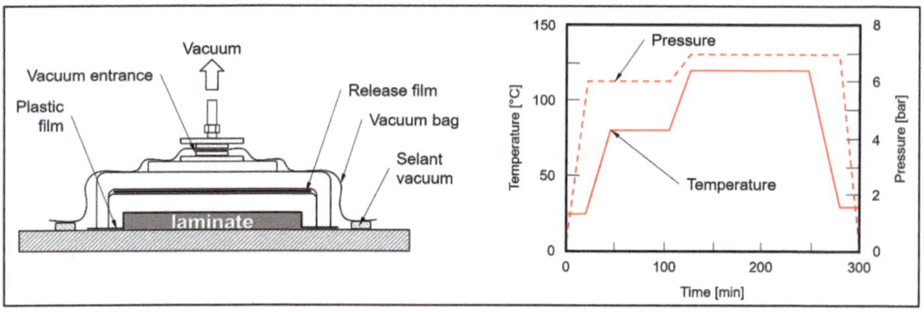

Fig. 3.42 Vacuum-bag setup for the impregnation and curing treatment of the composite lay-up. (Redrawn from Bothelo et al. 2006)

the time of curing, which is the most critical part of the production process. Thanks to the removal of air via a vacuum system, this procedure allows to press the multilayer and helps the spread of the resin and the elimination of excess residues and volatile components. Limits and possible problems at this stage of the process are essentially two: firstly, curing temperature can be comparable with the temper treatment temperature of the selected aluminum alloy, being the use of aged and stabilized alloys a possible remediation. Secondly, an optimal distribution of the resin can be achieved only if dimensions of the workpiece are not too big.

To overcome these limitations and to render the whole production process less expensive and applicable to relatively larger pieces too, the *vacuum-assisted resin transfer molding* (VARTM) process has been developed (Jensen et al. 2009). In this process, the resin is injected into the fiber preform and spreads through the metal and fiber layers. This is a fundamental step of the process, since a great deal of the final mechanical properties of the FML is dependent on the homogeneous spread of the resin within the fiber felt and at the interfaces with the metal foils. During the VARTM process, any excess resin is sucked away from the opposite side, with respect to the inlet duct of the vacuum bag (Fig. 3.43).

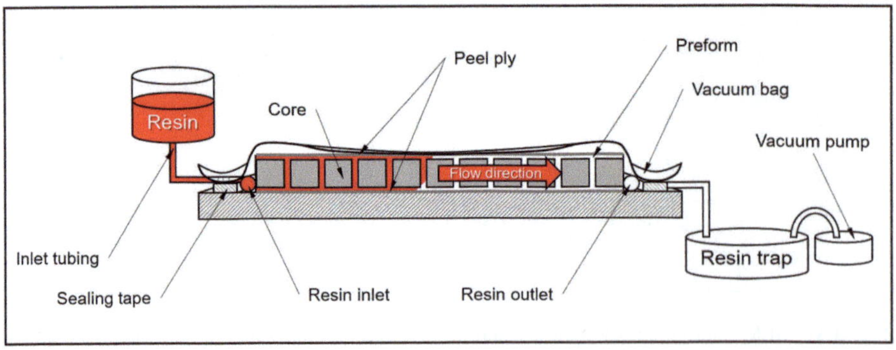

Fig. 3.43 Scheme of the fabrication set up for the VARTM. (Redrawn from Jensen et al. 2010)

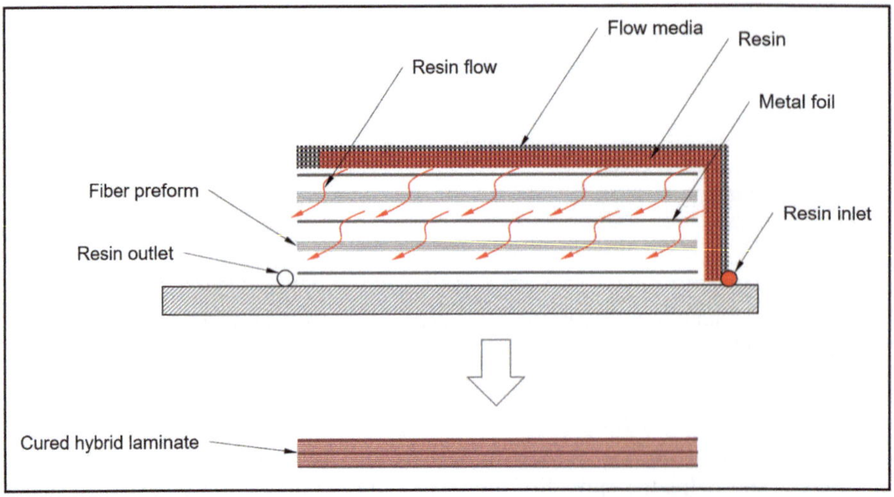

Fig. 3.44 Sequence of resin infiltration and curing for the VARTMFML. (Redrawn from Jensen et al. 2009)

To promote the homogeneous distribution of the resin within the composite, holes are drilled into the metal foils core, although this step requires a particular care, since an important influence of the hole features on the mechanical properties of the composite has been reported, with particular regard to fatigue life. Porous aluminum foils have been considered as a possible alternative approach (Fig. 3.44). In the process, named VARTM-infused *plasma coated laminates* (VARTMPCL), the fiber fabric is coated with a porous metal layer, obtained by depositing the Al-alloy via plasma-spraying (see Sect. 7.2.1). In Fig. 3.45 the main steps of the deposition process are illustrated. The fabric, fixed on a rotating drum, is coated with the metallic layer by repeated scans of the torch.

Fig. 3.45 Deposition of porous Al-alloy layer onto the fabric to produce FML via the VARTMPCL processing (Jensen et al. 2009, reproduced with permission of the Publisher)

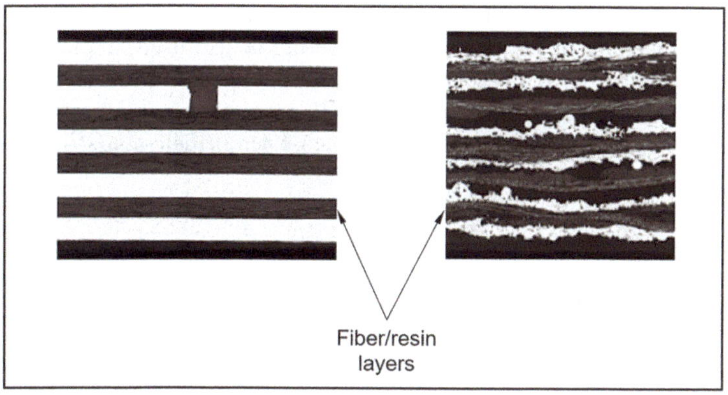

Fig. 3.46 Comparison of the microstructure of the VARTMFLM (left) and VARTMPCL (right) – 0.38 mm thick foils (Jensen et al. 2009, reproduced with permission of the Publisher)

The thickness and the other coating features can be tuned by selecting the torch deposition parameters and the rotating and translational speed of the drum. It is also possible to coat the fabric on both sides by just turning it over and replacing it back onto the drum and then going ahead with the deposition. The alternate layers of fiber felts and porous alloy layers are subjected to the VARTM process in the same way as the "hole-drilled" VARTMFML systems.

Figure 3.46 shows the microstructure of FML with the same stratigraphy and composition but with a difference consisting in the VARTM technique used for their production. Considering the rather different microstructures and, thereby, properties of the two materials, they may have different applications: the VARTMFML typically exhibits good mechanical properties, whereas the VARTMPCL products are preferable for their functional properties, like enhancing electrical and thermal conductivity, interesting, for instance, to improve the resistance of the composite to lightning strikes. From the structural point of view, the VARTMPCL materials suffer from the extreme irregularity of the aluminum layers, generally resulting in an early crack initiation. Although the intrinsic damage tolerance of these materials renders them capable to withstand also relatively high defect concentrations, still, safety criteria may suggest more conservative choices.

For a complete exploitation of FMLs, several post-production treatments have been developed to manufacture composite components with the right geometry required by any specific application. Joining techniques are very important to produce components with complex shapes and geometries. An alternative approach, largely used even in the preparation of the GLARE fuselage panels of the A380, is the *stretch forming*. The only precaution that has to be taken when stretch forming FML components is the delamination that may occur when excessive bending and straining are imparted to the component, favored by the elastic modulus mismatch between the metal and prepreg layers. Notwithstanding the possibility of forming FMLs using different approaches, joining still remains a fundamental aspect for obtaining complex piece geometries. Adhesives are used to join different parts, but in doing this, a continuity of the fiber felts should be preserved, in order to avoid an excessive weakening of the final pieces. Riveting is also used, exploiting the good deformability of the FML, and still sticking to the recalled precautions. The complexity and critical dependence of the FML properties, namely, the mechanical properties, which will be discussed in Sect. 3.4.4, on their actual microstructure, renders all production steps extremely critical.

3.4.4 Mechanical Properties

With reference to the values in Table 3.28, the generally anisotropic character of the mechanical properties of FMLs can be clearly seen, depending on the direction of the applied load with respect to the composite symmetry. The tensile and compression behavior of FMLs is influenced by the plasticity of the metallic component, displaying a nonelastic trend, depending on the load application direction: either longitudinal (L) or transverse (T). Although the metallic component behaves as a monolith and its properties cannot be modulated and adapted that much, the orientation of the fibers can be optimized, to avoid a too heterogeneous and anisotropic material to be obtained and to maximize the performances along the most demanding directions in terms of mechanical stress field.

The composite design requires a thorough analysis of its complex mechanical behavior. For this purpose the Stress-Strain Calculation Program (SSCP) provides excellent indications, at least in static loading conditions (Hagenbeek et al. 2003). The key factor for a good fatigue resistance of FMLs can be identified in the relatively thin layers building up the composite. This condition determines, within each layer, a biaxial plane stress, to be regarded as a beneficial factor for a slow crack propagation (Dieter 1988). In addition to this mechanism, crack bridging, i.e., the transfer of the load from the cracked metallic foil to the still undamaged fibers,

Table 3.28 Tensile and compressive mechanical properties of ARALL and GLARE materials along longitudinal (L) and transverse (T) directions (Laliberte et al. 2000)

Parameter	Direction	Material					
		A-2	A-3	G-2	G-3	G-4	G-5
Ultimate tensile strength [MPa]	L	717	821	1213	745	1027	683
	LT	317	780	317	745	607	683
Yield strength [MPa]	L	366	607	393	303	352	–
	LT	228	331	228	303	25	–
Young's modulus [GPa]	L	66	66	66	58	57	70
	LT	53	49	50	58	50	70
Ultimate strain [%]	L	2.5	2.2	4.7	4.7	4.7	–
	LT	12.7	8.6	10.8	4.7	4.7	–
Compressive yield strength [MPa]	L	255	345	413	310	365	283
	LT	234	360	236	310	285	283
Compressive elastic modulus [GPa]	L	65	64	67	59	60	–
	LT	53	50	52	59	54	–

A ARALL, *G* GLARE, *L* longitudinal, *T* transverse

contributes to enhance the fatigue resistance of the composite. An appropriate adhesion strength between fibers and metal foil completes the picture and determines the right balance among different mechanisms, all of them very much dependent on the layered structure of the FML. This aspect, i.e., the central role of the layered structure in determining the mechanical properties and in particular the fatigue behavior of FMLs, is clearly demonstrated by the curves in Fig. 3.47. In the case of FMLs, the crack propagation rate tends to decrease with increasing crack length, whereas in a bulk Al-alloy, the opposite situation is usually observed.

This behavior, exhibited by the laminate composites, is most interesting for aerospace applications, since the integrity of the component is preserved all through its useful lifetime. This can be achieved thanks to the effective action of the mentioned toughening mechanisms, capable to actively counteract the fatigue damages introduced into the material structure by service conditions. The damage tolerance of FMLs is proved by the fatigue data in Fig. 3.48. The behavior of ARALL is compared to that of bulk aluminum alloys. Interestingly, the number of cycles for the appearance of first fatigue crack (N_i) in the case of the ARALL material is lower or, at most, comparable with N_i of the bulk alloys. The maximum number of cycles (N_f), i.e., the fatigue life, of the FML is order of magnitudes larger than the fatigue life of all the reference Al-alloys (Lanciotti and Lazzeri 2009). The damage tolerance of FMLs has been also confirmed with reference to notched samples: the data for the FMLs are comparatively better than those for aluminum alloys (Wu and Yang 2005). This situation is particularly important for materials to be used for those parts more directly exposed to impacts, like the leading edges of the wings, the engine entrance ducts, etc., in which notch-like defects can be produced

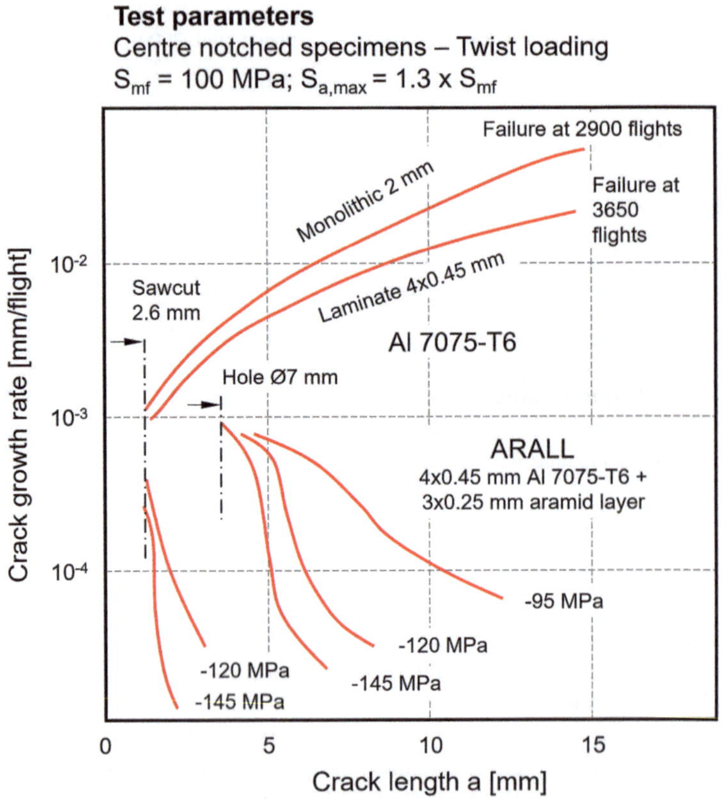

Fig. 3.47 Crack growth rate vs crack length for a bulk aluminum alloy (7075 – T6) and an ARALL composite. The maximum applied loading spectrum twist was truncated at 1.3 times the S_{mf}. Three different compression residual stress levels in the Al sheets are shown: −95 MPa, −120 MPa, and −145 MPa. (Redrawn from Buhl 1992)

by the bird strikes, hailstorms, etc. Other kinds of defects can be rather ascribed to structural features, like riveting, joint edges, and screws. Not always, the impact damages are immediately visible, and radiographic investigations are recommended, to prevent possible severe consequences, particularly as concerns the strength and stiffness of the structure. In addition to mechanical damages, changes in the FML properties may derive from environmental interactions, in the form of corrosion phenomena and water absorption. Water may enter the composite as air humidity, which may permeate the fibrous part of the composite.

Moreover, condensation on the cold metallic surfaces may result in liquid water infiltration, although the presence of outer aluminum foils and epoxy-sealed layers reduce significantly these phenomena with respect to the standard structural composites. Water absorption may produce the delamination of the prepreg layers and their detachment from the metallic foils with a consequent reduction in strength, stiffness, and fatigue life (Botelho et al. 2007, 2009). Surface corrosion may be

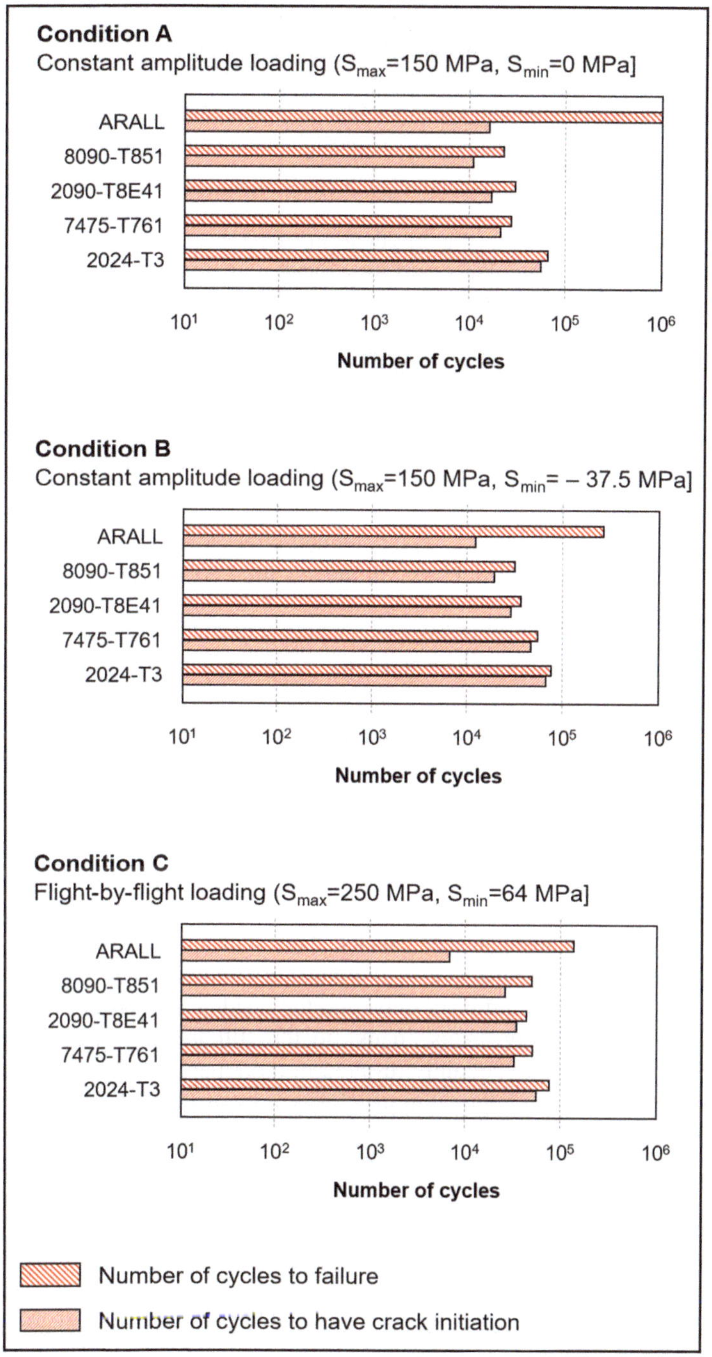

Fig. 3.48 Comparative fatigue data for bulk Al-alloys and an ARALL composite. For each graph the relevant fatigue test conditions (**a–c**) are indicated. (Redrawn from Buhl 1992)

avoided by anodizing the aluminum foils and by protecting the surface with a primer layer containing inhibitor substances (Wu and Yang 2005).

3.4.5 Applications

The initial focus, as concerns the application of FMLs was to use ARALL in the wing structures, with particular regard for a panel with a tank inspection hatch of the Fokker F-27 aircraft (Vermeeren 2003). This aircraft was selected for this study, since at the time, i.e., the early 1980s, a complete background information was available. The idea was discussed by the involved research parties, i.e., Fokker, Delft University, and ALCOA. The outcome of the study was a full-size panel (2.0×1.5 m^2), which achieved the target of 25% weight reduction with respect to the same part made in aluminum alloy. A series of products followed the early panel in ARALL 1, using different alloys:

- ARALL 1: with 7075 alloy layers, stretched after curing;
- ARALL 2: with 2024 alloy layers, in the as-cured condition.

In the following years, other products were made available:

- ARALL 3: same as ARALL 1 but with 7475 alloy in the place of 7075 alloy layers;
- ARALL 4: produced using a high-temperature adhesive, interesting for military applications.

An important indication emerging from the qualification tests carried out on these early products was the sensitivity of ARALL to the presence of drilled holes, that reduced significantly the strength of the material. Moreover, notwithstanding general excellent material properties as concerns fatigue crack growth, specific piece geometries displayed premature fatigue cracks. This is the case of the areas close to doublers bonded on the structure to increase its strength. However, these cracks regarded the outer aluminum layer only, and the residual strength of the component was still within the safety levels, thanks to the still integer fibers. Interestingly, similar cracks would have certainly resulted in a premature failure of an aluminum component, and in the end, the wing panel was very successful. In addition to wing panels, ARALL provided an effective solution to the excessive weight problems of the C-17 military transport aircraft that limited the loading capacity. For this reason, ARALL was chosen for the cargo door (5.6×9.7 m^2) with a relevant weight reduction of 26%. However, the complex production steps and joining necessary to obtain a complete door resulted too demanding, so that in the end, only 30 of these doors were actually produced. Furthermore, the cost-effective, highly performing aluminum alloys initially hindered the use of the FMLs in the fuselage. However, when GLARE 3 and 4 were made available, most of the limitations, which were there in ARALL materials and in GLARE 1 and 2, mainly because of the unidirectionality of the fibers, were overcome. The cross-ply structure of the fibers and the asymmetry in the properties, associated with the peculiar fiber distribution,

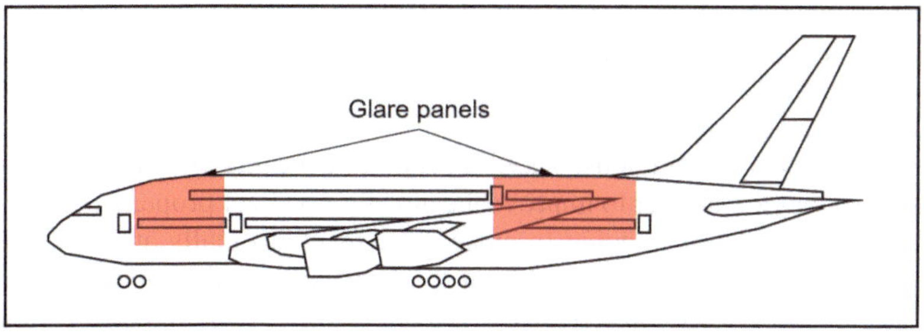

Fig. 3.49 The position of GLARE panels in the fuselage of the A380. (Redrawn from Wu and Yang 2005)

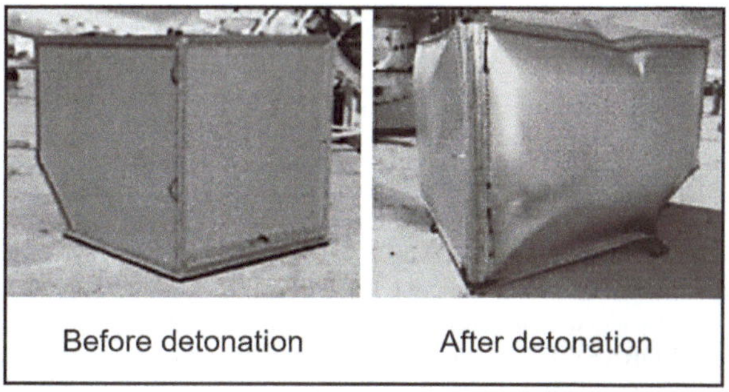

Fig. 3.50 The ECOS^{3}™ container before and after a blast test (Vlot and Gunnink 2001, reproduced with permission of the Publisher)

could be exploited for the fabrication of large fuselage parts. In this fundamental development for FMLs in aerospace applications, the studies and tests conducted by Airbus for several years, in some cases on full size components, have been fundamental. Figure 3.49 gives the distribution of GLARE panels used for the fuselage of the Airbus A380. This laminate composite is employed in association with other structural composites, namely, graphite-fiber reinforced polymer, with a significant overall weight reduction.

The combination of aluminum and glass fibers in GLARE induces in the composite an extremely high fire resistance. In fact, although fire resistance of aluminum is rather poor, when an aluminum alloy is coupled with the glass fiber felt, the situation is completely different. A GLARE sample, exposed to a fire source, might still exhibit the melting of the aluminum outer layer. However, the glass fibers felt remains in place, thus preventing the heat to propagate toward the inner layers of the composite, that in this way are preserved. Fire resistance, coupled with the excellent impact resistance of FMLs in general and particularly of GLARE, has fostered the use of these materials for cargo and luggage containers capable to resist possible explosions. Figure 3.50 shows the effect of a blast test on the ECOS^{3}™ container made of GLARE 3.

As a final example of application of FMLs, these materials are also used as patches for repairing cracks in polymer matrix composites. In this regard, the smaller thermal mismatch between aluminum and GLARE, as compared to aluminum with respect to polymer matrix composites, is exploited. An example of this application is the use of a GLARE patch to repair a crack in the fuselage of a C-5A Galaxy transport aircraft (Vermeeren 2003).

New systems are continuously investigated, as concerns both the metallic part and the prepreg felt, in order to improve the already excellent properties achieved so far. Processing is also paramount to make these materials really attractive also for large-scale, low-cost productions. One of the latest aspects that is under consideration is the use of thermoplastics rather than thermosetting polymers as binder materials. In the future, renovated and growing interest can be foreseen for FMLs featuring natural reinforcing fibers, like the mentioned BRALL and RAFAL with bamboo and ramie natural reinforcement, respectively.

References

Aleris Corporation (2015) Aerospace Aluminum A5028 AlMgSc – The Strong Lightweight. Available via DIALOG. https://www.aleris.com/wp-content/uploads/2016/02/AL-2342_012-Aktualisierung-BR-AlMgSc-2015-06-03-WEB.pdf. Accessed 04 January 2018.

APWORKS (2017) Material Data Sheet – Scalmalloy®. Available via DIALOG. http://www.apworks.de/en/scalmalloy/. Accessed 04 January 2018

APWORKS (2017) Scalmalloy ®. Available via DIALOG. http://www.apworks.de/en/scalmalloy/. Accessed 04 January 2018

ASM International (1991) ASM Handbook Vol. 2 – Properties and selection: Nonferrous Alloys and Special-Purpose Materials. ASM International, Materials Park, Ohio

ASM International (1992) ASM Handbook Vol. 3 – Alloy phase diagrams. ASM International, Materials Park, Ohio

ASM International (2003) ASM Handbook Vol. 13A – Corrosion: Fundamentals, Testing and Protection. ASM International, Materials Park, Ohio

ASM International (2015) ASM Subject Guide – Aluminum and Aluminum Alloys. ASM International, Materials Park, Ohio

Atrens A et al (2015) Review of Recent Developments in the Field of Magnesium Corrosion. Advanced Engineering Materials 17(4): 400–453

Belov et al (2002) Iron in Aluminium Alloys: Impurity and Alloying Element. CRC Press

Benedyk J C (2010) International Temper Designation Systems for Wrought Aluminum Alloys: Part II – Thermally Treated (T Temper) Aluminum Alloys. Light Metal Age: 16–22

Blawert C et al (2006) Anodizing Treatments for Magnesium Alloys and Their Effect on Corrosion Resistance in Various Environments. Advanced Engineering Materials 8 (6): 511–533

Boeing (2008) AERO Magazine 06. Available via DIALOG. http://www.boeing.com/commercial/aeromagazine/articles/qtr_4_06/index.html. Accessed 8 October 2017.

Botelho E C et al (2006) A Review on the Development and Properties of Continuous fiber/epoxy/aluminum Hybrid Composites for Aircraft Structures. Materials Research 9 (3): 247–256

Botelho E C et al (2007) Influence of Hygrothermal Conditioning on the Elastic Properties of Carall Laminates. Applied Composite Materials 14: 209–222

Botelho E C, Da Silva D A, Rexende M C (2009) Hygrothermal Aging Effect on Fatigue Behavior of GLARE. Journal of Reinforced Plastics and Composite Materials 28: 2487–2499

Buhl H (1992) Advanced Aerospace Materials. Springer Verlag, Berlin

Cahn R et al (2005) Structure and Properties of Nonferrous Alloys. Materials Science and Technology 8–9. Wiley & sons

Cardarelli F (2008) Materials Handbook 2nd edn. Springer Verlag, London
Carter C, Norton M (2007) Ceramic Materials – Science and Engineering. Springer
Czerwinski F (2011) Magnesium Alloys – Design, Processing and Properties. InTech
Davis J R (2001) Aluminum and Aluminum Alloys. ASM International
Davy H (1808) Electro-Chemical Researches, on the Decomposition of the Earths: With Observations on the Metals Obtained from the Alkaline Earths, and on the Amalgam Procured from Ammonia. Philosophical Transactions of the Royal Society of London 98: 333–370.
Dieter G E (1988) Mechanical Metallurgy. McGraw Hill
Djukanovic G (2017) Are Aluminium-Scandium Alloys the Future? http://aluminiuminsider.com/aluminium-scandium-alloys-future/. Accessed 1 January 2018
Donachie M J (2000) Titanium – A Technical Guide. ASM International, Materials Park, Ohio
Dziubinska A, Gontarz A (2015) A new method for producing magnesium alloy twin-rib aircraft brackets. Aircraft Engineering and Aerospace Technology: An International Journal 87(2): 180–188
Dziubinska A, Gontarz A (2016) A new technology for producing AZ31 magnesium alloy aircraft brackets with a triangular outline. Aircraft Engineering and Aerospace Technology: An International Journal 88(3): 452–457
Esmaily M et al (2017) Fundamentals and advances in magnesium alloy corrosion. Progress in Materials Science 89: 92–193
Federal Aviation Administration (2012) Aviation Maintenance Technician Handbook-Airframe Volume 1. Federal Aviation Administration
Filatov Y A et al (2000) New Al-Mg-Sc alloys. Materials Science and Engineering A 280: 97–101
Gasik M (2013) Handbook of Ferroalloys: Theory and Technology. Butterworth Heinemann
Gupta M, Gupta N (2017) The Promise of Magnesium Based Materials in Aerospace Sector. International Journal of Aeronautical Science & Aerospace Research 4 (1):141–149
Gupta M, Gupta N (2017) Utilizing Magnesium based Materials to Reduce Green House Gas Emissions in Aerospace Sector. Aeronautics and Aerospace Open Access Journal 1 (1): 1–6
Gutowski T G (1997) Advanced Composites Manufacturing 1st edn. John Wiley & Sons, New York
Hagenbeek M, Van Hengel C, Bosker O J, Vermeeren C A J R (2003) Static Properties of Fiber Metal Laminates. Applied Composite Materials 10: 207–222
Heinz A et al. (2000) Recent developments in aluminum alloys for aerospace applications. Mat Sci Eng A 280: 102–107
Horst F, Mordike B (2006) Magnesium Technology-Metallurgy, Design Data, Applications. Springer Verlag Berlin-Heidelberg
Hosford W F (2010) Mechanical Behavior of Materials. Cambridge University Press
Hull D (1981) An Introduction to Composites Materials. Cambridge University Press
Hull D, Bacon J (2011) Introduction to Dislocations. Elsevier
Jensen B J et al (2009) Fiber Metal Laminates made by the VARTM process. In: 17th International Conference on Composite Materials, British Composite Society, Edinburgh, 27–31 July 2009
Kaufmann G (2000) Introduction to Aluminum Alloys and Tempers. ASM International
Kettner M et al (2007) The InnMag Project – Processing Mg for Civil Aircraft Application. Advanced Engineering Materials 9(9): 813–819
Laliberte J F, Poon C, Straznicky P V(2000) Applications of fiber-metal laminates. Polymer Composites 21: 558–567
Lanciotti A, Lazzeri L (2009) Fatigue resistance and residual strength of riveted joints in FML. Fatigue & Fracture of Engineering Materials & Structures 32: 837–846
Lequeu P, Lassince P, Warner T (2007) Aluminum alloy development for the Airbus A380-Part 1 & 2. Adv Materials & Processes July: 33–35 & 41–44
Lequeu P, Smith K P, Daniélou A (2010) Aluminium-Copper-Lithium Alloy 20150 Developed for Medium to Thick Plate. Journal of Materials Engineering and Performance 19: 841–847
Lumely R (2011) Fundamentals of Aluminium Metallurgy. Woodhead Publishing Ltd

Lyon P, Syed I, Heaney S (2007) Elektron 21 – An Aerospace Magnesium Alloy for Sand Cast and Investment Cast Applications. Advanced Engineering Materials 9(9): 793–798

Magnesium Elektron (2012) Magnesium Alloy Welding Rod Datasheet. Available via DIALOG. https://www.magnesium-elektron.com/wp-content/uploads/2016/10/Magnesium-Alloy-Weld-Rod_0.pdf. Accessed 31 December 2017.

Magnesium Elektron (2014) Elektron ® 675 Datasheet. Available via DIALOG. https://www.magnesium-elektron.com/wp-content/uploads/2016/10/Elektron-675_0.pdf. Accessed 28 November 2017.

Magnesium Elektron (2014) Elektron ® ZREI Datasheet. Available via DIALOG. https://www.magnesium-elektron.com/wp-content/uploads/2016/10/Elektron-ZRE1_0.pdf. Accessed 31 December 2017)

Magnesium Elektron (2015) Elektron ® QE22 Datasheet. Available via DIALOG. https://www.magnesium-elektron.com/wp-content/uploads/2016/10/Elektron-QE22_2.pdf. Accessed 28 November 2017.

Magnesium Elektron (2016) Elektron ® 21 Datasheet. Available via DIALOG. https://www.magnesium-elektron.com/wp-content/uploads/2016/10/Elektron-21_1.pdf. Accessed 28 November 2017.

Marker T R (2013) Evaluating the flammability of various magnesium alloys during laboratory and full-scale aircraft fire tests. US Department of Transportation

McCafferty E (2010) Introduction to Corrosion Science. Springer Verlag, New York

Meyer G (2001) Die (ungleichen) Dydimium-Zwillinge. Chemie in unserer Zeit 35(2):116–123

Nature Materials (2016) No Easy Solutions for Aerospace (editorial). Nature Materials 15 (8): 803

Ostrovsky I, Henn Y (2007) Present state and future of magnesium application in aerospace industry. In: International Conference "New Challenges in Aeronautics" ASTEC'2007, Moscow, 19–22 August 2007

Parker R L (1967) Data of Geochemistry – Composition of the Earth's crust. US Government Printing Office

Pekguleryuz M O, Celikin M (2010) Creep resistance in magnesium alloys. International Materials Science Reviews 55: 197–217

Pepperhoff W, Acet M (2001) Constitution and Magnetism of Iron and its Alloys. Springer, Berlin Heidelberg

Perkguleryuz M O, Kainer K U, Arslan Kaya A (2013) Fundamentals of Magnesium Alloy Metallurgy. Woodhead Publishing

Polmear I J (2006) Light Alloys – From Traditional Alloys to Nanocrystals, 4th end. Butterworth-Heinemann

Polmear I J et al. (2017) Light Alloys – Metallurgy of the Light Metals. Butterworth-Heinemann

Porter D, Easterling K E, Sherif M Y (2009) Phase Transformations in Metals and Alloys, 3rd end. CRC Press

Prasad E N, Gokhale R, Wanhill R (2014) Aluminum – Lithium Alloys. Elsevier

Prasad E N, Wanhill R (2017) Aerospace Materials and Material Technologies. Springer

Rendigs K H (1994) Metallic structures used in aerospace during 25 years and prospects. In: 50 years of Advanced Materials or Back to the Future, Proceedings of the 15th International European Chapter Conference of the Society for the Advance of Material and Process Engineering, Toulouse, 1994

Röyset J, Ryum N (2005) Scandium in Aluminium Alloys. International Materials Reviews 50 (1): 19–44

Scandium International Mining Corp (2017) The Aluminum-Scandium Alloy Advantage. Available via DIALOG. http://www.scandiummining.com/i/pdf/Scandium-Alloy-Fact-Sheet.pdf. Accessed 01 January 2018

Sinmazçelik T et al (2011) A review: Fibre metal laminates, background, bonding types and applied test methods. Materials and Design 32: 3671–3685

Staley J T, Lege D J (1993) Advances in aluminum alloy products for structural applications in transportation. Journal of Physics IV, Colloque C7, 3: 179–190

Starke Jr E A, Sanders Jr T H, Palmer I G (1981) New Approaches to Alloy Development in the Al-Li System. Journal of Materials Aug 1981: 24–33
Starke Jr E A, Staley J T (1996) Application of modern aluminum alloys to aircraft. Progr Aerospace Sci 32: 131–172
The Aluminum Association (2015) International Alloy Designation and Chemical Composition Limits for Wrought Aluminum and Wrought Aluminum Alloys. The Aluminum Association Inc, Arlington
The CM Group (2012) The Global Mg Industry in 2011 – The Impact of Chinese Production, Costs and Shipments. IMA Annual Conference, San Francisco, 21 May 2012
Thompson G E et al (1999) Anodizing of Aluminium Alloys. Aircraft Engineering and Aerospace Technology 71: 228–238
Totten E G, MacKenzie S, Scott D (2003) Handbook of Aluminum Volume 1 – Physical Metallurgy and Processes. Marcel Dekker Inc.
Vermeeren C A J R (2003) An Historic Overview of the Development of Fiber Metal Laminates. Applied Composite Materials 10: 189–205
Vlot A, Gunnink J W (2001) Fibre Metal Laminates. Springer–Science+Business Media
Williams J C, Starke Jr E A (2003) Progress in structural materials for aerospace systems. Acta Materialia 51: 5775–5799
Wu G, Yang J M (2005) The Mechanical Behavior of GLARE Laminates for Aircraft Structures. Journal of Materials 57(1): 72–79
Wu R et al (2015) Recent progress in magnesium-lithium alloys. International Materials Review 60 (2):65–100
Wulandari W et al (2010) Magnesium: current and alternative production routes. Chemeca: Australasian Conference on Chemical Engineering. Available via DIALOG. http://ro.uow.edu.au/engpapers/1254. Accessed 07 Nov 2017
Xu W et al (2015) A high-specific-strength and corrosion-resistant magnesium alloy. Nature Materials 14: 1229–1237
Zhang J et al. (2013) Experimental study on strengthening of Mg-Li alloy by introducing long-period stacking ordered structure. Scripta Materialia 68: 675–678
Zhu Y M, Morton A J, Nie J F (2010) The 18R and 14H long-period stacking ordered structures in Mg-Y-Zn alloys. Acta Materialia 58(8): 2936–2947

Further Reading

ASM International (1999) ASM Specialty Handbook: Magnesium and Magnesium Alloys. ASM International, Materials Park, Ohio
ASM International (2008) ASM Handbook Vol. 15 – Casting. ASM International, Materials Park Ohio
Baker et al (2004) Composite Materials for Aircraft Structures 2nd edn. American Institute of Aeronautics and Astronautics Inc.
Davis J R (2001) ASM Specialty Handbook: Aluminum and Aluminum Alloys. ASM International, p 351–416
Rana S, Fangueiro R (2016) Advanced Composite Materials for Aerospace Engineering: Processing, Properties and Applications. Woodhead Publishing Limited
Zhang S, Zhao D (2012) Aerospace Materials Handbook. CRC Press

Chapter 4
Titanium and Titanium Alloys

4.1 Introduction

Titanium is a chemical element with atomic number 22, occupying the fourth group, fourth period of the periodic table of elements. It is classified as a transition metal, with a density of 4.51 g/cm^3 and a Young's modulus of 120 GPa. Titanium exhibits two polymorphs (see Sect. 4.2): α titanium (*hcp* structure), stable up to 882 °C, a temperature named β-transus. The other polymorph is indeed β-titanium (*bcc* structure), stable from 882 °C up to the melting point (1670 °C). The widespread use of titanium and its alloys can be ascribed to a number of interesting properties of these materials, like:

- Exceptional corrosion resistance, in many cases better than in stainless steels, owing to the spontaneous formation of a stable, adherent, and protective TiO_2-based surface layer.
- High specific strength. The tensile strength of some alloys is comparable to that of low-strength martensitic and austenitic stainless steels, like the A286 material (PHSS, see Sect. 5.4.3.3) and Co-based superalloys, such as the L-605 alloy (see Sect. 6.2).
- Low density, as compared to the densities of Ni-based superalloys and steels.
- Processing capabilities. Titanium can be forged, hot-rolled, and welded but also machined and cast. For this latter process, the best choice is investment casting (see Sect. 4.4). Finally, titanium can be processed by powder metallurgy techniques.

Aerospace applications played a major role in the development of titanium alloys, and the driving force for the massive production of titanium came directly from the highly demanding and rapidly growing aerospace industry in the late 1950s (Donachie 2000). Of course, the great variety of properties render titanium alloys an attractive choice for other fields too, like architectural, automotive, and biomedical. Titanium accounts for about the 0.6% of the Earth's crust, and it is the fourth

S. Gialanella, A. Malandruccolo, *Aerospace Alloys*, Topics in Mining, Metallurgy and Materials Engineering, https://doi.org/10.1007/978-3-030-24440-8_4

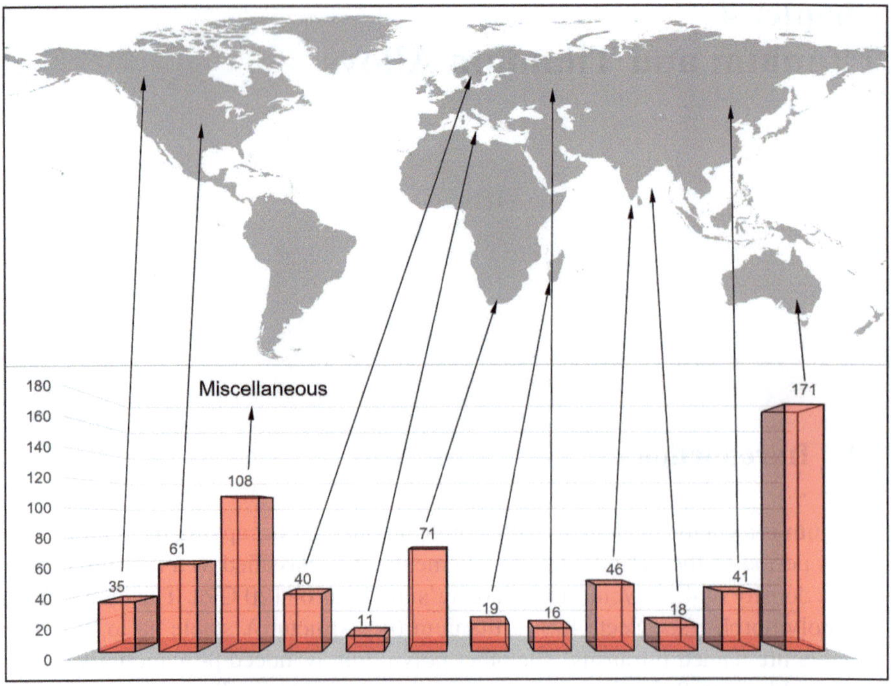

Fig. 4.1 The distribution of the global titanium ores deposits. Quantities are expressed in billions of metric tons. (Data from Leyens and Peters 2003)

most abundant structural metal, following iron, aluminum, and magnesium. Like other most important metals used for technological purposes, it is present on Earth in form of oxides. The most important minerals containing titanium are TiO_2 (rutile), TiO_2 (brookite), TiO_2 (anatase), $FeTiO_3$ (ilmenite), $CaTiO_3$ (perovskite), $MgTiO_3$ (geikielite), and $CaSiTiO_5$ (sphene). Of all these raw materials, the three polymorphs of titanium oxide are those exhibiting the higher concentration of titanium and therefore to be preferred in the extraction metallurgy process (Froes and Imam 2018). Figure 4.1 shows the geographical distribution of the main provenance regions for titanium ores.

The reactivity of titanium with oxygen and other common gaseous species, like hydrogen and nitrogen, and the stability of the resulting compounds are the main reasons why this metal started to be used only in relatively recent times. In 1791, the English clergyman and amateur mineralogist, William Gregor, proposed the existence of a new unknown element contained in a dark magnetic iron ore (ilmenite). Some years later, in 1795 the German chemist Martin H. Klaproth analyzed rutile from a Hungarian site and pointed out the presence of an unknown element, the same as the one reported by Gregor. The new element was then named titanium after the

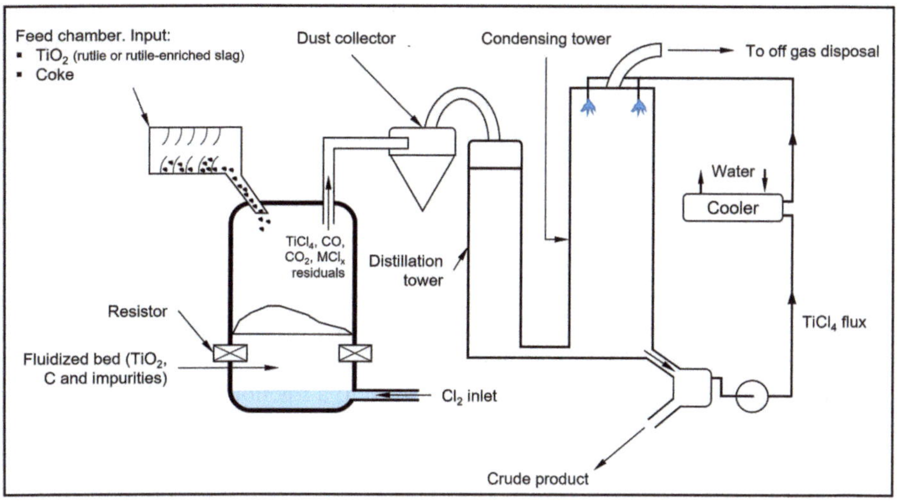

Fig. 4.2 Schematic of the Kroll's reactor plant for the extraction of primary titanium (Lüterling and Williams 2007)

Titans, the members of the second generation of divine creatures, according to the Greek mythology. Matthew A. Hunter was the first able to extract pure metallic titanium in 1910, starting from the thermal decomposition of $TiCl_4$ (titanium tetrachloride) induced by sodium in a steel reactor. This process was named the Hunter process after its developer, although William J. Kroll is unanimously recognized as the father of titanium extractive industry. In fact, in 1932 in Luxemburg he produced significant amounts of titanium, starting from $TiCl_4$ combined with calcium, using what can be regarded as the prototype of the modern pyrometallurgical Kroll's process. Afterward, the process was modified, by replacing the reducing agent: magnesium instead of calcium, leading to the modern Kroll's process.

The primary output of the Kroll's process is the so-called titanium *sponge*. The name derives from the porous and spongy appearance of titanium obtained from the purification process that removes magnesium chloride residuals through evaporation. The density of titanium *sponge* ranges from 1.2 to 3.0 g/cm^3. Therefore, before any usage, it is densified in a hydraulic press. Once densified, pieces of titanium *sponge* are assembled to form electrodes for the melting processes. Generally, multiple remelting steps are necessary, for the morphology of the sponge, for adjusting the chemical composition of the target alloy, and for eliminating, at the same time, any possible remaining contaminants (Lüterling and Williams 2007; Leyens and Peters 2003).

The Kroll's process consists of six main stages that can be described as follows, with reference to the fluidized bed reactor plant in Fig. 4.2, by far the most used for the extraction of primary titanium.

1. Chlorination. A mix of raw materials (titanium ores) and coke undergo chlorination in the reactor chamber to produce titanium tetrachloride ($TiCl_4$). The process is carried out at temperatures in between 800 °C and 1000 °C, reached thanks to the resistor bed supporting the raw material and carbon mix. The formation of $TiCl_4$ occurs according to the following reaction:

$$2TiO_2(s) + 3C(s) + 4Cl_2(g) \rightarrow 2TiCl_4(g) + 2CO(g) + CO_2(g) \tag{4.1}$$

   Carbon particles, residues of the ores and quartz fragments, and fly ashes are the main impurities typically present in the gas flux leaving the reactor chamber. Since their presence in the final product (*sponge*) would be extremely detrimental, they are eliminated using a cyclone dust collector.
2. Distillation. The tetrachloride flux is conveyed to a fractional distillation unit for its complete purification, since the starting grade, as deriving from the chlorinator directly, would not guarantee a sufficient purity level of the final product. The process is carried out in two steps. The first is meant to remove the impurities with low boiling points, e.g., CO and CO_2. The second step removes the high boiling point impurities, such as $SiCl_4$ and $SnCl_4$.
3. Condensation of $TiCl_4$. At the end of the distillation, the purified $TiCl_4$ is collected as crude product and stored under an inert gas atmosphere.
4. Reduction of $TiCl_4$. This is the actual Kroll's reaction. Titanium tetrachloride is reduced by means of molten magnesium. The reduction using magnesium is performed according to the reaction:

$$TiCl_4 + 2Mg \rightarrow Ti + 2MgCl_2 \tag{4.2}$$

   The reaction temperature is kept in between 712 and 920 °C. Magnesium is used in excess with respect to the reaction stoichiometry in order to ensure the complete elimination of chlorine from the products. Incidentally, the Hunter process, mentioned at the beginning of this section, differs from the Kroll's approach for the use of sodium as reducing agent. The sodium-based process was not only more expensive but involved also the risk of a higher flammability due to the presence of residues of this element, sodium, trapped into the final titanium *sponge*. As concerns the reducing reaction, its efficiency is limited by the following competitive reactions:

$$\begin{aligned} 3TiCl_4 + Ti &\rightarrow Ti + 4TiCl_3 \\ 2TiCl_4 + Ti &\rightarrow 3TiCl_2 \end{aligned} \tag{4.3}$$

5. Extraction of titanium from the *sponge* cake. The process aims at eliminating $MgCl_2$ (see Eq. 4.2) residuals. This can be achieved by acidic washing of the crushed sponge and exploiting the preferential solubility of the $MgCl_2$ in HCl-based solutions. This process requires the extraction of the *sponge* cake from the reactor. An alternative approach, which is carried out in situ, i.e., without removing the titanium *sponge* from the Kroll's reactor, is a preferable alternative.

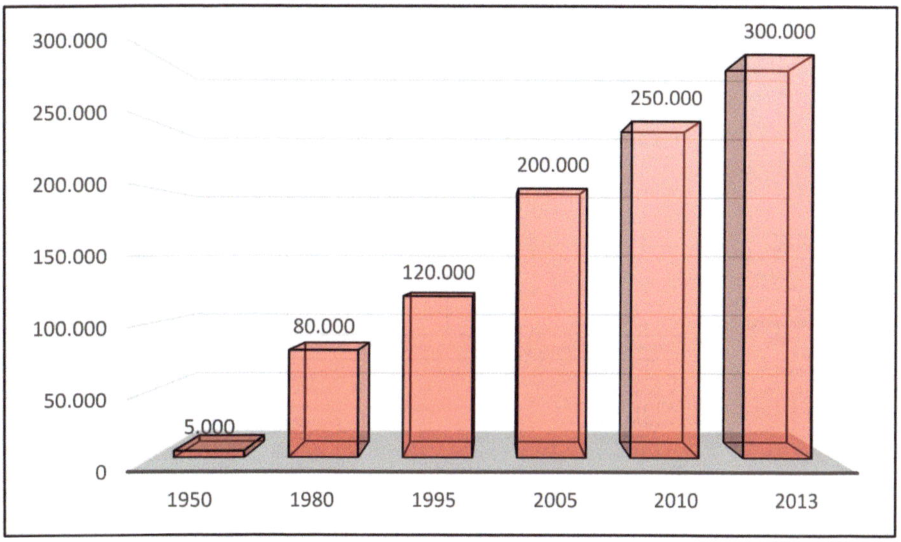

Fig. 4.3 Titanium sponge global production expressed in metric tons. (Data from Prasad and Wanhill 2017)

The evaporation of $MgCl_2$ is induced by the high vapor pressure of this chloride that evaporates from the *sponge* and, after recondensation, can be removed from the reactor. A vacuum condensation process can alternatively be used: the *sponge* is heated under vacuum without being removed from the Kroll's vessel. The vapor flux that is emitted from the sponge condensates in another vessel, from which it is recovered.

6. Crushing and sizing. This final stage is required for obtaining pieces of the *sponge* to be welded to form electrodes for the remelting processes (see Sect. 4.4).

The aerospace interest for titanium started after the World War II, between the late 1940s and the early 1950s. In that period, titanium and its alloys were rapidly recognized as key materials for aircraft engine parts. In the USA, the government-sponsored programs for the construction of large production plants of titanium sponge, such as those at TIMET (1951) and RMI (1958). The industrial production in Europe started in 1951 in the UK at the Metals Division of Imperial Chemical Industries that later became IMI and Deeside Titanium (see Fig. 4.3). The Soviet Union started the production of Ti sponge in 1954. The dramatic increase in the production of titanium led Soviet Union to become the world's largest producer in 1979. Innovation and alloy development in the USA rapidly progressed thanks to the attempts of alloying titanium with other elements such as Al, Mo, and Sn (see Sect. 4.3.1). Contemporarily, there was also an increase in the production of commercially pure (CP) titanium for non-aerospace applications, with a similar trend also in Japan (Leyens and Peters 2003; Lüterling and Williams 2007).

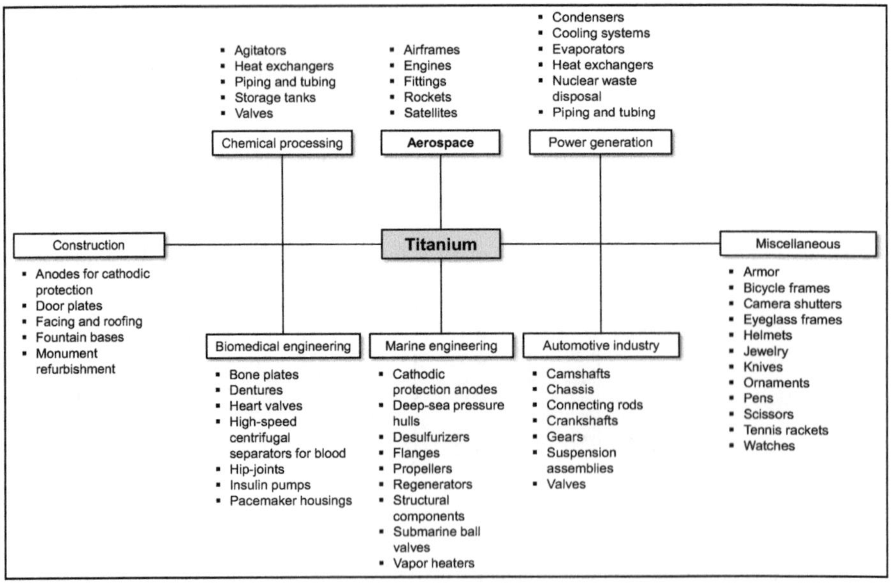

Fig. 4.4 Typical markets and main applications for titanium and titanium alloys (ASM International 1991; Donachie 2000)

The combination of mechanical properties and surface stability renders the use of titanium alloys interesting and ubiquitous in aircraft structures, aero-engines, and other aerospace applications. These have been for quite some time the main reference market for titanium alloy products, until recent, when titanium and titanium alloys are being massively used also in other fields, like power industry, civil engineering, and general consumer goods and commodities. The continuously growing trend of the primary titanium production (Fig. 4.3) is coherent with its expansion to new market shares (Fig. 4.4). The percentage of titanium used in other applications than aircraft structures and engines is increasing, and this trend seems to be growing further (see Fig. 4.5). At the same time, the link between the primary titanium production (*sponge* production) and the aerospace market has become more tenuous. Furthermore, the lower demand for titanium due to the reduction of the military budgets, both in the USA and in the Countries of the former Soviet Union, has led to an estimated 25% decrease in the world production capacity over the 1990–1995 interval. After this drop, the worldwide titanium *sponge* production resumed, mainly due to the rise in commercial airplanes sales and the increasing demand from other new markets (Lüterling and Williams 2007). In the early 1990s, a series of low-cost titanium alloys was introduced into the market. One example of these alloys is the Ti-6Al-1.7Fe-0.1Si (TIMETAL 62S), with comparable properties as those of Ti-6Al-4V. The lower cost of these alloys, mainly based on recycled titanium scraps, was attained also thanks to changes in the production and subsequent fabrication steps. The substitution of some alloying elements contributed to a further cost reduction. Notwithstanding the success of these low-cost alloys, the

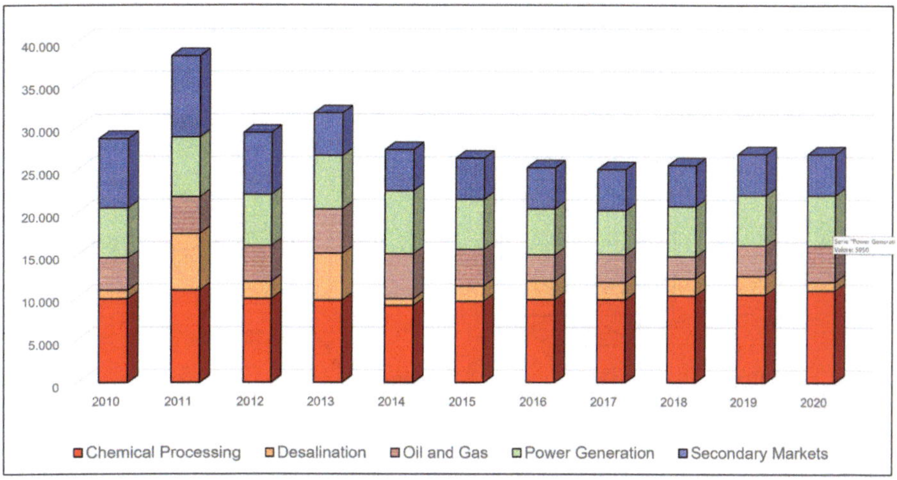

Fig. 4.5 Industrial titanium demand forecast expressed in metric tons. Aerospace applications are intended to belong to secondary markets. Data do not include shipment within China and Russia. (Redrawn from Cain 2016)

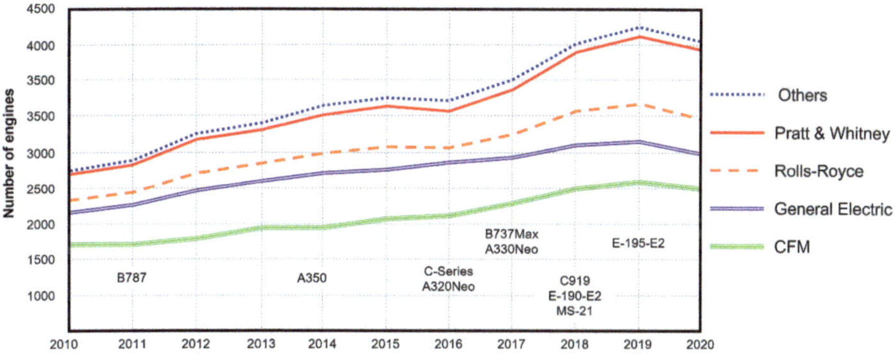

Fig. 4.6 Commercial jet engine deliveries and relevant forecast for the main Producers. (Redrawn from Leach 2016)

workhorse Ti-6Al-4V still remains the most used titanium alloy, with particular regard for aerospace applications (Donachie 2000).

As discussed in Chap. 1 (Fig. 1.11), a state-of-the-art commercial airliner, like the Boeing B787, has a significant number of components made of titanium and titanium alloys, more than in the past, as indirectly indicated by with the trend shown in Fig. 4.6, referring to the deliveries of commercial jet engines and forecast of the main Producers, from 2010 onward.

As in other fields, also titanium industry has experienced a consolidation of material specifications, with benefits for both producers and customers.

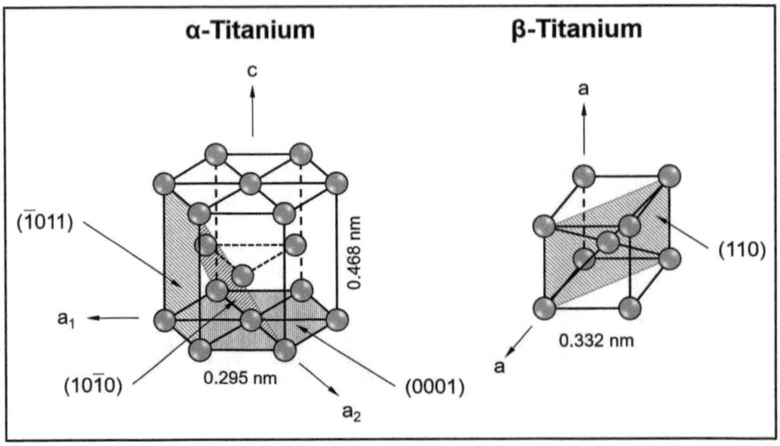

Fig. 4.7 Crystallographic structures of titanium polymorphs with the indication of slip planes (see also Table 4.1) and lattice parameters. The α phase has hexagonal close packed (*hcp*) structure, while the β phase is body centered cubic (*bcc*). (Redrawn from Leyens and Peters 2003)

This has contributed to widen the field of applications of titanium alloys, which can also be exploited in the production of innovative materials, such as titanium matrix composites and advanced alloys based on intermetallic phases, like gamma titanium aluminides (see Sect. 4.3.2).

4.2 Physical Metallurgy Aspects

The allotropic transformation, α-*hcp* to β-*bcc* (see Fig. 4.7) occurring at 882 °C in pure titanium, is also present in titanium alloys, with variations in temperature depending on their actual composition. Specific alloying elements can promote the stabilization of one polymorph preferentially, rather than the other. Moreover, through the balanced combination of stabilizing elements of the two allotropic phases, it is possible to have both of them in the same alloy in different concentrations. These metallurgical aspects will be recalled herewith, considering the importance they have in determining the properties of titanium alloys.

Some of the main features of α-Ti and β-Ti are as follows:

- *α phase*. This *hcp* (hexagonal close packed) phase, stable up to the β-transus temperature, is characterized by good mechanical properties with some limitations as concerns formability and toughness, deriving from the limited number of slip systems, just three in the *hcp* structure (see Table 4.1). The c/a ratio between the main axes of the crystallographic cell (see Fig. 4.7) is 1.587 for the α-Ti polymorph. This value, which is slightly lower than the ideal value for a hexagonal cell (1.633), can be varied when interstitially dissolved atoms are present

Table 4.1 Some characteristics of the crystallographic structures of the two polymorphs of pure titanium (Leyens and Peters 2003)

				Slip planes		Slip directions				
Polymorph	N	CN	P	Indices	Number	Indices	Number	Slip systems per unit cell	Lattice parameters	Atomic density in slip planes
α	6	12	74%	{0001}	1	<11$\bar{2}$0>	3	3	$a = 0.295$ nm $c = 0.468$ nm	91%
β	2	8	68%	{110}	6	<110>	2	12	$a = 0.332$ nm	83%

N number of atoms per unit cell, *CN* coordination number, *P* packing density (e.g., hydrogen, carbon, nitrogen, and oxygen) or when substitutional atoms with smaller than titanium radii are incorporated into the host lattice. An important example of this latter situation is provided by alloying with aluminum. The reduced c/a ratio induces a larger spacing between prism planes, and an increase in their packing density, so that they become possible slip planes, alternative to the basal planes. With reference to Fig. 4.7, prism planes are the $(10\bar{1}0)$ type planes; the pyramidal plane is $(10\bar{1}1)$, and the basal plane is (0001). According to the von Mises criterion, at least five independent slip systems are required for a homogeneous plastic deformation. This explains the limited deformability of polycrystalline α-Ti and why its limited ductility may result in the formation of mechanical twinning, as an additional plastic deformation mode together with the activation of secondary slip systems.

- *β phase*. This *bcc* polymorph is characterized by 12 slip systems, and this makes plastic deformation easier. However, the no-close packed character of the *bcc* lattice limits other properties, creep resistance in first place. Other interesting features of the β polymorph are given in Table 4.1.

The transition from the β phase to the α phase upon cooling can occur according to two mechanisms, depending on the cooling rate and the alloy composition: martensitic transformation or diffusion-controlled nucleation and growth.

A cooperative movement of atoms in a shear-type process accompanies the martensitic transformation. The hexagonal martensite of titanium is usually named α' and may exhibit two microstructures, featuring either massive or acicular grains. Massive martensite is also named *lath* or *packed* martensite. This type of martensite occurs in pure titanium, in low-alloyed materials, and in alloys in which the martensitic transformation occurs at relatively high temperatures. Massive martensite would usually feature a coarse-grained microstructure, with an average grain size in the 50–100 μm range. Within the coarse grains, small packets of parallel α plates or lathes with a thickness of 0.5–1.0 μm are present. Acicular martensite is characterized by an intimate mixture of α plates that generally contain a high density of dislocations. This is the so-called basketweave microstructure. The α plates may also contain twins. In titanium alloys, the α' martensite is supersaturated in β-stabilizing elements. When α' martensite is annealed in the α + β field (see Figs. 4.8 and 4.17), it decomposes into α + β mixture, either by the precipitation of β phase layers at the plate boundaries, or by precipitating β grains in regions with higher dislocation density. An increasing concentration of solute atoms hosted in martensite would tend to distort the lattice upon cooling, so that the hexagonal symmetry is lost and an orthorhombic martensite would form instead. The designation for this latter structure is α''. In Ti-X binary systems, the α'/α'' boundary concentrations are the following (in brackets, wt%): Mo (4), Nb (10.5), Ta (26.5), V (9.4), W (8). For the sake of completeness, another quenching structure should be mentioned as concerns β-Ti-alloys (see Sect. 4.3.1.4): the ω phase. The formation of the ω phase occurs for high solute concentrations, which inhibit the standard martensitic transformation. Upon quenching from the β field, the ω phase precipitates athermally, and it is characterized by a uniform dispersion

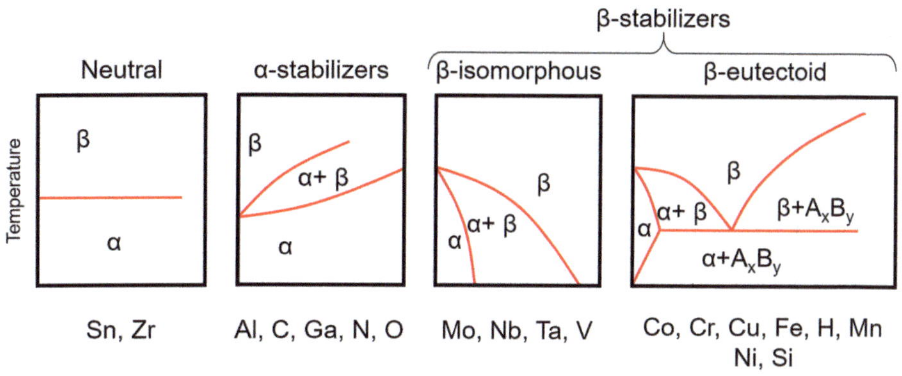

Fig. 4.8 Schematic of the influence of different alloying elements on the Ti binary systems. (Redrawn from Leyens and Peters 2003)

of extremely fine particles. The ω phase shows a trigonal symmetry in heavily β-stabilized alloys and a non-close-packed hexagonal symmetry in low-alloyed compositions, being the transition from hexagonal to trigonal continuous, as a function of the content of the alloying elements.

Alternatively, diffusion-controlled nucleation and growth of the α phase occurs when the titanium alloy is cooled from the β phase field to the α + β field (see Figs. 4.8 and 4.17 in Sect. 4.3.1) at a moderate rate. The nucleation of the α phase occurs at the β grain boundaries preferentially, evolving in a continuous layer of α phase. α grains may nucleate inside the β grains in form of parallel platelets, that continue their growth until they get in contact with other α colonies. This process is often defined as *sympathetic* nucleation and growth. α plates within the α colonies are separated by *retained* β, remnant of the high-temperature phase. The resulting microstructure features alternating α and β *lamellae*. An example of this structure can be obtained by slow cooling the Ti-6Al-4V alloy from the β field. The size of the α colonies, as well as their thickness, is reduced by increasing the cooling rates.

The new α plates that nucleate on existing ones tend to grow perpendicularly to that plate in order to minimize the overall elastic strain. This particular type of nucleation and growth usually leads to the formation of a Widmannstätten structure (see Fig. 4.23) (Boyer et al. 1994; Lüterling and Williams 2007).

The stability range of the two polymorphs, α and β, can be varied by means of suitable alloying elements. The alloying elements are usually classified into three main families; according to the polymorph, they are able to stabilize: α-stabilizers, β-stabilizers, and neutral elements. β-stabilizers can also be split in other two subgroups: isomorphous and eutectoid. A schematic representation of the influence of each one of them can be inferred from the relevant binary phase diagrams in Fig. 4.8. An important aspect of alloying is the appearance of a two-phase region (α + β),

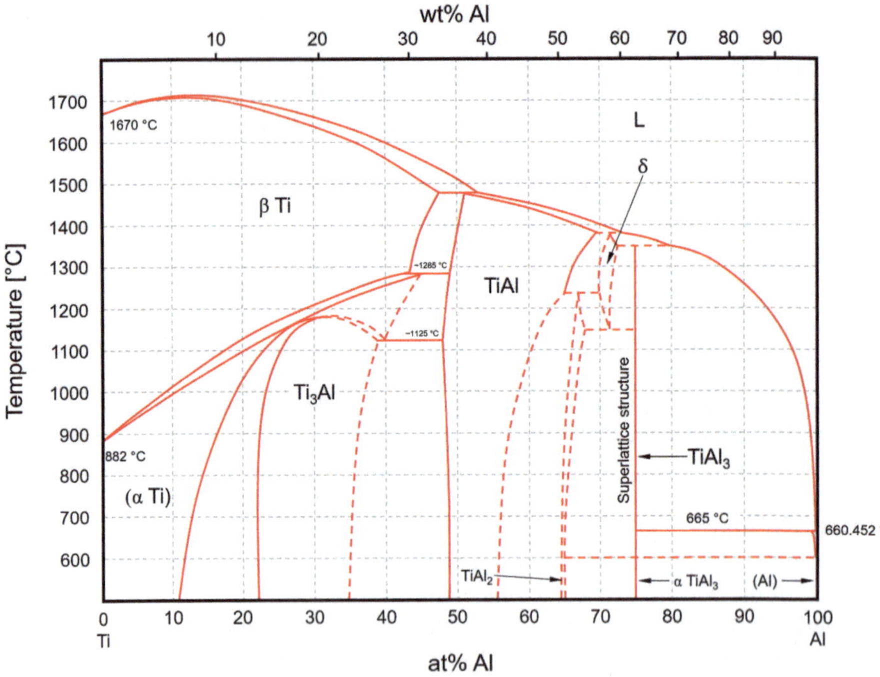

Fig. 4.9 Ti-Al phase diagram. Aluminum is an α-stabilizer element. (Redrawn from ASM International 1992)

extremely interesting for the development of titanium alloys. Elements like Al, C, N, and O are classified as α-stabilizers. Additions of aluminum stabilize the alpha polymorph of titanium also at temperatures higher than 882 °C, i.e., they raise the alloy β-transus (T_β). In general, aluminum is deliberately added to titanium and its alloys to obtain the stabilization effect and related improvements in the mechanical properties. In addition to the α and β phases, a number of intermetallics are present in the binary Ti-Al phase diagram. Among them, α_2-Ti_3Al and γ-TiAl are of specific technological interest, since on these ordered phases the advanced titanium aluminides alloys are based (see Sect. 4.3.2) (see Fig. 4.9).

The other α-stabilizers would generally originate from unwanted contamination, coming, for instance, from the working atmosphere, like nitrogen and oxygen, and production plant, i.e., carbon from graphite electrodes of nonconsumable arc melting (see Fig. 4.34). Alloy properties are very sensitive even to comparatively low concentrations of these elements, which in fact are used to qualify the grade of the alloy itself (see Sect. 4.3).

This aspect is particularly important for oxygen that, as shown by the binary Ti-O phase diagram (Fig. 4.10), has quite a high solid-state solubility and can be interstitially hosted in the titanium lattice, up to relatively high concentrations. This, of course, poses severe restrictions and control requirements on the different production steps, particularly to avoid air contamination (see Sect. 4.4). Obviously, the impurity

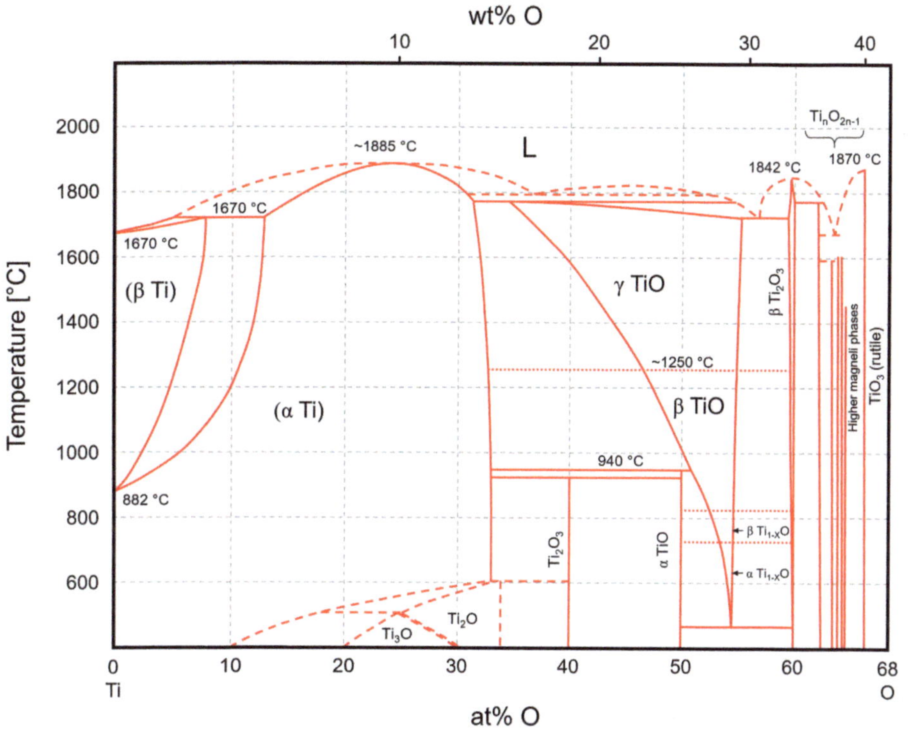

Fig. 4.10 Ti-O phase diagram. Oxygen is a α-stabilizer element. (Redrawn from Massalski et al. 1986)

content of the raw materials, i.e., titanium and alloying elements, is another important aspect to consider.

Substantially no effect on the β-transus is introduced by the neutral elements, like tin (Fig. 4.11) and zirconium, that are soluble both in α and β phases. They act as solid solution strengtheners and guarantee this mechanism also at high temperatures and in creep regime. This is an aspect that has been exploited for the development of high-performance alloys (see Sect. 4.3.1.3).

β-stabilizers are those elements that widen the stability temperature range of the high-temperature polymorph of titanium, the β phase, by lowering the β-transus. These alloying elements can be divided into two main categories: *isomorphous* and *eutectoid*. Isomorphous β-stabilizers, e.g., Nb and V (Fig. 4.12), have the same *bcc* crystalline structure as β-titanium. Moreover, they are characterized by a relatively limited solubility in the α phase and a complete solid-state solubility in β-titanium above given temperatures. β phase can be stabilized down to room temperature for sufficiently high concentrations of β-stabilizers. Mo, Ta, Nb, and V are the most important β-isomorphous elements, showing relatively high solubility in Ti. The melting temperature of a β-isomorphous-containing alloy may be either lower or higher than the melting temperature of pure titanium. This aspect, together with the

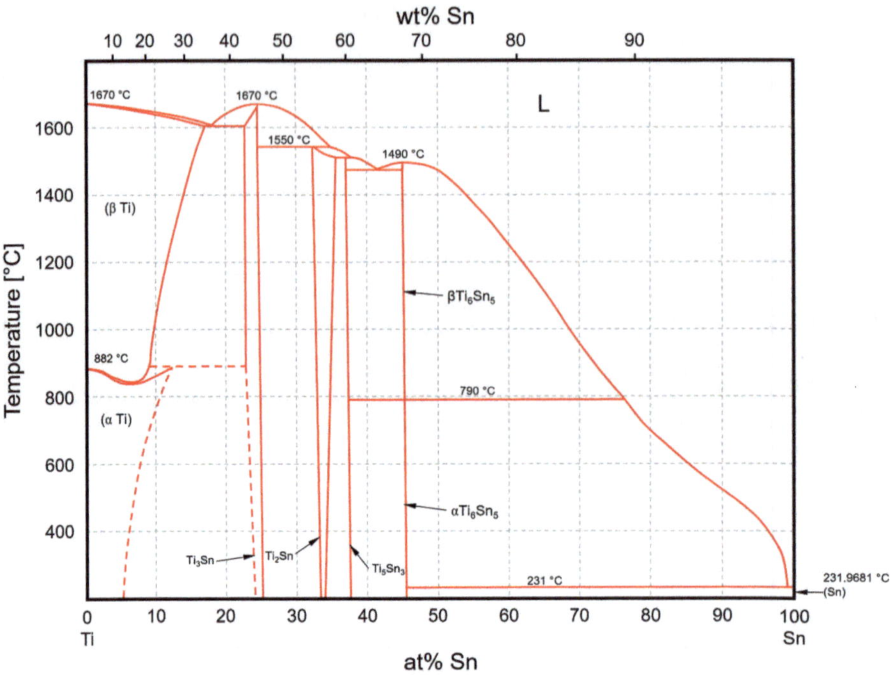

Fig. 4.11 Ti-Sn binary phase diagram. Tin additions do not change significantly the β-transus, and in fact tin is classified as *neutral* alloying element. (Redrawn from Massalski et al. 1986)

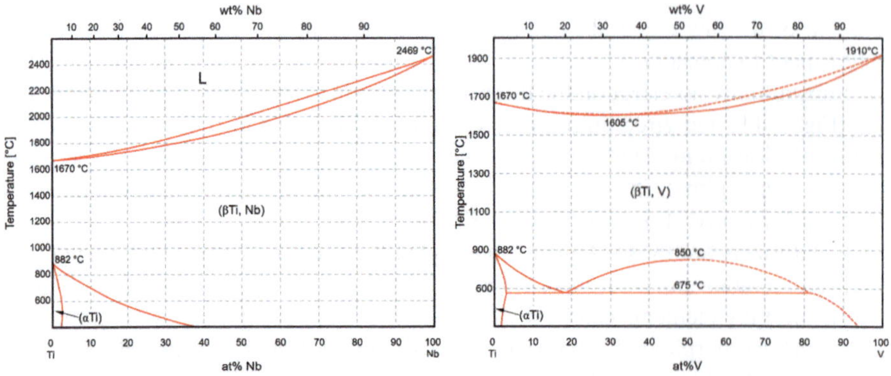

Fig. 4.12 Two examples of binary phase diagram involving titanium and two isomorphous β phase stabilizers: niobium and vanadium. See main text for the comments on the general features of these diagrams. (Redrawn from Massalski et al. 1986)

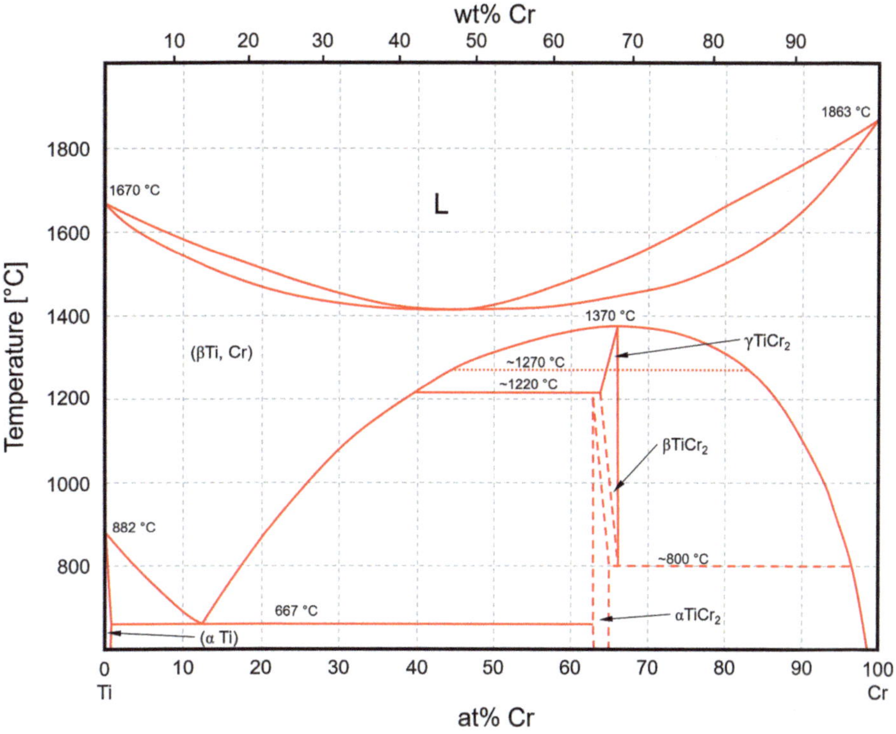

Fig. 4.13 Ti-Cr phase diagram. Chromium is a passive eutectoid β phase stabilizer. (Redrawn from Massalski et al. 1986)

presence of a liquid-solid two-phase field, has to be considered from the technological point of view, in relation to possible segregation effects during the solidification of the alloy from the melt.

The binary phase diagrams involving titanium and eutectoid β-stabilizers do not display a complete solid-state solubility range (Figs. 4.13 and 4.14). The solubility of these elements is limited also in the α phase, and the common feature to all relevant systems is the presence of an eutectoid transformation. Upon cooling, the decomposition of the β phase into a mixture of α phase and an intermetallic is observed. Depending on the kinetics of this transformation, these alloying elements can be classified as either *active* or *passive* β-stabilizers. For active β-stabilizers, like copper and silicon, the relevant decomposition reactions are β-Ti to α-Ti + Ti_2Cu and β-Ti to α-Ti + Ti_3Si. Elements that can be classified as passive β-stabilizers are, for instance, chromium (*bcc*), iron (*bcc*), and manganese (*bcc*). In this case, the decomposition reaction is more sluggish, if compared to active elements. In the Ti-Cr system (Fig. 4.13), the reaction is β-Ti to α-Ti + $TiCr_2$. TiFe and TiMn are other

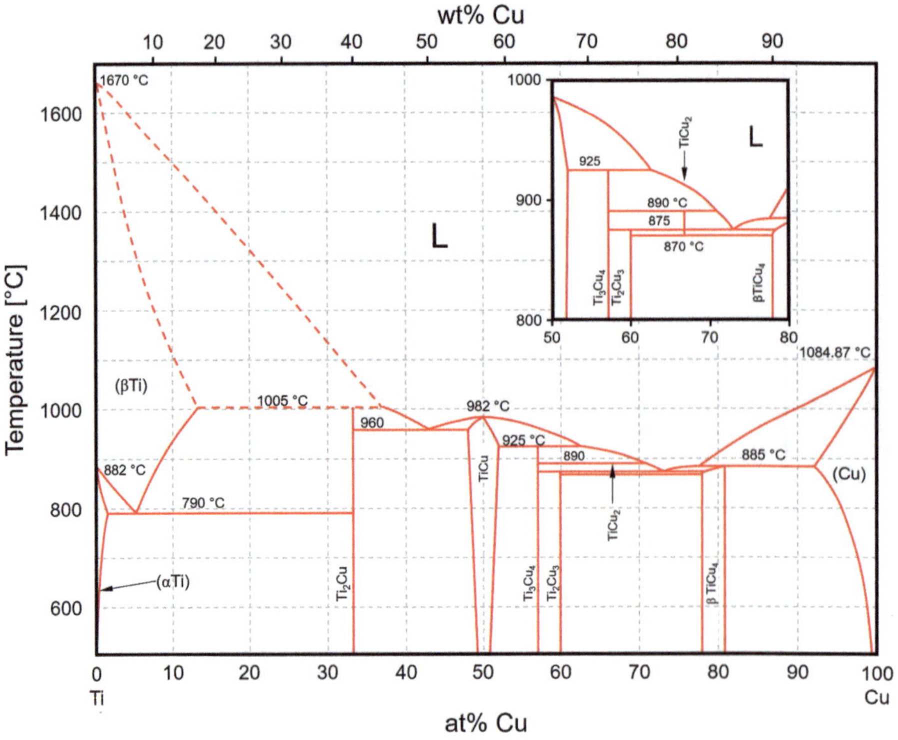

Fig. 4.14 Ti-Cu phase diagram. Copper is an active eutectoid β phase stabilizer. (Redrawn from Massalski et al. 1986)

intermetallics that may precipitate in the α phase in the Ti-Fe and Ti-Mn binary systems. As a general comment on the *active* or *passive* character of the eutectoid β-stabilizers, some clues can be inferred from the relevant binary phase diagrams, with the necessary precautions recommended when drawing conclusions on kinetic aspects from thermodynamic tools, like phase diagrams are. The stoichiometry of the intermetallic phases are an indirect indication of the compositional changes involved with the eutectoid transformations. The larger is the compositional change ($TiCr_2$, TiFe, TiMn, etc.), the longer time it will take for the transformation to be fully accomplished. A sluggish eutectoid transformation of the *passive* titanium alloying elements would be coherent with this aspect. Intermetallics with titanium-rich composition, closer to the eutectoid, result from the transformation involving *active* β-stabilizers. From the alloy design point of view, an interesting information is the concentration of each one of the β-stabilizer elements, needed to retain the high-temperature polymorph down to room temperature. This critical concentration (β_c) for some of the following transition metals is given among brackets (wt%): Co (7), Cr (8), Cu (13), Fe (4), Mn (6), Mo (10), Nb (36), Ni (9), Ta (45), V (15), and W (23) (Boyer et al. 1994). To complete the picture of the β-stabilizers, it is worth

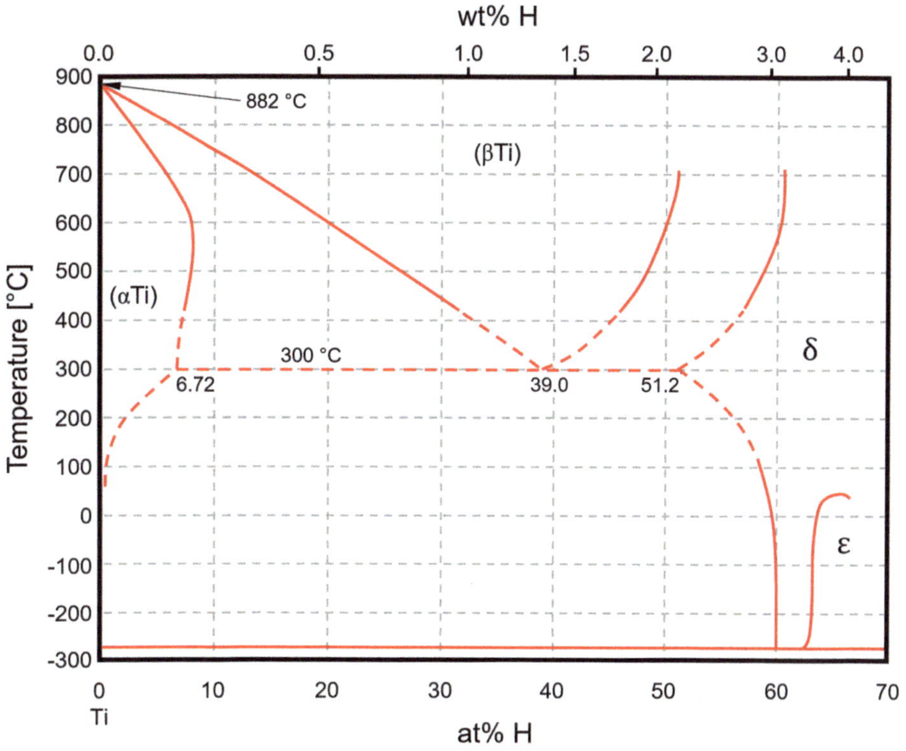

Fig. 4.15 Ti-H phase diagram. (Redrawn from Massalski et al. 1986)

mentioning hydrogen, an important gaseous element that is classified as a *passive* β-stabilizer (Fig. 4.15). The eutectoid reaction leads to the formation of the hydrogen rich titanium δ-hydride: β-Ti to α-Ti + δTiH_2. This hydride may cause an important embrittlement of the alloy; the more severe, the higher is its concentration. For the same reasons quoted with reference to carbon and the gaseous α-stabilizer, i.e., nitrogen and oxygen, hydrogen can possibly be present in titanium and its alloys out of contamination. This raises the need for a strict control on the processing and fabrication conditions, all through the production cycle, starting from extraction or recycling of titanium up to the final forming operations.

4.3 Main Groups of the Titanium Alloys

Several classifications can be found in the literature as concerns titanium alloys, generally coherent with each other, just introducing some naming differences. Titanium alloys can be grouped into two main families: *conventional* titanium alloys and *advanced* titanium alloys, as displayed by the diagram in Fig. 4.16.

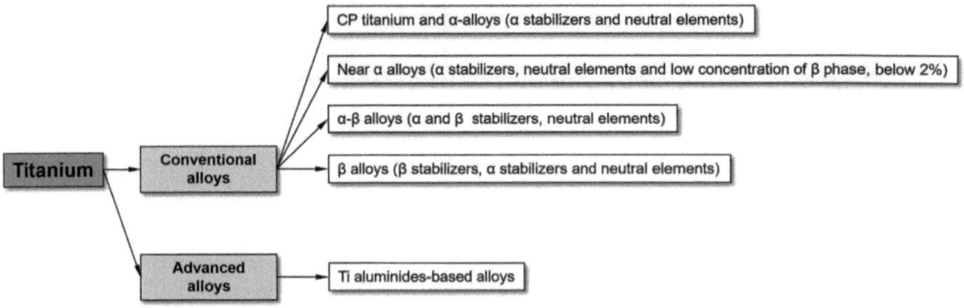

Fig. 4.16 The main classes of titanium alloys and relevant short descriptions as concerns their compositions

The classification of the *conventional* titanium alloys, is based on their phase composition, with reference to the α and β polymorphs. This is certainly an important aspect to be considered for materials selection purposes, although the actual microstructure and relevant properties of any specific Ti-alloy are very much dependent also on its thermomechanical history.

The *advanced* titanium alloys are based on the intermetallic aluminide phases: γ-TiAl and α_2-Ti_3Al. Several systems have been taken into account, single and multiphase alloys, although at the present time, the only important aircraft application has been achieved with TiAl-based alloys (see Sect. 4.3.2). In the following sections, both families of conventional and advanced titanium alloys will be presented, with particular regard to the metallurgical aspects and important characteristics that are interesting for the applications in the aerospace field.

4.3.1 Conventional Titanium Alloys

The main groups of *conventional* titanium alloys are α-, near-α, α + β, and β-alloys. Table 4.2 gives the compositions of selected titanium alloys, of the first three groups of the classification proposed in Fig. 4.16. In Sect. 4.3.1.4, similar information is provided in Table 4.4, for the main titanium β-alloys. Another information that is specified for each alloy is the relevant *beta transus*, above which the alloy is fully β. This temperature is most important for setting up the conditions of the thermal treatments and hot-working operations.

The classification adopted for the conventional titanium alloys can be visualized using a pseudo-binary phase diagram (Fig. 4.17), in which the alloy phase fields are given as a function of temperature and equivalent β-stabilizer concentration.

4.3.1.1 α-Alloys

These alloys are characterized by the absence of the ductile-to-brittle transition, which is typical of the *bcc* alloys. For this reason, α-alloys can be employed also for

Table 4.2 Composition and classification of some important conventional Ti-alloys

Alloy type	Alloy designation	Alloy composition [wt%]	β-transus [°C]
α	Grade 1	CP-Ti (0.2Fe, 0.18O)	890
	Grade 2	CP-Ti (0.3Fe, 0.25O)	915
	Grade 3	CP-Ti (0.3Fe, 0.35O)	920
	Grade 4	CP-Ti (0.5Fe, 0.40O)	950
	Grade 6	Ti-5Al-2.5Sn	1040
	Grade 7	Ti-0.2Pd	915
	Grade 12	Ti-0.3Mo-0.8Ni	880
	IMI 230	Ti-2.5Cu	–
	Ti-3-2.5	Ti-3Al-2.5V	935
Near-α	IMI 685	Ti-6Al-5Zr-0.5Mo-0.25Si	1020
	IMI 834	Ti-5.8Al-4Sn-3.5Zr-0.5Mo-0.7Nb-0.35Si-0.06C	1045
	Ti6–2–4-2-S	Ti-6Al2Sn-4Zr-2Mo-0.1Si	995
	Ti-811	Ti-8Al-1V-1Mo	1040
	TIMETAL 1100	Ti-6Al-2.7Sn-4Zr-0.4Mo	1010
α + β	Ti-6-4	Ti-6Al-4V (0.20O)	995
	Ti-6-4 ELI	Ti-6Al-4V (0.13O)	975
	IMI 315	Ti-2Al-2Mn	–
	IMI 550	Ti-4Al-2Sn-4Mo-0.5Si	975
	IMI 679	Ti-2Al-11Sn-4Zr-1Mo-0.25Si	–
	Ti-17	Ti-5Al-2Sn-2Zr-4Mo-4Cr	890
	Ti-6-2-2-2-2	Ti-6Al-2Sn-2Zr-2Mo-2Cr-0.25Si	–
	Ti-6-2-4-6	Ti-6Al-2Sn-4Zr-6Mo	940
	Ti-6-6-2	Ti-6Al-6V-2Sn	945

For the β-alloys the reader is referred to Table 4.4. The β-transus for each alloy is also indicated (Cotton et al. 2015; Leyens and Peters 2003; Lüterling and Williams 2007; Meetham 1981)

cryogenic applications, e.g., spacecraft tanks, for which Ti-5Al-2.5Sn is used. α-Alloys exhibit good strength, creep resistance up to intermediate temperatures, and toughness. Owing to their plain composition and microstructure, α-alloys are generally well weldable. Unalloyed titanium, also known as CP (commercially pure) titanium, is generally classified as a *conventional* α-alloy, since its structure at room temperature is fully *hcp*. There are several grades of CP-titanium (see Table 4.2), and the main difference among them consists in the oxygen and iron content. CP-titanium is characterized by an excellent corrosion resistance, as it can be inferred from the Pourbaix diagram for pure titanium (Fig. 4.18). The stability field of a protective passivation layer is very wide and the corrosion resistance of CP-Ti has made of this material an interesting choice for the fabrication of components of chemical and petrochemical industry. Although CP-Ti is more expensive than stainless steel, still, the products made of this alloy may have a lower lifecycle cost thanks to their better durability. Therefore, CP-Ti, whose tensile properties are

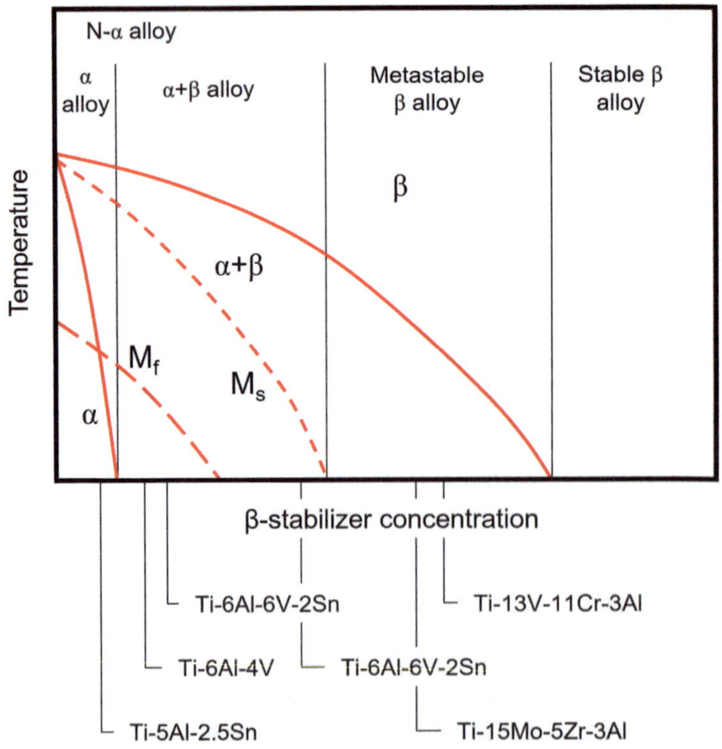

Fig. 4.17 Portion of binary phase diagram describing the classification and phase composition of the main titanium alloys, with the indication of some of them. (Redrawn from Boyer et al. 1994; Lüterling and Williams 2007)

listed in Table 4.3, is often used as substitute material of various grades of stainless steel (Lüterling and Williams 2007).

Grade 1 CP-Ti has the lowest strength level, but an excellent cold formability. Thanks to its properties, it can be used as cladding alloy for steel reactors and deep drawing applications. Grade 2 has a higher strength and it is the most used CP-Ti grade, while Grade 3 is quite exclusively used for the fabrication of pressure vessels. Grade 4 has the highest strength and it is used for fittings and mountings (Leyens and Peters 2003). α-Alloys were the first type of titanium alloys to be developed. The IMI 317 (Grade 6) belongs to this group and also alloys with limited concentrations of β-stabilizers, like the IMI 230, containing 2.5% of Cu (see Table 4.2). However, in this alloy the main role of copper is not much the stabilization of the β phase, but rather a contribution, through the formation of the ordered intermetallic phase Ti_2Cu, to precipitation strengthening. In α-alloys, alloying elements are those that can be solutioned to some extent into the *hcp* lattice, either as substitutional (Al, Sn, Zr) or interstitial (C, N, O) atoms. The main strengthening mechanisms for these alloys rely on solid solution hardening in the first place, with possible contribution from

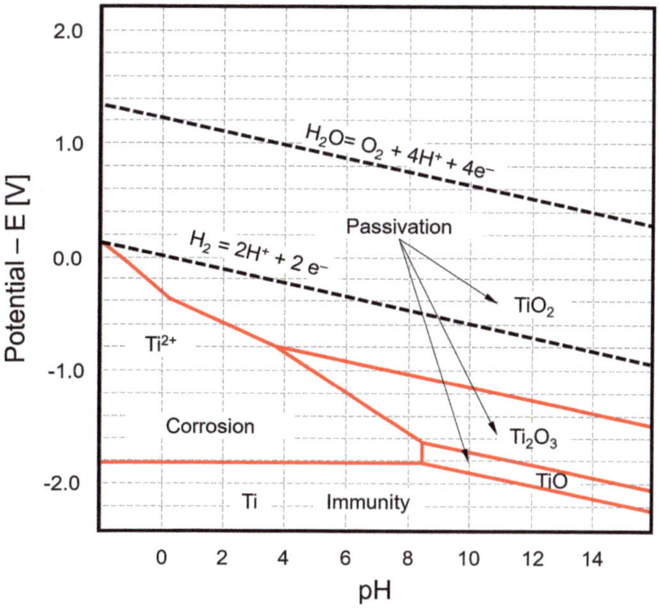

Fig. 4.18 Pourbaix diagram for pure titanium. (Redrawn from McCafferty 2010)

Table 4.3 Tensile properties of CP-titanium

Alloy	E [GPa]	UTS [MPa]	Yield strength [MPa]	Elongation [%]
Grade 1	105	240	170	24
Grade 2	105	345	275	20
Grade 3	105	445	380	18
Grade 4	105	550	480	15

E = Young's modulus; UTS = ultimate tensile strength (Lüterling and Williams 2007)

precipitation hardening. Substitutional strengthening relies on aluminum mainly, neutral elements (Sn and Zr in particular) and limited concentrations of β-stabilizers. Oxygen enters the hexagonal lattice interstitially, and its concentration determines the grade of the alloy (see Table 4.2), influencing its mechanical properties. The precipitation of coherent α_2-Ti_3Al particles contributes to precipitation hardening, in a similar way as the γ' strengthening in nickel-based superalloys (see Sect. 6.4). Dislocations that form in the alloy matrix are blocked by the precipitate boundaries, resulting in extensive piling-up. Alternatively, dislocations may cut through the precipitates and in this case their movement is hindered by the DO_{19} ordered structure of the intermetallic phase. For these reasons, although a generally good weldability and structural stability of α-alloys promoted their use for noncritical components, multiphase systems responded better to the production and service requirements of aerospace components (Leyens and Peters 2003; Lüterling and Williams 2007).

4.3.1.2 α + β Alloys

The first "aeronautic" α + β alloy was the IMI 315 (Ti-2Al-2Mn), used for the compressor blades of the Avon Rolls-Royce engine, starting from 1954 (Froes 2015). α + β alloys turned out to be excellent candidates for overcoming the limitations of single-phase α-alloys. When molybdenum was introduced as β-stabilizer, the Ti-7Al-4Mo and Ti-4Al-4Mo-2Sn-0.5Si (IMI550) alloys were developed, this latter with a particular compositional focus on improving creep resistance. The IMI550 alloy, introduced in 1958, together with the notorious and most used Ti-6Al-4V (IMI318), became the state-of-the-art reference alloys for aircraft applications at the beginning of 1960s. The Ti-6Al-4V alloy was developed in the USA in the early 1950s and it is characterized by an excellent balance of mechanical properties. Aerospace industry is its largest market, but not the only one. In fact, Ti-6Al-4V (also known as Grade 5) is largely used in biomedical applications, especially in its cleaner extra low interstitial (ELI) form: Ti-6Al-4V ELI (Donachie 2000; Leyens and Peters 2003; Lüterling and Williams 2007). Other examples of α + β alloys are Ti-17 and Ti-6-2-2-2-2 (Table 4.2). The main difference between α + β and near-α alloys, considered in the next section (Sect. 4.3.1.3), is essentially in the maximum fraction of β phase that can form. For instance, in the Ti-6Al-4V, the β phase content is equal to 15% at most. By adding β-stabilizers a β-transus reduction is observed (see Table 4.2). The α + β alloys can be strengthened and their microstructure modified by solution and aging treatments. The M_s and M_f curves in Figs. 4.17 and 4.19 represent the temperature below which the β phase starts and completes, respectively, the transformation to the α′ hexagonal martensite, if cooled down at a sufficiently high rate (see Sect. 4.2). Upon aging, quenched from beta field alloys show a continuous raise of the mechanical strength (Fig. 4.19), determined by the solid solution hardening of both polymorphs.

The introduction of the β phase is also important for the increased forgeabilty of the alloys. Hot forging, either in α + β or β field, becomes in fact the main processing approach for producing structural and engine components (Leyens and Peters 2003; Lüterling and Williams 2007).

The service temperature for α + β alloys usually does not exceed 440 °C. Above this limit, creep deformation rate becomes too high and reduces intolerably the useful lifetime of the involved components. This behavior is determined by too a high concentration of β phase, that is not only more deformable at room temperature, but features also a diffusivity two orders of magnitude higher than the hexagonal counterpart (Leyens and Peters 2003, Lüterling and Williams 2007).

Fig. 4.19 The effects of different thermal treatments on different alloy compositions. (Redrawn from Polmear 2006)

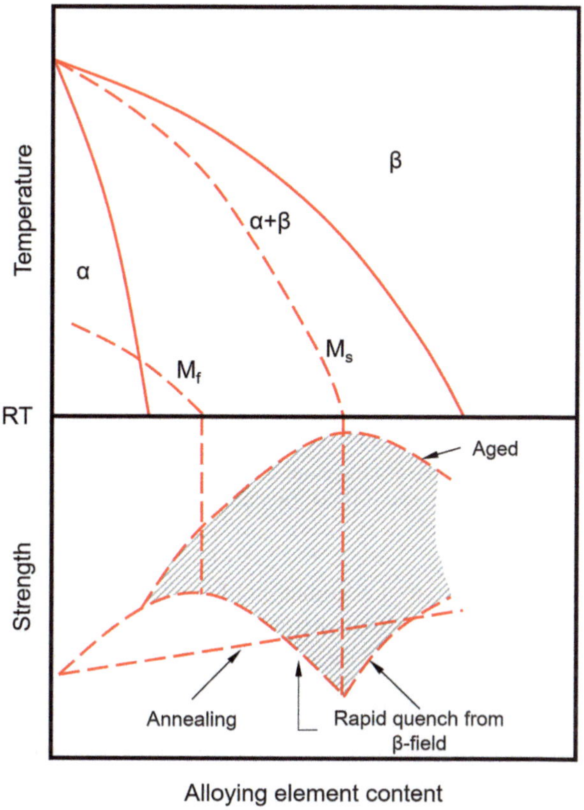

4.3.1.3 Near-α Alloys

An obvious approach to overcome creep limitations of the α + β alloys was to develop alloys with a lower concentration of the cubic titanium polymorph, the so-called near-α (N-α) alloys. These alloys are characterized by comparatively low concentrations of β-stabilizers in an α-alloyed chemistry. Therefore, the β phase that forms within the α phase matrix, would not generally exceed 10% (see Fig. 4.17) (Lüterling and Williams 2007). At 800 °C, the β phase content in the IMI834 top performing near-α alloy is 5 vol% only, whereas the value for the Ti-6Al-4V α + β alloy is three times higher (Leyens and Peters 2003; Lüterling and Williams 2007). For the stabilization of the β phase, refractory elements, like molybdenum and niobium, are preferentially used rather than vanadium, so that lower diffusivity alloys are obtained. At the same time, the grade of the raw materials can be improved, in order to reduce the content of unwanted impurities, like, in particular, iron. Comparatively high concentrations of neutral elements, like tin and zirconium, are also there. These elements enhance the room temperature strength of the alloys, thanks to a solid solution strengthening mechanism. Zirconium is also contributing,

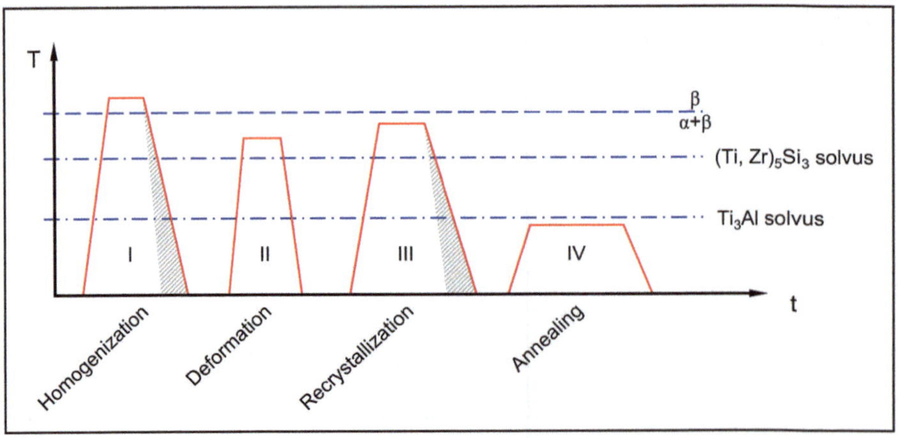

Fig. 4.20 Scheme of a thermomechanical treatment capable to induce a bimodal (duplex) structure in a near-α alloy, like IMI834. (Redrawn from Lüterling and Williams 2007)

with titanium, to the formation of silicide intermetallic compounds, like Ti_5Si_3 and $(Ti, Zr)_5Si_3$. These incoherent precipitates, that form preferentially along grain and α-β lamellae boundaries, stabilize the alloy microstructure, particularly at high temperatures. The excellent creep resistance of the *hcp* α phase, together with the combination of strengthening, thermal stability and toughness associated with the presence of a limited *bcc* β phase, allow using the near-α alloys up to 600 °C (Donachie 2000). This temperature still represents the upper limit for the applications involving *conventional* titanium alloys under high stress conditions, like it is very often the case of the aircraft engine applications. The grain structure of N-α alloys can be changed through the recrystallization obtained via cold-working and annealing. The very first alloy developed for high-temperature applications was the Ti-811 (Table 4.2). The high Al content used to obtain the N-α structure may result in the formation of the α_2-Ti_3Al intermetallic, which acts as a barrier to dislocation movement. The stability field of these phases is interesting for the definition of acceptable service conditions but also to design thermomechanical treatments, like the one illustrated in Fig. 4.20. The sequence of operations highlights the possibility to carry out the treatments both in α + β and β field, forming α′ martensite on cooling. The microstructures obtained from these treatments are intrinsically more stable, considering the higher values of the β-transus of these materials, in comparison with all the other groups of *conventional* Ti-alloys (Table 4.2). The alloy development process led to a couple of alloys that are still regarded as at the top of the performance ranking. The IMI 834 alloy (year of issue into the market: 1984. The alloy became later on: TIMETAL 834, when TIMET acquired IMI Titanium Ltd.) and the companion alloy Ti 1100 (1988), although no longer on the market, had comparable peak performances and its codename indicates clearly the maximum operating temperature, i.e., 1100 °F corresponding to 600 °C approximately. Although other alloys, mostly prototypal though, have been developed afterwards, the above limit still remains, depending on the constitutional aspects of these materials (Zhao et al. 2017). Subsequent research, concerning the improvement of high-temperature properties of these alloys, pointed out that the addition of silicon in small amounts (up to 0.1%) would significantly improve their creep resistance. Following a similar approach, RMI developed an alloy named Ti-6-2-4-2-Si (see Table 4.2), featuring a fine dispersion of titanium silicide intermetallic particles, acting as dislocation pinning centers, thus reducing the creep deformation rate. According to these results, all newly developed high-temperature alloys contained up to 0.5% of silicon. The IMI 685 was the first alloy employed after a β-annealing, a treatment resulting in a superior creep resistance (Leyens and Peters 2003).

The highly alpha stabilized N-α alloys may contain an excess of the α_2-Ti_3Al intermetallic, particularly after long-term exposure to high-temperature conditions. This has a twofold detrimental effect: increasing the alloy brittleness and, under particular environmental conditions, triggering hot-salt stress corrosion cracking (see Sect. 8.2). Therefore, real N-α alloys would generally feature an aluminum content evaluated on the basis of the *aluminum equivalent concentration*, or *ordering parameter* (Eq. 4.4, all concentration in wt%).

$$\%\mathrm{Al} + \frac{1}{3}\%\mathrm{Sn} + \frac{1}{6}\%\mathrm{Zr} + 10(\%\mathrm{O} + \%\mathrm{C} + 2\%\mathrm{N}) < 9\% \tag{4.4}$$

4.3.1.4 β-Alloys

β-Alloys contain a sufficient amount of β-stabilizers so to suppress the martensitic transformation during cooling the alloy from the beta field (Eylon et al. 1994). The as-quenched condition may be metastable, since the precipitation of the α-Ti phase may still occur upon aging the alloy or by exposing it at relatively high operational temperatures. Fully stable β-alloys do exist above a critical beta stabilizer concentration (Fig. 4.17). To evaluate the level of constitutional stability of a β-Ti-alloy, a parameter called the *molybdenum equivalent concentration*, Mo_{eq}, can be used. It is given by Eq. 4.5 (relevant element concentrations in wt%).

$$Mo_{eq} = 1.0Mo + 0.67V + 0.44W + 0.28Nb + 0.22Ta + 2.9Fe + 1.6Cr - 1.0Al \tag{4.5}$$

Values of Mo_{eq} in excess of 8 are necessary to stabilize the β phase upon quenching (β-metastable alloys). The limit for fully stable β-alloys is not so well clear-cut. In fact, even the nominally β-stable Ti-30Mo alloy turns out to be unstable at elevated temperatures under applied stresses (Bania 1994). In Table 4.4, ranked according to relevant Mo_{eq}, a list of the main β-alloys is provided, with the indication of the β-transus values and of the more common applications. In some respect,

Table 4.4 Commercial name, composition, and applications of main β-alloys sorted by increasing Mo equivalent (Leyens and Peters 2003)

Alloy designation	Composition [wt %]	Mo equivalent	β-transus [°C]	Main applications	Developer
13-13	Ti-13Nb-13Zr	3.6		Biomedical (orthopedic implants)	Smith & N. (1992)
Beta-CEZ	Ti-5Al-2Sn-2Cr-4Mo-4Zr-1Fe	5.1	890	High-strength applications at medium temperature	CEZUs (1990)
SP 700	Ti-4.5Al-3V-2Mo-2Fe	5.3	900	High-strength applications	NKK (1989)
Ti-17	Ti-5Al-2Sn-2Zr-4Mo-4Cr	5.4	885	High-strength applications at medium temperature	GEAE (1968)
VT 22	Ti-5V-5Mo-1Cr-1Fe-5Al	8	850	High-strength forgings	–
Ti-10-2-3	Ti-10V-2Fe-3Al	9.5	800	High-strength forgings	Timet (1971)
15-3	Ti-15V-3Cr-3Sn-3Al	12	760	Airframe castings, plates, sheets	USAF (1978)
Beta III	Ti-11.5Mo-6Zr-4.5Sn	12	745	High-strength applications	Crucible (1969)
TIMETAL 21S	Ti-15Mo-2.6Nb-3Al-0.2Si	13	807	Corrosion/oxidation resistance, TMCs[b]	Timet (1989)
8-8-2-3	Ti-8V-8Mo-2Fe-3Al	15	775	High-strength forgings	Timet (1969)
IMI 205	Ti-15Mo	15	727	Corrosion resistance	IMI (1958)
Beta-C	Ti-3Al-8V-6Cr-4Mo-4Zr	16	795	Fasteners, oils fields, springs	RMI (1969)
VT 35	Ti-15V-3Cr-1Mo-0.5Nb-3Al-3Sn-0.5Zr	16	–	High-strength airframe castings	–
TIMETAL LCB[a]	Ti-4.5Fe-6.8Mo-1.5Al	18	810	High-strength applications	Timet (1990)
TMZF	Ti-12Mo-6Zr-2Fe	18	743	Biomedical (orthopedic implants)	Howmedica (1992)
1–8-5	Ti-1Al-8V-5Fe	19	825	Fasteners	RMI (1957)
B120VCA	Ti-13V-11Vr-3Al	23	700	Airframe (SR-71), landing gears, springs	RemCru (1952)
TIMETAL 125	Ti-6V-6Mo-5.7Fe-2.7Al	24	704	High-strength fasteners	Timet (1990)
–	Ti-30Mo	30		Corrosion resistance	RemCru (1952)
–	Ti-40Mo	40		Corrosion resistance	RemCru (1952)
Alloy C	Ti-35V-15Cr	47	–	Burn-resistant alloy	Pratt & Whitney (1990)

[a]*LCB* low cost beta
[b]*TMCs* titanium matrix composites

β-alloys can be regarded as the most versatile class of titanium alloys, featuring the highest strength to weight ratios and very attractive combinations of toughness, strength, and fatigue resistance, even when produced in large cross-section parts. In the past, the β-alloy B120VCA had been extensively used for the supersonic SR-71 Blackbird project. The α phase can precipitate from the metastable β phase in form of very fine and extremely hard platelets with tensile strength exceeding 1400 MPa. For this reason, β-alloys can be hardened up to much higher yield strength values than α + β alloys. β-alloys can be processed at lower temperatures than those required for α + β alloys, and they are also more hydrogen-tolerant than the other conventional titanium alloys, with particular regard for α-alloys. This latter feature allows their use in environments where hydrogen pickup is an issue.

Metastable β-alloys have gained both interest and importance because of their combination of mechanical properties. They are characterized by a complex microstructure that enables the alloy designer to optimize not only strength but also toughness. Some examples coming from this family of β-alloys are Ti-6246, Ti-17, Beta-CEZ, SP-700, and Ti-10-2-3. Beta-CEZ is an alloy specifically designed for applications at moderate temperatures in gas turbine engines. The Japanese SP-700 is characterized by an excellent superplastic deformability, even at temperatures as low as 700 °C. This result is possible thanks to a very fine-grained duplex microstructure. The high-strength β-alloy Ti-10-2-3 combines strength, toughness, and high-cycle fatigue life. However, these performances can be achieved if the alloy has been properly processed, including an appropriate thermomechanical cycle. For instance, Fig. 4.21 shows a TTT diagram for the Ti-10-2-3 alloy, with the indications of the microstructures developed using different cooling rates. Heavily stabilized β-alloys contain higher fractions of β-stabilizing elements and consequently a lower volume fraction of α phase, resulting in a reduction of mechanical strength. Most of the commercial β-alloys have a Mo_{eq} in the β-metastable range, i.e., 10–24% (see Table 4.4) (Leyens and Peters 2003; Lüterling and Williams 2007).

The metastability of the β-alloys is responsible for their strengthening. Processing of β-alloys usually comprises a hot-working stage followed by heat treatments (see Sect. 4.3.1.5).

Specific characteristic properties of β-alloys are:

- Critical processing parameters, i.e., small processing window of some alloys.
- Good fatigue resistance.
- Good mechanical resistance up to 400 °C.
- Good workability and weldability.
- High cost due to the high concentration of alloying elements.
- High springback.
- Low elastic modulus (application: springs and biocomponents).
- Presence of impurities and segregation in the structure.

These properties are important selection criteria for suitable applications of these alloys that are mostly employed in structural parts at medium low temperatures apart a few exceptions (see Sect. 4.6).

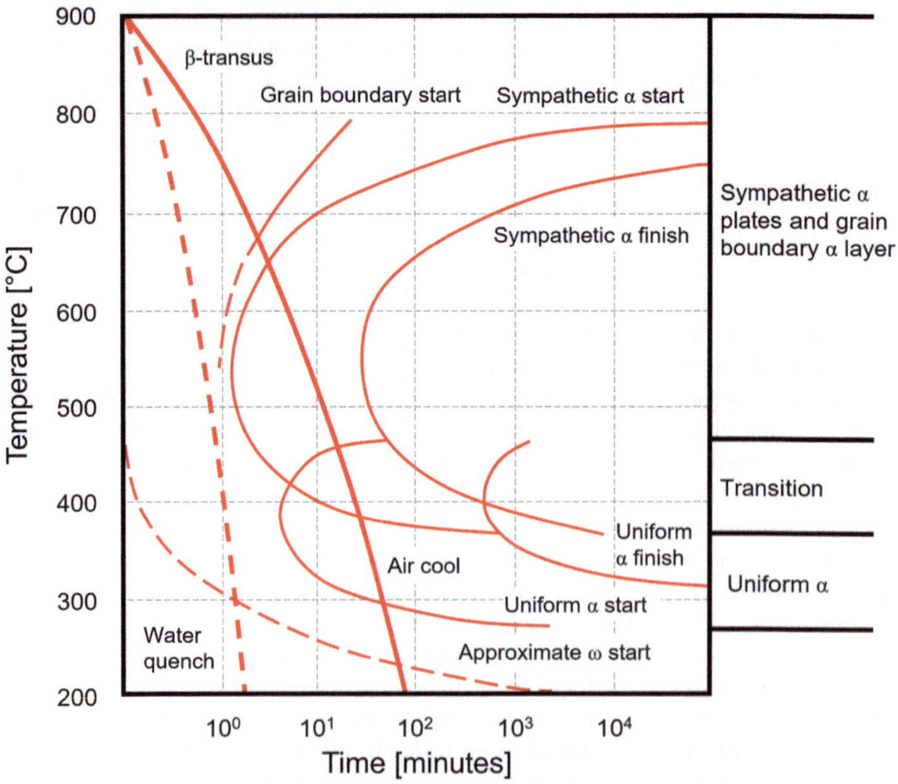

Fig. 4.21 TTT diagram for the high-strength Ti-10-2-3 alloy (redrawn from Leyens and Peters 2003)

4.3.1.5 Thermal Treatments and Properties of Conventional Titanium Alloys

Alloy composition is not the only parameter that determines its properties, very much dependent also on the microstructural features induced by suitable thermal treatments and deformation steps. The working principles behind these aspects are presented with reference to the Ti-6Al-4V alloy. The main aspects and alloy features that are involved in the treatment concern:

- α-Ti, α′-Ti, and β-Ti transformations.
- Intermetallic phase precipitation.
- Microstructure, e.g., grain morphology, grain size, composition.
- Phase composition.

The principal actions potentially interesting for a thermomechanical treatment protocol are summarized in Fig. 4.22. The atmosphere during the treatments is either inert or reduced down to relatively low pressure. In this way, air contamination is avoided or kept under control, considering that gaseous species, like oxygen and

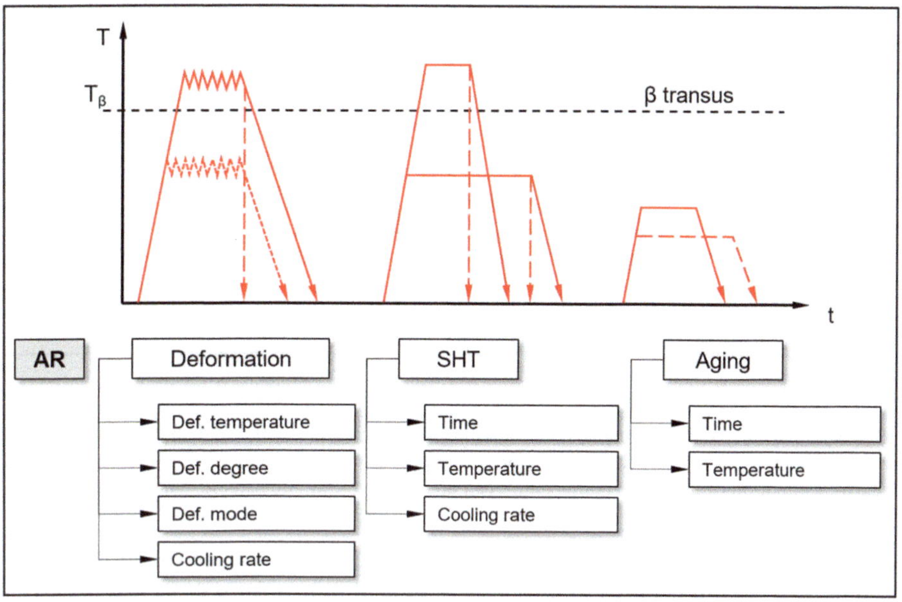

Fig. 4.22 Scheme of the four main phases of thermomechanical treatments that can be carried out on Ti-alloys. *AR* Arc-remelted state of the alloy, *SHT* solutioning heat treatment

nitrogen, have an active role in the stabilization of the α polymorph. Thanks to specific treatments, it is possible to obtain various microstructures and therefore properties best suited for any application. The Ti-6Al-4V alloy has an estimated $T_{\beta} = 995$ °C (see Table 4.2). In order to understand differences arising from process parameters variations, four different thermal treatments will be discussed.

Treatment 1 The first treatment comprises a β-solutioning followed by air cooling down to room temperature. The main steps and relevant transformations are:

- Solutioning: 1037 °C for 1 h in order to obtain a fully β phase alloy.
- Air cooling inducing the transformation: β-Ti → β-Ti + α-Ti.

Air cooling occurs with a relatively slow rate, and a *Wiedmanstätten* structure (Fig. 4.23a) is obtained from the diffusive decomposition of β phase into the mixture of α + β. A semiquantitative estimation of the phase composition can be inferred from the relevant pseudo-binary phase diagram (Fig. 4.23b). This grain morphology induces in the alloy a good fatigue resistance, since it tends to hinder crack propagation.

Treatment 2 The second treatment still involves a β-solutioning, followed this time by an aging step at intermediate temperature in the α + β range:

- Solutioning: 1037 °C for 1 h to obtain a fully β phase alloy.
- Aging: 704 °C for 2 h.

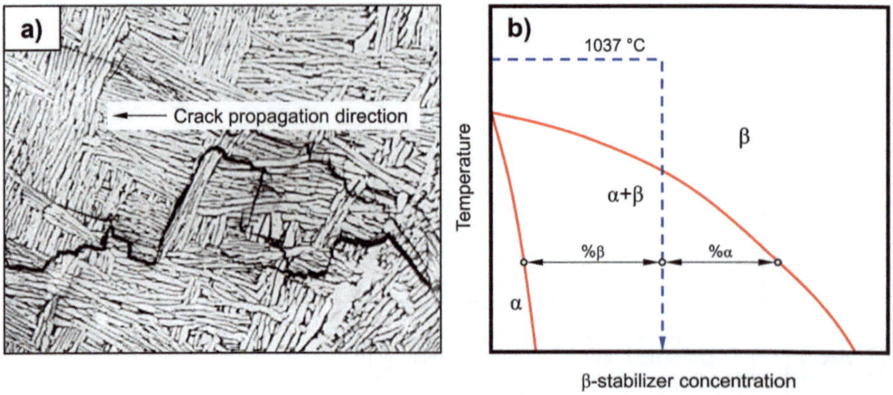

Fig. 4.23 (**a**) Typical Wiedmanstätten structure obtained from treatment 1 and its effect on the crack propagation (Polmear 2006, *reproduced with permission of the Publisher*); (**b**) a semiquantitative estimation of the phase composition can be obtained from the pseudo-binary phase diagram

During aging, the primary β phase decomposes into alternate α and β *lamellar* grains. The resulting alloy properties are not that different from those just discussed for treatment 1, although the aging step introduces a higher stability of the alloy microstructure in particular as concerns the β component.

Treatment 3 In the third treatment, a faster cooling rate from solutioning is experienced. In fact, after β-solutioning, the alloy is water-quenched and then aged in the α + β range:

- Solutioning: 1066 °C for 1 h in order to obtain a fully β phase alloy.
- Water cooling: upon rapid quenching, the α′ martensite is forming with acicular grain morphology.
- Aging: 550 °C for 24 h.
- Air cooling.

This treatment induces the formation of the so-called *basketweave* structure (Fig. 4.24), preserving the martensite grain morphology but with the α + β equilibrium phase composition. The alloy is characterized by a good mechanical strength, toughness, and fatigue resistance, enhanced by the fine-grained martensitic microstructure.

Treatment 4 The main difference of this treatment concerns the substitution of β-solutioning with an annealing in the α + β field, i.e., at a temperature below β-transus. The other stages of the temper treatment are similar to the equivalent ones discussed for the former treatment (n.3): water quenching and aging in the α + β range:

- Annealing: 954 °C for 2 h. In this case the alloy remains in the two-phase field also during the high-temperature annealing. So the phase mixture of β + α titanium is obtained.

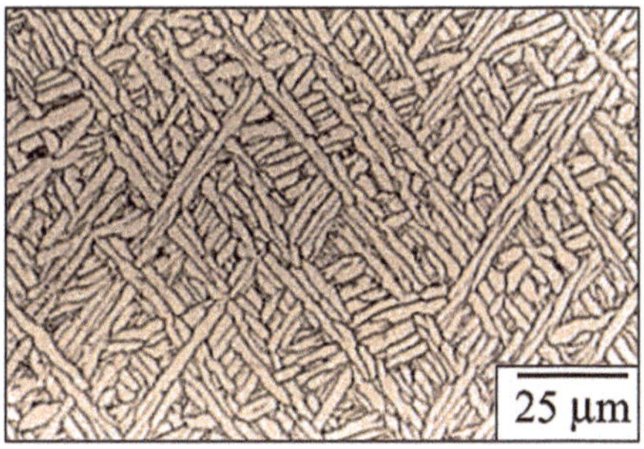

Fig. 4.24 The *basketweave* grain microstructure mentioned in the description of *Treatment 3* (Polmear 2006, reproduced with permission of the Publisher)

- Water cooling: the β phase present in the initial phase mixture, transforms into the α′ martensite:

  β-Ti + α-Ti → α′-Ti + α-Ti.
- Aging: 538 °C for 4 h. During aging the α′ martensite component undergoes a similar transformation as that induced by the *Treatment 3*. Therefore, the single-phase α′ grains turn into a phase mixture, β + α, exhibiting a *basketweave* microstructure, according to the following transformation:

  α′-Ti + α-Ti → α-Ti + β-Ti (both from martensite decomposition) + α-Ti (from the high-temperature annealing step, commonly named primary α phase).

This microstructure (Fig. 4.25) can be identified as *duplex*, since it features two different kinds of grain morphologies. Equiaxed coarse grains of the primary α phase, which form during the high-temperature annealing treatment and remain substantially unchanged during the following steps of the treatment. In between these equiaxed grains, a two-phase region still exhibits a *basketweave* microstructure, featuring finely intermixed laths of α and β grains, whose actual size depends of the cooling rate. The material is characterized by a good strength, good creep resistance, and toughness.

Tensile properties of selected alloys are given in Table 4.5, with the indication of their specific temper condition.

As an example of the dependence of mechanical properties on the specific microstructural features, some data for the Beta-CEZ alloy are given in Table 4.6.

A very important issue for the aerospace high-temperature applications of titanium alloys is their creep resistance. The Larson-Miller curves in Fig. 4.26 summarize the creep behavior of several Ti-alloys in terms of stress needed to induce 0.2% plastic deformation. The Larson-Miller parameter (LMP) is defined as:

$$\mathrm{LMP} = T \cdot [C_1 + \log_{10}(t)] \tag{4.6}$$

Where:

- T is the temperature expressed in Kelvin.
- t is, in the present context, the time needed to reach 0.2% plastic deformation.
- C_1 is a material constant that in case of metallic alloys can be taken equal to 20–22.

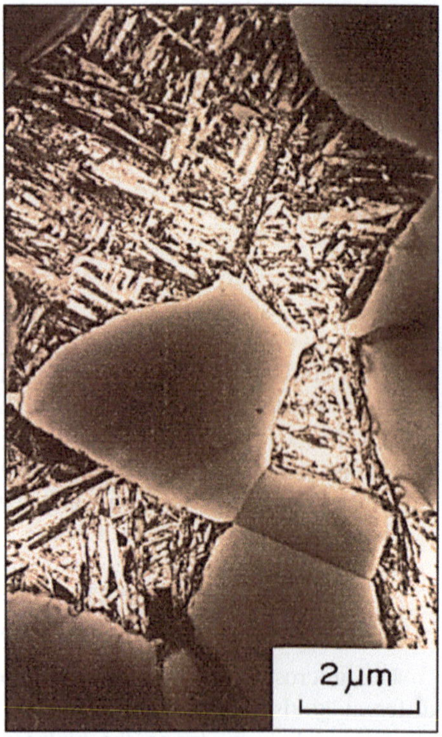

Fig. 4.25 The *duplex* grain microstructure obtained from the *Treatment 4* (Polmear 2006, reproduced with permission of the Publisher)

The LM curves are widely used for the evaluation of the creep performance and are particularly useful for comparative studies (Kutz 2002). Reliable indications can be obtained performing the test according to standard conditions and testing times not shorter than 1% of the expected service lifetime are recommended.

The creep curves plotted in Fig. 4.26 are fully coherent with the timeline of the development of conventional titanium alloys. The performances of the early α-alloys were ameliorated by the α + β alloys first and, subsequently, by state-of-the-art near-α alloys. In agreement with this evolution, the alloys of the β-group are not present in the Larson-Miller plot, since their higher concentrations of β phase render the relevant creep deformation rates exceedingly faster. In the plot, the dotted curves for two intermetallic phases, Ti_3Al and TiAl, are present. These ordered compounds have represented, and still are, a major breakthrough in the field of titanium alloys. High strength at room and high temperatures, creep resistance, and acceptable fracture toughness, particularly for the improved compositions and microstructures, are some of the interesting properties of titanium aluminide alloys, especially those based on the γ-TiAl intermetallic. The advanced Ti-aluminide alloys are now competing with the nickel-based superalloys for the fabrication of blading of low-pressure stages of commercial gas turbine engines (see Sects. 4.3.2 and 4.6).

Table 4.5 Tensile properties of selected Ti-alloys (ASM International 1991)

Alloy type	Alloy denomination	Tensile strength [MPa]	Yield strength [MPa]
α-Alloys	Grade 6	790	760
	Grade 12	480	380
Near-α	IMI 685	790	690
	IMI 834	1030	910
	Ti-811	900	830
α + β alloys	Ti-6-4[a]	900	830
	Ti-6-4 ELI[a]	830	730
	IMI 550	1100	960
	Ti-17[b,c]	1125	1055
	Ti-6-2-2-2-2[c]	1030	970
	Ti-6-2-4-6[b]	1170	1100
	Ti-6-6-2[a]	1030	970
β-Alloys	Beta-C[1,3]	900	830
	Beta III[a]	690	620
	B120VCA[b]	1170	1100
	Ti-10-2-3[a,c]	1170	1100
	Beta 21S	862	793
	Ti-15-3[b]	1000	965

[a]Annealed condition, may be solution treated and subsequently aged to increase the strength
[b]Solution treated and aged
[c]Mechanical properties and composition limits are subjected to negotiation with suppliers

Table 4.6 Tensile properties of Beta-CEZ alloy as a function of its microstructural features

Alloy	UTS [MPa]	Yield strength [MPa]	Elongation [%]
Beta-CEZ beta annealed	1280	1180	4.0
Beta-CEZ duplex	1275	1200	13.0
Beta-CEZ beta annealed (400 μm)	1335	1275	1.0
Beta-CEZ beta annealed (60 μm)	1350	1275	8.0

Among brackets the values of the average grain size of the two alloy tempers (Lüterling and Williams 2007)

As can be inferred from the thermogravimetric curves in Fig. 4.27, titanium aluminide alloys are also those performing better as concerns oxidation resistance. The Ti-6Al-4V alloy displays an extremely fast oxidation kinetics, and the close-to-linear trend indicates that the surface reaction products are not protective, so that oxidation may proceed and, in the long run, propagate to the whole alloy. The reaction kinetics is slower in the case of the two state-of-the-art N-α alloys: Ti-1100 and IMI834. However, the trend of the thermogravimetric curves still suggests a non-protective surface oxide growth. Similar comments can be applied to α_2-Ti_3Al and the derived Super-α_2 alloy (see Sect. 4.3.2). Definitely better is the behavior exhibited by the TiAl intermetallic compound that in fact at this temperature develops a protective mixed oxide layer, based on TiO_2 and Al_2O_3, which becomes protective at some stage (see Sect. 8.2.3).

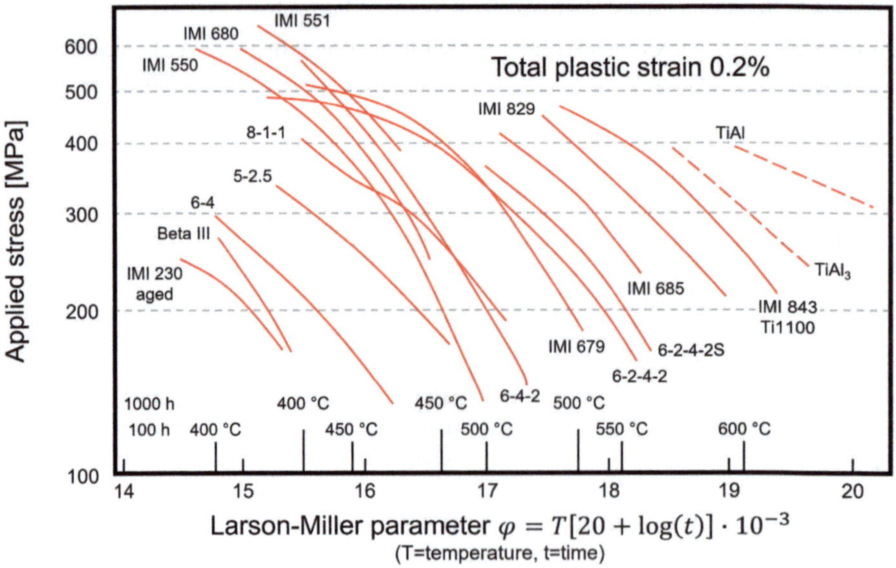

Fig. 4.26 The Larson-Miller parameter for several conventional Ti- alloys of the α, α-β, and near-α families, these latter being still the best-performing conventional alloys: IMI 834 and Ti 1100. (Redrawn from Polmear 2006)

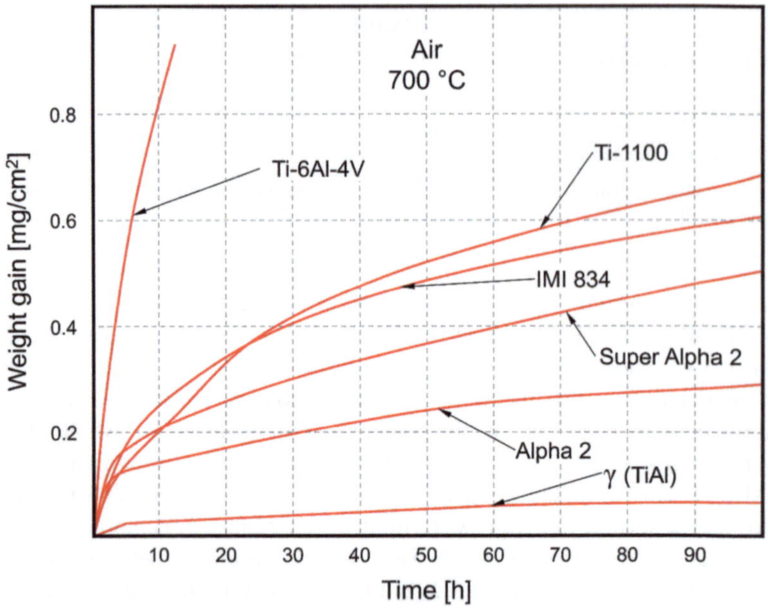

Fig. 4.27 Thermogravimetric curves for different Ti-alloys exposed to 1 atm air at 700 °C

Under particular conditions, i.e., high air pressure and elevated temperature, the kinetics of oxidation of titanium alloys may proceed at a very high rate, resembling rather a combustion process, and indeed the term "ignition" is often used in this regard. In a gas turbine engine, the following situations may be envisaged for the occurrence of the alloy ignition:

- Aerodynamic stall, with a localized increase of the part temperature, since heat removal is no longer effective.
- Impact with particles entering the engine and with the debris that come from outer stages of the compressor. In both cases, the alloy surface will be directly exposed to the outer atmosphere for the removal of the surface protective oxide layer.
- A similar damage can be induced by an occasional rubbing between static and rotary parts of the compressor.

In all these situations, oxidation can proceed and involves also the inner part of the component (Fig. 4.28) for the lack of any protective layer on the alloy surface. Moreover, the oxidation process is sustained by the enthalpy output associated with its exothermic character. In general, two stages can be identified during this catastrophic oxidation: initiation, when the process starts with the accidental damage of the protective oxide surface layer; propagation, accompanied by the spread of the oxidation process to the bulk component.

The oxidative reaction may structurally weaken the component and/or produce loose debris that may contribute to the damage of the other components. Not all titanium alloys exhibit the same behavior as concerns ignition and a testing rig (Fig. 4.29) can be used to qualify the different materials in this respect. The test specimen is kept under a high-pressure hot air flux and heated locally by a laser beam, which is meant to trigger the possible ignition. As displayed by the graph in Fig. 4.29, ignition is sustained by the alloy over specific temperature-pressure ranges and a ranking of titanium alloys, as concerns their resistance to ignition can be drawn from this approach:

- β alloys;
- TiAl and Ti_3Al-based alloys;
- α + β alloys;
- α and near-α alloys.

Ti-aluminide and β-alloys, particularly those fully β-stabilized, resist ignition better than the other groups of titanium alloys, for an intrinsically lower diffusivity. In titanium aluminide alloys, in fact, the comparatively higher concentration of slow diffusing aluminum and correspondingly lower concentration of titanium result in lower effective diffusion coefficients than in conventional titanium rich alloys. The significant presence of refractory elements has a similar effect in β-alloys.

The alloy C (Ti-35V-15Cr), whose ignition behavior is displayed by the diagram in Fig. 4.29, has been developed on purpose by Pratt & Whitney to resist to burning also when used in afterburners and in the hottest stages of the gas turbine compressor.

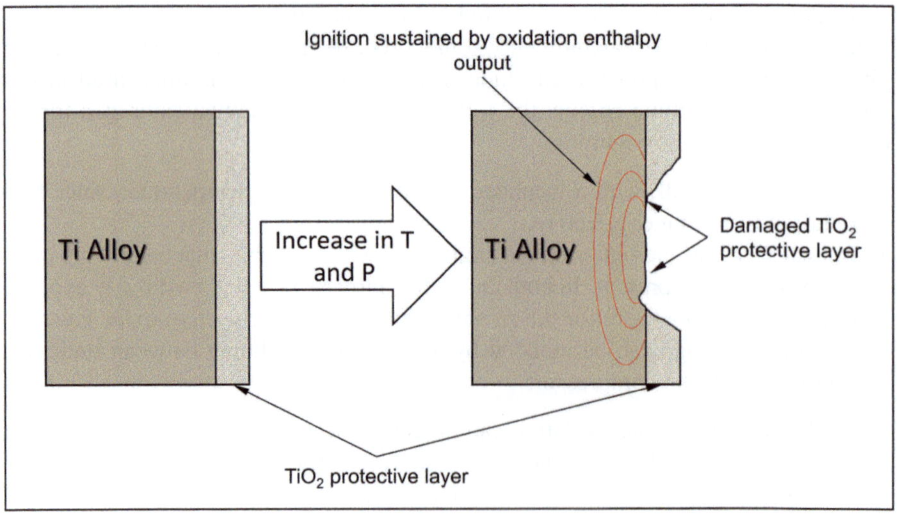

Fig. 4.28 Schematic of the ignition process involving Ti-alloys at high temperature and air pressure, once the protective oxide surface layer has been damaged

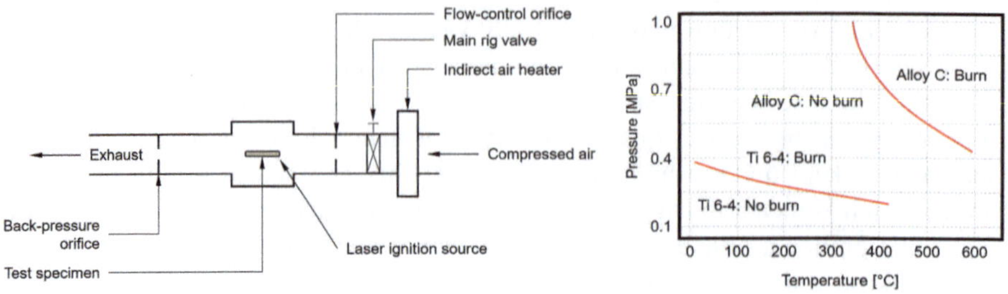

Fig. 4.29 Schematic of the test rig used to qualify the ignition behavior of an alloy. The graph shows the P-T fields in which ignition is sustained or not for the $\alpha + \beta$ alloy, Ti-6Al-6V, and the highly stabilized β-alloy, codenamed Alloy C. The plot refers to conditions with a flow in the 50–120 m/s. (Redrawn from Donachie 2000)

4.3.2 Advanced Titanium Alloys

In the search for alloys capable to overcome the limitations of the *conventional* titanium alloys, systems based on the intermetallic phases such as Ti_3Al and TiAl have attracted important research interests and efforts. Table 4.7 illustrates the potentials of the binary intermetallic Ti-Al phases, as compared to average reference properties of *conventional* Ti-alloys. A common feature to all intermetallic compounds is their ordered crystallographic structure (see Sect. 9.4). The ordered character refers to the selective occupation of the crystalline sites by each atomic species, as indicated in Fig. 4.30. This ordered atomic configuration induces

Table 4.7 Main properties of the binary intermetallics γ-TiAl and α_2-Ti_3Al compared to the average corresponding properties of *conventional* Ti-alloys

	Density [g/cm3]	Young's Modulus [GPa]	RT ductility [%]	RT fracture toughness [MPa√m]	Creep limit T [°C]	Oxidation limit T [°C]
Ti-alloys	4.5	110	10–25	35–60	600	600
Ti_3Al	4.3	145	2–10	25	800	650
TiAl	3.8	176	1–4	25	950	900

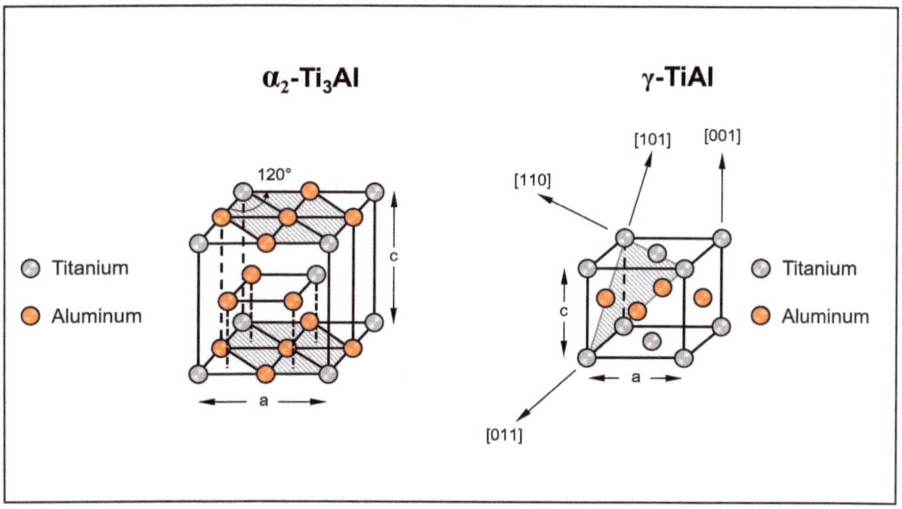

Fig. 4.30 Crystallographic structure of the α_2-Ti_3Al and γ-TiAl. The positions of aluminum and titanium atoms for a fully stoichiometric composition are also indicated

particular properties in the alloy. The mechanical behavior of an intermetallic alloy is in fact strongly dependent on the peculiar character of the chemical bonding and the relevant influence on the dislocation structure and dynamics.

As concerns Ti-aluminides, the early alloys that have been developed for aerospace applications are those based on the α_2-Ti_3Al phase, with hexagonal DO_{19} structure. This intermetallic was well known already as a strengthening precipitate phase in titanium α-alloys (Sect. 4.3.1). The composition of the alloys belonging to this family typically includes 23–25 at% Al, 11–18 at% Nb, up to 3 at% V, and approximately 1 at% Mo (Donachie 2000). The first generation of Ti_3Al-based alloys was named near-α_2, and it was characterized by a Nb concentration between 10 at% and 12 at%. Subsequent research suggested that an increased Nb content, i.e., 13–20 at%, would have a beneficial influence on the balance between room and high-temperature properties. The top product of this development process was the extensively investigated TIMET Super Alpha 2 alloy (Ti-25Al-10Nb-3V-1Mo at%), mentioned already as concerns oxidation resistance of Ti-alloys (see Fig. 4.27).

Table 4.8 Tensile properties of Super Alpha 2 alloy as a function of the microstructure (Lüterling and Williams 2007)

Structure	UTS [MPa]	Yield strength [MPa]	Elongation [%]
Bimodal	1235	1010	5.2
Lamellar	615	585	0.3
Equiaxed	900	835	1.8

The complex composition of Super Alpha 2 alloy was specifically designed to enhance the intrinsic properties of the binary intermetallic that otherwise suffered from poor room temperature ductility and fracture toughness (Table 4.7), quite a common drawback for ordered compounds (see Sect. 9.4). The tensile properties of Super Alpha 2 alloy are given in Table 4.8, with reference to different microstructures achievable with appropriate thermal treatments.

In Super Alpha 2 alloy, vanadium is used as "low-cost" replacement of niobium, although its addition also leads to a reduction in the oxidation resistance of the alloy. Molybdenum is used to enhance the Young's modulus, the creep resistance, and the high-temperature strength and, together with niobium and vanadium, contributes also to the stabilization of the β phase, providing a decisive contribution to the excellent ductility and fracture toughness of this alloy. Actually, in this alloy the cubic phase has an ordered CsCl structure, designated by β_0, or β_2 in analogy with α_2 (Leyens and Peters 2003; Proske et al. 1992).

The stabilization of the β cubic phase, in association with both, or just one, of the Ti-aluminide phases, is rather a constant feature of these alloys. In fact, the alloy design process resulted in important improvements in the composition of the Super Alpha 2 and other companion alloys. Interesting mechanical resistance and a reasonable workability were attained also thanks to specific processing routes, developed on purpose for these materials, like the pack-rolling process, capable to produce 0.2 mm thick sheets of Super-α_2 (Peters and Blank-Berwersdorff 1992). However, the oxidation and creep performances of α_2-Ti_3Al-based alloys at temperature in excess of 750 °C (Table 4.7) were not adequate for the long-term structural requirements of the hot parts of the compressor of a gas turbine engine. Therefore, alloys based on another Ti-aluminide phase, i.e., the tetragonal $L1_0$ γ-TiAl, were developed instead (Donachie 2000). Several properties of this intermetallic are particularly attractive in view of the structural applications at elevated temperatures, required by a gas turbine aerospace engine:

- Excellent oxidation resistance (see Fig. 4.27).
- Good creep properties, better than any other Ti-based alloy (see Fig. 4.26).
- High ignition resistance when compared to *conventional* titanium alloys.
- High specific yield strength.
- High specific Young's modulus.

The high concentration of aluminum lowers the alloy density below the typical values of the other Ti-alloys (Table 4.7). The ordered structure, which results in limited ductility and fracture toughness, contributes on the other hand to the intrinsic high strength and creep resistance. Oxidation resistance is also positively

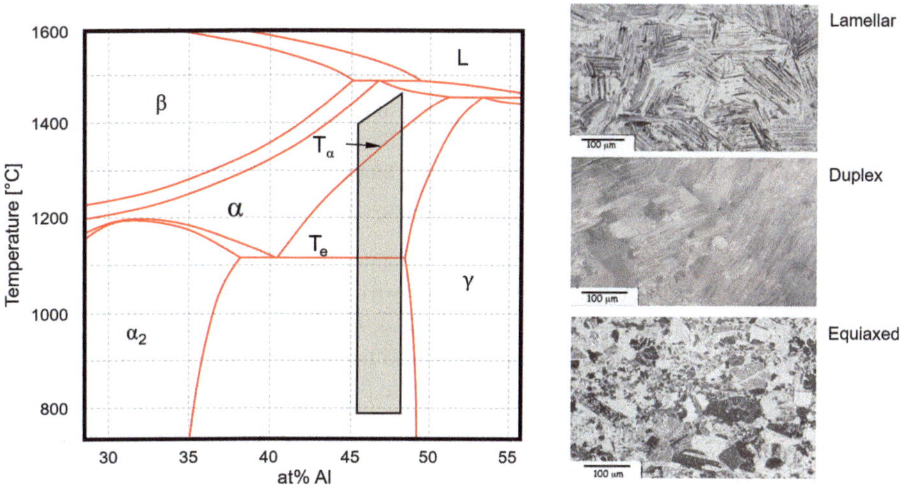

Fig. 4.31 Portion of the Ti-Al phase diagram interesting to discuss the phase stability and the alloy microstructures as a function of the thermal treatments of γ-TiAl-based alloys. The gray-shadowed area refers to the equivalent composition of these alloys and relevant temperature ranges for the treatments: T_e, the eutectoid transition temperature, and T_α, the *alpha transus*. The main microstructures that can be obtained with specific thermal treatments are also displayed on the right. They have been obtained with an alloy with composition: Ti-48Al-2Cr-2Nb-1B (at%) and the treatment parameters listed in Table 4.5 (micrographs properties of the authors)

influenced by the ordered structure, that limits the interdiffusion rate, and, most importantly, the high concentration of the low diffusive aluminum atoms is a prerequisite for the formation of a protective oxide layer. Single-phase alloys have been developed with compositions like Ti-(50 $\div$ 53)Al-(1 $\div$ 2)x_1, being x_1 one or a combination of the following β-stabilizers: Cr, Mn, or V. Multiphase alloys have been also considered, including the ordered intermetallics only (α_2-γ alloys), or the three-phase systems including the β-Ti polymorph (α_2-γ-β alloys). The general description of these multiphase compositions can be parametrized as Ti-(44 $\div$ 49) Al-(1 $\div$ 3)x_1-(1 $\div$ 4)x_2-(0.1 $\div$ 1)x_3. The term x_1 is the same as above, x_2 is one or a combination of isomorphous β-stabilizing elements like Mo, Nb, Ta, and W. The term x_3 is referred to B, C, or Si. Boron and carbon are present in the alloy in the form of several different compounds, like Ti_3AlC (P-type carbide with perovskite structure), Ti_2AlC (H-type carbide with hexagonal structure), and a variety of titanium borides, like TiB and TiB_2 with the combined role of alloy strengtheners, in both static and creep loading regimes, and grain boundary stabilizers (Liu et al. 2017; Wang et al. 2017). As for conventional Ti-alloys, the microstructure and, thereby, the mechanical properties of these advanced Ti-alloys can be tailored by suitable thermomechanical treatments, capable to induce different microstructures. In Fig. 4.31 the midsection of the Ti-Al phase diagram is displayed. The composition range for the γ-TiAl alloys is marked by the shadowed region. T_e, the eutectoid transition temperature, and T_α, the *alpha transus*, are two temperatures interesting for setting the heat treatment parameters of these alloys.

Table 4.9 Thermal treatments carried out on the alloy: Ti-48Al-2Cr-2Nb-1B (at%), resulting in the microstructures in Fig. 4.31.

Treatment T [°C]	Annealing time [h]	Cooling rate	Final microstructure
1200	4	Slow	Equiaxed
1300	3	Slow	Duplex
1380	1	Slow	Lamellar

For this alloy, $T_{\alpha} = 1340$ °C and $T_e = 1120$ °C, respectively

Figure 4.31 shows also the reference microstructures for these alloys, obtained from the thermal treatments described in Table 4.9. The alloy Ti-48Al-2Cr-2Nb-1B is a three-phase alloy, comprising in addition to the two aluminide phases (γ-TiAl, α_2-Ti_3Al) also a very limited amount of the cubic β phase, an important ingredient to optimize the fracture toughness and the ductility of the alloy. In the present discussion, the evolution of the β phase, and its ordered variants, is not considered herewith, requiring ternary phase diagram sections for a satisfactory description of the relevant transformations (Hashimoto et al. 1998) (Table 4.9).

The *equiaxed* microstructure exhibits grains of the three phases that form and are stabilized by the annealing at 1200 °C, above the eutectoid temperature. Upon cooling down to room temperature, the phase composition and relevant microstructure do not change significantly, and the only transformation that occurs is the ordering reaction from the α solid solution to the ordered intermetallic α_2-Ti_3Al. The *duplex* structured alloys are obtained from annealing treatments just below the α-*transus*. During this treatment too, the constituent phases are initially stabilized at the annealing temperature, but upon cooling down to room temperature, their concentration changes, as indicated by the lever rule. In particular, the concentration of the α_2-Ti_3Al decreases, whereas that of γ-TiAl increases. This determines the formation of colonies of alternate γ-α_2 lamellar grains of the two phases inside the grains that at the annealing temperature were fully α_2-Ti_3Al. These lamellar grains are typical also of the *fully lamellar* microstructure that forms from a fully α-stabilized alloy, obtained when annealing above α-*transus*. Upon cooling from high temperature, the following decomposition occurs: $\alpha \rightarrow \alpha + \gamma$. This decomposition is followed by the disorder-to-order transformation of the α phase, to form the α_2-Ti_3Al intermetallic: $\alpha + \gamma \rightarrow \alpha_2 + \gamma$. In this case no important changes in the lamellar grain morphology occur.

The resulting properties are very much dependent on the microstructure of the alloys, as exemplified by the stress-strain curves for the Ti-48Al-2Cr-2Nb (at%) alloy in Fig. 4.32, with two different microstructures induced by suitable thermal treatments, to be selected also in view of the possible subsequent steps of mechanical processing. In fact, extrusion, hot-rolling, and forging are usually conducted above T_e (eutectoid temperature), whereas for superplastic and conventional hot-forming of sheet materials, and residual stress relief heat treatments, moderate temperatures, below T_e, are more suitable.

The level of knowledge and technological development achieved with Ti-aluminide-based alloys has slowed down the research on other intermetallic alloys,

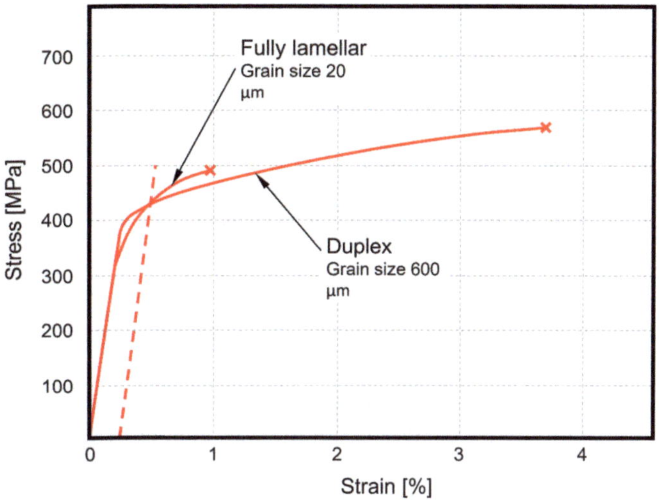

Fig. 4.32 Stress-strain curves for the Ti-48Al-2Cr-2Nb (at%) alloy, with different microstructures: duplex and lamellar. (Redrawn from Dimiduk et al. 1992)

having in principle very close target applications as those of γ-TiAl alloys. This is the case of the alloys based on the orthorhombic Ti_2AlNb intermetallic Heusler phase (Partridge and Shelton 2001). Development steps for these alloys are mainly concentrated on the design of new refined compositions and on the improvement of process related aspects (Chen et al. 2014).

4.4 Titanium Processing Technologies

Some aspects of the production cycle of titanium alloy components have been discussed already, with regard to primary titanium extraction (Sect. 4.1), types of available alloys, and some aspects of thermomechanical treatments, developed to induce specific structural properties (see Sect. 4.3). The production of a master alloy, involving melting technologies, is indeed particularly critical for titanium, owing to its elevated reactivity with elements like oxygen, nitrogen, and hydrogen, which might be present in the working atmosphere. Moreover, titanium tends to react with carbon, when nonconsumable electrodes of arc melting plants and graphite containers and liners are in contact with the molten alloy. Similarly, refractory ceramic crucibles and casting molds, used in this field, suffer from important thermochemical erosion phenomena, that may induce the release of brittle particles into the liquid alloy. To eliminate contamination from gaseous species, titanium alloy melting is conducted under vacuum or inert atmosphere, using argon or helium. To prevent contamination from the mold and lining materials, a system involving water-cooled copper containers is used. By far the most common approach is based on cold hearth

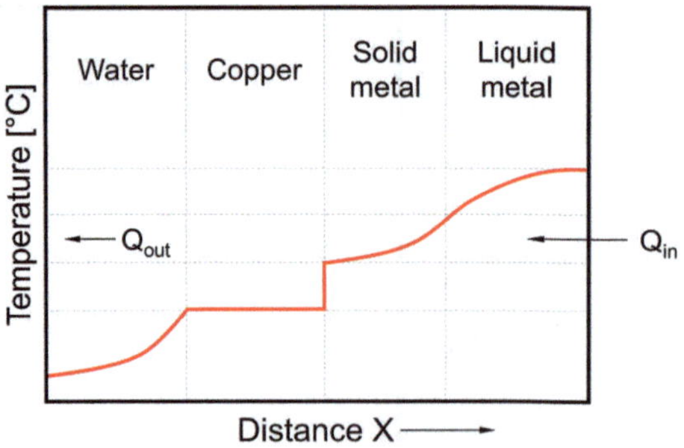

Fig. 4.33 Qualitative steady-state temperature profile for the skull-melting process. (Redrawn from Sears 1990)

melting (CHM), also known as “skull” melting. Cold heart melting has become an industrial process for the production of rotor grade titanium starting from 1985. A water-cooled copper mold assures a rapid extraction of the solidification enthalpy, so that a solidified alloy layer forms inside the mold, at the interface with the rest of the solidifying alloy (Fig. 4.33). Therefore, a direct contact between the liquid alloy and the copper mold is avoided, and the solidification of the inner part of the ingot can proceed in contamination-free conditions. At the end, the outer part of the resulting ingot can be eliminated with all the contaminant burden, coming from the refinement that the CHM has on the master alloy batch. The skull-melting principle is quite flexible, and it can be employed in different production plants, involving secondary melting for the production of ingots and powders, like in the case of a gas atomization stage.

Vacuum arc-remelting (VAR), using CHM, is classified for the production of ingots of CP-titanium and other titanium alloys, characterized by high purity and chemical homogeneity. These features are a prerequisite to guarantee adequate standard of the resulting mechanical properties. Differently from VAR applied to steels (see Sect. 5.3.6) and nickel base superalloys (see Sect. 6.4), in the production of titanium alloys it may be a primary melting process, whereas for special steels and superalloys, the primary melting is performed either by electric arc melting (EAF, see Sect. 5.3.1) or by vacuum induction melting (VIM, see Sect. 5.3.2). In addition to VAR, other CHM processes are becoming more and more important for titanium alloy melting (Lüterling and Williams 2007). The production path for titanium alloys includes single, double, or even triple VAR processing, to form, in the first place cylindrical ingots, i.e., the (consumable) electrodes of the process. The very first electrode is actually obtained by compacting and welding together pieces and pellets of titanium sponge and alloying elements. Since there are strong incentives in using titanium scraps also in VAR, this source of titanium can be incorporated into the starting electrode. Each electrode will have the desired alloy nominal composition. Compacted blocks have sufficient strength to remain intact during handling and melting. If needed, mechanical compaction by means of a hydraulic press can be used to strengthen the electrode. A twin cone mixer can be used to optimize the blending of the different components: titanium sponge fragments, scraps, and alloying elements (Lüterling and Williams 2007). The process is carried out using a cylindrical crucible under vacuum. Initially, an arc is sparked between the electrode and a small amount of the same alloy to be remelted, positioned at the bottom of the crucible. Ideally, the melting rate is constant all through the process. The main advantages of the VAR technology (also known as CEVAM or CEVAR, consumable electrode vacuum arc melting/vacuum arc remelting) are:

- Considerable reduction of dissolved gases, such as oxygen, nitrogen, and hydrogen, since under vacuum conditions these gases are released from the liquid metal and pumped away by the vacuum system.
- High degree of control on the solidification microstructure, thanks to the possibility of tuning and adapting the melting rate of the alloy to the specific process requirements.
- Minimization of segregations.
- Alloy chemical homogeneity across the ingot, improved by the successive remelting.

The main critical aspects concern the production of electrodes and the segregation of alloying elements within the electrodes. Remediation approaches and other details about the technique can be found in Sect. 5.3.6. However, the electrode issue can be solved by another melting process, involving the use of the non-consumable electrode vacuum melting (Fig. 4.34).

In this process, the electrode is not made of the same alloy that has to be melted but of another material, like graphite or tungsten. The working principle is similar to gas tungsten arc welding (GTAW). The non-consumable electrode rotates above the molten pool, and in the meantime its temperature is kept at acceptable values by a dedicated water cooling system. The technique ensures good flexibility and allows for the production of both powders and ingots. Moreover, since there is no consumption of the primary electrode and no need to prepare it, melting can even proceed from scraps directly, as shown by the scheme in Fig. 4.34. In this regard, the technique is an excellent choice for recycling purposes. Two other melting processes interesting for titanium alloys processing are electron beam melting (EBM) and plasma arc melting (PAM). In EBM plants, melting, carried out under vacuum (10^{-4}–10^{-6} torr) is induced by an electron beam bombardment of the alloy charge. The general scheme of an EBM equipment is given in Fig. 4.35. EBM furnaces employ one or more electron beam guns (power up to 1.2 MW).

To enhance the homogeneity of the final product and to reduce the density of defects and the incidence of ballistic effects, several melting stages (two in the scheme in Fig. 4.35) are usually employed. EBM can be used for the production of both primary electrodes, finished ingots, and slabs starting from sponge and scraps. Owing to the highly controlled working conditions needed by the electron beam guns, this technique guarantees elevated purity standards, although it is not recommended for large-scale production and as melting stage of powder atomization and rapid solidification plants. These latter processes, in fact, require comparatively higher gas pressures to facilitate the rapid solidification of the powders (Sears 1990).

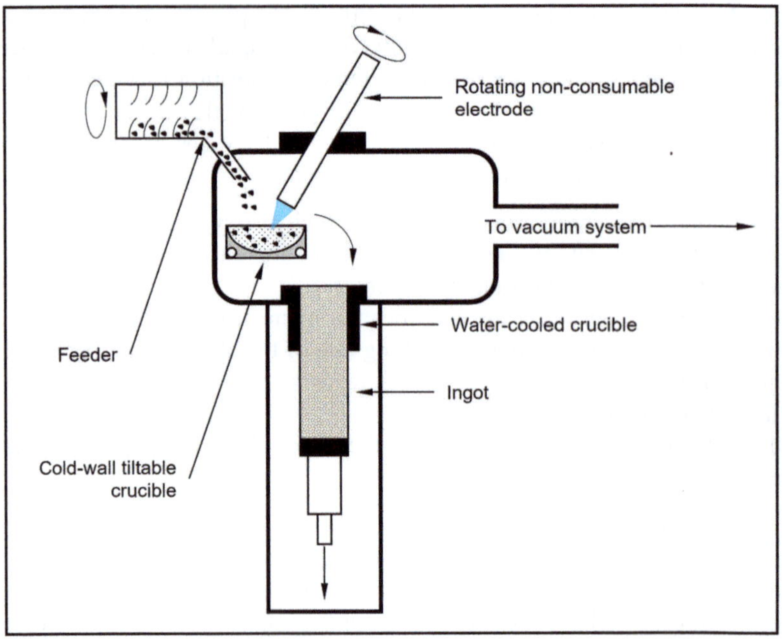

Fig. 4.34 Schematic of a non-consumable electrode vacuum arc remelting (VAR) equipment used for the production of titanium and titanium alloy ingots. (Redrawn from Sears 1990)

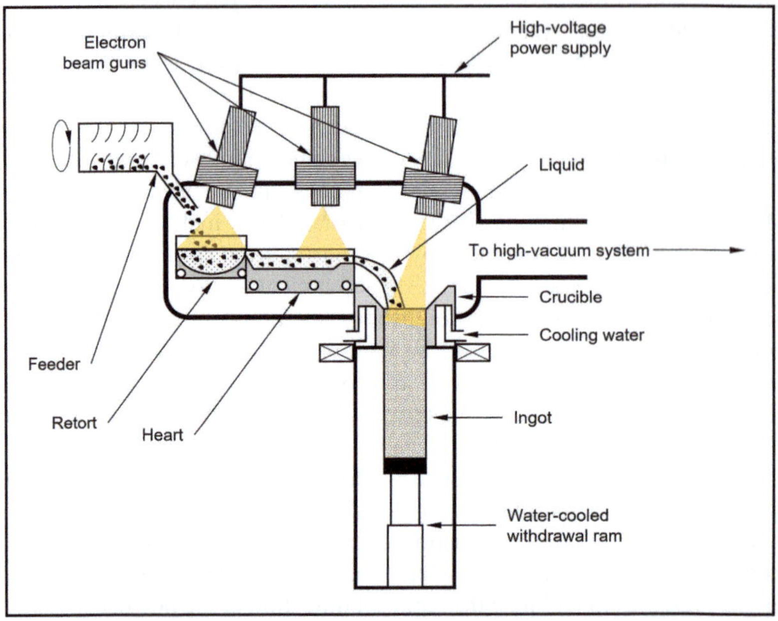

Fig. 4.35 Schematic of an electron beam melting (EBM) apparatus used for the production of ingots. (Redrawn from Sears 1990)

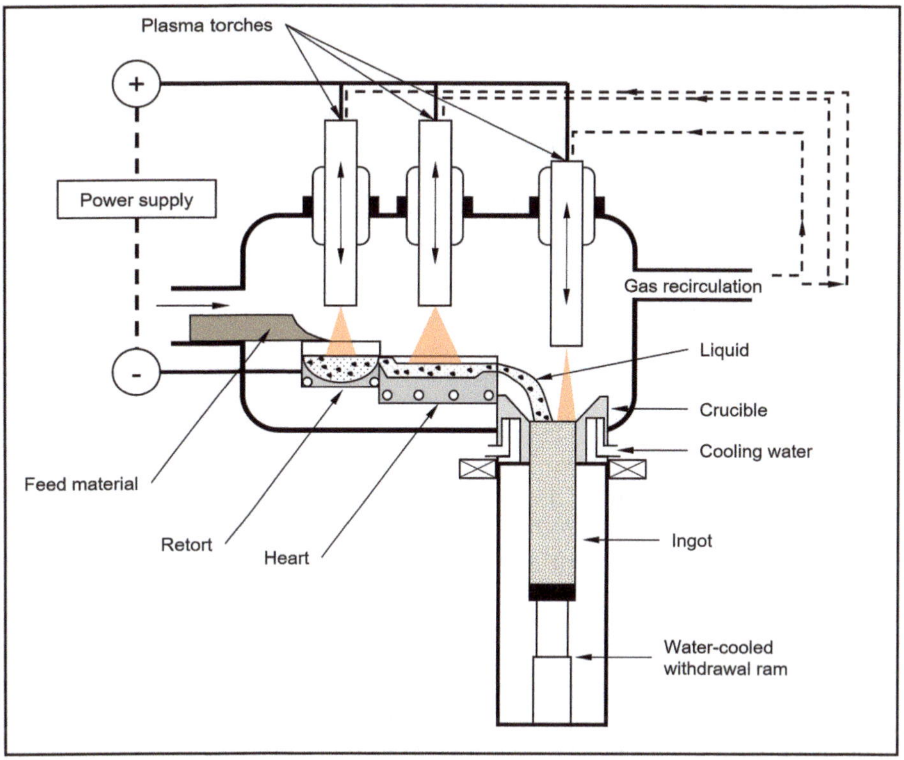

Fig. 4.36 Schematic of a plasma arc melting (PAM) apparatus used for the production of ingots. (Redrawn from Sears 1990)

Plasma arc melting (PAM) features an architecture which is similar to that of EBM but based on plasma torches to melt the alloy. This latter feature is illustrated by the scheme of the plant in Fig. 4.36. A series of plasma torches are present for the same reason as in EBM plants. The composition of the plasma gas and of the working atmosphere are chosen considering the high reactivity of titanium and the possibility of limiting any possible contamination of the alloy.

Titanium sponge fragments, scraps, and pieces of recycled ingots constitute the feed material for the PAM process. The molds used in both EBM and PAM exploit the skull-melting principle to eliminate contamination from the crucibles and retorts (Sears 1990). Technical data of real titanium-melting plants are listed in Table 4.10

Several post-melting processes may be used for turning the titanium master alloys into any final product. Rapid solidification (RS) is one option that finds application in the production of powders for powder metallurgy (PM) routes. RS is known to have had an impact on the metallurgy of several alloys for the changes, mostly far from equilibrium conditions, introduced by the very fast removal of the solidification enthalpy from the melt and cooling down of the solid phase (Das and Davis 1988).

Table 4.10 Technical specifications of real plants for melting and casting of titanium alloys (Sears 1990)

			Efficiency	
Technique	Type	Total power[a] [MW]	kWh/kg	kg/h
Consumable arc (VAR)	Direct melt	1.0	0.85	1200
Nonconsumable arc	Tilt-pour	0.6	2.0	225
Electron beam (EBM) – 4 guns	Cold hearth	2.0	1.1–2.2	450–1360
Electron beam (EBM) – 2 guns	Cold hearth	2.4	1.6–3.0	450–900
Plasma arc (PASER)	Direct melt	0.75	1.1	690
Plasma arc (PAM) – 4 torches	Cold hearth	2.1	2.5[b]	900[b]
Plasma arc (PAM) – 4 torches	Cold hearth	2.25	2.5[b]	900[b]

[a]Data based on arc or beam power only. The power value can vary greatly due to the type of feed material
[b]Predicted values

RS of titanium and its alloys is interesting as concerns the enhancement of the solubility limits and the obtainment of either micro- or nanocrystalline structures. On some occasion, this has resulted in improvements of the physical and mechanical properties of the alloys and has afforded the possibility of developing novel compositions (Sastry et al. 1983, Suryanarayana et al. 1991). Thanks to the higher concentrations of reactive elements that can be solutioned into the titanium alloy matrix through RS, the formation of a fine dispersion of oxide particles can thereafter be induced by post-processing thermal treatments. The oxide particles hinder the motion of grain boundaries, so that a micrometric grain size is retained in the consolidated bulk alloy. The oxide dispersion has also beneficial effects on the creep resistance of the alloy that behaves in this respect like an oxide dispersion strengthened (ODS) material (see Sect. 9.3) (Sastry et al. 1983; Suryanarayana and Froes 1990).

Powder metallurgy (PM) represents an important class of processing for titanium alloys, which is recently gaining more importance, even as concerns additive manufacturing. Powder metallurgy is regarded as an economically advantageous method for the production of near-net shape components. Mechanical properties of PM titanium products can reach values that may be even better than those of cast alloys and comparable to those of forged alloys. At the present time, powder metallurgy is experiencing a sort of renewed interest, in particular, thanks to innovative 3D printing techniques, offering the possibility of producing stiff and wear resistant components based on reinforced titanium PM alloys. The production of high-quality powders is of course paramount for PM processing, and, in this regard, the high surface area may critically enhance the reactivity of titanium powders. The expertise developed over the years in powder atomization and rapid solidification methods is very useful indeed for setting up the production of high-grade titanium alloy powders. Before consolidation, for instance, through hot isostatic pressing (HIP), the powders are usually encapsulated in austenitic steel cans and then degassed and evacuated. These steps are absolutely recommended in order to avoid surface gas absorption, with the risk of forming oxide defects on the alloy surface and inside the open pores. The HIP process conducted to consolidate a Ti-6Al-4V powder has a duration of 1–3 h, at a temperature between 920 and 970 °C, while the compaction pressure is about 2000 bar (Leyens and Peters 2003). Titanium powder production might benefit soon from the consolidation and widespread use of an alternative approach to the Kroll's extraction method: the so-called *Meltless Ti Process* (Dargusch and Keay 2009). The substantial difference with respect to the classical and still dominant Kroll's extraction method is that the *Meltless* approach, since producing sponge powders, instead of sponge cakes, requires PM approaches for the production of the bulk components. At the present time, the interest for this novel technique is sustained by a number of several positive aspects, for instance, an estimated 50% reduction in energy consumption and a comparable reduction of greenhouse gas emission. A similar reduction concerns also the emission of pollutant gases, like SO_2 and NO_x.

The development of γ-TiAl-based alloys has prompted research activities on the casting technology, aiming at obtaining components with enhanced mechanical properties. A particular research interest has been dedicated to the directional solidification of γ-TiAl alloys, using a technology in some respects similar to directional solidification of polycrystalline columnar grained or single crystal superalloys (see Sect. 6.5.2). A major difficulty of directional solidification applied to γ-TiAl alloys is controlling the formation of an optimal lamellar microstructure (see Fig. 4.31), resulting from the solid-state phase decomposition of the alloy. Moreover, the apparatus employed for this casting technology, considering the required solidification kinetics, cannot rely on skull melting, typical of titanium casting techniques. In fact, ceramic molds would be the obvious choice, considering the comparatively long solidification times caused by the involved low growth rates. This implies a prolonged contact between the molten alloy and the ceramic, with a consequent erosion of the mold that may lead to contamination of the melt by ceramic inclusions. The solidification parameters, in particular the growth rate, can be tuned to reduce this contamination and at the same time induce the formation of a better dendritic microstructure, with particular regard to improved creep resistance (Ding et al. 2011; Lapin et al. 2002; Lapin 2006).

Solidification using mold casting is just one of the possible processing approaches available, used for both conventional and advanced Ti-alloys, including the manufacturing of aerospace near-net shape components. Referring to investment casting, different procedures are available: centrifugal, tilt, and counter-gravity casting. Casting techniques are preferred for those alloys exhibiting poor workability. In all titanium alloys, this feature is to some extent dependent on the presence of the β phase, either as a stable constitutional component of the alloy or as a phase present at the metal working temperature only, like for α and near-α alloys. However, the high performances and elevated reliability levels required for aerospace components can be fully accomplished only through complex processing routes. For instance, the complete manufacturing sequence of the low-pressure blades, first mounted on the GEnX engine (Bewlay et al. 2016), made of the General Electric alloy: Ti-48Al-2Cr-2Nb (see Sect. 4.6), involves the following main processing steps:

- Master alloy ingot production, by casting.
- Blade production by melting and casting.
- Pre-HIP (hot isostatic pressing) solution heat treatment.
- HIP, to eliminate casting porosity.
- Recovery post-HIP heat treatment.
- Radiographic part control and inspection.
- Machining for part finishing and oxidation resistant coating deposition.
- Assembly, in the low-pressure turbine (LPT) stages.

The success of the Ti-48Al-2Cr-2Nb alloy in the LPT blades of commercial engines is now fully established, with remarkable advantages regarding the overall performances of the engine, also in terms of fuel saving. Recent developments in the directional solidification process have demonstrated the possibility to obtain a single crystal TiAl alloy with exceptional creep resistance (Chen et al. 2016). This trend and the promising results may suggest to extend the use of these alloys to other structural components of the gas turbine engines.

To complete the picture as regards the processing of titanium alloy components for aerospace applications, metal forming still maintains an important role, notwithstanding some critical aspect existing in the manufacturing of some titanium products, like sheets. The difficulties in processing titanium and its alloys are primarily due to some specific material features, such as strong springback effect, fret tendency when in contact with steel tools, high strength, and anisotropic sheet properties. For these reasons, deep drawing and stretch forming are not suitable for titanium alloys due to the low strain that can be achieved and the correspondingly high required stresses. Hot-forming is an important option. Warm calibration, i.e., creep forming, is used for sheet forming and hot-working, which is usually performed at around 850 °C (see Figs. 4.20 and 4.22). After hot-working, an oxide layer may form on the surface of the material at 500 °C. This layer should be removed by means of pickling processes, since an incomplete removal may result in a surface embrittlement and a consequent reduction in the fatigue strength and toughness of the alloy.

Superplastic forming (SPF), i.e., the possibility for some materials, under specific loading conditions, to undergo extremely large tensile strains without necking (Perez-Prado and Kassner 2009) has been considered for titanium alloy too. SPF for the manufacturing of aerospace components begun with two projects for the fabrication of supersonic aircrafts: the Concorde and the Rockwell B-1. An adequate knowledge on the superplastic behavior of the relevant materials was developed, including Ti-6Al-4V alloy. At that time, when both the Concorde and the B-1 were under development, composite sandwich structures were selected for several components. These structures were obtained by the combination of SPF and diffusion bonding (DB). DB is a welding technique, based on the solid-state interdiffusion of the two materials to be joined, pressed together at relatively high temperatures ($0.5{\cdot}T_m$ of the alloy approx.). The Ti-6Al-4V alloy, during isothermal tensile test performed at $0.5{\cdot}T_m$, shows a superplastic behavior, with a tensile elongation of

Table. 4.11 Superplastic properties of selected Ti-alloys (ASM International 1991)

Alloy type	Common alloy name	Test temperature [°C]	Strain rate [s^{-1}]	m	Elongation [%]
α + β	Ti-6-4	840–870	1.3×10^{-4} to 10^{-3}	0.75	750–1170
β	B120VCA	800	2×10^{-4}	0.50	229

1000% while developing no necking (Leyens and Peters 2003; Nieh et al. 1994). To rank the SPF behavior of material, the strain rate sensitivity factor, *m*, has been introduced, using the Eq. 4.7 (ASM International 1991):

$$m = \frac{\partial \ln \sigma}{\partial \ln \dot{\varepsilon}} \tag{4.7}$$

where:

- m is the strain rate sensitivity factor. The higher the factor, the greater the superplastic deformation exhibited by the alloy.
- σ is the flow stress, expressed in MPa.
- $\dot{\varepsilon}$ is the strain rate expressed in s^{-1}.

The large strain values associated with superplasticity can be explained by means of three mechanisms occurring at the same time: grain boundary diffusion, grain boundary sliding, and grain rotation. Nowadays, it is assumed that more or less all polycrystalline materials may exhibit a superplastic behavior, under appropriate temperature and strain rate conditions, if it has a sufficiently fine-grained microstructure and a good resistance to grain growth during SPF processes (Leyens and Peters 2003; Nieh et al. 1994). Since some Ti-alloys exhibit a superplastic behavior in their standard production state, no particular modifications in the alloy chemistry or in the high-temperature processing is needed (ASM International 1991). Details regarding the superplastic properties of the Ti-6Al-4V and B120VCA alloys are given in Table 4.11. Further details about SPF of materials can be found both in the further readings, and in the references at the end of this chapter.

Among the different plastic deformation processes, forging is by far the most used for titanium and its alloys, although rolling is used as well for the production of specific products. A comparison among the main plastic deformation processes, with relevant deformation rates, is given in Fig. 4.37. One of the advantages of forging is that it allows for a precise tailoring of the resulting alloy properties and microstructures.

Forging of titanium alloys can be conducted either in α/β or β field. In α/β forging, the alloy is heated up at a temperature 30–100 °C below its *β-transus*. The final cooling process is usually performed in air. In case of β forging, the alloy is heated above its *β-transus*, and the process has to be completed before the start of the β-to-α transition. β-forged alloys tend to be much more sensitive to process conditions, especially as concerns deformation temperature. For β forging, the cooling rate needs to be selected on the basis of the characteristics of the alloy: β-alloys can be

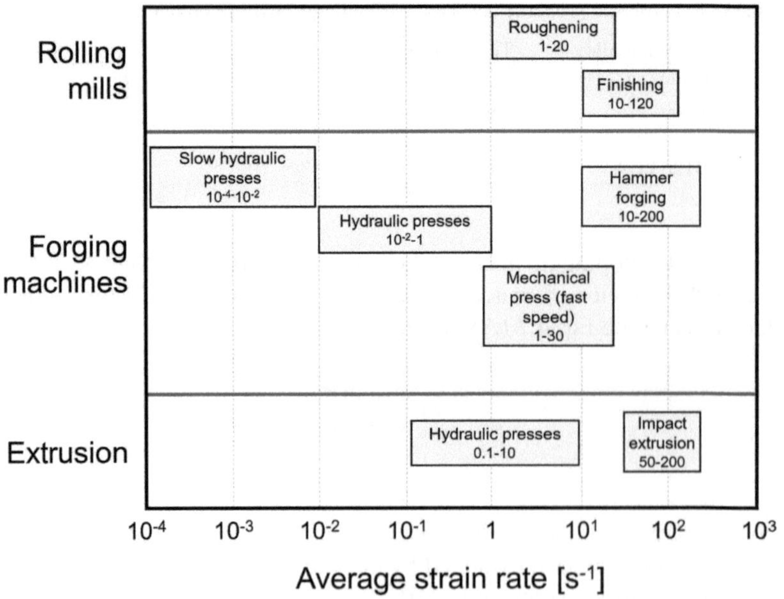

Fig. 4.37 Deformation processes, equipments, and relevant strain rates. (Redrawn from Leyens and Peters 2003)

slowly cooled down in air, while near-α alloys need a faster cooling, for which water quench can be used (Leyens and Peters 2003).

The main forging techniques are hammer forging and press forging. In the case of hammers, the following three main families can be considered:

- Gravity hammers.
- Counter-gravity hammers (counterblow hammers).
- Double action hammers.

The main features of the processes using hammers are:

- Ease of piece positioning during forging processes (especially gravity hammer).
- High fraction of scrap.
- High strain rates.
- Post-finishing and excess material removal are needed.
- Series of blows and heating of the workpiece.

Press forging is different from standard forging process. In fact, while in forging the force is applied impulsively, press forging works by slowly applying a continuous pressure, so the strain rate is lower if compared to that of hammer forging processes (see Fig. 4.37). The main features of the press forging are:

- Deformation rate is controlled.
- The piece can be subjected to multidirectional stresses.
- Material is processed in the right quantity without and excessive scrap at the end.

Hot forging is usually performed in Ni-based superalloy dies that are preheated at a temperature in a range between 760 and 1050 °C, rather than 300–500 °C, like in traditional forging processes. Isothermal forging is performed at same temperatures for die and workpiece, in inert atmosphere and with low deformation rates, to avoid overheating. For Ti-alloys, the temperature of the process can be in the 900–1050 °C range, depending on the forging process that the material has to undergo. Of course, also during the hot-forming steps, the control of the atmosphere to avoid surface reaction is paramount, requiring the use of environmental chambers, or inert gas flux, to protect the part surface.

4.5 Defects in Titanium Alloys and Diagnostics

Notwithstanding the sophisticated control and monitoring systems operated all along the production cycle of titanium casting and metal working, titanium alloy products still require quality controls and inspection tests both during their processing and at the end of the production cycle. These controls are needed to guarantee, with a high level of confidence, that the materials and the components have actually reached the targeted properties, in order to comply with the relevant performance and safety requirements. The main kinds of defects and inhomogeneities present in titanium products have been classified according to their formation mechanisms. This is particularly useful, since starting from defect analysis, the identification of possible faulty aspects of the process can be straightforwardly identified and, hopefully, fixed. The main kinds of defects that can be encountered in titanium alloys are:

- Type I (hard α) defects, or hard interstitial defects (HID), are mostly associated with the presence of oxygen. These defects may be originated during various steps of the alloy processing: sponge production, first melt electrode formation, and melting and remelting stages. Oxygen is able to contaminate the titanium sponge during its production. The same may happen during the production of the VAR electrodes, in particular during welding operations, handling, especially under fast oxidation conditions, contamination introduced by scraps, contaminated master alloy, air, or water leaks. These defects are typically constituted by highly interstitially alloyed titanium, featuring a particularly brittle behavior. Since these defects are generally due to contamination by gaseous, and therefore light, elements, like oxygen but also nitrogen, they are also alternatively identified as low-density inclusions (LDIs).
- Type II (α-stabilized) defects are regions in which an excess of α-stabilizers is present so to change locally the alloy phase composition. These defects can be produced by the ballistic drop in EBM or PAM plants from scrap source and segregation during the final stages of solidification during VAR processes.
- High-density inclusions (HDIs) are due to debris and fragments from nonconsumable electrodes and grains of unmelted refractory metals, e.g., isomorphous β-stabilizers. Another source of HDIs is the scraps and debris from machining tools.

- β-flecks. Like Type II defects, these defects too are generated by the segregation in the melt, but of β-rather than α-stabilizers. Alternatively, an excess of β phase can be trapped into the alloy if presumed α/β treatments actually occur in the β field, possibly for the effects of adiabatic heating during metal working operations.
- Voids and residual porosity from solidification volume contraction. These pores remain trapped into the alloy and should be eliminated. HIP is a typical and effective approach to reduce or to get completely rid of these detrimental defects, which may significantly reduce the actual bearing section of the component and act as stress concentrators (Lüterling and Williams 2007).

Several inspection methods are available for a full characterization of the defect structures that may be present in titanium alloys, starting from the as-cast conditions up to the real component and, if necessary, even after a prolonged period of service. Actually, these same methods can be used for other alloys and, if electrical conductivity is not required, like for the eddy currents, also for other materials, like ceramics and structural polymer matrix composites, all relevant to aerospace structures. Considering that investigations on real components are required, nondestructive methods (NDTs, nondestructive tests) are definitely preferred. NDTs are defined as *examination, evaluation, or test performed on any type of test object without changing or altering that object in any way, in order to determine the absence or the presence of conditions or discontinuities that may have an effect on the usefulness or serviceability of that object* (Hellier 2001). There are several NDT methods, and a possible classification can be made considering the main working principle the method is based on:

- Acoustic emission testing.
- Eddy current testing.
- Magnetic particle testing.
- Penetrant liquid testing.
- Radiographic testing (including X-ray, gamma rays, and neutron radiographic testing).
- Thermal/infrared testing.
- Ultrasonic testing.
- Visual and optical testing.

For bulk analyses, very effective approaches are based on radiographic inspections and ultrasonic tests. In the first case, using high-energy electromagnetic radiation, i.e., X- or gamma rays, if available, with tomographic instrumentation, 2D or 3D images of the inner structure of the ingot, bar, or component can be obtained. The main working principle of the radiographic approach is based on, concerns the different absorption coefficients of the radiation in the alloy, the inclusions, pores, etc. Radiographic testing provides information about the shape, size, and position of defects in the material/component. Usually, gamma ray analyses are performed on larger components, while X-rays are used for smaller ones. The energy associated with the electromagnetic radiation that has to be used for the analysis of the specific component must be calibrated according to several parameters, for instance, the thickness of the component to analyze and the material of which it is made.

Ultrasonic testing (UT) is also sensitive to bulk defects, since the ultrasonic wave is reflected with different amplitudes by sample regions with different densities. In this way, using different ultrasonic wave transducers, capable to sample the test specimen to different depths, the shape, position, and distribution of the defects can be mapped within the component. Since the method is essentially based on the reflection of an ultrasonic wave, defects that are paralleled to the wave propagation direction are difficult to be detected. An interesting development in this field, also for the imaging capabilities that affords, is the ultrasonic phased array technique. In this case, the transducer is composed of a matrix of several elements, so the ultrasonic wave can be dynamically focused at variable depths within the tested material. This can be achieved by sequentially pulsing the elements of the transducer using a specific combination of them in the array. Furthermore, multiple wave fronts are used in combination, in order to form a beam with a particular shape, optimized for the geometry of the region to analyze (Hellier 2001; Mix 2005).

A characterization of surface defects can be obtained using the distribution of the eddy currents induced into the alloy by a suitable electromagnetic probe, whose geometry is usually designed to be compliant with the shape and geometry of the component to be tested. The spatial distribution of the induced currents is determined in the first place by the geometry of the component itself and, most importantly for this methodology, by the value of the local electrical resistivity. This is the parameter that, being influenced by the defects localized in the subsurface regions, is providing a very powerful tool for their detection and characterization.

Surface etching is classified as a visual testing method, and it is based on the same working principles of the metallographic etching: the regions with different microstructures of the examined components, like rotors, blades, and airfoils, possibly machined for surface finishing, will be selectively attacked by a suitable chemical etching solution, inducing chromatic changes on the alloy surface. The approach is in principle very simple and can be applied directly onto the surface of the components, new or already serviced. Any chemical inhomogeneity in the alloy results in a specific reaction kinetics, that modifies the surface topography and can be examined by visual and microscopy inspection to reconstruct the spatial distribution of the surface defects. An alternative analysis uses dye penetrant liquids, whose absorption in the alloy component is dependent on the open porosity and surface microcracks. These defects can therefore be visualized, since retaining a larger concentration of the coloring agent (Hellier 2001; Mix 2005). Another technique, particularly interesting for analyzing the microstructure of a component, is based on the replication of the surface features using film-forming substances, like cellulose acetate, aldehyde acetate, etc. Once consolidated, the film is lifted off from surface, of which the film retains the morphology. The replica prepared in this way can then be examined with microscopy techniques, without the need to cut off real specimens from the component, an operation not always feasible. Moreover, using an appropriate sequence of chemical etchings, it is possible to remove from the sampled surface also traces of the underlying material (extraction replica), with the possibility of obtaining also chemical and structural information acting in a fully nondestructive way (ASM International 1992).

4.6 Main Aerospace Applications of Titanium Alloys

When dealing with titanium in aircraft structures, it is in a way mandatory to mention what is still the case of the most significant use of titanium for aircraft applications, i.e., the Lockheed SR-71 "Blackbird" (see Chap. 1, Fig. 1.8). An estimated 95% of its total mass was made of titanium alloys, including engine's components and most of the airframe and fuselage, using the B120VCA alloy (see Table 4.4) (Boyer 1994). Irrespective of the comparatively limited number of these aircrafts, the SR-71 is a clear demonstration of the great flexibility of titanium alloys, in principle usable in many parts of any aircraft. Although the SR-71 is an unprecedented example of titanium applications to aircraft structures, still a significant percentage, in the 35–50% range, of the whole weight of a fighter aircraft is titanium. Furthermore, titanium alloys have been, and still are, widely used in rocket manufacturing (Leyens and Peters 2003). Some examples coming from the former Soviet Union are the Vostok and Soyuz spacecraft and the unpiloted Luna, Mars, and Venera. For these particular applications, titanium and its alloys were employed for the fabrication of liquid-propellant rocket engines, skin, solid-fuel engines, fasteners, and high-pressure vessels (Moiseyev 2006).

When coming to material selection for commercial aircrafts, other parameters than top performances, like peak speed, a priority in the military field, must be considered.

Weight saving, which may also concern the replacement of part originally made of aluminum alloys, particularly when service temperatures in excess of 130 °C are possible, is one of the reasons for the success of Ti-alloys. In case of replacement of steels and superalloys, the lower density of titanium alloys is sufficient to compensate for the higher absolute strength of the replaced alloys. Weight saving and reduction of component size become essential criteria for selecting titanium alloys. Structural weight savings provide the possibility of increasing the payload, a most interesting parameter when dealing with valuable items, like satellites and other deployable objects carried by extra-atmospheric flights.

The main applications of titanium and its alloys in aircraft structures, aero-engines, and aerospace field are herewith presented, with the deliberate intention to highlight the extreme flexibility of these alloys. The alloys that will be considered are those from the groups discussed in the former sections, i.e., the *conventional* and *advanced* titanium alloys, the latter mostly based on γ-TiAl intermetallic.

Additional applications of titanium (containing) alloys will be considered in Chap. 9 (see Sect. 9.5.2), with reference to shape memory alloys, based on the NiTi intermetallic compound. However, even though these systems actually contain significant concentrations of titanium, they are rather characterized by the particular functional properties associated with the phenomenology of the shape memory effect (see Sect. 9.5.1).

Airframe Titanium alloys can have a central role in the production of the fuselage, whereas aluminum alloys are not suitable, due to aero-kinetic heating of the surface skin. Titanium sheets and rivets have been extensively used in the aft end of the McDonnell Douglas F-15 and F-15E "Strike Eagle" fighters. This latest model also

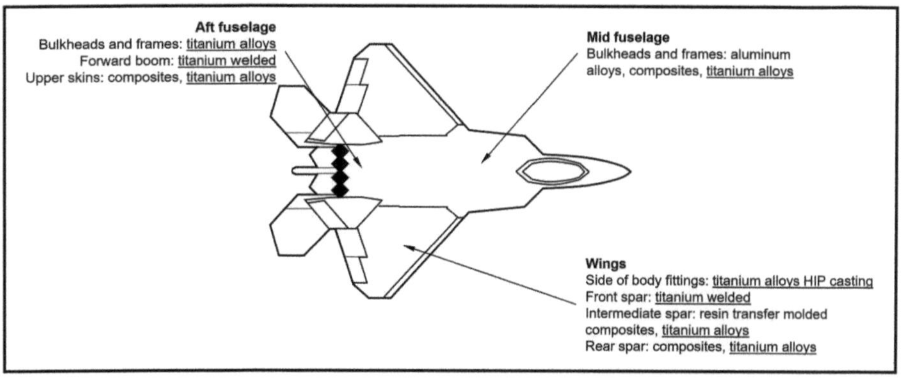

Fig. 4.38 Some of the parts made of titanium alloys in the Lockheed Martin-Boeing F22 Raptor advanced fighter aircraft. (Redrawn from Donachie 2000)

employs components made using advanced SPF-DB (superplastic forming and diffusion bonding) technique. In this way up to 726 part details and 10,000 fasteners could be eliminated with a consequent weight reduction and an enhanced maintainability. The Ti-6Al-2Zr-2Sn-2Mo-2Cr-0.25Si alloy was used in the airframe of the Lockheed Martin F-22 (see Fig. 4.38) and in the JSF (Joint Strike Fighter) project. This alloy is characterized by a moderate temperature capability and it is used in the fabrication of engine bay bulkheads, where fuselage temperatures reach the highest values. Several advantages derive from the use of this alloy in the mid-fuselage bulkhead of the F-22, resulting in a final weight of 150 kg only.

An important application of titanium alloys in aircraft structures is as reinforcing bands, strengthening the fuselage against fatigue cracks, that might propagate from the inner towards the outer skin layer, with obviously potentially catastrophic effects. The surface stability of titanium alloys, ensured by the self-passivating titanium oxide surface layer, is interesting for piping applications, particularly for corrosive fluids, like systems for air conditioning and de-icing. In this case, an additional burden from temperatures up to 200 °C must be considered. For these applications, the equivalent steel tubes would be 40% heavier. The easily deformable $\alpha + \beta$ Ti-3Al-2.5V alloy is a standard choice. Still corrosion resistance and weight saving recommend the use of CP-titanium for the aircraft floors of the onboard kitchens and toilets. The higher electrochemical compatibility of Ti-alloys, coupled with higher stiffness and strength, and also a better match in the coefficient of thermal expansion have recommended the use of these material as vertical and horizontal reinforcing and stabilizer structures for empennage tail, made of CFRP structural composites (Leyens and Peters 2003).

Landing Gear To meet volume constraints and, at the same time, the high-strength requirements, titanium alloys have been extensively used for the landing gear assembly of the Boeing 777 and 787 and Airbus A380. Boeing introduced on the B777 the metastable-β Ti-10V-2Fe-3Al alloy (Boyer 2010). In the landing gear, titanium alloy replaced steel (Fig. 4.39), with an estimated weight saving of 580 kg.

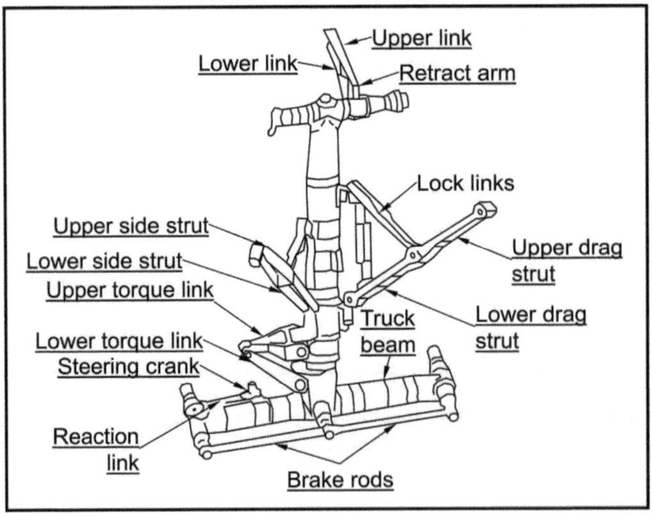

Fig. 4.39 Scheme of the main landing gear of the B 777 airplane. All underline-labeled parts are potential candidates for being manufactured using forgings made of Ti-5553 alloy. (Redrawn from Boyer 2010; Polmear 2006)

In the B787 a higher-strength alloy is used: Ti-5Al-5V-5Mo-3Cr. Using titanium alloys in landing gear structure, in addition to weight saving, reduces the relevant maintenance costs, mainly for an intrinsically better corrosion resistance.

Strictly related to the landing gear is the beam running between the wing and the fuselage, which actually supports the landing gear itself. Usually this part, as most of the airframe in commercial aircraft, is made of an aluminum alloy. In the Boeing aircraft models, B737, B747, and B757, due to a significantly larger applied load, a titanium alloy had necessarily to be employed. Indeed, the standard Al-alloy choice would not be feasible, since the section of the beam would not fit within the envelope of the wing. The alternative choice, i.e., a high-strength steel, would excessively increase the total weight of the aircraft. For the window frames of the pilot cockpit of commercial aircrafts, a forged titanium alloy is preferred to aluminum, suitable for the other widow frames. The reason for this is the requirement to resist bird strikes, an aspect that is not so critical in case of lateral fuselage surfaces.

Engine The front fan blades are other parts that have benefitted from the availability of advanced titanium alloys, although this is not the only class of materials that has been used for this part. Larger fan blades, up to 1 m long, required by larger by-pass ratio engines, would exhibit serious flutter problems, due to the shock waves generated by their tips when reaching a velocity close to that of sound. Several approaches have been implemented. In the first place, with the intention of increasing the structural stiffness of these components, mid-span shrouds were used to connect one blade to the other. However, this solution, which was certainly successful in solving the flutter vibration problem, presents some adverse aerodynamic effects. Novel design criteria were developed and lead to the so-called wide-chord fan blade design. These larger blades, lower in number, are intrinsically stiffer. Therefore, they do not require any structural reinforcement and guarantee a better engine efficiency. On the other hand, they would have been too heavy if made of solid titanium alloy forgings, as it was in case of shrouded blades. Lighter material systems were necessary and two solutions were actually proposed. After the pioneering attempts by Rolls-Royce, which were abandoned for excessive erosion problems, General Electric introduced in its workhorse engine, GE90, fan blades made of carbon fiber reinforced composites with strengthened leading edges made of titanium alloy. A full titanium alternative approach was adopted by Rolls-Royce and Pratt & Whitney. The problem of the excessive weight of a solid blade was overcome by adopting hollow components. Nowadays, titanium fan blades are produced by superplastic forming and solid-state diffusion bonding of titanium alloy sheet (Leyens and Peters 2003).

In Table 2.1 (Chap. 2), the parts of the compressor rotor of a commercial gas turbine engine are listed. Titanium alloys are the best choice for the compressor blades, as emerged since the early use of fan and compressor blades made of the $\alpha + \beta$ IMI550 alloy in the Pratt & Whitney JT3D gas turbine engine (1960). This resulted in remarkable improvements with respect to the same engine with steel blades: the take-off thrust increased by 42%; the SFC and total weight of the engine were reduced by 13% and 18%, respectively. The near-α alloys, which have been mainly developed to improve creep resistance, in order to widen the application field of titanium alloys, cannot overcome the 600 °C temperature. Therefore, the hottest parts of the compressor require the usage of steels and nickel-based superalloys. This aspect has become increasingly important with the raise of the air compression ratio reached in the last compressor stages in the state-of-the-art gas turbine engines (Fig. 2.8). The development of stronger titanium alloys allowed also using these materials for the compressor discs and shafts, thus reducing further the overall weight of the engines.

Until recent, compressor was the only internal part of a gas turbine engine in which titanium and its alloys could be safely used (see Table 2.1). In the latest General Electric gas turbine engine, GEnX, developed for commercial airliners and installed already on Boeing 747–8 and Boeing 787 aircrafts, the low-pressure turbine blades have been made of a γ-TiAl-based alloy: Ti-48Al- 2Cr-2Nb (at%) that, as discussed (Sect. 4.3.2), has adequate oxidation and creep resistance to be operated in the demanding turbine conditions. In this engine, thanks to this material and to the new design solutions, remarkable reductions in fuel consumption (20%), peak noise (50%), and NOx emissions (80%) compared with prior engines have been achieved. The use of a γ-TiAl alloy for low-pressure turbine blades would render this material suitable also for the blading of the hot part of the compressor, afterburners, and nozzle case in military engines, which at the present time are realized with steel and superalloys. However, since it is still too expensive, the use of the Ti-aluminide intermetallic alloy is so far restricted to the mentioned parts of the gas turbine engine (Linger 2009). Interesting to note, which concerns Ti-aluminide-based alloys, the very early applications involved alloys based on the α_2-Ti_3Al intermetallic (Sect. 4.3.2). However, although tested for components and parts of military aircraft prototype engines, like afterburner sheets and low-pressure turbine blades, the alloy never reached a sufficient level of technological development and reliability to justify its employment in commercial aircraft. In particular, oxidation resistance resulted still too poor to guarantee a sufficiently long lifetime.

A better success was achieved by replacing Inconel 625 with the metastable β-alloy TIMETAL B21S for the plug and nozzle assembly of the Rolls-Royce Trent 400 engine, mounted on Airbus 340 and Boeing 777 aircrafts, resulting in more than 160 kg weight reduction. The positive results, obtained in this case, have encouraged research efforts aiming at using titanium alloys also for the thrust reverser for the Boeing 737MAX. The use of a titanium alloy is intended in this specific case, not only to reach the expected weight reduction but also an implementation in the engine design, namely, longer fan blades, without the need of a proportional increase in the size of the engine nacelle. As reported for similar situations, the wider usage of titanium determines a significant reduction in the maintenance costs (Cotton et al. 2015). New concept design, like an integrally bladed disc, conventionally called "blisk," has contributed to push further the benefits associated with the use of lighter alloys for compressor components. The "blisk" is a single part in which disc and blades are bonded together by linear friction welding. In this way, the typical nucleation sites for fatigue fracture are eliminated, and, thereby, the relevant inspection intervals are extended. Moreover, friction welding can be used for replacing individual blades, if damaged during service (Leyens and Peters 2003).

Springs The elastic moduli of titanium and $\alpha + \beta$ Ti-alloys fall in the 100–110 GPa range that is definitely smaller than the elastic modulus of stainless steel (210 GPa). Therefore, one of the more interesting areas in which titanium alloys may replace steels is in spring applications. Indeed, considering the lower value of the Young's modulus, only half of the coils of the same size are needed to replace an equivalent steel spring. Moreover, the lower density of titanium, 60% approximately the density of the steel, results in up to 70% weight reduction for these parts. Titanium aircraft springs can be made of titanium alloys, and β-alloys, like the β-metastable Ti-15-3 (Ti-15V-3Cr-3Sn-3Al, Table 4.3) and Beta-C (Ti-3Al-8V-6Cr-4Mo-4Zr, Table 4.3), are particularly recommended. Interestingly, the low cost beta (LCB) alloy (metastable-β, with composition Ti-4.5Fe-6.8Mo-1.5Al, see Table 4.4) had been originally developed for car springs. Although this application was not feasible, since the material and production costs were still too high for the automotive field, this research made an alloy available to aerospace spring manufacturing, having half the price of another all-around alloy, the Beta-C (see above), and exhibiting very much the same mechanical properties.

Accessories Cargo handling systems, crew microphone holder (Boeing 737 and 757), access door clips, interior window clips, and other similar precision forged parts are all made of titanium alloys in commercial airliners starting from the early 1990s (Boyer et al. 1994). Structural parts, like brake torque tubes of aircrafts (military fighter and commercial), can be made of castings of metastable β-alloys, like the Ti-15-3 alloy.

Containers and Tanks A peculiar application of titanium alloys in aerospace structures is for high-pressure containers, ranging from fire extinguisher bottles to pressure tanks for space transportation systems (Boyer et al. 1994). In both cases, the very well-predictable behavior of titanium alloys and their low density are paramount selection criteria. A further plus of titanium alloys for aerospace fuel tanks is the possibility to be nondestructively tested before being launched with a level of statistical reliability not achievable with polymer matrix structural composites. This is an aspect which is sometimes overlooked, even though it is of absolutely important for these applications, in which any kind of critical defects may jeopardize the success of the entire mission.

Helicopters The high strength and fatigue resistance afforded by β-Ti-alloys have been very important for improving the performance of top-class helicopters (Boyer et al. 1994). From this point of view, the progressive increase in the all-up weight (AUW) of the Westland Lynx helicopter is an interesting case. The semirigid rotor head, the hub disc, the sleeve, and the mast were originally made of the $\alpha + \beta$ classical Ti-6-4 alloy, which was no longer adequate to support an increase in the AUW from 38.6 tons up to the 56.0 tons. The Ti-6-4 alloy was replaced by a metastable β-alloy: Ti-10-2-3 (Ti-10V-2Fe-3Al, see Table 4.4) that exhibited enhanced high-cycle fatigue, lower stiffness, and superior strength. Other helicopter companies adopted similar alloys and technological development process for their products.

References

ASM International (1991) ASM Handbook Vol. 2 – Properties and Selection: Nonferrous Alloys and Special-Purpose Materials. ASM International

ASM International (1992) ASM Handbook Vol. 17 – Nondestructive Evaluation and Quality Control

ASM International (1992) ASM Handbook Vol. 3 – Alloy phase diagrams. ASM International

Bania P J (1994) Beta Titanium Alloys and Their Role in the Titanium Industry. Journal of Materials 46 (7): 16–19

Bewlay BP et al (2016) TiAl Alloys in Commercial Aircraft Engines. Materials at High Temperatures 33 (4–5): 549–559

Boyer R (1994) Aerospace Applications of Beta Titanium Alloys. Journal of Materials 46(7): 20–23

Boyer R (2010) Application of Titanium and Its Alloys. Journal of Materials 62(5): 21–24

Boyer R, Collings E W, Welsch G (1994) Materials Properties Handbook: Titanium Alloys. ASM International

Cain K J (2016) Industrial Titanium Demand Forecast 2016. In: Titanium USA 2016, the 32nd annual conference and exhibition produced by the International Titanium Association (ITA), J. W. Marriot Desert Ridge Resort, Scottsdale, 25–28 September 2016

Chen G et al (2016) Polysynthetic twinned TiAl Single Crystals for high-temperature Applications. Nature Materials 15: 876–881

Chen W et al (2014) Development of Ti_2AlNb Alloys: Opportunities and Challenges. Advanced Materials & Processes 172(5): 23–27

Cotton J et al (2015) State of the Art in Beta Titanium Alloys for Airframe Applications. Journal of Materials 67 (6):1281–1303

Dargusch M S, Keay S M (2009) Meltless Ti-Al New Light Metals Industry. In: Dargusch M S, Keay S M (eds) Light Metals Technology, Trans Tech Publications Ltd, p 135–138

Das S K, Davis L A (1988) High Performance Aerospace Alloys via Rapid Solidification Processing. Materials Science and Engineering 98: 1–12

Dimiduk D M et al (1992) Development of Intermetallic materials for Aerospace Systems. Materials Science and Technology 8(4): 367–375

Ding X F et al. (2011) Microstructure evolution of directionally solidified Ti-45Al-8.5Nb-(W, B, Y) alloys. Journal of Alloys and Compounds 509(9): 4041–4046

Donachie M J (2000) Titanium – A Technical Guide, 2nd edn. ASM International

Eylon D et al (1994) Issues in the Development of Beta Titanium Alloys. Journal of Materials 46 (7): 14–15

Froes F H (2015) Titanium: Physical Metallurgy, Processing and Applications. ASM International, Materials Park, Ohio

Froes F H, Imam A (2018) Titanium: A Historic and Current Perspective-Part I. Advanced Materials and Processes 176: 19–25

Hashimoto K et al. (1998) Alloy design of gamma titanium aluminides based on phase diagrams. Intermetallics 6(7–8): 667–672

Hellier C (2001) Handbook of Nondestructive Evaluation. McGraw-Hill

Kutz M (2002) Handbook of Materials Selection. John Wiley and Sons

Lapin J (2006) Creep behavior of a cast TiAl-based alloy for industrial applications. Intermetallics 14(2): 115–122

Lapin J, Ondrúš L, Nazmy M (2002) Directional Solidification of Intermetallic Ti-46Al-2W-0.5Si Alloy in Alumina Moulds. Intermetallics 10 (10): 1019–1031

Leach W (2016) Titanium Demand and Trends in the Aero Engine Market. In: Titanium USA 2016, the 32nd annual conference and exhibition produced by the International Titanium Association (ITA), J. W. Marriot Desert Ridge Resort, Scottsdale, 25–28 September 2016

Leyens C, Peters M (2003) Titanium and Titanium Alloys – Fundamentals and Applications. Wiley VHC Verlag, Weinheim

Linger D (2009) Titanium Utilization & Vision for The Next Decade: An Aircraft Engine OEM Perspective. In: Titanium 2009, the 25th annual conference and exhibition produced by the International Titanium Association (ITA), Hilton Waikoloa Village, Hawaii, 13–16 September 2009

Liu B G et al. (2017) Structural stability and the alloying effect of TiB polymorphs in TiAl alloys. Intermatallics 90: 97–102

Lütjering G, Williams J C (2007) Titanium, 2nd edn. Springer Berlin Heidelberg, Berlin

Massalski T B et al (1986) Binary Alloys Phase Diagrams Vol. 2. ASM International

McCafferty E (2010) Introduction to Corrosion Science. Springer Verlag

Meetham G W (1981) The Development of Gas Turbine Materials. Applied Science Publishers Ltd, London

Mix P E (2005) Introduction to Nondestructive Testing: A Training Guide. John Wiley & Sons

Moiseyev V N (2006) Titanium Alloys-Russian Aircraft and Aerospace Applications. Taylor & Francis

Nieh T G et al. (1994) Superplasticity in Metals and Ceramics. Cambridge University Press

Partridge A, Shelton E F J (2001) Processing and Mechanical Property Studies of Orthorhombic Titanium-Aluminide-based Alloys. Air & Space Europe 3 (3–4): 170–173

Pérez-Prado M T, Kassner ME (2009) Superplasticity. In: Kassner M E (ed) Fundamentals of Creep in Metals and Alloys, 2nd edn. Elsevier, p 137–149

Peters J M, Blank-Berwersdorff M (1992) Titanium Aluminide Foil. Materials and Design 13 (2): 83–92

Polmear I J (2006) Light Alloys – From Traditional Alloys to Nanocrystals, 4th edn. Butterworth-Heinemann

Prasad E N, Wanhill R (2017) Aerospace Materials and Material Technologies Volume 1: Aerospace Materials. Springer

Proske G et al. (1992) The Microstructure and Mechanical Properties of the Intermetallic Compund Super Alpha 2. Materials Science and Engineering A 152: 310–316
Sastry S M L et al (1983) Rapid Solidification Processing of Titanium Alloys. JOM 35 (9): 21–28
Sears J W (1990) Current Processes for the Cold-Wall Melting of Titanium. Journal of Minerals, Metals and Materials Society 42 (3):17–21
Suryanarayana C et al. (1991) Rapid Solidification Processing of Titanium Alloys. International Materials Review 36 (1): 85–123
Suryanarayana C, Froes F H (1990) The Current Status of Titanium Rapid Solidification. Journal of Materials 42 (3): 22–25
Wang L et al. (2017) Influence of Alloy Composition and Thermal History on Carbide Precipitation in γ-based TiAl Alloys. Intermetallics 89:32–39
Zhao Z B et al. (2017) Effect of heat treatment on the crystallographic orientation evolution of a near-α titanium alloy Ti60. Acta Materialia 131: 305–314

Further Reading

Boyer R R (1996) An overview on the use of titanium in the aerospace industry. Materials Science and Engineering: A 213 (1–2):103–114
Gillo G (2011) Superplastic Forming of Advanced Metallic Materials – Methods and Applications. Woodhead Publishing Ltd
Joshi V A (2006) Titanium Alloys – An Atlas of Structures and Fracture Features. Taylor & Francis
Moore H D (1981) Materials and Processes for NDT Technology. The American Society for Nondestructive Testing
Seong S et al. (2009) Titanium – Industrial Base, Price Trends and Technology Initiatives. RAND Corp., CA (USA)

Chapter 5
Steels

5.1 Introduction

A very simple definition for steels is "iron-carbon-based alloys". In the common language, iron and steel are sometimes used to describe the same material, and this is mainly due to the fact that people tend to associate each metallic object to the most notorious and better known metal, i.e., iron, a chemical element with the atomic number equal to 26.

The name directly comes from the Latin word *ferrum*, from which the Italian term (*ferro*) derives. Concerning the English term, iron derives from the Scandinavian term *iarn*. On the Earth, the most abundant sources of this metal are the Fe_3O_4 (magnetite) and Fe_2O_3 (hematite)-containing ores. Pure iron is only available in meteoritic and particular native form. Native iron, also known as telluric, can be found in basaltic rocks, as small grain inclusions, with a maximum Ni content of 3 wt %, and resulting from a thermal reduction occurring within the silicic matrix. Nickel is also present in meteoritic iron, in which higher concentrations of this metal have been measured, ranging from 4 to 26 wt% (Buchwald and Mosdal 1985). Although rather scarce, it seems that the first iron that mankind actually used was the meteoritic type. It was regarded as a magic "metal from the stars," and this marked the vocabulary of iron-related activities. Iron metallurgy in Latin languages, like the French *sidérurgie*, is named after *sidus-sideris*, i.e., star. Examples of both meteoric- and telluric-based iron objects have been found in Greenland. The main source of meteoric iron in Greenland was the Cape York meteorite that collided with the Earth approximately 10,000 years ago, with fragments falling out over a wide area (see Fig. 5.1). The early reported examples of iron production from ores date back to 2000 BC in India, and by 1200 BC, iron production processes were widespread both in China and in the Near East. This was the time when iron metallurgy appeared also in Mediterranean area, and it is taken as the starting time for the Iron Age (Ashby 2005; Tylecote 2002). Nowadays, the reference material has become steel, named after the Sanskrit word *stakati*, (*Stahl* in German, *stalin* in Russian), identifying a

S. Gialanella, A. Malandruccolo, *Aerospace Alloys*, Topics in Mining, Metallurgy and Materials Engineering, https://doi.org/10.1007/978-3-030-24440-8_5

Fig. 5.1 Part of the Cape York meteorite at the American Museum of Natural History in New York (USA). The polished part from which the *Widmanstätten* structure can be clearly seen is several centimeters in width. (Photo property of the authors)

class of iron alloys featuring carbon as component. Carbon content can strongly vary depending on the steel grade. The lowest concentrations are those of the low-carbon steels, containing no more than 0.06 wt% C.

Until relatively recent times, in the Western metallurgy, the production of steel was mainly the result of the skills of blacksmiths, heating bloomery iron in charcoal fires. Blooming was an important stage, as concerns both the alloy refinement and the optimization of the carbon content. The situation was subsequently ameliorated thanks to several improvements. Pushed by the shortage of wood and legislation restrictions, the process to eliminate sulfur from the fossil coal was developed, resulting in the production of coke coal, by the end of the seventeenth century AD in England. The coke coal boosted any metallurgical production, with particular regard for iron metallurgy, since the mechanically stronger fuel afforded the possibility to realize larger furnaces and raw material stacks. The process lead in the end to blast furnaces, and cast iron was the main product of these plants. Thereafter, the production cycle required the treatment of cast iron to reduce the excessive carbon content. This was achieved thanks to the development of the puddling furnaces first and then of the Bessemer converter, unanimously recognized as the beginning of the modern steelmaking in 1856 and of the so-called open-heart process shortly afterward (Hosford 2012). Steel is nowadays present in almost every technological field, including aerospace industry, of which some relevant aspects are considered in this chapter.

5.2 Iron and Steel Metallurgical Base Concepts

Steels are characterized by a variety of properties and microstructures, which can be achieved through solid-state transformations and processing. Like titanium (see Chap. 4), iron shows allotropic forms, three in this element, stable at increasing temperatures: α (*bcc*), γ (*fcc*), and δ (*bcc*). Iron exhibits also a magnetic transition, occurring at 768 °C (the Curie temperature) above which it becomes paramagnetic, still retaining a *bcc* structure. This form is called β-iron. At atmospheric pressure, the α polymorph (*bcc*) transforms into the close packed *fcc* γ-iron above 911 °C. This phase is stable up to 1392 °C, above which iron turns back to the *bcc* structure of δ-iron. This phase is stable up to the melting temperature (1538 °C). The stability of the different polymorphs can be justified in terms of the distribution of the density of electronic levels, with some complication coming from the partially filled d-bands and relevant magnetic properties (Ashcroft and Mermin 1988). For instance, if nonmagnetic iron was stable at room temperature, it would have had an *hcp* instead of the *bcc* structure (Pepperhoff and Acet 2001).

The polymorphism of iron is maintained in steels, with stability fields now dependent on temperature and composition too. This can be clearly seen from the steel prototype phase diagram, i.e., the Fe-C binary phase diagram in Fig. 5.2.

The combination of alloying elements, including undesired impurities, that might be present in lower-grade alloys and heat treatments lead to a variety of microstructures and, thereby, properties. Some alloying elements will dissolve into the iron matrix; others will form precipitates and secondary phases.

Concerning steels, there are two main criteria for the classification of alloying elements. The first is based on the relevant role in the stabilization of austenite; the second regards the tendency for the alloying element to form carbides. It is worth noting that carbide formation needs the concentration of the alloying element to overcome a threshold value, having to do with its solubility in iron. According to these criteria, in Appendix 4 the most important alloying elements and their effects on steel properties are given. The microstructure of steel products resulting from any composition can be modified by specific heat and thermomechanical treatments, as described in full details in the cited literature at the end of the chapter.

5.2.1 Annealing

Annealing treatments are conducted after either high-temperature (casting, hot-rolling, forging) or cold-forming processes. The main effects induced by the annealing regard hardness reduction, grain refinement, increased microstructural stability, and better machinability. The driving force for these changes is the energy stored into the material as a consequence of previous processing steps, like casting (internal solidification stresses) and deformation processes, dealing with wrought products. The most important categories of annealing treatments are summarized in Table 5.1.

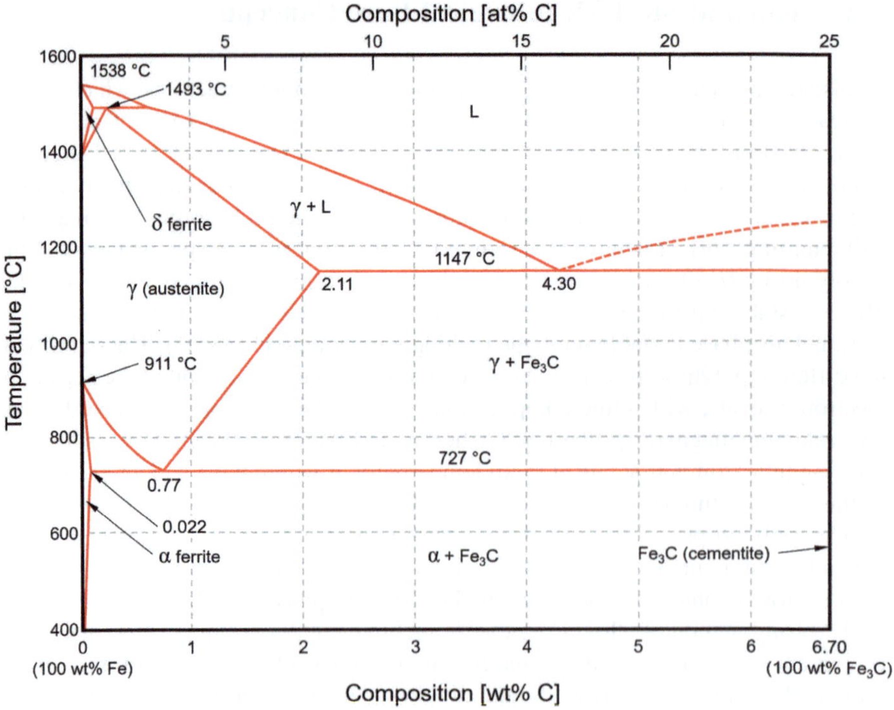

Fig. 5.2 Fe-C phase diagram, iron rich corner. (Redrawn from Callister and Rethwisch 2012; Hosford 2012)

5.2.2 *Quenching*

Quenching involves a sufficiently fast cooling of a steel from its austenitizing temperature field, capable to induce the formation of a non-equilibrium structure, called *martensite*. The martensite formation occurs via an athermal transformation that starts once the alloy has been taken below the so-called martensite start, M_s, temperature, and it is accomplished below the martensite finish temperature, M_f. Due to its athermal character, the transformation requires a continuous cooling to proceed. Since the rate of change propagates at the speed of sound, the transformation would typically take place in a relatively short time, so that no carbon diffusion throughout the crystal lattice or any other compositional change may occur. Carbon remains trapped into the crystalline lattice, which is *fcc* in the initial state (austenite), and transforms into body-centered tetragonal (*bct*), owing to the deformation associated with the excessive concentration of carbon into the iron lattice. The cooling rate necessary to attain the martensite varies according to the chemical composition of the steel. Optimal values can be determined from the continuous cooling transformation (CCT) and time-temperature-transformation (TTT) diagrams (Totten 2007). An excessively high cooling rate will cause tension cracks in the material.

Table 5.1 A summary of the most important annealing treatments for steels (Campbell 2008; Totten 2007)

Name	Aim	Description
Diffusion annealing	Elimination of chemical inhomogeneities in alloy composition, in particular along grain boundaries, to reduce liquation phenomena	Diffusion annealing is carried out in the austenite stability region between 1100 and 1300 °C, to reduce inhomogeneity in alloying elements distribution that may result in liquation. Therefore, microhardness gradients are eliminated as well. Finally, the overall hardness of the steel decreases with respect to the initial value
Normalization	Grain size refinement, better microstructure, improved machinability, and enhanced response to hardening processes. Normalization is generally carried out for producing a uniform structure made of ferrite and pearlite	It is carried out at a temperature which is not much higher than the ferrite-to-austenite transformation temperature, this for avoiding grain coarsening phenomena. Normalization temperature usually lays 20–50 °C above the ferrite-to-austenite transformation temperature. Cooling is usually performed in air. The higher the cooling rate, the finer the final microstructure
Softening	Production of a globular structure, known as pearlite, in steels with more than 0.5% C. Production of a uniform fine microstructure with finely dispersed carbides after quenching process	Softening is performed by heating the materials up to a temperature slightly above its eutectoid temperature and holding the condition for a reasonable long time. Cooling after softening can be performed in air as well but only when the temperature has been decreased below 600 °C
Solution annealing	Dissolution of particular phases precipitated during high-temperature processes, such as hot-rolling. Since solution annealing is the most used heat treatment for austenitic stainless steels, phases that have to be dissolved are the Cr-rich $M_{23}C_6$ carbides (M = Cr, Fe, Mo) or the Mo- and Cr-rich deleterious σ phase	The alloy is heated above the solvus temperature of the secondary phases (either carbides, or intermetallics) and cooled down sufficiently fast, in order to avoid re-precipitation. Stainless steels like AISI 303, 304, and 304L grades are usually solution annealed in a temperature range in between 1010 and 1120 °C. Mo-containing stainless steels such as AISI 316, 316L, 317, and 317L are treated between 1040 and 1120 °C. For Ti-stabilized austenitic stainless steels such as AISI 321 grade, the temperature range is between 955 and 1065 °C. Complete dissolution of carbides has to be carried out, and this usually takes considerable time to happen. As a general rule, the higher the treatment temperature, the faster carbide dissolution takes place.

(continued)

Table 5.1 (continued)

Name	Aim	Description
		Duration of the treatment depends on the component size and on the furnace type. The maximum annealing temperature and duration are limited by the need to avoid abnormal grain growth. Concerning the cooling media, the type used is related to the desired cooling rate. For instance, water is a stronger cooling medium if compared to oil, and oil is stronger than air. Most common cooling medium is water; however oil and solutions containing polymers are also used
Spheroidizing	Production of a spheroidized (globular) structure with maximum ductility and minimum hardness	In order to obtain spheroidizing, the steel has to be heated at a temperature just below its ferrite-to-austenite transformation for a sufficiently long time. Another possibility is to perform spheroidizing by cyclic oscillation above and below the ferrite-to-austenite transformation temperature. The treatment produces small cementite (Fe_3C) spheroids in a ferrite matrix. It is often performed after a normalization annealing
Stabilization annealing	Maximization of intergranular corrosion resistance of stabilized austenitic stainless steels	The material is heated up to the temperature range 845–955 °C in order to promote the formation of the thermodynamically stable MC carbides, in place of the less stable $M_{23}C_6$ carbide. MC carbide precipitation temperature is higher than that of the Cr-rich $M_{23}C_6$. Furthermore, $M_{23}C_6$ carbide formation leads to Cr depletion in the stabilized austenitic stainless steels. This phenomenon is a particular type of intergranular corrosion named *sensitization*. The formation of MC rather than $M_{23}C_6$ leads to a lower decrease in Cr content of the bulk and therefore a better corrosion resistance, particularly along grain boundaries
Stress relief annealing	Elimination of macroscopic stresses in the material due to cold-working	The process temperature strongly depends on the preliminary deformation the material has been subjected to. The higher the degree of deformation, the lower is the process temperature necessary for stress relief. Two

(continued)

Table 5.1 (continued)

Name	Aim	Description
		different phenomena lead to stress relief: recovery and recrystallization. The material is slowly cooled down from the annealing temperature (typically in the 425–550 °C range for austenitic stainless steels) in order to reduce stresses. However, this can cause sensitization of the steel, i.e., $M_{23}C_6$ carbide precipitation, with the consequent removal of chromium from the alloy and consequent worsening of the corrosion resistance. Fast cooling causes residual stresses leading to a structure which is more prone to stress corrosion cracking. However, it is generally preferable to have a small sensitization rather than catastrophic failures caused by stress corrosion cracking (SCC – see Sect. 8.1.5) that can indeed be reduced by the present treatment

On the other hand, a too low cooling rate will not allow martensite to form at all or just to a limited extent. There are two main types of quenching processes: normal and thermochemical quenching. The first one involves the treatment of the bulk volume and the second the surface. This means that using processes, like surface carburizing for low-carbon steels, an external layer only is transformed into martensite. In this way, the surface properties, e.g., hardness, would be improved, without excessively embrittling the bulk. Since martensite has a high hardness and a low toughness, quenching is usually followed by tempering treatments (see Sect. 5.2.3). After quenching, retained austenite can be observed, i.e., a residual fraction of the initial austenite phase that does not undergo any transformation. If required, the retention of austenite can be avoided by means of a cryogenic, or subzero temperature, treatment (Totten 2007). Concerning steels that will be discussed in this chapter, martensitic transformation, with material-specific differences, regards maraging steels (Sect. 5.4.1), AerMet steels (Sect. 5.4.2), martensitic stainless steels (Sect. 5.4.3.2), and martensitic PH stainless steels (Sect. 5.4.3.3).

5.2.3 Tempering

Tempering treatments are meant to optimize the properties of the metastable martensite, as concerns particularly its low toughness. Depending on the material composition, part size, and residual stress distribution, tempering can be performed over either low (e.g., 100–200 °C) or relatively higher (e.g., 450–700 °C) temperature intervals, however, still well below the ferrite-to-austenite transition temperature. The tempering process can be split into four main stages (ASM International 1992; Totten 2007):

- Stage I ($T < 250$ °C). At this stage, a reduction of the carbon content in the martensite takes place with the formation of ε carbide ($Fe_{2.4}C$). The martensite is still a supersaturated solid solution and may undergo further transformations at higher temperatures.
- Stage II (200 °C < T < 300 °C). During the II stage, retained austenite decomposes into ferrite and cementite.
- Stage III (250 °C < T < 350 °C). The third stage of tempering implies the replacement of transition carbides, like the ε carbide, and low-temperature martensite substitution by ferrite and cementite. Cementite precipitates heterogeneously, as needles and platelets at the inter-lath boundaries (Campbell 2008). The formation of cementite results in a further reduction of the carbon level in the martensite, and, thereby, the *bct* structure will lose its tetragonal symmetry.
- Stage IV (350 °C < T < 700 °C). This stage is typical of highly alloyed steels. During the fourth stage of tempering, further equilibrium carbides, like M_7C_3 and $M_{23}C_6$, precipitate, resulting in the so-called secondary hardening.

A detrimental phenomenon associated with tempering is temper embrittlement, involving a rise in the ductile-to-brittle transition temperature and, consequently, a reduction in the low-temperature fracture toughness. Temper embrittlement is associated with two phenomena, depending on the temperature range: tempered martensite (irreversible) embrittlement and the temper embrittlement. Tempered martensite embrittlement typically occurs over the 230 °C–400 °C temperature range, corresponding to II and III tempering stages. It is observed in plain carbon steels containing phosphorous, and low-alloy steels subjected to heat treatments to reach high-strength levels. In steel materials relatively free from nitrogen and phosphorous, a decrease in toughness is expected due to the formation of cementite, which is a preferential path for fracture propagation. In case the steel contains a high amount of phosphorous (e.g., 0.03 wt% or more), the embrittlement results in an intergranular failure along initial austenite grain boundaries, for the decrease in their cohesive strength (Campbell 2008; Totten 2007).

Temper embrittlement occurs at higher temperatures (450 °C–600 °C). It is also known as *two-step* embrittlement, since it is observed either on slow cooling through the critical temperature range or upon aging the alloy in the same temperature interval. Since a heating cycle above 600 °C followed by rapid cooling can eliminate temper embrittlement, differently from tempered martensite embrittlement, this phenomenon is reversible and is characterized by an intergranular fracture morphology. Impurities, like As, P, Sb, and Sn, play an important role in this phenomenology, since affecting the ductile-to-brittle transition temperature of the alloy. The effect of these elements can be enhanced further by the presence of Cr, Mn, Ni, and Si (Campbell 2008; Totten 2007).

Tungsten and molybdenum (see Appendix 4) slow down temper embrittlement, by reducing the alloy diffusivity. Temper embrittlement can also be prevented if the critical temperature interval is crossed on cooling at a sufficiently fast rate (Campbell 2008; Totten 2007).

Strictly related to the above treatments is *austempering*, not involving modifications of the martensite but still interesting in several respects. This thermal treatment, developed by Eduard Devenport and Edgar Bain (US Patent 1,924,099), can be performed on both cast irons (ADI, austempered ductile irons) and steels and involves an isothermal transformation of the alloy at a temperature above its martensite start temperature (M_s) but below that of pearlite transformation. The alloy is cooled down to this temperature interval after homogenization in the austenitic field. The most used cooling medium is a molten salt bath, and the main improvements introduced by this treatment are increased ductility, strength, and toughness; reduction in the distortions associated with heat treatment; and possibility to rapidly obtain hardness values in the 35–55 HRC range (ASM International 2016).

5.2.4 Age Hardening

Age (or precipitation) hardening treatments produce an increase of the steel mechanical properties due to the solid-state precipitation of secondary phases. Age-hardened materials are of particular interest for aerospace applications, and precipitation treatments are interesting not only as concerns steels but also aluminum and magnesium alloys (see Chap. 3), titanium alloys (see Chap. 4), iron- and nickel-based superalloys, and also the latest generation of cobalt-based superalloys (see Sect. 6.2.2). The main steel families that exhibit precipitation hardening are AerMet, maraging, and PH (PH = precipitation hardening or precipitation hardenable) steels. Details on age-hardening treatments of maraging, AerMet, and PH stainless steels can be found in Sects. 5.4.1, 5.4.2, and 5.4.3.3, respectively.

5.2.5 Steel Designation

The most common steel designation system is based on their chemical composition, although other classifications do exist, based on different intrinsic and technological aspects:

1. Crystallographic structure, e.g., ferritic, martensitic, austenitic, duplex, bainitic, and pearlitic
2. Heat treatment condition and thermomechanical processing, like air cooled, solution annealed, etc.
3. Manufacturing methods: vacuum arc remelting, electric slag remelting, etc.
4. Main application fields: stainless steels, tool steels, structural steels, etc.

5. Oxidation protocol used for their production: *killed* and *semi-killed*
6. Finishing method: hot-rolled, cold-rolled, and continuously cast
7. Product shapes: bars, coils, plates, and sheets
8. Required strength level according to ASTM standards
9. Quality indications and classification: commercial quality, forging quality, etc.

A detailed description of each classification system can be found in specific references (ASM International 1993; Totten 2007). In the USA, the AISI (American Iron and Steel Institute) system is the most common, together with UNS (unified numbering system), that is widely accepted also in the rest of North America. In Europe, the EN (Euronorm) and the DIN (Deutsche Institut für Normung) standard, including the Werkstoff number (WN), are common designation systems. For instance, the austenitic stainless steel AISI 316 is the 1.4401 (WN) and the X5CrNiMo17-12-2 according to the EN designation system (ASM International 1993; Totten 2007).

5.3 Standard and Special Production Techniques

This section is dedicated to a brief overview on the principal steel production techniques. The aim is to provide a general information on the production and processing of particular steel grades, especially those employed in the aerospace field that will be discussed in Sects. 5.4 and 5.5. Steel metallurgy is either primary or secondary, depending on the feedstock type used in the first melting process. Iron-rich ores are used as feedstock for extractive metallurgy, using a blast furnace. Alternatively, steel scraps can be used as feedstock in the alternative metallurgy processing, for which the electric arc furnace (EAF) is the most common production plant. The output of an EAF melting may be an alloy whose chemical composition differs from that of the final product. This is the case of highly alloyed steels, that need a further refining step to set the correct chemical composition. They are usually characterized by a carbon content in the 1–2 wt% range, lower than a typical EAF product. Therefore, to remove the excess carbon, an oxidation is required. Other refining approaches can be employed as concerns the tuning of the concentration of other alloying elements, like Al, Cr, Mo, Ti, etc. The final chemical composition of the products and the desired cleanliness level are achieved through refining processes (e.g., argon oxygen decarburization, AOD), before casting and subsequent solidification, following a standard production route. Plain carbon steels and low-alloyed grades can be attained directly from the EAF thanks to a careful selection of the scraps and to specific practices during the process. Therefore, the composition of the scraps charge must be calibrated according to the steel grade that has to be produced and to the melting and refining steps that are needed. Concerning the steel grades that are interesting to the aerospace industry, the main processes discussed in this section are:

- Arc melting
- Vacuum induction melting

- Argon oxygen decarburization, vacuum oxygen decarburization (VOD and VODC)
- Electroslag remelting
- Vacuum arc remelting

These same processes are also used in the production of superalloys and, as concerns vacuum processes, in the production of titanium alloys (see Chaps. 6 and 4, respectively).

5.3.1 Arc Melting

The EAF is one of the most important plants in foundries and steel production factories. It was first introduced at the beginning of the twentieth century, and nowadays it is worldwide used by more than 80% of steel producers. The rest will use induction furnaces (around 17%) and other melting practices (ASM International 2008; Schifo and Radia 2004; Toulouevski and Zinurov 2013).

As shown in Fig. 5.3, the EAF is composed of three main parts: a vessel, a roof, and the electrodes. The steel vessel has an internal refractory layer either made of bricks or of monolithic material. The roof is movable, in order to allow feeding the scrap charge. Its inner surface is also protected by a refractory layer. The EAFs used for steel production are in general alternating current furnaces, featuring three graphite-based electrodes. Direct current furnaces have only one electrode. The electrodes are mounted vertically through the roof of the furnace and float above the solid steel scrap, or the molten bath, depending on the specific step of the production cycle and on the distance between the electrodes and the charge/melt, which varies during the melting process. The temperature is increased until charge melting is achieved thanks to the energy transfer through the electric arc sparking between the charge and the electrodes. Usually, due to the high volume of the scraps, the whole charge is added to the furnace in several steps. For a 50t EAF, there are usually three charging steps. After the first portion is introduced into the furnace chamber, the roof is repositioned onto the vessel, and the electrodes are lowered into the charge. As melting initiates, the voltage is gradually increased and the process continues until the charge is totally molten. At this stage, since the volume occupied by the initially solid, now molten scraps is lower, it is possible to add a second batch of metallic scraps. This process is repeated until the nominal volume capacity of the furnace is reached. In addition to the metal scraps, it is possible to introduce into the furnace other materials, such as DRI (direct reduced iron) and substances, like mixtures of oxides and fluorine compounds, such as CaF_2, which promote the formation of the slag and optimize its viscosity. The slag is an important by-product of the melting process, playing an important role in embedding impurities and undesired substances present in the melt. Since the slag, mainly consisting of metal oxides, has a lower density than liquid steel, it can float on the liquid alloy surface. The chemical equilibrium between the slag and the liquid alloy is an important parameter, influencing the partition between liquid and slag of several elements (Turkdogan 2010). After the complete melting and homogenization, the

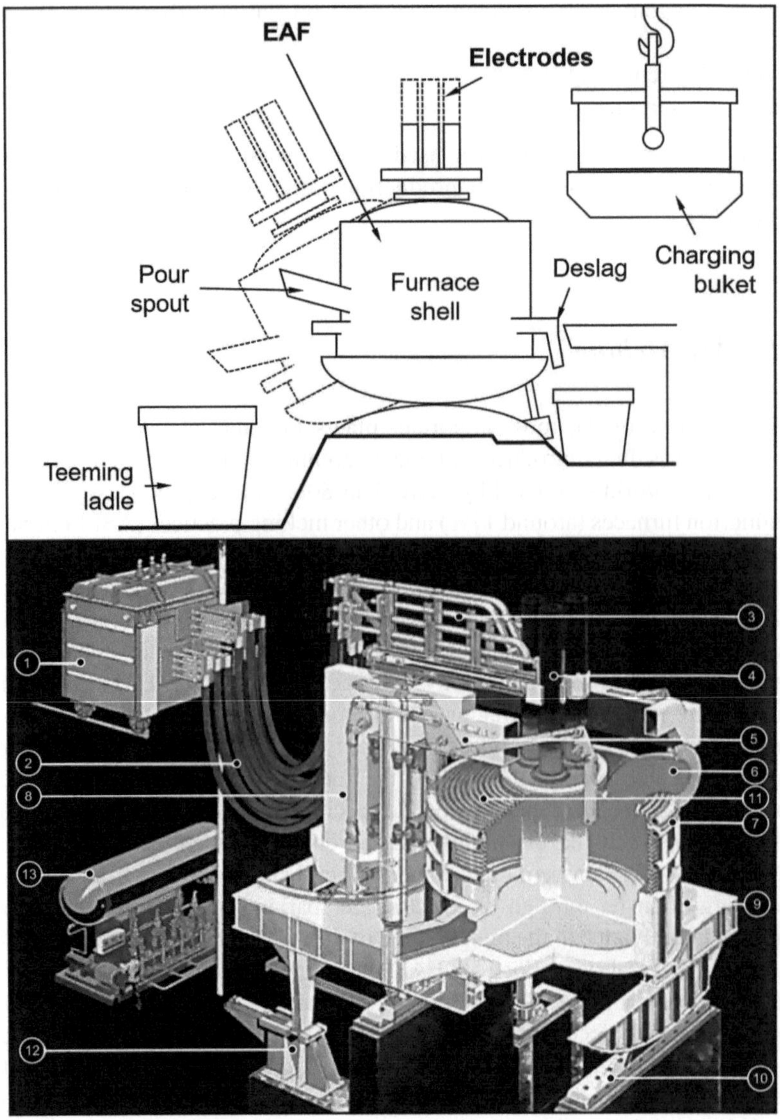

Fig. 5.3 Schematic of an electric arc furnace with traditional architecture and detailed components list. (Redrawn from ASM International 2008; Donachie and Donachie 2002): (*1*) transformer; (*2*) flexible cable connection; (*3*) electrode arms; (*4*) electrodes and electrodes clamping system; (*5*) arms; (*6*) cooled off-gas duct; (*7*) cooled panes (furnace shell); (*8*) structure; (*9*) basculating structure; (*10*) rack; (*11*) cooled roof; (*12*) basculating device; (*13*) hydraulic group (Madias 2013, reproduced with permission of the Publisher)

liquid steel is poured into a ladle and sent to the next production step, e.g., a conversion process. The slag is removed, either completely or partially, thus allowing its reformation and promoting further reactions to occur during the conversion process (see Sects. 5.3.3 and 5.3.4).

5.3.2 Vacuum Induction Melting (VIM)

The development of the vacuum induction melting (VIM) process started around the mid-nineteenth century, and its industrial use started in the early 1950s. The driving force for the development of VIM was the need of producing materials with a particular compositional cleanliness and controlled mechanical properties. One example is the production of steels for rolling bearings. It has been observed that the improvement in the steelmaking process, using VIM, results in significantly better performances of the material, particularly as concerns contact fatigue life (Meetham 1981). Compared to traditional air-melting processes (e.g., EAF plus AOD, see below), VIM ensures a considerable reduction in both nitrogen and oxygen content in the final products. Also other impurity elements, like Sb and Pb, can be eliminated through their evaporation during melting, favored by the low-pressure environment. As compared to EAF + AOD, VIM affords a better control of the alloying element content and thereby improved stress rupture and fatigue properties.

Nonetheless, some drawbacks are there in vacuum induction melting too:

- The composition of the molten charge is essentially determined and fixed by the choice of the charge materials, and no compositional changes, such as C, Si, S reductions, can be implemented.
- The process is definitely more expensive as compared to EAF + AOD.
- The compositional control of elements with high vapor pressure, like manganese, might be critical.

Figure 5.4 shows the general schematic of a VIM furnace. It consists of a melting vessel inside a steel shell kept under vacuum using a high pumping rate vacuum system. The heart of the furnace is the crucible, and the induction heating is obtained by a magnetic field that induces a current into the metal charge (ASM International 2008). The induction coils, wrapped around the melting vessel, are made of water cooled copper tubing.

One important process parameter is the induction coil frequency, to be varied according to the composition and the mass of the alloy charge. For aerospace applications, VIM is often a first melting step, followed by further remelting stages, such as ESR and VAR, as illustrated by the scheme in Fig. 5.5.

5.3.3 Argon Oxygen Decarburization (AOD)

Argon oxygen decarburization (AOD) is a refining process that was originally developed for the production of stainless steel and successively extended to other metal alloys:

- Carbon steels.
- Superalloys (see Chap. 6).

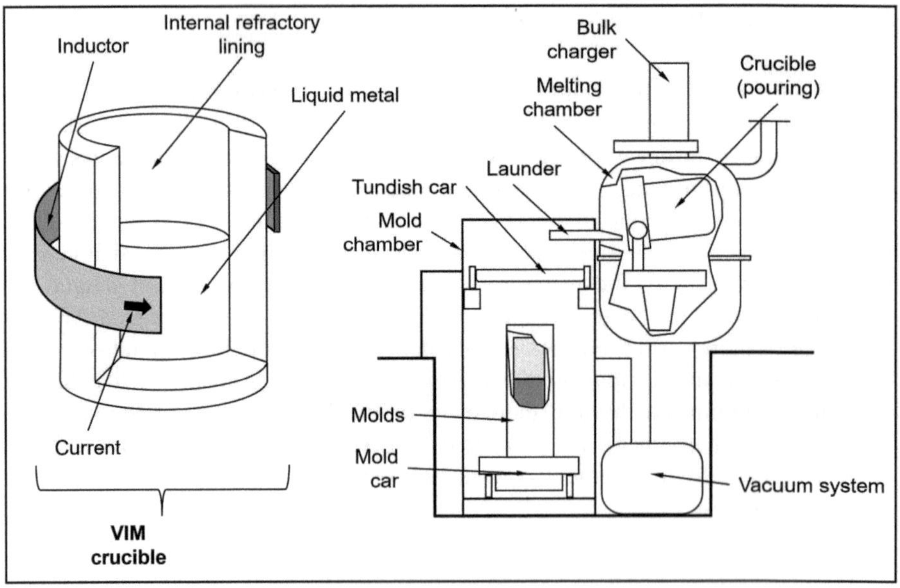

Fig. 5.4 Schematic of a VIM crucible and of a top-opening, double-chamber vacuum induction melting furnace. (Redrawn from ASM International 2008)

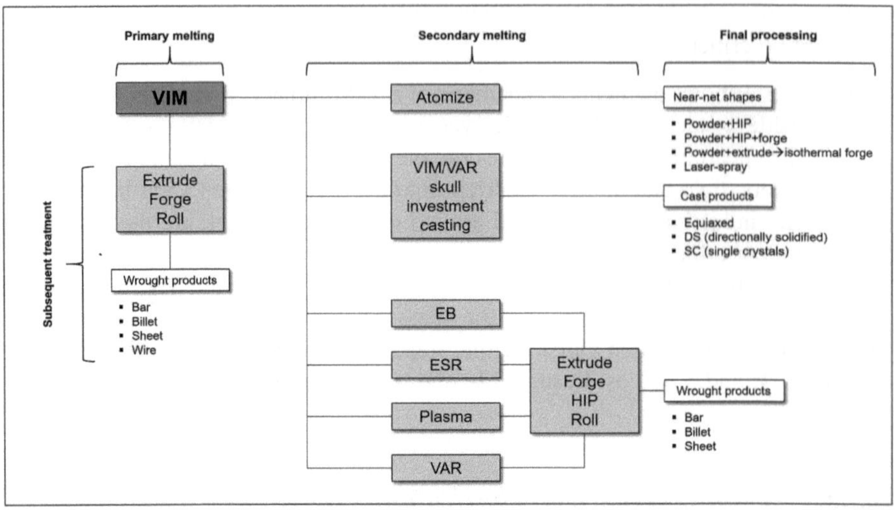

Fig. 5.5 Possible processing routes used for VIM casting products (ASM International 2008)

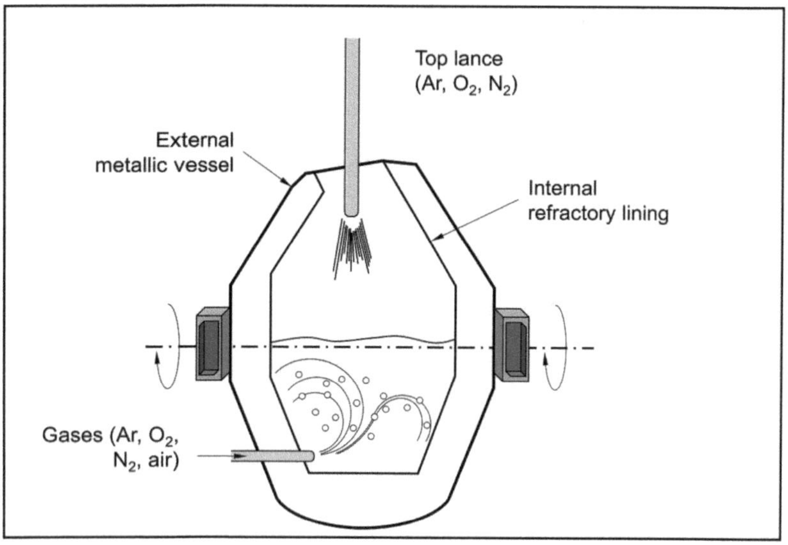

Fig. 5.6 AOD converter with the oxygen top blowing (OTB) system. (Redrawn from The AISE Foundation 1998)

- Refractory alloys (see Sect. 9.2).
- HSLA (high-strength low-alloy) steels.
- Low-alloy steels.
- Tool steels.

The process is conducted using the AOD converter, i.e., a processing vessel comprising a refractory-lined steel shell mounted on a tiltable trunnion ring. Concerning the refractory material, most AOD converters use dolomite refractories (ASM International 2008, Fig. 5.6).

The AOD converter looks very much as a Bessemer converter and has tuyeres, located in the lower sidewalls, which are used for the injection of argon, nitrogen, and oxygen. The most recent AOD converters use a combined blowing system, called OTB (oxygen top blowing), in which gases are injected into the molten bath also through a top lance (see Fig. 5.6). The inert component of the gas flux is used for stirring the molten mass with a consequent better chemical and thermal homogeneity of the liquid. In the AOD process, the heat transferred to the liquid alloy is supplied by exothermic oxidation reactions taking place into the liquid alloy and involving the injected oxygen, so that no external heating source is required. The temperature is controlled by adjusting the amount of oxygen, alloying elements, and solid steel scraps introduced into the converter during the refinement. The reactions occurring during the refining process involve a preliminary reduction of the oxides, formed in the melt, and subsequent release of carbon monoxide. Two decarburization reactions, involving chromium, are:

$$\frac{1}{4}\mathrm{Cr_3O_4} + \mathrm{C} \rightarrow \frac{3}{4}\mathrm{Cr} + \mathrm{CO(g)} \tag{5.1}$$

and

$$\mathrm{Cr_2O_3} + 3\mathrm{C} \rightarrow 3\mathrm{CO(g)} + 2\mathrm{Cr} \tag{5.2}$$

The relevant equilibrium constants, K, can be expressed as:

$$K = \frac{(a_{\mathrm{Cr}})^{\frac{3}{4}} \cdot (P_{\mathrm{CO}})}{(a_{\mathrm{Cr_3O_4}}) \cdot (a_{\mathrm{C}})} \text{ and } K = \frac{(a_{\mathrm{Cr}}) \cdot (P_{\mathrm{CO}})^3}{(a_{\mathrm{Cr_2O_3}}) \cdot (a_{\mathrm{C}})^3} \tag{5.3}$$

where:

a_{Cr} is the activity of chromium.
P_{CO} is the partial pressure of the carbon monoxide.
$a_{\mathrm{Cr_3O_4}}$ and $a_{\mathrm{Cr_2O_3}}$ are the activities of chromium oxides.
a_{C} is the activity of carbon.

Equation 5.3 suggests that a reduction in the partial pressure of CO leads to an increase in the equilibrium concentration of chromium with carbon in the molten steel (Fig. 5.7).

Similar reactions may involve other oxygen reactive elements, which are indeed considered in the refinement of steels. The carbon content at meltdown can typically vary between 0.5% and 3.0%, depending on the composition of the furnace charge (ASM International 2008). The AOD refining process implies a stepwise injection of oxygen and inert gases, whose ratio is varied during the process (see inset in Fig. 5.8). The schematic of a typical stainless steel refining cycle for a 30 t vessel charge is shown by Fig. 5.8.

According to these curves, as the oxygen to inert gas, Ar or N, ratio increases, the carbon content decreases, owing to its progressive oxidation. Other processing parameters for the AOD decarburization refinement of an AISI 304 stainless steel are given in Tables 5.2 and 5.3.

During the AOD refinement, the chemical composition of the bath is checked at regular intervals and when the target composition is reached, the alloy is either cast, e.g., ingot or continuous casting products, or poured into a ladle furnace for further refining treatments. The charge composition control during AOD process is extremely accurate and involves weight control of the heat and sampling of the liquid steel.

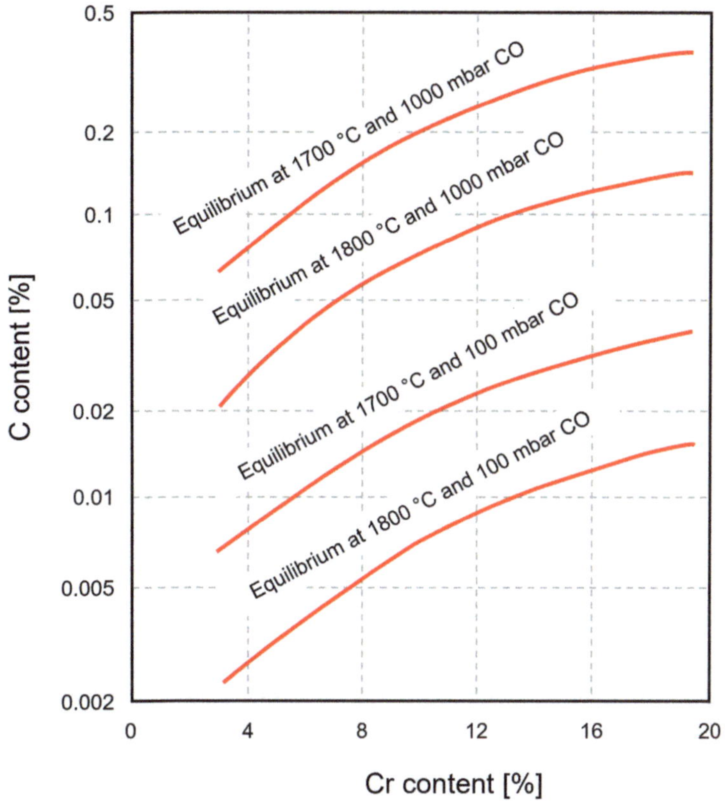

Fig. 5.7 C-Cr equilibrium curves in the AOD converter. The data show that a reduction in the partial pressure of CO corresponds to lower carbon levels at higher chromium concentrations, also at lower temperatures. (Redrawn from ASM International 2008)

The dilution of oxygen with inert gases results in an increased carbon removal efficiency, resulting in carbon concentrations below 0.01%. At the same time, the desulfurization process can reduce the sulfur content down to 0.01% or less (ASM International 2008).

5.3.4 Vacuum Oxygen Decarburization (VOD and VODC)

Vacuum oxygen decarburization is another refining process conducted under vacuum in a converter vessel (*VODC* vacuum oxygen decarburization converter) or in a ladle (VOD). The conversion process is not that different from the AOD (-OTB) process, although operating under a vacuum environment requires a thorough control of the bottom blowing of the reaction gases (Fig. 5.9).

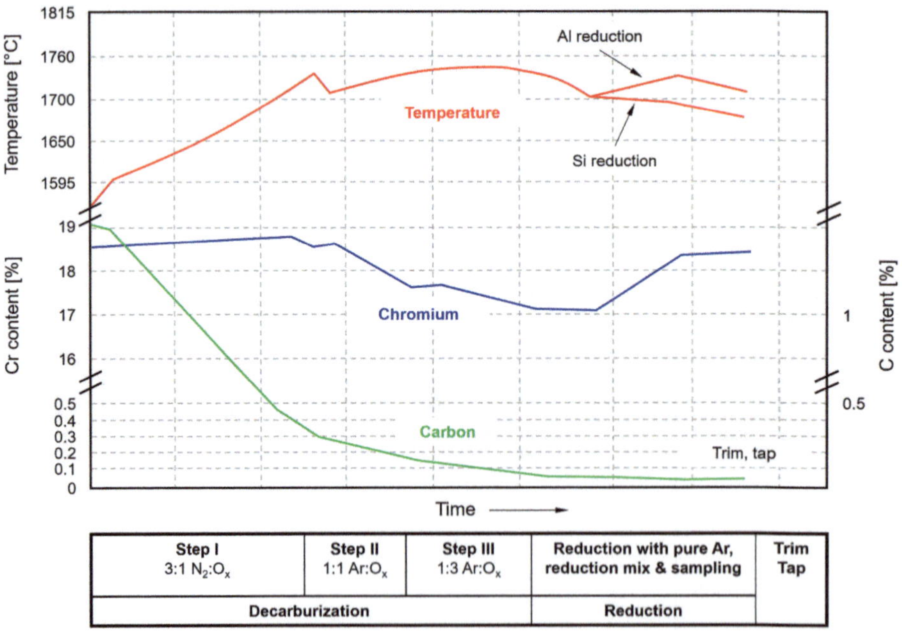

Fig. 5.8 Schematic of a stainless steel refining for a small AOD converter (30 t charge) showing the relationship between C and Cr contents as a function of time and temperature. (Redrawn from ASM International 2008)

Table 5.2 Processing parameters for AOD refinement of an AISI 304 austenitic stainless steel (ASM International 2008)

	Gas ratios			Partial pressure [atm]					
Carbon level	O_2	N_2	Ar	Ar	N_2	CO	CRE %[a]	CRR [ppm/min][b]	Temperature [°C]
3.0–0.7	4	1	–	–	0.14	0.86	80	1200	1510
1–0.25	3	1	–	–	0.20	0.80	65	830	1700
0.25–0.12	1	1	–	–	0.53	0.47	45	430	1700
0.12–0.04	1	–	3	0.83	–	0.17	30	133	1730
0.04–0.01	1	–	8	0.96	–	0.04	15	26	1730

Total gas flow rate = about 1.5 m^3/min/ton
[a]*CRE* carbon removal efficiency
[b]*CRR* carbon removal rate

VODC ensures low chromium losses during refining and higher carbon removal rates thanks to the very low CO partial pressures (Fig. 5.10). Sulfur levels can be kept in the 10–30 ppm range for any steel grade, and the resulting hydrogen content is generally below 2 ppm. A common feature to all the converter treated products, including VODC, is gas inclusions, which remain trapped into the liquid during

Table 5.3 Composition control of an AOD-refined high-strength low-alloy steel (ASM International 2008)

Element	Target [%]	Mean [%]	Standard deviation [%]
Aluminum	0.05	0.043	0.009
Carbon	0.27	0.264	0.006
Chromium	0.50	0.56	0.07
Manganese	0.80	0.83	0.03
Molybdenum	0.28	0.28	0.01
Nickel	–	0.13	0.03
Phosphorous	–	0.019	0.002
Silicon	0.50	0.49	0.04
Sulfur	–	0.002	0.001

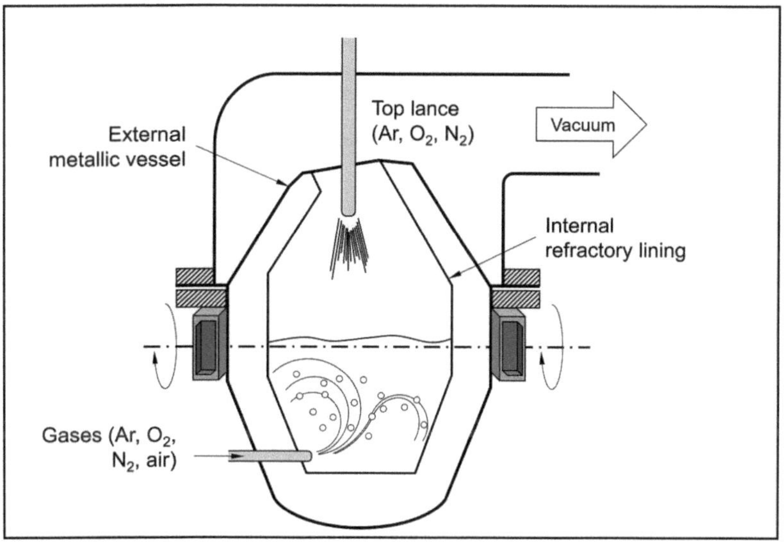

Fig. 5.9 Schematic of a VODC with the oxygen top blowing (OTB) system. (Redrawn from The AISE Foundation 1998)

tapping into the teeming ladle. To overcome or reduce the incidence of this phenomenon, if critical, a successive ladle treatment may be employed.

5.3.5 Electroslag Remelting (ESR)

Electroslag remelting (ESR) is a refining process (Fig. 5.11a), highly recommended if better alloy homogeneity, cleanliness, and control of the solidification structures are paramount. The early design of ESR dates back to the 1930s and successively patented in 1945. Nowadays, it is a fully established process, with a steel production

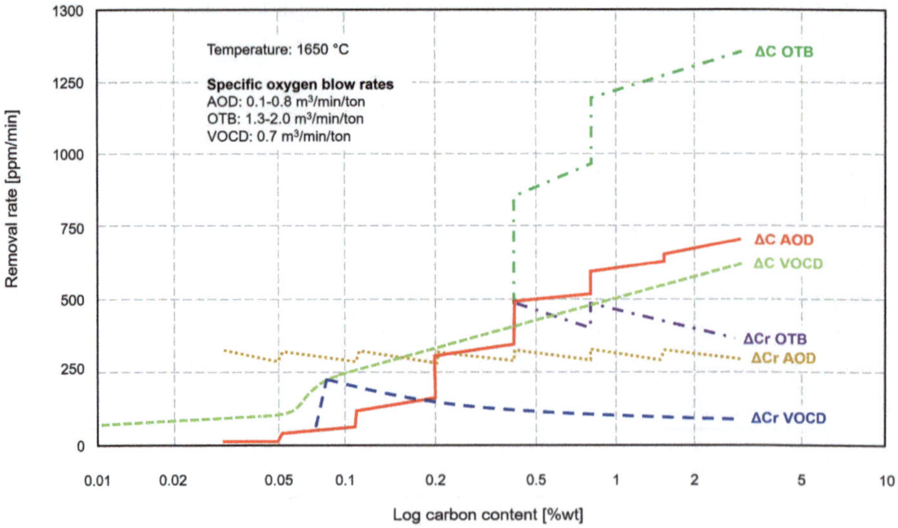

Fig. 5.10 Chromium oxidation rates and carbon removal rates for the VODC, OTB, and AOD processes. (Redrawn from ASM International 2008)

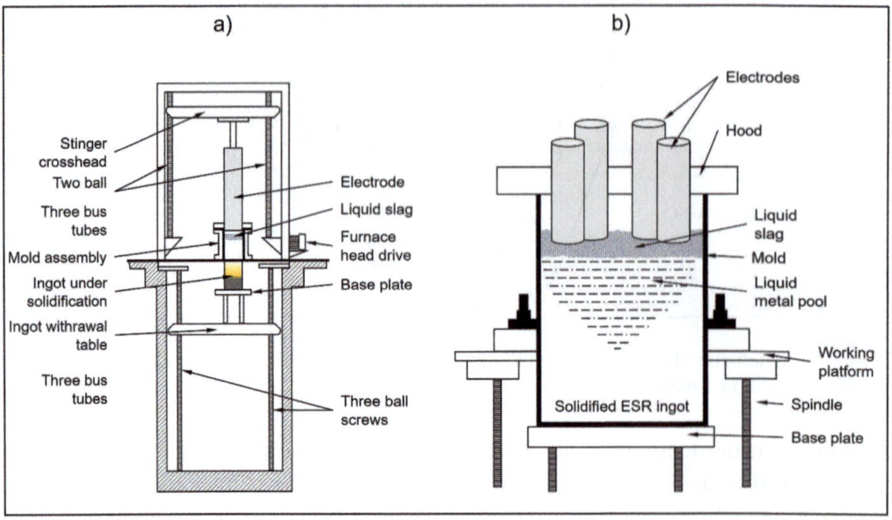

Fig. 5.11 (**a**) ESR furnace with retractable bottom baseplate. A fixed mold design is a possible alternative. (**b**) Schematic of an ESR furnace with multiple electrodes and detail of the ingot solidification. (Redrawn from ASM International 2008)

capacity of more than 1 million tons per year worldwide (ASM International 2008). The ESR process is not restricted to steels, and it is actually employed in the processing Ti-alloys (see Chap. 4) and Ni-based superalloys (see Chap. 6) as well. As concerns steels, ESR is particularly suited for HSLA steels, stainless steels, tool steels, bearing, and heat-resistant steels. ESR requires a consumable electrode, which is melted by immersion into the molten slag. Melting temperature is reached thanks to an electric current flowing from the base plate through the slag and the electrode (Fig. 5.11b). ESR process is used for the production of cylindrical ingots, with diameters of 50–75 cm. A variant of the process, named ESRR (electroslag rapid remelting), is used for the production of smaller ingots with slab, square, or round cross section that can be directly processed, using conventional rolling mills and presses.

Possible alternatives to the ESR process are VAC-ESR and pressurized ESR. In the first, remelting is conducted under a low gas pressure, possibly slightly reducing, with additions of nitrogen and hydrogen. The conventional ESR instead is performed in air, with a worse control of the impurity level in the final products. In pressurized ESR, a gas, e.g., nitrogen, overpressure promotes its own diffusion into the alloy and modifies the resulting mechanical properties (see Appendix 4). Most ESR plants have AC furnace, using multiple electrodes for the production of single ingots (Fig. 5.11b). At the beginning of the remelting process, the slag induces the melting of the tip of the electrode, and metal droplets falling through the liquid slag are collected into a water-cooled copper mold. The solidification rate is constant since the mold is subjected to a constant heat extraction through the cooling water circuit. The metal undergoes a refinement process, and it is purified from unwanted substances, like oxide particles and other kinds of inclusions that are trapped into the slag. The main parameters that control the metal/slag reaction are:

- Superheating of both slag and metal that promotes metal/slag reactions and consequent cleaning up of the liquid metal.
- Slag chemical composition, which can be tuned according to specific applications. General purpose slag has composition: 70% CaF_2, 20% CaO, and 10% Al_2O_3. A slag made of 30% CaO and 70% CaF_2 has a high thermal conductivity, recommended when alumina is not allowed and there is a risk of hydrogen pickup (ASM International 2008).
- Metal-slag interaction. Metal droplets fall into the mold passing through the liquid slag. The large specific surface area of the metal/slag interface enhances the kinetics of the reactions between the liquid alloy and the slag.

The inclusions that remain trapped into the solidifying ingot are usually very fine grained, and their distribution is rather homogeneous, considering the kinetics of the remelting process, featuring a comparatively limited thickness of the liquid pool. Concerning the slag chemical composition, it can be adjusted in order to improve the dephosphorization and desulfurization processes. The final desulfurization that can be attained for a given a slag composition is usually evaluated by means of the following reactions:

$$\mathrm{FeS} + \mathrm{CaO} \rightarrow \mathrm{CaS} + \mathrm{FeO} \tag{5.4}$$

$$CaS + \frac{3}{2}O_2(g) \rightarrow CaO + SO_2(g) \tag{5.5}$$

These two reactions describe the transfer of sulfur from the metal to the slag and from the slag to the process gas, collected and removed from the working atmosphere. In this way, the desulfurization capacity of the slag remains substantially unaltered with time. When the index of basicity (also known as IB) of the slag, evaluated as the CaO/SiO_2 ratio, is larger than 3, more than 80% of sulfur can be removed from the slag during remelting. Reactions between the liquid alloy and oxygen cannot be avoided when the process is carried out in air. Therefore, oxidation of the alloy occurs to some extent, either by a direct reaction of the electrode surface above the liquid slag or through the oxidation of elements, such as iron and manganese, at the slag surface. These oxides are subsequently reduced by alloy elements that have high affinity to oxygen, e.g., aluminum and silicon, resulting in the unwanted depletion of these elements in the ingot. In order to limit this drawback, a continuous deoxidation of the slag can be carried out by means of aluminum additions that can even lead to an oxygen content below 20 ppm in unalloyed steels (ASM International 2008). ESR ingots are characterized by a very clean and smooth surface: this helps reducing the production costs, since there is no need for grinding prior to hot-working. The homogeneity of ESR ingots, as concerns macro-segregation, and the uniform distribution of inclusions result in improved mechanical properties, with particular regard to yield strength and toughness. Uniformity in transverse and longitudinal direction after hot-working is another important technological advantage. Concerning process parameters, the control of the molten pool depth and shape has direct influence on the solidification structures, which are dependent on the temperature gradient at the solid/liquid interface. A relatively high-temperature gradient at the solidification front leads to the development of directional dendritic structures. Micro-segregation in ESR ingots increases with increasing dendrite arm spacing (DAS). Minimum micro-segregation can be achieved through a dendrite growth direction as closer as possible to the ingot solidification axis. This latter condition cannot be achieved in all cases, particularly in larger ingots. ESR ingots are prone to the formation of specific defects, like white spots, freckles, and tree-ring pattern, all described in the next section (Sect. 5.3.6), since the same kind of defects can be found in VAR ingots.

5.3.6 Vacuum Arc Remelting (VAR)

Vacuum arc remelting (VAR) has been specifically developed for the production of superalloys (see Chap. 6), and it was first used in the late 1950s for aerospace materials. The use of the VAR has been thereafter extended to other alloys (Mitchell 2005):

- Steels.
- Titanium alloys (see Chap. 4).
- Refractory alloys (see Sect. 9.2).

VAR can be used for refinement and purity improvement of air-melted or VIM ingots, and it is also a fundamental step in the superalloy-specific *triplex process*, i.e., VIM-ESR-VAR (see Sect. 6.3.2) (ASM International 2008; Cobb 2010). The refined alloys, produced by vacuum arc remelting, are characterized by improved mechanical properties, in particular fracture toughness and fatigue resistance. The main reasons for this improvement can be ascribed to the following factors or to a combination of them:

- Elimination of porosity.
- No macro-segregations, thanks to micro-fusion and *layer-by-layer* solidification of the ingot.
- Reduction of impurity content, thanks to their preferential evaporation under vacuum, as observed for lead and tin. Unfortunately, other alloying elements, like manganese, with high vapor pressure, intentionally added to the master batch for their beneficial effects, may undergo the same process so that their concentration is unwillingly reduced, thus requiring an adequate compensation in setting the composition of the initial batch.
- Reduction of micro-segregations.
- Removal of dissolved gaseous elements, e.g., N_2, O_2, CO, and H_2, that are easily outgassed from the melt.
- Reduction of incidence of oxides and consequent alloy purity improvement.

Oxides, and also nitrides, are eliminated from the melt by thermal dissociation or for the reducing effect of carbon and subsequent elimination via gas phase (ASM International 2008). Very stable oxides, like Al_2O_3, and titanium carbonitrides are removed by floatation, favored by their lower density (ASM International 2008).

The overall design of a VAR furnace has not undergone particular changes over the years (scheme in Fig. 5.12). The mold is placed into a water-cooled vessel, similarly to ESRs. In Fig. 5.12 it is possible to see that the electrode is driven into the crucible by a ram, while the electrode itself is being consumed, and its position is continuously controlled in order to maintain the optimal arc gap, i.e., the distance between the electrode tip and the molten bath. Reproducibility of the process and homogeneity of the produced ingots strongly depend on a close control of the process parameters, which has been considerably improved in fully automatized modern plants.

VAR ingots exhibit a microstructure influenced by both the temperature gradient of the liquid-solid interface and the local solidification rate. A relatively high-temperature gradient at the solidification front is a necessary requirement to obtain an oriented dendritic primary structure, whose growth direction, always perpendicular to the solidification front, is parallel to the direction of the temperature gradient.

Vacuum pressure typically ranges in between 0.1 and 1.0 Pa, although remelting can also be carried out under an inert gas pressure that can be as high as 10 kPa, in order to reduce the evaporation losses of the more volatile alloying elements. On the other hand, an increase in the working pressure corresponds to a decrease in the

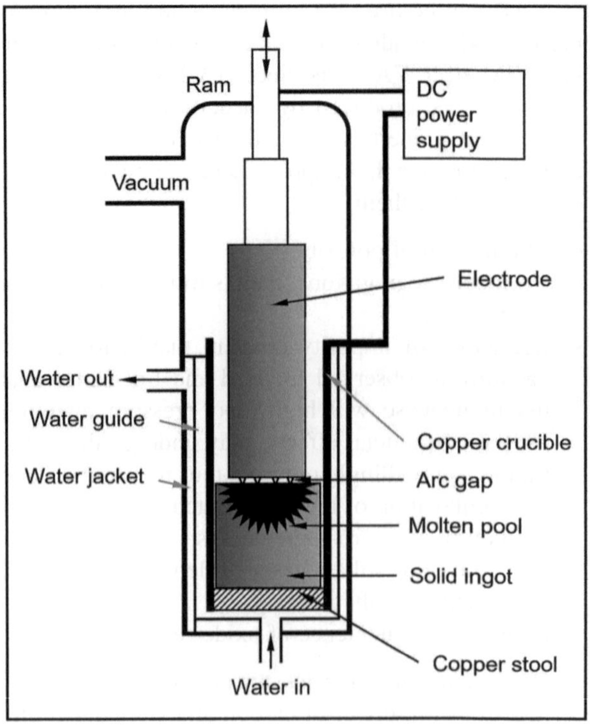

Fig. 5.12 Schematic of a VAR furnace with its key components. (Redrawn from ASM International 2008)

melting rate, due to the thermal losses induced by the residual gas pressure (Boxman et al. 1996). The peak direct current can be as high as 40 kA, remelting rate can reach 19 kg/min, and a typical ingot's diameter can be up to 1.5 m.

Figure 5.13 shows the scheme of the molten metal pool-electrode region of a VAR ingot during remelting. The electrical current flowing through the electrode determines the depth of molten pool, mainly influenced by the thermal buoyancy and Lorentz stirring. An excessive depth of the liquid inhibits the migration of the impurity oxides toward the outer surface and favors instead their segregation along the centerline of the ingot. ESR and VAR ingots share similar defect structures, which may importantly affect the quality and properties of the relevant products. The main kinds of defects are:

- Freckles. They are regions of the alloy containing an excess of carbide-forming elements or carbides, resulting from segregations within the liquid pool, induced by its excessive depth. The situation may be worsened by turbulences induced by external causes, such as stray magnetic fields.
- "Tree rings." They are series of comparatively small equiaxed grains that form within the elongated dendritic microstructure for a locally higher grain nucleation rate, due to thermal and compositional fluctuations inside the melt (Xu et al. 2002).

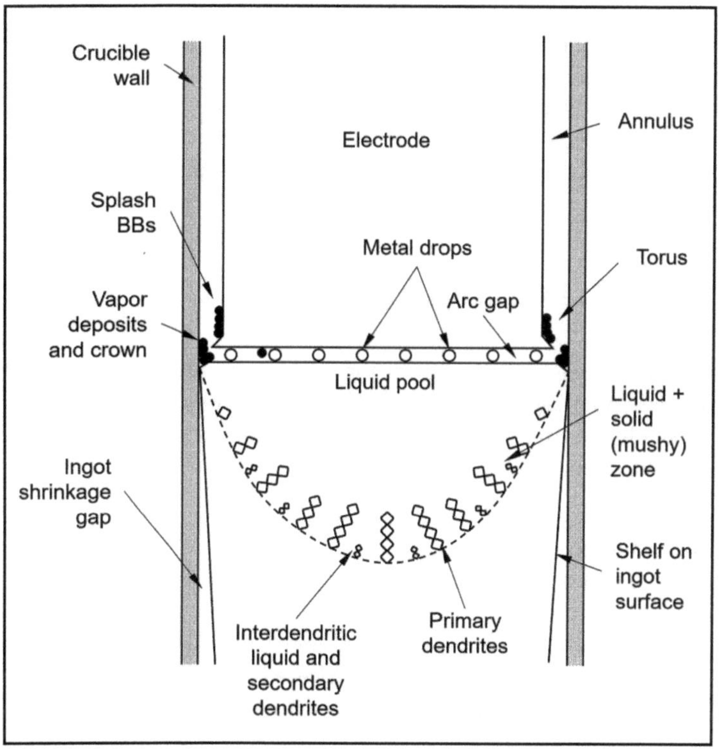

Fig. 5.13 Schematic of the liquid pool region of a VAR ingot during remelting process. (Redrawn from ASM International 2008)

- White spots. They can be easily recognized on the external surface of the ingot, as white-etched spots, featuring compositional fluctuations, in particular a depletion of alloy elements, originating from unmolten dendrites that fall into the metal pool directly from the electrode, thus retaining the original electrode composition. Solid pieces, contributing to the formation of white spots, may detach from other parts of the VAR too, like the *crown* (see below) that forms on the inner wall of the mold, along the ingot-electrode gap (see Fig. 5.13).

Freckles and white spots act as stress concentrators and tend to reduce the fatigue life of structural components. In parts, like compressor discs of the final stages and turbine superalloy disc (see Table 2.1, Sect. 5.5.3, and Chap. 6), these defects have been identified as one of the main causes of premature failure. To reduce the incidence of these defects, in particular of the white spots, a careful control of the following remelting conditions and parameters is required:

- High melting rates, capable to rapidly melt any solid particle that may fall into the molten metal pool.

- High-quality electrodes, with no chemical and microstructural heterogeneities, like cracks and solidification shrinkage voids, from which a preferential particle detachment may more easily occur.
- Reduction of the electrode-ingot gap, so to limit the length of the arc and the relevant tendency to form the so-called *crown*, which is a possible source of solid fragments contributing to the formation of *white spots*.

5.4 Steels for Aerospace Applications

Selected kinds of steels, interesting for the aerospace industry, are described herewith, also mentioning criteria for their selection. Together with the following main families of steels, i.e., maraging, AerMet, and stainless steels, others are used in aerospace applications, like:

- Advanced carburizing steels (e.g., Pyrowear 53).
- Carburizing/nitriding steels (e.g., X30).
- Conventional carburizing steels (e.g., AMS 6265).
- Conventional nitriding steels, such as Nitralloy N.
- High fracture toughness steels such as AF1410 and HP 9-4-30.
- Medium-carbon low-alloy steels (e.g., AISI 4130, AISI 4140, AISI 4340, and 300M) (Campbell 2006; Flower 1995).

5.4.1 Maraging Steels

The total use of steel for structural components in aircrafts has progressively decreased, due to the development and availability of aluminum (Chap. 3) and titanium (Chap. 4) alloys. However, there are still components requiring high tensile strength, wear resistance, and stiffness that render steels still the best-candidates, particularly when ultrahigh tensile strengths are required. Since high strength is often accompanied by a reduced ductility (Megson 2013), the development of a new class of steels began at the end of the 1950s at the INCO research laboratories, to overcome this limitation (Würzinger et al. 2004). This class of steels is named *maraging*, i.e., martensite aging or martensite age hardening, from the main strengthening mechanism. It involves the age hardening of the low-C, Fe-Ni lath martensite structures present in the steel after quenching. The main phase transformations occurring in these alloys are illustrated in the phase diagrams in Fig. 5.14. Maraging steels are characterized by chemical compositions rich in nickel, cobalt, and molybdenum and relatively poor in carbon, as shown by Table 5.4. In fact, intermetallic compounds precipitate during the age-hardening treatments and do not involve any solid-state reaction with carbon (ASM International 1993). Except 18Ni (180), all maraging steels are based on the 18Ni-Co-Mo type, and the available

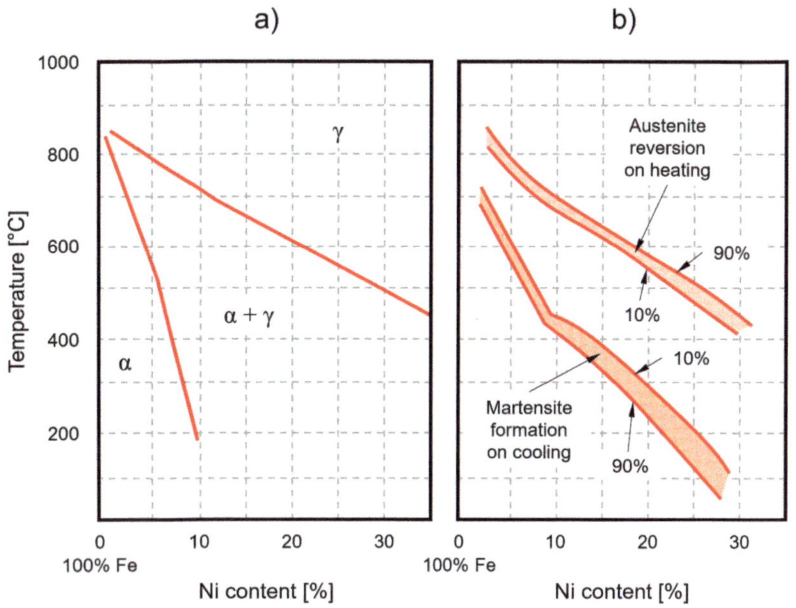

Fig. 5.14 (**a**) Equilibrium phase relationships in the Fe-Ni system; (**b**) metastable phase relationships in the Fe-Ni system. (Redrawn from ASM International 1993)

Table 5.4 Nominal composition (wt%) of maraging steels. Fe concentration balance (ASM International 1993; Novotny and Maurer 2007)

Grade	C	Ni	Mo	Co	Ti	Al	Nb
18Ni(200)	–	18.0	3.3	8.5	0.2	0.1	–
18Ni(250)	0.03 max	18.0	5.0	8.5	0.4	0.1	–
18(Ni300)	0.03 max	18.0	5.0	9.0	0.7	0.1	–
18(Ni350)	0.05 max	18.0	4.2	12.2	1.6	0.10	–
18Ni(Cast)	–	17.0	4.6	10.0	0.3	0.1	–
Co-free 18Ni(200)	–	18.5	3.0	–	0.7	0.1	–
Co-free 18Ni(250)	0.03 max	18.5	3.0	–	1.4	0.1	–
Co-free 18Ni(300)	0.03 max	18.5	4.0	–	1.85	0.1	–
Low-Co 18Ni(250)	0.03 max	18.5	2.6	2.0	1.2	0.1	0.1
12-5-3(180)	–	12.0	3.0	–	0.2	0.3	–

commercial grades are 18Ni(200), 18Ni(250), 18Ni(300), 18Ni(350), 18Ni(cast), and 12-5-3(180) (Prasad and Wanhill 2017). The shortage of cobalt during the late 1970s and 1980s, with the consequent cobalt price fluctuations, led to the development of several low-Co grades and Co-free grades of maraging steels: Co-free 18Ni (200), Co-free 18Ni(250), low-Co 18Ni(250), and Co-free 18Ni(300). Designation of maraging steels refers to their nominal yield strength, which is the value (in ksi) among brackets. For example, the standard 18Ni(350) steel has a yield strength of 350 ksi (2420 MPa) (ASM International 1993).

Table 5.5 Effect of the most important alloying elements in maraging steels (ASM International 1993; Prasad and Wanhill 2017)

Name	Symbol	Features and alloying effects
Aluminum	Al	It is usually kept below 0.2–0.3% in order to avoid a decrease in ductility
Chromium	Cr	Chromium increases the strain hardening coefficient of martensite as well as the corrosion resistance
Cobalt	Co	Cobalt rises the martensite start temperature (M_s) so that it is possible to increase the concentration of hardening elements without increasing the retained austenite content. Furthermore, its presence leads to an increase in the volume fraction of Ni_3Mo and Ni_3Ti precipitates
Molybdenum	Mo	Molybdenum, in addition to forming the reinforcing Ni_3Mo phase, promotes the plasticity and ductility in the aged condition. This is so, since molybdenum is able to reduce the diffusion coefficient of several elements, particularly at the grain boundary, with a consequent reduction in precipitation of second phases during age hardening
Nickel	Ni	Nickel can be regarded as the most important alloying element in maraging steels. Its presence leads to the formation of a lath martensitic structure
Titanium	Ti	The main effect of titanium is to harden the material after heat treatment. If enough carbon is present in the alloy, it can react with titanium leading to the formation of the deleterious, since embrittling TiC

In Table 5.5 the main effects of the most important alloying elements present in maraging steels are indicated. Although not included in the table, other alloying elements too, like Cu, Nb, Si, Ta, and V, may contribute to the steel hardening (Sha and Guo 2009).

Figure 5.14a shows the Fe-rich region of the Ni-Fe equilibrium phase diagram and indicates that the equilibrium phases at low temperature for high nickel content are austenite (γ) and ferrite (α). Figure 5.14b shows the metastable phase diagram, including the austenite-to-martensite transformation upon cooling and the reverse transformation on heating. From the diagram in Fig. 5.14b, it turns out that there are no phase transformations occurring until the martensite start temperature is reached (i.e., martensite formation on cooling region in Fig. 5.14b). One interesting feature of maraging steels is the possibility of obtaining a fully martensitic structure even with a very slow cooling rate and also when treating large sections. In other words, maraging steels would exhibit some hardenability anyway. M_s can be tuned through a suitable choice of the alloy chemical composition that does not introduce any change in the character of the martensitic transformation, which remains independent from the cooling rate. Nickel and the other alloying elements, except cobalt, tend to lower M_s. Cobalt raises M_s, and it can be exploited to compensate for the addition of the other alloying elements in order to maintain a martensitic transformation still fully accomplished at room temperature. In fact, the M_s of maraging steels lays in the range of 200 °C–300 °C. The resulting martensite is relatively soft with hardness of 30–35 HRC, so that it can be machined and deformed quite easily (Campbell 2006). The martensitic transformation is not sufficient to reach the peak properties, which require instead an aging treatment, to induce precipitation strengthening mechanisms (see Chap. 6).

Table 5.6 Characteristics of intermetallic phases in maraging steels (Moshka et al. 2015; Tewari et al. 2000)

Intermetallic phase	Unit cell	Composition/stoichiometry	Lattice parameters [Å]
H	Hexagonal	$Ni_3(Ti, Mo)$	$a = 5.101$
			$c = 8.307$
Laves	Hexagonal	Fe_2Mo	$a = 4.755$
			$c = 7.767$
Ni_3Mo	Orthorhombic	Ni_3Mo	$a = 4.224$
			$b = 4.448$
			$c = 5.064$
S	Hexagonal	A_8B	$a = 7.04$
			$c = 2.48$
μ	Rhombohedral	Fe_7Mo	$a = 4.751$
			$c = 8.988$
			$(\gamma = 120°)$
X	Hexagonal	A_3B	$a = 2.55$
			$c = 8.30$

Several studies have been performed for the identification of the main reinforcing intermetallic phases in maraging steels; it has been observed that the precipitation process occurs preferentially at dislocations and within the martensite laths, leading to the formation of a fine and uniform distribution of coherent precipitates (see Sect. 6.4.1.1) (Moshka et al. 2015; Pardal et al. 2005; Tewari et al. 2000). Thanks to their nanometric size and coherent interface with the alloy matrix, the otherwise embrittling intermetallic precipitates ensure a beneficial strengthening effect (Leitner et al. 2011; Yang et al. 2015). Phases observed in maraging steels are given in Table 5.6. The alloying element providing the highest contribution to hardening is molybdenum, forming the metastable Ni_3Mo phase. Coherency strains tend to limit the growth of Ni_3Mo, and, therefore, further aging results in the transformation of Ni_3Mo into the equilibrium Fe_2Mo precipitate. Titanium promotes further hardening with the formation of Ni_3Ti, and another intermetallic precipitate, FeTi, may form in the nickel-depleted regions, after the formation of the Ni_3Mo and Ni_3Ti precipitates (ASM International 1993). During the hardening treatments, detrimental phases may also form, like elongated nonmetallic inclusions (NMIs), visible on the fracture surfaces (Fig. 5.15).

Energy dispersive X-ray (EDX) analysis has been used to identify the composition of these NMIs, of which two main types have been recognized: blocky titanium nitrides (Fig. 5.15) and titanium sulfides (Fig. 5.16) (Würzinger et al. 2004). Low carbon concentration (Table 5.4) is also meant to inhibit the formation of other NMIs phases, like TiC. These and similar phases may have deleterious consequences on the toughness, ductility, and strength of the alloy (ASM International 1993; Würzinger et al. 2004).

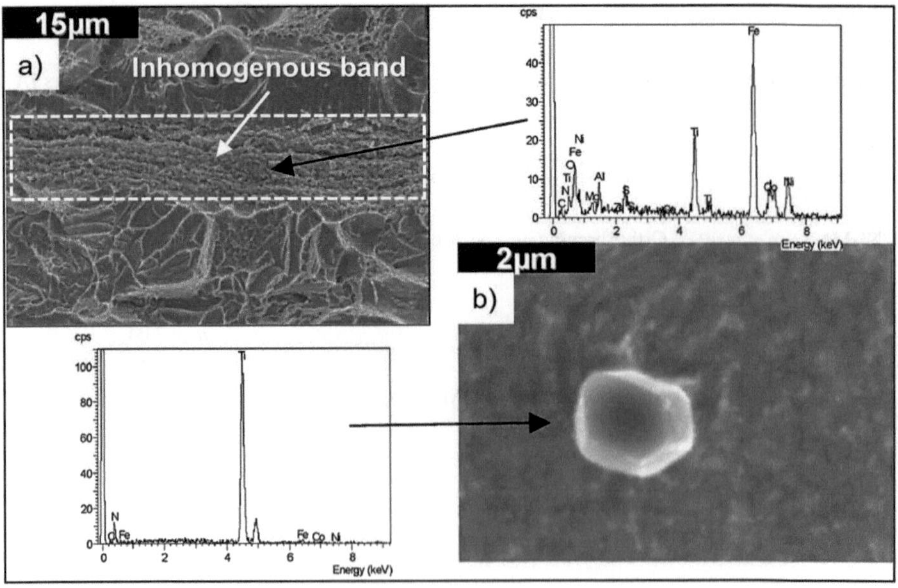

Fig. 5.15 (**a**) NMIs inhomogeneous band; (**b**) blocky titanium nitride. (Würzinger et al. 2004, reproduced with permission of the Publisher)

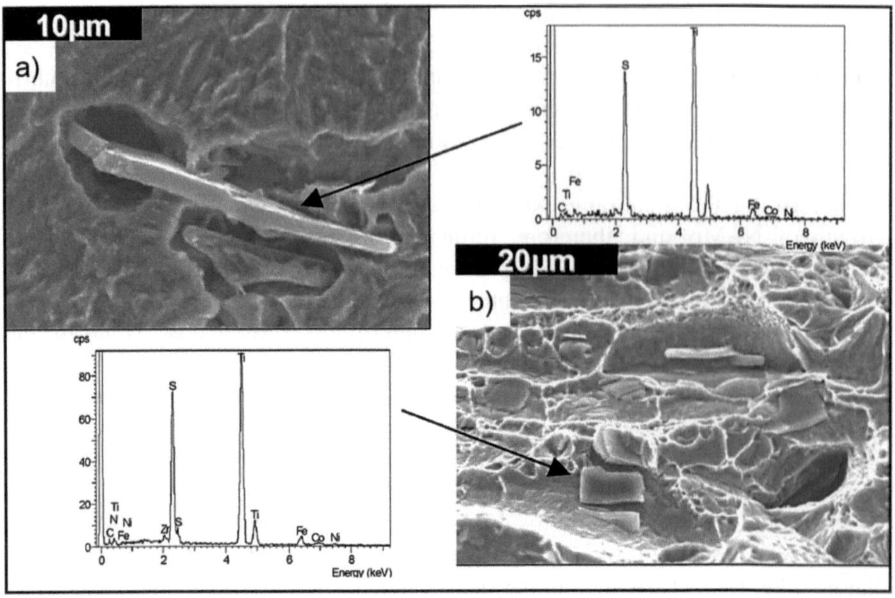

Fig. 5.16 SEM investigation on inclusions showing (**a**) and (**b**) to be titanium sulfides. (Würzinger et al. 2004, reproduced with permission of the Publisher)

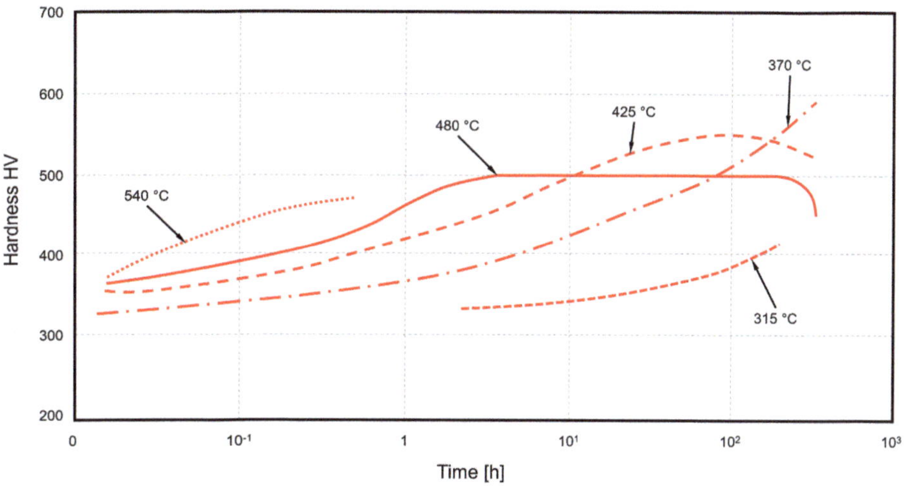

Fig. 5.17 Vickers hardness of 18Ni(250) maraging steel versus aging time for various aging temperatures. (Redrawn from ASM International 1993)

The melting process used for maraging steels depends on the target grade of the final product and relevant applications. There are two main options: either EAF or VIM, followed in both cases followed by VAR (ASM International 1993). EAF/VAR route is mainly used for the production of tool steel grades (Würzinger et al. 2004). Maraging steels that belong to the family of cast grades can be melted and cast in air, in view of their adequate fluidity and castability (Prasad and Wanhill 2017). In Ti- and Al-containing grades, the affinity of these elements for nitrogen and oxygen may result in an increase in the inclusion content (Würzinger et al. 2004).

The relatively high titanium and molybdenum concentrations render the steel more prone to micro-segregations during solidification. Therefore, premium grades, used for the more demanding applications, where homogeneity and high mechanical properties are paramount, are prepared using the so-called triple process, EAF melting, VIM, and final VAR, since in this way, it is possible to reach very low concentrations of gaseous, i.e., N_2, H_2, and O_2, and of other impurities, like Mn, C, P, and S (ASM International 1993).

As concerns the post-processing heat treatments of the maraging steels, before age hardening, the steel is usually solution annealed in the austenite field; 1 h is an average time for 25 mm section parts. Subsequently, the material is cooled down to room temperature before aging. Properties and microstructure are only slightly affected by the cooling rate after solution annealing. Age hardening requires heat treatment at temperatures in the 455 °C–530 °C range, lasting from 3h up to 12 h, depending on the grade and desired properties (Fig. 5.17). Specific applications, such as die casting, require an aging temperature of about 530 °C. If it is maintained for too a long time, coarsening of precipitates leads to a sharp softening and to the martensite to austenite reverse transformation. The refinement of coarse-grained microstructure can be obtained in maraging steels thanks to a thermal cycling

Table 5.7 Mechanical properties and heat treatments of standard maraging steel grades (ASM International 1993)

Grade	Heat treatment	Tensile strength [MPa]	Yield strength [MPa]	% Elongation in 50 mm	% Reduction in area	Fracture toughness [MPa√m]
18Ni (200)	A	1500	1400	10.0	60	155–240
18Ni (250)	A	1800	1700	8.0	55	120
18Ni (300)	A	2050	2000	7.0	40	80
18Ni (350)	B	2450	2400	6.0	25	35–50
18Ni (Cast)	C	1750	1650	8.0	35	105

Heat treatment A: solution treatment 1 h at 820 °C. Age hardening 3 h at 480 °C
Heat treatment B: solution treatment 1 h at 820 °C. Age hardening 12 h at 480 °C
Heat treatment C: annealing 1 h at 1150 °C. Age hardening 1 h at 595 °C. Solution treatment 1 h at 820. Age hardening 3 h at 480 °C

Table 5.8 Mechanical properties of three variants of the heat-treated 18Ni(250) steel (ASM International 1993)

Grade	Tensile strength [MPa]	Yield strength [MPa]	% Elongation in 25 mm	% Reduction in area	Charpy V-notch impact toughness [J][a]	Fracture toughness [MPa√m][a]
18Ni (250)	1870	1825	12.0	64.5	37	138
Co-free 18Ni (250)	1895	1825	11.5	58.5	34	127
Low-Co 18Ni (250)	1835	1780	11.0	63.5	43	149

Heat treatment: solution treatment 1 h at 815 °C. Age hardening 5 h at 480 °C
Samples: 63.5 × 88.9 mm^2 billets originated from Ø200 mm VIM/VAR billet
[a]Orientation of the test: L-S (longitudinal-short transverse orientation)

between M_f and a temperature which is significantly higher than the precipitate solutioning temperature. For instance, an ASTM grain size of −1/1 in 18Ni(300) has been refined by Saul et al. (1970) by means of three cycles between room temperature and 1025 °C. Prolonged aging treatments induce the formation of the equilibrium phases, i.e., austenite and ferrite (Saul et al. 1970; ASM International 1993).

The mechanical properties of various maraging steel grades are given in Table 5.7, with the indication of the relevant heat treatments.

The Table 5.8 shows the main mechanical properties of three variants of the 18Ni (250) maraging steel: standard composition, Co-free, and low-Co bearing grade (See Table 5.4).

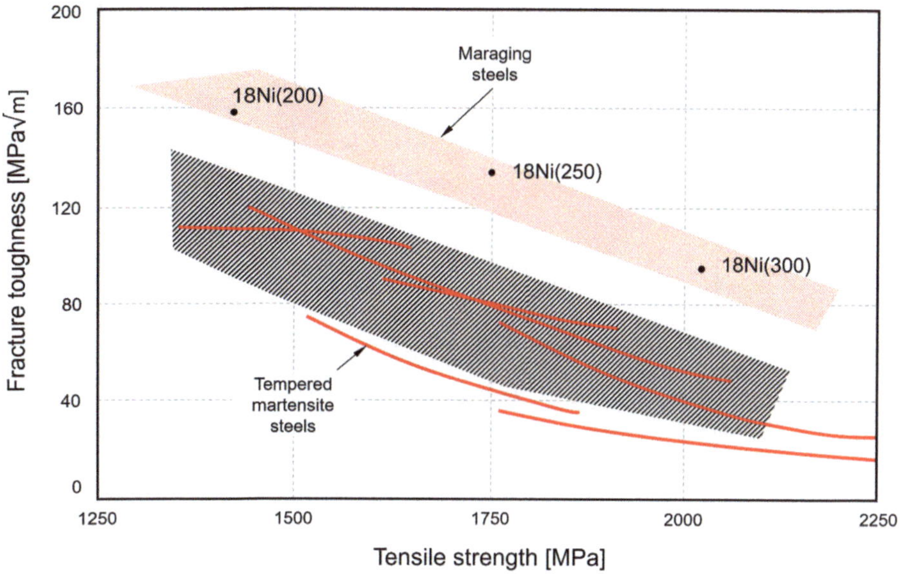

Fig. 5.18 Plane-strain fracture toughness of several steel materials: maraging steel family and ultrahigh strength steels as a function of tensile strength. (Redrawn from ASM International 1993)

A comparison of the plane-strain fracture toughness of maraging steels and conventional tempered martensite steels is given in Fig. 5.18 as a function of their tensile strength.

The evolution with temperature of tensile strength, yield strength, ductility, and hardness of several maraging steels is given in Fig. 5.19.

From the plots in Fig. 5.19, it turns out that the maraging steels can be used for long times at temperatures up to 400 °C, considering that the tensile and yield strengths at this temperature are 80% of the room temperature values approximately. Concerning fatigue properties, the behavior of maraging steels and other high-strength steels is shown by Fig. 5.20.

Fatigue crack growth rate is comparable to the rates observed in conventional steels. An overview of most important technological processes applied to maraging steels is given in Table 5.9.

The corrosion resistance of maraging steels is slightly better than that of tempered martensitic steels: in industrial and marine atmospheres, it is about one and a half of that observed in conventional steels. Maraging steels have essentially the same behavior of conventional steels when in contact with static or flowing seawater. Nickel and cobalt improve their high-temperature oxidation resistance, which is indeed better than that of conventional martensitic alloy steels (ASM International 1993). Maraging steels do not undergo passivation (see Sect. 8.1.6) and are subjected to stress corrosion cracking (SCC, see Sect. 8.1) in most aqueous environments: the higher the yield strength, the higher the susceptibility to SCC (Flint et al. 2010).

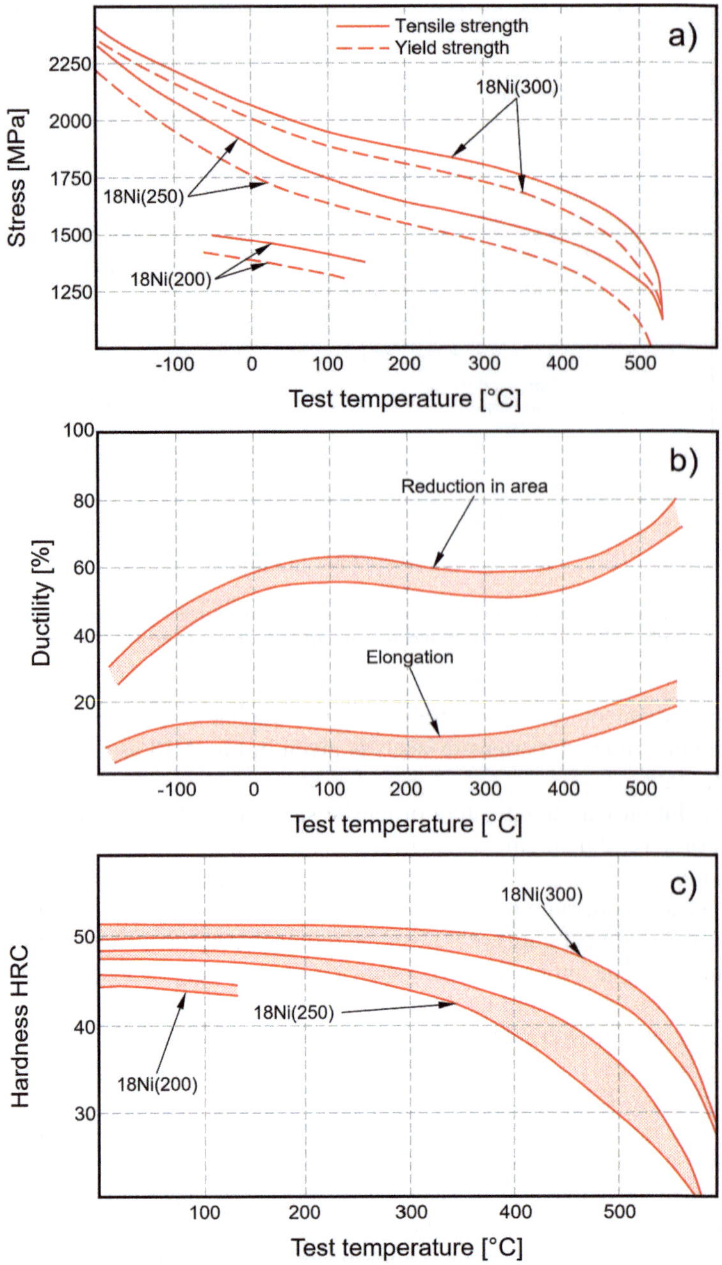

Fig. 5.19 Effect of the temperature on the mechanical properties of 18Ni maraging steel family: (**a**) tensile and yield strength, (**b**) ductility, and (**c**) Rockwell hardness C. (Redrawn from ASM International 1993)

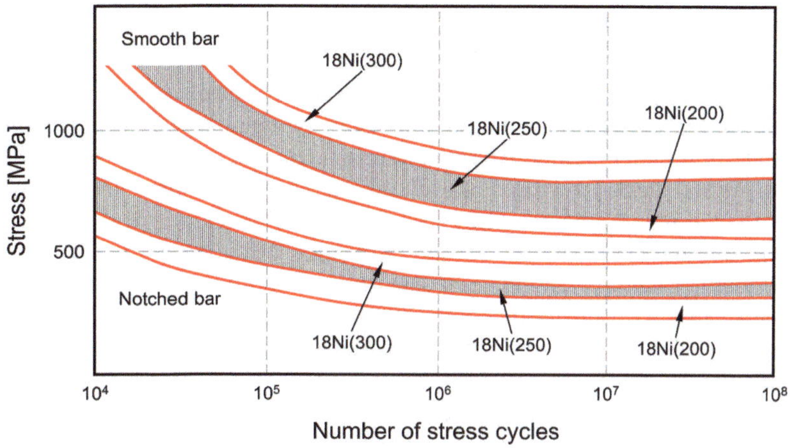

Fig. 5.20 Rotating-beam fatigue properties of 18Ni maraging steels family. (Redrawn from ASM International 1993)

Table 5.9 Overview table of the main processing routes used for maraging steels (ASM International 1993; Campbell 2008)

Process	Features
Hot-working	Hot-working can be carried out by conventional forging and steel mill techniques. Precipitation of TiC at the grain boundary must be avoided; therefore a slow cooling process in the range of 750–1025 °C after hot-working is undesirable. The same holds for hot-working processes above 1260 °C. Furthermore, Ti-containing grades have greater hot strength if compared with conventional steels; therefore they need either higher working temperatures or higher hot-working loads. However, for mechanical properties maximization, maraging steels should be hot-worked at temperatures as lower as possible
Cold forming	When in un-aged condition (i.e., solution annealed or as-transformed), maraging steels can be cold-worked by any conventional technique. They show very low work-hardening rates and can accept heavy reductions ($\leq$50%) with only slight increase in hardness
Machining	Maraging steels can be machined by means of conventional techniques both in solution-annealed and age-hardened condition. Machinability of maraging steels is similar, and sometimes better, as that of conventional steels of comparable hardness
Welding	Maraging steels show an excellent weldability. They can be readily welded by means of GTAW (gas tungsten arc welding) in both the age-hardened and annealed condition
Powder metallurgy	Maraging steels produced via powder metallurgy of elemental powders show reasonable ductility combined with a yield strength around 1600 MPa. Gas-atomized alloy powders canned and extruded are able to generate a material with excellent ductility and yield strength up to 1900 MPa. Contamination of the surface of the powder during atomization and subsequent collection is one of the major issues of this process

5.4.2 AerMet Steels

There has been a growing interest for ultra-high strength steels over the last decades, in particular for those alloys exhibiting secondary hardening, due to the precipitation of carbides during tempering. After the development of commercial maraging steel grades, AerMet steels have been introduced. AerMet steels show both outstanding toughness and ultra-high strength. These steels were developed at the McDonnell-Douglas, as candidate materials for the fabrication of the landing gear of the F/A 18 E/F military aircraft. The most demanding requirement for this material was a toughness value at least twice that of the 300M steel (AISI 4340M) but still retaining, or even improving, its strength. AerMet 100 had met these requirements, considering its minimum toughness of 110 MPa√m and 1930 MPa ultimate tensile strength. After its very first application for the F/A 18 E/F fighter, AerMet 100 was used as replacement for other steels, such as AF1410 (Air Force 14Co 10Ni), HP-9-4-30, and AISI 4340. As compared to maraging steels and titanium alloys, AerMet steels show better fatigue life and excellent combination of toughness and strength (Novotny and Maurer 2007). They can be considered as a family of secondary hardening steels based on the Ni-Co-Cr-Mo alloying system (Prasad and Wanhill 2017). In addition to AerMet 100, other varieties of this family have been developed over the years, with increasingly higher strength. This is the case of the 488 °C (875 °F) age-hardened AerMet 100, the AerMet 310 ($UTS_{min} = 2140$ MPa), and the AerMet 340, this latter featuring the highest strength among all of the AerMet alloys developed so far ($UTS_{min} = 2344$ MPa) (Novotny and Maurer 2007). The nominal compositions of the three AerMet steels are listed in Table 5.10.

Like maraging steels, AerMet alloys have an age-hardened Fe-Ni-Co-C lath martensite matrix. They attain their peak strength thanks to the precipitation of M_2C carbides, where M can be either a Mo/Cr combination or just one of the two elements (Novotny and Maurer 2007). The important role of carbides in strengthening AerMet steels is coherent with the higher carbon concentration in these alloys (Table 5.10). The M_2C carbides replace M_3C (i.e., Fe_3C, cementite), thus removing an important grain boundary embrittling agent. Moreover, the fine-grained M_2C carbides, being better distributed within the alloy matrix, contribute to increasing both hardness, enhanced by age-hardening thermal treatments (Novotny and Maurer 2007), and fracture toughness (Graves 1994). The melting process used for the production of AerMet steels is generally a combination of VIM and VAR, suitable to achieve the required mechanical properties and low impurity levels. The amount of inclusions and the impurity concentration are kept to a minimum thanks to the double vacuum process, as concerns particularly silicon and manganese that tend to reduce the grain boundary cohesion of the austenite phase. After ingot production, the AerMet steels are processed to obtain different products: billets, plates, sheets, strips, die forgings, bars, and wires (Campbell 2008). The maximum starting temperature for forging (primary breakdown) is 1230 °C. Finishing operations can be carried out starting from 980 °C. Air cooling and subsequent annealing must be performed on workpieces after forging; a normalization treatment is generally needed to restore the properties of the dead zones, not involved in the plastic

Table 5.10 Nominal compositions (wt%) of AerMet alloys (Novotny and Maurer 2007)

Grade	C	Cr	Ni	Mo	Cu	Co	V	Ti	Al	B	Zr	Ca	Fe
AerMet 100	0.23	3.10	11.50	1.20	–	13.50	–	0.015 max	0.015 max	–	–	–	Balance
AerMet 310	0.25	1.40	11.00	1.40	–	15.00	–	0.015 max	0.015 max	–	–	–	Balance
AerMet 340	0.33	2.25	12.00	1.85	–	15.60	–	0.015 max	0.015 max	–	–	–	Balance

Table 5.11 Standard heat treatment sequence for AerMet alloys (Prasad and Wanhill 2017)

	AerMet 100	AerMet 310	AerMet 340
Normalization	Austenitization: 900–925 °C. Soaking time: at least 1 h each 25 mm of thickness. Final cooling: air	Austenitization: 970 °C. Soaking time: at least 1 h each 25 mm of thickness. Final cooling: air	Austenitization: 968 °C. Soaking time: at least 1 h each 25 mm of thickness. Final cooling: air
Stress relief	Stress relief is performed after normalization annealing. Processing temperature = 675 °C		
Annealing	Annealing is performed at 677 °C with a soaking time of 16 h followed by air cooling. The process leads to a maximum hardness equal to HRC 40		
Quenching/hardening	Austenitization: 870–900 °C. Soaking time: at least 1 h each 25 mm of thickness. Cooling process either in air or oil in order to reach 65 °C in 1–2 h. Cryogenic treatment for 1 h at −75 °C and final warming to room temperature	Austenitization: 900–925 °C. Soaking time: at least 1 h each 25 mm of thickness. Cooling process either in air or oil in order to reach 65 °C in 1–2 h. Cryogenic treatment for 1 h at −75 °C and final warming to room temperature	Austenitization: 955–988 °C. Soaking time: at least 1 h each 25 mm of thickness. Cooling process either in air or oil in order to reach 65 °C in 1–2 h. Cryogenic treatment for 1 h at −75 °C and final warming to room temperature
Age hardening	Age hardening can be carried out between 475 and 488 °C as a function of the final desired properties. Age-hardening processes carried out below 468 °C result in a material with very low toughness values		

deformation during the metal forming process (Prasad and Wanhill 2017). Joining processes, e.g., welding, are not a problem for AerMet alloys, and no preheating is generally required (Campbell 2008). AerMet steels can undergo machining, although the process might be rather challenging and requires tungsten carbide tools (Prasad and Wanhill 2017).

As concerns age-hardening treatments, the optimum temperature ranges from 450 °C to 600 °C (Graves 1994; Shi et al. 2016). In Table 5.11 the main indications for the heat treatments of each grade of AerMet alloys are provided.

AerMet alloys are very sensitive to the age-hardening treatments, as concerns microstructure and resulting properties. For this reason, specific standard treatments have been developed.

Figure 5.21 confirms that the optimum combination of strength and fracture toughness is achieved with an age-hardening treatment carried out at 482 °C, resulting in carbide precipitation. This transformation is accompanied by the formation of a reverted austenite film at the martensite lath boundaries. The combination of M_2C carbides and the reverted austenite film present along the boundaries of martensite laths are responsible for the optimum balance of mechanical strength and toughness/ductility (Shi et al. 2016). The heat treatments of AerMet steels should be performed under vacuum, inert atmosphere, or salt bath due to the decarburization tendency of the material, a feature that they share with other high-strength steels with relatively high carbon concentration (Prasad and Wanhill 2017).

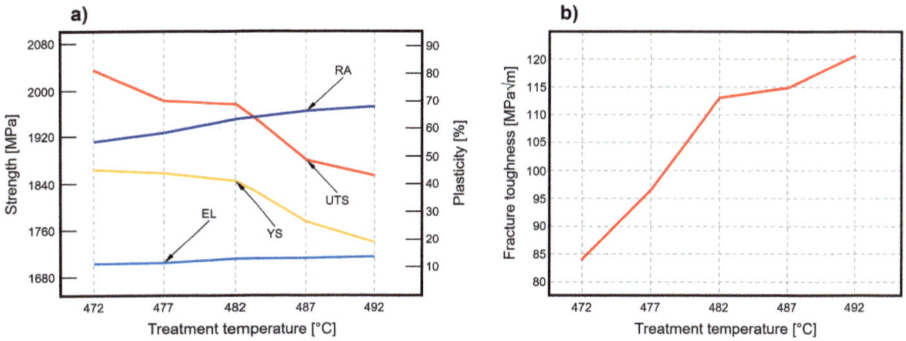

Fig. 5.21 (**a**) Tensile properties; (**b**) fracture toughness of AerMet 100 as a function of the tempering temperature. *RA* % reduction in area, *UTS* ultimate tensile strength, *YS* yield strength, *EL* % elongation to fracture. (Redrawn from Shi et al. 2016)

Table 5.12 Mechanical properties of AerMet alloys (Novotny and Maurer 2007)

	482 °C aged AerMet 100	454 °C aged AerMet 100	AerMet 310	AerMet 340
UTS [MPa]	1978	2082	2172	2427
Yield strength [MPa]	1696	1779	1896	2165
% Elongation	16.0	14.0	14.5	11.0
% Reduction in area	67.0	64.0	63.0	55.0
Charpy V-notch impact test [J]	47.5	40.7	27.1	14.9
Fracture toughness [MPa√m]	132	109	71.4	35.2
Density [g/cm^3]	7.889	7.889	7.972	7.861
Fatigue strength 10^7 cycles [MPa]	945	944.6	1034.2	986
Specific strength [(MPa cm^3)/g]	251	264	272	309

The reference values of the main mechanical properties of aged AerMet alloys are listed in Table 5.12.

AerMet ductility is generally higher than that of maraging steels. As concerns hardness, AerMet 100 steel exhibits values in between 53 and 54 HRC, 55 HRC for AerMet 310, and 56.5 HRC for AerMet 340 (Prasad and Wanhill 2017). As concerns surface stability, AerMet steels do not contain a sufficient Cr concentration for developing a protective self-passivating layer. For this reason, applications, including aerospace ones, usually require a coating or a plating process, necessary to prevent in-service corrosion. Like other high-strength materials, AerMet steels can be subjected to SCC (stress corrosion cracking, see Sect. 8.1.6), with the exception of AerMet 100 steel, which is generally regarded as better SCC resistant (Campbell 2008). The application of AerMet steels in the aerospace field is extended to a wide range of components, as described in detail in Table 5.24 (Sect. 5.5.7).

5.4.3 Stainless Steels

The main alloying element in stainless steels is chromium, since the specific feature of these materials, i.e., corrosion resistance, is determined by a passivation surface layer based on chromium oxide. The first time chromium was identified as a new chemical element was more than 200 years ago, being extracted from $PbCrO_4$ (crocoite). The name chromium derives from the Greek word *khrôma-atos*, which means color (Totten 2007). At the present time, the main source of metallic chromium is $FeCr_2O_4$ (chromite). The process to extract chromium from this compound was used for the first time in 1798 by the German chemist Tobias Lowitz. The low-C ferrochromium processing (aluminothermic process), needed in view of the low-carbon stainless steel production, was patented in 1891 in Germany by Hans Goldschmidt.

Even Michael Faraday and James Stodart had been involved in this metallurgical research on new alloy steels. In their paper "Experiments on the Alloy Steel, made with a view to its improvements," they described possible approaches for producing better cutting tools and alloys less prone to oxidation (Stodart and Faraday 1820). In 1821, the French Pierre Berthier found out that chromium-alloyed iron had a better corrosion resistance against acids (Cobb 2010). In the following years, the production of stainless steels started quite simultaneously in different countries in Europe and also in the USA (Totten 2007). Lion Guillet, professor of metallurgy and metal processing at the Conservatoire National des Arts et Métiers, published in 1904 a series of papers on Fe-Cr alloys characterized by carbon and chromium concentrations comparable to modern ferritic and martensitic stainless steels, such as AISI 421, AISI 446, AISI 410, AISI 420, and AISI 440C (Cobb 2010). In 1913, the English metallurgist Harry Brearley, head of the Brown-Firth Research Laboratories in Sheffield, coordinated the production of the first heat of an alloy containing 0.24% C, 0.44% Mn, 0.22% Si, and 12.86% Cr. At the same time in the USA, Frederick Becket and Christian Dantsizen were starting the production of alloys containing from 0.007% to 0.015% C and from 14% to 16% Cr, obtained thanks to the low-C ferrochromium precursor and produced via the aluminothermic process (Totten 2007). The well-known 18–8 austenitic stainless steel first appeared in 1920, although its wide-scale production just started a few years later. Nowadays, stainless steels are classified into five main groups, according to their microstructure and composition: austenitic, martensitic, ferritic, duplex, and PH (precipitation hardened). The remarkable corrosion resistance, together with a very wide range of mechanical properties, is the main reason for the massive use of these alloys in many different fields: food industry, cryogenic applications, cutlery, engine valves, bearings, medical applications, power generation, and aerospace industry, of course. This paragraph presents the three main types of stainless steels used in the aerospace field, austenitic, martensitic, and PH, highlighting their most important features, particularly as concerns their properties and applications for aircraft components.

5.4.3.1 Austenitic Stainless Steels

Benno Strauss and Edward Maurer at the Krupp Research Laboratories developed the first commercial austenitic stainless steel in between 1909 and 1912. This very year, a stainless steel with a "modern" composition (18% Cr −8% Ni) codenamed AISI 304 was developed. At the beginning of their era, austenitic stainless steels were prone to intergranular corrosion, i.e., sensitization (see Sect. 8.1.4), because of the precipitation of the Cr-rich $M_{23}C_6$ carbides, occurring quite often upon cooling the alloy from the melt. Two remediation approaches were adopted. The first was a reduction in the carbon content, in the so-called L-type austenitic stainless steels, such as AISI 304L and AISI 316L. The second approach was the addition of specific alloying elements featuring a higher affinity to carbon than chromium. This is the case of titanium and niobium, which resulted in the production of the stabilized grades AISI 321 and AISI 347, respectively. Thanks to these successful developments, austenitic stainless steels are nowadays regarded as benchmark alloys, as far as mechanical strength coupled with oxidation and corrosion resistance are concerned (Cobb 2010; Demeri 2013; Totten 2007). Austenitic stainless steels can be classified as Fe-based alloys containing from 16 to 26 wt% chromium and 8–35 wt% nickel. Other alloying elements are also present but in lower percentage. Austenitic stainless steels belong to the AISI 2xx and 3xx groups (see Sect. 5.2.5). According to the international classification, the AISI 2xx are Cr-Mn-Ni- or Fe-Cr-Mn-Ni-N-based alloys. The other AISI 3xx series have a general composition referring to the Fe-Cr-Ni-C system.

Austenitic stainless steels are essentially nonmagnetic in the annealed condition and can be hardened by cold-working. Their chemical composition is designed so to suppress the austenite to ferrite phase transformation on cooling, thus keeping the steel fully austenitic also at room temperature (ASM International 1993; Demeri 2013; Hosford 2010; Totten 2007). The chemical compositions of the main austenitic stainless steels are given in Table 5.13, while the "family tree" of austenitic stainless steels is illustrated in Fig. 5.22.

Chromium is essential for the formation of the protective passivation layer. Other elements are added to improve the quality of the protective film, but no other element can fully replace chromium in this role. Nickel is a γ-stabilizer, and therefore its presence is absolutely necessary for these steels. It promotes re-passivation, particularly in chemically reducing environments. Furthermore, it enhances the mechanical properties and the workability. Manganese can replace nickel partially: the 2xx series contains manganese from 5.5 to 15.5 wt% and nickel up to 6 wt%. Concerning the formability, it decreases upon decreasing the nickel and chromium content. Other elements, such as niobium, tantalum, titanium, and also carbon, may introduce some limitations due to their tendency both to form precipitates and to contribute to grain refinement (see Appendix 4 for further details). For these reasons, the formability of the AISI 321 and AISI 347 is lower than that of the AISI 302 and AISI 304. Nitrogen has several positive effects on the properties of austenitic stainless steels: it increases mechanical strength, enhances pitting corrosion resistance, and retards the formation

Table 5.13 Chemical composition of the principal austenitic stainless steels. Fe concentration balance (ASM International 1993)

Type	C	Mn	Si	Cr	Ni	P	S	N	Others
AISI 201	0.15	5.5–7.5	1.0	16.0–18.0	3.5–5.5	0.06	0.03	0.25	–
AISI 202	0.15	7.5–10	1.0	17.0–19.0	4.0–6.0	0.06	0.03	0.25	–
AISI 205	0.12–0.25	14.0–15.5	1.0	16.5–18.0	1.0–1.75	0.06	0.03	0.32–0.40	–
AISI 301	0.15	2.0	1.0	16.0–18.0	6.0–8.0	0.045	0.03	–	–
AISI 302	0.15	2.0	1.0	17.0–19.0	8.0–10.0	0.045	0.03	–	–
AISI 303	0.15	2.0	2.0–3.0	17.0–19.0	8.0–10.0	0.20	0.15 min	–	Mo = 0.6
AISI 303Se	0.15	2.0	2.0–3.0	17.0–19.0	8.0–10.0	0.20	0.06	–	Se = 0.15 min
AISI 304	0.08	2.0	1.0	18.0–20.0	8.0–10.5	0.045	0.03	–	–
AISI 304H	0.04–0.10	2.0	1.0	18.0–20.0	8.0–10.5	0.045	0.03	–	–
AISI 304L	0.03	2.0	1.0	18.0–20.0	8.0–10.5	0.045	0.03	–	–
AISI 304LN	0.03	2.0	1.0	18.0–20.0	8.0–10.5	0.045	0.03	0.10–0.16	–
AISI 304N	0.08	2.0	1.0	18.0–20.0	8.0–10.5	0.045	0.03	0.10–0.16	–
AISI 309	0.20	2.0	1.0	22.0–24.0	12.0–15.0	0.045	0.03	–	–
AISI 309S	0.08	2.0	1.0	22.0–24.0	12.0–15.0	0.045	0.03	–	–
AISI 310	0.25	2.0	1.5	24.0–26.0	19.0–22.0	0.045	0.03	–	–
AISI 310S	0.08	2.0	1.5	24.0–26.0	19.0–22.0	0.045	0.03	–	–
AISI 316	0.08	2.0	1.0	16.0–18.0	10.0–14.0	0.045	0.03	–	Mo = 2.0–3.0
AISI 316F	0.08	2.0	1.0	16.0–18.0	10.0–14.0	0.20	0.10 min	–	Mo = 1.75–2.5
AISI 316L	0.03	2.0	1.0	16.0–18.0	10.0–14.0	0.045	0.03	–	Mo = 2.0–3.0
AISI 316LN	0.03	2.0	1.0	16.0–18.0	10.0–14.0	0.045	0.03	0.10–0.16	Mo = 2.0–3.0
AISI 316N	0.08	2.0	1.0	16.0–18.0	10.0–14.0	0.045	0.03	0.10–0.16	Mo = 2.0–3.0
AISI 317	0.08	2.0	1.0	18.0–20.0	11.0–15.0	0.045	0.03	–	Mo = 3.0–4.0
AISI 317L	0.03	2.0	1.0	18.0–20.0	11.0–15.0	0.045	0.03	–	Mo = 3.0–4.0
AISI 321	0.08	2.0	1.0	17.0–19.0	9.0–12.0	0.045	0.03	–	Ti = 5 × %C min
AISI 330	0.08	2.0	0.75–1.5	17.0–20.0	34.0–37.0	0.04	0.03	–	–
AISI 347	0.08	2.0	1.0	17.0–19.0	9.0–13.0	0.045	0.03	–	Nb = 10 × %C min

AISI 348	0.08	2.0	1.0	17.0–19.0	9.0–13.0	0.045	0.03	–	Co = 0.2
									Nb = 10 × %C min Ta = 0.10
Nitronic 50	0.06	4.0–6.0	1.0	20.5–23.5	11.5–13.5	0.040	0.03	0.2–0.4	Mo = 1.5–3.0
									Nb = 0.1–0.3
									V = 0.1–0.3
Nitronic 60	0.10	7.0–9.0	3.5–4.5	16.0–18.0	8.0–9.0	0.040	0.03	0.08–0.18	–
253MA	0.05–0.10	0.8	1.4–2.0	20.0–22.0	10.0–12.0	0.040	0.03	0.14–0.20	Al = 1.0
									Ce = 0.03–0.08

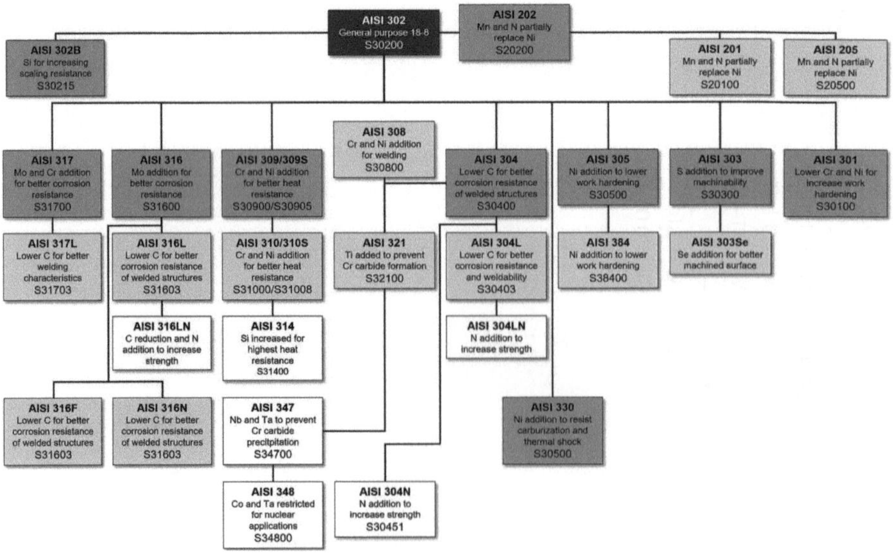

Fig. 5.22 Austenitic stainless steels family tree, AISI, and UNS designation (ASM International 1993)

of the detrimental Cr-Mo-based σ phase. In association with molybdenum, nitrogen enhances also the halide pitting corrosion resistance. Molybdenum is interesting not only for its solid solution strengthening of the alloy but also because it promotes the formation of the protective passivation layer (see Appendix 4).

The resistance to stress corrosion cracking (SSC, see Sect. 8.1.6) is improved by the addition of silicon and copper. Selenium and sulfur are added to improve the steel machinability. The majority of stainless steel, including austenitic ones, is produced using scrap based metallurgy processes. The typical melting practice implies electric arc furnace (EAF, see Sect. 5.3.1) and argon oxygen decarburization (AOD see Sect. 5.3.3). The casting of austenitic stainless steels is based on continuous casters and ingot casting processes. Aerospace applications often require products subjected to double melting processes: the first is conducted via EAF and AOD; then the starting ingot becomes the electrode for ESR (see Sect. 5.3.5) or for VAR (see Sect. 5.3.6). Double melting leads to a much cleaner material, in terms of inclusions and homogeneity, and produces ingots with improved hot-workability (ASM International 1993; ASM International 2008).

Austenitic stainless steels are typically supplied in the annealed condition, since they are not strengthened by thermal treatments but rather by cold-working. Cooling from the annealing temperature (see Sect. 5.2.1) must be sufficiently rapid to avoid the trajectory to cross the carbide transformation curves in the relevant TTT diagram (Fig. 5.23); otherwise a precipitation of carbides occurs (e.g., $Cr_{23}C_6$), with potentially negative effects on the corrosion resistance of the steel, determined by chromium depletion in the alloy (see Sect. 8.1.4).

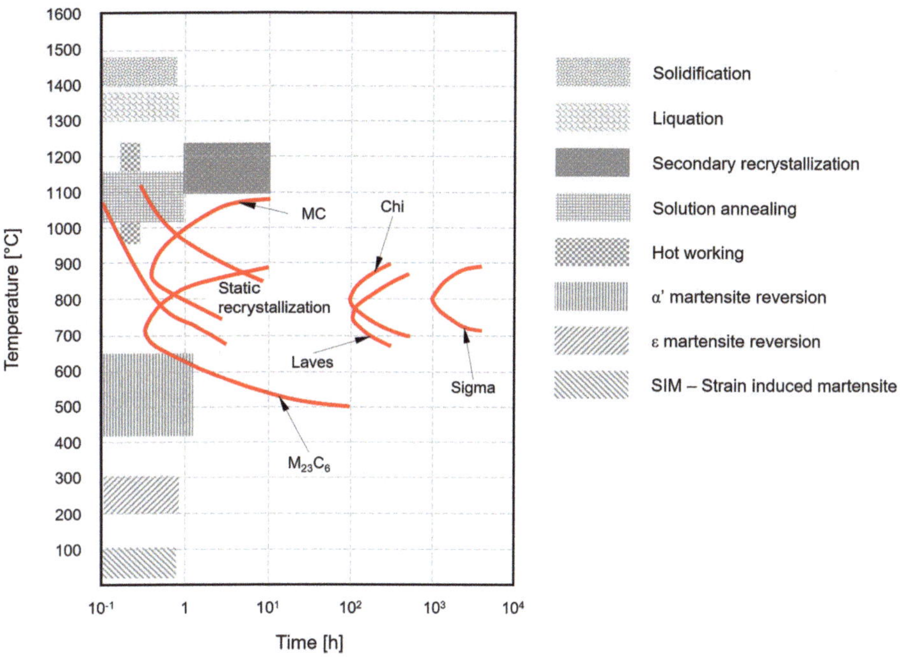

Fig. 5.23 Main thermal treatments and relevant transformations occurring in austenitic stainless. (Redrawn from Totten 2007)

Austenitic stainless steels do not form martensite on cooling because their M_s (martensite start) is well below room temperature. However, martensite can still be induced by mechanical strain (see Sect. 9.5.1), the so-called strain-induced martensite (SIM). Two types of martensite can occur, α' (*bcc* and ferromagnetic) and ε (*hcp*, non-ferromagnetic), depending on the composition, stacking fault energy, deformation rate, and temperature range (Kaieda and Oguchi 1985; Totten 2007). The transitions from γ to α' and from γ to ε involve 2.57% volume increase or 0.8% shrinkage, respectively. The temperature below which martensite can be induced by mechanical strain is defined as M_d (see also Sect. 9.5.1). It is given by empirical expressions, based on the chemical composition of the alloy and on the percentage of strain-induced martensite, like Eq. 5.6, referring to a 30% true tensile strain needed to obtain 50% α' martensite (Angel 1954):

$$M_d(30/50) = 413 - 13.7(\%Cr) - 9.5(\%Ni) - 8.1(\%Mn) \\ - 18.5(\%Mo) - 9.2(\%Si) - 462(\%C + \%N) \tag{5.6}$$

The annealing of stress induced martensite may cause its reversion to austenite for treatments carried out at temperatures and times lower than those required by recrystallization treatments (Demeri 2013, Totten 2007). The most common heat treatments for these steels are solution annealing, stabilization annealing, and stress

Table 5.14 Annealing treatments for austenitic stainless steels (Totten 2007)

Treatment	Temperature range
Solution annealing	1010–1120 °C. For grades that are more subjected to abnormal grain growth and secondary recrystallization (e.g., stabilized grades): 1065 °C max
Stabilization annealing	845–955 °C
Stress relief annealing	425–550 °C

Table 5.15 Typical mechanical properties at room temperature of some austenitic stainless steels used in the aerospace field (ASM International 1993)

AISI type	Tensile strength [MPa]	Yield strength [MPa]	Elongation [%]	Reduction in area [%]
304	640	235	76.0	82.0
304L	660	405	78.0	81.0
310	585	340	50.0	76.0
316	595	275	60.0	–
321	675	430	55.0	79.0
347	670	340	57.0	76.0

relief annealing (see Sect. 5.2.1). In Table 5.14 the description for the most important heat treatment parameters for austenitic stainless steels is provided.

Important room temperature mechanical properties of some austenitic stainless steels used in the aerospace field (see also Table 5.25) are listed in Table 5.15. These steels do not exhibit any ductile-to-brittle transition, and for this reason, they are extensively used in cryogenic applications, even down to temperatures below −200 °C. Actually, the yield and ultimate tensile strengths of austenitic steels increase substantially as the testing temperature decreases. Obviously, cold-working greatly increases the yield strength while reduces the alloy ductility (ASM International 1993).

High-temperature plastic deformation is generally not an issue for austenitic stainless steels. Most grades can be subjected to forging operations in a wide range of temperatures above 925 °C, with no equipment damage and degrading quality of the material. Chemical composition differences affect the generally good weldability of austenitic stainless steels. The carbon content plays an important role too, since it influences carbide precipitation in the HAZ (heat-affected zone). AISI 304L is known to minimize this deleterious phenomenon when compared with standard AISI 304. Grades with improved machinability, the so-called free-machining steel grades, contain phosphorous, sulfur, and selenium, that, on the other hand, reduce weldability since their presence can cause severe hot cracking of the metal. The weldability of AISI 318, AISI 321, AISI 347, and AISI 348 is guaranteed by alloying with Ti, Nb, and Ta that preventing the formation of chromium carbides avoids sensitization phenomena (see Sect. 8.1.4).

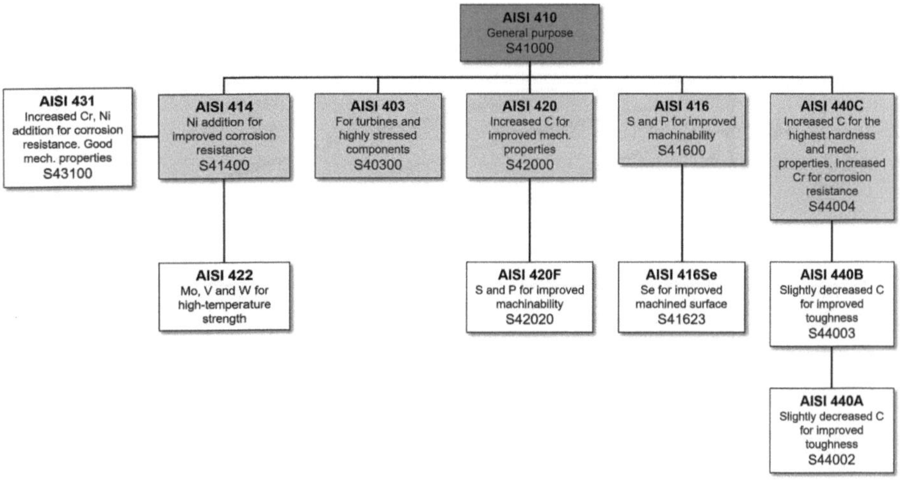

Fig. 5.24 Martensitic stainless steels. The family tree of the martensitic stainless steels. AISI and UNS designation (ASM International 1993)

In addition to sensitization, austenitic stainless steels are prone to SCC in chloride containing environments. In order to decrease the susceptibility to SCC, a high nickel content is recommended. In high-temperature applications, austenitic stainless steels with 10–20% Ni and 20–25% Cr can be used, still exhibiting acceptable performances as concerns oxidation and mechanical properties.

5.4.3.2 Martensitic Stainless Steels

The very first commercial martensitic stainless steel was probably developed by Harry Brearley in 1913. The present family of stainless steels is based on the Fe-Cr-C system, where Cr ranges in between 11.5 and 18 wt %, while carbon content is kept between 0.1 and 1.2 wt %, with little or no additions of other alloying elements (Totten 2007). Martensitic stainless steels belong to the AISI 4xx series (see Sect. 5.2.5) and are ferromagnetic and resistant to corrosion in relatively mild environments. Their corrosion resistance is lower if compared to that of austenitic and ferritic grades. Most commonly used martensitic grades are those with low-carbon content and up to 12 wt% Cr. AISI 410 can be regarded as the reference, general purpose martensitic stainless steel. It is possible to classify the standard martensitic stainless steel grades according to the scheme in Fig. 5.24.

The application of martensitic stainless steels in the aerospace field is limited to specific components that benefit from the combination of specific properties of these alloys. Of course, tribological (i.e., wear resistance) features together with remarkable tensile strength, as well as their corrosion resistance, are the main reasons for their adoption (see Table 5.26 in Sect. 5.5.7). Martensitic stainless steels can be

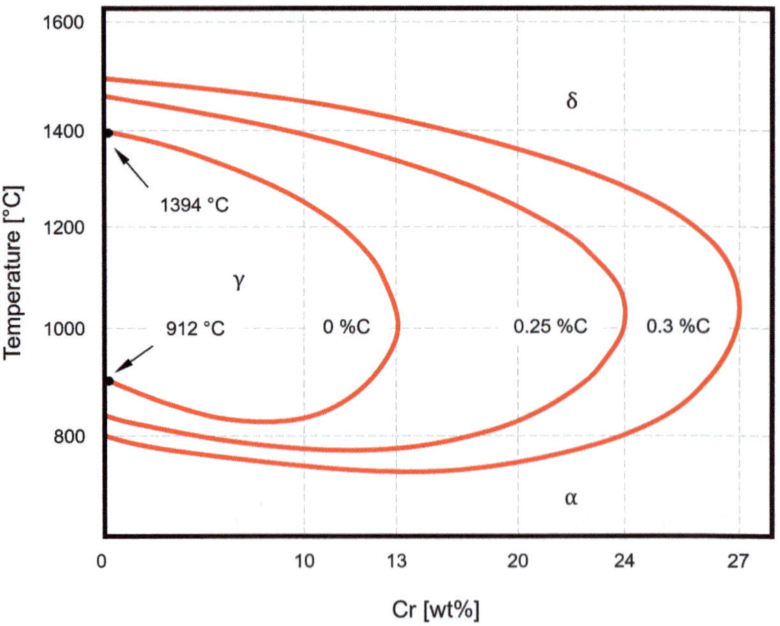

Fig. 5.25 Projection of the ternary Fe-Cr-C diagram on the temperature vs. %Cr plane. (Redrawn from Totten 2007)

subdivided into three main subgroups, according to their carbon content and relevant applications: low-C steels for gas turbines, medium-C steels for cutlery, and high-C grades for wear-resistant applications. Martensitic stainless steels with high carbon content are those able to reach the highest strength. The carbon percentage influences the microstructure of the alloy. Low-C alloys typically show the formation of martensitic needlelike microstructure; medium-C alloys usually have a fine martensitic structure, while high-C grades show ultrafine martensitic structure with primary carbides. Ni-containing grades (e.g., AISI 431) and high-C grades (e.g., AISI 440C) may show the presence of a high volume fraction of retained austenite after quenching, due to a very low M_f (martensite finish), that may fall in the cryogenic range (Totten 2007). Cr is kept as high as possible in order to attain the maximum corrosion resistance. The concentrations of chromium and carbon must be balanced to ensure a martensitic structure after a quenching hardening treatment (see Sect. 5.2.2). As chromium level is increased, carbon has to be increased as well, in order to stabilize the austenite field (see Fig. 5.25). If other elements other than C and Cr are present, their concentration does not usually exceed 2–3 wt%. This is the case of nickel, a γ-stabilizing element (see Fig. 5.26). Martensitic steels containing nickel (e.g., AISI 414 and AISI 431) will also feature lower C content. Therefore, they will exhibit higher toughness and better corrosion resistance. A similar effect can be obtained with nitrogen, another γ-stabilizer. Its influence on the γ phase stability

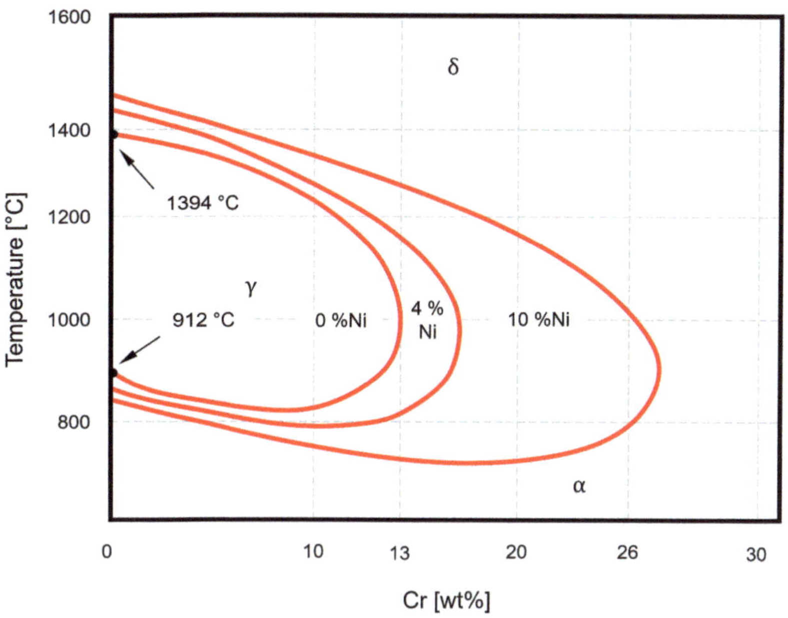

Fig. 5.26 Projection of the ternary Fe-Cr-C diagram on the temperature vs. %Cr (in wt%) plane and influence of Ni content. (Redrawn from Totten 2007)

field is shown by Fig. 5.27. The chemical compositions of the most common martensitic stainless steels are listed in Table 5.16.

High chromium contents have also influence on the tempering response of the material, since chromium slows down the softening, allowing for the removal of residual stresses without any appreciable decrease in the strength and hardness.

Some martensitic grades contain up to 1% Mo for the improvement of both corrosion resistance and mechanical properties. Nb, Si, V, and W are added in order to modify the tempering response of the material after the hardening process. Sulfur and selenium are added for improving machinability (ASM International 1993; Campbell 2008; Totten 2007). Martensitic stainless steels may be subjected to the precipitation of several carbides: MC, M_2C, M_3C, M_7C_3, and $M_{23}C_6$. σ phase precipitation is usually not an issue for this class of stainless steels (Totten 2007).

Concerning melting and casting practices, martensitic stainless steels can be produced via EAF + AOD route or, in some cases, via arc melting, AOD refining, and final VAR process. This approach is very much recommended for those applications requiring homogeneity and cleanliness, in order to ensure a safe fatigue life. This aspect can be particularly interesting for bearing systems for aerospace uses (e.g., AISI 440C). Martensitic stainless steels can be produced via ingot casting or continuous casting, depending on the chemistry and on the final product specifications.

Martensitic stainless steels can be heat-treated in the same way as plain carbon, low alloy, and alloy steels. This implies the presence of at least three main stages: austenitization, quenching, and tempering. Heat treatment of these alloys is very

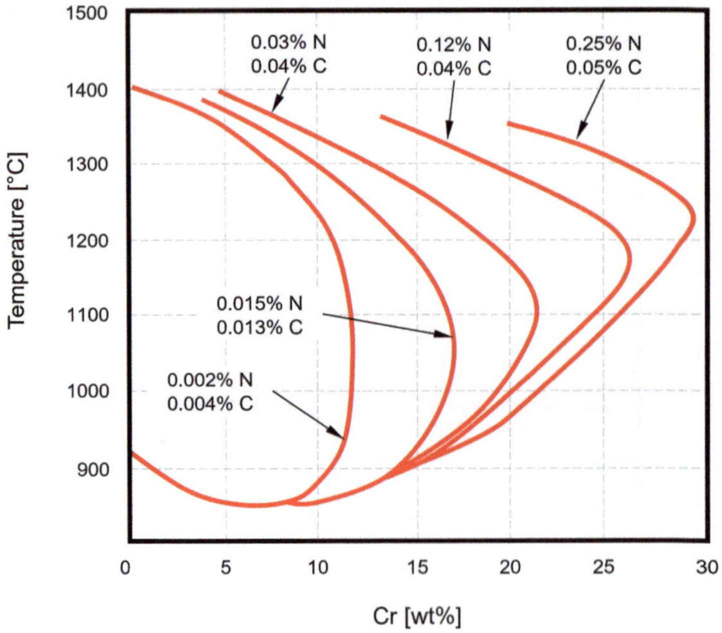

Fig. 5.27 Shift of the γ loop with carbon and nitrogen content. (Redrawn from Totten 2007)

Table 5.16 Chemical composition of martensitic stainless steels using the AISI code. Fe concentration balance (ASM International 1993)

Type	C	Mn	Si	Cr	Ni	P	S	Others
403	0.15	1.0	0.5	11.5–13.0	–	0.04	0.03	–
410	0.15	1.0	1.0	11.5–13.5	–	0.04	0.03	
410S	0.08	1.0	1.0	11.5–13.5	0.60	0.04	0.03	–
414	0.15	1.0	1.0	11.5–13.5	1.25–2.50	0.04	0.03	–
416	0.15	1.25	1.0	12.0–14.0	–	0.06	0.15 min	Mo = 0.60
416Se	0.15	1.25	1.0	12.0–14.0	–	0.06	0.06	Se = 0.15 min
420	0.15 min	1.0	1.0	12.0–14.0	–	0.04	0.03	–
420F	0.15 min	1.25	1.0	12.0–14.0	–	0.06	0.15 min	Mo = 0.60
422	0.20–0.25	1.0	0.75	11.5–13.5	0.5–1.0	0.04	0.03	Mo = 0.75–1.25
								W = 0.75–1.25
								V = 0.15–0.30
431	0.20	1.0	1.0	15.0–17.0	1.25–2.50	0.04	0.03	–
440A	0.60–0.75	1.0	1.0	16.0–18.0	–	0.04	0.03	Mo = 0.75
440B	0.75–0.95	1.0	1.0	16.0–18.0	–	0.04	0.03	Mo = 0.75
440C	0.95–1.20	1.0	1.0	16.0–18.0	–	0.04	0.03	Mo = 0.75

Table 5.17 Typical austenitizing and tempering temperatures for martensitic AISI stainless steels used in the aerospace field (Totten 2007)

Type	Austenitizing [°C]	Tempering [°C]
403	925–1010	150–370
410	925–1010	150–370
431	980–1065	205–370
440C	1010–1065	150–370

important, since the material is purchased either in the annealed or in the fully treated (i.e., hardened and tempered) condition. Table 5.17 presents some typical values for the austenitizing and tempering temperatures for martensitic stainless steels used in the aerospace field.

The high chromium content leads to a significant shift of the time-temperature-transformation curves to the right (see Fig. 5.28) (Verhoeven 2007).

For this reason, martensitic stainless steels are said to be highly hardenable, using as cooling media either air or oil. These alloys exhibit a much lower thermal conductivity if compared with plain carbon and low-alloy steels. This results in higher-temperature gradients over the cross section and higher probability of warpage and cracking, even on heating. Moreover, metal carbides tend to dissolve slowly in the austenite, and carbide density is affected by the austenitizing temperature in such a way that an increase in this temperature determines a decrease in the carbide density. An example of this behavior is given in Fig. 5.29 where the carbide density vs. austenitizing temperature is given for the AISI 440C material (Verhoeven 2007).

Since these grades harden when cooled off the mill after hot-processing, the material requires an annealing at 650–760 °C for about 4 h. This treatment differs from full annealing that is usually carried out in between 800 °C and 880 °C and followed by a slow cooling in the furnace down to about 450 °C, before air cooling. Oil quenching is preferred for avoiding carbides precipitation, while air cooling is indicated to avoid part distortion. Low-C grades (e.g., AISI 410) show a relatively high M_s and M_f temperatures, 350 °C and 250 °C, respectively, and are fully martensitic at room temperature. High-C grades may retain a relatively high concentration of austenite down to room temperature, thus requiring cryogenic treatment to avoid this phenomenon (*vide infra*).

The evolution of retained austenite as a function of the austenitizing temperature is given in Fig. 5.30 for the AISI 440C steel. Figure 5.31 shows that for the AISI 440C at the austenitizing temperature of 1100 °C, a significant increase of the C-saturation line occurs. This requires a higher Cr concentration in order to obtain, upon quenching, a martensite with a Cr content above the threshold value of 13 wt%, necessary to be in the self-passivation range. This higher Cr concentration results in a lower both M_s and M_f. Consequently, a higher amount of retained austenite is present in the quenched alloy. Therefore, the completion of martensite transformation requires cryogenic quenching treatments, down to temperatures as low as −96 °C (ASM International 1993; Campbell 2008; Totten 2007; Verhoeven 2007).

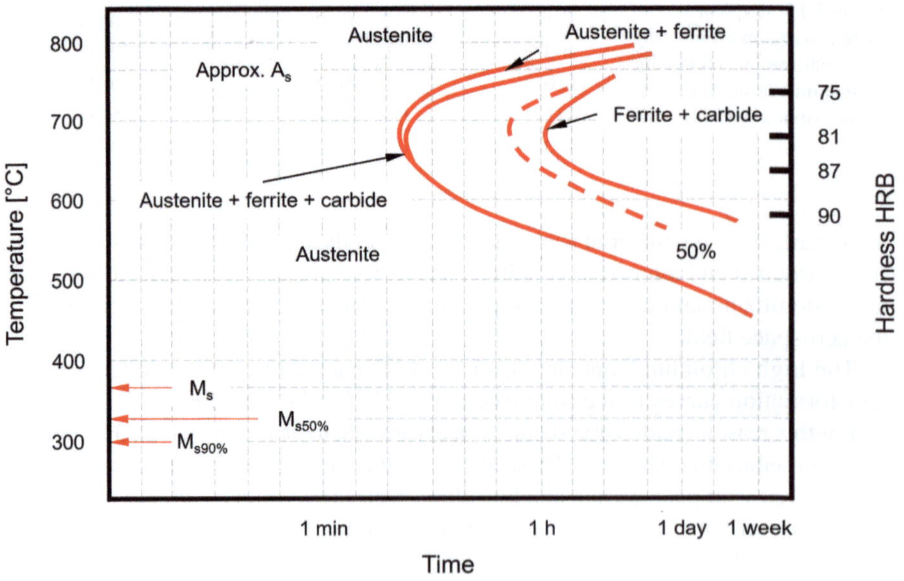

Fig. 5.28 Isothermal transformation diagram for AISI 410 stainless steel. (Redrawn from Verhoeven 2007)

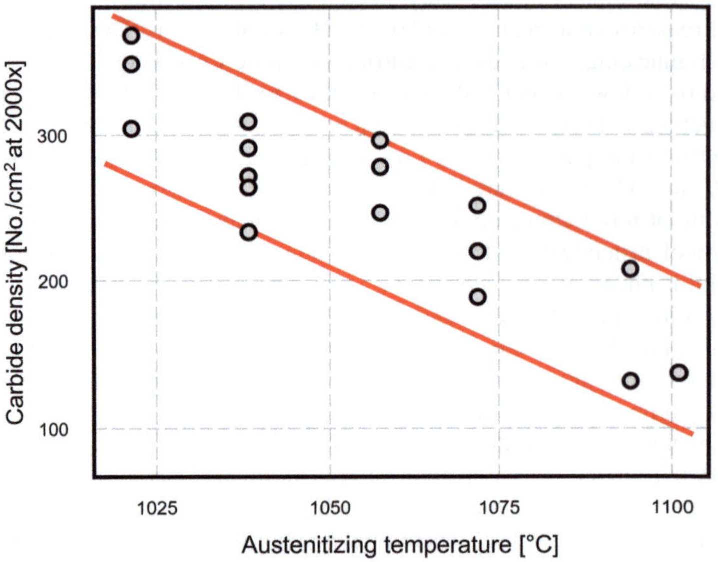

Fig. 5.29 Carbide density as a function of the austenitizing temperature for AISI 440C. (Redrawn from Verhoeven 2007)

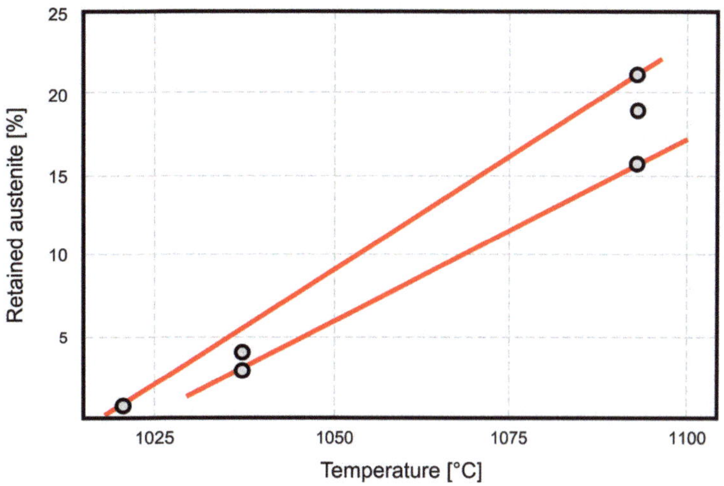

Fig. 5.30 Evolution of retained austenite as a function of the austenitizing temperature for room-temperature quench for the AISI 440C. (Redrawn from Verhoeven 2007)

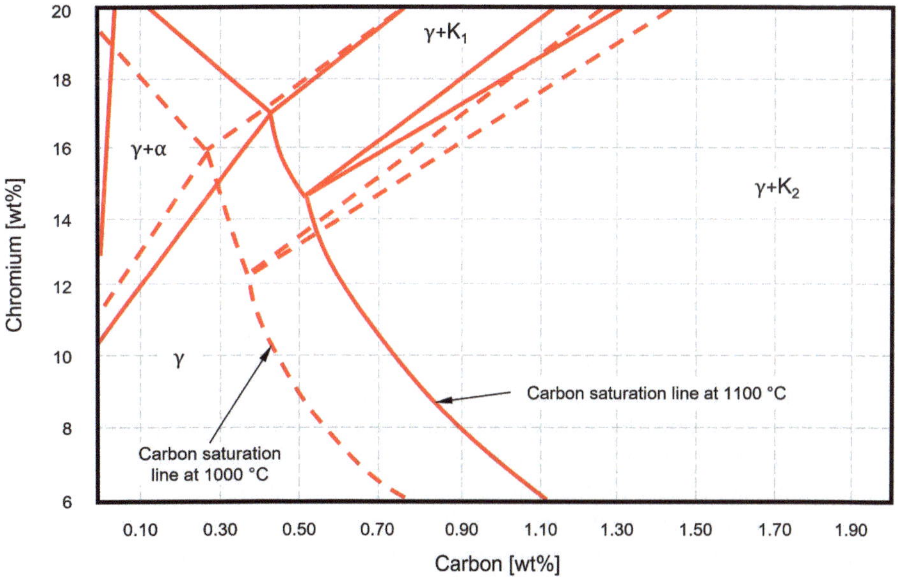

Fig. 5.31 The Fe-Cr-C isothermal sections at 1000 and 1100 °C for AISI 440C. (Redrawn from Verhoeven 2007)

Table 5.18 Mechanical properties of several martensitic stainless steels (Campbell 2008)

AISI type	Condition	Tensile strength [MPa]	Yield strength [MPa]	Elongation in 50 mm [%]	Reduction in area [%]	Hardness
403	Annealed	517	276	35.0	70.0	82 HRB
	Tempered	765	586	23.0	67.0	97 HRB
410	Oil quenched from 980 °C and tempered at 540 °C	1089	1006	13.0	70.0	–
414	Annealed	793	621	20.0	60.0	235 HB
	Cold drawn	896	793	15.0	58.0	270 HB
	Oil quenched from 980 °C and tempered at 650 °C	1006	800	19.0	58.0	–
420	Annealed	655	345	25.0	55.0	195 HB
	Annealed and cold drawn	758	690	14.0	40.0	228 HB
431	Annealed	862	655	20.0	55.0	260 HB
	Annealed and cold drawn	896	758	15.0	35.0	270 HB
	Oil quenched from 980 °C and tempered at 650 °C	834	738	20.0	64.0	–
440C	Annealed	758	448	14.0	25.0	97 HRB
	Annealed and cold drawn	862	690	7.0	20.0	260 HB
	Hardened and tempered at 315 °C	1965	1896	2.0	10.0	580 HB

Mechanical properties of selected martensitic stainless steels are summarized in Table 5.18 for different temper conditions.

Concerning processing, martensitic stainless steels afford several options. Due to their mechanical properties and structures, hot-working must be performed when they are in the austenitic field, since the material is close to a ductility maximum. Hot-working temperature, typically in excess of 1100 °C, for both forging and rolling processes, must be selected according to the chemical composition of the specific steel grade. Other process parameters, such as strain rate, temperature during hot-working, are crucial for the fabrication of flawless material.

Martensitic stainless steels can be machined, but the process has to be carried out with some precaution. In comparison with carbon steels, the required power for the process is higher, the cutting speed must be lower, and particular care must be taken to ensure correct cooling and adequate lubrication.

The weldability of martensitic stainless steels depends on their hardenability, considering that the metallurgical changes occurring during hardening may also take place during welding, regardless of the initial condition of the alloy: hardened, annealed, or tempered. The HAZ (heat-altered zone) is therefore characterized by a reduced toughness and becomes more susceptible to cracking. Pre-heating is recommended, and it is usually performed at temperatures in the 205 °C–315 °C range (ASM International 1993).

The high carbon concentration in martensitic stainless steels results in a lower corrosion resistance of these alloys if compared to austenitic and ferritic grades, since $M_{23}C_6$ (M = Cr, Fe) carbide may form, thus reducing chromium concentration in the alloy, even below the sensitization limit (Cr 13 wt%). Martensitic stainless steels are also prone to SCC (see Sect. 8.1.6) when exposed to marine environments, particularly in the presence of high chloride concentrations (ASM International 1993; Verhoeven 2007).

5.4.3.3 PH Stainless Steels

Until the 1930s there were three classes of stainless steels only: austenitic, martensitic, and ferritic. The first precipitation hardenable stainless steel was developed in Luxemburg in 1929 thanks to the work of William Kroll, who is nowadays notorious mainly for the invention of the first commercial extraction process for titanium (see Sect. 4.1) and zirconium. Precipitation hardening stainless steels were introduced into the market more or less at the same time on both sides of the Atlantic. In Sheffield, Firth Vickers was working on FV520B, while in 1948 Armco (USA) first introduced in the market the well-known 17-4PH, that was followed by 17-7PH and subsequently by 15-7Mo. Other PH stainless steels, named A286, AM350, and AM355, were developed by Allegheny Ludlum Steel Company. Finally, the Carpenter Steel Company developed the Custom 450, Custom 455, and Custom 630 (Cobb 2010; Farrar 2004).

Precipitation hardening stainless steels (PHSS or simply PH) are a family of alloys that exhibit a secondary hardening for the precipitation of compounds that are not carbides. PH stainless steels can be defined as a Cr-Ni steels, containing specific alloying elements, that induce precipitation hardening. Indeed, PH steels were introduced for both overcoming the limits of solution strengthening of the austenitic stainless steels and the low toughness and ductility of high-C martensitic grades. The PH steels retain good mechanical properties in association with an excellent corrosion resistance, so to be attractive for several aircraft components with quite a broad range of requirements, as concerns service temperatures; mechanical strength, also in creep and fatigue regime; and corrosion resistance (see Table 5.27 in Sect. 5.5.7). This section is devoted to the description of the main families of PH stainless steels, with a particular focus for those features interesting in aerospace applications. These alloys can be grouped into three main classes, according to the parent phase from which precipitation hardening occurs: martensitic, semiaustenitic, and austenitic. Martensitic PHSS are also called *single-treatment* alloys, since their peak strength condition is attained after a single low-temperature aging treatment, carried out after a solutioning annealing. The microstructure is austenitic at the annealing temperature and becomes martensitic on cooling, so that a martensite matrix is attained at the end of the final aging treatment. Semiaustenitic PHSS are also called *double-treatment* alloys. Their starting structure is austenitic, and the two-stage heat treatment includes austenite conditioning (heating at a temperature that allows martensite formation on cooling) and final age hardening after intermediate cooling. Austenitic PHSS feature an austenitic structure in both the solution-annealed and age-treated conditions. Strength improvement of these alloys is achieved by means of single or double precipitation hardening. The chemical composition of PHSS is the key factor for the development of their final microstructure and mechanical properties. Classification and chemical composition of the main PH stainless steels are shown in Table 5.19.

Figure 5.32 shows the Cr-Ni composition of a martensitic (17-4PH), semiaustenitic (17-7PH), and austenitic (A286) PHSS. 17-4PH lies in the region where rapid cooling produces martensite, and precipitation hardening occurs after a final aging treatment. 17-7PH requires a double treatment, since the first is needed to produce an austenite matrix with minor concentrations of δ-ferrite, while the second is needed for the martensitic transformation of the alloy matrix. A286 instead lies in the stable austenite region, and this implies that precipitation hardening occurs in a fully austenitic matrix (Campbell 2008; Everhart 1971; Verhoeven 2007; Totten 2007).

Precipitation hardening involves the formation of intermetallics, carbides, nitrides, and carbonitrides. Cu-, Ni-, and Al-rich intermetallics, with composition of the Ni_3M type (M = Al, Mo, Nb, Ti), are the most common phases (Campbell 2008; Chandler 1996; Farrar 2004; Mouritz 2012). Titanium and niobium tend to form also the very stable MCN and MN carbonitrides and nitrides (Miller and Burke 1991; Yeli et al. 2017). CrN precipitates have been reported to form in 17-4PH steel, along dislocation lines and microstructural defects preferentially (Yeli et al. 2017). However, the complex transformation intervening during precipitation treatments of the 17-4PH is still attracting major research efforts to be fully characterized.

The EAF + AOD process is commonly used for PHSS. After melting and refining, the material can be either continuously or ingot cast, depending on the grade and on the subsequent production stages, i.e., forging or hot-rolling. For specific applications, including aerospace, ESR, ESRR, or VAR processes can be used for improving the quality of the material in terms of microstructural homogeneity and segregation reduction. The 17–4PH steel can be produced with EAF + ESR/ESRR; for the martensitic 13-8Mo, a combination of VIM and VAR processes is a common choice. Different production routes are recommended for 5-5PH martensitic PH stainless steel: EAF + AOD, EAF + AOD + ESR, or EAF + AOD + VAR. Finally, the semiaustenitic 17-7PH is EAF melted and AOD refined.

Figure 5.33 shows the typical microstructure of 17-4PH and A286 in different conditions of treatment. Figure 5.33a shows the martensitic microstructure of the 17-4PH steel. The microstructure of the semiaustenitic 17-7PH would look rather similar, since the Al- and Ni-rich intermetallic precipitates, forming in the martensite, would not change that much the situation as compared to 17-4PH. Figure 5.33b shows the microstructure of the austenitic A286, after solution annealing and aging treatments: the austenitic grains can be clearly observed.

Table 5.19 Chemical composition of some important PHSS

Grade	Designation	Type	%C	%Mn	%Ni	%Cr	%Mo	%Si	%Others
17-10P	–	A	0.07	0.75	10.0	17.0	–	0.5	P = 0.3
17-14CuMo	AISI 653	A	0.12	0.75	14.0	16.0	2.5	0.5	Cu = 3.0 Nb = 0.4 Ti = 0.3
A286	AISI 660	A	≤0.08	≤2.0	24.0–27.0	13.5–16.0	1.0–1.5	≤1.0	Al = 0.35 B = 10–100 ppm Ti = 1.90–2.35 V = 0.1–0.5
HNM	–	A	0.3	3.5	9.5	18.5	–	–	–
13-8Mo	–	M	≤0.05	≤0.2	7.5–8.5	12.25–13.25	2.0–2.5	≤0.1	Al = 0.90–1.35 N = 0.01
15-5PH	–	M	≤0.07	≤1.0	3.5–5.5	12.5–14.0	–	≤1.0	Cu = 2.5–4.5 Nb = 0.15–0.45
17–4PH	AISI 630	M	≤0.07	≤1.0	3.0–5.0	12.5–17.5	–	≤1.0	Cu = 3.0–5.0 Nb = 0.15–0.45
Custom 450	–	M	≤0.05	≤1.0	5.0–7.0	14.0–16.0	0.5–1.0	≤1.0	Cu = 1.25–1.75 Nb ≥ 8 × %C
Stainless W	AISI 635	M	≤0.08	≤1.0	6.0–7.5	16.0–17.0	–	0.5	Al = 0.4 Ti = 0.4–1.2
15-7Mo	AISI 632	SA	≤0.09	≤1.0	6.50–7.75	14.0–16.0	2.0–3.0	≤1.0	Al = 0.75–1.50
17-7PH	AISI 631	SA	≤0.09	≤1.0	6.50–7.75	16.0–18.0	–	≤1.0	Al = 0.75–1.50
AM355	AISI 634	SA	0.13	0.95	4.3	15.5	2.75	–	N = 0.1

A austenitic, *M* martensitic, *SA* semiaustenitic
Fe concentration balance (Campbell 2008; Chandler 1996; Mouritz 2012; Totten 2007; Verhoeven 2007)

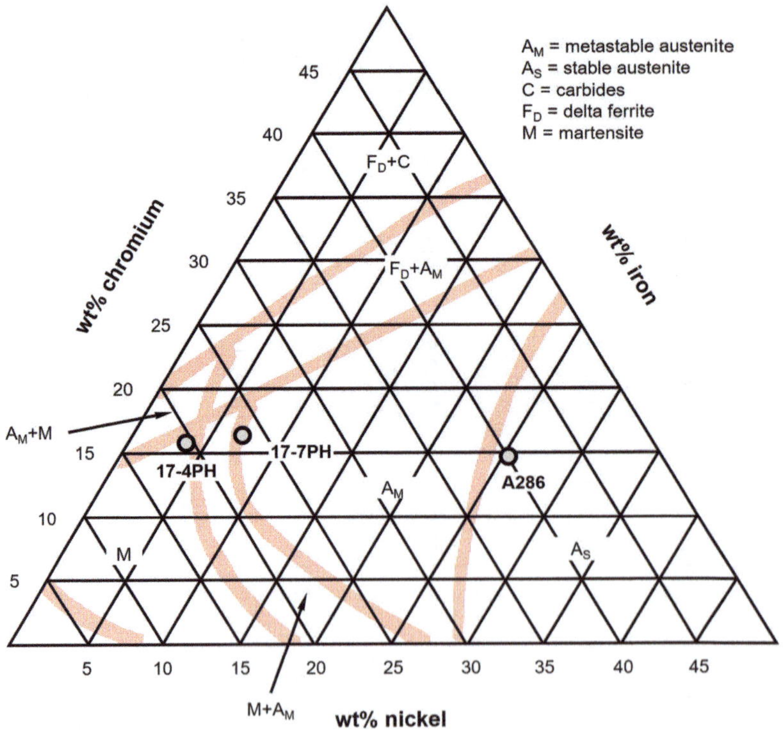

Fig. 5.32 Position of A286, 17-4PH, and 17-7PH steel in the relevant portion of the Fe-Cr-Ni ternary phase diagram. The diagram indicates the position of the three PH stainless steels after cooling from the 1100 °C austenitizing temperature. (Redrawn from Verhoeven 2007)

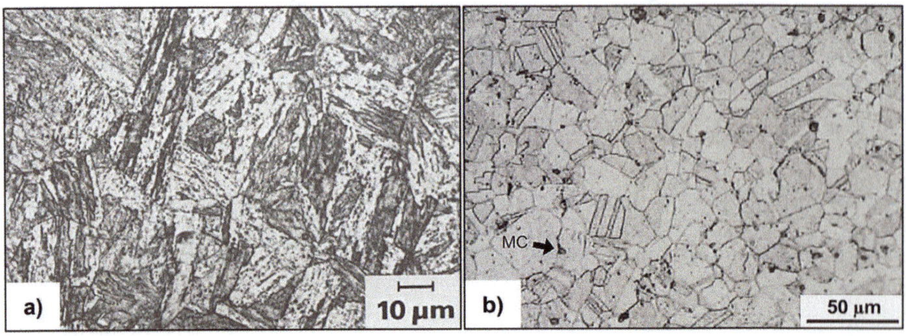

Fig. 5.33 (**a**) 17–4PH microstructure after 1 h aging at 470 °C (Viswanathan et al. 1988, reproduced with permission of the Publisher); (**b**) A286 microstructure after annealing at 930 °C and subsequent aging at 730 °C. (De Cicco et al. 2004, reproduced with permission of the Publisher)

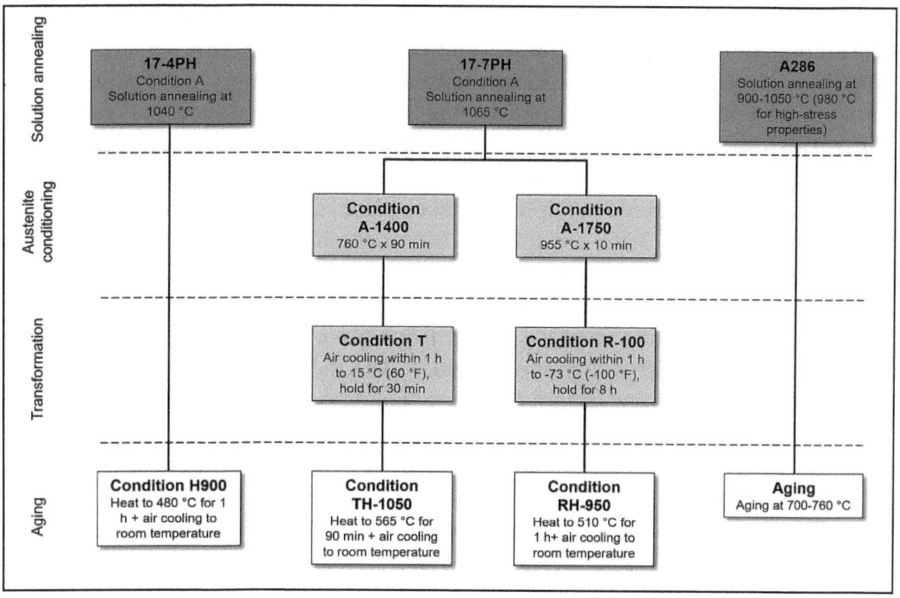

Fig. 5.34 Heat treatments for three types of PHSS: martensitic 17-4PH, semiaustenitic 17-7PH, and austenitic A286 (Campbell 2008; Chandler 1996; Totten 2007)

Since there are important constitutional differences among the three families of PHSS, heat treatments conducted for their microstructural refinement and property optimization must be designed accordingly. Regarding the heat treatment code, the solution treated condition is designed by "A" (see scheme in Fig. 5.34), while the aged condition is identified by "H" followed by the treatment temperature expressed in Fahrenheit degrees. Figure 5.34 shows a few examples of heat treatments for the main representatives of each type of PH steel: martensitic, semiaustenitic, and austenitic.

Martensitic PHSS are solution annealed and transform on cooling, leading to a fully martensitic structure. Afterward, aging is performed to induce the desired phase precipitation, without the need for austenite conditioning, which is necessary instead for semiaustenitic PHSS. The solution annealing (condition A) of the martensitic 17-4PH (see Fig. 5.34) is conducted at 1040 °C, usually followed by air cooling, resulting in the formation of a fully martensitic structure, i.e., no retained austenite, since M_f is above room temperature, although some δ-ferrite may be present. Age hardening, to obtain peak properties, is performed at 480 °C (900 °F), the so-called H900 condition. Another common heat-treated condition is H1095 (590 °C for 4 h). Whereas H900 is used to obtain peak hardness and strength, the H1095 improves ductility and toughness, through a suitable increase of the aging temperature. In both these main aging conditions, the strengthening phases are NbN-, CrN-, and Cu-rich precipitates (CRPs), these latter forming to different extent in the H900 and H1050 materials and providing the main strengthening contributions.

Semiaustenitic PHSS are also subjected to a solution annealing process (condition A, see Fig. 5.34). With reference to 17-7PH (Fig. 5.34), solution annealing is performed at 1065 °C, resulting in the formation of thermodynamically unstable austenite. Austenite conditioning, necessary to induce the subsequent martensite formation, can be performed at various temperatures, that will influence the kinetics of the transformation. At high temperatures (e.g., 955 °C, condition A1750), a reduced carbide precipitation takes place, and this leads to the need of a subzero treatment (−73 °C, condition R100) to obtain the formation of martensite. Otherwise, austenite conditioning at 760 °C for 90 min (condition A1400) is sufficient and does not require subzero cooling for the conversion into martensite, since Cr and C depletion from the matrix takes place as a consequence of precipitation, mainly of the $M_{23}C_6$ type of carbides. The material is then cooled down to about 15 °C and held for 30 min leading to condition T. In case this procedure is followed by aging at 565 °C for 90 min with subsequent cooling process down to room temperature, the final condition is named TH1050. During age-hardening treatment of 17-7PH, Al in the martensitic matrix combines with Ni, thus strengthening the alloy through the precipitation of the Ni_3Al and NiAl intermetallics (Totten 2007). It is worth reminding that in these alloys too, like in austenitic stainless steels (see Sect. 5.4.3.1), martensite can also be induced by mechanical strain (SIM, see Sect. 5.4.3.1).

The heat treatments of austenitic PHSS usually involve a solution annealing followed by an aging. Figure 5.34 shows the typical sequence used for solution annealing and age hardening of the austenitic A286 steel. Solution annealing is performed at 900–1050 °C. For high stress rupture properties, the recommended treatment temperature is 980 °C, and cooling is performed either in oil or water, although no martensitic transformation would actually occur. Aging treatment is carried out in the 700–760 °C temperature range and leads to the precipitation of carbides and γ' phase. For γ' to form, the chemical composition of the alloy should contain a sufficient amount of aluminum and titanium, as it is in iron- and nickel-based superalloys (see Sects. 6.3.1 and 6.4.3 and Appendix 6 for further details on γ') (Campbell 2008; Chandler 1996; Totten 2007; Verhoeven 2007; Yeli et al. 2017). A two-step aging treatment can also be used, to control the size and the distribution of the γ' precipitates, thus increasing the alloy yield strength (see Sect. 6.6). This kind of aging treatment is also used to control the morphology and size of grain boundary carbides, particularly the MC type.

Both martensitic and semiaustenitic PHSS after aging are able to reach tensile strength that exceeds 1300 MPa, although they also exhibit a significant ductility decrease when shifting from the annealed to the aged condition. Austenitic PHSS are not characterized by that great improvement in tensile strength after aging, reaching maximum values of around 950 MPa, again with a reduced ductility. However, in austenitic steels this effect is markedly lower as compared to martensitic and semiaustenitic PH steels. In Table 5.20 the mechanical properties of selected PHSS, as resulting from different heat treatments, are quoted.

In addition to heat treatments, cold-working can be used to modify further the microstructure and, thereby, the mechanical properties of semiaustentic PHSS.

Table 5.20 Properties of some PHSS as resulting from the specified aging treatments (Campbell 2008)

Type	Treatment	Tensile strength [MPa]	Yield strength [MPa]	Elongation [%]	Hardness
13-8Mo	Aged at 510 °C	1520	1410	6.0–10.0	45 HRC min.
15-5PH	Aged at 480 °C	1310	1170	–	–
17-4PH	Aged at 480 °C	1310	1170	5.0–10.0	40 HRC min.
Custom 450	Aged at 480 °C	1240	1170	3.0–5.0	40 HRC min.
Custom 455	Aged at 510 °C	1530	1450	≤4.0	44 HRC min.
15-7Mo	Aged at 480 °C	1650	1590	1.0	46 HRC min.
17-7PH	Aged at 510 °C	1450	1310	1.0–6.0	43 HRC min.
AM355	Aged at 540 °C	1170	1030	12.0	37 HRC min.
A286	Aged at 730 °C	860–965	655	4.0–15.0	24 HRC min.

Table 5.21 Mechanical properties of the semiaustenitic 17-7PH resulting from heat treatments and cold-working processes (Totten 2007)

Condition	Treatment	Tensile strength [MPa]	Yield strength [MPa]	Elongation [%]
A	1040–1080 °C, air or water cooled	<1030	<380	18.0
TH1050	A + 90 min at 955 °C air cooled +60 min at room temperature + 90 min at 565 °C	1170–1240	960–1030	5.0–6.0
RH950	A + 10 min at 955 °C + 8 h at −73 °C + 60 min at 510 °C	1370–1450	1240–1310	4.0–5.0
C	A + 50% cold-working	>1400	>1200	1.0
CH900	C + 60 min at 480 °C	>1650	>1580	1.0

Table 5.21 illustrates how some heat treatments and cold-working processes may determine specific mechanical properties of the 17-7PH steel (Campbell 2008; Everhart 1971; Totten 2007).

Hot-working is another important processing option for PHSS, that can be forged and hot-rolled. Heating cycles in furnaces prior to hot-working must be carefully calibrated according to the grade and to the final quality requirements for any specific product. PHSS can effectively be joined by resistance and fusion welding techniques. The main critical aspects to consider are the same as for the other families of stainless steels, i.e., the HAZ and relevant metallurgical issues. The machinability of PHSS is good in the annealed and decreases in the aged condition, especially for martensitic and semiaustenitic grades, due to the strength improvement that results from the thermal treatments. PHSS show better corrosion resistance than traditional high-strength steels, as a direct consequence of their chemical compositions and effect on the self-passivation behavior. However, PHSS are rather prone to hydrogen embrittlement (see Sect. 8.1.7), especially in alloys with increasing yield strength. Like low-carbon steels, PH stainless steels exhibit a dependence of the corrosion resistance from strength level and microstructure: a decrease in the corrosion resistance is observed at increasing yield strengths. Although the mechanical properties of several PHSS are comparable to those of martensitic stainless steels, their corrosion resistance is definitely better. The 17-4PH has a corrosion resistance that is very close to that of the austenitic AISI 304L. The corrosion resistance of Custom 450 is similar to that of 17-4PH, although somehow better as concerns pitting corrosion (see Sect. 8.1.2). This is due to the Mo content of the Custom 450 alloy, higher than that of 17-4PH steel (see Table 5.19). Among the austenitic PHSS, A286 displays a good corrosion resistance thanks to its chromium-rich austenitic structure. Improvements in SCC resistance can be attained by aging the alloy at the maximum treatment temperature, compatible with the achievement of an adequate strength level (Savoie et al. 2007; Campbell 2008; DeCicco et al. 2004; Farrar 2004).

5.5 Steel-Based Aerospace Components

The Flyer I (see Sect. 1.1) had some parts made of steel, which was indeed used since the beginning of the aerospace era, although for structural parts only. Steel was the standard choice for bracing wires and wing struts, being wood preferred for the fuselage and other important parts, like propellers, wing ribs, etc. This situation lasted until the mid-1930s, when wood started to be replaced by other materials due to the emerging limitations associated with its thermal stability at high speeds and for moisture absorption. The material universally used for most engineering structures was ready to replace wood and indeed steel tubes started to be used for the fabrication of the main fuselage parts, as well as for the wing spars (Cutler 2005). Afterward, the percentage of steel used in aircraft structures and engines was progressively reduced, passing from about 40% down to 5–15%, depending on the aircraft, as a consequence of the increased availability of light alloys (see Fig. 1.11, Chaps. 3 and 4) (Prasad and Wanhill 2017). On average, helicopters and aircrafts account for 5–10 wt% of their structure to be made of steel. Designers use steels in applications where there is no alternative valuable choice, especially for components that are regarded as highly critical, both for safety and performances. The reason for the limited structural aerospace applications of steel is mainly due to the comparatively high density of these alloys, being on average 3.5 times more than that of carbo-epoxy composites, 1.5 times that of titanium and its alloys, and 2.5 times the density of aluminum alloys.

However, there are still aircraft parts for which steel is an excellent choice, as it will be illustrated in the upcoming sections on:

- Bearings.
- Brake systems.
- Engine parts.
- Gearings.
- Undercarriage (landing gear).
- Wings (root attachments, carry trough, slot track, fasteners) and parts of engine pylons.

A detailed list of general aerospace applications of steels in provided at the end of this section.

5.5.1 Bearings

Bearings are rolling components usually comprising a cage, inner and outer raceways, and balls (or rollers) (Fig. 5.35). Bearing systems in aerospace must withstand rather severe conditions. Moreover, they have to guarantee reliable performances and predictable lifetimes, particularly those in very important positions, like the engine shaft bearings. Steels are still the preferred choice, either low or highly alloyed materials (Flower 1995). Helicopters' bearings are usually made of low-alloy steels that have been case-hardened. For applications where high strength is required in combination with corrosion resistance, a suitable choice is the AISI 440C martensitic stainless steel. Other examples of bearing steels are SAE S2100, AISI M50, M50NiL, and X30 steels (Flower 1995).

Two main factors may affect the performances of bearings:

- Grain flow, i.e., forging flow orientation with respect to the rolling contact surface. In this respect, a parallel, rather than perpendicular, orientation is recommended since it increases the fatigue life.
- Melting procedures, influencing the homogeneity, microscopic cleanliness (e.g., carbide and inclusion distribution) of the product.

Bearing life is limited by the high rotational and vibrational levels reached under service conditions and relevant temperature rise. Depending on the specific applications, they may reach peak values in the 350–400 °C in case of most demanding conditions, like turbine bearings (Flower 1995). Of course, much higher flash temperatures may be reached at asperity contact regions, although for comparatively much shorter times. Another important element is the corrosive environment, due to the lubricants, that may turn out to be an important lifetime limiting factor. The development of more performing engines would pose more challenging requirements to bearing systems too, in terms of higher rotational contact speeds and load capacity, longer life, and lower weight. Steel processing is continuously evolving, so that inclusion-free and cleaner materials can be produced, with generally positive

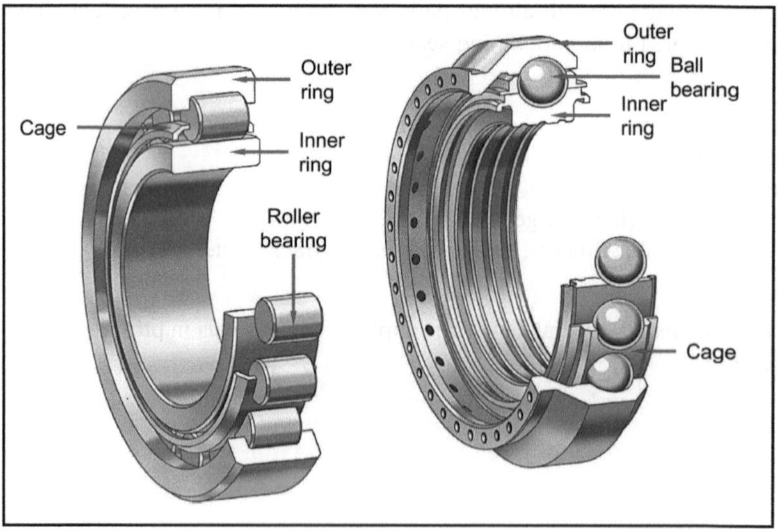

Fig. 5.35 Roller-bearing and ball-bearing structures. (Soares 2015, reproduced with permission of the Publisher)

effects on the expected lifetime of the bearing (Scott 1969). In this regard, the main specific requests for bearing materials are:

- Adequate rolling fatigue resistance.
- Corrosion resistance.
- High hardness, which may guarantee an adequate resistance to wear.
- Structural stability at the working stress and temperature conditions.

5.5.2 Brake Systems

In the early days of the aircraft era, airplanes had no brakes, and, once landed, during taxing they just exploited the frictional forces resulting from the contact between the aircraft wheels and tail and the grassy ground of the early runways. The very first aircraft braking systems were piston brakes derived from road vehicles. They became absolutely necessary when hard, concrete or asphalt, runways started to be used. These brakes were made of pads and drums, which looked very similar to brake systems of contemporary cars. However, this system is characterized by some disadvantages that were not compatible with the increasing needs in terms of braking performances of the newly developed aircrafts. In particular, the thermal expansion of both brake fluid and drum occurring during braking leads to the fading of the brake action, thus reducing brake performances and increasing the space needed to stop the aircraft. To overcome this drawback, expander tube brakes were introduced and became the standard for any kind of aircrafts over the 1930s–1950s period.

Expander tube brakes were then replaced progressively by diverse kinds of new disc brake systems: single disc, dual and multiple discs, and segmented-rotor disc brakes. Single disc brakes were installed on aircrafts with nonretractable landing gear (Jetley 2007). The system is built in such a way that the disc rotates together with the wheel assembly. Meanwhile, when the braking pressure is applied, a stationary caliper resists the rotation producing a frictional torque against the disc (Federal Aviation Administration 2012). Increasing flight speeds called for retractable landing gear systems, and for this reason, wheels had to become smaller and lighter, to save space and weight. The need for limiting the wheel and brake assembly size, still preserving an adequate stopping power, promoted the design of dual and, afterward, multiple-disc systems. A dual disc brake system is composed of two discs for each wheel, with lining on each side of the disc. A frictional force is developed when the linings are pressed against the discs. The multiple-disc concept is an obvious extension of the above brakes. An alternate stack of stators, usually made of steel, and bronze alloy discs provide a very effective frictional system, capable to dissipate large amounts of kinetic energy. Multiple-disc brakes were designed for large and heavy aircrafts and can still be found in cargo aircrafts. However, in multiple-disc brakes, heat dissipation is not optimized, and they have high-warp tendency, considering the very large frictional heat concentration that can be reached in these highly dense packed systems. This has suggested the development of the so-called segmented-rotor disc design. Segmented-rotor disc brakes are multiple-disc braking systems with an improved design, capable to overcome the cooling limitations of multiple-disc brakes, with obvious benefit in terms of brake efficiency and part durability (Jetley 2007). Their use is widely spread in air carrier and high-performance aircrafts and commercial airliners (Federal Aviation Administration 2012). The brake unit is constituted by a set of discs that include alternating rotors and stators built with slots or spacings among them, resulting in a better heat dissipation (Federal Aviation Administration 2012). Rotor discs rotate with the wheel, while stator discs are stationary with respect to the landing gear. When the brake fluid hydraulic pressure increases, pistons press stator discs against rotors, thus generating a friction between the two sets of discs, leading to a reduction in the rotational speed and eventually to stop the aircraft (Young 2018). The force generated by the braking system can be determined according to Eq. 5.7 (Young 2018):

$$F = \mu_b \left(W - \cos \gamma_g - L \right) \tag{5.7}$$

where:

F is the braking force.

μ_b is the braking coefficient. It is experimentally determined and depends on the airplane's weight and speed: when the speed decreases, the braking coefficient increases. The braking coefficient depends also on maximum slip that occurs between tire and runway, maximum braking torque, and tire-runway interface conditions (dry or wet runway, runway surface condition and profile, tire pressure and wear).

W is the airplane weight.
γ_g is the runway slope angle expressed in rad.
L is the instantaneous lift.

The lift is calculated according to Eq. 5.8.

$$L = \frac{1}{2}\rho S C_L v^2 \tag{5.8}$$

where:

L is the instantaneous lift force, expressed in N.
ρ is the air density expressed in kg/m^3.
S is the wing reference area expressed in m^2.
C_L is the lift coefficient (dimensionless).
v is the speed in m/s.

Thanks to the small angle approximation, $\cos\gamma_g$ can be taken equal to 1. Equation 5.7 can therefore be rearranged using the definition of instantaneous lift given by Eq. 5.9:

$$F = \mu_b\left(W - \frac{1}{2}\rho S C_L v^2\right) \tag{5.9}$$

There are of course different braking regimes, the most severe one being the emergency braking, i.e., the braking action necessary to abort take-off once the aircraft has reached a sub-maximum speed. In this case, in fact, the whole burden of stopping the aircraft relies on the wheel braking system, which cannot be backed by any additional stopping contribution, like thrust reversers, if available. Severe braking regimes may also result in the melting down of disc surface, with a consequent significant reduction in the braking power, the so-called *brake fade* (Young 2018). Although, disc replacement is in general necessary after an emergency braking event, still it is important that even these extreme conditions are not producing any catastrophic effect that may jeopardize the passenger and crew safety. Other elements that may influence the brake system performances are the additional and uneven loads on wheel, that may be acting in case of crosswinds and consequent compensation through wheel steering. The kinetic energy generated by the aircraft engines will also vary depending on the altitude at which the airport is located. Higher altitude leads to an increased landing and take-off speed owing to the less dense atmosphere (Jetley 2007). Material requirements for brake applications concern wear and heat resistance, heat dissipation capacity, and adequate toughness to withstand rapidly changing loading conditions. Steels exhibit properties that render them suitable candidates for this undercarriage subsystem, as proved by the important role they have played in different types of aircraft brake systems. Stators are steel plates having the lining material (usually a Cu-based alloy) on both sides, either riveted or deposited. Rotors are made of steel, and they may have a frictional surface

Table 5.22 Chemical composition (wt%) of AMS 6320 (Jetley 2007)

C	Mn	Cr	Mo	Si	S	P	V
0.28–0.33	0.45–0.65	1.0–1.5	0.40–0.60	0.55–0.75	0.04	0.04	0.2–0.3

made of sintered metal or cermet systems, to improve brake efficiency (Federal Aviation Administration 2012). Wear resistance directly depends on the hardness and, therefore, on the microstructure and on possible thermal treatments. Road vehicles, mainly cars and heavy duty vehicles, have cast iron or austenitic stainless steel discs. Stainless steel, like the AISI 410 steel, is the typical choice for motorcycles (Sarip et al. 2013). For aircraft applications one possibility is heat-resistant low-alloy steels. These steels have chemical compositions that allow heat treatment to be carried out to induce structural transformations in order to meet the severe requirements of aircrafts service conditions. One example of heat-resistant low-alloy steel used in the production of brake discs is the AMS 6320. Its standard chemical composition is given in Table 5.22.

The standard heat treatment for this steel is austempering (see Sect. 5.2.3). Austempering consists in heating the steel up to the relevant austenitic field. After homogenization, the steel is cooled down to a temperature above its M_s and held until isothermal transformation into bainite has occurred (Jetley 2007).

The increased dimension of the multiple-disc brakes rendered particularly critical the weight issue, considering the densities of the metallic alloys used for discs and stators. Carbon-carbon (C–C) composites turned out to be excellent candidates for the substitution of steels and bronze alloys. C–C composites afford an excellent combination of high strength, retained also at the highest peak temperatures, and tribological properties, once the oxidation and humidity problems have been addressed and solved. The early important aircraft application of C–C composites was in the brakes of the Concorde supersonic airliner in the mid-1970s (Stimson and Fisher 1980). By replacing a steel counterpart, this choice introduced an estimated 600 kg weight saving, being the density of C–C composites equal to 2.3 g/cm^3, i.e., one third of the density of steels. Another important material property for brake applications, in which discs behave by all means as heat sinks, is their heat capacity, which is 2.5 higher in C–C composites than in steels. Since then, these brake systems became the main choice of the aircraft industry, particularly for large airliner, considering the remarkable weight saving that a C–C composite design introduces with respect to the use of steel and bronze alloy (Gatt et al. 2017; Windhorst and Blount 1997). In the Boeing 737 airliner, the weight reduction is 300 kg approximately, involving a significant fuel saving over the whole lifetime of the aircraft. The better performances of the C–C brakes, capable to manage higher landing and take-off energies, are fundamental to keep up with the tendency to develop larger and, thereby, heavier airliners, like, for instance, the Airbus 380 and the Boeing 747-8. Moreover, these brakes would also allow shorter landing and take-off distances. Interestingly, C-C brakes have been developed and deployed on commercial aircrafts first, with respect to military ones, which actually on several occasions have benefited from this technological transfer. For instance, the brakes of the F-16 fighter are derived from the brakes of the Boeing 747-400; those of the C-130 Hercules are a derivative of the brake systems of the Airbus A320 and Boeing 777LR. Although out of the reach of the monograph, it is worth mentioning that at the present time, new materials, i.e., the C/SiC composites, originally developed for top cars, are now under investigation in view of their possible usage for aircraft brakes too (Fan et al. 2016; Narottam and Boccaccini 2012).

5.5.3 Gas Turbine Jet Engine Parts

AISI 403 and AISI 403 Cb stainless steels have been the base materials for compressor blades until the 1980s, when a new steel for these components, Custom 450 (see Table 5.19), was introduced. Custom 450 (also known as GTD-450), features better tensile strength, stress corrosion resistance and both higher fatigue corrosion and high-cycle fatigue resistance, than AISI 403 (Boyce 2012). Other steels used for the fabrication of compressor blades of GTs are the FV458, FV46 7, FV520, and 15-5 PH stainless steel and the Rex 706 steel (Meetham 1981). As concerns compressor discs, despite steel-based discs are still in production, the use of this material has been reduced over the years mainly for the increasing temperature requirements. Concerning the combustor, some parts, which are not directly in contact with high-temperature gases, are made of AISI 309 stainless steel (Boyce 2012). Both ferritic and austenitic stainless steels have been used for turbine components, in particular for turbine discs. The austenitic PH stainless steel A286 has been successfully used for several years in aircraft applications (see Sect. 5.4.3.3), and starting from 1965, this same material has been used in the fabrication of discs for land-based gas turbines. Occasionally, A286 has been replaced by the M152 steel (12% Cr and 2–3% Ni), exhibiting a higher tensile and an intermediate rupture strengths, this latter roughly in between the rupture strength of Cr-Mo-V steel (1%Cr-1.25%Mo-0.25%V) and that of A286 (Boyce 2012). Stainless steels are used in the exhaust ducts of the engine too, where they are exposed to relatively high gas temperatures, in particular when the afterburner is enabled. For these parts, an important request is not to transfer excessive heat to the surrounding aircraft structure. A possible approach is to protect the hot parts with insulating "blankets" (Prasad and Wanhill 2017).

5.5.4 Gearboxes

Gearboxes are particularly critical in helicopters, since they have to match the rotational speed of the turbine with that of the rotors (Flower 1995). Concerning aircrafts, gearing systems are important too and are used to drive accessory units

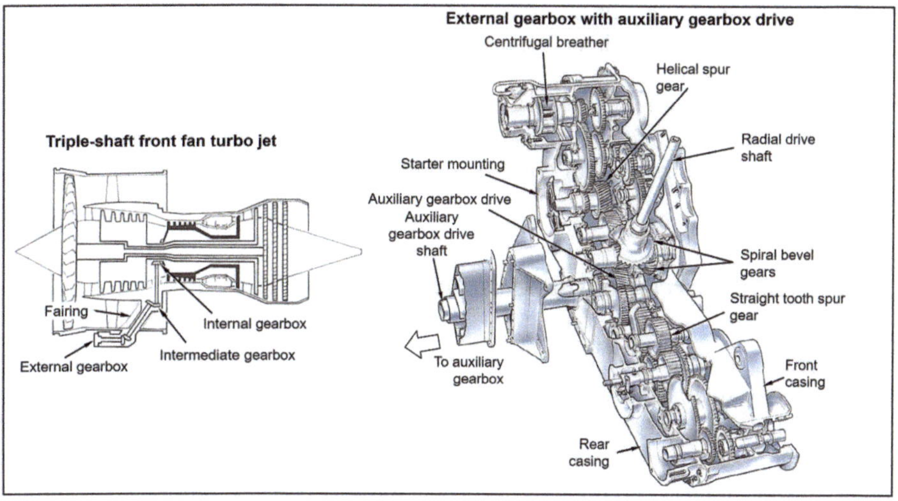

Fig. 5.36 Position of gearing systems in a triple-shaft front fan turbo jet and assembly of an external gearbox with auxiliary gearbox drive. (Rolls-Royce 1996, images courtesy of Rolls-Royce plc)

connected to a rotating engine shaft, as illustrated in Fig. 5.36. Accessory units have to provide the power for aircraft hydraulic, electrical and pneumatic systems, pumps and other control systems, all of great importance, both for performance and safety reasons. Therefore, they require an extreme reliability, also as concerns materials properties (Rolls-Royce 1996).

Materials used for gearing purposes must meet the following main requirements:

- Adequate resistance to surface degradation phenomena, such as wear, pitting, and scuffing.
- Corrosion resistance in oil environments over a temperature range of 90 °C–130 °C.
- Sufficient strength for withstanding both static and fatigue loads.
- Toughness for resisting against shock loads.

For these reasons, gearing steels are usually subjected to case-hardening, either nitriding or carburizing. Typical nitriding steels for these applications are Nitralloy N and 32CDV13 (see Sect. 5.4). Two examples of conventional carburizing steels are AMS 6265 and BS S156, replaced in some engines by advanced carburizing steels, like VASCO X-2M and Pyrowear 53.

The performances of gearing materials may be affected by metallurgical factors, the most important being:

- Core microstructure. Since surface hardening may alter the properties of the surface and substrate, it is important to check for an adequate toughness of the core.
- Grain size. Fine and uniform grain size is preferred, and usually an ASTM grain size of 6–8 is required for matching fatigue strength and ductility requirements.
- Steel cleanliness in terms of type, size, and shape of inclusions.

5.5.5 *Undercarriage Applications*

As regards undercarriage components, the landing gear is of course particularly important, considering its central role in aircraft operations, and steels have been a standard choice over the years (Fig. 5.37). Landing gears are subjected to repeated high dynamic loads on landing: civil aircrafts are designed for 30,000–60,000 flights, while military aircrafts are designed for an expected 5000–10,000 flights, which are spread over a 20–30 year operational time. The landing gears have also to sustain the full weight of the aircraft, i.e., masses up to several hundreds of tons (e.g., 380 t for Boeing 747), during land operations, e.g., taxing, and parking. Another important requirement is a design that reduces not only the weight but also the volume of the undercarriage in the retracted condition, since this volume, and weight too, competes with the volume available to the payload.

High-strength low-alloy steels are excellent candidates for landing gear applications. A few examples of this kind of alloys are the high-strength steels (see labels in Fig. 5.37): S99 (WN 1.6745), AISI 4340, AISI 4340M (alternatively known as 300M), and AMS 6487. In addition to conventional steels, the family of advanced AerMet steels, like AerMet 100 (see Sect. 5.3.2), are also used for undercarriage applications.

5.5.6 *Wings and Engine Pylons*

Steels are used for other important structural aircraft components, like wing root connections, slat tracks (i.e., part of the leading edge of the wings), and joints/fittings of engine pylons (Cantor et al. 2001). Wing roots and parts of the engine pylons are particularly critical, and the use of steels is still recommended. Wing roots connect wings to the fuselage and are subjected to diverse stress conditions, particularly static, dynamic, and fatigue loads. Slat tracks are moving beams that allow the relative displacement between the wing and the front part of the wing itself that enhances the lift during take-off and landing (see Fig. 3.1). Owing to their high strength and damage tolerance, maraging steels are an excellent choice for these parts. Fittings exhibit quite a broad range of odd design and relevant stress conditions, depending on the specific parts of the aircraft they are connected to. Fasteners are relatively small items, although often in critical positions for the safety and the

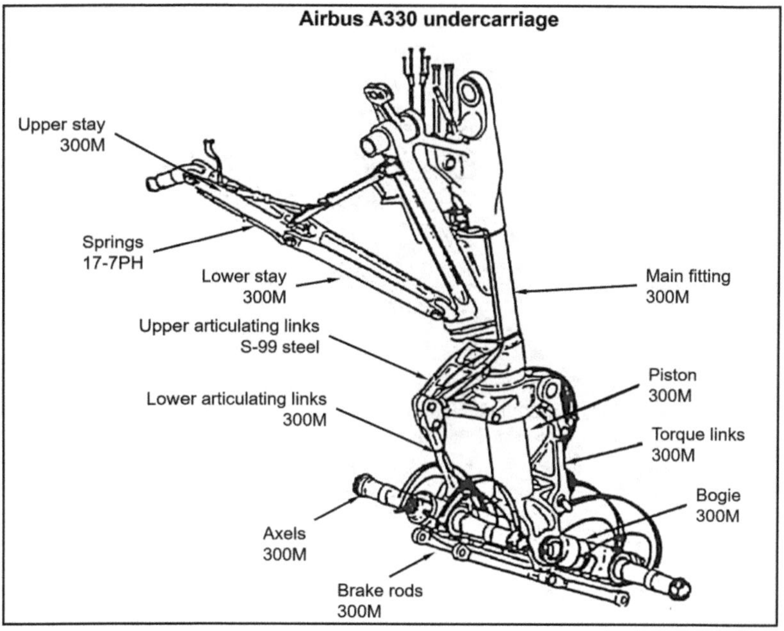

Fig. 5.37 Airbus A330 undercarriage with the indication of steel components. (Flower 1995, reproduced with permission of the Publisher)

efficient operation of aircrafts. Typical steels used for the fabrication of fasteners are 17-4PH and A286 (Speck 2015). Dealing with fasteners, an important issue is the way in which they are connected to the main part, bolting and riveting being the most common approaches. In this context, AerMet 100 can be a good candidate (Speck 2015). All parts of the joint are to be selected in order to have compatible mechanical properties, with no excessive increase of internal stresses, and electrochemical compatibility.

5.5.7 Aerospace Applications of Steels: An Overview

Steels introduced in this chapter are used for several aircraft components. They are selected on the basis of safety, critical loading, and environmental conditions. Maraging steels are the best choice for a large variety of components, as shown by the information in Table 5.23.

AerMet steels are employed in the fabrication of components that require particularly high strength, toughness, and ductility. They have the advantage over titanium alloys, of which AerMet steels can be regarded as direct competitors, of reduced dimensions of the parts, due to their higher density (Novotny and Maurer 2007). This aspect may be a decisive selection criterium in case space saving is an issue.

Table 5.23 Aircraft components made of maraging steels (ASM International 1993; Llewellyn and Hudd 1995; Megson 2013)

Component/part	Component/part
Aircraft arrest hooks	Hydraulic hoses and bolts
Aircraft forgings	Landing gear components
Bearings	Missile cases
Cannon recoil springs	Slat tracks
Cargo systems	Transmission shafts
Fan shafts of commercial jet engines	Undercarriage pivot brackets
Fasteners	Wing lateral panels and wing roots attachments

Table 5.24 Aircraft components made of AerMet steels (Steel Service Aerospace Corp. 2016; Novotny and Maurer 2007; Prasad and Wanhill 2017)

Component/part	Steel
Actuators	AerMet 100, AerMet 310
Armor	AerMet 100, AerMet 310
Connecting rods	AerMet 340
Crank shafts	AerMet 340
Drive shafts	AerMet 100, AerMet 310, AerMet 340
Fasteners	AerMet 100, AerMet 310
Jet engine shafts	AerMet 100, AerMet 310
Landing gear components	AerMet 100, AerMet 310
Ordnance	AerMet 100, AerMet 310
Springs	AerMet 340
Structural tubing	AerMet 100, AerMet 310

The main applications of AerMet alloys are summarized in Table 5.24.

Depending on the atmosphere, the temperature range for austenitic stainless steel applications can reach 900–1100 °C (Bhadeshia and Honeycombe 2006; Young 2015). Therefore, AISI 321 is an ideal candidate for the production of some aerospace components, such as piston engine exhaust manifolds, exhaust ducts, and flanges. In fact, this steel retains stable and adequate mechanical properties over a temperatures interval between 400 and 820 °C also resisting a to high-temperature oxidation. Other applications of austenitic stainless steels are listed in Table 5.25.

The components for which martensitic stainless steels are typically used are listed in Table 5.26. Other possible applications for martensitic stainless steels are axels cams, shafts and springs, blading in steam, and gas turbines, operating at temperatures up to 540 °C (ASM International 1993; Campbell 2008; Meetham 1981; Prasad and Wanhill 2017).

The 58 HRC hardened condition is the standard for the AISI 440C used in the production of rolling bearings operating in corrosive environments, where there is no other way to protect the material from corrosion (ASM International 1993; Campbell 2008; Harnoy 2003; Totten 2007).

Table 5.25 Aircraft components made of austenitic stainless steels (Boyce 2012; Prasad and Wanhill 2017)

Component/part	Steel
Exhaust components and medium-temperature engine structural parts	AISI 309, AISI 310, AISI 316, AISI 316L, AISI 321
Fuel tanks	AISI 304, AISI 304L
Jet engine parts	AISI 321, AISI 347
Parts of the combustor that are not directly in contact with high temperatures	AISI 309
Pins for compressor variable vanes	Nitronic 60
Piston engine exhaust manifolds and exhaust ducts and flanges	AISI 321

Table 5.26 Aircraft components made of martensitic stainless steels (ASM International 1993; Boyce 2012; Campbell 2008; Harnoy 2003; Meetham 1981; Prasad and Wanhill 2017)

Component/part	Steel
Ball bearings	AISI 440C
Bolts	AISI 410
Compressor blades	AISI 403, AISI 410, AISI 414
Spray nozzles	AISI 440
Valves and valve needles	AISI 410, AISI 431

Table 5.27 Aircraft components made of PH stainless steels (Boyce 2012; Speck 2015)

Component/part	Steel
Afterburner parts	A286
Aircraft bulkheads	15-7Mo
Compressor blades	A286, 15-5PH
Door surrounds	15-5PH, 13-8Mo
Fasteners	A286, 13-8Mo, 17-4PH
Landing gear components	15-5PH, 13-8Mo
Nose gear door	15-5PH, 13-8Mo
Pylon and engine mounting parts	15-5PH
Rockets and missiles propulsion systems	15-5PH
Slat tracks	17-4PH, 15-5PH
Springs	A286, 17-7PH, 15-7Mo
Tail cones	A286
Turbine wheels	A286
Valves	15-5Mo, 13-8Mo
Welded and brazed honeycomb panels	15-7Mo
Wing flap tracks	17-4PH, 15-5PH

Thanks to their unique combination of mechanical properties and corrosion resistance, PH stainless steels are a class of alloys that is massively used in several parts and componets of aircraft structure and engine, the main ones being listed in Table 5.27.

References

Angel T J (1954) Formation of martensite in austenitic stainless steels. ISIJ International (Japan) 177: 165–174

Ashby M F (2005) Materials – A brief History. Philosophical Magazine Letters 88 (9–10): 749–755

Ashcroft N W, Mermin N D (1988) Solid State Physics. HRW International Editions, Hong Kong

ASM International (1992) ASM Handbook Vol. 3 – Alloy phase diagrams. ASM International, Materials Park, Ohio

ASM International (1993) ASM Handbook Vol. 1 – Properties and Selection: Iron, Steel and High Performance Alloys. ASM International, Materials Park, Ohio

ASM International (2008) ASM Handbook Vol. 15 – Casting. ASM International, Materials Park, Ohio

ASM International (2016) ASM Handbook Vol. 4E – Heat treating of Nonferrous Alloys. ASM International, Materials Park, Ohio

Bhadeshia H. K. D. H, Honeycombe R. W. K. (2006) Steels – Microstructure and Properties, 3rd edn. Butterworth-Heinemann, Oxford

Boxman R L et al (1996) Handbook of Vacuum Arc Science and Technology: Fundamentals and Applications. Noyes Publications, Park Ridge, New Jersey

Boyce M P (2012) Gas Turbine Engineering Handbook, 4th edn. Butterworth-Heinemann, Oxford

Buchwald V F, Mosdal G (1985) Meteoric Iron, Telluric Iron and Wrought in Greenland. Man & Society 9: 3–49

Callister W, Rethwisch D (2012) Fundamentals of Materials Science and Engineering: An Integrated Approach, 4th edn. John Wiley & Sons Ltd

Campbell F (2006) Manufacturing Technology for Aerospace Structural Materials. Elsevier

Campbell F (2008) Elements of Metallurgy and Engineering Alloys. ASM International

Cantor B et al (2001) Aerospace Materials. Institute of Physics Publishing

Chandler H (1996) Heat Treater's Guide – Practices and Procedures for Nonferrous Alloys. ASM International

Cobb H M (2010) The History of Stainless Steel. ASM International

Cutler J (2005) Understanding Aircraft Structures, 4th edn. Wiley-Blackwell

De Cicco H, Luppo M, Gribaudo L et al (2004) Microstructural Development and Creep Behavior in A286 Superally. Materials Characterization 52: 85–92

Demeri M (2013) Advanced High-strength Steels: Science, Technology and Applications. ASM International

Donachie M, Donachie S (2002) Superalloys, a Technical Guide. ASM International

Everhart J L (1971) Engineering Properties of Nickel and Nickel Alloys. Plenum Press, New York

Fan s et al (2016) Progress of Ceramic Matrix Composites Brake Materials for Aircraft Applications. Reviews on Advanced Materials Science 44 (4): 313–325

Farrar J (2004) The Alloy Tree – A guide to Low-Alloy Steels, Stainless Steels and Nickel-base Alloys. Woodhead Publishing Limited and CRC Press

Federal Aviation Administration (2012) Aviation Maintenance Technician Handbook-Airframe Volume 2. Federal Aviation Administration

Flint G, Oldfield J, Dautovich D P (2010) Corrosion of Iron Nickel Alloys and Maraging Steel. Shreir's Corrosion: 1789–1801

Flower H (1995) High Performance Materials in Aerospace. Springer-Science+Business Media

Gatt A et al (2017) Hydro-Mechanical Coupling in Unstable Aircraft Braking Systems. In: Harvie J M, Bawersad J (ed) Shock & Vibration, Aircraft/Aerospace, Energy Harvesting, Acoustics & Optics, Volume 9 – Proceedings of the 35th IMAC, A Conference and Exposition on Structural Dynamics, Garden Grove, California, 2017

Graves J (1994) Effect of Heat Treatment on the Microstructure and Properties of AerMet 100 Steel. Army Research Laboratory. Available via DIALOG. https://apps.dtic.mil/dtic/tr/fulltext/u2/a284387.pdf Accessed 30 March 2019

Harnoy A (2003) Bearing Design in Machinery. Marcel Dekker Inc

Hosford W F (2010) Physical Metallurgy. CRC Press, Boca Raton

Hosford W F (2012) Iron and Steel. Cambridge University Press, Cambridge

Jetley S (2007) Martempering to Improve Wear Properties of Aircraft Brake Steel Rotors. Journal of Industrial Technology 23 (2)

Kaieda Y, Oguchi A (1985) Strain-induced transformation and plastic deformation behavior of a 17Cr-7Ni-1Al steel at high hydrostatic pressure. Journal of Materials Science 20: 1847–1858

Leitner H et al. (2011) Strengthening behavior of Fe–Cr–Ni– Al–(Ti) maraging steels. Materials Science and Engineering A, Structural Materials Properties, Microstructure and Processing 528 (15): 5264–5270

Llewellyn D T, Hudd R C (1995) Steels: Metallurgy and Applications 3rd edn. Butterworth-Heinemann

Madias J (2013) Electric Arc Furnace. In: Seetharaman S (ed) Treatise on Process Metallurgy Volume 3: Industrial Processes. Elsevier, p 271–300

Meetham G W (1981) The Development of Gas Turbine Materials. Applied Science Publishers Ltd, London

Megson T (2013) Aircraft Structures for Engineering Students, 5th edn. Butterworth-Heinemann, Oxford

Miller M K, Burke M G (1991) Characterization of copper precipitation in a 17/4PH steel: a combined APFIM/TEM study. In: Proc. 5th Intern Conf Environ Degrad Mater Nucl Power Syst React, LaGrange Park, 1991

Mitchell A (2005) Solidification in Remelting Processes. Materials Science and Engineering A 413–414: 10–18

Moshka O et al (2015) Adressing the Issue of Precipitates in Maraging Steels – Unambiguous Answer. Materials Science and Engineering A 638: 232–239

Mouritz A (2012) Introduction to Aerospace Materials. Woodhead Publishing Limited

Narottam B, Boccaccini A (2012) Ceramic and Compostites Processing Methods. John Wiley & Sons Ltd

Novotny P, Maurer G (2007) Ultra High Strengh Steels vs Titanium Alloys. Advanced Materials and Processing 165(11): 37–40

Pardal JM et al (2005) Modeling of precipitation hardening during the aging and overaging of 18Ni–Co–Mo–Ti maraging 300 steel. Journal of Alloys Compounds 393(1–2):109–113

Pepperhoff W, Acet M (2001) Constitution and Magnetism of Iron and its Alloys. Springer, Berlin Heidelberg

Prasad E N, Wanhill R (2017) Aerospace Materials and Material Technologies Volume 1: Aerospace Materials. Springer

Prasad E N, Wanhill R (2017) Aerospace Materials and Material Technologies Volume 2: Aerospace Material Technologies. Springer

Rolls-Royce (1996) The Jet Engine 5th edition. Rolls-Royce plc, Derby

Sarip S et al (2013) Analysis of the Transient Thermomechanical Behaviour of a Ligthweight Brake Disc for a Regenerative Braking System. In: SAE-China, FISITA (eds) Proceedings of the FISITA 2012 World Automotive Congress Volume 8: Vehicle Design and Testing (II). Lecture Notes in Electrical Engineering, vol 196. Springer, Berlin, Heidelberg, p 1075–1088

Saul G, Roberson J A, Adair A M (1970) The effects of thermal treatment on the austenitic grain size and mechanical properties of 18 Pct Ni maraging steels. Metallurgical and Materials Transactions B 1 (2): 383–387

Savoie M, Esnouf C, Fournier L et al (2007) Influence of Ageing Heat Treatment on Alloy A-286 Microstructure and Stress Corrosion Cracking Behavior in PWR primary water. Journal of Nuclear Materials 360; 222–230

Schifo J F, Radia J T (2004) Theoretical/Best Practice Energy Use In Metalcasting Operations. U.S. Department of Energy Industrial Technologies Program, Washington, DC

Scott D (1969) The Effect of Steelmaking, Vacuum Melting and Casting Techniques on the Life or Rolling Bearings. Vacuum 19(4):167–169

Service Steel Aerospace Corp. (2016) AerMet ® 100 Alloy Data Sheet. Available via DIALOG. https://www.ssa-corp.com/documents/Data%20Sheet%20AerMet100.pdf. Accessed 16 April 2017

Sha W, Guo Z (2009) Maraging steels – Modeling of microstructure, properties and applications. CRC Pess

Shi X, Zeng W et al (2016) Study of the Microstructure and Mechanical Properties of AerMet 100 Steel at the Tempering Temperature around 482 °C. Journal of Alloys and Compounds 679: 184–190

Soares C (2015) Gas Turbines: A Handbook of Air, Land and Sea Applications, 2nd edn. Butterworth-Heinemann

Speck J A (2015) Mechanical Fastening, Joining and Assembly. CRC Press

Stimson I L, Fisher R (1980) Design and Engineering of Carbon Brakes. Philosophical Transactions of the Royal Society A Mathematical Physical and Engineering Sciences 294 (1411): 583–590

Stodart J, Faraday M (1820) Experiments on the Alloys of Steel, made with a View to its Improvement. Philosophical Magazine 56 (257): 26–35

Tewari R et al (2000) Precipitation in 18 wt% Ni Maraging Steel of Grade 350. Acta Materialia 48 (5): 1187–1200

The AISE Steel Foundation (1998) The Making Shaping and Treating of Steel, 10th edn. The AISE Foundation, Pittsburg

Totten G E (2007) Steel Heat Treatment – Metallurgy and Technologies, 2nd edn. CRC Press

Toulouevski Y, Zinurov I (2013) Innovation in Electric Arc Furnaces, 2nd edn. Springer

Turkdogan E T (2010) Fundamentals of Steelmaking. Maney Publishing

Tylecote R F (2002) A History of Metallurgy, 2nd edn. Maney Publishing

Verhoeven J D (2007) Steel Metallurgy for the Non-Metallurgist. ASM International

Viswanathan U et al (1988) Effects of Aging on the Microstructure of 17-4 PH Stainless Steel. Materials Science and Engineering A 104: 181–189

Windhorst T, Blount G (1997) Carbon-Carbon Composites: a Summary of Recent Developments and Applications. Materials & Design 18 (1): 11–15

Würzinger P, Rabitsch R, Meyer W (2004) Production of Maraging Steel Grades and the Influence of Specified and Nonspecified Elements for Special Applications. Journal of Materials Science 39 (24): 7295–7302

Xu X et al (2002) Tree-ring Formation during Vacuum Arc Remelting of INCONEL 718: Part I. Experimental Investigation. Metallurgical and Materials Transactions A 33 (6): 1795–1804

Yang S et al (2015) Phase transformation and its effect on mechanical properties of C300 weld metal after aging treatment at different temperatures. Journal of Iron and Steel Research (International) 22(6): 527–533.

Yeli G, et al (2017) Sequential Nucleation of Phases in a 17-4PH Steel: Microstructural Characterization and Mechanical Properties. Acta Materialia 125: 38–49

Young D (2015) High temperature oxidation and corrosion of metals. Elsevier, Oxford

Young T M (2018) Performance of the Jet Transport Airplane. John Wiley & Sons Ltd

Further Reading

ASM International (1994) ASM Specialty Handbook – Stainless Steels. ASM International, Materials Park, Ohio

Faivre D, Frankel R (2016) Iron Oxides: from Nature to Applications. Wiley-VHC, Weinheim

Gale W F, Totemeyer T C (2004) Smithells Metals Reference Book 8th ed. Butterwoth-Heinemann

Chapter 6
Superalloys

6.1 Introduction

In the first half of the twentieth century, stainless steels were employed in high-temperature structural applications, including some components of the newly developed jet engines (see Sects. 2.1 and 2.2). Materials used in hot parts of jet engines must retain elevated performances and structural stability for prolonged periods under service conditions involving high mechanical stresses and corrosive agents. The increasing performances of the gas turbine engines evidenced some limitations of steels. This fostered the development of a new class of metallic alloys, named *superalloys* in view of their remarkable structural properties retained even up to relatively high temperatures. The main features of superalloys can be summarized as follows:

- Capability to withstand loads at operating temperatures that are important fractions of their incipient melting point, expressed in terms of the ratio T_{op}/T_m, i.e., operating temperature versus incipient melting temperature ratio, with temperatures expressed in Kelvin. In the hot spots of gas turbine components, like the blades of the turbine high-pressure stage, values in excess of 0.70 can be achieved.
- Mechanical stability over long times. For high-temperature structural applications, materials with low creep rates are desirable. Other mechanical properties, such as ultimate tensile strength, tensile yield strength, and fracture toughness, are important too and should be preserved over the time, to guarantee safe performances of the structural components.
- Surface stability in aggressive environments. The highly corrosive environments generated by the combustion of jet fuels and ingestion of contaminants into the gas turbine engine (see Sect. 8.2.4) may induce important gas phase high-temperature corrosion phenomena, enhanced by the possible condensation of salt deposits, either solid or liquid (see Sect. 8.2.5).

S. Gialanella, A. Malandruccolo, *Aerospace Alloys*, Topics in Mining, Metallurgy and Materials Engineering, https://doi.org/10.1007/978-3-030-24440-8_6

The early superalloys were stainless varieties of the existing steels, subsequently officially named *superalloys*. Patents for Al and Ti additions to the Nichrome-type alloys were issued in the 1920s. The superalloy industry really emerged with the introduction of a Co-based alloy, initially used in dentistry: the Vitallium. It was selected as a candidate material capable to satisfy the high-temperature requirements of the emerging gas turbine aircraft engines. At the same time, other alloy systems attracted research interests. This led to the widening of the superalloy family tree. Nowadays, a widely accepted definition of superalloys refers to a group of austenitic Fe, Co, or Ni (group VIII A)-based alloys that are extensively used for applications involving operating temperature exceeding 600 °C (Donachie and Donachie 2002; Reed 2006). Interestingly, although all of them are austenitic, i.e., they are face-centered cubic (*fcc*), nickel only exhibits this crystalline structure from room temperature up to the melting point. The crystallographic structure of cobalt at room temperature and below is hexagonal close packed (*hcp*). Upon heating, the structure changes martensitically into the *fcc*-Co polymorph, stable up to the melting temperature. The transition temperature for pure Co from the low- to the high-temperature polymorph, i.e., *austenite start* (A_S, see Sect. 9.5.1), is 430 °C. The reverse transformation occurs, with some hysteresis, at the *martensite start* temperature (M_S), equal to 370 °C. Iron is body-centered cubic (*bcc*) up to 911 °C (see Sect. 5.1). Above this temperature, the stable phase is the *fcc* austenite (γ). Therefore, the common austenitic phase on which all superalloys are based needs to be stabilized in the cobalt and iron alloys, using suitable stabilizers elements.

The austenitic phase exhibits several interesting features:

- The close-packed atomic arrangement would generally feature intrinsically lower diffusivities, than, for instance, a *bcc* structure.
- A lower diffusivity results in a good stability of the precipitate phases and of the alloy microstructure. Creep resistance too is beneficially influenced by a low diffusivity (see Sect. 6.5.1).
- The hexagonal close-packed structure, although sharing the same packing factor of the *fcc* structure, still has the important limitation of an intrinsic brittleness due to the limited number of slip systems (see Sects. 3.3 and 4.2), just 3 as compared to the 12. Actually, also the *bcc* structure has 12 slip systems, but none of them comprises close-packed planes, like *fcc*. This feature has important consequences on the stability of the dislocation in *bcc* structures and, thereby, on the ductility of the alloy that turns out to be generally not so high as in *fcc* structures.
- The close-packed structure and the large number of slip systems of austenite are prerequisites for better mechanical properties, as concerns also toughness, ductility, and mechanical resistance.

The austenitic structure displays along close-packed directions elevated stiffness that can be modulated by selecting alternative directions on purpose. This is the case of turbine blades of the high-pressure stage. To enhance their thermal fatigue resistance, their main axis is aligned parallel to the more compliant <001> direction (see Sect. 6.5.1).

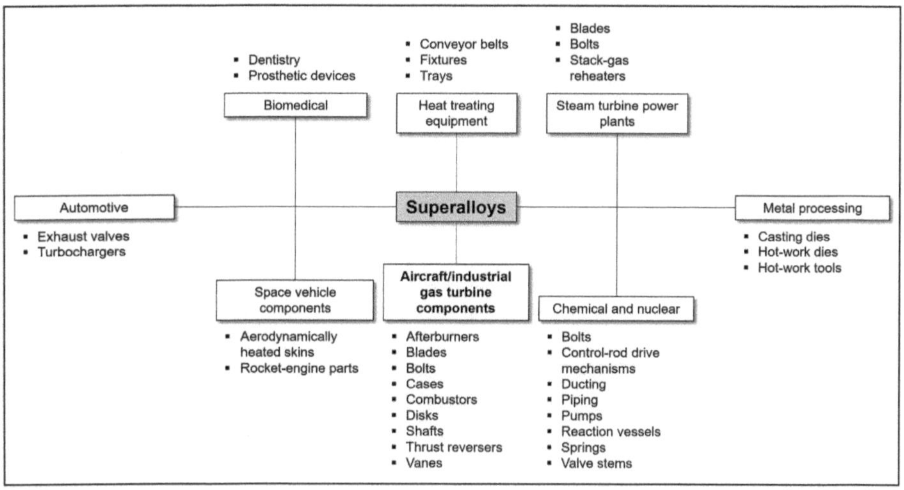

Fig. 6.1 A scheme with the main application fields of superalloys. (Donachie and Donachie 2002)

One of the main drives for the development of superalloys has been their use in aircraft GTEs. Thanks to their properties, superalloys represent nowadays the highest percentage of all materials used in the fabrication of gas turbine engines. However, the aerospace industry is not the only field where superalloys find successful applications. An overview of the most important applications of superalloys is shown in Fig. 6.1 (Donachie and Donachie 2002; Reed 2006).

Considering the application of superalloys in the production of gas turbine components, material capabilities alone would not be sufficient to bear the extreme conditions of the hottest stages. The situation is well represented by the curves of the Larson-Miller parameter in Fig. 6.2 (see also Sect. 4.3.1.5), for different classes of alloys interesting for high-temperature structural applications. The nickel-based superalloys SRR99, CMSX-4, and the RR3000 all show a creep resistance that is definitely better than other titanium alloys, intermetallic alloys, like the TiAl and Ti_3Al, in addition to iron and nickel aluminide alloys. The refractory alloys only, based on niobium, tungsten, and rhenium, can do better (see Sect. 9.2), although the elevated density is certainly a detrimental parameter as far as aerospace applications are concerned.

Indeed, specific processing routes have been developed on purpose to produce high-pressure hollow, air-cooled blades with single-crystalline superalloy, having remarkable mechanical properties in association with extremely high surface stability against corrosive attack. Superalloys have played a key role in the development of high-performance gas turbine engines since the 1950s. At that time, 10% only of the total weight of an aircraft engine was made of superalloys; the percentage reached 50% in 1985.

An obvious approach to increasing the efficiency of a jet engine is to raise its operating temperature (see Fig. 2.6, Brayton cycle). An important limiting factor are

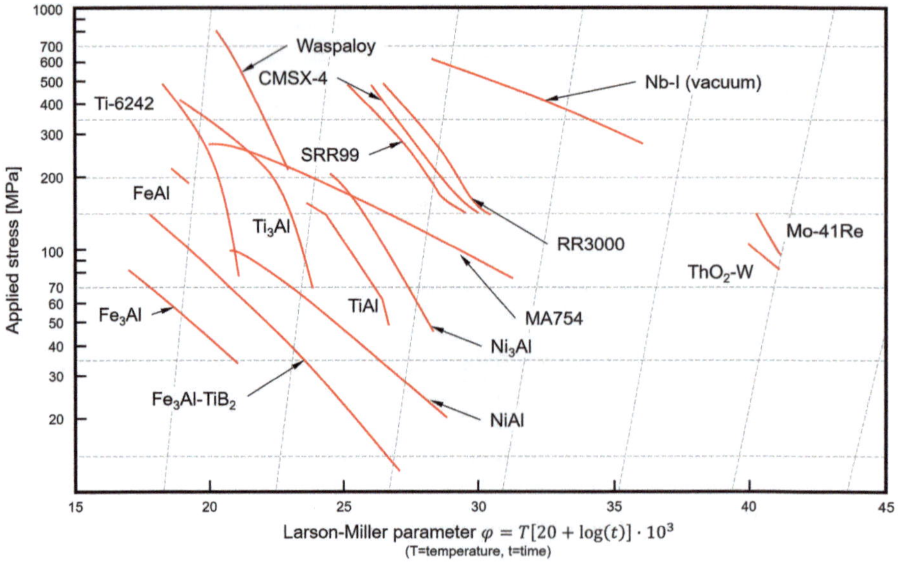

Fig. 6.2 Larson-Miller parameter plot for several high-temperature alloys. (Redrawn from Reed 2006)

the materials used for the most demanding parts of the engine (see point 3 in the Brayton cycle in Fig. 2.7). In this regard, the high-pressure blades all through the development history of GTEs have been a most challenging component as concerns material capabilities, which can be ranked, in the first place, with reference to the turbine entry temperature (TET, see Figs. 1.14, 2.8, and 2.20). Indeed, the operating temperatures are affecting the strength of the materials but also the deformation rate in the relevant creep regime, largely determined by diffusive phenomena (see Sect. 6.5.1). Still, diffusion is paramount for the corrosive phenomena occurring at high temperatures, contributing to reduce the service time of turbine blades. The success of the superalloys is based on the combination of high-temperature structural properties with very good resistance to high-temperature corrosion (see Sect. 8.2). This aspect is well illustrated by Fig. 6.3, in which the behavior of several alloy groups is comparatively considered (Geddes et al. 2010). The damages, particularly on the blade surfaces, induced by creep and oxidation, can be important nucleation sites for micro-cracks and flaws. In order to avoid catastrophic propagation to the rest of the component, a further requirement for superalloys used in the turbine blading is an excellent damage tolerance and high fracture toughness, particularly under a fatigue regime.

Several superalloy parts of a GTE, in addition to high-pressure stage turbine blades, would usually operate at temperatures in excess of 600 °C, a fundamental limit for titanium alloys (see Chap. 4). The success of the superalloys is to be ascribed to the wide range of properties that can be attained through a suitable selection of rather complex compositions and relevant processing routes, including

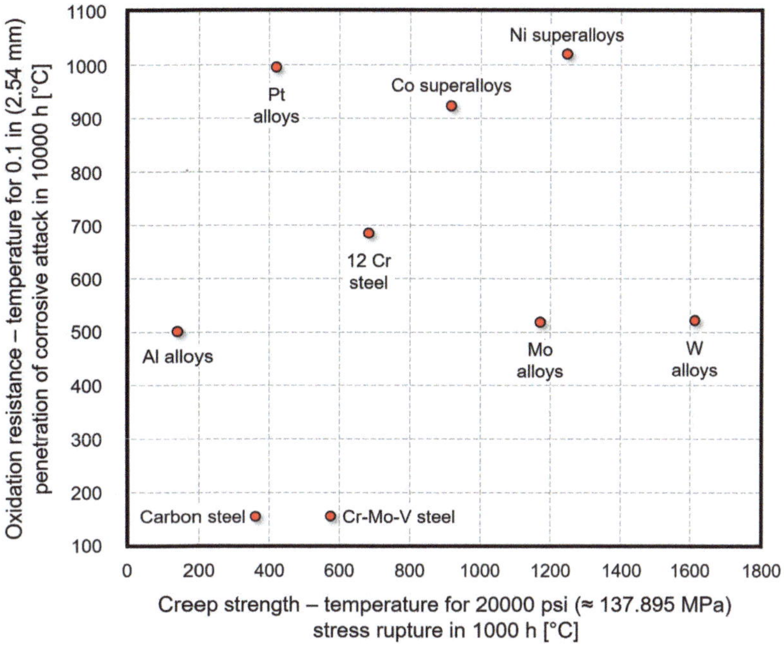

Fig. 6.3 Interplay between environmental resistance and creep strength for several families of metallic alloys. (Redrawn from Geddes et al. 2010)

the post-processing thermal treatments. All these aspects contribute to optimize the diverse strengthening mechanisms (Fig. 6.4). Appendix 5 illustrates the main specific effects of the most important alloying elements in each superalloy family. Appendix 6 presents a set of information concerning the phases that may form in different superalloys. The superalloy design has been importantly guided by the need for better high-temperature corrosion resistance. The general strategy of forming a protective scale of reaction products, which for all superalloy families are based on chromium and aluminum oxides (see Sect. 8.2), has been refined through the selection of minority elements, like yttrium, scandium, cerium, lanthanum, etc. (the so-called reactive elements), that resulted in an improved barrier effect of the surface reaction layer (see Sect. 7.3.3).

At the present time, the environmental advantages that can be achieved through alloy density reduction and consequent improved engine efficiency are becoming major drives for further developments (see Chap. 2 and particularly Sect. 2.6).

As concerns processing, the early steps of the production route, common to the three families of superalloys, are melting and refining. A possible process flowchart for the production of superalloy components is given in Fig. 6.5. During primary melting of the raw material or scrap, the chemical composition of the alloy is set. Common primary melting routes for superalloys are VIM (see Sect. 5.3.2) and AOD (see Sect. 5.3.3) of material coming from EAF melting (EAF + AOD or arc AOD process, see Sects. 5.3.3 and 5.3.1, respectively). Once melting and conversion

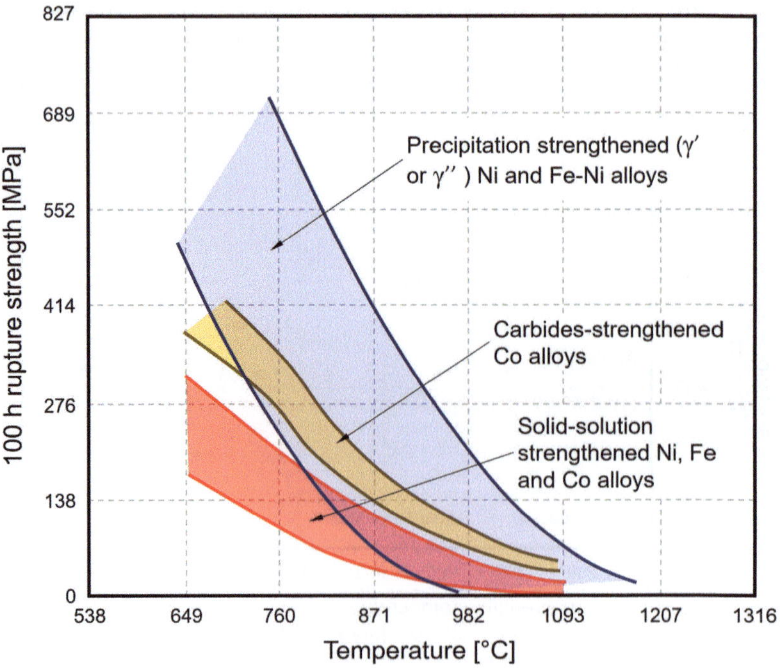

Fig. 6.4 Effect of different strengthening mechanisms on creep-rupture life of cobalt-, iron-, and nickel-based superalloys. (Redrawn from ASM International 2000)

processes are accomplished, the melt is sent to the casting unit. Castings of superalloys produced via arc AOD process are poured ingots, electrodes for secondary melting, or continuously cast products. VIM is especially employed for the production of electrodes for secondary melting, when higher purity standard is an issue. ESR or VAR secondary processes, applied to superalloys produced via arc AOD, can compensate to some extent for the lower quality of the produced material by increasing the cleanliness and improving the solidification structures. As concerns secondary melting, superalloys are generally processed either through electroslag (see Sect. 5.3.5) or vacuum arc (see Sect. 5.3.6) remelting.

For obtaining finished or semifinished products, casting is definitely the most common and suitable process, when maximum strength levels are required, being typically associated with an optimum microstructural control. High volume fractions of ordered precipitates, i.e., γ' and γ'', as well as the control of carbide concentration and morphology can be attained using multistep post-processing thermal treatments (see Sect. 6.6). Concerning the grain size of a cast alloy, it is generally coarser than that obtained from forging processes, unless rapid chill casting is used. Investment casting allows to control the grain alignment, using directional solidification techniques (see Sect. 6.5), resulting in exceptional high-temperature strength and creep resistance, when the alloy component is subjected to stresses along the grain growth direction (Geddes et al. 2010). Hot-rolling and

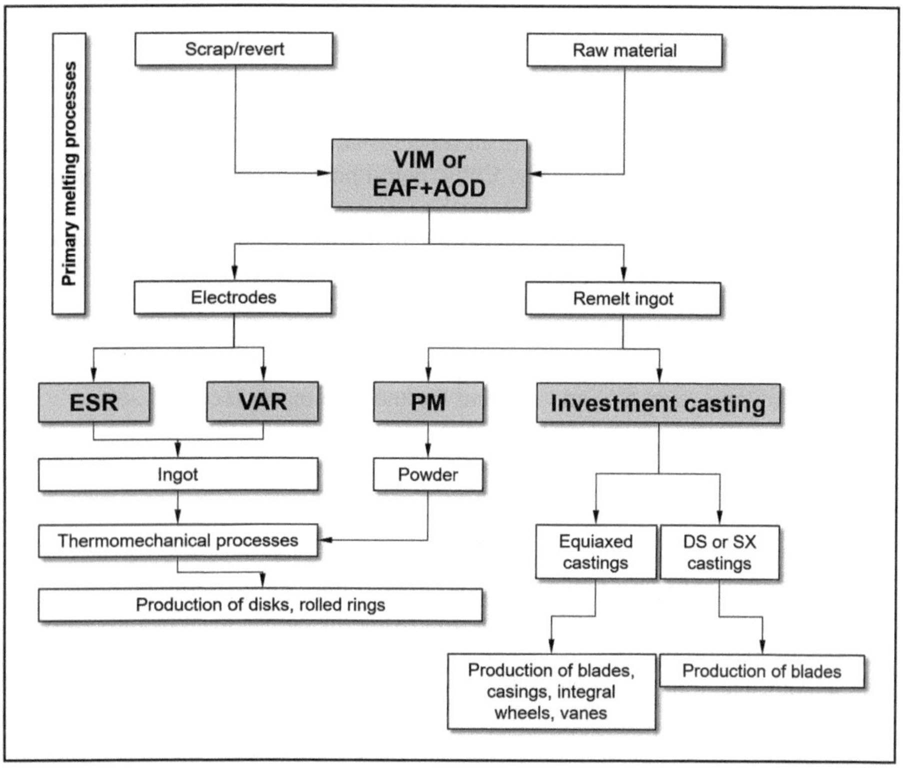

Fig. 6.5 Process flowchart for the fabrication of superalloy components. (Campbell 2006; Prasad and Wanhill 2017)

forging are other interesting approaches to the production of superalloys components. Concerning hot-rolling, material temperature, as well as process parameters, like rolling speed and in-line cooling, must be carefully tuned in order to avoid the formation of both surface and internal defects. Selecting the right concentrations of aluminum and titanium is paramount when using forging to produce high toughness alloys. These alloys may be very difficult to forge when the sum of the two elements overcomes 6% approximately, with a consequent formation of an excessive fraction of the γ'-$Ni_3(Al,Ti)$ strengthening phase. A solution would be a forging process carried out above γ'-*solvus*. Owing to the proximity of the γ'-*solvus* to the incipient melting temperature, temperature control of the component is paramount, in order to avoid unwanted local melting. The input for forging process can be either ingot metallurgy (usually triplex process) or powder metallurgy (PM) products. The choice of the production route is mainly a function of the chemistry of the selected superalloy. For instance, Alloy 718 and Waspaloy are generally produced via ingot metallurgy. The introduction of alloys like Astroloy and René 95 made it evident that conventional ingot metallurgy was not suitable, thus leading to the development of PM processing (Reed 2006). These products are also interesting for the specialized thermal treatments that have been developed on purpose (see DMHT in Sect. 6.6) to enhance further their already excellent properties through a pinpointed development of specific microstructural features. Effective strengthening, due to intermetallics and carbides, the ability to retain hardness also at high temperature, and the work hardening tendency are all aspects that render the superalloys forming via machining not always that easy. Cubic boron nitride (CBN) tools are generally required.

Additional complications are faced when machining parts with an intrinsic anisotropy due to directionally solidified (DS) or single-crystal (SC) casting (Geddes et al. 2010). In the next sections, the main aspects concerning the cobalt- and iron-based superalloys will be introduced. Nickel-based superalloys will follow, including the description of specific processing routes, like directional and single-crystal solidification (see Sect. 6.5) and relevant thermal treatments (see Sect. 6.6).

6.2 Cobalt-Based Superalloys

Cobalt is a transition metal (fourth period, ninth group) with atomic number 27. Its name derives from the Latin *Cobaltum*, following the names of malicious spirits of the Nordic mythology, the *Kobolds*. There was the belief that *Kobolds*, inhabitants of the provenance mines of the cobalt ores, were used to steal good minerals, containing more precious metals like gold and nickel, substituting them with other useless ones, among which cobalt itself. Indeed, Co-containing minerals were often mistaken for those containing more noble metals. Moreover, cobalt-containing minerals released toxic vapors during smelting (Fontani et al. 2015).The element was identified by Torbern Bergman in 1780, although it had been already unknowingly isolated by Georg Brandt in 1742. In the 1920s, intensive research in the field of magnetic materials for the electrical industry fostered a major interest for cobalt alloys, in particular as concerns the production of permanent magnets. Several alloys emerged from these studies, that explored the effects of alloying cobalt with aluminum, titanium, and nickel, resulting in the production of new alloys, such as Alcomax, Alnico, and Titanal (Tylcote 2002). Nowadays, the largest suppliers of cobalt ores are located in the so-called “Copper Belt” region in Africa. This area comprises the Democratic Republic of Congo (the former Zaire) and Zambia. The distribution of the world’s cobalt mining capacity is shown in Fig. 6.6. The main refiners of cobalt ores, like Australia, Belgium, Canada, China, Finland, Japan, Madagascar, Norway, Russia, and Zambia, purchase raw material concentrate of this strategic and critical metal and ship it to their own production sites (ASM International 2000).

Cobalt extraction is generally subsidiary to that of other metals, like copper and nickel. Therefore, the extraction process varies according to the ores in which Co is contained. The main ones are listed in Table 6.1.

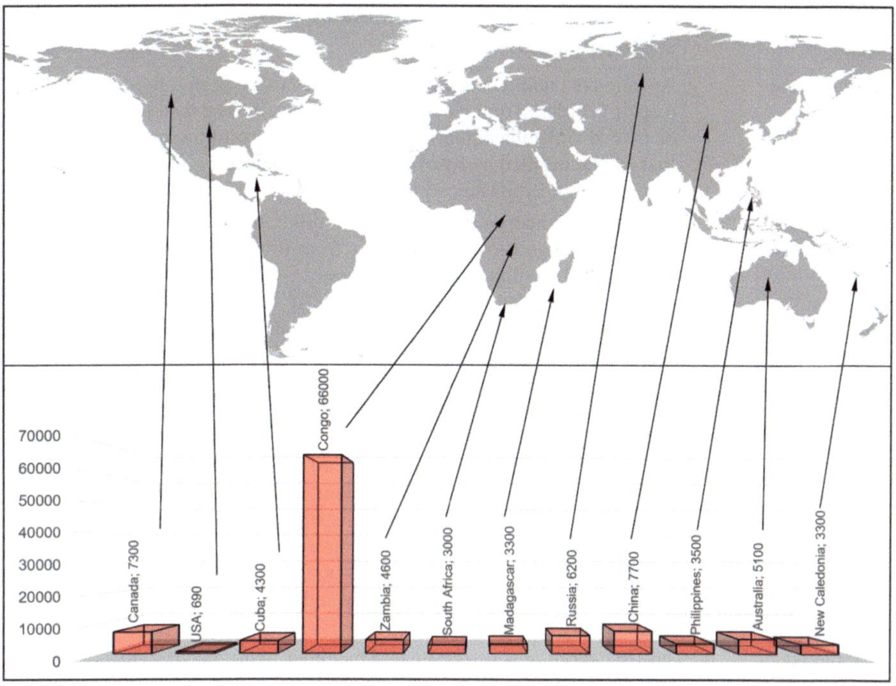

Fig. 6.6 Cobalt world mine production 2016. Data are expressed in metric tons of Co content. (Data from US Geological Survey 2017)

Table 6.1 The main cobalt ores and relevant main geographical origin (ASM International 2000)

Ore	Composition	Geographical origin
Asbolite	Mixed Mn-Fe oxides + Co	Congo, New Caledonia, Zambia
Carrolite	$CuCo_2S_4$, $CuSCo_2S_3$	Congo, Zambia
Cattierite	CoS_2	Congo
Cobaltite	CoAsS	Idaho (USA), Ontario (Canada)
Heterogenite	$2Co_2O_3CuO{\cdot}6H_2O$	Congo
Laterites	Weathered igneous rock containing limonite, Mg oxide, silicates and 2–3% Ni+Co	Australia, Brazil, Moa Bay (Cuba), Russia
Linnaeite	Co_3S_4 (+Cu, Fe, Ni)	Congo, Mississippi Valley (USA), Zambia
Nickel cattierite	(Co, Ni)S_2	Congo
Pyrrhotite	Ni, Cu, (Co) sulfides	Ontario (Canada)
Skutterudite	(Co, Ni)As_3	Bou-Azzer (Morocco), Ontario (Canada)

Table 6.2 An overview of the cobalt extraction and refining processes (ASM International 2000)

Source	Process details
Arsenide ores	Arsenic-containing concentrate undergoes roasting in a fluidized bed reactor in order to remove from 60% to 70% As. The resulting product can be treated with either chlorine or sulfuric acid, resulting in a leach solution that can be purified by hydrometallurgical processes. Cobalt is usually recovered by carbonate precipitation or electrolysis
Cu-Co oxide concentrates	The mineral undergoes crushing and milling before the metal-bearing fraction undergoes concentration by froth-flotation to separate the valuable mineral from the unwanted gangue. Co-rich oxides are sintered in pellets and then fed to electric smelting furnaces. The concentrate is then mixed up with coal and lime and is melted in a reducing atmosphere. This leads to the production of Co-Cu-Fe alloys. The light white alloy obtained from this process contains approximately 40% Co. The compound is then subjected to further hydrometallurgical refining to produce 99% purity cobalt
Cu-Co sulfide concentrates	The ores are treated by a sulfatizing roast in a fluidized bed furnace leading to the conversion of Cu and Co sulfides into soluble oxides. If present, Fe is turned into hematite. The calcine is subsequently leached with sulfuric acid from the spent Cu recovery electrolyte, which partially enters the Co recovery circuit. Purification is then performed prior to Co precipitation, as hydroxide, which is then redissolved. Final Co refining is carried out by electrolysis
Laterite ores	Ni-Co laterite ores can be treated by either hydrometallurgical or pyrometallurgical processes. Hydrometallurgical processes can use ammonia or sulfuric acid leach solutions. Pyrometallurgical processes such as ferronickel or matte smelting require the entire ore to be melted to separate the metal from the other residual components
Ni-Co sulfide concentrates	These compounds can be treated by either roasting or flash smelting. The resulting Co- and Ni-containing matte can be recovered hydrometallurgically or treated by pressure leach with an ammonia solution. The Sherritt Gordon process uses a feed matte that contains approximately 0.4% Co and 30% S, and the pressure leaching process is carried out in an ammonia solution. The resulting compound contains in addition to Co, also Cu and Ni. After Cu has been separated, the remaining solution undergoes a further separation by selective reduction processes leading to the production of a Co powder with a purity of 99%. Temperature, pressure, and pH value are critical process parameters

The main extraction processes for cobalt are electrolytic, hydrometallurgical, pyrometallurgical, vapormetallurgical extraction, or a combination of them. Depending on the mineralogy of the ores, Table 6.2 provides synthetic information on the relevant preferred processes.

Cobalt is used in several products, as shown by the block diagram in Fig. 6.7. Cemented carbide tools, corrosion- and wear-resistant alloys, high-speed tool steels, orthopedic implants, and, of course, superalloys are the main metallurgical products involving cobalt. Cobalt-based compounds are interesting for quite a broad range of products, like catalysts for chemical industry, pigments, food industry, and batteries.

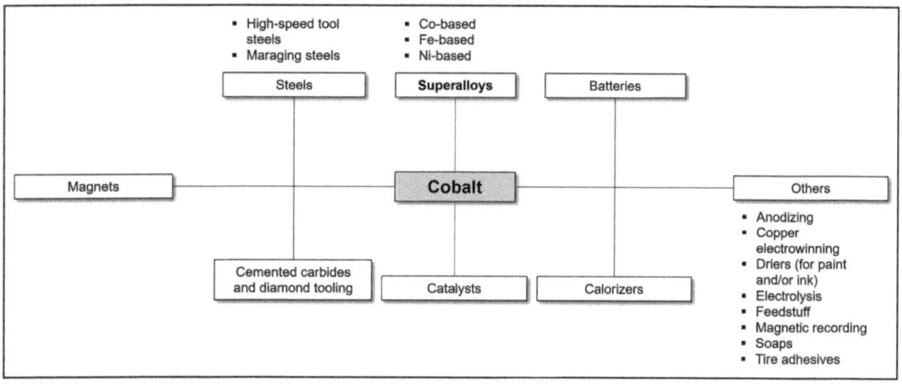

Fig. 6.7 Most important uses of cobalt. (ASM International 2000)

Cathodes for lithium ion batteries are made of the cobalt mixed oxide, $LiCoO_2$ (Fergus 2010). The main applications of cobalt are summarized by the scheme in Fig. 6.7.

The development of cobalt-based superalloys dates back to the early 1900s. In 1907, Elwood Haynes patented (No. 873,745 US patent) a Co-Cr binary alloy and for the production of this and other cobalt-based alloys created the Haynes Stellite Company in Kokomo, Indiana (Kracke 2010). Several commercial cobalt-based alloys nowadays available derive from Co-Cr-Mo and Co-Cr-W ternary systems that were investigated first by Haynes. Both high strength and corrosion resistance characterized the Co-Cr-based alloys developed by Haynes. The mechanical strength was improved further thanks to the introduction of molybdenum and tungsten. Haynes named this family of alloys Stellites, due to their star-like shiny luster. Later on, in the 1930s and 1940s, Co-based alloys for high-temperature applications featuring corrosion resistance were developed further, leading to the production of a Co-Cr-Mo alloy with moderately low C content, used for biocompatible investment casting applications: the Vitallium (Sjöberg 2010), a still existing tradename (ASM International 2000). In the same period, General Electric (GE) was particularly concentrated in the innovation process of high-strength aerospace alloys. A casting process, originally used for dentistry implants, was extended to the production of turbochargers components, using the newly developed Stellite 21 (Table 6.3), directly derived from the prototype alloy Vitallium. This alloy exhibited, in addition to high-strength level, interesting tribological, namely, wear resistance, properties. Cobalt-based alloys can be regarded as the first set of alloys deliberately designed for aerospace gas turbine components, exploiting the well-established carbide-based strengthening mechanisms, which were boosted by the stronger cobalt-rich alloy matrix. Stellites are by all means the prototypes of modern cobalt-based superalloys, developed and still used for tribological applications, also as wear-resistant coatings, deposited onto cutting tools and turbine blades to enhance their erosion resistance against particulate matter carried by the hot gas flux (Antony 1983; Riddihough

Table 6.3 Nominal composition of cobalt-based superalloys. Concentrations are expressed in wt% (ASM International 2000; Donachie and Donachie 2002; Geddes et al. 2010)

Alloy	Other designations	Cr	Ni	Co	Mo	W	Al	Fe	C	Others	Form
AiResist 13	AR-13	21.0	–	62.0	–	11.0	3.4	–	0.45	Ta = 2.0; Y = 0.1	Cast
AiResist 213	AR-213	20.0	0.5	64.0	–	4.5	3.5	0.5	0.2	Ta = 6.5; Y = 0.1; Zr = 0.1	Cast
AiResist 215	–	19.0	0.5	63.0	–	4.5	4.3	0.5	0.35	Ta = 7.5; Y = 0.1; Zr = 0.1	Cast
FSX-414	–	29.0	10.0	52.5	–	7.5	–	1.0	0.25	B = 0.01	Cast
FSX-418	–	29.5	11.0	52.0	–	7.0	–	2.0 max	0.25	B = 0.012; Mn = 1.0 max; Si = 1.0 max Y = 0.2	Cast
FSX-430	–	29.5	10.0	51.0	–	7.5	–	–	0.4	B = 0.027; Zr = 0.9	Cast
Jetalloy 1570	–	20.0	28.0	46.0	–	–	–	2.0	0.2	Ti = 4.0	Cast
Jetalloy 1650	–	19.0	27.0	36.0		12.0			0.2	B = 0.02; Ta = 2.0; Ti = 3.8	
Haynes 188	HA188 HS-188	22.0	22.0	37.0	–	14.5	–	3.0 max	0.10	La = 0.9	Wrought
Haynes 21	HS21 Stellite 21	27.0	3.0	64.0	5.0	–	–	1.0	0.25	–	Cast
Haynes 25	L-605 Stellite 25	20.0	10.0	50.0	–	15.0	–	3.0	0.10	Mn = 1.5	Wrought
J-1650	–	19.0	27.0	36.0	–	12.0	–	–	0.2	B = 0.02; Ta = 2.0; Ti = 3.8	Cast
MAR-M 302	–	21.5	–	58.0	–	10.0	–	0.5	0.85	B = 0.005; Ta = 9.0; Zr = 0.2	Cast
MAR-M 322	–	21.5	–	60.5	–	9.0	–	0.5	1.0	Ta = 4.5; Ti = 0.75; Zr = 2	Cast

MAR-M 509	MM-509	23.5	10.0	54.5	–	7.0	–	–	0.6	Ta = 3.5; Ti = 0.2; Zr = 0.5	Cast
MAR-M 918	–	20.0	20.0	52.0	–	–	–	–	0.05	Ta = 7.5; Zr = 0.1	Cast
MP159	–	19.0	25.0	36.0	7.0	–	0.2	9.0	–	Nb = 0.6; Ti = 3.0	Wrought
MP35-N	MP35N	20.0	35.0	35.0	10.0	–	–	–	–	–	Wrought
NASA Co-W-Re	–	3.0	–	67.5	–	25.0	–	–	0.4	Re = 2.0; Ti = 1.0	Cast
S-816	–	20.0	20.0	42.0	4.0	4.0	–	4.0	0.4	Mn = 1.2; Nb = 4.0; Si = 0.4	Cast
Stellite 23	–	24.0	–	65.5	–	5.0	–	1.0	0.4	Mn = 0.3; Si = 0.6	Cast
Stellite 27	–	25.0	32.0	35.0	5.5	–	–	1.0	0.4	Mn = 0.3; Si = 0.6	Cast
Stellite 30	–	26.0	15.0	50.5	6.0	–	–	1.0	0.5	Mn = 0.6; Si = 0.6	Cast
Stellite 31	X-40	22.0	10.0	57.5	–	7.5	–	1.5	0.50	Mn = 0.5; Si = 0.5	Cast
Stellite B	–	30.0	1.0	61.5	–	4.5	–	1.0	1.0	–	Wrought
UMCo-50	Haynes 150	28.0	–	49.0	–	–	–	21.0	0.12	–	Wrought
X-45	–	25.5	11.0	56.0	–	7.0	–	2.0 max	0.25	B = 0.01; Mn = 1.0 max	Cast
V-36	–	25.0	20.0	42.0	4.0	2.0	–	3.0	0.27	Mn = 1.0; Nb = 2.0; Si = 0.4	Cast
WI-52	–	21.0	–	63.5	–	11.0	–	2.0	0.45	Ta+Nb = 2.0	Cast

1970). Stellite 21, based on the Co-Cr-Mo system, was followed by other two important alloys: S-816 (Co-Ni-Cr) and X-40 (Co-Ni-Cr-W) for gas turbine blades and vanes, respectively (see Table 6.3).

The leading position of cobalt alloys was rapidly overtaken by nickel-based superalloys, and this situation is lasting until present for different reasons, in the first place, the price, lower for nickel-based alloys mainly for a lower impact of the raw material costs. The top properties achieved by the nickel-based superalloys are by far better than those of cobalt alloys, except for a newer, recently developed group of alloys (see Sect. 6.2.2). Eventually, on the eve of the potential extensive spread of cobalt alloys in the aerospace market, the end of the Colonialism era led to consider, in some way, unreliable supply of several raw materials, including cobalt, whose main provenance was, and still is, the African state of Congo, former Zaire (see Fig. 6.6). This set of reasons, and the contemporary achievements attained with the design of novel nickel alloys, reduced the interest for cobalt alloys. Nonetheless, for some specific applications in gas turbine engines, they are still the preferred choice.

Cobalt-based superalloys feature incipient melting temperature larger than that of nickel-based alloys, considering that cobalt has a melting temperature that is 40 °C higher than that of nickel. The diffusivity of alloying elements and the stacking fault energy (SFE) are comparatively lower (Neumeier et al. 2016). These two features favor good creep performances, considering that a lower SFE is a prerequisite to achieve higher concentrations of dislocations and, thereby, higher frequency of dislocation locks. Carbides, on which the main strengthening mechanisms of these alloys are based, are stable up to higher temperatures than γ' precipitates, the main strengthening agents of nickel superalloys, on which they rely for their exceptional mechanical properties. Moreover, cobalt alloys exhibit a good weldability and excellent hot corrosion resistance in particular conditions, i.e., sulfate-based deposits (see Sect. 8.2.4).

Concerning the composition, Co superalloys usually contain more than 60 wt% of cobalt. Nickel and iron are present as stabilizers of the austenitic, high-temperature polymorph of cobalt. Chromium is usually present in the concentration range of 20–30%, while molybdenum, tungsten, and other refractory metals sum up to a maximum value of about 10%, whereas carbon is present in concentrations not exceeding 1 wt% approximately. The chemical compositions of some cobalt-based superalloys are given in Table 6.3.

The right combination of alloying elements in cobalt alloys is intended to attain both solid solution and carbide strengthening and surface and γ phase stabilization. Cobalt-based superalloys do not generally require vacuum melting, and the process can be carried out by air induction melting, except for those alloys, like MAR-M 509, MAR-M 302, and MAR-M 918, containing Al, Ta, Ti, and Zr. For these formulations, VIM is required (Campbell 2006).Vacuum melting of other Co alloys would lead to enhancement of their strength and ductility, thanks to an improved cleanliness and microstructural control. The charge that should undergo the primary melting process is generally the mixing of virgin material with selected scraps. The amount of virgin material is usually around 40%. Concerning argon oxygen

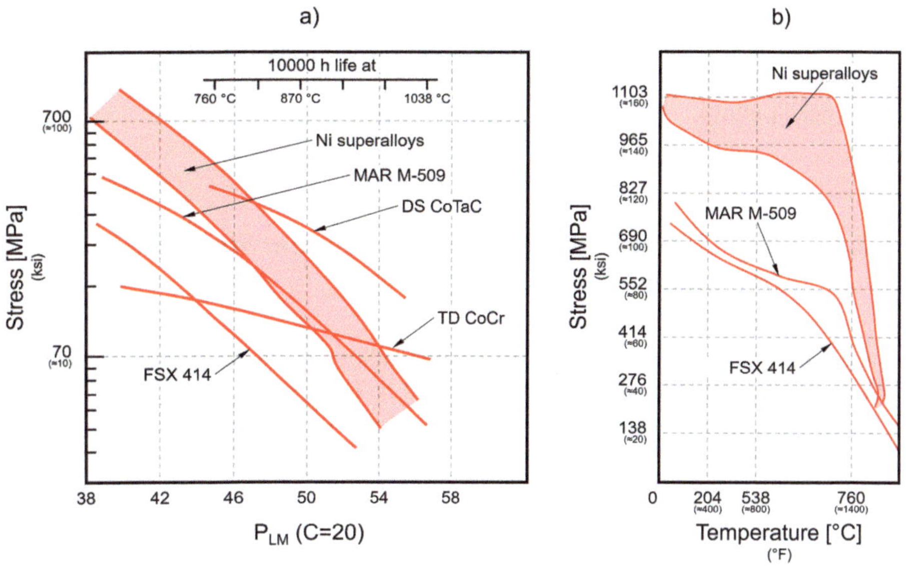

Fig. 6.8 A comparison between Ni and Co superalloys. (**a**) Creep properties (stress rupture) where P_{LM} is the Larson-Miller parameter; (**b**) ultimate tensile strength. (Redrawn from Sims et al. 1987)

decarburization (AOD), the process can be successfully used for the reduction of the C, Si, and S content in Co-based superalloys, as well as for the reduction of the dissolved gases. The control on sulfur is performed through the use of artificial slags, rich in CaO, or other basic oxides, capable to reduce sulfur levels down to 10 ppm, or less (ASM International 2008). Co-based superalloys are frequently remelted by means of ESR or VAR.

According to the graphs in Fig. 6.8, adapted from the classical textbook *Superalloys II* (Sims et al. 1987), cobalt alloys still have interesting stability and associated structural properties at high temperatures, although not achieving the top mechanical strength and creep resistance of nickel-based superalloys. This defines the typical application fields for these alloys, generally used in components that operate under low stresses but still need excellent hot corrosion resistance. Incidentally, in the Larson-Miller plot in Fig. 6.8a, a set of data for two cobalt alloys, not strictly classified as "superalloys," are shown. The first is an oxide dispersion strengthened (ODS see Sect. 9.3) alloy named TD-Co-Cr. TD means "thoriated", i.e., strengthened by a dispersion of thorium oxide (ThO_2) particles. Incidentally, it is worth making clear the pure historical interest of TD alloys, all abandoned for the radioactive character of the oxide particles containing thorium. As regards the DS-CoTaC alloy, it is classified as an in situ composite and consists of a metallic matrix, a Co-Cr alloy in this case, reinforced by tantalum carbides (TaC), that form directly from the melt during the alloy solidification. The process is conducted following the directional solidification principles (see also Sect. 6.5). Therefore, the carbides rather resemble to long fibers that provide exceptional structural properties to the component. Both ThO_2 particles and TaC elongated grains can guarantee remarkable strengthening mechanisms up to temperatures much higher than those accessible to the metallic counterparts, as a further confirmation of the typical application fields of these cobalt alloys.

6.2.1 Main Constituent Elements and Relevant Strengthening Mechanisms

Each Co-based superalloy contains different alloying elements (see Table 6.3), whose main roles in the alloy are summarized in Table 6.4.

Carbide dispersion and solid solution are the main strengthening mechanisms in Co-based superalloys that, together with the high-temperature corrosion requirements, determine the main families of constituent elements of these alloys: γ (*fcc*)-phase stabilizers and strengthener, surface stabilizers, and carbide formers. Nickel and iron, added to stabilize the high-temperature *fcc* polymorph of cobalt down to room temperature, tend to reduce the corrosion resistance of the alloy. This is among the main limiting factors to the upper concentration of these elements.

Chromium definitely is one of the most important alloying elements in cobalt superalloys, in view of its positive influence on the high-temperature corrosion resistance, through the formation of a Cr_2O_3-based protective layer on the alloy surface. Moreover, chromium is an important carbide former. Other elements that have beneficial effects on the corrosion resistance of Co-based superalloys are aluminum and the so-called reactive elements (see Sec. 6.1). Yttrium is particularly effective when added to alumina-forming alloys, and in fact, the AiResist alloys may contain up to 4.3 wt% aluminum (see Table 6.3). Lanthanum is another reactive element, particularly suited for chromia-forming alloys, like Haynes 188.

Table 6.4 Alloying elements in Co-based superalloys and relevant role in the alloy – see also Appendix 5 for a comparative overview, referring to all superalloy families (Geddes et al. 2010)

Expected effect	Alloying elements
Carbide formation	C, Cr, Mo, Nb, Ta, W
Austenite stabilization	Fe, Ni
Grain boundary refinement	B, C, Zr
Hot corrosion resistance	La, Th, Y
Intermetallics and/or other hardening precipitates	Al, Mo, Ta, Ti, W
Carbonitride formation	C, N
Oxidation resistance	Al, Cr
Rupture strength	B, Zr
Solid solution strengthening	Cr, Mo, Nb, Ni, Ta, W
Sulfidation resistance improvement	Cr
Workability improvement	B, C, Fe

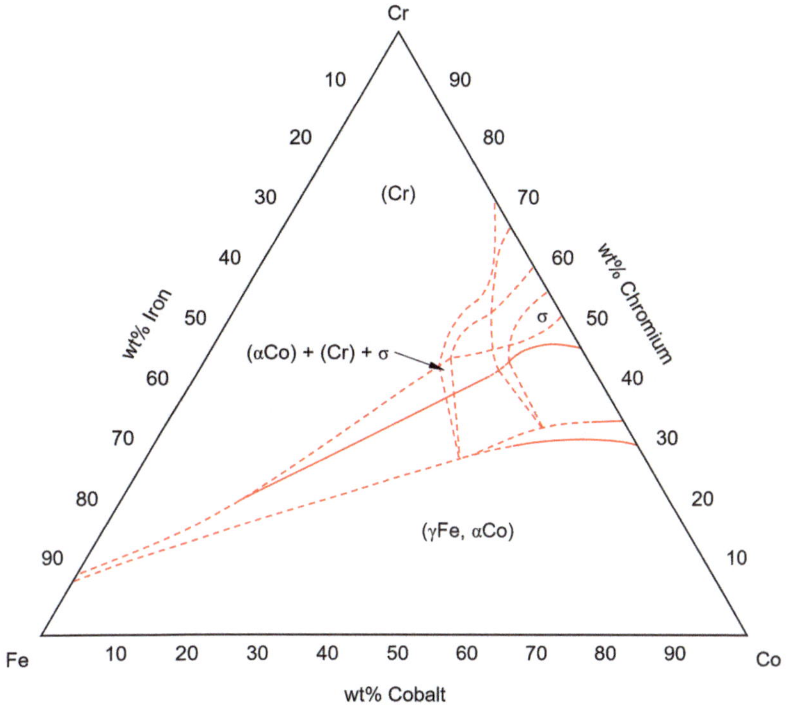

Fig. 6.9 Isothermal section at 1200 °C of the ternary Fe-Co-Cr phase diagram. The occurrence of the σ phase is possible at concentrations in excess of 30 wt% of chromium. (ASM International 2000)

A maximum concentration of 30 wt% chromium is not exceeded in commercial cobalt superalloys (Table 6.3), since above this limit the formation of the σ phase becomes increasingly likely, at the operating conditions of the gas turbine engines. σ phase is a topologically close-packed phase (TCP, see Sect. 6.4.2), and its stability field is visible in the isothermal section at 1200 °C of the Co-Cr-Fe phase diagram in Fig. 6.9. The relevant degrading effects of TCP phases are discussed in Sect. 6.4.2.

The σ phase is a Cr-rich phase, whose formation may be due to carbide transformation, with the preferential replacement of carbon atoms by nickel and cobalt.

Other intermetallic phases may form as indicated in Table 6.5, as alternatives to carbides.

Thus, Mo, Nb, Ta, W, and Zr are other constituents of carbides that can be found in a cobalt alloy in association with Cr carbides. The optimum concentrations of carbides can be attained thanks to the comparatively high carbon content of cobalt alloys, as compared to other technological alloys:

- Co-based superalloys have typical concentration in the 0.25–1.00% range (MAR-M 322).
- Ni-based superalloys have typical concentration in the range 0.05–0.20% range.

Table 6.5 Carbides and intermetallics in Co-based superalloys; see Appendix 6 for a comparative overview on intermetallic phases that may form also in other superalloy families (ASM International 2000; Geddes et al. 2010)

Element	Carbides	Intermetallics	Effects
B	–	–	Increase stress-rupture strength
C	MC, M_7C_3, M_6C, $M_{23}C_6$	–	Strengthening through carbides precipitation
Cr	M_7C_3, $M_{23}C_6$	–	Oxidation and sulfidation resistance; solid solution strengthener
La	–	–	Oxidation resistance
Mo	M_6C	Co_3M	Solid solution strengthener
Nb	MC, M_6C	Co_3M	Solid solution strengthener
Ta	MC, M_6C	Co_2M	Solid solution strengthener
W	M_6C	Co_3M	Solid solution strengthener
Y	–	–	Oxidation resistance
Zr	MC	–	Increase stress-rupture strength

M metal

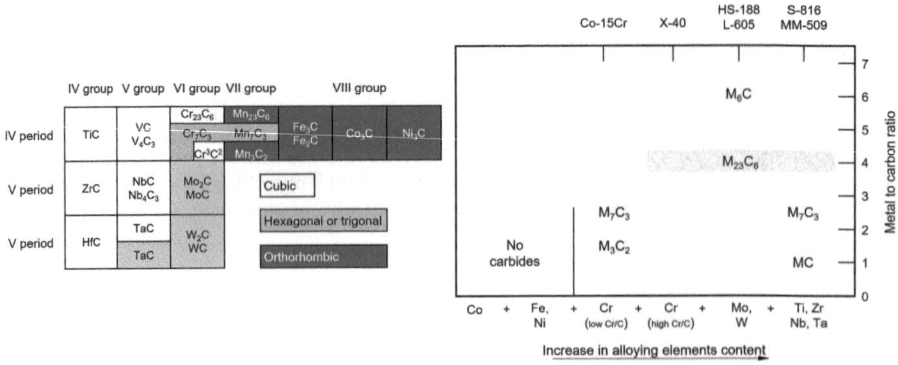

Fig. 6.10 Portion of the periodic table of the elements with the indication of the main carbide formers in Co-based superalloys and relevant structure (left). In the plot (right), the typical stoichiometry of the main carbides forming in different alloys, as a function of the content of the carbide forming elements and relative electronegativity difference with cobalt. (Redrawn from Sims et al. 1987)

- Stainless steels show typical concentration in the 0.02–0.2% range (although some tool steels and some martensitic stainless steels such as AISI 440C may contain up to about 1.0% C).

The formation of carbides is ruled by the difference in electronegativity of each carbide-forming element, as shown by the graph in Fig. 6.10. In particular, elements sitting in the columns of the periodic table to the left of cobalt, being more electronegative than cobalt, would preferentially tend to form carbides.

Typical refractory metal carbides are MC and M_6C types, generally with complex compositions.

MC are primary carbides, characterized by cubic structure, equiaxed microstructure, and incoherent interface, and tend to form directly from the melt. Changes in the composition and thereby in the crystallographic structure and relevant morphology may occur during service, according to the reactions with the γ matrix of the alloy:

$$MC + \gamma \rightarrow M_6C \text{ at } T > 1000^{\circ}C$$

$$\text{Example } TaC + (Co, Ni, Cr, C) \rightarrow (Co, Ni)_4(Cr, Ta)_2 + (Co, Ni, Cr, C) \quad (6.1)$$

M_6C carbides also tend to form from the melt, particularly in W- and Mo-rich alloys. Similarly to MC carbides, they may undergo transformations, reacting with the γ matrix of the superalloy:

$$M_6C + \gamma \rightarrow M_{23}C_6 \text{ at } T > 800 \div 900^{\circ}C$$

$$\begin{aligned}\text{Example}: (Co, Ni)_3Mo_3C + \gamma(Co, Ni, Cr\ ) \\ \rightarrow Cr_{21}Mo_2C_6 + \gamma(Co, Ni, Cr, Mo)\end{aligned} \quad (6.2)$$

Concerning chromium, there are three types of carbides that form in cobalt superalloys, for which the binary Cr-C phase diagram in Fig. 6.11 provides relevant indications. These carbides are:

- Cr_3C_2, forming from a peritectic reaction.
- Cr_7C_3 that forms during alloy solidification from the melt. In this case, the carbide is classified as primary.
- $Cr_{23}C_6$, usually obtained from other carbides' transformation.

Like refractory metal carbides, also Cr carbides are not strictly binary but feature more complex compositions that may evolve during the alloy lifetime, with changes similar to those described already for refractory metal carbides, including the formation of the TCP σ phase.

Carbides, particularly primary ones, produce several beneficial effects on the alloy stability and microstructure, with positive consequences on their mechanical properties: blockage of grain boundary sliding, grain size stabilization against recrystallization, and hindering of the dislocation movement (Orowan mechanism, see Sects. 3.2.3.3 and 6.4.3).

The evolution of carbides, upon exposure to operating conditions, would generally determine a downgrading of the mechanical properties of the alloy, since both carbide and solid solution strengthening mechanisms are reduced (see Table 6.6). The formation of secondary carbides, and their coarsening and coalescence along grain boundary regions, leads to an embrittlement and to reduction in the alloy's fracture toughness. At the same time, the depletion of the alloy matrix of the solid

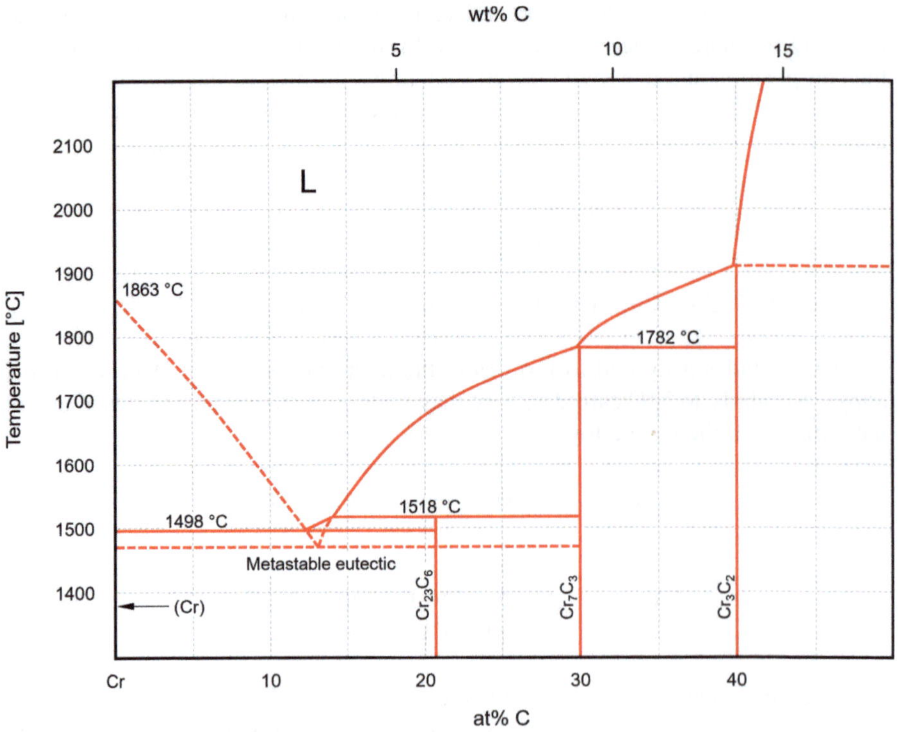

Fig. 6.11 Chromium-carbon binary phase diagram with the indication of the main carbide phases interesting also for cobalt alloys. (Redrawn from Massalski et al. 1986)

solution strengthening elements results in a constitutional weakening of the alloy (see also Sect. 6.4.2 on the weakening effects of TCP phases). To counterbalance this effect, cobalt alloys in which the precipitation of ordered intermetallic as strengthening agents occurs have been designed at different stages and lately (see Sect. 6.2.2) with extremely promising outcomes. Refining heat treatments of cobalt alloys are an additional factor which can be exploited for improving their performances. A complete overview on the main aspects of these thermal treatments can be found in Chandler (1996).

6.2.2 Latest Developments of Co Superalloys

Considering the enormous success of nickel-based superalloys, mainly relying on the very effective strengthening due to the precipitation of the γ' phase with an $L1_2$ structure (see Appendix 6), a few cobalt alloys have been developed in which a similar mechanism has been implemented. Since an analog of the Ni_3Al intermetallic, the prototype of the γ' phase in Ni alloys, is missing in the Co-Al phase diagram,

Table 6.6 Tensile properties of selected Co-based superalloys at three temperatures. E% = fracture strain (ASM International 2000)

Alloy	UTS at RT [MPa]	Yield strength at RT [MPa]	E% at RT [%]	UTS at 650 °C [MPa]	Yield strength at 650 °C [MPa]	E% at 650 °C [%]	UTS at 870 °C [MPa]	Yield strength at 870 °C [MPa]	E% at 870 °C [%]
AiResist 13	600	530	1.5	475	385	4.5	290	275	21.0
AiResist 213	1200	625	14.0	960	455	28.0	315	220	55.0
FSX-414	740	440	11.0	485	215	15.0	310	165	23.0
Haynes 25	1005	475	51.0	745	330	60.0	415	250	45.0
Haynes 188 (sheet)	945	465	53.0	710	275	59.0	415	250	64.0
Haynes 188 (plate)	985	470	56.0	755	265	73.0	450	250	77.0
MAR-M 302	930	690	2.0	785	450	–	450	310	11.0
MAR-M 322	825	625	3.2	655	415	6.0	550	345	12.0
MAR-M 509	785	570	4.0	560	370	7.0	350	290	20.0
MP159 (annealed)	850	400	60.0	–	–	–	–	–	–
MP159 (48% cold-worked)	1585	1415	12.0	–	–	–	–	–	–
MP159 (48% cold-worked + aged)	1895	1825	8.0	–	–	–	–	–	–
MP35-N (25% work strengthened)	1172	1034	28.0	–	–	–	–	–	–
MP35-N[a]	1282	1207	24.0	–	–	–	–	–	–
UMCo-50 (as-cast)	550	315	8.0	–	–	–	–	–	–
UMCo-50 (wrought)	925	610	10.0	–	–	–	–	–	–
X-40	745	525	9.0	515	260	12.0	325	–	–

[a] 25% work strengthened +538 °C aged and air cooled

the early attempts were made with alloy formulations resulting in the precipitation of the ordered Co_3Ti intermetallic compound, indeed with the same $L1_2$ structure of the γ' phase in Ni superalloys (Sims et al. 1987). These early attempts were subsequently abandoned for several reasons: the comparatively low solution temperature of Co_3Ti, ranging from 815 to 900 °C, depending on the actual alloy composition. Moreover, the lattice mismatch between Co_3Ti and the austenitic alloy matrix resulted to be larger in absolute terms than 1%, a value recommended both for an adequate stability of the precipitates against recrystallization and for better high-temperature properties (Rogister et al. 1967; Viatour et al. 1973). Moreover, the maximum volume fraction of the reinforcing intermetallic phase that can be obtained is just 20%, due to the limited solubility of Ti in Co (12 at% approx.). This value is definitely low, if compared with 70% precipitate percentage attained in state-of-the-art nickel superalloys (see Sect. 6.4) (Murray 1982, 1987). A further drawback of the Co-Ti system emerged when Ni additions have been attempted to produce Ni-Co superalloys. The formation of plate-like η-Ni_3Ti ($D0_{24}$ structure) precipitates was observed, whose cellular structure influenced negatively the fracture toughness of the alloy (Cui et al. 2006). All these factors cooled down, but not eliminated, the interest for cobalt alloys. The next step in the development of new alloys consisted in the stabilization, through ternary additions, of a more suitable than the Co_3Ti ordered intermetallic as strengthening phase of the austenitic γ-matrix. The goal was achieved with a new generation of cobalt alloys, based on the ternary Co-Al-W system, having as strengthening phase the ternary $L1_2$ ordered γ' phase: $Co_3(Al,W)$. The mechanical properties of these alloys turned out to be extremely promising for structural applications at high temperatures, although some critical features required some amendment. In fact, the newly developed alloys had a relatively high density, considering that the investigated compositions have tungsten concentrations in the 15–30 wt% (5–11 at%) range, corresponding to a maximum density of the alloy of 9.4 g/cm^3 (a reference alloy composition in at% is Co-9Al-8W). This density value is definitely higher than the density of the conventional MAR-M 509, i.e., 8.8 g/cm^3 (Bauer et al. 2010; Sato et al. 2006; Suzuki and Pollock 2008). Moreover, the γ' solvus of 990 °C only and a narrow compositional stability range of the two-phase γ-γ' field represent other important limitations to any further alloy development. Alloys with slightly better properties were developed more or less at the same time, with more complex compositions, still featuring a γ-γ' structure and a relatively high concentration of nickel. The best performing material was the alloy with composition Co-28.6Ni-10.9Ti-8.0Cr-1.3Al-1.5Mo-0.6W wt% (Co-27.6Ni-12.9Ti-8.7Cr-2.6Al-0.8Mo-0.2W at%) (Cui et al. 2006). So far, the latest step of the alloy development is represented by γ-γ' alloys of the Co-Ti-Cr system. The alloy Co-11Ti-15Cr (at%, alloy codename CTC-1), reinforced by the $L1_2$ Co-Ti14.1-Cr11.3 γ' phase, has a density of 8.1 g/cm^3, i.e., 14% lower that the density of the Co-9Al-8W alloy. Moreover, this material has a γ' solvus in excess of 1100 °C and the possibility of precipitating up to 60 vol.% of this reinforcing phase.

Figure 6.12 displays the evolution of the yield strength with temperature of various Co-based superalloys. The Co-11Ti-15Cr (code-named CTC-1) alloy results to be stronger than the other alloys in the graph over the entire temperature range. The relevant curve displays a yield strength peak, typical of the γ' strengthened

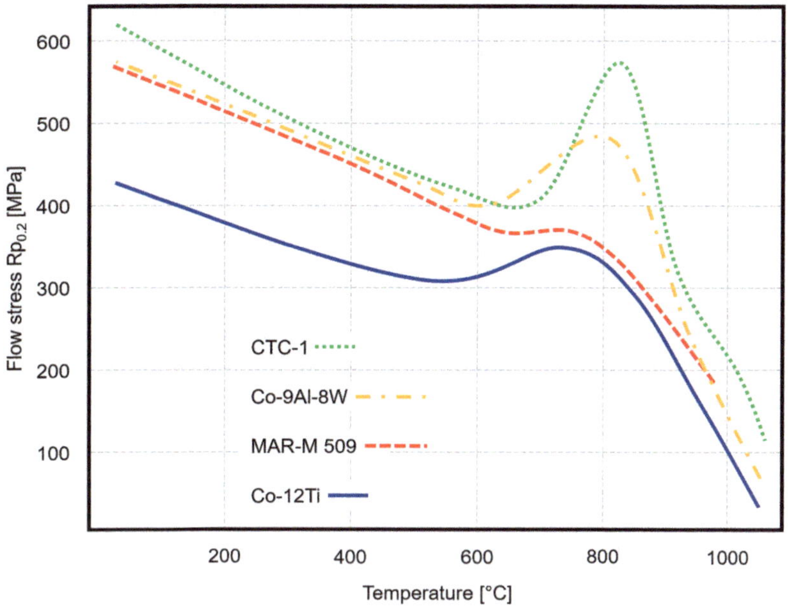

Fig. 6.12 Yield strength as a function of the temperature for the Co alloys listed in the graph legend. The composition of alloy CTC-1 is Co-11Ti-15Cr at %; that of the Co-based superalloy Mar-M 509 is given in Table 6.3. (Redrawn from Zenck et al. 2017)

alloys (see Sect. 6.4), that is not exhibited by the conventional, carbide, and solid solution strengthened MAR-M 509 Co superalloy (see Sect. 6.2.2). In the same graph, the yield strength curves for the alloys Co-9Al-8W and Co-12Ti (Sato et al. 2006) are drawn for a comparison.

6.3 Iron-Based Superalloys

As concerns composition and microstructural features, iron-based superalloys are somehow intermediate between austenitic stainless steels and nickel-based superalloys. Therefore, it is not surprisingly that Fe-based superalloys are characterized by the same production and processing routes considered for austenitic stainless steels (see Sect. 5.4.3.1). The era of austenitic stainless steels began in the 1910–1915, fostered by the increasing needs, as concerns structural stability at higher temperatures, of the aerospace industry. This led to the birth of early Fe-based superalloys with peculiar chemistry and improved properties. Initially, these materials, in addition to iron and nickel, contained a sufficient amount of chromium for improving the oxidation resistance. Afterward, small amounts of aluminum, niobium, and titanium started to be added, in order to produce a coherent, creep-resistant phase, named γ' (Sims 1984). Most of Fe-based superalloys contain high concentrations of nickel, which sometimes, like in the most popular Alloy 718, is still the majority element.

This aspect justifies the nomenclature occasionally adopted in the literature and in the present text too of Ni-Fe superalloys. Iron is present in concentrations up to 75 wt %. Nickel is not only meant to stabilize the austenitic *fcc* crystallographic structure but also contributes to the formation of strengthening ordered precipitates, like in nickel-based superalloys (see Sect. 6.4.3) and in the newly developed cobalt-based superalloys (see Sect. 6.2.2). The compositions of the main Fe-based superalloys are given in Table 6.7.

Chromium is an essential component of iron-based superalloys, interesting for carbide formation and, although to a limited extent, solid solution hardening. Moreover, chromium provides a protective Cr_2O_3-rich native oxide scale, which is an essential protection barrier against corrosion at high temperatures (see Sect. 8.2). The different effects of other alloying elements are given in Table 6.8.

6.3.1 Strengthening Mechanisms and Classification of Iron-Based Superalloys

This class of superalloys is perhaps the one that is nowadays receiving the lowest research interest worldwide, if compared to nickel and, more recently, cobalt-based superalloys. Nonetheless, iron-based superalloys are still very important for aerospace applications, in view of the very well-established knowledge available on the main products that are used for an extremely wide range of components, present in almost all parts of a gas turbine engine. Some examples are: compressor discs and blading, casings, heat exchangers, discs, and blading of the intermediate and low-pressure stages of the turbine, exhaust system, and cans. Fe-based superalloys are used in the wrought condition only, since in this way a better control of the grain size and morphology can be attained (Geddes et al. 2010). One of the reasons for the long-lasting success of these alloys is that they are relatively cheaper, if compared to the other superalloy families. The lower cost is mainly due to the high Fe content. Good mechanical strength at room and high temperatures, including also a good creep resistance, and wear, oxidation, and corrosion resistance are other properties that not surprisingly have considerably widened the fields of application of iron-based superalloys since the early times of their development. Furthermore, they are characterized by high fracture toughness and ductility. Just to mention a few applications of iron-based superalloys, often in association with steels (see Chap. 5): steam turbine power plants (as bolts and stack of gas reheaters), reciprocating engines (exhaust valves, hot plugs, turbochargers), metal processing (hot work tools, forging dies), and pollutant abating systems, like scrubbers for large diesel engines.

Precipitation hardening is the strengthening mechanism peculiar of Fe- and Ni-based superalloys, and in some of them, two hardening phases may form (see Fig. 6.13). One is the γ' phase (Appendix 6 and Sect. 6.4.2), having an ordered $L1_2$ structure based on the face-centered cubic lattice and reference stoichiometry

Table 6.7 Chemical compositions of selected Fe-based superalloys

Alloy	Type	Cr	Ni	Co	Mo	Ti	Al	Fe	C	Others
19-9DL	SSH	–	9.0	–	1.3	0.3	–	66.8	0.3	Mn = 1.1
										Nb = 0.4
										Si = 0.6
										W = 1.3
Alloy 718	PH	19.0	53.0	–	3.0	0.9	0.5	18.5	0.08 max	Cu = 0.15 max
										Nb = 5.1
D-979	PH	15.0	45.0	–	4.0	3.0	1.0	26.44 min	0.05	B = 0.01
										Mn = 0.75 max Si = 0.75 max
										W = 4.0
Discaloy	SSH	14.0	26.0	–	3.0	1.7	0.25	55.0	0.06	–
Hastelloy X	SSH	22.0	49.0	1.5 max	9.0	–	2.0	15.8	0.15	W = 0.6
Haynes N-1655	SSH	21.0	20.0	20.0	3.0	–	–	32.2	0.15	La = 0.2
										N = 0.15
										Nb = 1.0
										W = 2.5
										Zr = 0.002
Haynes 556	SSH	22.0	20.0	20.0	3.0	–	0.3	29.0	0.4	La = 0.02
										N = 0.2
										Nb = 0.1
										Ta = 0.9
										W = 2.5
Incoloy 800	SSH	21.0	32.5	–	–	0.38	0.38	45.7	0.05	–
Incoloy 800H	SSH	21.0	33.0	–	–	0.15–0.6	0.15–0.6	45.8	0.05–0.1	–
Incoloy 800HT	SSH	21.0	33.0	–	–	0.25–0.6	0.25–0.6	46.0	0.06–0.1	Cu = 0.4
										Mn = 0.8
										Si = 0.5

(continued)

Table 6.7 (continued)

Alloy	Type	Cr	Ni	Co	Mo	Ti	Al	Fe	C	Others
Incoloy 801	SSH	20.5	32.0	–	–	1.13	–	46.3	0.05	–
Incoloy 802	SSH	21.0	33.0	–	–	–	–	44.8	0.35	–
Incoloy 901	PH	12.5	43.0	–	6.0	2.7	–	36.2	0.10 max	–
Incoloy 903	PH	0.1 max	38.0	15.0	0.1	1.4	0.7	41.0	0.04	Nb = 3.0
Incoloy 907	PH	–	38.0	13.0	–	1.5	0.03	42.0	0.01	Nb = 4.7
										Si = 0.15
Incoloy 909	PH	–	38.0	13.0	–	1.5	0.03	42.0	0.01	Nb = 4.7
										Si = 0.4
Incoloy 925	PH	20.5	44.0	–	2.8	2.1	0.2	29.0	0.01	Cu = 1.8
Inconel 706	PH	16.0	42.0	–	–	1.8	0.2	37.0 min	0.01	Nb = 2.9
										Si = 0.05
Inconel 783	PH	3.0	29.0	34.0	–	0.1	5.4	25.5	–	Nb = 3.0
MA-956	ODS PM	20.0	–	–	–	–	4.5	75.0	0.05	Y_2O_3 = 0.5
Pyromet CTX-1	PH	0.1 max	37.7	16.0	0.1	1.7	1.0	39.0	0.03	Nb = 3.0
REX-78	SSH	14.0	18.0	–	4.0	0.6	–	60.0	0.01	B = 0.015
Thermo-span	PH	5.5	25.0	29.0	–	0.85	0.45	34.5	0.004	Nb = 4.8
										Si = 0.35
V57	PH	14.8	27.0	–	1.25	3.0	0.25	48.6	0.08 max	B = 0.01
										V = 0.5 max
W-545	PH	13.5	26.0	–	1.5	2.85	0.2	55.8	0.08 max	B = 0.05

PH precipitation hardening, *SSH* solid solution hardening, *ODS* oxide dispersion strengthened, *PM* powder metallurgy (ASM International 1993; Donachie and Donachie 2002; Geddes et al. 2010)

Table 6.8 Alloying elements in iron-based superalloys and relevant effects. See also Appendix 5 for a comparative overview, concerning the compositions of all families of superalloys (Geddes et al. 2010)

Desired effect	Element required
Carbide formation	Cr, Mo, Nb, Ta, Ti, W
Fcc matrix stabilization	C, Ni, W
Grain boundary refinement	B, C
Hot corrosion resistance improvement	La, Y
Intermetallics and/or other hardening precipitates	Al, Nb, Ni, Ta, Ti
Carbonitride precipitation	C, N
Oxidation resistance	Cr
Retarded formation of η phase	Al, B, Zr
Rupture strength improvement	B
Solid solution strengthening	Cr, Mo
Sulfidation resistance improvement	Cr
Workability improvement	B, C

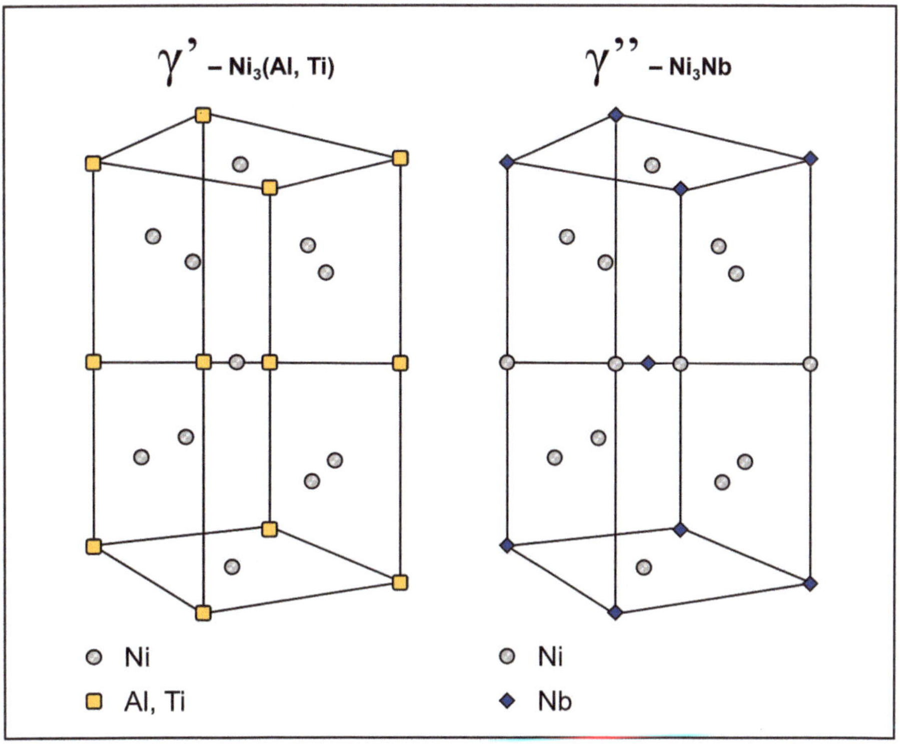

Fig. 6.13 Crystallographic structures of the γ′ $Ni_3(Al,Ti)$ and of the γ″ Ni_3Nb intermetallics. (Redrawn from Reed 2006)

$Ni_3(Al,Ti)$, although in real alloys much more complex compositions have been experimentally measured (see Sect. 6.4.3).

The second precipitate phase is the body-centered tetragonal (*bct*) ordered structure, $D0_{22}$, designated γ'', and with a reference composition Ni_3Nb. The ordered atomic occupation in both γ' and γ'' has important effects on the dislocation structure, inducing the formation of the so-called super-dislocations (see Sect. 6.4.1) and influencing their dynamics.

This results in the peak of the yield strength as a function of the temperature, similar to the trend illustrated for cobalt alloys in Fig. 6.12. Several classifications of the iron-based superalloys are available. Sims et al. (1987) proposed that they can be grouped into four families:

- **γ'- $Ni_3(Al,Ti)$ strengthened alloys**. This group comprises Ni-rich compositions (Ni% > 40 wt%), such as Incoloy 901. Other important members of the group are Fe-rich alloys, with a lower content of nickel, ranging in between 25 and 35 wt%. An example belonging to this group of alloys is V57. Mostly for historical reasons, the Tinidur alloy is also to be mentioned (Fe-30Ni-15Cr-1.8Ti-0.08C). Friedrich Krupp developed it in 1936, and as suggested by its denomination, an elevated strength was achieved by adding to an austenitic steel, a relatively high concentration of titanium and aluminum. Due to its poor weldability, another alloy replaced it for specific applications: Cromadur (Fe-12Cr-18Mn) (Kracke 2010). Although the actual strengthening mechanism, as it happened for precipitation-hardened Al-alloys (see Sect. 3.2.3.3), was fully understood at a later stage, through transmission electron microscope observations, its remarkable effect on the alloy performances was an important boost for the development of iron- and nickel-based superalloys. An elevated concentration of titanium is a common feature to all these alloys. This choice is also determined by the necessity to attain a better matching between the lattice parameter of the austenitic γ matrix and the γ' ordered precipitates, in order to retain a coherent interface (see Sect. 6.4.1), as required to achieve an effective strengthening effect. On the other hand, a critical side effect of the elevated titanium content is that the exposure at too elevated temperatures, in excess of 650 °C, results in the rapid formation of the hexagonal η-Ni_3Ti phase, to be accurately avoided considering its detrimental effects on the alloy properties, particularly a reduction in ductility and fracture toughness.
- **γ'- $Ni_3(Al,Ti)$ strengthened alloys with compositions on the Fe-rich range of the Ni-Co-Fe system**. In these alloys, mechanical strength is enhanced by increasing the fraction of the precipitate phase, whose formation is favored by the high aluminum concentrations (>1 wt%). The limiting application temperature is still 650 °C, although an insufficient oxidation resistance rather than the formation of embrittling phases sets the application limits. The need to overcome the 650 °C temperature has driven the development of iron superalloys in which the strengthening is due to other phases than ordered precipitates, as described further on, concerning the fourth group of iron alloys. Members of this group are Incoloy 903, 907, 909, and Pyromet CTX-1 (see Table 6.7). These alloys are able

to combine high strength up to 650 °C with a low thermal expansion coefficient obtained thanks to the elimination of the ferrite-stabilizing elements. A small amount of titanium is added to favor, during the aging treatments, the formation of γ'-$Ni_3(Ti,Al)$. In the Pyromet alloys, the stable *hcp* plate-like η phase may also form (see Appendix 6). Table 6.7 shows that these alloys are essentially Cr-free and therefore with a relatively limited oxidation resistance (ASM International 1993).

- **γ''- Ni_3Nb strengthened alloys**. In this group of alloys, the prevailing, although not only, strengthening precipitate phase is γ''-Ni_3Nb, whose formation is ensured by appropriate concentrations of niobium: 5.1 wt% in Alloy 718, 3 wt% in Inconel 706, just to mention a few important examples, considering that Alloy 718 is still the most used superalloy ever. In this group of alloys, the prevailing precipitate phase depends also on the concentration of aluminum. In fact, γ-$''$-Ni_3Nb would preferentially form in alloys with lower concentrations of aluminum (<0.2 wt%). Aluminum, a component of the γ'-$Ni_3(Al,Ti)$, contributes to the precipitation of this phase if present with a concentration in excess of 0.5 wt%. Similarly to the previous group of alloys, a limiting temperature of 650 °C is set by the possible formation of orthorhombic δ-Ni_3Nb, evolving from the *bct* γ''-Ni_3Nb. The formation of the δ-Ni_3Nb has similar detrimental effects on the alloy properties as the formation η-Ni_3Ti phase (see also Appendix 6).
- **Alloys strengthened by carbides and solid solution**. These alloys, all featuring a good oxidation resistance, can be divided into two main groups, according to the strengthening mechanism. The first group includes alloys strengthened by the precipitation of carbides, nitrides, and carbonitrides. The higher stability of these secondary phases raises the maximum application temperature up to 800 °C. However, since the active strengthening mechanisms are not so effective as in case of ordered precipitates, particularly as concerns the ultimate tensile strength, lower applied stresses are recommended.

  The second group of alloys is reinforced by solid solution, a reinforcing mechanism that allows the materials to be used up to temperatures in excess of 1000 °C. The main representatives of this family are described herewith. Hastelloy X (see Table 6.7) is characterized by an exceptional stress corrosion cracking resistance in specific environments, as well as by a remarkable oxidation resistance. Furthermore, it possesses good ductility also after prolonged exposure at temperatures at 650–870 °C (Haynes International 2017a). Haynes N-155, also known as Multimet, whose first production dates back to the late 1940s, (Table 6.7), can be used in high-stress applications at temperatures up to 800–1100 °C. Incoloy 800, in its three versions 800, 800H, and 800HT, has a similar composition as the austenitic stainless steel AISI 330 (see Table 5.13). They are usually supplied in the solution treated condition followed by quenching. Quenching is necessary for giving the alloy a stable austenitic structure. Second phases that may form in these alloys are titanium carbides and nitrides, as well as chromium carbides. Alloy 800 was first developed by INCO in the 1950s aiming at the production of a heat- and corrosion-resistant alloy with a relatively low nickel content, to save this strategic material. Incoloy 800 is not

Table 6.9 Typical heat treatments for precipitation hardening of Fe-based superalloys. The treatment time intervals, in the 1–4 h range, depend on the plant features and part size and shape (ASM International 2000; Chandler 1996)

Alloy	Solution treatment temperature [°C]	Aging temperature [°C]	Cooling process
Incoloy 903	845	620–720	Air or furnace cooling
Incoloy 907	980	775–620	Air or furnace cooling
Incoloy 909	980	620–720	Air or furnace cooling
Incoloy 925	1010	620–730	Air or furnace cooling
Alloy 718	980	620–718	Air or furnace cooling
Pyromet CTX-1	845–870	720	Air cooling

prone to the formation of the detrimental σ phase, even after long exposure times at 650–870 °C. Furthermore, it shows good resistance to carburization, oxidation, and sulfidation (see Sect. 8.2), combined with creep and rupture strength up to temperatures of about 800 °C. It was found that high-temperature properties could be improved thanks to a better control of the carbon content, and this led to the development of the 800H composition (Everhart 1971; Farrar 2004). Alloy 800HT was developed next as a consequence of both the need for tight control on aluminum and titanium content and to limitations in the grain size.

γ' precipitation in Fe-based superalloys occurs for relatively low concentrations of aluminum only, so that titanium turns out to be the main constituent of the precipitate. For this reason, long-term exposure at temperatures above 650 °C would result in the conversion of the γ'-Ni_3Ti into the detrimental η phase (see Appendix 6), which has an *hcp* structure and causes a decrease in the strength of the alloy. Examples of thermal treatments used for precipitation hardening of Fe-based superalloys are given in Table 6.9.

An alternative classification of iron-based superalloys can be found in Geddes et al. (2010). Three groups of alloys are proposed:

- Precipitation hardenable alloys.
- Low coefficient of thermal expansion (CTE) alloys.
- Modified stainless steels, featuring solid solution hardening as primary strengthening mechanism.

The first family includes γ'- and γ''-strengthened alloys, irrespective of the γ phase actual composition, i.e., Fe- or Ni-rich compositions. The second group includes alloys like Incoloy 903, Incoloy 907, Incoloy 909, Inconel 783, and Thermo-span, featuring the lowest CTE among all superalloys, with the additional beneficial aspect that the CTE remains relatively constant even at higher temperatures (Geddes et al. 2010). The low CTE results from the so-called Invar effect, observed in diverse materials, including austenitic iron and nickel, ordered and random, as well as amorphous alloys. In 1897 Charles Edouard Guillame discovered austenitic iron alloys, having a nickel content around 35 at%, with a particularly low thermal expansion over a wide range of temperatures. Materials exhibiting the Invar effect also show anomalies in other physical properties such as atomic volume, Curie temperature, elastic modulus, heat capacity, and magnetization. It has been found that the Invar effect is related to magnetism. In practice, the contraction associated with the loss of ferromagnetism upon heating the alloy counterbalances the normal thermal expansion. In general chromium content is strongly reduced in these alloys, since it depresses the Curie temperature (Table 6.7). Above the Curie temperature, the material shows paramagnetic behavior, and therefore the thermal expansion is no more compensated by any magnetic effect. A lower chromium content renders these alloys more sensitive to corrosion phenomena (Van Schilfgaarde et al. 1999).

Finally, the third group, of modified stainless steels, is primarily strengthened via solid solution hardening and, to a minor extent, via the precipitation of carbides and ordered intermetallics. 19-9DL and Inconel 800H are two examples of alloys belonging to this family. The first is essentially an 18–8 austenitic stainless steel (see Table 5.13) with higher C content. The second has 21% of Cr and 33% of Ni with additions of Ti and Al to promote the γ' precipitation (Geddes et al. 2010).

6.3.2 *Alloy 718 and Properties of Iron-Based Superalloys*

Dealing with iron-based superalloys, a few words have to be spent on the most popular alloy of this family the family: the γ''-Ni_3Nb strengthened Alloy 718, also known as INCO 718, IN 718, or Inconel 718. It is regarded as one of the most versatile and successful superalloys ever. Alloy 718 is used not only in the aerospace field but also for the fabrication of components used in other industries, such as oil and gas, power generation, and critical applications in hot corrosive environments. Alloy 718 was developed to meet the needs of the jet engine market and to overcome limitations of existing alloys, such as the PHSS A286 and the Ni-based superalloy René 41. Alloy 718 was patented on July 24, 1962, by Herb Edelstein, and it has been used in the aerospace field ever since, both for commercial and military applications. The Rolls-Royce Trent 1000 and XWB (see Sect. 2.6) engines use this alloy for several components. The GTE of the next generation of the defense military aircraft, the JSF, is equipped with components made of Alloy 718. Its versatility is demonstrated also by the high number of component elements (almost 20), melt practices (seven different schemes having multiple varieties), and hot-working processing, such as hot-rolling and forging employed in the production and processing of this alloy (Kracke 2010). Concerning the melt practices, critical components made of Alloy 718 are produced via the triplex process (VIM-ESR-VAR – see Sect. 5.3.6) and usually would not feature freckles or whites spots,

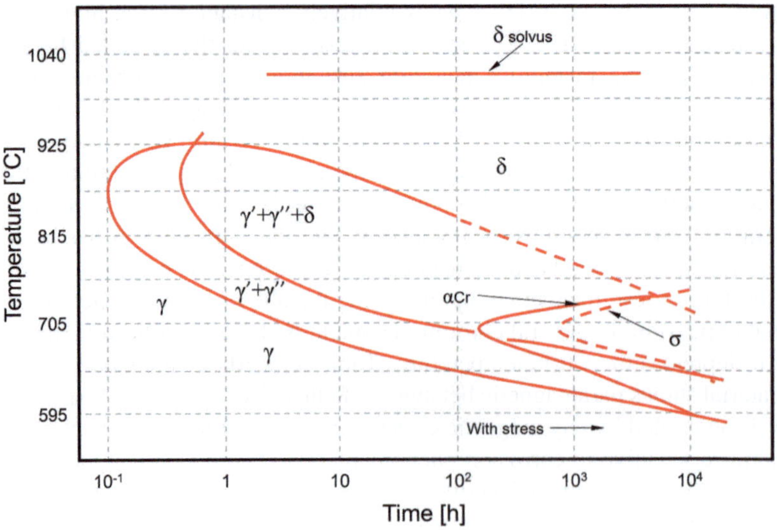

Fig. 6.14 TTT diagram for Alloy 718. (Redrawn from ASM International 2000)

possibly associated with the remelting practice (see Sect. 5.3.6). During solidification, niobium, molybdenum, and titanium tend to segregate in the inter-dendritic regions, where Nb-rich Laves phases (see Sect. 6.4.2) may form. A solution treatment is required for dissolving these phases and homogenizing the alloy composition. γ'' precipitates are regarded as the main strengthening agents in Alloy 718. It has been observed that an increasing niobium content results in a higher *solvus* temperature for the δ- Ni_3Nb phase, with obvious consequences on the design of the solution treatment. γ'' can transform into γ' and δ on increasing temperature and exposure times. The formation of large amounts of δ phase at the expense of γ'' leads to a heavy drop in the mechanical properties of the alloy. The TTT diagram for Alloy 718, with the relevant stability field of the alloy phases, is shown in Fig. 6.14. MC and TiN do not change with thermal cycles at low temperatures and are considered as inert phases, differently from γ'', γ', δ, and Laves phases. The use of Alloy 718 in critical applications promoted the development of two other variants: HS718 (high strength) and DA718 (direct age). Both alloys employ processing improvements, such as cleaner input stock for the melting process and a dedicated chemistry. In fact, both HS718 and DA718 are characterized by a higher niobium content in the 5.0–5.5% range.

DA718 is characterized by a production process that avoids solution treatment and uses a double aging treatment followed by water quenching. The double aging process implies a first treatment at 718 °C followed by a second at 621 °C. As-quenched material is characterized by retained strain that increases the γ'' and γ' precipitation kinetics (Radavich 1989; Schafrik et al. 2001).

One of the many reasons for the success of Alloy 718 are its exceptional mechanical properties, whose evolution with temperature, as concerns UTS, yield

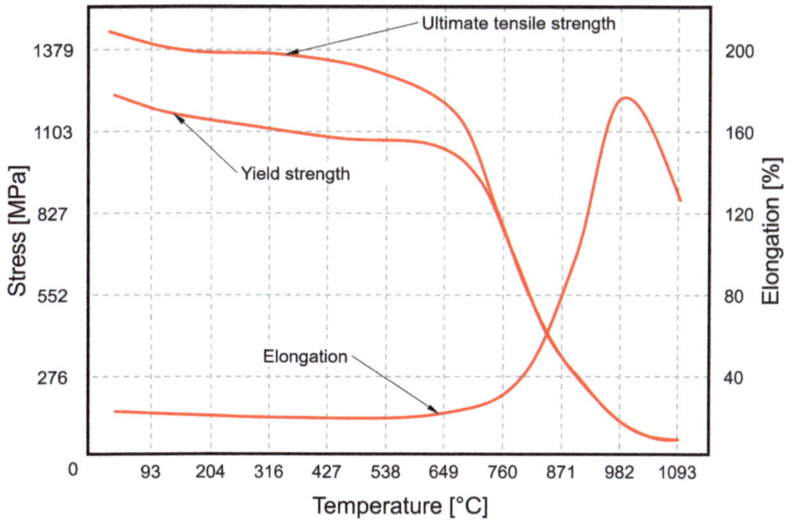

Fig. 6.15 Evolution of tensile properties as a function of the temperature for a hot-rolled, annealed, and aged 13 mm round bar of Alloy 718. (Redrawn from ASM International 2000)

strength, and elongation, is given in Fig. 6.15. Table 6.10 shows how this alloy is ranked with respect to other iron-based high-strength superalloys (ASM International 1993).

6.4 Nickel-Based Superalloys

Nickel is a transition metal (fourth period, tenth group) with atomic number 28. The melting temperature for pure nickel is 1453 °C; its Young's modulus is 207 GPa, with a density of 8.902 g/cm^3 (ASM International 2000), relatively high and due to the combined effect of atomic mass and close packing bonds induced by the outer shell *d* electrons, typical of transition metals (Reed 2006).

Apart from nickel of meteoritic origin (see also Sect. 5.1), the earliest reported example of an artificial alloy containing nickel was the Pai-Thong (white copper). This Chinese alloy was obtained by adding zinc to nickel-copper ores. Owing to its aspect, nickel was used for the production of several decorative pieces, imported to Europe by the East India Company in the seventeenth century. In the early eighteenth century, attempts of smelting a newly discovered, presumed, copper ore were performed in the German region of Saxony. Although unaware, miners, had actually found niccolite, a nickel arsenide containing traces of Co, Fe, S, and Sb. The result of the extraction attempts was a white metal, too hard to be effectively hammered. For this reason, German miners thought the material was somehow cursed, blaming for this *Nickel*, a deceptive demon in the German mythology. Therefore, the name given to the new alloy was the German *Kupfer-Nickel*, literally "copper-demon," which was abandoned soon, since not regarded as technologically

Table 6.10 Mechanical properties of selected Fe-based superalloys at room and high temperatures (650 °C and 870 °C). E% = fracture strain (ASM International 1993)

Alloy	UTS at RT [MPa]	Yield strength at RT [MPa]	E% at RT [%]	UTS at 650 °C [MPa]	Yield strength at 650 °C [MPa]	E% at 650 °C [%]	1000 h rupture strength at 650 °C [MPa]	1000 h rupture strength at 870 °C [MPa]
19-9 DL	815	570	43.0	517	360	30.0	–	–
Alloy 718	1350–1530	1105–1365	16.0–21.0	1130–1235	860–1090	19.0–24.0	405–600	–
Discaloy	1000	730	19.0	720	630	19.0	275	–
Haynes 556	815	410	48.0	590	225	52.0	275	55
Haynes N-155	815	400	40.0	545	295	32.0	295	70
Incoloy 800	595	250	44.0	405	150	51.0	165	30
Incoloy 801	785	385	30.0	540	305	26.0	–	–
Incoloy 802	690	290	44.0	525	200	25.0	170	69
Incoloy 901	1205	895	14.0	960	760	13.0	–	–
Incoloy 903	1310	1105	14.0	1000	895	18.0	510	–
Incoloy 907	≈1365	≈1110	≈12.0	≈1035	≈895	≈10.0	–	–
Incoloy 909	1310	1020	16.0	1025	870	24.0	345	–
V57	1170	830	26.0	895	745	22.0	–	–

Table 6.11 Main Ni-containing ores used in the modern production of primary nickel, with relevant geographic provenance (ASM International 2000)

Type	Ni-containing ore	Chemistry	Impurities	Deposit
Laterite (hydrated magnesium-iron-silicates)	Garnierite	$(Ni, Mg)_6Si_4O_{10}(OH)_8$	Co	Central America
			Cr_2O_3	Cuba
			MgO	Indonesia
				New Caledonia
				Philippines
Laterite (limonitic-type)	Goethite	$(Ni, Fe)_2O_3 \cdot H_2O$	Co	South America
			Cr_2O_3	
			Fe	
Sulfide ore	Pentalandite	$(Ni, Fe)_9S_8$	Al_2O_3	Kola Peninsula (Russia)
			CaO	Norlisk (Siberia, Russia)
			$CuFeS_2$ (chalcopyrite)	South Africa
			Fe_7S_8 (pyrrhotite)	Sudbury (Ontario, Canada)
			MgO	Voisey's Bay (Labrador, Canada)
			SiO_2	Western Australia

promising. Nickel was identified by the Swedish mineralogist Axel Fredrick Cronstedt in 1751, after 5 years of work using an instrument that previously was utilized only by jewelers and metal workers: the blowpipe. Cronstedt named the metal simply nickel after the Swedish word: *kopparnickel*, the translation of the original German *Kupfer-Nickel* (ASM International 1991; Fontani et al. 2015; Tylcote 2002).

Massive nickel production did not start until 1865, when an ever-increasing market request for this metal, determined a major upscaling of the production plants. In Table 6.11 the main nickel ores are recalled, with the indication of the main mining sites that have been identified over the years.

Among the different Ni-bearing minerals, sulfides and laterites (oxides and silicates) are the most important ones and represent the majority of the ores from which primary nickel is nowadays extracted.

The industrial production of nickel was initially based on the so-called Orford process, a technique introduced in 1890. The process involves the gravity segregation of nickel from a mixture of sodium sulfide and Cu-Ni-containing matte fused together in a smelter with the help of a suitable coke charge. The separation occurs during solidification when Cu-Na sulfide segregates above 99.8% pure Ni sulfide. This intermediate product can be refined further using, for instance, a Bessemer

Table 6.12 Processes used for the refinement of Ni concentrates (ASM International 2000)

Process type	Description
Hydrometallurgical	Pure nickel cathode sheets are produced thanks to electrorefining and electrowinning, two electrolytic processes. Electrorefining uses a nickel electrolytic refining cell consisting of a thin nickel starting sheet (cathode) and a slab of crude nickel that must be refined (anode). Both are immersed into an aqueous electrolyte, and a direct current supply is fed into the cell. Nickel, together with some impurities, undergoes dissolution in the electrolyte. The solution obtained from this process is pumped out from the cell, and then impurities are chemically removed. The electrolyte is purified and returns to the cell depositing as pure nickel on the cathode. Electrowinning is also known as electroextraction and consists in the electrodeposition of metallic nickel
Pyrometallurgical	Nickel concentrates are calcined using a roasting furnace. Then, the compounds undergo smelting in a reverberatory furnace or in a similar equipment, using natural gas or coal for fuel before being blown with air in a converter. Further refining occurs in order to increase the purity level. In particular, refinement is used to reduce the content of elements that may have detrimental effects on the performances of the final alloy. Removed elements are, for instance, As, Bi Cu, Fe, P, Pb, Sb, Sn, and Zn
Vapometallurgical	The extraction of pure nickel from nickel oxide implies the use of a gas-to-metal transformation according to the carbonyl process. Nickel oxide is reduced with hydrogen, and the resulting nickel selectively reacts with CO to form gaseous Ni-carbonyl. Subsequently, Ni-carbonyl is thermally decomposed to obtain pure nickel, in the form of pellets or powders with a very high purity degree

process, in order to obtain metallic nickel. The Orford process remained the only used for the industrial production of nickel until 1948, when the development of new and more efficient techniques led to its progressive replacement (ASM International 1991; Tylcote 2002). Instead of being smelted using sodium sulfide, the Cu-Ni matte is allowed to slowly solidify in order to obtain the crystallization of the two separate compounds, Ni_3S_2 and Cu_2S, both solidified in form of blocks. Blocks are then crushed and separated by means of froth flotation or magnetic methods. Ni_3S_2 is oxidized to NiO and then reduced to metallic Ni (Tylecote 2002).

Several processes can be employed for the extraction of nickel from laterite ores. One possibility is to dry up and preheat the ore, commercially known as ferronickel, under reducing conditions, followed by melting in an electric arc furnace. The crude metal is then refined and cast into ferronickel *pigs* (ASM International 2000; Davis 2000). Hydrometallurgical, pyrometallurgical, and vapormetallurgical were, and still are, the main processes to refine the nickel concentrates (Table 6.12) (ASM International 1991). The main market volumes of nickel, accounting for 47.1% of the total, belong to China (33%), the USA (7.3%), and Japan (6.8%) (Nakajima et al. 2018). From an industrial point of view, the most important use of nickel, accounting for more than 50% of the total usage, is as alloying element in ferrous alloys, i.e., steels (see Fig. 6.16).

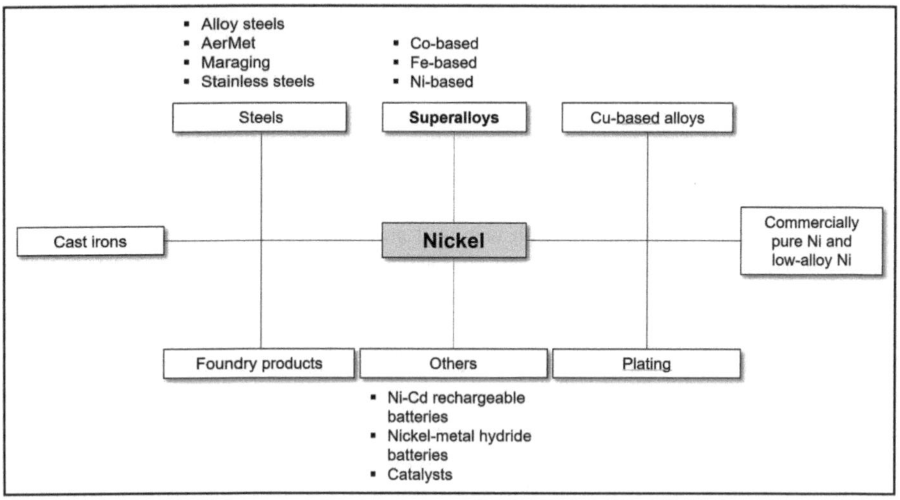

Fig. 6.16 Most important applications of nickel. (Davis 2000)

Nickel-based superalloys, in view of their strength and temperature capabilities under different loading regimes, static, creep, and fatigue, in association with corrosion resistance (see Figs. 6.2, 6.3, 6.4, and 6.8), are the most used materials in gas turbine engines. This results from a combination of sophisticated alloy design methodologies, capable to optimize rather complex compositions, and microstructural refinement, achieved with processing techniques developed on purpose for specific part manufacturing. An excellent example regards single-crystal turbine blades, with operating peak temperatures that can be raised up to 70% of their incipient melting temperatures in Kelvin.

Thanks to vacuum melting, including VIM and VAR technologies (see Sects. 5.3.2 and 5.3.6), new alloy compositions were developed. Elwood Haynes, known for the development of the cobalt-based superalloys (see Sect. 6.2), investigated the Ni-Cr alloys too, used for wear- and oxidation-resistant applications and in 1907 patented a Ni-Cr binary alloy used in cutting tools for machining automotive components. Albert Leroy Marsh was experimenting Ni-Cr alloys for electrical applications, resulting in the patent of alloys of the Nichrome series. Furthermore, the research effort made by Marsh paved the way for the development of Ni-Cr-Ti alloys: the Nimonic series. The work on Ni-Mo alloys led to the development of the Hastelloy family, known for their remarkable corrosion resistance. In 1919 the International Nickel Company Research Laboratory in Bayonne (NJ, USA) started a development project on high-strength Monel 400, used for steam turbine blades. The chemical composition was tuned by increasing the Al content up to 5%, in order to enhance age hardening, an important step forward in the development of the modern nickel-based superalloys. The composition of this alloy was patented in 1926 (US Patent 1,572,744) and became later known as Monel K-500. Age hardening was then patented by the International Nickel Company in 1930 (US Patent

1,755,554). Finally, Nimonic 80 and Nimonic 80A were developed during the 1940s. Nimonic 80 was developed in 1946 by Mond Nickel Company Ltd.: the first steps of a success story that is still going on until the present times (ASM 1991; Kracke 2010; Sims 1984).

Nickel-based superalloys generally contain at least 50 wt% of Ni, together with different combinations of Cr, Al and Ti (all together summing up to 8 wt%), and Co and refractory elements, like Mo, W, Nb, and Re, in lower concentrations. Ni-based superalloys can be produced in both wrought and cast forms. The chemical compositions of some important wrought Ni superalloys are given by Table 6.13.

Wrought alloys are usually produced by remelting VIM ingots by means of VAR, ESR, or electron beam cold heart remelting (EBCHR) processes. This leads to the production of secondary ingots, which undergo deformation processing. The secondary processing is needed because Ni-based superalloys are very sensitive to chemical and microstructural inhomogeneities and to the presence of inclusions. Furthermore, large ingots produced via VIM are often characterized by macro-segregations and shrinkage cavities. This is mainly due to the intrinsic low thermal conductivity of the solidifying alloy, which limits the heat transfer out of the ingot. Defects are therefore originated by large-scale solute segregations associated with dendritic solidification under low thermal gradients. Special processing routes, such as powder metallurgy (PM) or isothermal forging, or a combination of the two, can be employed for the production of wrought highly alloyed compositions. For instance, Astroloy is an alloy that can be processed in this way (ASM 2000; Prasad and Wanhill 2017), and, at the present time, three generations of PM superalloys are present on the market. The René 95 alloy belongs to the first generation of PM superalloys. Their most important features are the fine grain size (in the 10 μm range) and high volume fraction of γ' precipitates (around 55%), a combination used to achieve maximum tensile strength. Therefore, sub-solvus heat treatments are employed in order to prevent modifications in both the γ' volume fraction and grain size that might degrade the resulting alloy properties. The second generation of PM superalloys, like René 88DT, features lower γ' volume fraction and a larger grain size: they have been developed with task of increasing the resistance against fatigue crack propagation. The LSHR and ME3 alloys belong to third-generation PM superalloys. They are characterized by a generally higher volume fraction of γ', as compared to second-generation alloys, and recently developed heat treatments (e.g., DMHT, see Sects. 6.4.3 and 6.6) are used on these alloys in order to induce dual microstructure and, consequently, improved properties within the same component (Reed 2006).

The chemical composition of wrought Ni-based superalloys has to comply with the hot-workability requirements. The development of cast Ni-based superalloys has been supported by two casting technologies, developed on purpose for high-temperature components, in particular the blading of the high-pressure gas turbine: directional solidification (DS) and single-crystal (SC) casting (see Sect. 6.5).

The chemical compositions of selected cast Ni-based superalloys are listed in Table 6.14, whereas the compositions specifically designed for DS and SC solidification processes are provided in Tables 6.15 and 6.16.

Table 6.13 Chemical analysis of selected wrought Ni-based superalloys

Alloy	Strengthening mechanism	Ni	Cr	Co	Mo	Al	Ti	Fe	Mn	Si	C	B	Zr	Others
Astroloy	PH	55.0	15.0	17.0	5.3	4.0	3.5	–	–	–	0.06	0.03	–	–
Hastelloy C-22	PH	51.6	21.5	2.5	13.5	–	–	5.5	1.0	0.1	0.01	–	–	V = 0.3 W = 4.0
Hastelloy C-276	SSH	59.0	15.5	2.5	16.0	–	–	5.5	1.0	0.1	0.01	–	–	V = 0.3 W = 3.7
Hastelloy S	SSH	67.0	15.5	–	14.5	0.3	–	1.0	0.5	0.4		0.009	–	La = 0.05
Hastelloy W	SSH	61.0	5.0	2.5	24.5	–	–	5.5	–	–	0.12	–	–	V = 0.6
Haynes 214	SSH	75.0	16.0	–		4.5	–	2.5	–	–		–	–	Y = 0.01
Haynes 230	SSH	57.0	22.0	–	2.0	0.3	–	–	0.5	0.4	0.1	–	–	La = 0.02 W = 14.0
Inconel 617	SSH	54.0	22.0	12.5	9.0	1.0	0.3	–	–	–	0.07	–	–	Nb = 3.6
Inconel 625	SSH	61.0	21.5	–	9.0	0.2	0.2	2.5	0.2	0.2	0.05	–	–	Nb = 1.0
Inconel 700	PH	Bal.	14.0	29.5	3.0	3.0	2.5	–	–	–	0.1	–	–	–
Inconel X750	PH	73.0	15.5	–	–	0.7	2.5	7.0	0.5	0.2	0.04	–	–	–
LSHR[a]	PH	Bal.	13.0	21.0	2.7	3.5	3.5	–	–	–	0.03	0.03	0.05	Nb = 1.5
														Ta = 1.6
														W = 4.3
M-252	PH	55.0	20.0	10.0	10.0	1.0	2.6	–	0.5	0.5	0.15	0.005	–	–
ME3[a]	PH	Bal.	13.1	18.2	3.8	3.5	3.5	–	–	–	0.03	0.03	0.05	Nb = 1.4
														Ta = 2.7
														W = 1.9
Nimonic 105	PH	53.0	15.0	20.0	5.0	4.7	1.2	–	0.3	0.3	0.16	0.008	0.1	–
Nimonic 115	PH	60.0	14.3	13.2	–	4.9	3.7	–	–	–	0.15	0.16	0.04	–
Nimonic 263	PH	51.0	20.0	20.0	5.9	0.5	2.1	–	0.4	0.3	0.06	0.001	0.02	–
Nimonic 75	SSH	76.0	19.5	–	–	–	0.4	3.0	0.3	0.3	0.1		–	–
Nimonic 80A	PH	76.0	19.5	–	–	1.4	2.4	–	0.3	0.3	0.06	0.003	0.06	–
Nimonic 86	SSH	65.0	25.0	–	10.0	–	–	–	–	–	0.05		–	Ce = 0.03
														Mg = 0.015

(continued)

Table 6.13 (continued)

Alloy	Strengthening mechanism	Ni	Cr	Co	Mo	Al	Ti	Fe	Mn	Si	C	B	Zr	Others
Nimonic 90	PH	59.0	19.5	16.5	–	1.5	2.5	–	0.3	0.3	0.07	0.003	0.06	–
René 41	PH	55.0	19.0	11.0	1.0	1.5	3.1	–	–	–	0.09	0.005	–	–
René 88DT[b]	PH	Bal.	16.0	13.0	4.0	2.1	3.7	–	–	–	0.03	0.015	0.03	Nb = 0.7
														W = 4.0
René 95[c]	PH	61.0	14.0	8.0	3.5	3.5	2.5	–	–	–	0.15	0.01	0.05	Nb = 3.5 W = 3.5
Udimet 500	PH	54.0	18.0	18.5	4.0	2.9	2.9	–	–	–	0.08	0.006	0.05	–
Udimet 520	PH	57.0	19.0	12.0	6.0	2.0	3.0	–	–	–	0.05	0.005	–	W = 1.0
Udimet 700	PH	55.0	15.0	17.0	5.0	4.0	3.5	–	–	–	0.06	0.03	–	–
Udimet 710	PH	55.0	18.0	15.0	3.0	2.5	5.0	–	–	–	0.07	0.02	–	W = 1.5
Udimet 720	PH	55.0	17.9	14.7	3.0	2.5	5.0	–	–	–	0.03	0.033	0.03	W = 1.3
Unitemp AF2-1DA6	PH	60.0	12.0	10.0	2.7	4.0	2.8	–	–	–	0.04	0.015	0.1	Ta = 1.5 W = 6.5
Waspaloy	PH	58.0	19.5	13.5	4.3	1.3	3.0	–	–	–	0.08	0.006	–	–
Custom age 625 Plus	PH	61.0	21.0	–	8.0	0.2	1.3	5.0	–	–	0.01	–	–	Nb = 3.4

PH precipitation hardening, *SSH* solid solution hardening, *PM* powder metallurgy (ASM International 2000; Donachie and Donachie 2002; Reed 2006; Zhang and Zhao 2012)

[a]Third-generation PM superalloy

[b]Second-generation PM superalloy

[c]First-generation PM superalloy

Table 6.14 Chemical compositions of selected cast Ni-based superalloys. (ASM International 2000; Donachie and Donachie 2002; Kracke 2010; Zhang and Zhao 2012)

Alloy	Ni	Cr	Co	Mo	W	Nb	Al	Ti	C	B	Zr	Others
Alloy 713C	Bal.	12.5	–	4.2	–	2.0	6.1	0.8	0.12	0.012	0.10	–
C 130	Bal.	21.5	–	10.0	–	–	0.8	2.6	0.04	–	–	–
C1023	Bal.	15.5	10.0	8.0	–	–	4.2	3.6	0.15	0.006	–	–
Hastelloy S	Bal.	16.0	–	–	15.0	–	0.40	–	0.01	0.009	–	Fe = 3.0
												Mn = 0.55
												La = 0.02 Si = 0.65
IN-939	Bal.	22.4	19.0	–	2.0	1.0	1.9	3.7	0.15	0.009	0.10	Ta = 1.4
Inconel 625	Bal.	21.5	–	8.5	–	4.0	0.2	0.2	0.06	–	–	Fe = 2.5
Inconel 700	Bal.	14.0	29.5	3.0	–	–	3.0	2.5	0.1	–	–	–
M-22	Bal.	5.7	–	2.0	11.0	–	6.3	–	0.13	–	0.6	–
MAR-M 200	Bal.	9.0	10.0	–	12.5	1.8	5.0	2.0	0.15	0.015	0.05	–
MAR-M 200 Hf	Bal.	9.0	10.0	–	12.5	1.0	5.0	2.0	0.14	0.015	–	Hf = 2.0
MAR-M 421	Bal.	15.8	9.5	2.0	3.8		4.3	1.8	0.14	0.015	0.05	–
MAR-M 432	Bal.	15.5	20.0	–	3.0	2.0	2.8	4.3	–	0.015	0.05	Ta = 2.0
MM 002	Bal.	9.0	10.0	–	–	–	5.5	1.5	0.15	0.015	0.05	Hf = 1.5 Ta = 2.5
Nimocast 100	Bal.	11.0	20.0	5.0	–	–	5.0	1.5	0.20	0.015	0.03	–
Nimocast 80	Bal.	19.5	–	–	–	–	1.4	2.3	0.05	–	–	Fe = 1.5
NX 188	Bal.	–	–	18.0	–	–	8.0	–	0.04	–	–	–
René 41	Bal.	19.0	10.5	9.5	–	–	1.7	3.2	0.08	0.005	0.01	–
René 77	Bal.	15.0	18.5	5.2	–	–	4.25	3.5	0.08	0.015	–	–
René 80	Bal.	14.0	9.5	4.0	4.0	–	3.0	5.0	0.17	0.015	0.03	–

(continued)

Table 6.14 (continued)

Alloy	Ni	Cr	Co	Mo	W	Nb	Al	Ti	C	B	Zr	Others
René 80 Hf	Bal.	14.0	9.5	4.0	4.0	–	3.0	4.7	0.15	0.015	–	Hf = 0.8
Udimet 500	Bal.	18.5	16.5	3.5	–	–	3.0	3.0	0.08	0.006	–	–
Udimet 700	Bal.	14.3	14.5	4.3	–	–	3.0	3.0	0.08	0.015	0.02	–
Udimet 710	Bal.	18.0	15.0	3.0	1.5	–	2.5	5.0	0.13	–	0.08	–
UDM 56	Bal.	16.0	5.0	1.5	6.0	–	4.5	2.0	0.02	0.07	0.03	V = 0.5
Waspaloy	Bal.	19.0	12.3	3.8	–	–	1.2	3.0	0.06	0.005	0.01	Mn = 0.45

Table 6.15 Chemical compositions of selected DS Ni-based superalloys. (ASM International 2000)

Alloy	Generation	Ni	Cr	Co	Mo	W	Al	Ti	C	B	Zr	Others
MAR-M 002	1	Bal.	8.0	10.0	–	10.0	5.5	1.5	0.15	0.015	0.03	Hf = 1.5 Ta = 2.6
MAR-M 200 Hf	1	Bal.	8.0	9.0	–	12.0	5.0	1.9	0.13	0.015	0.03	Nb = 1.0 Hf = 2.0
MAR-M 247	1	Bal.	8.0	10.0	0.6	10.0	5.5	1.0	0.15	0.015	0.03	Hf = 1.5 Ta = 3.0
PWA 1422	1	Bal.	9.0	10.0	–	12.0	5.0	2.0	0.14	0.015	0.1	Nb = 1.0 Hf = 1.5
René 80 H	1	Bal.	14.0	9.0	4.0	4.0	3.0	4.7	0.16	0.015	0.01	Hf = 0.8
CM 186 LC	2	Bal.	6.0	9.0	0.5	8.4	5.7	0.7	0.07	0.015	0.005	Hf = 1.4 Re = 3.0 Ta = 3.4
CM 247 LC	2	Bal.	8.0	9.0	0.5	10.0	5.6	0.7	0.07	0.015	0.01	Hf = 1.4 Ta = 3.2
PWA 1426	2	Bal.	6.5	10.0	1.7	6.5	6.0	–	0.10	0.015	0.10	Hf = 1.5 Re = 3.0 Ta = 4.0
René 142	2	Bal.	6.8	12.0	1.5	4.9	6.15	–	0.12	0.015	0.02	Hf = 1.5 Re = 2.8 Ta = 6.35

Table 6.16 Chemical compositions of selected SC Ni-based superalloys. (ASM International 2000; Prasad and Wanhill 2017)

Alloy	Generation	Ni	Cr	Co	Mo	W	Al	Ti	Ta	Others
AF 56 (SX 792)	1	Bal.	12.0	8.0	2.0	4.0	3.4	4.2	5.0	–
AM1	1	Bal.	8.0	6.0	2.0	6.0	5.2	1.2	9.0	–
AM3	1	Bal.	8.0	6.0	2.0	5.0	6.0	2.0	4.0	–
CMSX-11B	1	Bal.	12.5	7.0	0.5	5.0	3.6	4.2	5.0	Nb = 0.1 Hf = 0.04
CMSX-11C	1	Bal.	14.9	3.0	0.4	4.5	3.4	4.2	5.0	Nb = 0.1 Hf = 0.04
CMSX-2	1	Bal.	8.0	5.0	0.6	8.0	5.6	1.0	6.0	–
CMSX-3	1	Bal.	8.0	5.0	0.6	8.0	5.6	1.0	6.0	Hf = 0.1
CMSX-6	1	Bal.	10.0	5.0	3.0		4.8	4.7	2.0	Hf = 0.1
PWA 1480	1	Bal.	10.0	5.0	–	4.0	5.0	1.5	12.0	–
PWA 1483	1	Bal.	12.8	9.0	1.9	3.8	3.6	4.0	4.0	–
René N4	1	Bal.	9.0	8.0	2.0	6.0	3.7	4.2	4.0	Nb = 0.5
RR 2000	1	Bal.	10.0	15.0	3.0	–	5.5	4.0	–	–
SC 16	1	Bal.	16.0	–	3.0	–	3.5	3.5	3.5	–
SRR 99	1	Bal.	8.0	5.0	–	10.0	5.5	2.2	3.0	–
CMSX-4	2	Bal.	6.5	9.0	0.6	6.0	5.6	1.0	6.5	Hf = 0.1 Re = 3.0
MC2	2	Bal.	8.0	5.0	2.0	8.0	5.0	1.5	6.0	–
PWA 1484	2	Bal.	5.0	10.0	2.0	6.0	5.6	–	9.0	Hf = 0.1 Re = 3.0
René N5	2	Bal.	7.0	8.0	2.0	5.0	6.2	–	7.0	Hf = 0.2 Re = 3.0
SC 180	2	Bal.	5.0	10.0	2.0	5.0	5.2	1.0	8.5	Hf = 0.1 Re = 3.0
CMSX-10	3	Bal.	2.0	3.0	0.4	5.0	5.7	0.2	8.0	Nb = 0.1 Hf = 0.03 Re = 6.0
René N6	3	Bal.	4.2	12.5	1.4	6.0	5.75	–	7.2	C = 0.05 Hf = 0.15 Re = 5.4
TMS-75	3	Bal.	3.0	12.0	2.0	6.0	6.0	–	6.0	Hf = 0.1
EPM-102	4	Bal.	2.0	16.5	2.0	6.0	5.6	–	8.25	B = 0.004 C = 0.03 Hf = 0.15 Re = 5.95 Ru = 3.0 Y = 0.001
TMS-138	4	Bal.	3.2	5.8	2.9	5.9	5.8	–	5.6	Hf = 0.1; Re = 5.0

(continued)

Table 6.16 (continued)

Alloy	Generation	Ni	Cr	Co	Mo	W	Al	Ti	Ta	Others
										Ru = 2.0
TMS-138A	4	Bal.	3.2	5.8	2.9	5.6	5.7	–	5.6	Hf = 0.1 Re = 5.8 Ru = 3.6
TMS-162	5	Bal.	3.0	5.8	3.9	5.8	5.8	–	5.6	Hf = 0.1 Re = 4.9 Ru = 6.0
TMS-173	5	Bal.	3.0	5.6	2.8	5.6	5.6	–	5.6	Hf = 0.1 Re = 6.9 Ru = 5.0
TMS-196	5	Bal.	4.6	5.6	2.4	5.0	5.6	–	5.6	Hf = 0.1 Re = 6.4 Ru = 5.0
TMS-238	6	Bal.	4.6	6.5	1.1	4.0	5.9	–	7.6	Hf = 0.1 Re = 6.4 Ru = 5.0

The design of new alloy compositions, the refinement of their microstructure through controlled thermal treatments, and the prediction of the relevant properties under different operating conditions involve some metallurgical aspects, which will be shortly revised in Sect. 6.4.1, with particular regard to solid interfaces and dislocation dynamics. Sections 6.4.2 and 6.4.3 will illustrate the main selection criteria of alloying elements in nickel-based superalloy, including the way to avoid the formation of topologically close-packed (TCP) phases, like the σ phase, discussed already with reference to cobalt-based superalloys (see Sect. 6.2.2).

The large variety of superalloy compositions affords a wide range of mechanical properties, regarded as valuable selection criteria in view of any specific structural application. A summary of the main mechanical properties of wrought and cast Ni-based superalloys is provided in Tables 6.17 and 6.18, respectively.

6.4.1 Relevant Metallurgical Issues

The strengthening mechanisms interesting for structural alloys to be used in the hottest part of a gas turbine involve phenomena occurring at different length scales: from nanometer up to millimeter. Precipitation strengthening in nickel-based superalloys is mainly based on the formation of ordered the γ'-$Ni_3(Al,Ti)$ precipitates (see Sect. 6.3 and Appendix 6). Their ordered atomic arrangement determines peculiar dislocation structures and, thereby, dynamics that will be considered and discussed in details. Beforehand, the main kinds of interfaces that can form between crystalline phases will be recalled, considering their central role in the strengthening

Table 6.17 Mechanical properties of selected wrought Ni-based superalloys. E% is the fracture elongation

Alloy	Type	UTS at RT [MPa]	Yield strength at RT [MPa]	E% at RT [%]	UTS at 760 °C [MPa]	Yield strength at 760 °C [MPa]	E% at 760 °C [%]	1000 h rupture strength at 650 °C [MPa]	1000 h rupture strength at 870 °C [MPa]
Astroloy	PH	1415	1050	16.0	1160	910	21.0	770	170
Hastelloy S	SSH	845	455	49.0	575	310	70.0	–	25
Inconel 625	SSH	965	490	50.0	550	415	45.0	370	50
Inconel X750	PH	1200	815	27.0	–	–	–	470	50
Nimonic 80A	PH	1000	620	39.0	600	505	17.0	420	–
Nimonic 90	PH	1235	810	33.0	655	540	12.0	455	60
René 41	PH	1420	1060	14.0	1105	940	11.0	705	115
René 95	PH	1620	1310	15.0	1170	1100	15.0	860	–
Udimet 500	PH	1310	840	32.0	840	730	39.0	760	110
Udimet 710	PH	1185	910	7.0	1020	815	25.0	870	200
AF2-1DA6	PH	1560	1015	20.0	1290	995	16.0	885	–
Waspaloy	PH	1275	795	25.0	650	675	28.0	615	110

PH precipitation hardening, *SSH* solid solution hardening. Tensile properties of all alloys are obtained from bar samples (Donachie and Donachie 2002; Geddes et al. 2010)

Table 6.18 Mechanical properties of selected cast Ni-based superalloys. E% refers to the elongation at rupture. RT is referred to room temperature condition, meaning 21 °C

Alloy	UTS at RT [MPa]	Yield strength at RT [MPa]	E% at RT [%]	UTS at high temperature [MPa]	Yield strength at high temperature [MPa]	E% at high temperature [%]	1000 h rupture strength at 815 °C [MPa]	1000 h rupture strength at 980 °C [MPa]
Alloy 713C	850	740	8.0	860[a]	740[a]	10.0[a]	305	70
M-22	730	685	5.5	780[a]	685[a]	4.5[a]	385	130
MAR-M200	930	840	7.0	325[b]	–[b]	–[b]	415	125
Nimocast 80	730	520	15.0	–	–	–	–	–
Udimet 500	930	815	13.0	895[a]	725[a]	13.0[a]	240	–
Udimet 710	1075	895	8.0	–	–	–	325	76
CMSX-2[c]	1185	1135	10.0	1295[d]	1245[d]	17.0[d]	–	170
MM 002	1035	825	7.0	550[e]	345[e]	12.0[d]	–	125
MAR-M200 Hf	1035	825	5.0	540[e]	345[e]	10.0[d]	–	125
PWA-1480[c]	–	–	4.0	685[e]	495[e]	20.0[d]	–	–
UDM 56	945	850	3.0	945[d]	725[d]	5.0[d]	–	125

Donachie and Donachie (2002) and Geddes et al. (2010)
[a]= 538 °C
[b]= 1093 °C
[c]= Single crystal [001]
[d]= 760 °C
[e]= 980 °C

mechanisms in alloys, including those regarding the mechanical behavior of Ni-based superalloys and also the latest generation of Co-based (see Sect. 6.2.2) and Fe-based superalloys (see Sect. 6.3.1 and Appendix 6).

6.4.1.1 Solid Phase Interfaces

Interfaces, like the twins and grain boundaries separating the same phase (although with different crystallographic orientations), are called homo-phase interfaces. Hetero-phase interfaces are present instead, between a precipitate phase and its hosting parent phase (i.e., the matrix). In this regard, grain boundaries separating grains of two or more different phases would belong to this category. Since many properties of engineering materials are determined by the structure of these solid-solid interfaces, a classification has been developed. It is based on the relationship, if any, between the crystallographic structures of the two phases across the interface. Three main types of interfaces can be classified as coherent, semicoherent, and incoherent.

A coherent interface is defined as a plane across which there is a continuity in the crystallographic structures of the two phases (Fig. 6.17). This implies a correspondence in the atomic occupation in the crystallographic planes facing each other across the interface. Coherent interfaces can form if the two phases have the same crystallographic structure. In this case, a coherent interface may form when equivalent crystallographic planes face each other along specific crystallographic directions. The same may happen between phases with different structures but still having planes with an equivalent atomic occupation geometry. A typical example is the coherent interface which may form between the basal plane of the *hcp* structure and

Fig. 6.17 Scheme of a coherent interface, featuring a small misfit elastic strain

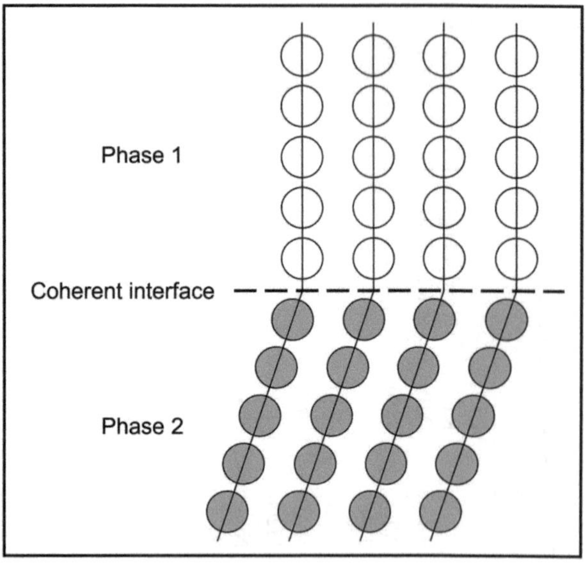

the diagonal plane of the *fcc* structure, both close-packed planes. The coherency orientation relationship can be written as:

$$\{111\}_{\text{cubic}}//\{0001\}_{\text{hexagonal}} \tag{6.1}$$

$$< 110 >\ _{\text{cubic}}// < 11\bar{2}0 >\ _{\text{hexagonal}} \tag{6.2}$$

According to Eqs. 6.1 and 6.2, coherent interfaces are identified by parallel crystallographic planes (Eq. 6.1), mating across the interface itself, and directions (Eq. 6.2), indicating the orientation of the crystallographic plane. In Ni-based superalloys, coherent interfaces separate the ordered γ'-$Ni_3(Al,Ti)$ precipitates and the austenitic γ matrix. Indeed, the difference between the lattice parameters of the two phases is specifically designed, through an appropriate compositional selection, to remain below the coherency limit over the whole temperature range interesting for their applications, irrespective of any differential expansion induced by different CTEs in the two phases. A coherent interface may tolerate differences in the interatomic distances of the facing planes up to 0.05, i.e., 5%, which is regarded as the coherency limit. Since both the γ matrix and the γ' precipitate have cubic structure, lattice strain is usually quantified in terms of the lattice misfit, δ, that can be defined, in terms of the lattice parameters, a_γ and $a_{\gamma'}$ of the two phases: γ matrix and γ' precipitate (Eq. 6.3) (Geddes et al. 2010; Reed 2006).

$$\delta = \frac{a_{\gamma'} - a_\gamma}{a_{\gamma'} + a_\gamma} \times 2 \tag{6.3}$$

Since it depends on the lattice elastic strain, the lattice misfit influences the interface surface energy. In case of coherent interfaces, this term is comparatively smaller than the surface energy term (Γ_{ch}) due to the different chemical compositions across the interface itself, which is indeed prevailing. Therefore, the surface energy of a coherent interface ($\Gamma_{coherent}$) is relatively low at room temperature, and it can be expressed as $\Gamma_{coherent} = \Gamma_{ch}$, with values usually lower than 200 mJ/m^2. Considering Ni-based superalloys, there exists a coherency orientation relationship between the planes of the γ matrix and the γ' precipitate, the so-called cube-cube orientation relationship (Eqs. 6.4 and 6.5).

$$\{100\}_\gamma//\{100\}_{\gamma'} \tag{6.4}$$

$$< 100 >\ _\gamma// < 100 >\ _{\gamma'} \tag{6.5}$$

In Sect. 6.3, γ''-Ni_3Nb has been introduced as another strengthening phase in iron-based superalloys (Fig. 6.13). The *bct* $D0_{22}$ structure of the γ'' phase features an ordered atomic arrangement with a symmetry similar to the ordered *fcc* $L1_2$ structure of the γ' phase. Also γ'' precipitates can form coherent interfaces with the γ matrix in iron-based superalloys. Because of the tetragonal structure of the γ'' cell, two coherency relationships are needed to describe the situation:

$$\{100\}_{\gamma}//\{100\}_{\gamma''} \tag{6.6}$$

$$< 100 >_{\gamma}// < 100 >_{\gamma''} \tag{6.7}$$

and

$$\{001\}_{\gamma}//\{001\}_{\gamma''} \tag{6.8}$$

$$< 001 >_{\gamma}// < 001 >_{\gamma''} \tag{6.9}$$

The coherency relationships influence the morphology and dimensions of the precipitates. In γ'-strengthened superalloys, the typical precipitate morphology is cube like, whereas with γ'' phase, owing to the anisotropy of the tetragonal structure, disc-shaped precipitates would usually form. The interest for coherent interfaces in superalloys strengthened by the precipitation of ordered intermetallics is twofold. In the first place, a low surface energy term is a prerequisite for a better stability of the alloy microstructure against recrystallization and precipitate coalescence phenomena, promoted by diffusion through chemical potential gradients (see Sect. 6.6 on thermal treatments). Secondly, coherent interfaces have an important role in the strengthening mechanisms, as concerns both static and creep loading regimes (see Sects. 6.4.3 and 6.5.1 on precipitation hardening and creep mechanisms).

At higher coherency strains, resulting from larger differences in the interatomic distances between precipitate phase and alloy matrix, in excess of the coherency limit (5%), a complete coherency at the interface is lost. This determines the formation of coherency dislocations (Fig. 6.18). A surface energy term due to the lattice elastic strain at the interface (Γ_{st}), including also the dislocation contribution, raises the overall interface surface energy, and the interface is now semicoherent. The surface energy of a semicoherent interface is given by both a chemical and strain contributions ($\Gamma_{semicoherent} = \Gamma_{ch} + \Gamma_{st}$) with larger values than in coherent interfaces: from 200 to 500 mJ/m^2 (Smallman and Ngan 2014).

Two phases are separated by an incoherent interface (Fig. 6.19) when there is neither continuity, nor correspondence between the relevant crystallographic structures. The surface energy contribution is mainly due to the volume mismatch between precipitate and matrix. Carbides, topologically close-packed (TCP) phases in steels and superalloys, are typical examples of incoherent precipitates. The surface energy of an incoherent interface ($\Gamma_{incoherent}$) can reach values as large as 1000 mJ/m^2 (Smallman and Ngan 2014).

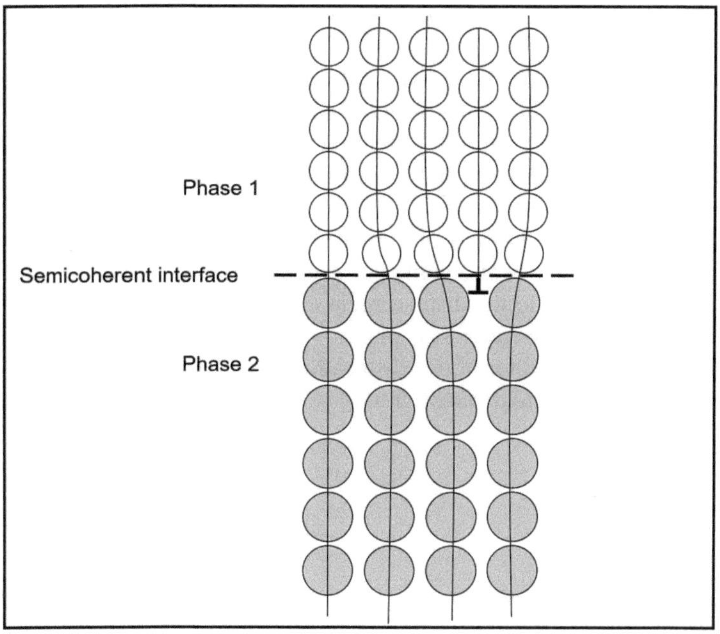

Fig. 6.18 Scheme of a semicoherent interface. Note the presence of a misfit dislocation

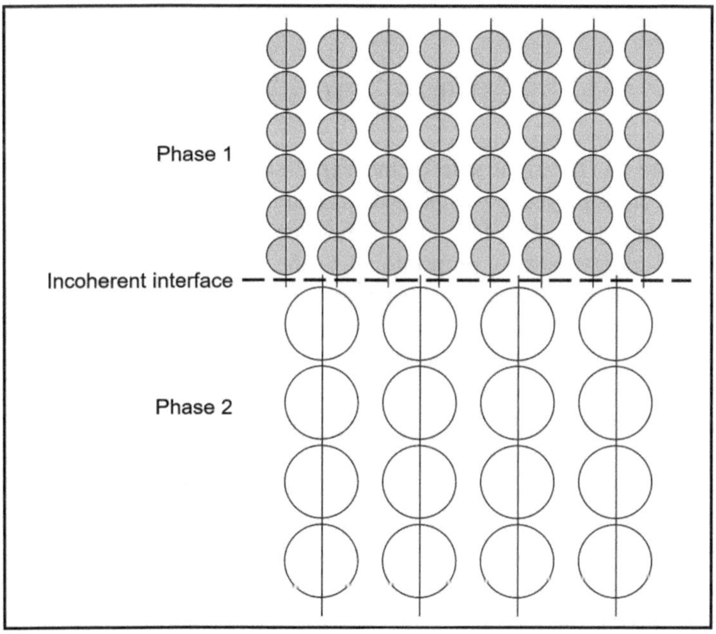

Fig. 6.19 Scheme of an incoherent interface

6.4.1.2 Dislocations Structure and Dynamics in Ordered Alloys

From a microscopic standpoint, plastic deformation in crystalline materials can be described in terms of dislocation movement and interactions with the microstructural elements present in the materials, like precipitates, grain boundaries, and other defects (Callister and Rethwisch 2012; Hull and Bacon 2011). The dislocation vector notation provides a very useful tool, when dealing with these aspects. In *fcc* metals, the elementary atomic displacement, i.e., the movement of each single atom (of the dislocation line) from one to another lattice site, so that the symmetry of the crystal is not broken, corresponds to the displacement along the cube diagonal, <110> direction, lying in the cube diagonal planes {111}, by half of its length. The corresponding Burgers vector of this dislocation, describing an elementary displacement, and named *complete dislocation* is:

$$\frac{a}{2}\langle 110\rangle \tag{6.10}$$

where

- a is the lattice parameter of the cubic cell. Sometimes a is omitted in this expression.

The movement of the atoms belonging to a complete dislocation line does not occur in a straight line but rather with a zigzag movement. In other terms, a dislocation dissociation occurs that can be expressed as:

$$\frac{a}{2}\langle 110\rangle \rightarrow \frac{a}{6}\langle 211\rangle + \frac{a}{6}\langle 12\bar{1}\rangle \tag{6.11}$$

Dislocations on the right side of the expression are called *partial* (Shockley), since they do not represent a complete displacement between two equivalent atomic positions. Nonetheless, their formation is favored since they reduce the lattice strain energy. This energy term is proportional to the squared modulus of the Burgers vector, and the above dissociation reaction confirms that the complete dislocation has a larger energy than that of the sum of the relevant partials (Reed-Hill 1973). *fcc* crystals result from an ABC stacking of {111} planes. The atomic displacement due to the movement of a partial dislocation takes the atoms to nearest neighbor positions, thus changing the correct stacking sequence, since atoms in the B layer become C-atoms and so forth. A repulsive force is acting on the two partials, which tends to increase their separation. An equilibrium distance is achieved when the repulsion energy is counterbalanced by the stacking fault energy (SFE). The defect structure consisting of two partials separated by a stacking fault band is called *extended dislocation*, and it is, by all means, a planar defect (Fig. 6.20).

The dislocation structure becomes slightly more complex in case of the ordered $L1_2$ structures. Slip systems are still the same as in an austenitic random alloy, although the minimum displacement required to preserve the lattice symmetry and

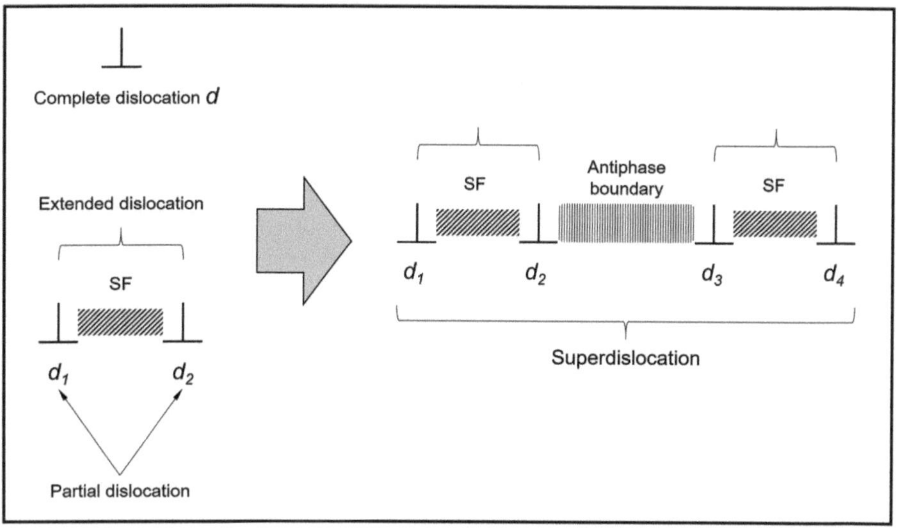

Fig. 6.20 Dislocation structures: dissociation of a complete dislocation into the extended dislocations. Structure of a super-dislocation, typical of ordered intermetallics, like γ'-$Ni_3(Al, Ti)$

the correct atomic occupation of lattice sites is now doubled. The relevant Burgers vector (Eq. 6.12) is equal to the entire cube diagonal, and it is named *super-dislocation*:

$$a \langle 110 \rangle \tag{6.12}$$

Similarly to the movement of a partial dislocation that creates stacking faults in an *fcc* lattice, the movement of an elementary complete dislocation $\left(\frac{a}{2}\langle 110 \rangle\right)$ within an ordered lattice induces the formation of a planar defect, featuring a wrong atomic occupation. This defect is named *antiphase boundary* (APB). In analogy with complete dislocations in *fcc* lattice, super-dislocations exhibit the tendency to dissociate. The dissociation scheme is slightly more complex, as shown by Fig. 6.20. The two complete dislocations, possibly dissociated into the relevant Shockley dislocations according to Eq. 6.11, are now separated by a band of antiphase boundary. The width of the APB is the results of the repulsive interaction between the two complete dislocations and their attraction, which limits the positive APB energy term.

An important effect determined by the super-dislocation dynamics is the anomalous dependence of the yield strength on temperature. In fact, as long as the leading and trailing dislocation glide together on {111} slip planes, under the effect of a shear stress, the width of the APB present in between the two component dislocations remains substantially constant. Upon increasing the temperature, a cross-slip mechanism, involving the leading super-partial dislocation, occurs, as a result of a mechanical torque, the so-called Yoo torque, acting on each component of the super-

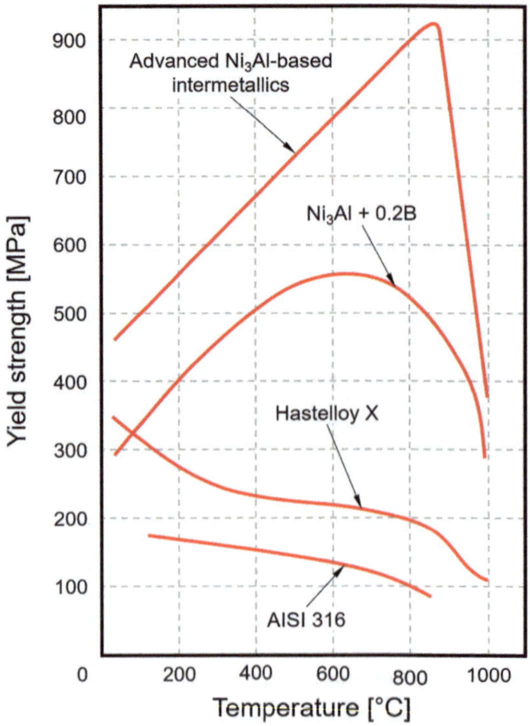

Fig. 6.21 Evolution of the yield strength with temperature for an austenitic stainless steel (AISI 316), an Fe-based superalloy (Hastelloy X), the γ'-Ni_3Al binary intermetallic with 0.2% B, and the advanced Ni_3Al-based intermetallics. (Redrawn from Jozwik et al. 2015)

dislocation (Yoo 1987). In particular, the cross-slip from {111} octahedral planes to {100} cube planes becomes more and more probable at increasing temperature. Once the leading dislocation has moved to a {100} plane, the trailing one remains blocked at the intersection between the two planes. These dislocation structures are named Kear-Wilsdorf locks, after the names of their discoverers (Dimiduk 1991; Kear and Wilsdorf 1962). Any further movement of the leading dislocation becomes a difficult process and requires a widening of the APB and, thereby, an increase of the associated energy. This is the reason for the increased strength of the γ'-Ni_3(Al, Ti) phase with temperature, a phenomenon that is paramount as concerns precipitation hardening of the nickel and of the new-generation cobalt superalloys (see Sect. 6.2.2) and has been one of the main driving forces for the development of novel alloy compositions. Figure 6.21 shows the yield strength curves for the binary intermetallic compound, Ni_3Al, which can be regarded as the prototype of the gamma prime phase, compared to *advanced* Ni_3Al alloys obtained from the binary compound with specific additions (see Sect. 9.4). These curves are displayed together with the yield strength curves for Hastelloy X, an iron-rich Fe-based superalloy (see Table 6.7), and the austenitic stainless steel AISI 316 (see Table 5.13). This latter displays a monotonous softening with temperature. A similar trend is there in the Hastelloy X material, mitigated though in the high-temperature range by the presence of ordered γ'-Ni_3(Al, Ti) precipitates. Since they become stronger at higher temperatures, a positive derivative change in the yield strength is observed, although not to the same extent as in pure γ' materials, i.e., γ'-Ni_3Al and *advanced* Ni_3Al-based intermetallic alloys.

The peak in the yield strength, more or less intense, depending on the actual composition of the alloy, is reached when the thermally activated unlocking of the trailing dislocations occurs. At this stage, the activation of additional slip systems results in a general softening of the alloy.

The yield strength dependence on temperature is strictly related to another "anomaly," observed in the creep behavior of the Ni_3Al-based (Hemeker et al. 1991) and also of other intermetallic alloys (Smallman et al. 2002). In fact, over an intermediate temperature range, close to the peak temperature of the yield strength vs temperature curve (see Fig. 6.21), the dislocation structures that have been considered under static loading conditions become important also to explain the creep deformation behavior. The primary creep is sustained by the gliding of super-dislocations (see Fig. 6.20) on the $\{111\}$ octahedral planes. The coordinated movement of these dislocation couples tends to saturate at the end of the primary creep regime for the cross-slip that produces sessile dislocations on cube planes, on which they are substantially blocked, i.e., the Kear-Wilsdorf locks (Hull and Bacon 2011). This results in a slowing down of creep deformation. Subsequently, the cross-slipped dislocations start both to bow out, generating more dislocation segments, and to glide on the cube planes. The consequent increase in the density of mobile dislocations on cube planes leads to an increase in the creep rate, that is, exactly the inverse creep rate effect.

6.4.2 *Selection Criteria for Alloying Elements in Nickel-Based Superalloys: The γ Matrix*

A common feature to superalloys in general and, in particular, to nickel-based ones is the complexity of their compositions, required to satisfy the demanding operating conditions typically faced in gas turbine engines and other structural applications at high temperatures and in potentially corrosive environments. Adopting a composite-like approach, a nickel-based superalloy can be regarded as made of the following main components:

- The alloy matrix, i.e., a Ni-based solid solution with austenitic structure. Its intrinsic strength and properties depend on the concentrations of alloying elements in solution.
- Reinforcing phases. γ'-$Ni_3(Al,Ti)$ and γ''-Ni_3Nb ordered precipitates are the main reinforcement of these alloys. In case of polycrystalline alloys, carbides are very important too, with beneficial effects on the grain boundary stability, hindering recrystallization and grain boundary slip. Direct carbide strengthening is based on their interaction with dislocation lines, mainly through the Orowan bowing mechanism (see Sects. 6.4.3 and 6.6). Carbides, indeed interesting for

polycrystalline alloys, are not even present in alloys developed for single-crystal castings, although carbon is still having a role in the latest formulations of Ni-based superalloys (see Table 6.16).

Moreover, the request for an intrinsic resistance against high-temperature corrosion determines the selection of specific alloying elements capable to produce protective native mixed oxide scales. Although this requirement is mitigated by the possibility of depositing suitable overlay coatings (see Sect. 7.3.1), still the design of an alloy composition stable against diffusion-assisted phenomena, including high-temperature corrosion, has beneficial effects also on creep resistance, for which diffusion plays a central role. Controlling diffusion is also interesting as regards the stability of the microstructure of the superalloys when exposed to service conditions. The retention of an optimized microstructure depends also on preventing the formation of unwanted phases, like the topologically closed packed (TCP) ones, including the σ phase introduced already dealing with Co-based superalloys (see Sect. 6.2.1 and Appendix 6). As concerns the nickel solutes, i.e., those elements that enter the γ matrix of the nickel-based superalloys, they should exhibit the following three main features: a good solid solubility in nickel, a high hardening coefficient, and they should also contribute to the improvement of creep resistance of the superalloy. The solubility of alloying elements in nickel can be assessed on the basis of the difference of their atomic diameters with respect to that of nickel itself.

In Fig. 6.22, the diameters of several atoms, some of which potentially interesting for being solid solution elements in nickel-based superalloys, are shown. The solid

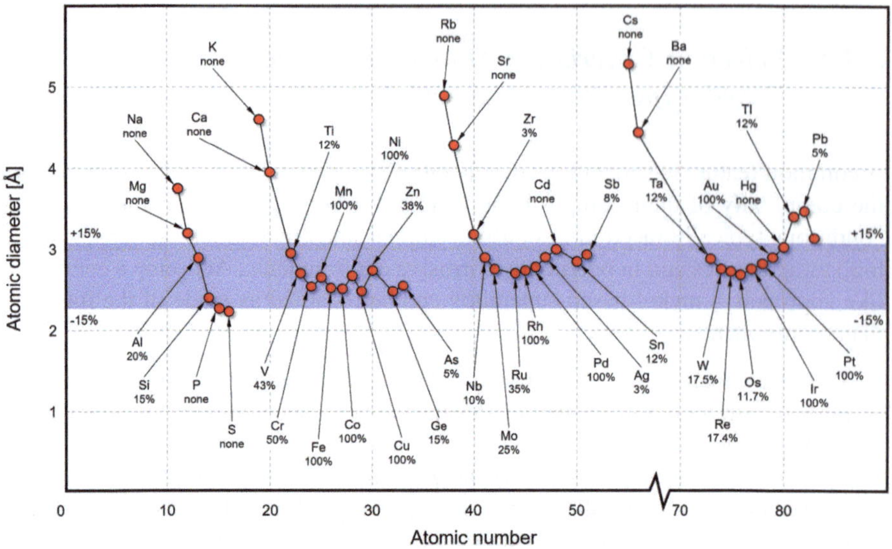

Fig. 6.22 Atomic diameter of several elements. The shadowed area marks a range of variability of ±15% with respect to the atomic diameter of nickel. For each element the maximum solid solubility in nickel is also quoted. (Redrawn from Jena and Chaturvedi 1984; Reed 2006)

Table 6.19 Some parameters relevant to solid solution hardening of nickel (Jena and Chaturvedi 1984)

Element	Melting T [°C]	%Δ in the lattice parameter of nickel per at% solute atom	%Δd[a] with respect to nickel	Groups separation with respect to nickel	Change in yield strength per at% of solute [MPa]
Co	1493	0.005	0.2	1	+0.3
Cr	1907	0.032	0.2	4	+4.9
Cu	1085	0.025	2.5	1	+1.6
Fe	1538	0.031	1.3	2	+3.3
Mo	2623	0.116	9.5	4	+14.2
Ti	1670	0.095	16.0	6	+22.7
W	3422	0.135	10.0	4	+16.5

[a]d indicates the atomic diameter

solubility limit of each element is also indicated. In general, a better solubility is exhibited by those elements whose atomic diameter differs by no more than 15% from the atomic diameter of nickel, i.e., if it falls within the shadowed band in Fig. 6.22. For structural applications, it is important not only the concentration that each solute element can reach but also how effective this element is in solid solution strengthening the alloy matrix. This mechanism relies on the increase in the resistance to dislocation movement, determined by the lattice distortions and changes in the shear modulus.

The effects of changes in lattice parameter of nickel, introduced by alloying elements, on the yield strength are more pronounced when there is a larger separation between the groups of the periodic table between nickel and the solute.

With reference to Table 6.19, titanium introduces a lattice expansion equal to 0.095% and a corresponding hardening of 22.7 MPa per at%. Molybdenum, although exhibiting a slightly larger change in the lattice parameter (0.116% per at %), has a strengthening factor of 14.2 MPa per at% only. This can be ascribed to the different position with respect to the group of nickel of the two elements; 6 and 4 are for Ti and Mo, respectively. A larger separation between elements in the periodic table corresponds to a valence difference, which influences the stacking fault energy (SFE) of the alloy. In particular, solute atoms with a larger valence difference with respect to nickel lead to a greater reduction in the SFE of the solid solution. This stabilizes the dislocations on the slip planes, inhibiting cross-slip. This condition leads to the deformation behavior shown by Fig. 6.23, referring to a classical study on a Ni-Co binary system (Cahn and Haasen 1996). The shear stress vs shear strain curves for pure nickel and for the Ni-Co solid solutions, with the composition values indicated in Fig. 6.23, exhibit an initial *easy glide* regime, whose maximum strain can reach larger values when cobalt concentration is higher. The larger deformation that can be reached for higher Co concentrations is in agreement with the possibility of stabilizing a higher density of dislocations, since they are more stable in case of a solute-richer alloy. The same is true for *strain hardening*, which is enhanced by the raise in the dislocation density and consequent entanglement. Eventually, when the

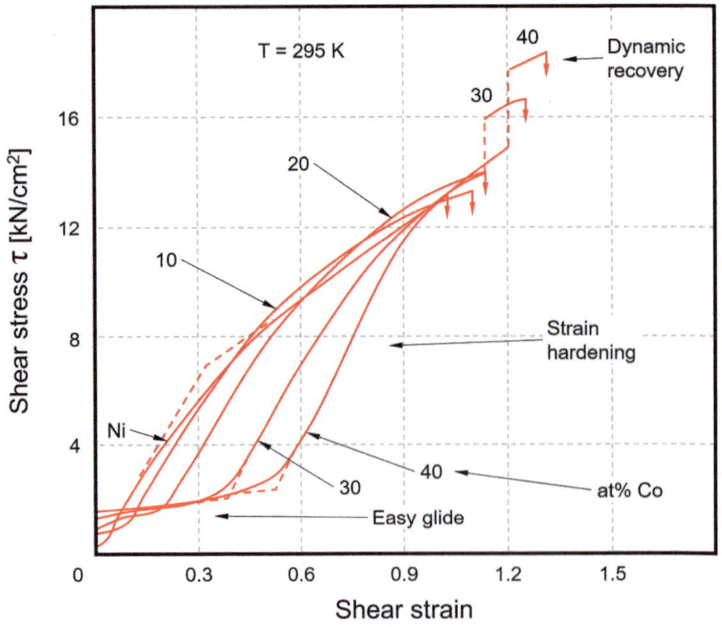

Fig. 6.23 Deformation behavior of pure nickel and Ni-Co solid solutions. Compositions are indicated in the plot. The extent of the three deformation regimes (easy glide, strain hardening and dynamic recovery) is influenced by the SFE values determined by the solute, i.e., cobalt, concentration. (Redrawn from Cahn and Haasen 1996)

dislocation density becomes unsustainably high, an abrupt softening of the alloy occurs. This is due to its *dynamic recovery*, resulting from the recombination and annihilation of dislocations.

Low SFE and high elastic modulus are important selection parameters also as concerns creep resistance, a fundamental property of nickel-based superalloys, particularly in view of their applications in gas turbines. Considering that creep deformation mechanisms and relevant deformation rates are very much dependent on diffusion, an excellent selection parameter for solute elements that may beneficially affect the creep resistance of the alloy is the melting temperature (see Fig. 6.24). In fact, the higher is the melting temperature of the pure element, the lower is its solid-state diffusivity in nickel and, thereby, the relevant creep deformation rate. Therefore, refractory elements, like Mo, Rh, Ta, and W, are good candidates to improve the creep resistance of the γ matrix of the superalloy and, to some extent, also to prevent other diffusion-assisted phenomena, like high-temperature oxidation and corrosion (see Sect. 8.2) and alloy recrystallization.

However, upper limits do exist for the concentration of refractory elements. If present to an excessively high extent, these elements may result in the formation of the so-called topologically close-packed (TCP) phases (see also Appendix 6). The denomination comes from the fact that the space filling principle is the optimization of the atomic packing factor with coordination polyhedral (the Frank-Kasper

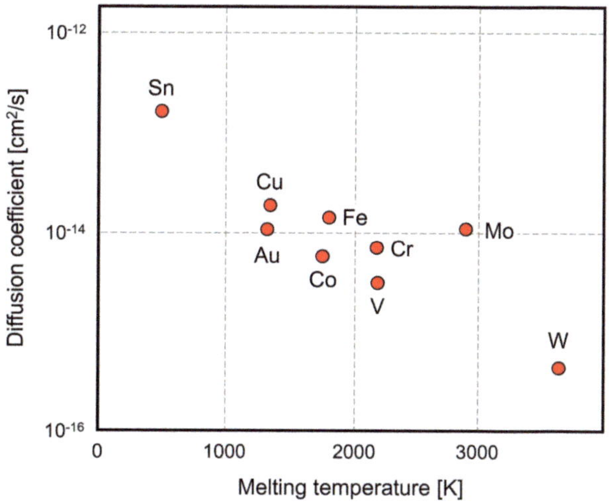

Fig. 6.24 Diffusivity vs melting temperature for solute species in nickel. (Redrawn from Jena and Chaturvedi 1984)

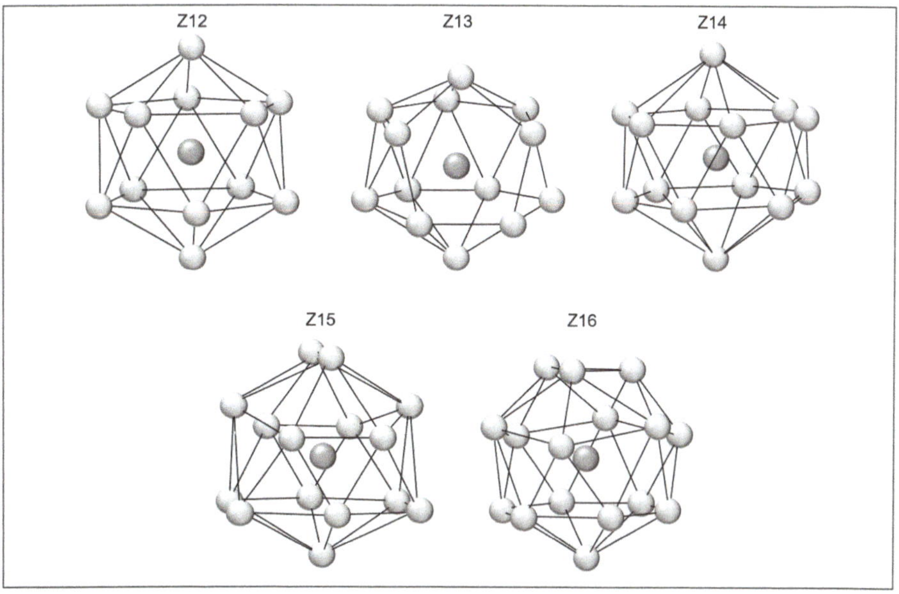

Fig. 6.25 Frank- Kasper coordination polyhedra typical of TCP phases. (Redrawn from Seiser et al. 2011)

polyhedra, see Fig. 6.25), allowing for coordination numbers, i.e., number of nearest neighbors atoms, equal to 12, 13, 14, 15, and 16 and resulting in very tight packing (Barrett and Massalski 1980). Concerning the standard geometrically packed phases like body- and face-centered cubic (*bcc* and *fcc*), hexagonal close packed (*hcp*), etc., they feature different coordination polyhedra that can be thought of as composed of suitable combinations of tetrahedral and octahedral voids. This is shown in

Table 6.20 Types and relevant number of coordination polyhedral of the main TCP structures together with geometrically packed ones. <Z> is the average coordination number for each one of the listed structures. f_{12} is the fraction of the 12-fold polyhedra featuring each structure, including *bcc*, *fcc*, and *hcp* (Seiser et al. 2011)

				Coordination polyhedra				
Structure	Atoms in basis	<Z>	f_{12}	Z12	Z13	Z14	Z15	Z16
fcc/hcp	1	12.00	1.00	(1)	0	0	0	0
C14	12	13.33	0.67	8	0	0	0	4
C15	24	13.33	0.67	16	0	0	0	8
C36	24	13.33	0.67	16	0	0	0	8
M	52	13.36	0.54	28	0	8	8	8
μ	13	13.39	0.54	7	0	2	2	2
R	53	13.39	0.51	27	0	12	6	8
δ	56	13.43	0.43	24	0	20	8	4
P	56	13.43	0.43	24	0	20	8	4
χ	58	13.10	0.41	24	24	0	0	10
σ	30	13.47	0.33	10	0	16	4	0
A15	8	13.50	0.25	2	0	6	0	0
bcc	1	14.00	0.00	0	0	(1)	0	0

Table 6.20, in which the most common TCP phases are listed together with the number and types of relevant Frank-Kasper polyhedra, ranked on the basis of the fraction content (f_{12}) of the 12-fold coordination polyhedra. The situation for *bcc*, *fcc*, and *hcp* structures is also displayed for a comparison. Since they are not made up of Frank-Kasper polyhedra, the relevant structures in the table are indicated among brackets. TCP phases can reach comparatively high atomic packing and rather complex geometries, as can be seen in Fig. 6.26 for the R, μ and χ (or C36 Laves) phase. Another important feature of these structures is the lack of active slip systems, which justifies their high intrinsic brittleness that may degrade the overall fracture toughness of the superalloy. Fracture nucleates at the stress concentration points introduced by the hard TCP grains and may propagate more easily through the brittle intermetallic grains (intersigmatic fracture). However, this is not the only effect these phases are able to produce. Their compositions, particularly rich in refractory elements, definitely richer than the alloy matrix, can be achieved thanks to a diffusive flow, occurring during the alloy exposure at high-temperature conditions, not only during service but also during the pre-service thermal treatments, or welding operations, particularly if the process conditions are not fully under control.

A consequence of these heat-induced transformations is the formation in the alloy matrix, nearby the TCP phase grains, of regions in which the depletion of the refractory elements has significantly reduced solid solution strengthening. This sort of constitutional softening, influencing both mechanical strength and creep resistance of the superalloys, is another important effect of the formation of TCP phases.

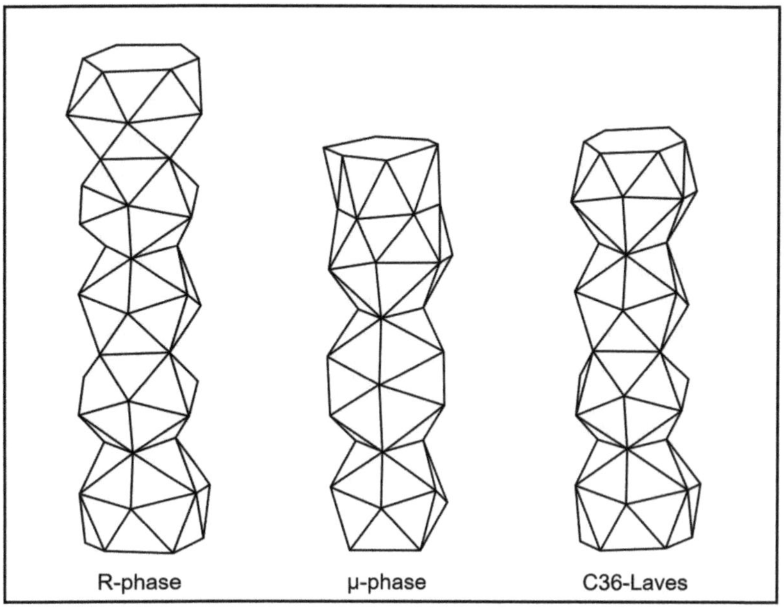

Fig. 6.26 The complex structures of the TCP phases are obtained from the combination of the coordination polyhedra (Fig. 6.25, Table 6.20), as exemplified in the figure by the R-phase, μ-phase, and C36-Laves phase. (Redrawn from Barrett and Massalski 1980)

Therefore, all through the evolution of novel compositions of superalloys, particular attention has been devoted to the design of the composition with a reduced tendency to form TCP phases. A major help came from alloy design and phase prediction programs implemented within PHACOMP – involving phase computation programs. These programs evaluate the average number of holes in the d-bands to define the stability range of the TCP phases, considering their formation in the austenitic γ matrix of the superalloy (Seiser et al. 2011). The improvement implemented in the new PHACOMP was the introduction of the average energy level of the d-orbital, obtained from electronic structure calculations. In this way, both atomic size and electronegativity differences are accounted for in the definition of the stability range of the TCP intermetallic phases, including a better temperature sensitivity. Thanks to the results of these calculations, notwithstanding the simplification of the approach if compared to the complexity of the superalloy compositions, reliable criteria for new alloy design have been identified. The starting point, that to some extent provided the first important guideline, was based on the observations that the detrimental σ phase in Co-Cr and Cr-Fe alloy forms at a fixed value of the valence electron concentration. Therefore, this can be assumed as a cut-off parameter for the newly developed alloys. According to this criterion, the main TCP phases that can form in superalloys have been predicted and identified, and the main ones are listed in Table 6.21.

Table 6.21 Main TCP phases that may form in superalloys, with a selection of stoichiometric compositions and relevant crystallographic parameters (Krakow et al. 2017; Kuo et al. 1985; Rae and Reed 2001; Rae et al. 2000)

Phase	Structure and space group	Lattice parameters [nm]	Possible stoichiometry
A15	Cubic, $Pm\bar{3}m$, 221	$a = 0.456$	Cr_3Si
C14 Laves	Hexagonal, $P6_3/mmc$, 194	$a = 0.515$	$MgZn_2$
		$c = 0.848$	
C15 Laves	Cubic, $Fd\bar{3}m$, 227	$a = 0.708$	$MgCu_2$
C36 Laves	Hexagonal, $P6_3/mmc$, 194	$a = 0.527$	$MgNi_2$
		$c = 1.33$	
M	Orthorhombic, Pnma, 62	$a = 0.930$	$Nb_{10}Ni_9Al_3$,
		$b = 1.62$	$Nb_{48}Ni_{38}Al_{13}$
		$c = 0.493$	
μ	Rhombohedral, $R\bar{3}m$, 166	$a = 0.476$	Mo_6Co_7,
		$c = 2.56$	Mo_6Fe_7
P	Orthorhombic, Pnma, 62	$a = 1.70$	$Mo_{42}Cr_{18}Ni_{40}$
		$b = 0.475$	
		$c = 0.907$ nm	
R	Rhombohedral, $R\bar{3}$, 148	a = b = 1.09	$Mo_{31}Cr_{18}Co_{51}$
		$c = 1.93$	
σ	Tetragonal, $P4_2/mnm$, 136	a = b = 0.880	$Cr_{46}Fe_{54}$
		$c = 0.456$	

The limits of the PHACOM approaches became more evident when dealing with new generations of superalloys, like those containing particularly high concentrations of rhenium. To overcome these limitations, the latest approaches are based on the development of phase stability maps, including two main coordinates: valence electron concentration and size factor difference (Seiser et al. 2011).

Several mechanisms may lead to the formation of the TCP phases. One is the transformation of carbides, a phenomenon favored by the similarities of their crystallographic structures. TCP phases originate from the degeneration of carbides, during which carbon is replaced by refractory elements. TCP phases have also been observed in high-temperature creep studies of single-crystal superalloys for the local compositional changes determined by the diffusive processes occurring under these loading regimes (Reed et al. 2007).

The formation of the TCP phases is not only due to a prolonged exposure to high operational temperatures but also to a too slow cooling from high-temperature stages, i.e., solutioning, of microstructural refinement heat treatments.

This risk was particularly important for early generation superalloys, in which the compositions were not yet sufficiently optimized in this respect. Figure 6.27 displays this aspect for a general-purpose nickel-based superalloy in conventionally cast and wrought conditions. The inhomogeneity, that is typical of castings, makes these products rather prone to the formation of the TCP phases, unless a careful control of the heat treatment conditions is implemented.

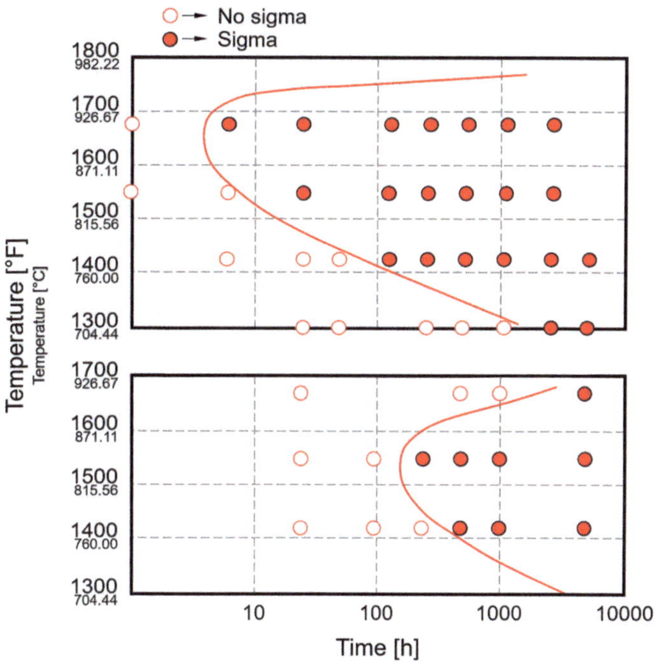

Fig. 6.27 A classical study showing TTT curves for a cast (upper) and wrought (lower) nickel-based superalloys, with the indication of the formation fields for the TCP σ phase. (Redrawn from Sims et al. 1987)

Another important selection rule for the components of the γ matrix of a nickel-based superalloy concerns the improvement in both outer and inner surface stability. Concerning the outer surface, specific elements tend to promote the formation of a native protective oxide layer, usually several tens of micrometer in thickness, whose main task is to avoid, or limit, further oxidation. Important elements for outer surface stability are Cr, Al, Ce, Cs, Mo, La, Mo, Sc, Ta, W, and Y (see also Sect. 6.1 and Appendix 5).

The first, and maybe the most important element, is chromium, which contributes to the formation of a surface protective layer including Cr_2O_3 as fundamental constituent. Its stability is generally guaranteed up to 950 °C, the actual temperature depending on the alloy and, thereby, on the native oxide layer composition. Above 950 °C, chromium oxide tends to become unstable and to evolve in a gaseous phase, according to:

$$Cr_2O_3(s) + \frac{3}{2}O_2(g) \rightarrow 2CrO_3(g) \text{ at } T > 950^{\circ}C \tag{6.13}$$

Aluminum forms a number of oxide phases, the most interesting in this context being the α-Al_2O_3 phase that, unlike Cr_2O_3, is stable also at temperatures in excess of 950 °C. The presence of cobalt improves the resistance to hot corrosion under salt

(Na_2SO_4) deposit, due to a lower diffusivity of sulfur in cobalt than in nickel, with a two order of magnitude difference. Moreover, the melting temperature of an eutectic mixture, like Co_4S_3-Co, is higher than that of Ni_3S_2-Ni: 877 °C vs 645 °C, and thus the corrosive attack is sustained by solid, rather than liquid, phase diffusion (see Sect. 8.2.5).

Refractory elements, like molybdenum, tantalum, tungsten, and rhenium, reduce both inward and outward diffusion rates, for oxygen and metallic alloying elements, respectively. Moreover, molybdenum and tungsten improve the adhesion of the surface oxide layer to the substrate, through the formation of intermetallic precipitates at the interface that produce a sort of "pegging" effect (see Sect. 7.3.3). A similar effect can be achieved with the addition of rare earths (La, Ce) and reactive elements (Cs, Sc, Y) (see Sect. 6.1). On the other hand, these elements may react with sulfur to form eutectics with low-melting temperatures and in this regard may reduce the alloy resistance to hot corrosion in salt deposit (see Sect. 8.2.5).

The main category of inner surfaces is the grain boundaries of the alloy matrix and of secondary phases, like carbides. Actually, some carbides are specifically localized along the alloy GBs and usually would stabilize them against sliding and migration under elevated stresses at high temperatures. Some elements, possibly impurities, are not soluble in nickel and therefore segregate along grain boundaries with detrimental or beneficial effects, depending in the first place on the element but also on its concentration. Elements that affect the stability of the inner surfaces are S, P, Bi, Pb, Te, and Th. These elements are classified as embrittling impurities and can reduce grain boundary strength, even if present in very little concentration. Manganese is able to compensate the S-induced embrittlement through the formation of MnS, which segregates at the grain boundary regions, thus sequestrating the detrimental element. Boron has a positive effect on grain boundary stability, since it increases the grain boundary cohesion, enhances the ductility, and improves oxidation resistance, being able to reduce the diffusional phenomena at the grain boundary. Its concentration is usually kept below 1 wt%, since higher values may result in the formation of coarse borides along the alloy grain boundaries that, in the long run, may reduce the grain boundary cohesion. Also hafnium and zirconium reduce diffusional phenomena at the grain boundaries, including the segregation of alloying elements. Table 6.22 recalls the main alloy elements for the γ phase and relevant role in a Ni-based superalloy.

Table 6.22 Alloying elements and relevant roles in Ni-based superalloys (Geddes et al. 2010)

Role	Element required
Carbides formation	Cr, Hf, Mo, Nb, Ta, Ti, W
Grain boundary refinement and stregthening	B, C, Hf, Zr
High-temperature corrosion resistance	La, Th
Oxidation resistance	Al, Ce, Cr, La, Y
Rupture strength improvement	B
Sulfidation resistance improvement	Co, Cr, Si
Workability improvement	B, C, Fe

6.4.3 Selection Criteria for Alloying Elements in Nickel-Based Superalloys: The γ' Precipitates

The main strengthening mechanism active in nickel-based superalloys is precipitation hardening, based on the formation of γ' precipitates. They have an ordered $L1_2$ structure (see Sects. 6.3.1 and 6.4.1.1 and Appendix 6), based on the *fcc* lattice. Although in real alloys γ' precipitates have rather complex compositions, the binary Ni_3Al intermetallic compound can be taken as a benchmark. The choice of γ' precipitates has essentially been determined by their thermal stability, the possibility of forming coherent interfaces, and the existence of common slip systems with the γ-matrix. These selection criteria have restricted the set of possible candidate phases. Figure 6.28 shows the unit cells of the austenitic *fcc* γ phase and the *fcc* cell with the ordered atomic occupation pattern for the $L1_2$ and $D0_{22}$ structures, corresponding to the γ' and γ'' precipitate phases, respectively (see also Fig. 6.13). The tetragonal $D0_{22}$ structure, introduced dealing with iron-based superalloys (see Sect. 6.3), can actually be regarded as consisting of two stacked *fcc* cells, with a clear atomic occupation similarity with the other ordered γ' phase. In fact, assuming a binary composition for γ'', e.g., Ni_3Nb, the only difference with respect to γ' is that the ordered atomic sequence exhibits, in the site at the center of the base and central crystallographic planes, the alternate occupation of Ni and Nb atoms, respectively. The structural similarity can be straightforwardly extended to slip planes too, as evidenced by Fig. 6.28.

The small coherency strains and relevant low surface energies of the interfaces among the ordered precipitates and the γ matrix are important prerequisites for the

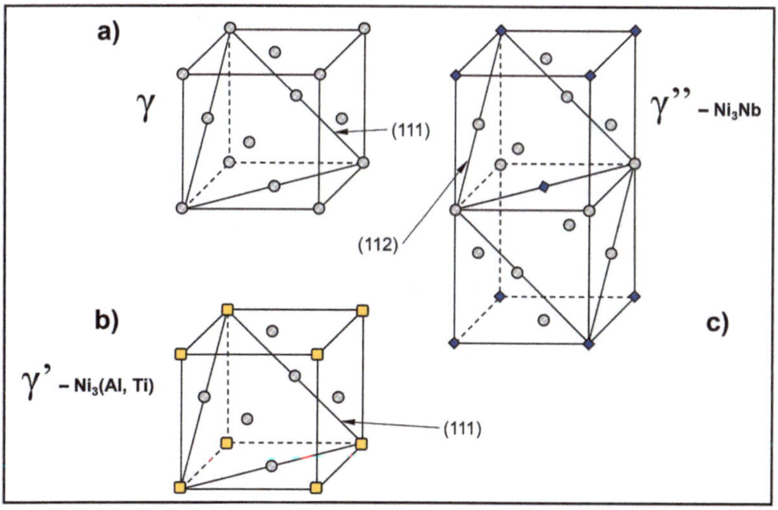

Fig. 6.28 Structural correspondences among the unit cells of the (**a**) fcc γ; (**b**) $L1_2$ γ', and (**c**) $D0_{22}$ γ'' phases. (Redrawn from Jena and Chaturvedi 1984)

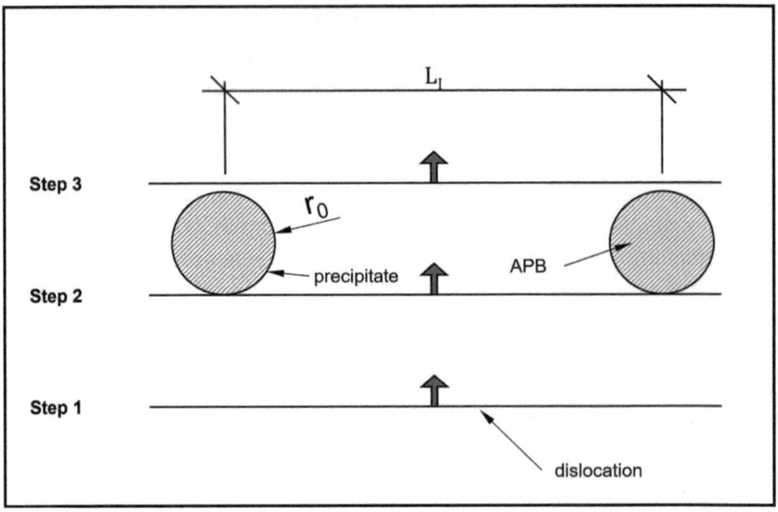

Fig. 6.29 Scheme illustrating the interaction among a dislocation line, length L_I, and ordered γ' precipitates of average radii r_0

alloy stability against recrystallization. The uniform spatial distribution and an optimized cuboidal shape of the precipitates are additional factors, which can be improved through suitable thermal treatments (see Sect. 6.6).

For the exploitation of the anomalous dependence of the yield strength of γ' precipitates on temperature (Sect. 6.4.1.1), it is important that the average size and reciprocal separation of the precipitates are sufficiently small, so that the prevailing interaction mechanism among them and the incoming dislocations is *cutting*. Dislocations generated in the γ matrix, being just *partials* for the ordered structure of γ' precipitates (see Fig. 6.20), when cutting through them require an additional energy term, needed to create the antiphase boundaries (*vide infra*). Therefore, either they are blocked at the matrix-precipitate interface or they are hindered in their movement through the precipitates.

To quantify this contribution, a simple model can be considered (Fig. 6.29). Ordered γ' precipitates are cut through by a dislocation line of length L_1, equal to the average distance between pairs of these precipitates, assumed to be of spherical shape, for the ease of computation, and with an average size of r_0.

The force per unit length, *t*, of a dislocation line with a Burgers vector *b*, moving under the effect of a shear stress, τ, is given by:

$$t = \tau b \tag{6.14}$$

As noted above, the dislocation that forms in the alloy matrix is an incomplete dislocation (*partial*) for ordered γ' precipitates. This implies that when the dislocation moves across an ordered γ' precipitate, it creates an antiphase boundary (APB) with a specific surface energy (Γ_{APB}). By equating the work required to move the dislocation line from one to the other end of the precipitates (see Fig. 6.29), to the energy term required for the creation of the APB across the precipitates, it is possible to obtain:

$$\pi r_0^2 \Gamma_{APB} = L_1 \tau b 2 r_0 \tag{6.15}$$

The Eq. 6.16 is obtained for precipitation strengthening associated with the *cutting* interaction mechanism:

$$\tau_{CUT} = \frac{\pi r_0 \Gamma_{APB}}{2bL_1} \tag{6.16}$$

The expression for τ_{CUT} shows clearly the importance of the APB energy term (Γ_{APB}) that is being actually adopted as a guideline for the design of novel alloys, including γ' precipitates with an optimized composition, increasing their strengthening efficiency. As shown in Table 6.23, the real compositions of the γ' precipitates turn out to be rather complex, definitely more complex than those of the just binary, Ni_3Al, or, at most, ternary $Ni_3(Al, Ti)$ intermetallics, usually taken as reference compositions for γ'.

To infer how different alloying elements tend to replace Ni and Al in the relevant atomic positions, it is possible to refer to the pseudo-ternary phase diagram in Fig. 6.30. Cobalt and copper are preferential substitutes of nickel; molybdenum, chromium and iron replace indifferently both nickel and aluminum; the other elements in the diagram, replace aluminum.

Still referring to the dislocation-precipitate model in Fig. 6.29, a simple calculation provides the contribution of dislocation *cutting* to the superalloy strengthening. Considering γ' precipitates with an average size $r_0 = 250$ nm and an average distance $L_1 = 25$ nm, being $b = 0.25$ nm and $\Gamma_{APB} = 0.1$ J/m^2, τ_{CUT} can be calculated using Eq. 6.16:

$$\tau_{CUT} = \frac{\pi r_0 \Gamma_{APB}}{2bL_1'} = \frac{\pi \cdot 2.5 \cdot 10^{-7} \cdot 0.1}{2 \cdot 0.25 \cdot 2.5 \cdot 10^{-8}} = 6000\ \text{MPa} \tag{6.17}$$

A competitive interaction mechanism to *cutting* is *bowing*, once the dislocation line is pinned by the precipitates, as illustrated in Fig. 6.31. This situation can be faced either in early superalloys, with a comparatively low fraction of γ' phase, or in case of overaged microstructures in which the ripening of the precipitates, for instance, after long-term exposure to the operating conditions, determines an increase in their average distance. The strengthening by the Orowan (*bowing*) mechanism is on average lower than *cutting*. It is therefore interesting to evaluate the conditions corresponding to the transition from one to the other dislocation interaction mechanism.

Table 6.23 Average compositions (at%) of selected nickel-based superalloys and actual compositions of the relevant γ' precipitates (Sims et al. 1987)

		Chemical composition [at%]									
Alloy	Component analyzed	Ni	Al	Ti	Cr	Co	Mo	W	Nb	V	C
Alloy 713C	Alloy	68.77	12.03	0.92	13.96	–	2.58	–	1.24		0.50
	γ'	73.36	17.95	1.26	3.73	–	1.21	–	2.49		–
MAR-M200	Alloy	59.60	10.97	2.47	10.24	10.95	–	4.02	0.64		0.74
	γ'	64.19	18.01	3.42	4.02	6.71	–	3.16	0.49		–
Udimet 500	Alloy	49.62	6.22	2.33	20.84	18.11	2.50	–	–		0.37
	γ'	50.00	–	4.00	1.60	4.00	1.30	–	–		–
Udimet 520	Alloy	55.35	4.45	3.58	20.87	11.63	3.57	0.31	–		0.24
	GB γ'	73.89	–	9.49	4.31	7.22	1.80	0.31	–		–

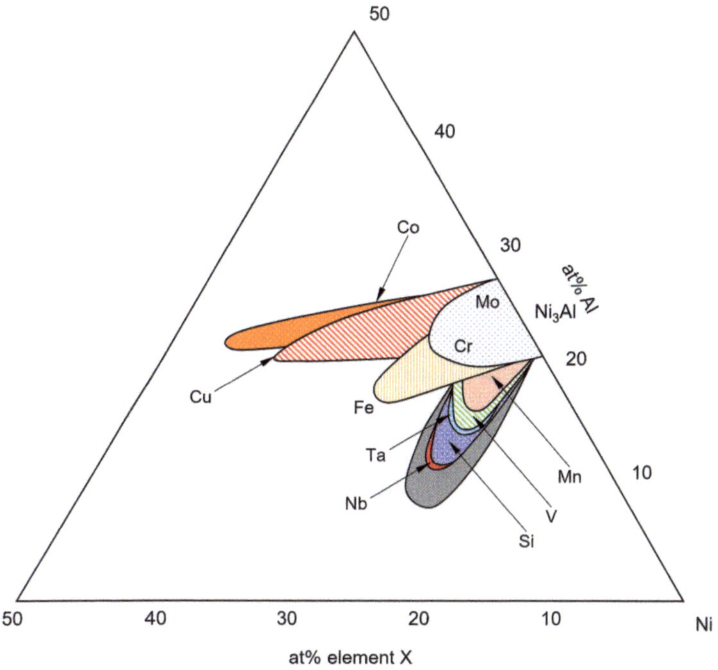

Fig. 6.30 Pseudo-ternary phase diagram, Ni-Al-X, indicating the tendency of ternary alloying elements to preferentially replace Ni, Al, or both atomic in the binary Ni_3Al reference compound. (Redrawn from Sims et al. 1987)

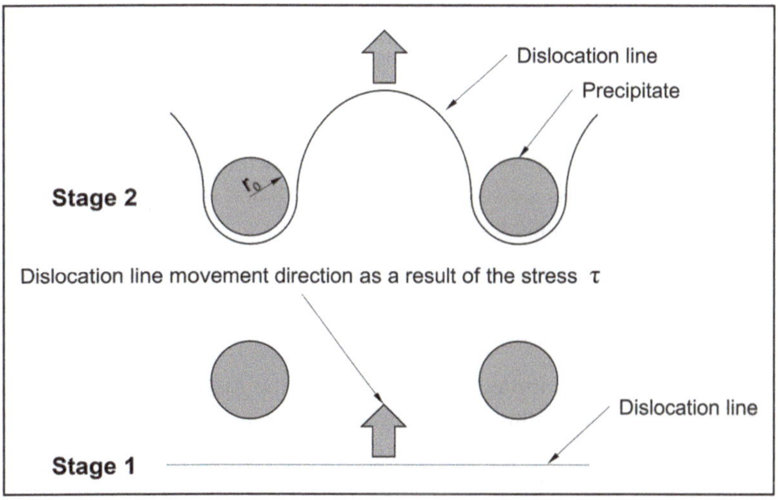

Fig. 6.31 Scheme illustrating the *bowing*, or Orowan, interaction mechanism of a dislocation line with a couple of precipitates. (Redrawn from Hull and Bacon 2011)

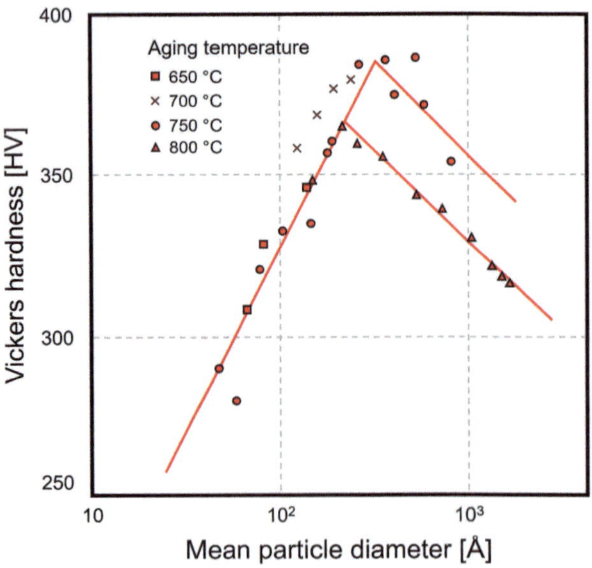

Fig. 6.32 Hardness of a Ni-Cr-Al superalloys as a function of the average particle size of the strengthening γ' precipitates. (Redrawn from ASM International 2000)

The strengthening due to bowing (τ_{BOW}) of dislocations can be expressed as (Hull and Bacon 2011):

$$\tau_{BOW} = \frac{2G_\gamma b}{L_1} \tag{6.18}$$

where:

$G_\gamma$is the shear modulus of the γ matrix.
b is the Burgers vector.
L_1 dislocation line length (equal to the distance between precipitates).

The transition from *cutting* to *bowing* of dislocations will occur when the condition: $\tau_{CUT} = \tau_{BOW}$ is satisfied. From Eqs. 6.18 and 6.16, a critical size of the γ' precipitates can be evaluated:

$$r_0^* = \frac{2Gb}{\pi\Gamma_{APB}} \tag{6.19}$$

Above this size, dislocations, once blocked at the matrix-precipitate interface, start bending and remain within the γ matrix. The situation is illustrated by the plot in Fig. 6.32 of the hardness as a function of the average size of γ' precipitates of a Ni-Cr-Al Ni-based superalloy (ASM 2000). The trend reminds the age-hardening curves discussed for heat-treatable aluminum alloys in Sect. 3.2.3.3 (Fig. 3.17), with a similar evolution in the dislocation-precipitate interaction mechanisms. The only important difference is that in case of aluminum alloys, metastable and non-equilibrium structures are involved, whereas in case of nickel-based superalloys, the equilibrium γ' phase is present.

The phase and microstructural stabilities are paramount in view of the high-stress field and temperatures structural components of a gas turbine engine have to stand. Adequate stability conditions can be achieved through a suitable selection of the alloy composition by refining the microstructure and tuning the lattice misfit, δ (see Sect. 6.4.1.1), a crucial parameter for the long-term in-service behavior of nickel-based superalloys. If γ′ particles are small in size and with low lattice misfit, they tend to have a spherical shape, i.e., an isotropic geometry not affected by any directionality in the elastic strain field at the interface with the alloy matrix. A cubic morphology is generally regarded as the more desirable for γ′ precipitates in Ni superalloys. In fact, this geometry exhibits an intrinsic stability, resulting from a combination of low-energy interfaces, thanks to a low lattice misfit, and for being substantially planar. This aspect and relevant implications will be discussed further on, since they affect also the response of superalloys to the thermal treatments carried out for their microstructural refinement (see Sect. 6.6).

As concerns the interface geometry of the precipitates, it can be noted soon, with reference Fig. 6.33, that the recrystallization and precipitate coarsening propagate starting from the corners of the cubes, the regions with the higher curvature. Since crystal lattice misfit plays an important role in determining the precipitate stability

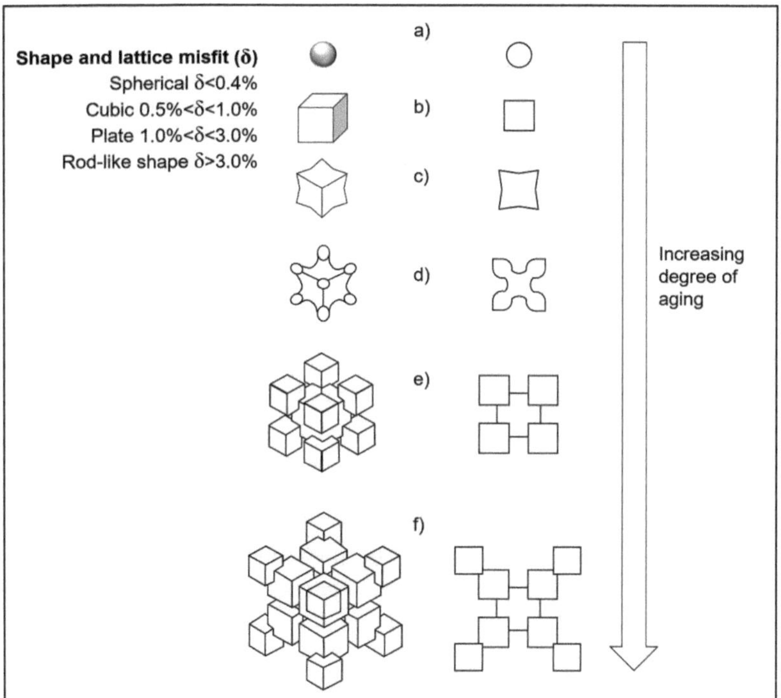

Fig. 6.33 Shapes that the ordered γ′ precipitates can assume during microstructural refinement thermal treatments or due to the exposure to real working conditions. (Redrawn from Ricks et al. 1983; Reed 2006)

and, thereby, the performances of the superalloy, research efforts have been invested in developing effective tools and models, based on PHACOMP, Thermo-Calc, DICTRA, and other similar modeling approaches, resulting in the optimization of this parameter through the design of appropriate alloy compositions (Reed 2006).

For a fast prediction of the effects of different alloying elements on the lattice parameter of the γ matrix and γ′ precipitates, a general expression of the Vegard's law, capable to treat multicomponent systems, has been developed. The lattice parameters of the γ and γ′ phases would change according to the expressions:

$$a_\gamma = a_\gamma^\circ + \sum_i \mu_{i,\gamma} X_{i,\gamma} \tag{6.20}$$

$$a_{\gamma'} = a_{\gamma'}^\circ + \sum_i \mu_{i,\gamma'} X_{i,\gamma'} \tag{6.21}$$

In Eqs. 6.20 and 6.21, a°_γ and $a^\circ_{\gamma'}$ are the lattice parameter values of pure nickel and of the binary Ni_3Al intermetallic compound, respectively (see Fig. 6.34a). The values obtained from Eqs. 6.20 and 6.21 depend on the molar fraction of the alloying elements (X_i) and on the relevant Vegard's coefficients (μ_i) given in Fig. 6.34b.

The total fraction of the γ′ precipitates has been, and still is, a very important parameter in the development of novel superalloy compositions, considering the effect this parameter has on the mechanical properties of the superalloys. Still referring to a very simple model (Fig. 6.35), the dependence of precipitation strengthening on the precipitate fraction is derived.

The number of precipitates (n_L) that are interacting with a dislocation line of length L is given by:

$$n_L = 4\pi r_0^2 L \cdot n \tag{6.22}$$

where n in the concentration of precipitate, i.e., the number of precipitates per unit volume. From the definition of volume fraction of precipitate phase, f:

$$f = \frac{V_{\text{tot.precipitate}}}{V_{\text{alloy}}} \tag{6.23}$$

An expression for n can be obtained by dividing f by the volume of a single precipitate. Again, for the ease of calculation, spherical precipitates with r_0 radius are assumed. Therefore, Eq. 6.22 becomes:

$$n_L = 4\pi r_0^2 L \cdot \frac{f}{\frac{4}{3} n \pi r_0^3} = \frac{3\text{Lf}}{r_0} \tag{6.24}$$

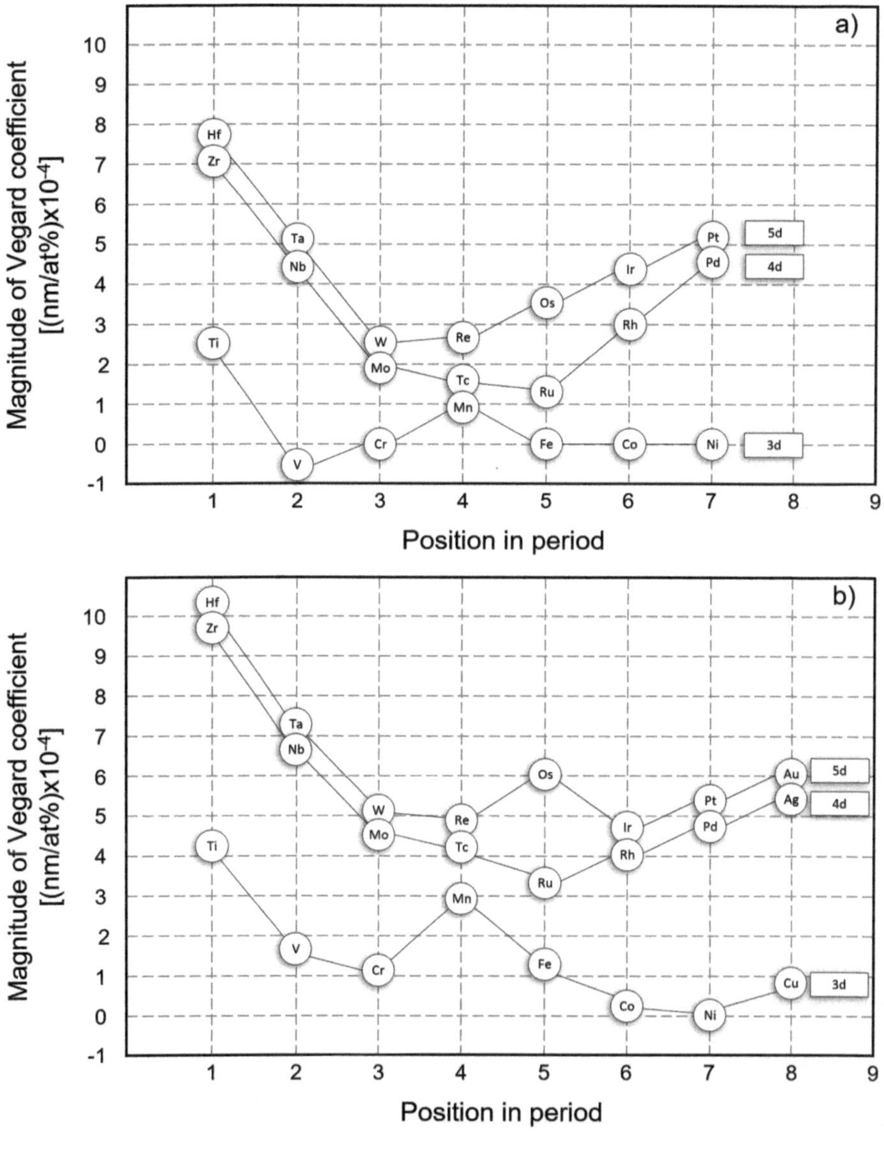

Fig. 6.34 Coefficients to be used in the generalized Vegard's law to evaluate the lattice parameters of (**a**) the ordered $L1_2$ γ' precipitates, having as reference composition Ni_3Al, and (**b**) a nickel-based *fcc* solid solution (the γ matrix of the superalloy). (Redrawn from Reed 2006)

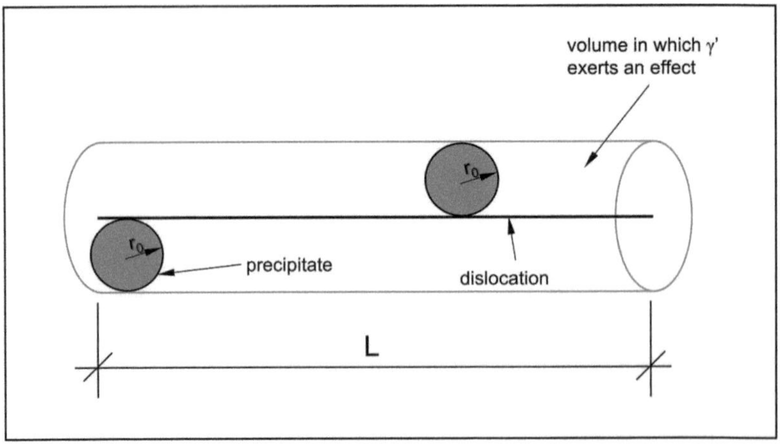

Fig. 6.35 A dislocation line interacting with two precipitates. Schematic of the model employed to derive the dependence of the strengthening effect on the volume fraction of precipitates

The average distance between couples of precipitates, L_I, can be written as:

$$L_I = \frac{L}{n_L} = \frac{r_0}{3f} \tag{6.25}$$

Therefore, the Eq. 6.16 for the strength enhancement of the alloy due to dislocations cutting through ordered precipitates can be expressed as a function of the volume fraction of precipitate phase:

$$\tau_{\mathrm{CUT}} = \frac{\pi r_0 \Gamma_{\mathrm{APB}}}{2\mathrm{b}\mathrm{L}_I} = \frac{3\pi f \Gamma_{\mathrm{APB}}}{2b} \tag{6.26}$$

Equation 6.26 confirms the expected increasing trend of the alloy strengthening with the volume fraction of ordered precipitates. A practical evidence of this result is represented in Fig. 6.36.

In agreement with these results, novel superalloys, particularly those for single-crystal turbine high-pressure blades (see Sect. 6.5), feature an increasing fraction of γ' phase and a corresponding improvement in the mechanical properties, including creep resistance (Fig. 6.37). However, the volume fraction of γ' phase has leveled off at 70%, in state-of-the-art materials, since higher values turned out to induce worse structural properties. This is an interesting aspect, which emerged in the pioneering paper by Murakumo et al. (2004) and was successively confirmed by other investigations (Murakumo et al. 2004).

Single-crystal superalloys guarantee the best creep performances, as far as vacancy diffusion creep is the dominating mechanism (see Sect. 6.5). However, for sufficiently high applied stress and elevated temperatures, diffusion-assisted

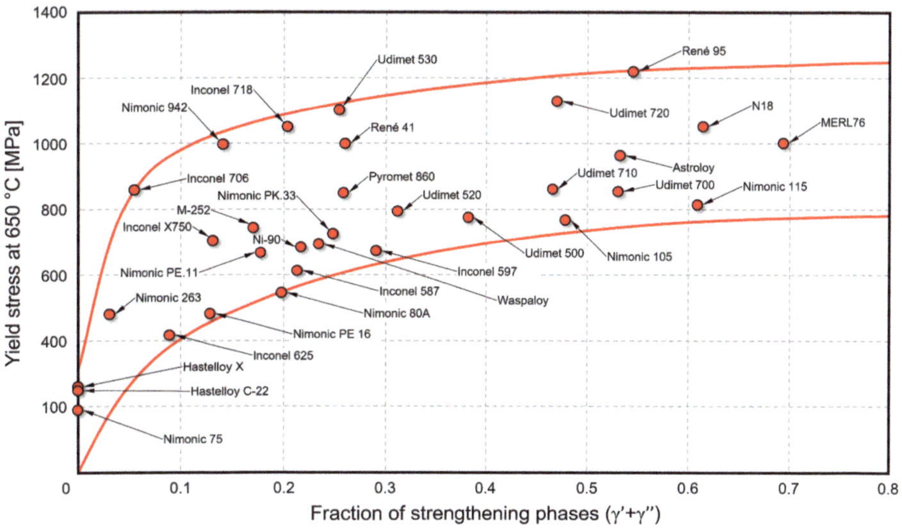

Fig. 6.36 Improvement of yield stress with increasing volume fraction of γ′ and γ″ phases in state-of-the-art nickel-containing superalloys at 650 °C. (Redrawn from Reed 2006)

dislocation motion may provide an important contribution to creep deformation. In this situation, the best way to improve the creep performances of the alloy, without excessively reducing other important structural properties, like fracture toughness and fatigue life, is to restrict creep to the so-called γ channel. In the γ austenitic matrix, which is a minority phase in the latest-generation superalloys and it is localized between the ordered γ′ precipitates, dislocations may form under the effect of the applied load. The results of creep tests and experiments demonstrate the fundamental role played by the γ/γ′ interfaces in preventing dislocations to enter the γ′ precipitates and confirm that an optimized microstructure still requires a minor fraction of γ matrix in which creep is confined and limited by the surrounding γ′ precipitates.

To complete the picture of the microstructural aspects of nickel-based superalloys, a few notes on the other secondary phases that can be found in the alloy γ matrix in association with the main reinforcing phase, i.e., γ′, like carbides and borides with different stoichiometries (see also Appendix 6). The main carbides are the same as those of cobalt-based superalloys (see Sect. 6.2.1). They may hinder dislocation movement, directly, through an Orowan-like interaction, and indirectly, through the lattice strain they introduce into the alloy. Other important roles that the carbides share with borides concern the optimization of grain boundaries, which are stabilized by these phases against sliding and grain recrystallization. For these reasons, the optimization of carbon and boron concentrations, together with the concentrations of carbide- and boride-forming elements, has been investigated with particular regard for polycrystalline alloys, usually obtained using casting technologies. As concerns borides, the main stoichiometry that is encountered in Ni

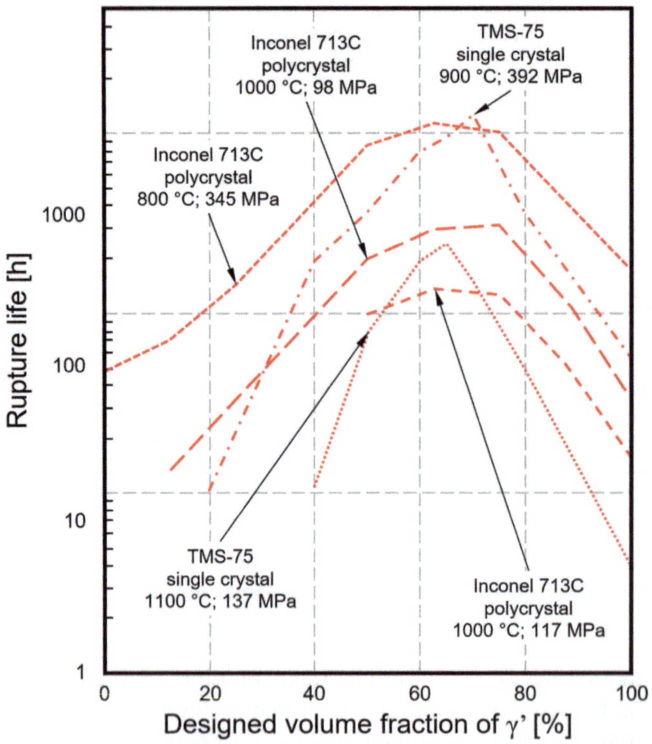

Fig. 6.37 Dependence of the creep life of TMS-75 single-crystal Ni-based superalloy and Inconel 713 C polycrystalline alloy on different volume fractions of γ' phase. (Redrawn from Murakumo et al. 2004)

superalloys is the M_3B_2, being M one or a combination of the following main metals: Cr, Hf, Mo, Ta, and Ti. The superalloys obtained by powder metallurgy (PM) routes are another class of products for which grain boundary engineering is an issue. As concerns aerospace gas turbines engines, turbine discs have recently attracted a remarkable technological interest of PM processing. The control of grain boundary structure has been improved with the development of specifically designed production steps. Differential thermal treatments have been applied to optimize the microstructure of different regions of the same critical components, like turbine discs. A suitable gradient of grain size, depending on the specific local requirements and part specifications, can be induced thanks to this approach (Reed 2006). The dual microstructure heat treatment (DMHT) technology, developed and patented by the NASA (see Sect. 6.6), has been applied to the third-generation PM superalloys, such as LSHR and ME3 (see Table 6.13) to induce different grain size and morphology in the bore and the rim of the same turbine disc (Reed 2006, see also Sects. 6.6 and 6.7).

6.5 Single-Crystal and Directionally Solidified Superalloys

The continuous progress in the aircraft GTEs has also been driven by the will to increase their efficiency and, consequently, to reduce fuel consumption, with obvious advantages not only from the economical but also the environmental point of view. Raising the operating temperatures is an effective strategy for achieving this goal (see Fig. 2.6). Over the years, the TET (see Sect. 2.2 and Fig. 2.20) has greatly increased, and this improvement can be ascribed to two main factors: more efficient design of the cooling system of turbine blades and vanes and improved superalloys chemistry and relevant casting processes. In this respect, high-pressure turbine blades, discs, and vanes are critical components because, in addition to the high temperature, they are subjected also to a combination of corrosive environment and mechanical stresses. In the early days of the GTEs history, the use of wrought alloys was predominant over castings, which where tolerated when strictly necessary only (castings had a bad reputation due to some failures associated with casting defects). However, higher-strength requirements continued to increase over the years, and, for the fabrication of some components, casting became the only choice because high-strength alloys cannot be subjected to forging processes successfully (Giamei 2013). Also for cast superalloys, some limiting factors emerged, related to fatigue and creep resistance, surface stability against oxidation, and corrosion phenomena.

Columnar-grained superalloys, with crystalline grains oriented parallel to the longitudinal axis or principal stress axis, were thought to be the ideal choice for components subjected to the axial loads produced by high-speed rotation (Giamei 2013). The directional solidification (DS) casting process was implemented to obtain columnar-grained, still polycrystalline, superalloy components. Further technological developments afforded the possibility to produce single-crystal (SC) components that still represent the state-of-the-art achievement of this process. Pratt & Whitney Aircraft (PWA) was the pioneering company in the development of both DS and SC superalloy products. Indeed, the history of both DS and SC alloys started at Pratt & Whitney Advanced Materials Laboratory in the mid-1960s (ASM International 1993; Giamei 2013). CGDS (columnar-grained directionally solidified) structures have been produced, allowing unidirectional heat flow during the solidification process of the alloy. The heat of solidification is extracted along one single direction, which is parallel to the main axis of the component. Although the application of the DS process to already-existing polycrystalline alloys, e.g., the MAR-M200 alloy, led to important improvements in their performances, yet only with the design of specific alloy compositions, the improved microstructure could be better exploited (Donachie and Donachie 2002). The same SC casting process is, in some respect, the natural evolution of the DS technology, and the optimization of the alloy composition became even more critical. The research and development at PWA proved that the removal of grain boundary strengthening elements, such as B, C, Hf, and Zr, resulted in the increase in the incipient melting temperature of the alloy (see Table 6.16 for the chemical composition of SC superalloys). Actually, a minor amount of carbon may still have beneficial effects, since it helps reducing sulfur content, and this is the reason why a few SC alloy formulations contain limited amounts of this element (Table 6.16). Tantalum turned out to be a strongly positive addition (see PWA 1480 alloy – Table 6.16), since it induces strengthening and oxidation resistance and improves also the alloy castability. Rhenium additions are responsible for reducing the alloy diffusivity, with beneficial effects on a slower precipitate coarsening and improved creep resistance (see PWA 1484 alloy – Table 6.16).

Notwithstanding the excellent performances achieved by DS and SC components (see Sect. 6.5.1), still the directional solidification technology introduced specific processing issues:

- Higher tendency to inclusion formation.
- Separately nucleated grains.
- Grain misorientation.
- Tendency of freckle grains formation (nucleation of grains as a consequence of inverse segregation caused by dendrite erosion).

They were addressed not only with the alloy design optimization but also with the evolution of the production apparatuses and solidification procedures (see Sect. 6.5.2).

Turbine blades made of a DS superalloy were first tested in the small PT6 engine, typically used for helicopters. Since the components did not fulfill the operational requirements, the test was suspended. Afterward, a replacement set of SC turbine blades was subjected to the same trial, fully successful this time. Progress in DS and SC superalloys brought very important positive consequences in terms of achievable engine performances, both as concerns the thrust increase and the improvement in the time between overhaul (TBO). At the beginning of the 1980s, SC alloys started to be employed in the engines of large commercial airliners, as well as in military aircrafts (Giamei 2013). The first reported commercial application of single-crystal turbine high-pressure blades dates back to 1981, on an Airbus aircraft, and these components are still to be regarded as the only structural parts featuring a single-crystalline microstructure (ASM International 1993). There are two generations of DS alloys and six of the SC ones (see Tables 6.15 and 6.16). The two generations of DS Ni-based superalloys differ for the presence of Re in their composition. In particular, the first-generation alloys do not contain this element, which is present instead, in concentrations of about 3 wt%, in the second generation of DS alloys. Furthermore, the first-generation alloys derived essentially from wrought products, following early attempts, made in the 1950s, to produce them by vacuum melting and casting processes. These alloys were developed with the intention of improving strength through the elimination of the deleterious Laves and σ phases, as well as by increasing the γ' volume fraction. In these alloys, Mo and W additions were meant to improve high-temperature capabilities, while Hf in concentrations around 2 wt% was used for increasing rupture stress ductility.

As concerns SC superalloys, the first generation has the same features as those of the first DS superalloy generation, while the second generation of SC superalloys was characterized by better oxidation resistance coupled with higher-temperature capabilities, about 30 °C more than the average peak temperatures of the first-generation alloys. This result was possible thanks to the addition of Re (3 wt% approx.) that contributed to the general increase of the refractory elements content. Moreover, the γ' volume fraction exceeded 60%. The third generation of SC superalloys was characterized by lower Cr concentrations and even higher Re content for improving both environmental resistance and creep strength. The fourth generation was essentially characterized by the addition of platinum-group metals, Ru in particular. A further increase in the refractory elements content aimed at improving the alloy strength. High-temperature low-stress resistance was also improved thanks to a higher negative γ/γ' misfit. The fifth generation of SC superalloys was characterized by even higher Ru concentrations, higher lattice misfit, and additional strengthening obtained through additions of Mo, Re, and Ru itself. Ru also improved the phase stability thanks to a reduction in the alloy diffusivity, having an indirect beneficial effect on the superior creep resistance than the alloys of the four previous generations. Finally, sixth-generation superalloys show structural properties that are in some respect comparable to those of fifth-generation SC superalloys but with a better oxidation and hot corrosion resistance (Prasad and Wanhill 2017). Still concerning gas turbine components, starting from high-pressure blading, there is a growing trend to apply the SC casting technology to a wider range of aero-engine components, which undergo complex loading conditions involving a combination of multiaxial and centrifugal stresses. At the same time, SC and DS blading has been considered for other than aerospace field: industrial natural gas-fueled and large land-based gas turbine engines used for electricity production (ASM International 1993).

6.5.1 Properties of DS and SC Superalloys

As shown in Fig. 6.38, the main benefits of SC and DS over polycrystalline superalloys derive from a combination of improved creep behavior, excellent intrinsic corrosion and thermal fatigue resistance.

The improved corrosion resistance displayed by these materials depends on several factors. The main ones are the absence, or the reduction, of grain boundaries, a controlled and homogeneous composition, both giving the alloy an intrinsic stability against inward and outward diffusion of chemical species. Thermal fatigue life benefits from the compliant <001> direction, along which the components are typically oriented, thanks to the directional solidification processing. In this way, as shown by Fig. 6.39, the thermal stress, σ_{TH}, associated with the temperature gradient along the hot spot regions, remains below critical values. The search for better creep performances has been the main driving force for developing the DS and SC technology, considering in particular the central role of time-dependent deformation in limiting the useful lifetime of the hot parts of gas turbine engines. Owing to the operational conditions, the creep is mainly diffusive in character, involving vacancy and, in general, point defect diffusion.

These aspects have been treated in a comprehensive approach by Ashby (1972) and Frost and Ashby (1982), in their pioneering works, who grouped together in stress-temperature maps, the main deformation modes in polycrystalline materials. In these maps the stress (tensile or shear), normalized by a suitable elastic modulus,

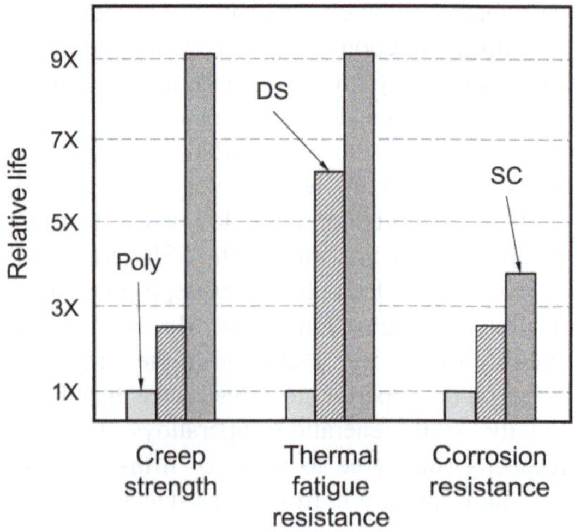

Fig. 6.38 Relative lifetime for polycrystalline, directionally solidified (DS) and single-crystal (SC) components, with regard to creep, thermal fatigue, and corrosion resistance. (Redrawn from Gell et al. 1987)

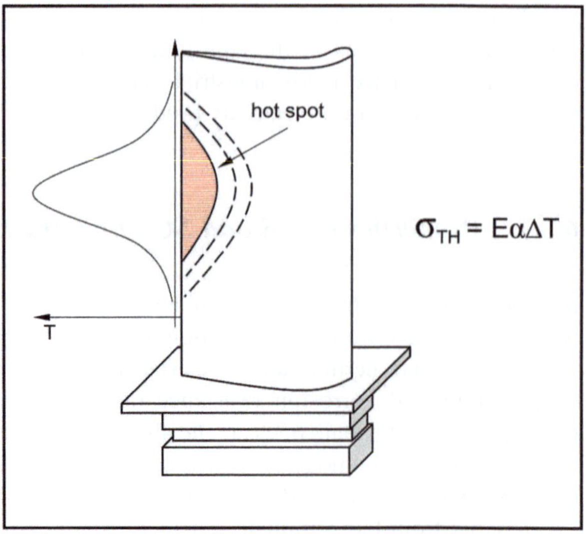

Fig. 6.39 Schematic of the temperature distribution along the axis of a gas turbine high-pressure blade. The expression for the thermal stress, σ_{TH}, indicates that the lower is the elastic modulus, E, the lower is the stress in the alloy resulting from a temperature gradient. α is the coefficient of thermal expansion (CTE) of the alloy

is plotted against the homologous temperature (T/T_m). According to this approach, there are six main different deformation mechanisms that have been acknowledged as essential to describe the behavior of real materials:

- Defectless flow.
- Dislocation glide.
- Diffusion creep along grain boundaries (Coble creep).
- Diffusion creep within grain boundaries (Nabarro-Herring creep).
- Dislocation creep.
- Twinning.

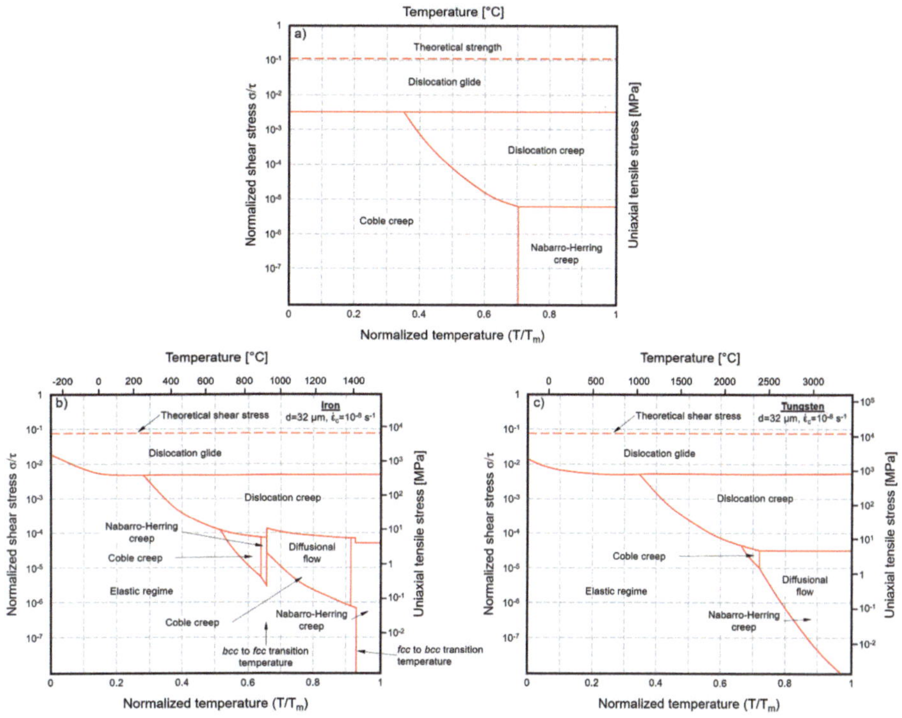

Fig. 6.40 Ashby deformation-mechanism maps: (**a**) general scheme; (**b**) deformation map for iron; (**c**) deformation map for tungsten. (Redrawn from Ashby 1972)

In Fig 6.40a the deformation map including the above mechanisms, but twinning, is shown. Twinning in general accounts for minor deformations, although it may be very important when dealing with some classes of materials, like shape memory alloys (see Sect. 9.5). In these particular alloys, twinning accounts for large strains (up to 8%), induced by applying comparatively low stresses, as compared to the yield stress. In Sect. 9.3, dealing with the applications of oxide dispersion strengthened (ODS) alloys, more extended Ashby deformation maps are considered, including also the effects of irradiation-induced deformations.

Apart from defectless flow, occurring in ideal conditions when the applied stress exceeds the theoretical shear strength of the material, all the other mechanisms involve different kinds of defects. The Ashby deformation maps are built starting from constitutive equations that depend on the dominant deformation mechanism and the specific microstructural features of the material. This is exemplified in the series of maps for polycrystalline iron and tungsten (Fig. 6.40b and c), respectively, and nickel (Fig. 6.41).

The transition borderline from one regime to the other is determined by equating the deformation rate due to each mechanism. Other important specifications quoted for each map are the minimum deformation rate (e.g., $\dot{\varepsilon} = 10^{-8}\ s^{-1}$) and the specific microstructure, usually the average grain size (e.g., $d = 32$ μm, see Fig. 6.40). When

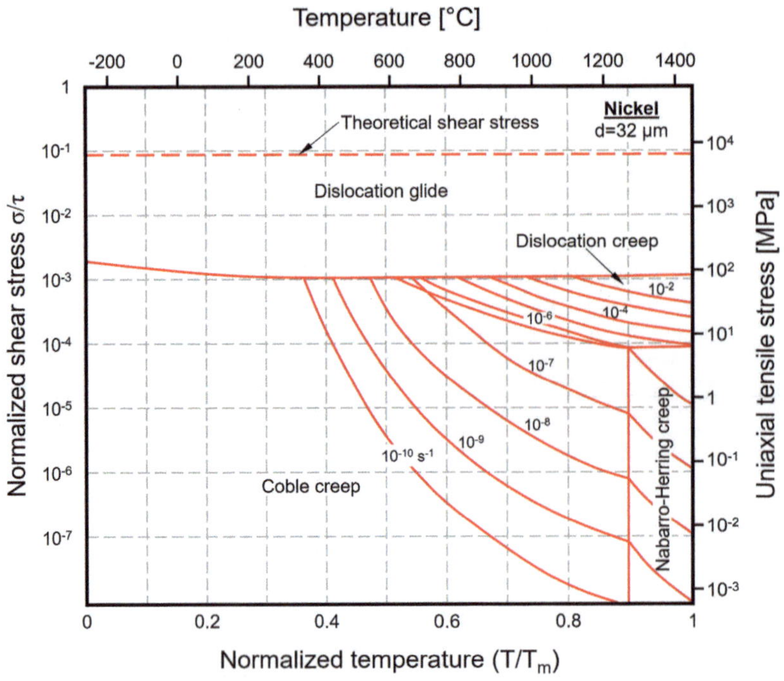

Fig. 6.41 Ashby deformation-mechanism maps for nickel, with constant deformation rate curves. (Ashby 1972)

the minimum deformation rate for a map is quoted, then an elastic regime region appears. This is used to indicate conditions that are not suitable to induce plastic deformation, at least not with the specified rate. Alternatively, a set of curves corresponding to constant deformation rates are introduced in the map (Fig. 6.41), so that the stress-temperature data point can be calculated under different loading conditions.

Dislocation glide, i.e., the movement of dislocation on slip planes along specific slip directions, may result in extensive plastic deformations in real materials, particularly metals and metallic alloys. The actual deformation depends on several factors, including the stacking fault energy, whose effect on dislocation stability and, thereby, plastic deformation has been considered in Sects. 6.4.1.2 and 6.4.2. At higher temperatures, dislocation climb, a typical diffusion-assisted phenomenon, may occur. This is the mechanism for *dislocation creep*. Also point defects may provide their contribution to the high-temperature deformation of materials by diffusing through the grains (*Nabarro-Herring creep*) and along the grain boundaries (*Coble creep*). These latter mechanisms will be considered next, and a simplified expression for their constitutive equations is derived.

The approach can be applied also to nonmetallic, polycrystalline materials as shown in Fig. 6.42, in which the deformation maps for germanium, magnesium oxide, and uranium oxide are shown (Ashby 1972).

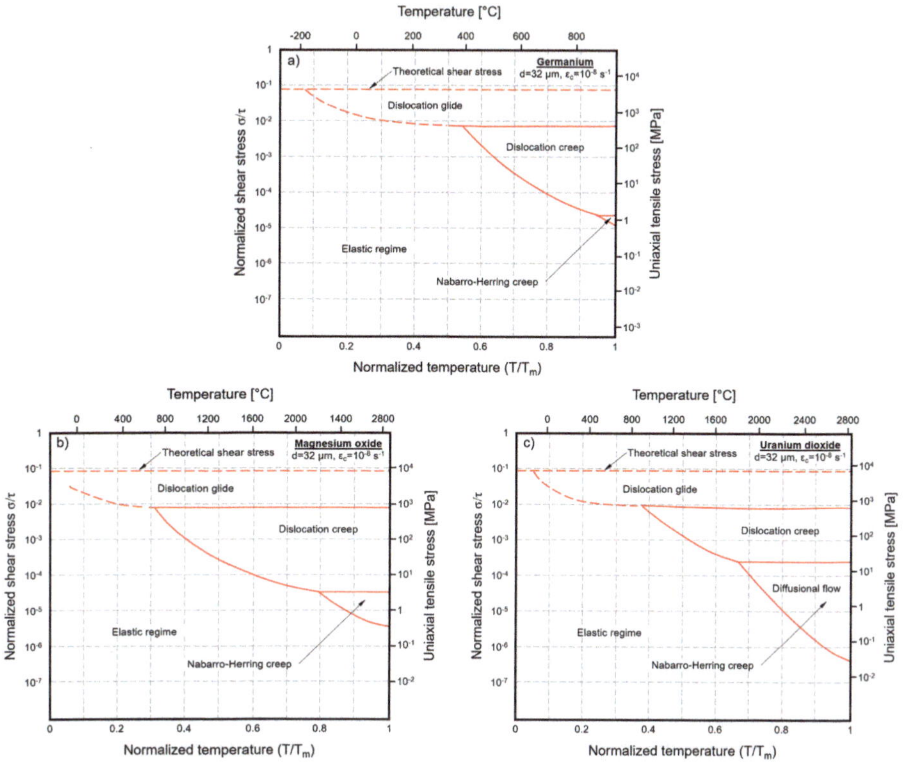

Fig. 6.42 Ashby deformation-mechanism maps for nonmetallic materials: (**a**) germanium; (**b**) magnesium oxide; (**c**) uranium oxide. (Ashby 1972)

The stress and temperature conditions faced by the high-pressure turbine blades would typically fall in the diffusive creep regions, as displayed by the Ashby map in Fig. 6.43, referring to the nickel-based superalloy MAR-M200., featuring a comparatively coarse grain size, i.e., 10 mm.

Grain coarsening results in a significant reduction of the deformation rate resulting from diffusive creep mechanisms (Coble and Nabarro-Herring creep), with a prevailing contribution from dislocation creep. To get a better understanding of the important effect that the grain coarsening has on diffusive creep, let us consider the schemes in Fig. 6.44, showing qualitatively the trajectories of the diffusive flow of atoms and counterflow of vacancies in a polycrystalline alloy under an applied stress and at sufficiently high temperature. The driving force for these flows is the concentration gradient of excess vacancies, i.e., those vacancies created by the atomic displacement from the equilibrium position induced by the applied stress. These excess vacancies give a net contribution to the vacancy concentration along the grain boundary regions perpendicular to the applied stress.

In the long run, this process leads to the accumulation of diffusing atoms in the grain boundary regions normal to the applied stress. These atoms leave behind

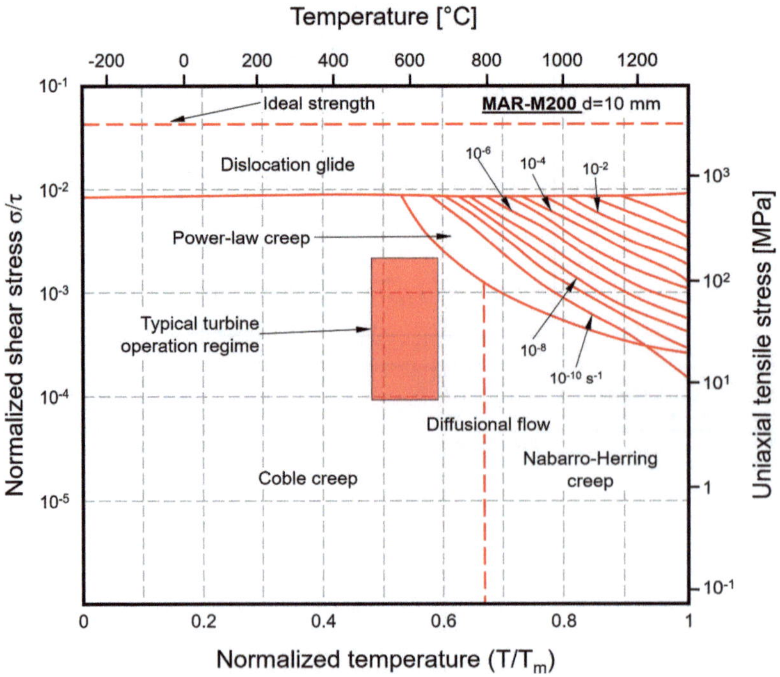

Fig. 6.43 Ashby deformation maps for the nickel-based superalloy, MAR-M200. Note the migration of the constant deformation rate curves to the dislocation creep region, determined by the comparatively coarse (10 mm) grain size. (Redrawn from Ashby and Frost 1982)

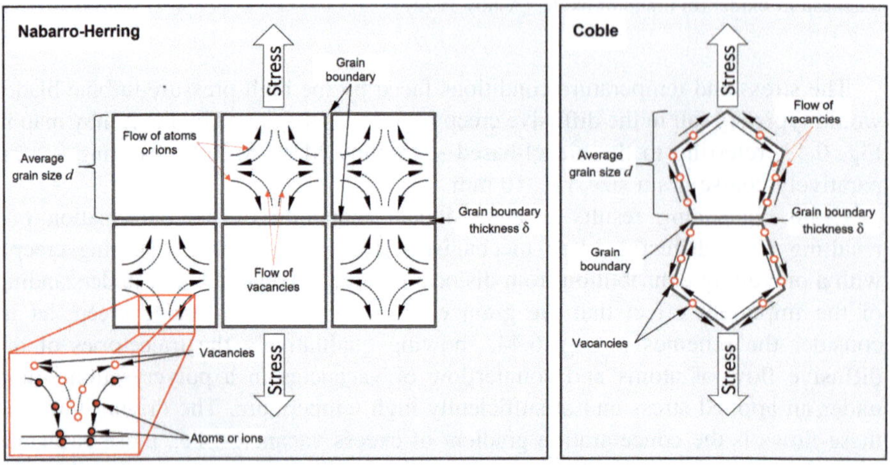

Fig. 6.44 Scheme of a polycrystalline sample with the diffusion mechanisms activated by the applied stress: bulk and grain boundary diffusion for the Nabarro-Herring and Coble creep, respectively. Diffusive flows of atoms and vacancies result in the accumulation of atoms in the grain boundary regions normal to the applied stress

vacancies, piling up along the grain boundaries parallel to the applied stress. A permanent deformation will then appear, accompanied by the reduction of the load-bearing section of the sample due to the creation of micropores for the coalescence of vacancies. This specific damaging mechanism may take the creeping sample to a tertiary creep stage, resulting in its rapid failure. The simple microstructural model depicted in Fig. 6.44 can be used to evaluate the deformation rate under diffusive creep conditions, with particular regard to the dependence of strain rate on the average grain size, *d*. Considering stationary conditions, the change of the grain size due to Nabarro-Herring creep occurs at a rate that can be expressed in terms of the Fick's first law:

$$\dot{d} = D_l\left(\frac{C_{\text{ex}}}{d}\right) \tag{6.27}$$

where:

$\dot{d}$ is the rate of change of the grain size.
d is the average grain size.
D_l is the lattice diffusivity, i.e., the diffusivity inside the grains.
C_{ex} is the concentration of excess vacancies, i.e., those vacancies produced by the atomic displacement due to the applied stress (Nabarro and de Villers 1995).

From (6.27) the strain rate for the Nabarro-Herring creep ($\dot{\varepsilon}_{\text{NH}}$) can be evaluated as:

$$\dot{\varepsilon}_{\text{NH}} = \frac{\dot{d}}{d} = D_l\left(\frac{C_{\text{ex}}}{d^2}\right) \tag{6.28}$$

A similar expression can be obtained for the strain rate due to the Coble creep. For this deformation mechanism, it has to be considered that diffusion occurs along the grain boundaries with a much larger diffusivity (D_{GB}). On the other hand, the effective diffusion path is now limited by the grain boundary section ($d\delta$), whereas the diffusion path for the Nabarro-Herring creep was the whole grain section (d^2). Therefore, the same functional relationship for the strain rate derived for the Nabarro-Herring creep can be assumed for Coble's, although scaled by the ratio between the relevant effective diffusion area:

$$\dot{\varepsilon_C} = D_{\text{GB}}\frac{C_{\text{ex}}}{d^2}\frac{d\delta}{d^2} = D_{\text{GB}}\frac{C_{\text{ex}}}{d^3}\delta \tag{6.29}$$

where:

δ is the grain boundary thickness.
D_{GB} is the diffusivity along grain boundary.

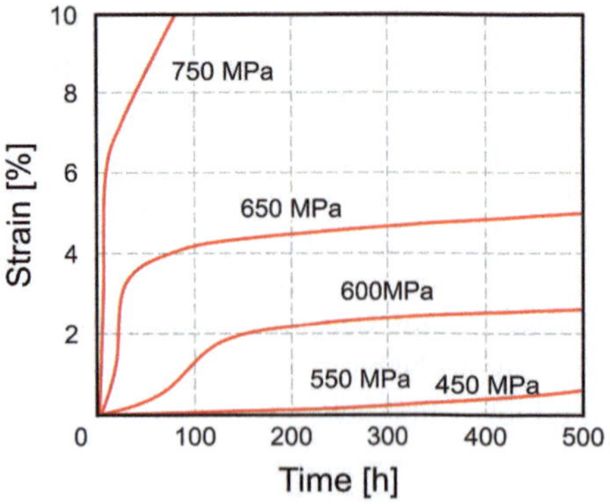

Fig. 6.45 Creep curves at 750 °C for the CMSX4 nickel-based superalloy, at the constant stress values indicated in the graph. (Redrawn from Reed 2006)

From these two expressions for the diffusive creep strain rates and from the exponents that the average grain size exhibits, the reason why DS and SC components have been developed turns out to be rather clear. The lack of grain boundaries normal to the main load axis and the extremely large grain size, as compared to conventionally cast alloys, are very effective in reducing the diffusive creep strain.

Obviously creep still remains an issue for superalloys, in particular as concerns materials of the latest generations, which have been developed with the deliberate intention to improve the engine performances acting on the turbine entry temperature (see Fig. 1.14 and Chap. 2). In this regard, it is important to consider that once the creep strain accumulated in service by a high-pressure turbine blade goes above a few percent, the involved tolerances are so strict that the part replacement becomes necessary, even if no other damages are present. Moreover, real conditions, featuring thermal gradients and corrosive and turbulent gases, influence the creep mechanisms in not always an easily predictable way. These are aspects that suggest some precautions, when transferring results from laboratory experiments to real components, even in view of the complex phenomenology reported for the creep behavior of nickel-based superalloys.

Primary creep is observed at lower temperatures (700–850 °C) for sufficiently high stresses, capable to induce the nucleation, typically occurring in the γ matrix, and movement of dislocations. In the low-stress range, *complete* dislocations in the austenitic γ phase are observed (see Sect. 6.4.1.2, Fig. 6.20). The movement of these a/2<110> dislocations, mostly by glide, is limited to the γ channels. The strain curve is monotonous with time, showing a trend like that in Fig. 6.45 for the 450 MPa applied stress. In the same graph, other curves are displayed, corresponding to higher applied stresses. The primary creep strain increases with increasing stress. Moreover, a transition to a secondary creep regime, following an inflection in the strain versus time curve, is observed. Transmission electron microscopy (TEM) investigations have demonstrated that this behavior is determined by the interaction of the dislocations with the ordered γ' precipitates. Thanks to dislocation decomposition and subsequent recombination, with other dislocations, complex dislocation structures would form, in particular, dislocation reactions like the one given in Eq. 6.30:

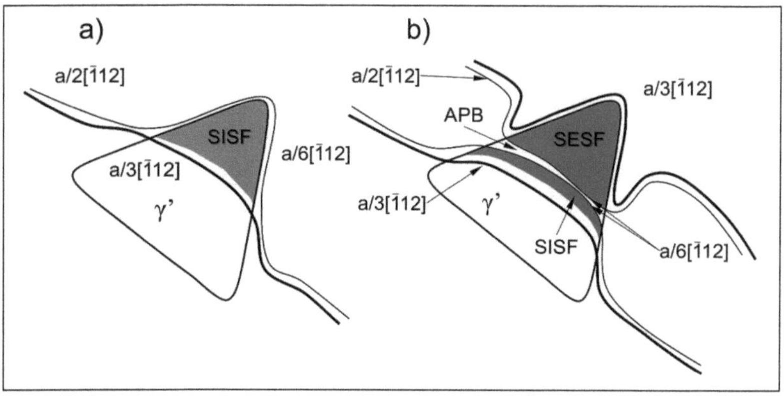

Fig. 6.46 γ' cutting by paired a/2 <112> dislocations. See main text for the detailed description of the involved dislocation reactions. (Redrawn from Reed 2006)

$$\frac{a}{2}[011]+\frac{a}{2}\left[\bar{1}01\right]\rightarrow\frac{a}{3}\left[\bar{1}12\right]+\frac{a}{6}\left[\bar{1}12\right] \quad (6.30)$$

As shown in Fig. 6.46a, the leading dislocation $a/3\left[\bar{1}12\right]$ can cut through the γ' precipitate if the shear component of the applied stress is sufficiently high. In doing this, it leaves behind a superlattice intrinsic stacking fault (SISF).

When a second similar decomposition reaction occurs, new dislocation components may interact with those that entered already the ordered precipitates. The formation of a super-dislocation may now occur, a line defect capable to move inside the precipitate conservatively (Fig. 6.46b). In the process another planar defect is formed, the counterpart of the SISF, i.e., a superlattice extrinsic stacking fault (SESF). The dislocation cutting and, thereby, the primary creep regime can continue as long as new dislocations are not blocked at the γ/γ' interface, preventing the conservative cutting through the precipitate to go ahead. At this stage, the secondary creep regime would initiate. The consequent restriction of the dislocations' movement and their accumulation inside the γ channel only, with a consequent increase of their concentration, may drive the superalloy to the tertiary creep regime that can be directly accessed from the primary one at higher temperatures. The density of dislocations that can now move also by diffusion-assisted climb and cross-slip in the γ channel can be taken as an indication of the increasing damage of the alloy. The strain rate, after an incubation time needed to the dislocation density to reach a critical value, goes up abruptly leading the superalloy rapidly to unacceptable strain values, as shown by Fig. 6.47a. At higher temperatures, a mechanism typical of single-crystal superalloys is observed. Under the effect of the applied stress, a progressive coalescence of the γ' precipitates occurs, resulting in plate-shaped rafts, so that the mechanism is called "rafting." This kind of *directional recrystallization* determines the formation of rafted structures, oriented with the basal planes orthogonal to the applied stress direction in case of negatively misfitting alloys, which is the case in most commercial SC superalloys. If positive lattice misfit

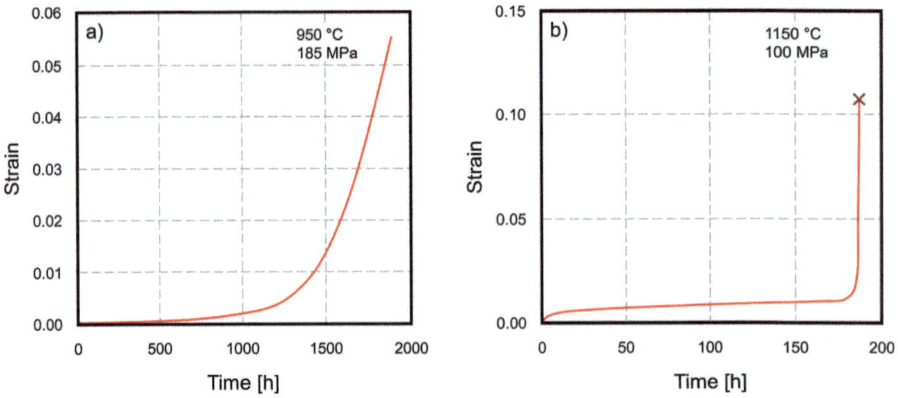

Fig. 6.47 Creep curves for the CMSX4 nickel-based superalloy. (**a**) 950 °C and 185 MPa, typical tertiary creep behavior; (**b**) 1150 °C and 100 MPa, the deformation behavior of the alloy is dominated by the rafting mechanism. (Redrawn from Reed 2006)

is present, rafting determines the formation of needlelike rafts, oriented parallel to the stress direction. Like for the tertiary creep, also rafting creep (Fig. 6.47b) displays an initial incubation time, required for the recrystallization of the γ' precipitates and for the accumulation of dislocation at the γ/γ' interface. Controversial aspects are still standing, as concerns the effect of rafting on the creep behavior of superalloys, reported to be beneficial in some respects. It is likely that an important contribution to weakening the rafted superalloy comes from the increase in dislocation density within the γ phase, similarly to the tertiary creep regime.

6.5.2 *Principles and Technological Aspects of Directional Solidification*

DS and SC superalloy components generally have complex geometry, surface topography, and a hollow internal structure, essential for an efficient cooling of the component. The main production approach is the investment casting, or lost-wax, process. This method implies the use of a "wax" model that ensures an adequate surface finishing and dimensional accuracy, in view of the final machining and polishing steps. The cooling of the blade is more and more crucial, considering the operating temperatures that can be reached in the hot stages of the turbine. To increase the heat exchange efficiency between the blade wall and the compressed air circulating through it, complex patterns of the cooling internal channels and film-cooling holes have been designed, much more complex than in early blades, for which extensive mechanical and laser drilling were used, with obvious limitations on the geometries that could be realized (see Fig. 6.48). The cooling patterns are typically obtained by inserting ceramic cores inside the wax model. Differently from the wax model, which is eliminated from the mold before its sintering and

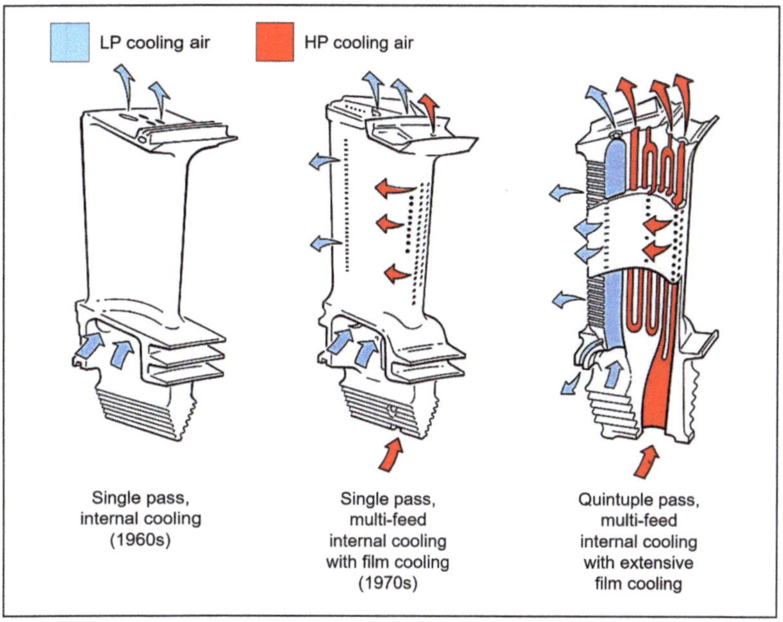

Fig. 6.48 Comparison between the cooling systems in turbine blades. On the left and in the middle, single-pass internal cooling; the straight channel and the film-cooling holes were originally just drilled in the solid blade. The cooling efficiency was limited by the limited contact surface area between the air flux and the alloy. On the right, a multi-pass serpentine cooling system, obtained by inserting in the casting mold ceramic cores, removed after the alloy solidification. (Rolls-Royce 1996, images courtesy of Rolls-Royce plc)

subsequent casting of the alloy, the ceramic cores remain in place until the alloy solidification is complete. Afterward, they are removed by ultrasonic crushing and chemical leaching (see Fig. 6.49).

The investment shell usually comprises multiple, up to 20 or so, chambers for producing several parts in each single run. For obtaining the investment shell, the multiple-wax model is dipped repeatedly into a ceramic slurry, containing a mixture of $ZrSiO_4$, Al_2O_3, and SiO_2 powders, suspended in colloidal silica or ethyl-silicate alcohol. To improve the strength of the shell, the dipping step in the slurry may be alternated with fluidized bed coating and stuccoing, using the same ceramic powders contained in the slurry. Once the wanted wall thickness is reached, the mold is sent to the consolidation thermal treatments. The first step is usually conducted in a steam autoclave and is needed to melt the wax model away and to promote the evaporation of the less volatile residues of the solvent of the slurry.

Subsequently, a proper firing treatment is carried out for a complete consolidation and sintering of the ceramic mold that is now ready for the casting. The production of reliable shells for investment casting to be used for the directional solidification process has been a challenging issue to solve for the full development of this technology. This is due to the more demanding conditions, as concerns the duration

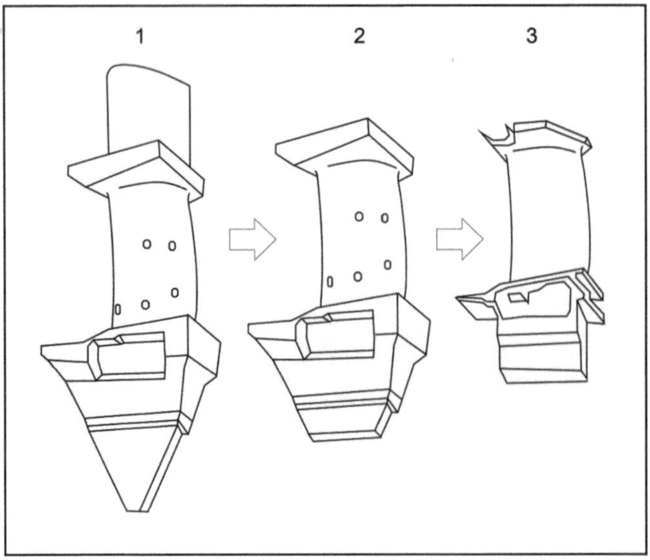

Fig. 6.49 Schemes of the different stages of the production of the high-pressure stage turbine blade of a gas turbine aero-engine: (1) wax model; (2) as-cast component; (3) finished blade after machining

of the process and the thermal gradients across the advancing liquid-solid interface, if compared to other conventional casting and solidification technologies.

For nickel-based superalloys, casting occurs under vacuum usually at a temperature of 1550 °C. The microstructure obtained in the solidified pieces is strictly dependent on the adopted solidification mode. In the early applications of investment casting to the production of gas turbine blades, polycrystalline components were obtained using the "power down" method, consisting in switching off the furnace after casting and leaving the solidification to occur spontaneously. The master alloy melting typically occurs in a disposable crucible made of refractory fibers. The furnace usually consists of an upper melting chamber, a central mold chamber maintained at the correct temperature by a graphite resistance heating element, and a lower withdrawal chamber (Fig. 6.50) (Reed 2006).

The directionality in the way heat is removed during solidification has been achieved with the Bridgman method using vacuum furnaces like the one illustrated in Fig. 6.50. To obtain the directional solidification, the mold that sits on a Cu cooled plate (water cooled in the scheme in Fig. 6.50) is extracted from the bottom part of the furnace. Typically, thermal gradients in the 4000 °C/m range are achieved. The time needed to pass from 1400 to 1000 °C is typically reached within 20 min (Reed 2006). The directionality of the solidification is thus assured, and in this way DS components can be obtained.

To obtain a SC component, it is necessary to eliminate from the solidifying cast all grains but one. The most used technology in this field exploits a particular design of the mold, also known as grain selector, which in the lower part, in contact with the

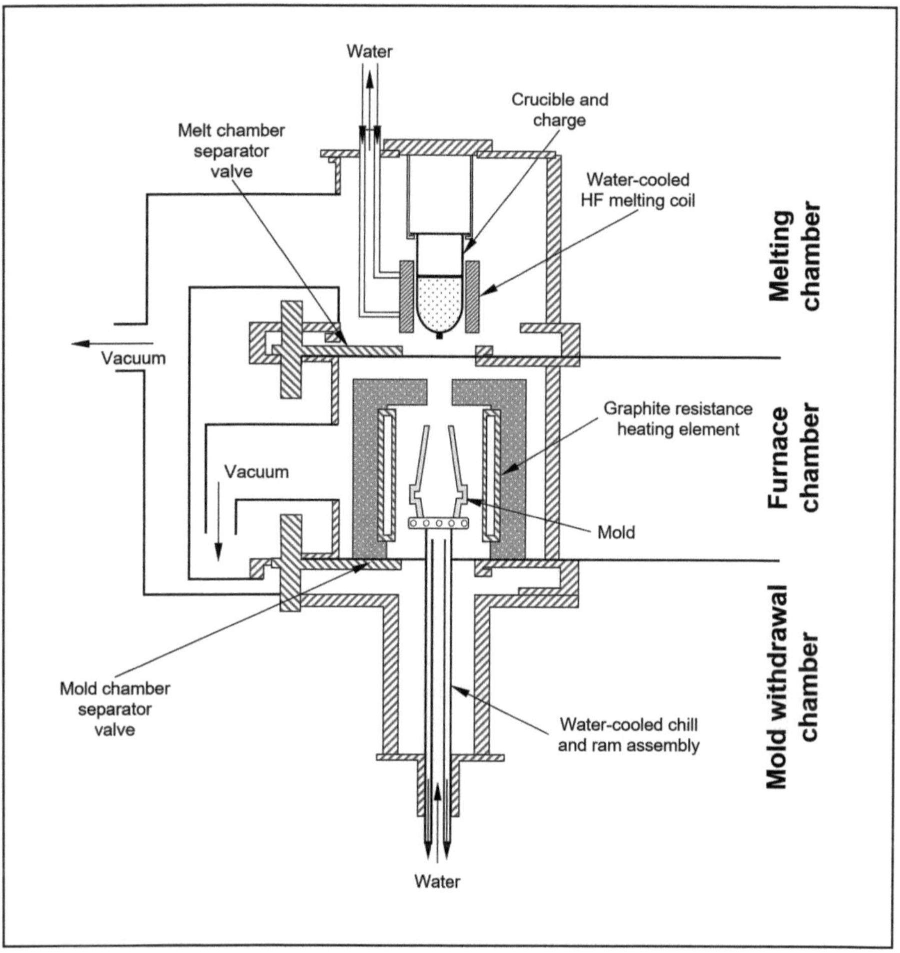

Fig. 6.50 Layout of an investment casting plant for the production of SC components, using the Bridgman method. (Redrawn from Reed 2006)

cooled plate where the solidification begins, is characterized by a spiral (see Fig. 6.51) or angular geometry. This particular design allows only one of the grains that nucleate on the cold plate at the bottom of the mold to enter its upper part, i.e., the part having the desired component shape.

A spontaneous growth direction does exist for nickel-based superalloys, parallel to the cube axis <001>. If, for any reason a different texture is interesting, the solidification can be initiated by a "seed" that is placed at the bottom of the mold, having the required crystallographic orientation, which will preferentially grow in the rest of the solidified component. Considering the microstructure of the base block of the SC blade shown in Fig. 6.51, it turns out that in case of directionally solidified

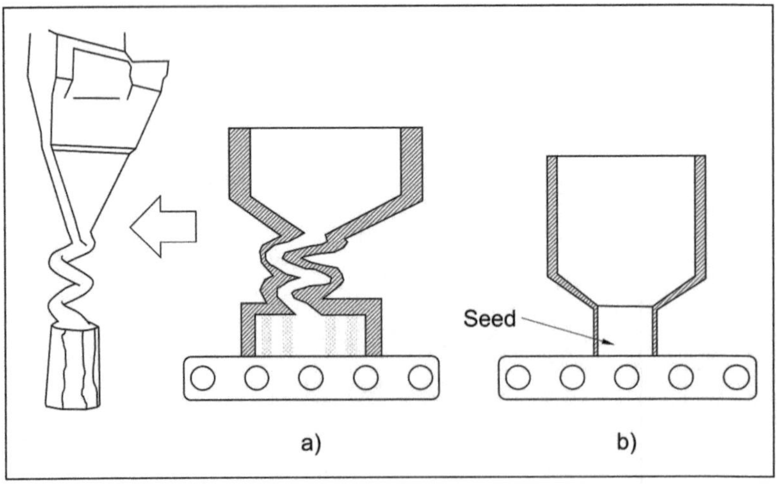

Fig. 6.51 (**a**) Working principle of the grain selector, used to obtain SC component, with sketch of the real part; (**b**) a "seed" with a specific orientation can be used to induce that orientation, possibly different from the spontaneous one, in the solidified component

components, no grain selector is necessary. The uniaxial orientation of the grains is simply caused by the growth direction induced by the heat extraction direction.

The spontaneous growth direction for nickel-based superalloys (actually, in general, for any *fcc* alloy) is the <001>, since it is perpendicular to the cube planes {001}, featuring the lowest atomic surface density for an *fcc* structure (see Fig. 6.52). Therefore, in case of a controlled solidification process, in which the solid phase would actually form one atomic layer at the time, as it occurs in the Bridgman solidification method, the <001> growth direction is naturally favored. The SC and DS solidification of superalloy components leads to interesting properties: low porosity, crystallographic texture, homogeneous microstructure and composition. Low porosity castings are favored not only by the limited shrinkage associated with the solid-liquid volume mismatch but also by the low-pressure atmosphere in the solidification environment, preventing any significant gas trapping in the melt. As shown in Sect. 6.6, the homogeneity in microstructure and composition is an important prerequisite to carry out the refinement thermal treatments needed to optimize the alloy properties. In fact, particularly the newest alloy compositions, featuring high fractions of γ' phase, exhibit a comparatively narrow solutioning temperature range. The main challenge of DS and SC technology is to obtain over the full component length the same level of chemical homogeneity that is usually attained within each single grain of a polycrystalline alloy, i.e., over a few tens of micrometers (Porter et al. 2009). Ruling out the equilibrium solidification, interesting for geological or planetary solidification processes, but totally unsuitable for a technologically relevant process, two alternative solidification regimes are in principle feasible. For the treatment of these solidification regimes, it is practical to refer

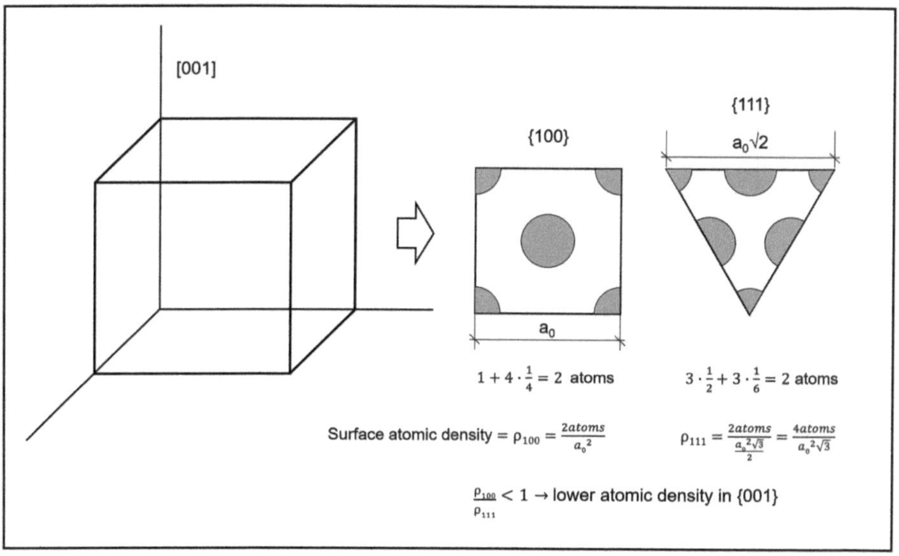

Fig. 6.52 Comparison between the atomic surface density, ρ_{hkl}, for the lowest and highest atomic packing planes in an *fcc* structure: the cube planes {001} and diagonal planes {111}

to the pseudo-binary phase diagram in Fig. 6.53, featuring a partition coefficient of the solute atoms between solid and liquid given by:

$$K = \frac{C_S}{C_L} \tag{6.31}$$

where:

C_S is the concentration of the solute in the solid.
C_L is the concentration of the solute in the liquid.

K in the present context is <1. Referring to the pseudo-binary phase diagram in Fig. 6.53, when dealing with the solidification of multicomponent superalloys, is still an acceptable choice, since superalloys solidify in a single phase (γ) and γ′ forms through solid-state precipitation afterward.

In the first solidification regime, the composition of the liquid phase is perfectly homogenized by mixing, e.g., by convection and stirring, two phenomena typically present in conventional casting processes. On the other hand, diffusion, occurring in the solid state, is not sufficiently fast to induce any meaningful homogenization of the alloy composition. In agreement with these assumptions, Fig. 6.54 shows the transient concentration profile together with the complete profile observed at the end of the solidification process that can be assumed to proceed from left to right as indicate by the arrow in Fig. 6.54.

The composition in the solid is not uniform, and a eutectic fraction is present in the final segment of the profile. This is coherent with the phase diagram in Fig. 6.53

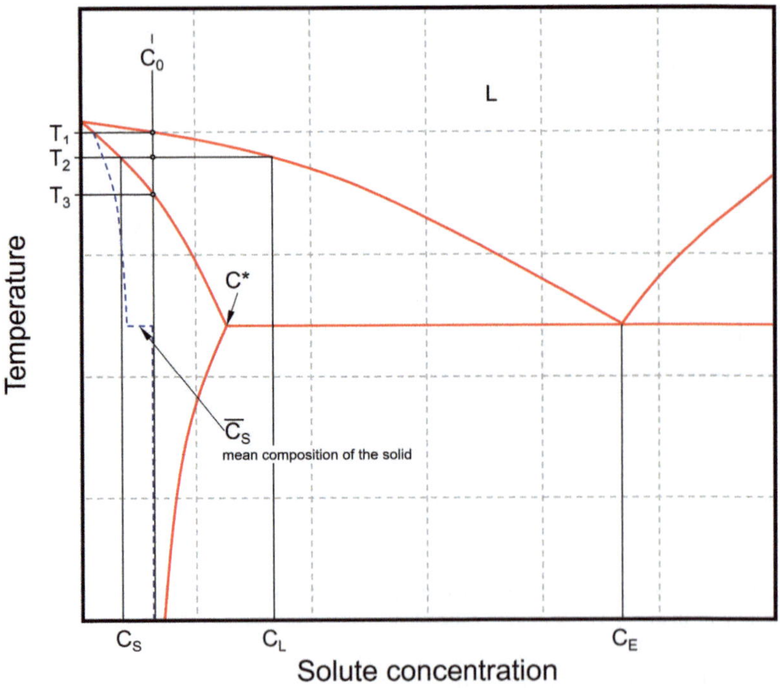

Fig. 6.53 Schematic of the pseudo-binary phase diagram used for the discussion of different solidification regimes

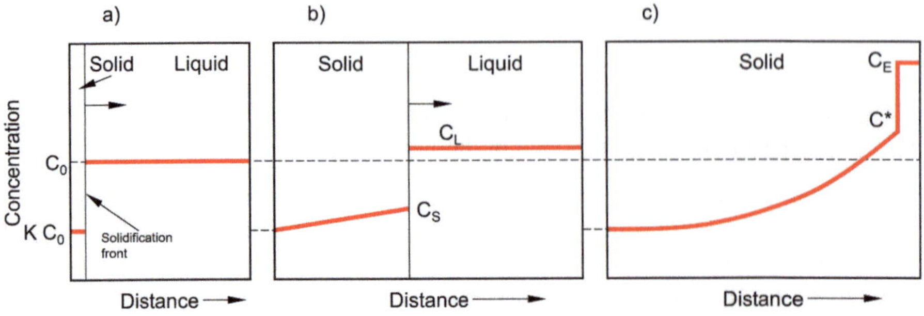

Fig. 6.54 Transient concentration profile in the liquid and solid phase during solidification occurring with perfect mixing in the liquid and no mixing in the solid. (**a**) $T = T_1$; (**b**) $T_2 < T < T_1$; (**c**) T = eutectic temperature (see Fig. 6.53)

and with the continuous solute enrichment of the liquid all through the solidification range. With reference to the single-phase region (no eutectics), the concentrations of both liquid and solid can be determined by solving the equation:

$$df_S(C_L - C_S) = f_L dC_L \tag{6.32}$$

where:

f_S is the solid fraction.
f_L is the liquid fraction.
C_L is the concentration of the solute in the liquid.
C_S is the concentration of the solute in the solid.

Assuming the same molar volume for the liquid and the solid phase, it is possible to rearrange Eq. 6.32 using the expressions:

$$\begin{cases} f_S + f_L = 1 \rightarrow f_S = 1 - f_L \rightarrow df_S = -df_L \\ K = \dfrac{C_S}{C_L} \rightarrow C_S = KC_L \rightarrow C_L - C_S = C_L(1-K) \end{cases} \tag{6.33}$$

So that Eq. 6.32 becomes:

$$\frac{-df_L}{f_L} = \frac{dC_L}{C_L(1-K)} \tag{6.34}$$

After the integration, being C_0 the composition of the molten alloy, expressions for C_L and C_S can be obtained as:

$$-\int_1^{f_L} \frac{df_L}{f_L} = \frac{1}{1-K}\int_0^{C_L} \frac{dC_L}{C_L} \rightarrow \ln\left(\frac{1}{f_L}\right) = \frac{1}{1-K}\ln\left(\frac{C_L}{C_0}\right) \rightarrow \frac{1}{f_L} = \left(\frac{C_L}{C_0}\right)^{\frac{1}{1-K}} \rightarrow \begin{cases} C_L = C_0\left(\frac{1}{f_L}\right)^{1-K} \\ C_S = KC_0\left(\frac{1}{f_L}\right)^{1-K} \end{cases} \tag{6.35}$$

The concentration profile in the solid phase can be easily homogenized through thermal treatments, if it refers to crystalline grains with average size not larger than a few hundreds of micrometers. Therefore, this concentration profile is not suitable when dealing with very large grain size, in the range of several centimeters, interesting for SC and DS products.

The solidification regime that provides a constant concentration in the solidified alloy is the one in which the solid is not having any significant solute homogenization via mixing phenomena in the liquid, being diffusion the dominant solute transport phenomenon in the liquid. The concentration profile, at the beginning of the solidification process, is shown in Fig. 6.55a. The progressive solute enrichment of the liquid at the interface with the solid results in a progressive increase of the solute concentration in the solid too (Fig. 6.55b). At some stage the composition of the solid phase equals the C_0 composition of the initial melt (Fig. 6.55c). From this point onward, the composition in the solid remains constant up to the end of the process, when a limited eutectic solidification would in general occur. Figure 6.55d

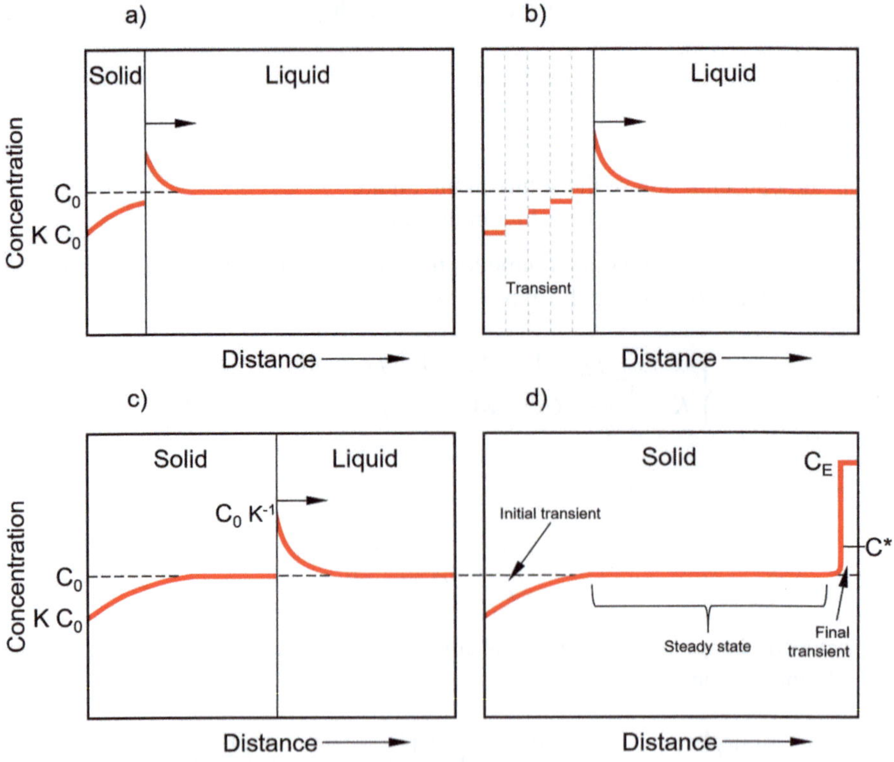

Fig. 6.55 Concentration profiles at different stages of the alloy solidification occurring with no homogenization in the solid and diffusion in the liquid phase. (**a**) Initial stage; (**b**) transient; (**c**) steady-state solidification; (**d**) solidification complete, with the formation of the final eutectics

shows the complete resulting concentration profile in the solid. Assuming a constant advancement velocity (v) for the solid-liquid interface, the expression for the solute concentration in the two phases over the steady-state solidification regime (stationary solidification) can be evaluated. By equating the amount of solute that is leaving the solid on its formation to the amount that is diffusing away from the solid-liquid interface into the liquid, with a rate given by the first Fick's law, the resulting expression can be written as:

$$(C_L - C_S)v = -D_L \frac{\partial C_L}{\partial x} \tag{6.36}$$

D_L being the diffusivity of the solute in the liquid phase. For the stationary state, where $C_S = C_0$, by introducing the variables:

$$y = C_L - C_0 \text{ and } dy = dC_L \tag{6.37}$$

and assuming the following boundary condition:

$$y_{x=0} = \frac{C_0}{K} - C_0 = C_0\left(\frac{1-k}{K}\right) \tag{6.38}$$

For the steady-state solidification stage, Eq. 6.36 becomes:

$$(C_L - C_0)v = -D_L\frac{\partial C_L}{\partial x} \tag{6.39}$$

Introducing the variables from Eq. 6.37, the Eq. 6.39 becomes:

$$\frac{dy}{y} = -\frac{v}{C_L}dx \tag{6.40}$$

By integrating Eq. 6.40:

$$\int_{C_0}^{y}\left(\frac{1-K}{K}\right)\frac{dy}{y} = -\int_0^x \frac{v}{D_L}dx \rightarrow \ln\left(\frac{y}{C_0\left(\frac{1-K}{K}\right)}\right) = -\frac{v}{D_L}x \tag{6.41}$$

Equation 6.42 provides the concentration profile in the liquid phase, during the stationary solidification stage:

$$C_L = C_0\left[1 + \left(\frac{1-K}{K}\right)\exp\left(-\frac{v}{D_L}x\right)\right] \tag{6.42}$$

The constitutional temperature profile (T_L, see Fig. 6.56) depends on the diffusion coefficient in the liquid and on the advancing velocity (v) of the solid-liquid interface.

The processing parameters involved in the former expressions strongly influence the features of the solidifying alloy. Therefore, major efforts have been devoted to the development of processing equipment and relevant procedures, capable to assure high reliability and productivity associated with flexible operational parameters (Pollock and Tin 2006; Pollock 2016). This resulted in the development of processes capable to comply with the specific processing requests of different products, like small and large size high-pressure blades for aircraft and terrestrial heavy duty gas turbines, respectively.

Two parameters are regarded as particularly critical in the directional solidification practice; they are thermal gradient in the liquid phase at the solid-liquid interface and growth rate, affecting very much the microstructure of the resulting products. Figure 6.57 gives the evolution of these two parameters with reference to three casting technologies, all of them used in the production of superalloys. The "power

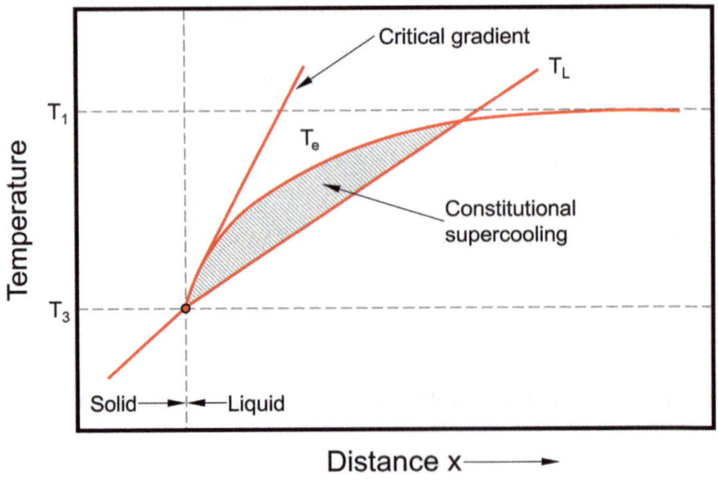

Fig. 6.56 Temperature profile in the solidifying liquid and constitutional supercooling. The liquid ahead of the solidification front has the T_L profile. T_e indicates the equilibrium liquidus temperature adjacent to the solid-liquid interface. Constitutional supercooling is obtained when the $T_L$line lies below the critical temperature gradient

down" approach has been mentioned already as the first used in investment casting of Ni-based superalloys. The "withdrawal" corresponds to the Bridgman method (see Fig. 6.50), and its performances have been remarkably improved by the liquid metal cooling (LMC) method. The main novelty of this latter technology relies on the substitution of the heat removal strategy from the investment casting shell and solidifying alloy. The combined system, involving water-cooled chill plate, in direct contact with the casting mold, associated with a radiative cooling system and boosted by a gas flow, has been replaced by a metal bath (tin or aluminum), extracting the heat by conduction in a more efficient and controllable way. In this way, the variation range of the operational parameters has significantly been expanded. This has resulted in the production of alloys with improved microstructural features, like a reduced fraction of inter-dendritic eutectics and the size and shape of the γ' precipitates, which render more effective the post-processing thermal treatments (see Sect. 6.6).

6.6 Heat Treatments

Some specific aspects concerning the heat treatments of superalloys are discussed in this section. Generally, these treatments are classified into four main groups: stress relief, annealing, solutioning (or solution annealing), and age hardening (or just aging); and a short description of each one of them is given in Table 6.24.

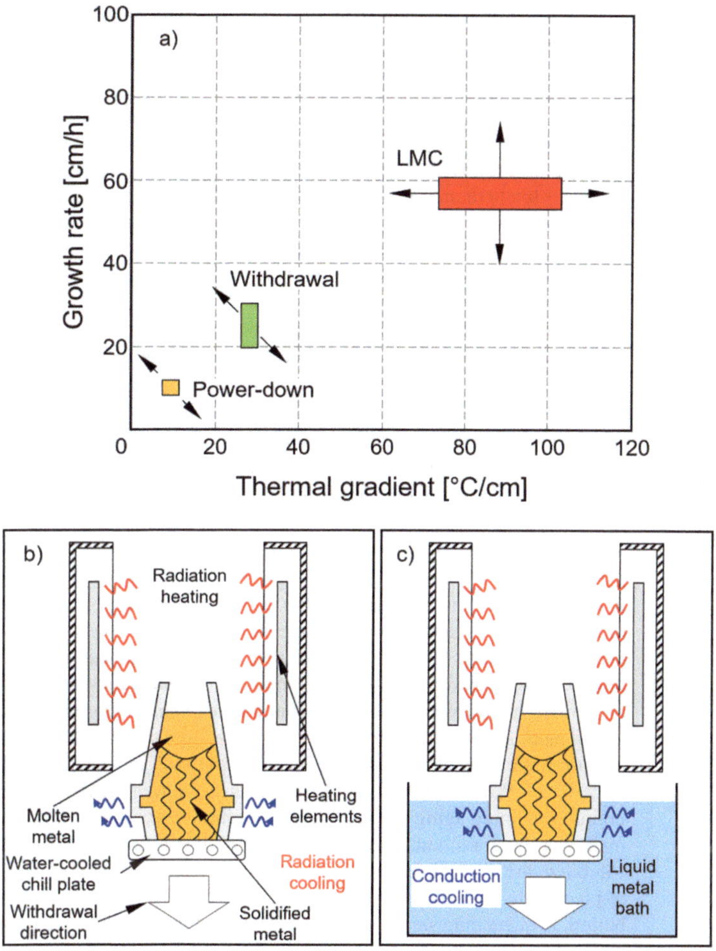

Fig. 6.57 (**a**) Variation ranges of the thermal gradients and growth rates accessible with different setup for directional solidification processes (see main text for figure description); (**b**) schematic illustration of the conventional Bridgman process; (**c**) schematic illustration of liquid metal cooling (LMC) process. (Redrawn from Reed 2006)

The first two types of treatments are dealing with the microstructural refinement of the alloy, with the elimination of defects and residual stresses. The focus of this section is rather on the second pair, i.e., solution and aging treatments, that still modify the microstructure of the superalloy but with the primary aim of inducing the precipitation and refinement of the strengthening phases, like γ', carbides, borides, etc. (see Appendix 6). In this respect, some aspects of these treatments are similar to those considered already with regard to other precipitation and age-hardenable alloys, like aluminum alloys (see Sects. 3.2.3.2 and 3.2.3.3) and steels (see Sect. 5.4).

Table 6.24 Thermal treatments performed on superalloys – a summary (ASM International 2000, 2016)

Name	Description
Stress relief	Stress relief is often a compromise between the relief of residual stresses in the material and effects that can be deleterious to corrosion resistance and/or high-temperature properties. The stress relieving temperature typically falls within the upper temperature range for precipitation hardening and is generally below the recrystallization or annealing temperature. For Ni-based superalloys, it is typically used to reduce/eliminate residual stresses in work-hardened no-age-treatable alloys, but avoiding the recrystallization of the microstructure. A typical temperature range for Ni alloys is in between 425 °C and 870 °C, depending on the chemical composition. Stress equalizing can be considered as a subgroup of stress relieving. In this latter case, a low-temperature treatment is used for homogenizing stresses after cold-working operations, without decreasing the mechanical strength gained via cold-working
Annealing	Annealing treatment of superalloys implies a full recrystallization and complete softening when dealing with no-age-hardenable wrought alloys. When applied to precipitation hardenable alloys, the treatment is essentially a solution annealing (see below). The most important outcomes of annealing are ductility increase, hardness reduction, solutioning of second phases. Annealing is in general used as preliminary to forming, machining, and welding operations
Solution annealing	A solution annealing treatment can be regarded as the first step of the standard heat treatment of superalloys. Treatments performed at lower temperatures are generally indicated for improving the resistance to notch-rupture sensitivity, to improve fatigue resistance thanks to smaller grain size, better tensile properties at elevated temperatures. No significant carbide solutioning or grain growth is taking place at low temperature. Higher temperature treatments lead to optimum creep-rupture properties thanks to coarser grain size, as well as the dissolution of carbides. When followed by age hardening, the treatment results in large grains containing aging phases (γ', γ'') and high concentrations of carbides mainly at the grain boundary. The main goal of solution treatments is to solubilize the γ' precipitates and carbides that after solutioning may re-precipitate with an optimized shape, more uniform size, and spatial distribution. A quenching step is usually employed after solution annealing. Cooling media depend on the alloy composition and can be water, oil, air, and also some inert gas cooling
Age hardening	Aging treatment is meant to induce the precipitation of one or more strengthening phases from a supersaturated alloy matrix. Important factors in this regard are combination of desired strength and ductility, type and number of precipitating phases, precipitate size

Dealing with the thermal treatments of superalloys and also as concerns the stability of precipitate phases, it is paramount to consider the effect of their surface energy term and its dependence on the curvature radius of the precipitate itself. Let us consider a two-phase alloy, made of α phase (parent phase) in which the β phase precipitates. To show how the stability of the phase, β, depends on the size and shape of the precipitates, in Fig. 6.58 two β-precipitates are shown: a spherical precipitate having a finite radius r and another precipitate with a planar interface, corresponding to an infinite radius.

Fig. 6.58 Scheme used to discuss the stability of the β phase in the α matrix, as a function of the curvature radius

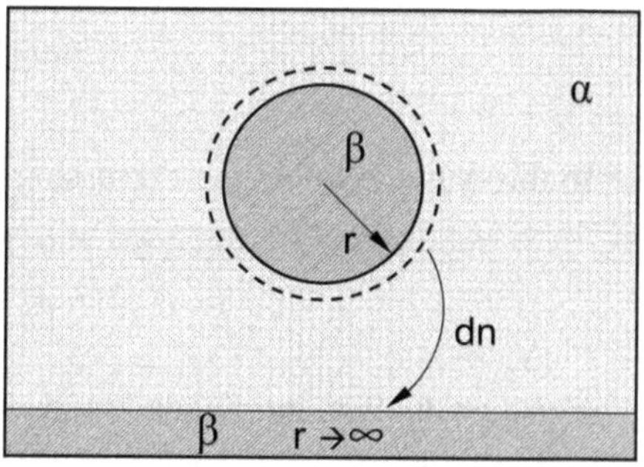

The difference in free energy between the two precipitates is $\Delta G_{\Gamma S}$, due to the different contribution from the surface energy term, being the composition of the two precipitates the same. The variation of the free energy of the system for a transfer of *dn* moles of β from the spherical to the infinite radius precipitate can be expressed as:

$$\mathrm{dG} = \Delta G_{\Gamma S} \mathrm{dn} \tag{6.43}$$

Assuming constant the free energy of the infinite radius precipitate that, being much larger than the other, is substantially unaffected by the infinitesimal mass change, the free energy change of the system is entirely due to the decrease in the surface area (dA) of the spherical precipitate:

$$\mathrm{dG} = \Gamma_S \mathrm{dA} \tag{6.44}$$

By equating the two expressions for the energy change:

$$\Delta G_{\Gamma_S} \mathrm{dn} = \Gamma_S \mathrm{dA} \rightarrow \Delta G_{\Gamma_S} = \frac{\Gamma_S \mathrm{dA}}{\mathrm{dn}} \tag{6.45}$$

In case of a spherical precipitate, simple expressions for its surface area and volume can be introduced in order to evaluate the relevant free energy terms:

$$A = 4\pi r^2 \rightarrow \mathrm{dA} = 8\pi \mathrm{rdr} \tag{6.46}$$

As concerns the number of moles contained in the spherical precipitate, the value is given by:

$$n = \frac{4\pi r^3}{3V_m^\beta} \tag{6.47}$$

where V_m^β is the molar volume of the β phase. Differentiating Eq. 6.47:

$$dn = \frac{4\pi r^2 dr}{V_m^\beta} \tag{6.48}$$

In this way the following expression for $\Delta G_{\Gamma_S}$is obtained:

$$\Delta G_{\Gamma_S}(r) = \frac{2}{r}\Gamma_S V_m^\beta \tag{6.49}$$

Equation 6.49 is the Gibbs-Thomson equation and has very important consequences on the free energy equilibrium curve of the α-β alloy.

As shown in Fig. 6.59, the extra energy term $\Delta G_\Gamma(r)$ determines the rise of the minima of the free energy curves for the β phase, depending on the different average size of the precipitates. This has an effect on the equilibrium concentration in the α matrix that can be evaluated, at a constant temperature, using the common tangent construction. As the radius of the precipitate decreases, the tangent to the free energy curve of α moves toward solute-richer concentrations.

A concentration gradient tends to be established in the alloy and activates a diffusive flow, resulting in the so-called Ostwald ripening of the precipitates. With reference to the scheme in Fig. 6.60, showing two β-precipitates with different radii, the diffusive flow, triggered by the concentration gradient, results in the progressive shrinking of the smaller one and coarsening of the larger one. The homogeneous distribution of the size and planar interfaces of the precipitates are essential features to obtain an intrinsically stable alloy microstructure. This latter feature is certainly

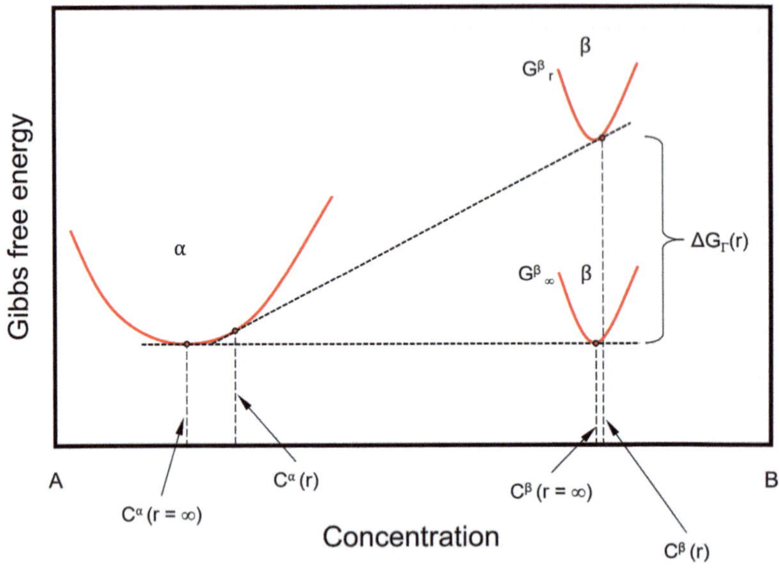

Fig. 6.59 Gibbs free energy stability curves for the α-β alloy binary, modified by the surface energy terms for a finite radius of the β-precipitate phase

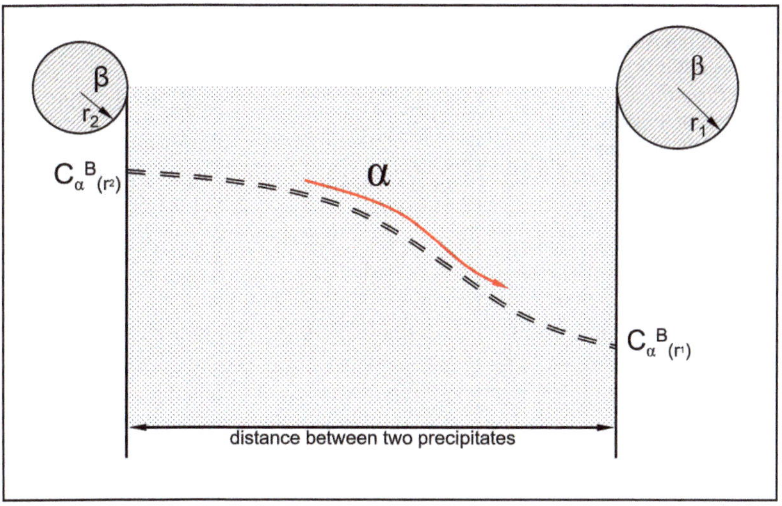

Fig. 6.60 Concentration profile in the α phase in between two β precipitates. The different equilibrium concentrations at the α-β interface are due to the Gibbs-Thomson effect. The arrow indicates the direction of the diffusive flow

one of the main aspects that the design of a successful thermal treatment should take into account.

Alloy-specific parameters, like γ′ solvus, incipient melting, and liquidus temperature, are essential information to set the right conditions of the treatments. The pseudo-binary phase diagram in Fig. 6.61 shows the general trend of the superalloy characteristic temperatures as a function of an increasing concentration of γ′ phase, observed with the development of nickel-based superalloys, leading to the γ′-richer compositions of the latest-generation SC superalloys.

The complexity of the treatment is dependent on the target microstructure and alloy properties, but also on the size of the component, ingot, bar, or semi-product, to be treated. This is an important aspect for the correct evaluation of the time necessary to reach any set temperature and for ensuring a controlled quenching rate.

Examples of real thermal cycles are shown by Fig. 6.62, concerning two conventionally cast superalloys: Inconel 700 and Udimet 700 featuring different fractions of γ′ (20% and 45%, respectively – see Fig. 6.61; for the alloy compositions see Table 6.14).

The treatments of both Inconel 700 and Udimet 700 are discussed hereafter with reference to the relevant temperature-transformation-time diagram in Fig. 6.62. For the Inconel 700 superalloy, with a 20% γ′ phase, the heat treatment features two main stages, involving the transformations shortly described by the scheme in Fig. 6.63.

The Udimet 700 alloy has a higher percentage of γ′ phase. Therefore, a multiple-stage treatment is recommended, in order to obtain the full precipitation of the γ′ and carbide phases and relevant microstructural refinement. The heat treatment described

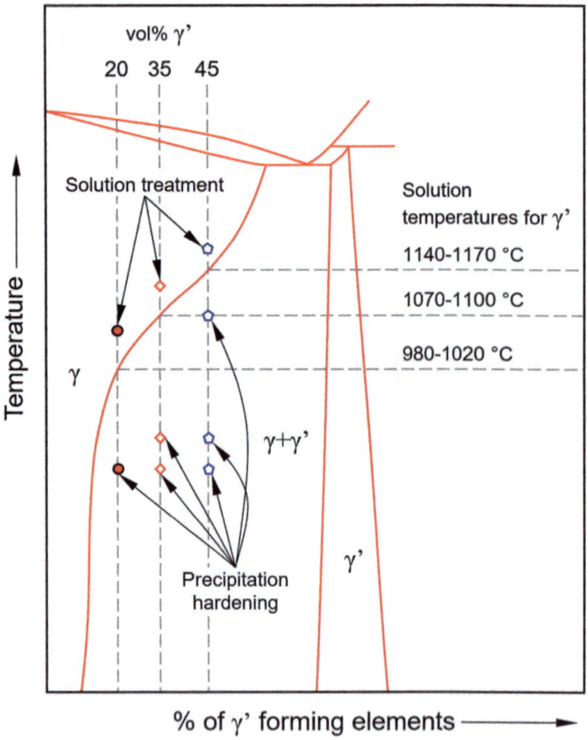

Fig. 6.61 Pseudo-binary phase diagram referring to different superalloys, featuring different concentrations of γ'- phase. (Redrawn from Guttmann 1981)

by Fig. 6.64 involves four main steps. The treatment sequence proposed for the Udimet 700 alloy comprises two solutioning treatments. A two-step solutioning, involving two different temperatures, is meant to obtain a complete solutioning of those γ' precipitates already present in the alloy, possibly featuring an inhomogeneous size distribution for the uncontrolled precipitation conditions and thereby a different stability (see Eq. 6.49). The precipitates are subsequently stabilized by a two-step aging. As detailed in the scheme in Fig. 6.64, the microstructural refinement during the treatment involves also carbide phases and the grain morphology of the superalloy.

Based on these principles, a broad range of heat treatment have been developed, according to the alloy chemistry and target properties (ASM International 2016; Donachie and Donachie 2002). Table 6.25 contains the main steps and parameters for a number of heat treatments used for different groups of superalloys.

Also DS and SC alloys, notwithstanding the extreme control of the investment casting process, still require post-processing treatments for the optimization of their microstructure. The main aspects that are managed through these treatments concern the optimization, usually the reduction, of the average size of the γ' precipitates and the elimination of the inter-dendritic eutectic regions. These latter are unavoidable features of the as-solidified alloy and result from the solute enrichment in the liquid phase, close to the solid-liquid interface (Figs. 6.54 and 6.55). The average size of

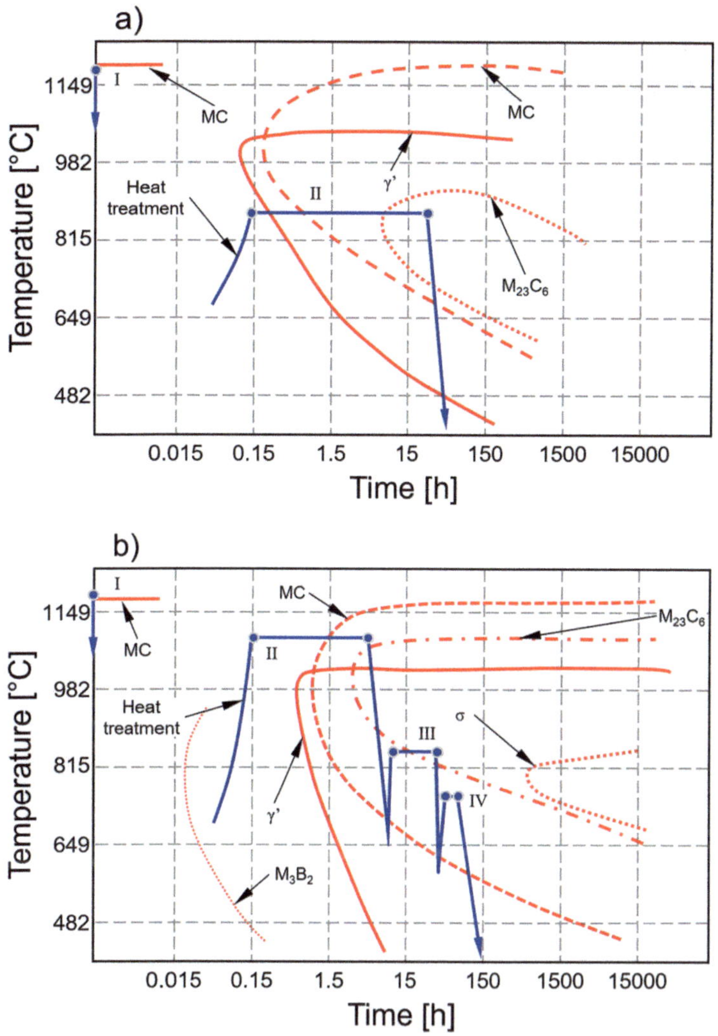

Fig. 6.62 TTT curves referring to the heat treatments for the microstructural refinement of two polycrystalline superalloys with different fractions of γ′ phase: (**a**) Inconel 700, 20% γ′; (**b**) Udimet 700, 45% γ′. (Redrawn from Guttmann 1981)

the γ′ precipitates tends to exceed the optimal value, since an excessive coarsening may be favored by the long-term exposure to comparatively high temperatures, required by the directional solidification process. The typical processing duration involved with the directional solidification process is, in fact, significantly longer than that of conventional casting processes. The refinement of the γ′ phase is a fundamental aspect considering that it is the majority phase in superalloys belonging to the latest generations (see Tables 6.15 and 6.16). The increasing fraction of γ′ in

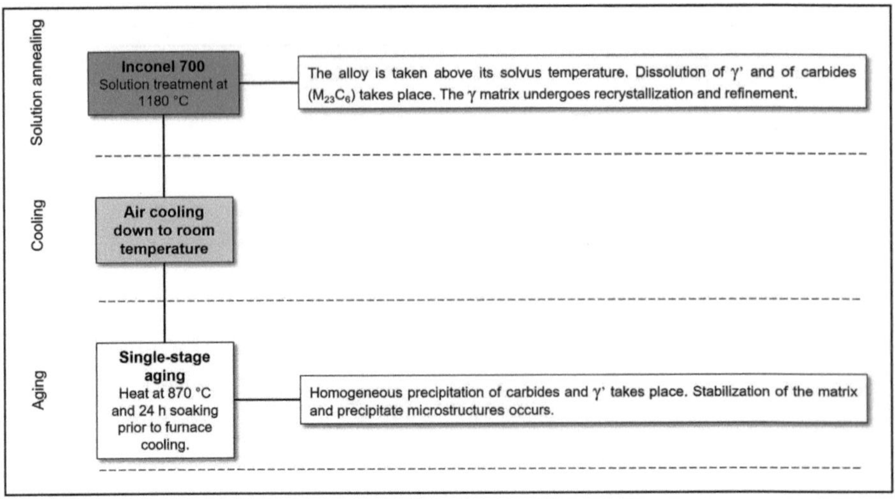

Fig. 6.63 Heat treatment steps for Inconel 700. Related TTT curves are given in Fig. 6.62a. (Guttmann 1981)

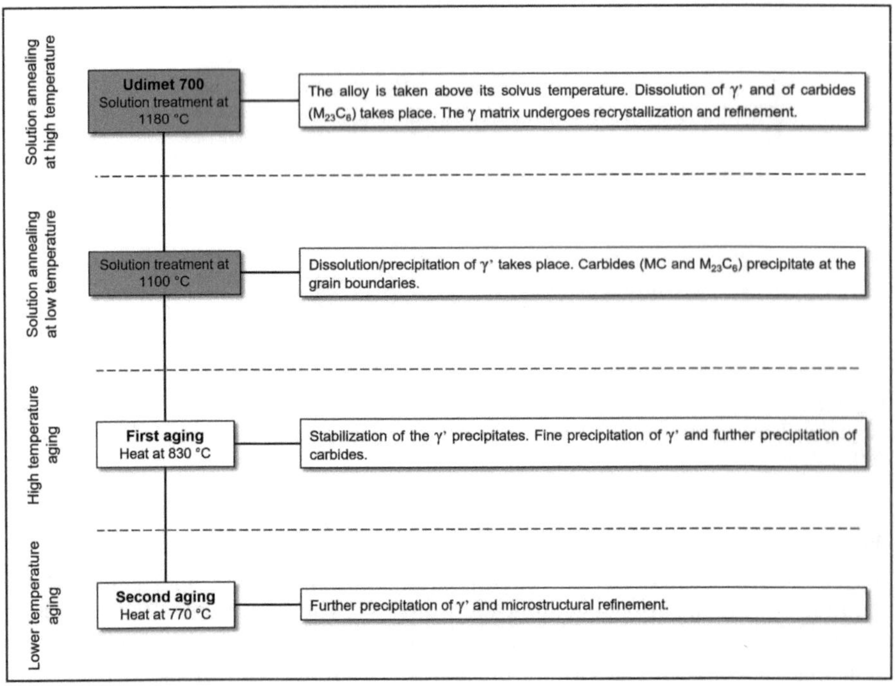

Fig. 6.64 Heat treatment steps for Udimet 700. Related TTT curves are given in Fig. 6.62b. (Guttmann 1981)

Table 6.25 Examples of thermal treatments for selected polycrystalline superalloys (ASM International 2000, 2016; Donachie and Donachie 2002)

		Stress relief	Annealing	Solution annealing			Aging		
Group	Alloy	Temperature [°C]	Temperature [°C]	Temperature [°C]	Time [h]	Cooling media	Temperature [°C]	Time [h]	Cooling media
Co-based	Haynes 188	–	–	1175	0.5	Rapid air cooling	–	–	–
	Haynes 25	Approx. 55 °C below annealing temperature (1)	1230	1230	1	Rapid air cooling	Aging occurs at elevated temperatures during service		
	S-816	Approx. 55 °C below annealing temperature (1)	1205	1175	1	(4)	760	12	Air
Fe-based	19-9DL	675 (2)	980	–	–	–	–	–	–
	Alloy 718	(3)	955	980–1065	1	Air or water	718	8	Furnace cooling
	Discaloy	(3)	1035	1010	2	Oil	650	20	Air
							730	20	Air
	Incoloy 907	–	–	980	1	Air	620	8	Air
							775	12	Furnace cooling
	Incoloy 909	–	–	980	1	Air	620	8	Air
	Incoloy 925	–	–	1010	1	Air	620	8	Air
							730	8	Furnace cooling
	Inconel 706	–	–	925–1010	–	–	620	8	Air
							845	3	Air
	N-155	Approx. 55 °C below annealing temperature (1)	1230	1175	1	Water	815	4	Air

(continued)

Table 6.25 (continued)

		Stress relief	Annealing	Solution annealing			Aging		
Group	Alloy	Temperature [°C]	Temperature [°C]	Temperature [°C]	Time [h]	Cooling media	Temperature [°C]	Time [h]	Cooling media
Ni-based	Astroloy	(3)	1135	1175	4	Air	845	24	Air
	Hastelloy S	–	–	1065	0.5	Rapid quench	–	–	–
	Inconel 617	–	–	1175	2	(4)	–	–	–
	Inconel 625	870	980	1150	2	(4)	–	–	–
	Nimonic 80A	(3)	1080	1080	8	Air	705	16	Air
	Nimonic 90	–	–	1080	8	Air	705	16	Air
	René 41	–	–	1065	0.5	Air	760	16	Air
	Udimet 500	(3)	1080	1080	4	Air	845	24	Air
	Waspaloy	(3)	1010	1080	4	Air	760	16	Air
							845	24	Air

(1) If further fabrication is needed, full annealing is recommended
(2) 675 °C is the nominal temperature, the 650–705 °C range is allowed
(3) Since intermediate temperatures lead to aging, full annealing is recommended
(4) Rapid cooling below 540 °C is necessary to allow adequate quench and to avoid precipitation in the intermediate temperature range. For thin sections rapid air cooling is enough; oil or water quenching is used for heavier sections that do not undergo cracking while cooling

the alloy results in an increasing γ' solvus and, correspondingly, involves a reduction of the temperature range available to the solutioning. This aspect renders an accurate control of the treatment conditions particularly critical.

In Table 6.26 a series of heat treatments adopted for conventionally cast, DS and SC superalloys are illustrated.

The importance of an optimized alloy microstructure is illustrated in Fig. 6.65, with the creep life of the PWA1480 SC alloy as a function of the average size of the ordered γ' precipitates, exhibiting a maximum value close to 0.4 μm (Fig. 6.65a). In Sect. 6.5.1, it has been demonstrated that the creep behavior of SC superalloys is very much dependent on mechanisms involving dislocations and, thereby, their interaction modes with γ' precipitates. If the ordered precipitates have too a small average size, dislocations may climb over them. Therefore, the obstacle provided by the precipitates to dislocation movement is not that effective as it would be in case dislocations were forced to cut through them, as shown by direct observations and relevant modeling (Fig. 6.46). This is the condition corresponding to the peak in the lifetime of the alloy, since the trailing dislocations remain blocked at the interface between the γ matrix and the γ' precipitates as long as their concentration becomes very high. At this stage, the alloy enters the high deformation rate tertiary creep regime, unless, as it is typically the case, the component is replaced. A reduction in the creep life of the alloy is observed when too coarse precipitates render the bowing (Orowan) mechanism relatively more important. The optimum solution temperature of the PWA1480 SC alloy is restricted to a narrow range, which, through the optimization of the precipitate shape and size, affords the most effective strengthening of the alloy (Fig. 6.65b). The softening of the alloy, observed for solution treatments performed at lower temperatures, is due to residual of γ' precipitates formed during the solidification and cooling of the alloy during primary investment casting solidification process. The size of these precipitates is too large to provide the best mechanical behavior. On the other hand, if an excessive solution temperature is used, defects in the microstructure of the alloy are introduced as a consequence of unwanted melting. Since the subsequent solidification would occur in an uncontrolled regime, the formation of defects and freckles would be unavoidable.

Concerning superalloy heat treatments, it is worth recalling the DMHT (dual microstructure heat treatment) (see Sect. 6.4.3) developed by NASA for turbine disc obtained via powder metallurgy. A sub-solvus heat treatment is conducted in the bore region, while the rim is subjected to a super-solvus heat treatment. This is achieved thanks to a particular apparatus, which is able to keep bore and ring at two different temperatures at the same time, while the whole disc is kept inside the treatment furnace. The bore region is encapsulated by a thermal insulating jacket, so that it remains at lower temperatures with respect to the disc rim. The thermal gradients present in the part during the heat treatment are enhanced by the heat sink placed in contact with the bore inside the insulation jacket. The heat sinks are usually solid metal cylinders. Thanks to this particular architecture of the furnace, finer grain size and larger γ' volume fractions can be attained in the bore region, while the rim, being subjected to a super-solvus heat treatment, is characterized by

Table 6.26 Thermal treatments for different kinds of nickel-based superalloys (Donachie and Donachie 2002)

		Step 1		Step 2		Step 3		Step 4	
	Alloy	Temp. [°C]	Time [h] + cooling	Temp. [°C]	Time [h] + cooling	Temp. [°C]	Time [h] + cooling	Temp. [°C]	Time [h] + cooling
Polycrystalline conventional castings	IN-939	1160	4 – rapid air cooling	1000	6 – rapid air cooling	900	24 – air	700	16 – air
	René 41	1065	3 – air	1120	0.5 – air	900	4 – air	–	–
	René 77	1163	4 – air	1080	4 – air	760	16 – air	–	–
	René 80	1220	2 – gas furnace quench	1095	4 – gas furnace quench	1050	4 – air	845	16 – air
	Udimet 500	1150	4 – air	1080	4 – air	760	16 – air	–	–
	Waspaloy	1080	4 – air	845	4 – air	760	16 – air	–	–
DS castings	MAR-M 247	1230	2 – gas furnace quench	980	5 – air	870	20 – air	–	–
	MAR-M 200Hf	1230	4 – gas furnace quench	1080	4 – air	870	32 – air	–	–
	René 80H	1190	2 – gas furnace quench	1080	4 – air	870	16 – air	–	–
SC castings	CMSX-2	1315	3 – gas furnace quench	980	5 – air	870	20 – air	–	–
	PWA 1480	1290	4 – gas furnace quench	1080	4 – air	870	32 – air	–	–
	René N4	1270	2 – gas furnace quench	1080	4 – air	900	16 – air	–	–

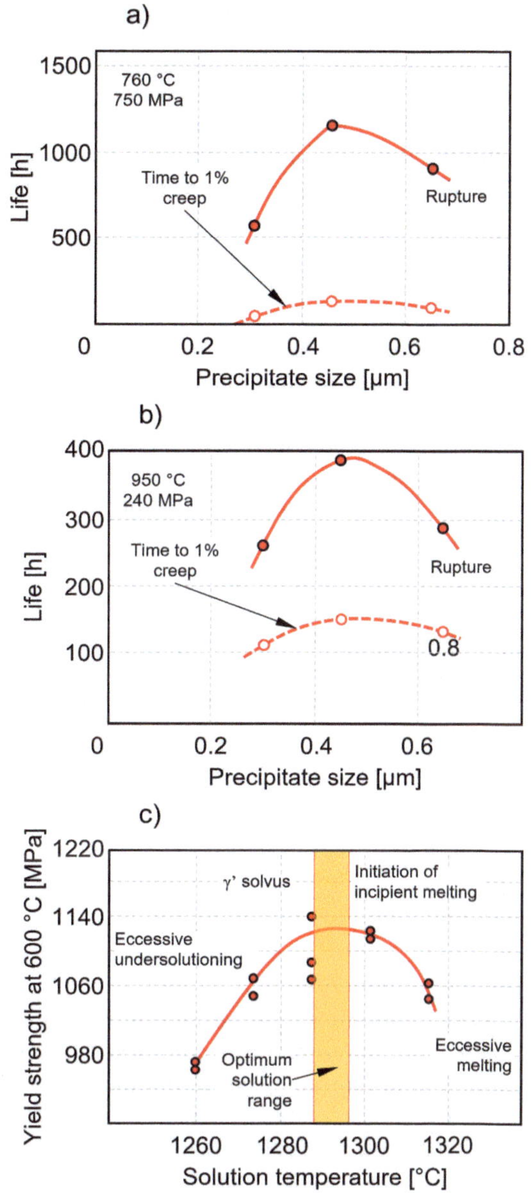

Fig. 6.65 Creep curves of the PWA1480 alloy for 1% deformation and rupture, as a function of the average size of the γ′ precipitates, at (**a**) 760 °C and (**b**) 950 °C. (**c**) Yield strength at 600 °C as a function of the solutioning temperature. The reason for the decay of the alloy strength, at temperatures above and below the optimum temperature range, is indicated in the graph

larger grain size and lower volume fraction of γ′ precipitates. The disc with this dual microstructure exhibits excellent performances when exposed to the complex working condition of state-of-the art GTEs (Reed 2006).

6.7 Applications of Superalloys in the Aerospace Field

All through this chapter, the properties of the superalloys belonging to the three main groups, cobalt-, iron- and nickel-based, have been discussed, and on some occasion, a direct reference to specific applications has been made, like in case of directionally solidified turbine blades and PM turbine discs. Considering the importance of these alloys in the manufacture of aerospace gas turbine engines, an overview of their specific applications is presented in this section, using the structure of a GTE as guideline, moving from the compressor to the low-pressure stages of the gas turbine (see Fig. 2.3).

As concerns compressor, in the inner stages where the highest air compressed temperatures are reached, nickel- and iron-based superalloys are used for stationary blades (vanes). Nickel-based superalloys are employed also for the blades and the discs of the last compressor stages, closer to the combustor. In modern engines the compression ratio can reach values close to 50 (see Fig. 2.8 and Sect. 2.6), and the relevant gas temperatures require such choice. Forged blades and discs are used, although the development of reliable powder metallurgy processes for the turbine discs may extend this technology to compressor parts too. The blades, as those used in the turbine, can be also produced by investment casting, although not requiring directionally solidified microstructure.

For the combustor chamber, wrought cobalt superalloys are an excellent choice, for their high-temperature stability and corrosion resistance, coupled with ease of manufacturing, considering the good weldability of these materials. This is an interesting feature both for the initial assembling of the engine and also for possible repair and maintenance. The inner surface of the combustion chamber requires protection by thermal barrier coatings (see Sect. 7.3.2), so that the hottest gases in the whole engine are not in direct contact with the inner surfaces of the metallic wall. An essential constituent of the TBC system is the bond coat, needed to assure a good matching of the mechanical and thermophysical properties of the outer ceramic part of the TBC and the underlying cobalt alloy. The composition of the bond coat has to be selected accordingly.

TBC coatings are also used for the high-pressure blades and vanes in the turbine. These parts are directionally solidified or single-crystal components. TBCs have to sustain more demanding conditions than in the combustor chamber, considering that the exhaust gases impinge directly onto the stationary and rotating blades, whereas in the combustor, although hotter, the flux is nearly tangent to the metal parts. Moreover, the rotating blades are stressed by centrifugal forces. For these reasons, the coating reliability is paramount, and this is one of the main reasons why electron beam physical vapor deposition (EB PVD) is used in the place of PS, as discussed in Sect. 7.2.2.

Investment-cast nickel-based superalloys are used for the blading and discs of the intermediate and low-pressure stages of the turbine. Considering the less demanding conditions, equiaxed alloys are used for the disc, either produced by forging or by PM approach. The turbine disc can be regarded as one of the most critical components in a

GTE. This becomes particularly evident when analyzing the incidence of the turbine disc on both the weight and the cost of a modern aircraft engine. For example, in the Rolls-Royce Trent 800, the turbine disc accounts for about the 20% of the total engine weight while representing about 10% of its value. In a military engine, such as the EJ200 (produced by Eurojet Turbo Gmbh) mounted on the Eurofighter Typhoon jet, the turbine disc accounts for about the 5% of the total engine weight, but for about 25% of its entire value. The main task of turbine discs is to provide fixturing to the turbine blades, that receive kinetic energy from the gas stream coming from the combustor (see Fig. 2.3). The disc-blade assembly employs this energy to operate the front fan and the compressor by means of single or multiple shafts, depending on the engine's architecture (see Figs. 2.10 and 2.11) (Reed 2006). The design of a turbine disc is optimized, so to satisfy the requirements of dimensional stability, weight minimization, mechanical resistance, and integrity, in order to sustain general yielding and low-cycle fatigue cracking in the region near the bore. Furthermore, turbine blades may transfer part of the heat from the hot gas flux so that the temperature in the disc's rim can reach peak values close to 650 °C. In these conditions, centrifugal forces due to the high rotational speed, even higher than 10,500 rpm, can have severe consequences on the material integrity (Reed 2006). Creep and high-temperature dwell crack propagation resistance are assuming an increasing importance in newly developed engines. In this regard, PM started to be used for the production of this critical component and the possibility to have locally optimized microstructure, and, thereby, properties in the bore and the rim of the same turbine disc have gained great interest (see Sect. 6.6). The use of PM was initially limited to GTEs used in the military field, like in the Pratt & Whitney F100 engine employed for the propulsion of the McDonnell Douglas F-15 "Eagle" jet fighter (early operational flight in 1974). Subsequently, the experience developed in the military field was transferred to civilian transportation, and PM discs are used in the engines for the Boeing 747 and 777 airliners, the General Electric CF6-80 and GE90, respectively (Reed 2006).

A list of superalloy components used in aerospace applications is presented herewith with the same order used in this chapter for the description of each family of superalloys. Concerning Co-based superalloys, they are used for those components subjected to working conditions where hot corrosion is a major concern at moderate-to-high temperatures and for low-stress structural applications. Moreover, Co-based superalloys have superior hot corrosion resistance in environments containing sulfur, lead oxides, and other compounds produced from the combustion of jet fuels. Superior stress-rupture properties and hot corrosion resistance make cobalt alloys a suitable choice for vanes and other stationary components (Geddes et al. 2010; Mouritz 2012). With reference to the alloys presented in Sect. 6.2 (see also Table 6.3), a list of components made of Co-based superalloys is given in Table 6.27.

Iron-based superalloys (see Sect. 6.3) provide a very wide combination of properties that are interesting for the fabrication of several components in the aerospace field and also for other structural, possibly at high temperature, applications. In Table 6.28, the main applications of these superalloys are listed.

Table 6.27 Uses of selected Co-based superalloys (ASM International 2000; Geddes et al. 2010; Mouritz 2012; Zhang and Zhao 2012)

Alloy	Application
AiResist 213	Gas turbine components
AiResist 215	Gas turbine engines (stator vanes)
FSX-414	Gas turbine stator vanes
FSX-418	Gas turbine stator vanes
FSX-430	Gas turbine stator vanes
Haynes 25	Bearing material (balls and races)
	Combustion chamber
	Flame holder components
	Liners
Haynes 188	Afterburner components
	Combustion cans
	Military and commercial GTE components
	Transition ducts
MAR-M 302	Buckets in GTE
	Nozzle guide vanes
	Turbine vanes
MAR-M 322	Turbine blades
	Turbine vanes
MAR-M 509	Gas turbine engines blades and vanes
MP35-N	Fasteners for aerospace applications
MP159	Aircraft landing gear
	Airframes
	Fasteners used in GTE
	Space shuttle propulsion system
NASA Co-Re-W	High-temperature space applications
S-816	Gas turbine engines blades, bolts, and springs
WI-52	Gas turbine engines stator vanes
X-40	Gas turbine engine stator vanes
X-45	Gas turbine engine stator vanes

All through the chapter, Ni-based superalloys have been described as materials possessing an exceptional combination of properties and high-temperature capabilities, and for this reason, their use is indicated for the fabrication of the most demanding components for aerospace turbine engines. In Table 6.29 the most important applications of Ni-based superalloys (see Sect. 6.4) restricted to the aerospace field are provided.

Table 6.28 Applications in aerospace and related high-temperature fields of selected Fe-based superalloys (Geddes et al. 2010; Haynes International 2017a, b; Schafrik et al. 2001)

Alloy	Application	Additional information
19-9DL	Afterburners components	High strength, oxidation, and corrosion resistance up to about 670 °C. Relatively low cost
	Casings	
	Industrial gas turbine blades	
	Turbine and supercharger wheels	
Alloy 718	Compressor blades and discs, casings, fasteners, locking lugs, other compressor components (rear shaft, hub tie clamps, and supports)	It is the most widely used Fe-based superalloy. The creep-rupture strength is maintained up to 650–700 °C
	Turbine discs and wheels	
Discaloy	Bolts	–
	GTE parts	
Hastelloy X	GTE combustors (combustor cans, flame holders, spray bars, transition ducts)	Hastelloy X is characterized by good oxidation resistance and good high-temperature strength. It also shows exceptional resistance to stress corrosion cracking
	GTE fabricated parts	
	GTE parts (afterburners, tailpipes)	
Haynes 556	GTE sheet components (afterburners, combustors, transition ducts)	The alloy is similar to N-155, but the compositional control has been improved. It has improved oxidation and hot corrosion resistance, high-temperature ductility, fatigue resistance, thermal shock resistance, and weldability
	Heat treating	
	Waste incineration	
Haynes N-155	Afterburner parts, exhaust manifolds, combustion chambers, nozzles, turbine blades	The alloy can be used in applications involving high stress at temperatures up to 800–1100 °C. This alloy is also used in the fabrication of heat treatment equipment, where high-temperature strength and corrosion resistance are two essential features
Incoloy 800	Heat exchangers	Excellent carburization and oxidation resistance. Good strength
	GTE parts	
Incoloy 901	GTE rotors	–
Incoloy 903	GTE rings and casings	Low and constant thermal expansion coefficient up to 430 °C. High thermal fatigue and thermal shock resistance
	Rocket components	
Incoloy 907	GTE seals	Low and constant thermal expansion coefficient and improved notch-rupture properties at elevated temperatures in comparison with Incoloy 903
	GTE shafts	
	GTE casings	
Incoloy 909	GTE casings	Low and constant thermal expansion coefficient. In comparison with Incoloy 903 and 907, the addition of Si leads to improved tensile properties and notch-rupture properties, while heat treatments can be shorter and processing becomes less restrictive
	GTE shafts	
	GTE shrouds	
	GTE vanes	
	Turbine shafts	
	Turbine supports and other components	

(continued)

Table 6.28 (continued)

Alloy	Application	Additional information
MA-956	GTE combustion chamber	This is an ODS alloy (see Sect. 9.3) with a ferritic matrix. Exceptional resistance to carburization, hot corrosion and oxidation at temperatures above 1100 °C
N-155	GTE sheet components (afterburners, combustors, transition ducts). Extensive use in older aircraft GTEs	It has been developed before Haynes 556. It can be used in high-stress applications up to 820 °C and in moderate-stress applications up to 1090 °C
V-57	GTE rotors	–
	GTE bolts	
REX-78	Blading in the Whittle engine (see Sect. 2.1)	–
W-545	GTE parts	–

Table 6.29 Applications in the aerospace field of selected Ni-based superalloys (Geddes et al. 2010; Mukherji et al. 2011; Reed 2006)

Alloy	Application	Additional information
Astroloy	Forgings for high temperatures	–
CMSX-4	Turbine blades	Second-generation SC superalloy
CMSX-10	Turbine blades	Third-generation SC superalloy
Hastelloy B	Old rocket engines and GTEs	Good strength up to 1095 °C, oxidation resistance up to 650 °C
Hastelloy S	GTE components subjected to low stress	Relatively low coefficient of thermal expansion. Good fatigue resistance and high thermal stability
Hastelloy W	Combustors	–
	GTE fabricated parts	
Haynes 230	Combustion environments	Properties in terms of oxidation resistance, strength, and stability up to 1150 °C
	High-temperature bellows	
Inconel 617	Combustion cans in GTE	Exceptional combination of oxidation resistance, metallurgical stability and strength at high temperature
	Ducting an transition liners in GTE	
Inconel X750	GTEs	High-stress relaxation resistance. High creep-rupture and tensile properties maintained up to 700 °C
	High-temperature springs and bolts	
	Rocket engines	
	Aircraft structures	
LSHR	Turbine discs	Third-generation PM superalloy
M-252	GTE blades, parts, sheets	–
ME3	Turbine discs	Third-generation PM superalloy

(continued)

Table 6.29 (continued)

Alloy	Application	Additional information
Nimonic 75	GTE sheet metal fabrications	–
Nimonic 80A	GTE blades, bolts, discs, rings	Properties in terms of corrosion resistance, tensile and creep-rupture strength up to 815 °C
Nimonic 86	GTE sheets (afterburners and combustion chamber)	The addition of Ce leads to scaling and oxidation resistance up to 1050 °C. Good formability and weldability
Nimonic 90	GTE blades and discs	High resistance against hot corrosion and oxidation. High creep and stress-rupture properties up to 920 °C
PWA 1480	Turbine blades	First-generation SC superalloy
PWA 1484	Turbine blades	Second-generation SC superalloy
René 41	GTE blades and other parts	–
René 88DT	Turbine discs	Second-generation PM superalloy
René 95	Discs for compressor and turbine in GTEs	First-generation PM superalloy
René N5	Turbine blades	Second-generation SC superalloy
TMS 162	Turbine blades	Fifth-generation SC superalloy
TMS 196	Turbine blades	Fifth-generation SC superalloy
Udimet 720	Discs for compressor and turbine in GTEs	High metallurgical stability. Good impact, oxidation, and sulfidation resistance
Unitemp AF2-1DA	Turbine discs	–
Waspaloy	Airframe assemblies	Used for the fabrication of highly loaded components with service temperature up to 650 °C. High strength and oxidation resistance at high temperature
	Fasteners	
	GTE components (casings, rotor discs, shafts, spacers, rings)	

References

Antony K C (1983) Wear-resistant Cobalt-base Alloys. JOM 35 (2): 52–60

Ashby M F (1972) A first Report on Deformation-mechanism Maps. Acta Metallurgica 20 (7): 887–897

Ashby M F, Frost H J (1982) Deformation Mechanisms Maps. Pergamon Press

ASM International (1991) ASM Handbook Vol. 2 – Properties and Selection of Nonferrous Alloys and Special Purpose Materials. ASM International, Materials Park, Ohio

ASM International (1993) ASM Handbook Vol. 1 – Properties and Selection: Iron, Steel and High Performance Alloys. ASM International, Materials Park, Ohio

ASM International (2000) ASM Specialty Handbook – Nickel, Cobalt and Their Alloys. ASM International, Materials Park, Ohio

ASM International (2008) ASM Handbook Vol. 15 – Casting. ASM International, Materials Park, Ohio

ASM International (2016) ASM Handbook Vol. 4E – Heat treating of Nonferrous Alloys. ASM International, Materials Park, Ohio

Barrett C S, Massalski T (1980) Structure of Metals 3rd edn. Pergamon Press

Bauer A et al. (2010) Microstructure and Creep Strength of different γ/γ'-strengthened Co-base superalloy variants. Scripta Materialia 63 (12): 1197–1200

Cahn R W, Haasen P (1996) Physical Metallurgy 4th edn. North Holland

Callister W, Rethwisch D (2012) Fundamentals of Materials Science and Engineering: An Integrated Approach, 4th edn. Wiley John & Sons

Campbell F C (2006) Manufacturing Technology for Aerospace Structural Materials. Elsevier

Chandler H (1996) Heat Treater's Guide: Practices and Procedures for Nonferrous Alloys. ASM International, Materials Park, Ohio

Cui C et al (2006) A New Co-Base Superalloy Strengthened by γ' Phase. Materials Transactions 47 (8): 2099–2102

Davis J R (2000) The Nickel Industry: Occurrence, Recovery, and Consumption. In: ASM Specialty Handbook – Nickel, Cobalt and Their Alloys. ASM International, Materials Park, Ohio, p 3–6

Dimiduk D (1991) Dislocation Structures and Anomalous Flow in L12 Compounds. Journal de Physique III 1 (6): 1025–1053

Donachie M J, Donachie S J (2002) Superalloys – A Technical Guide. ASM International

Everhart J (1971) Engineering Properties of Nickel and Nickel Alloys. Springer US

Farrar J (2004) The Alloy Tree – A guide to Low-Alloy Steels, Stainless Steels and Nickel-base Alloys. Woodhead Publishing Limited and CRC Press

Fergus JW (2010) Recent developments in cathode materials for lithium ion batteries. Journal of Power Sources, 195(4): 939–954

Fontani M et al (2015) The Lost Elements: The Periodic Table's Shadow Side. Oxford University Press

Frost J H, Ashby M F (1982) Deformation-Mechanism Maps: The Plasticity and Creep of Metals and Ceramics. Pergamon Press

Geddes B et al (2010) Superalloys: Alloying and Performance. ASM International, Materials Park, Ohio

Gell M et al (1987) Advanced Superalloy Airfoil. JOM 39 (7): 11–15

Giamei A. F. (2013) Development of Single Crystal Superalloy: A brief History. Advanced Materials and Processes 171 (9): 26–30

Guttmann V (1981) Phase Stability in High Temperature Alloys. Applied Science Publishers LTD, London

Haynes International (2017a) HASTELLOY® X alloy. Available via DIALOG. http://www.haynesintl.com/alloys/alloy-portfolio_/High-temperature-Alloys/HASTELLOY-X-alloy/HASTELLOY-X-principal-features.aspx. Accessed 27 Dec 2018

Haynes International (2017b) MULTIMET® alloy. Available via DIALOG. https://www.haynesintl.com/alloys/alloy-portfolio_/High-temperature-Alloys/MULTIMET-alloy/principal-features.aspx. Accessed 28 Dec 2018

Hemeker K J et al (1991) An Investigation of the Mechanisms that Control Intermediate Temperature Creep of Ni_3Al. Acta Metallurgica et Materialia 39 (8): 1901–1913

Hull D, Bacon J (2011) Introduction to Dislocations 5th edn. Butterworth-Heinemann

Jena A K, Chaturvedi M C (1984) The Role of Alloying Elements in the Design of Nickel-base Superalloys. Journal of Materials Science 19 (10): 3121–3139

Jozwik P et al (2015) Applications of Ni_3Al Based Intermetallic Alloys – Current Stage and Potential Perspectives. Materials 8 (5): 2537–2568

Kear B H, Wilsdorf H G F (1962) Dislocation Configurations in Plastically deformed Cu_3Au Alloys. Transactions of the Metallurgical Society of AIME 224: 382–386

Kracke A (2010) Superalloys, the most successful Alloy System of modern Times – Past, Present and Future. In: Ott E A et al (eds) Superalloy 718 and Derivatives – Proceedings of the 7th International Symposium on SUPERALLOY 718 and DERIVATIVES, Pittsburgh, 2010

Krakow R et al (2017) On the Crystallography and Composition of topologically close-packed phases in ATI 718Plus ®. Acta Materialia 130: 271–280

Kuo K H et al (1985) Tetrahedrally close-packed Phases in Superalloys: new Phases and Domain Structures observed by high-resolution Electron Microscopy. Journal of Materials Science 21 (8): 2597–2622

Massalski T B et al (1986) Binary Alloys Phase Diagrams Vol. 2. ASM International

Mouritz A (2012) Introduction to Aerospace Materials. Woodhead Publishing Limited

Mukherji D et al (2011) Beyond Ni-based Superalloys: Development of CoRe-based Alloys for Gas Turbine Applications at Very High Temperatures. International Journal of Materials Research 102(9): 1125–1132

Murakumo T et al (2004) Creep Behaviour of Ni-base Single-crystal Superalloys with various γ' Volume Fraction. Acta Materialia 52 (12): 3737–3744

Murray J L (1982) The Co-Ti (Cobalt-Titanium) System. Bulletin of Alloy Phase Diagrams 3: 74–85

Murray J L (1987) Phase Diagrams of Binary Titanium Alloys. ASM International

Nabarro F R N, de Villers F (1995) The Physics of Creep and Creep-resistant Alloys. CRC Press

Nakajima K et al (2018) Global Distribution of Material Consumption: Nickel, Copper and Iron. Resource, Conservation and Recycling 133: 369–374

Neumeier S et al (2016) Diffusion of Solutes in fcc Cobalt investigated by Diffusion Couples and First Principles Kinetic Monte Carlo. Acta Materialia 106: 304–312

Pollock T M (2016) Alloy Design for Aircraft Engines. Nature Materials 15 (8): 809–815

Pollock T M, Tin S (2006) Nickel-Based Superalloys for Advanced Turbine Engines: Chemistry, Microstructure, and Properties. Journal of Propulsion and Power 22 (2): 361–374

Porter D, Easterling K E, Sherif M Y (2009) Phase Transformations in Metals and Alloys, 3rd end. CRC Press

Prasad E N, Wanhill R (2017) Aerospace Materials and Material Technologies Volume 1: Aerospace Materials. Springer

Radavich J F (1989) The Physical Metallurgy of Cast and Wrought Alloy 718. In: Loria E A (ed) Superalloys 718 Metallurgy and Applications. TMS, p 229–240fi

Rae C M F et al (2000) Topologically close-packed Phases in an experimental Rhenium-containing Single Crystal Superalloy. Superalloys 2000: 767–776

Rae C M F, Reed R C (2001) The Precipitation of Topologically close-packed Phases in Rhenium-containing Superalloys. Acta Materialia 49 (19): 4113–4125

Reed R C (2006) The Superalloys – Fundamentals and Applications. Cambridge University Press, Cambridge (UK)

Reed R C et al. (2007) Damage Accumulation during Creep Deformation of a Single Crystal Superalloy at 1150 °C. Materials Science and Engineering A 448 (1–2): 88–96

Reed-Hill R E (1973) Physical Metallurgy Principles 2nd edn. PWS-KENT Publ., Boston

Ricks R A et al (1983) The Growth of γ Precipitates in Nickel-Base Superalloys. Acta Materialia 31: 43–53

Riddihough M (1970) Stellite as a wear-resistant material. Tribology 3 (4): 211–215

Rogister C et al (1967) Improvement of Heat Resisting Cobalt Base Alloys by Precipitation Hardening. Cobalt 34: 3–9

Rolls-Royce (1996) The Jet Engine 5th edition. Rolls-Royce plc, Derby

Sato J et al (2006) Cobalt-Base High-Temperature Alloys. Science 312 (5770): 90–91

Schafrik R E et al (2001) Application of Alloy 718 in GE Aircraft Engines: Past, Present and Next Five Years. In: Fifth International Symposium on Superalloys 718, 625, 706 and Derivatives, Embassy Suites Hotel, Pittsburg, Pennsylvania, 17–20 June 2001

Seiser B et al (2011) TCP Phase Predictions in Ni-based Superalloys: Structure Maps Revisited. Acta Materialia 59 (2): 749–763.

Sims C T (1984) A History of Superalloy Metallurgy for Superalloy Metallurgist. Superalloys 1984: Proceedings of the Fifth International Symposium on Superalloys, Seven Springs Mountain Resort, Champion, Pennsylvania, 7–11 Oct 1984

Sims C T et al (1987) Superalloys II. John Wiley & Sons, New York
Sjöberg G (2010) Casting Superalloys for Structural Applications. In: Ott E A et al (ed) Superalloy 718 and Derivatives – Proceedings of the 7th International Symposium on SUPERALLOY 718 and DERIVATIVES, Pittsburgh, 2010
Smallman R E et al (2002) Inverse Creep in Intermetallics. Materials Science and Engineering A 329–331: 852–855
Smallman R E, Ngan A H W (2014) Modern Physical Metallurgy 8th edn. Butterworth-Heinemann
Suzuki A, Pollock T M (2008) High-temperature Strength and Deformation of γ/γ′ two Phase Co-Al-W-base alloys. Acta Materialia 56 (6): 1288.1297
Tylcote R F (2002) A History of Metallurgy. Maney Publishing
U.S. Geological Survey (2017) Mineral Commodity Summaries. Available via DIALOG. https://minerals.usgs.gov/minerals/pubs/commodity/cobalt/mcs-2017-cobal.pdf. Accessed Jun 2018
Van Schilfgaarde M et al (1999) Origin of the Invar Effect in Iron-Nickel Alloys. Nature 400: 46–49
Viatour P et al (1973) Stability of the Gamma Prime Co_3Ti Compound in Simple and Complex Cobalt Alloys. Cobalt 3: 67–74
Yoo M H (1987) Stability of Superdislocations and Shear Faults in L12 ordered Alloys. Acta Metallurgica 35 (7): 1559–1569
Zenck C H et al (2017) A novel Type of Co-Ti-Cr-base γ/γ′ Superalloys with low Mass Density. Acta Materialia 135: 244–251
Zhang S, Zhao D (2012) Aerospace Materials Handbook. CRC Press

Further Reading

Durand-Charre M (1997) The Microstructure of Superalloys. CRC Press
Gessinger G H (2013) Powder Metallurgy of Superalloys. Butterworth-Heinemann
Kazantseva N V et al (2019) Superalloys – Analysis and Control of Failure Process. CRC Press
Ott E et al (2010) Superalloy 718 and Derivatives – Proceedings of the 7th International Symposium on SUPERALLOY 718 and DERIVATIVES, Marriott Pittsburgh City Center, Pittsburgh, Pennsylvania, 10–13 Oct 2010
Srivastava R R et al (2014) Resource recycling of superalloys and hydrometallurgical challenges. Journal of Materials Science 49 (14): 4671–4686

Chapter 7
Coatings

7.1 General Aspects

Coatings are extensively used in gas turbine engines, as illustrated by Fig. 7.1. They are designed to improve the capabilities of the different components of the engine, with specific regard to surface stability against high-temperature corrosion and thermally activated transformations. Particularly for the compressor, erosion and fretting resistance are other aspects that can be tackled using suitable coating systems.

Coatings are also important to ensure and retain high engine efficiency by reducing the mechanical tolerances and gas leakages, for which abradable and seal coatings are used.

As concerns surface stability, the main degradation phenomena are high-temperature corrosion, involving oxidation and hot corrosion (see Chap. 8), wear, and erosion. High-temperature corrosion is particularly important in the hottest stages of the turbine and requires the use of overlay and diffusion coatings, capable to form on their own surface protective layers, mainly made of native oxides. Wear is rather a complex phenomenon, involving both mechanical stresses, possibly cyclic, like fretting, and chemical reactions, like oxidation. For this reason, the surface treatments and coating systems used for these purposes require specific design for each part to be protected. Erosion is quite an issue for the front part of the engine, i.e., fan and early compressor stages, considering the ingestion of particulate matter and sands from the outer atmosphere. Since the erosion rate depends on the direction of the impinging particles, composite cermet coatings are the best candidates, considering the wide range of impact angles faced in real conditions. For ground-based gas turbines, generally using not so refined fuels as in aerospace engines, erosion is also observed in the high-pressure stages of the turbine as an effect of the combustion soot carried by the high-velocity exhaust gas flow coming out from the combustion chamber.

S. Gialanella, A. Malandruccolo, *Aerospace Alloys*, Topics in Mining, Metallurgy and Materials Engineering, https://doi.org/10.1007/978-3-030-24440-8_7

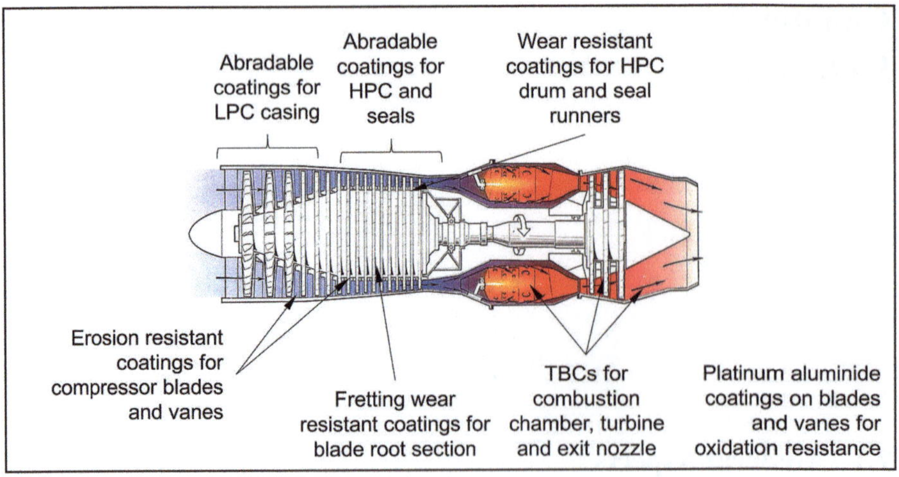

Fig. 7.1 Coating systems in a gas turbine aero-engine, with the indication of the main surface degrading mechanisms (Rajendran 2012; Rolls-Royce 1996, image courtesy of Rolls-Royce plc)

Clearance control and seal coatings are obviously interesting for retaining an excellent engine efficiency over its entire lifetime. These systems are there to optimize the mechanical matching between rotary and stationary parts. Wear-resistant and abradable coatings are generally facing one another to achieve this. Still, in some way related to efficiency improvement are the thermal barrier coatings, because they allow increasing operating temperatures of the engine, beyond the limits of the component alloys.

For a correct design and selection of an appropriate coating system, it is paramount to know the variation range of the operating parameters. Reference values of the more critical of these parameters for gas turbine aero-engines are listed in Table 7.1. They are classified with reference to the dominant character of the attack: mechanical, chemical, and thermal. These parameters are of course fundamental not only to identify and select the most suitable coating for any specific application but also the relevant deposition techniques, discussed in Sect. 7.2.

7.2 Coating Deposition Techniques for Gas Turbine Engines

Several techniques have been developed over the years and employed to deposit coatings onto components of gas turbine engines in order to protect them from the attack and damaging phenomena described in the previous section. Two of the most important requirements for these techniques are the good reproducibility of the coating features and the flexibility as concerns the geometry of the substrate on which the coating is deposited. These features are particularly important for aero-engine applications and

Table 7.1 Indication of the main working conditions and relevant variation range interesting for the design of coating systems for aerospace (and terrestrial) gas turbine engines (Hocking et al. 1989)

Mechanical		
Load or aggressive phenomenon	Maximum value (terrestrial)	Damage
Centrifugal stresses	170 MPa	Fatigue
Non-stationary stresses (gas T and P)	30 MPa (16 MPa)	Micro-cracking
Non-stationary stresses (gas flux)	600 m/s (300 m/s)	Coating removal, erosion
Shocks	–	Coating removal
Chemical		
Oxidizing atmosphere	–	Oxidation and corrosion, compositional changes, localized attack
Combustion products	Up to 0.3 wt% S	Corrosion, compositional changes
Air pollution	Marine salt, industrial pollution, fly ash, particulate matter	Corrosion, erosion, surface roughening
Thermal		
Exhaust gas	1400–2000 °C	Phase and microstructural transformations
Components	1000–1300 °C	Phase and microstructural transformations
Thermal gradients vs time	100 °C/s	Thermomechanical fatigue
Thermal gradients vs space	200 °C/mm	Thermomechanical fatigue

represent a prerequisite for the achievement of adequate safety and reliability levels. All coatings are deposited, or directly form, like diffusion coatings, on engine parts sometimes having rather complex shapes and surface topography. Nonetheless, the thickness and adhesion strength have to be uniform all over the coated surface. The same uniformity requirement can be extended to the main coating characteristics, like composition, microstructure, and internal stress fields. The deposition techniques herewith discussed are very well established and optimized to produce coatings that for gas turbine engines have thickness typically ranging from tens to hundreds of micrometers, i.e., much larger than the average thickness of thin films used in other fields, like microelectronics and optical and aesthetic coatings, just to mention a few major categories. The different operating conditions present in different parts of the engine require different coating systems and, consequently, different deposition techniques. The main ones used for gas turbine aero-engines belong to the following groups: thermal spraying deposition, chemical vapor deposition (CVD), and physical vapor deposition (PVD). In the next sections, the main features of the aforementioned techniques will be discussed with particular reference to the most important process parameters and resulting properties.

7.2.1 Thermal Spraying

Quite a large number of thermal deposition techniques are available, the most important and interesting for the aerospace field being (Fauchais et al. 2014):

- Flame spraying
- High-velocity oxy-fuel flame spraying
- Detonation gun and super detonation gun
- Electric arc spraying
- Plasma spraying

All these coating deposition techniques are characterized by a specific heat source, after which they are named, needed to melt the feedstock to be deposited. Combustion is the approach to generate heat in the first three techniques of the list, with differences in the fuel and carrier gases. The energy transferred by an electrical discharge is exploited by the electric arc deposition system, whereas in plasma spraying, the energy input to the feedstock powders comes from a high-temperature ionized gas. The droplets are produced by torches or guns and carried by a gas stream, sometimes comprising also an auxiliary carrier gas flow, towards the part to be coated and splat onto it. Depending on the deposition system, the feedstock can be in the form of powders, wires, or rods. In all cases, the complete deposition of a thermally sprayed coating requires the controlled movement of the spraying nozzle over the surface of the component with an appropriate overlapping in order to achieve the complete coverage of the surface. The temperature rise in thermal spraying techniques is meant to promote the melting of the source material in order to improve the coating formation process. In this regard, the bonding mechanisms, their control, and their optimization are very important indeed. Initially, the interaction occurs essentially between the substrate and the impinging particles. Subsequently, once the first layer of the coating has formed, the incoming particles will interact and try to adhere to this layer. Mechanical bonding is one of the main adhesion mechanisms, and surface roughness of the substrate and of the already formed layers is therefore paramount. Diffusion and chemical bonding are other possible interaction mechanisms (Fauchais et al. 2014).

The thermal spraying techniques are line-of-sight processes. Therefore, those parts only that can be directly reached by the particle jet will have the chance to be coated. To reach the final thickness, repeated passes are generally required, and this can be achieved by combining together controlled relative displacements and rotations of the "source" of material droplets/particles and the substrate. For the controlled movement of the spraying nozzle, depending on the complexity of the geometry and topography of the surface to be coated, options ranging from a plain x-y system up to more sophisticated robotic arms can be selected. At the same time, the part is also displaced and rotated relatively to the particle flux, in order to obtain a homogeneous coverage of the substrate and uniform thickness of the deposit. An optimum surface finishing of the as-deposited coating, typically obtained using sand blasting or mechanical machining, may have a very important effect on the coating

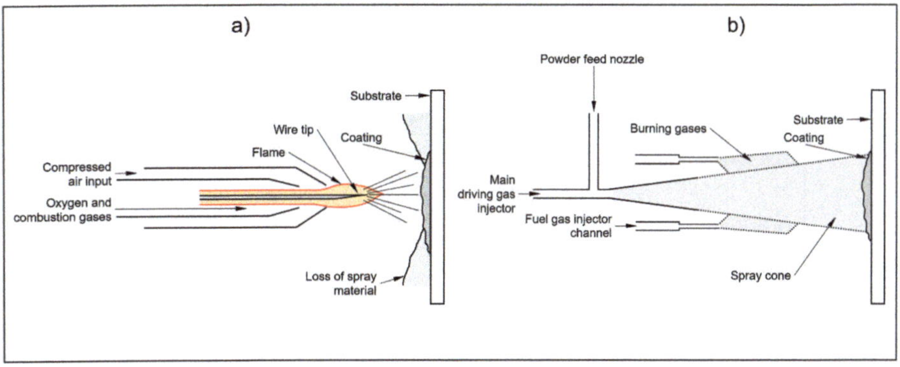

Fig. 7.2 Schematic of two flame spraying systems: (**a**) wire and (**b**) powder fed. (Redrawn from Hocking et al. 1989)

performances and durability. Of course, each deposition technique has specific features, important to be considered for the selection of the most suitable for each coating system.

7.2.1.1 Flame Spraying

It is regarded as a relatively cheap technique (Fig. 7.2). The torch is operated in air and employs an oxy-acetylene combustion gas mixture, capable to reach temperatures in excess of 2500 °C and flame velocities of 100 m/s or so. Methane and hydrogen can alternatively be used; the choice is made on the basis of the features of the material to be deposited and of the target properties of the coating (Hermanek 2001).

The feedstock material can be in the form of powders, rods, or wires. It is important to note that in case of rod and wire sources (Fig. 7.2a), the material is completely melted by the flame. This is not necessarily the case for powder particles (Fig. 7.2b) that are fed into the flame and can melt down only if they reach the hot part of the plume for a sufficiently long time. The main features of the flame spraying process are:

- Coatings are characterized by a relatively high porosity.
- Complex geometries can be easily coated, but not narrow channels, which cannot effectively be accessed by the material flow.
- Deposition can only involve materials that do not decompose during the melting or, in general, heating up.
- High deposition rate.
- Poor adhesion of the coating on the substrate.
- Relatively low cost of process and equipment.
- Low-temperature rise of the substrate.

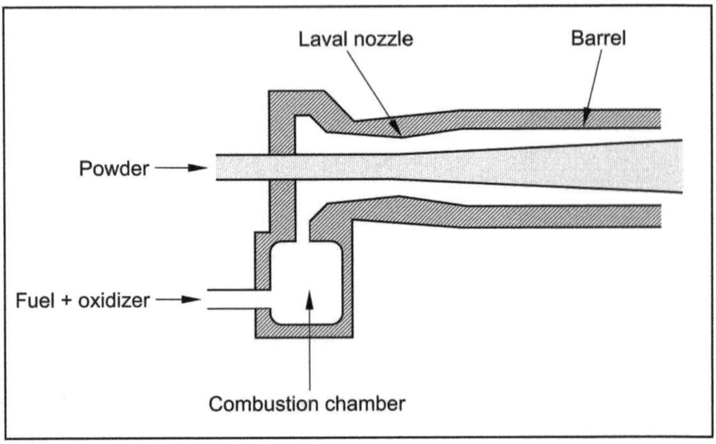

Fig. 7.3 Schematic of the HVOF system, based on the Jet Kote gun. (Redrawn from Thorpe and Richter 1992)

The main systems for which flame spraying is used are metal-ceramic (CERMET) mixed coatings and wear-resistant coatings, based on carbide phases, e.g., WC and Cr_3C_2, embedded into a metallic matrix, for which Co-based alloys have been a widespread and common choice, so far.

7.2.1.2 High-Velocity Oxy-fuel Flame Spraying

HVOF can be regarded as a more energetic version of the flame spraying process, as can be inferred also from the denomination. A higher velocity of the powder particles can be reached since the combustion is triggered inside a pressurized chamber (Fig. 7.3). As a consequence of this improved combustion, HVOF supersonic speeds can be reached, even as high as 2000 m/s, whereas the particle velocity in flame spraying is typically in the 100–200 m/s range and anyway never above the speed of sound. To sustain combustion, the HVOFs are fed with elevated flow rates of both fuel and oxygen, which are typically one order of magnitude higher than in flame spraying equipment. Thanks to the elevated velocity of the powder particles, high-density coatings can be obtained. As an alternative to powder, HVOF can also have a wire source that is progressively melted down by the hot gas flow, and the droplets are carried away towards the substrate.

7.2.1.3 Detonation Gun (D-Gun) and the Derived Super-Detonation Gun (Super D-Gun)

Both are based on controlled, alternate explosions of a mixture of fuel, generally acetylene, and oxygen, mixed with the powder of the coating material. This material

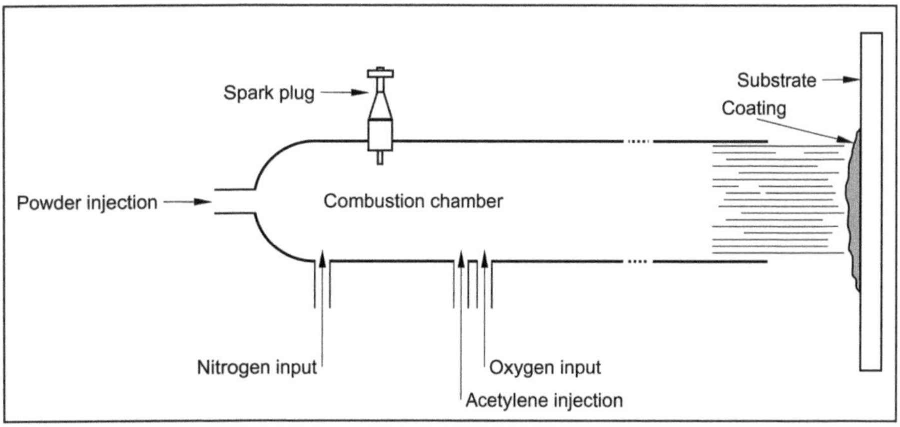

Fig. 7.4 Schematic of a detonation gun apparatus. (Redrawn from Hocking et al. 1989)

charge is thereby melted and ejected from the D-Gun barrel pointed toward the substrate (Fig. 7.4) (Hermanek 2001).

After each detonation, sparked by a plug located close to the bottom of the barrel, the inner chamber is purged with a nitrogen gas flux, to eliminate any combustion residue. A new filling of combustion gases and powders is prepared next, ready for another shot. The sequence continues with a frequency of 4–8 explosions per second, with particle velocities up to 3000 m/s, until the deposition is fully accomplished. Thanks to the high velocity of the material flux, extremely high compaction, good adhesion, and elevated densities, i.e., low residual porosity, are typical features of the coatings obtained with this technique. Other interesting features of the D-Gun products are good mechanical and tribological properties and presence of internal compressive stresses due to a sort of shot pinning effect induced by the particles arriving and hitting the already deposited layers. An appropriate setting of the process parameters allows for an optimum tuning of the temperature and deposition rate, in order to avoid overheating and unwanted phase transformations to occur within the coating.

7.2.1.4 Electric Arc Spraying

It is a suitable deposition technique to obtain electrically conductive coatings. The source features two wires, also acting as consumable electrodes, between which an arc is struck to melt the coating material (Fig. 7.5). The molten material reaches the substrate carried by a flux either of compressed air or, in case of particularly reactive alloys, inert gas. The gas velocities and consequently the flux of molten particles can reach comparable velocities as those of a flame spray torch, i.e., 100–200 m/s. Therefore, the deposited coatings will generally feature a relatively high porosity and may contain oxide grains, although the complete melting of the feeding

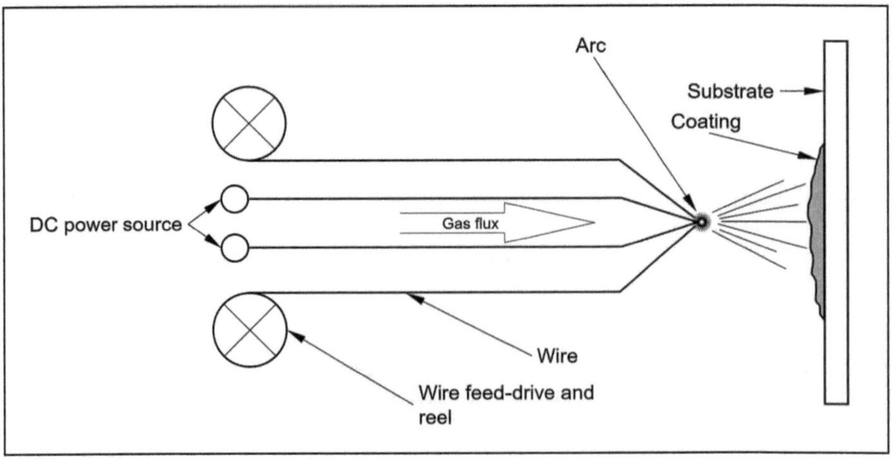

Fig. 7.5 Main parts of an electric arc spraying technique (schematic). (Redrawn from Hocking et al. 1989)

materials results in better adhesion and higher deposition rates than in the case of flame spraying, notwithstanding a certain fraction of metallic droplets may solidify during the flight. The carrier gas is mostly unaffected by the arc heating, and consequently it does not heat up the substrate. For this reason, electric arc spraying can be used also to deposit coatings on low melting or heat-sensitive substrate materials, for instance, polymers.

7.2.1.5 Plasma Spraying

It is a coating technique that uses a plasma torch (Fig. 7.6) as heat source. As for the previous technique, an arc is also present in plasma spraying. The arc is not meant to melt directly the material to be deposited but to ionize a plasma gas, in which the particles of the coating material are injected, melted, and carried to the workpiece. Several plasma gas mixtures are used: Ar, Ar–H_2, Ar–He, Ar–He–H_2, N_2, and N_2–H_2, just to mention the most common ones. In the design and selection of these mixtures, two main parameters are considered: a primary heavy gas (Ar or N_2) carrier, to assure an adequate flow and particle entrainment, and a secondary gas, like H_2 and He, which is generally introduced to improve the heat transfer. In the case of metallic powders, the above gas selections are also very effective in preventing oxidation. At 1 atm pressure, a stable plasma torch is attained for temperatures in excess of 7500 °C approximately. Under particular operating conditions, temperatures as high as 20,000 °C can be achieved, so that the technique is potentially capable to melt any kind of powder, once it is injected into the plasma-flame dart. This allows different types of coatings to be deposited: metallic, particularly wear- and corrosion-resistant coatings, ceramic coatings, cermets, and thermal barrier coatings.

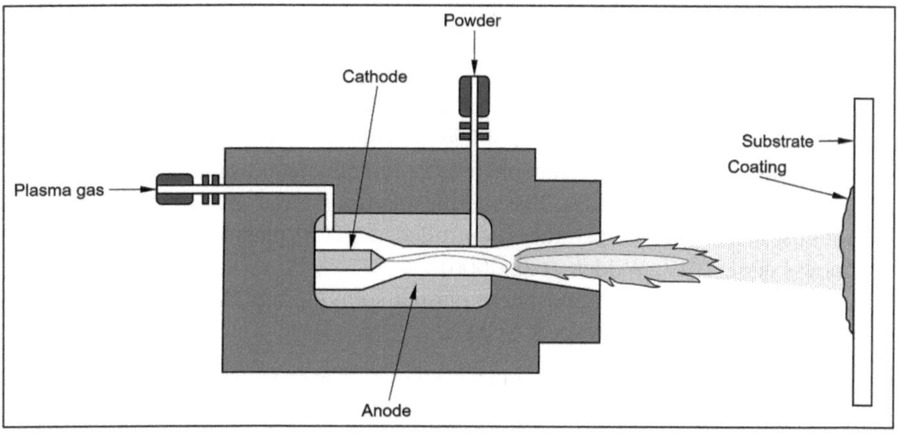

Fig. 7.6 Main parts of a plasma torch used in plasma spraying coating deposition. (Redrawn from Reed 2006)

Plasma spraying coatings are used in many technological fields: furnaces for thermal treatments, petrochemical plants, biomedical applications, ground-based gas turbine plants, and aero-engines (Clarke et al. 2012; Driver 2012; Heimann 2006). Thermal barrier coatings (TBCs), originally developed for increasing the ignition temperature of diesel engines of heavy duty vehicles, have been paramount for improving the efficiency of terrestrial and aircraft gas turbine engines (see Sect. 7.3.2).

Depending on the deposition environment, plasma spraying (PS) process can be classified as atmospheric (or air) PS (APS) and vacuum PS (VPS), alternatively defined as low-pressure PS (LPPS). APS is by far the most used technique. Deposition is carried out in air at atmospheric pressure inside a chamber, mainly meant to reduce the environmental noise and to contain the dispersion of the airborne powders. The main reasons why powder particles may not reach the target or not get embedded into the forming coating can be either the unwanted escape from the plasma dart or their bouncing off after hitting the surface of the substrate target. According to the torch conditions, for any specific powder, the deposition rate can be optimized in order to control and improve the features of the coating. In general, the high melting efficiency that is achieved with the plasma torches coupled with a relatively high particle velocity that can reach the supersonic limit of 300 m/s results in the formation of coatings that are relatively dense and well adherent to the substrate. However, notwithstanding the inert or slightly reducing plasma gas, the presence of spurious oxides in a metallic coating deposited via APS cannot be fully avoided for the interaction of the molten alloy powders with the air in the atmosphere. Even to prevent this effect, the VPS has been implemented. In this case, the whole process occurs inside a sealed chamber connected to a vacuum system, typically a mechanical roughing pump. Once the limiting pressure has been achieved, the deposition starts. Alternatively, the vacuum chamber is backfilled with a low pressure (50–250 mbar) of inert gas, i.e., argon, which still assures cleaner and more controlled conditions than in APS. Thanks to the reduced gas pressure used in VPS, the plasma flame can be longer, and higher particle velocity (500–800 m/s) and higher deposition efficiency can be attained. This latter aspect is favored also by a reduced fraction of losses in the sprayed powder. The possibility of controlling the processing atmosphere is a prerequisite to implement, if available, additional operating PS modes, like the plasma-transferred arc (PTA) process and the atmosphere and temperature controlled (ATC) plasma spraying process (Firdaouss et al. 2017; Wilden et al. 2006).

Vacuum and controlled atmosphere plasma spraying are characterized by some important advantages with respect to APS:

- Possibility to deposit refractory and hard metals. These materials would otherwise tend to oxidize, and vacuum avoids or, at least, limits this effect.
- In VPS, including PTA, the substrate may be heated before starting the deposition, with limited risks of oxidizing it. This operation is useful both for removing any impurity from the substrate surface and to induce compressive stress fields into the coating.

On the other hand, VPS is more expensive, owing to the higher costs of the spraying equipment and running expenses, considering that each deposition requires high volumes of inert gas. Another important restriction to the use of VPS is the limited possibilities to move the component inside the vacuum chamber, although these latter aspects have been profitably managed with the introduction of rotating stages and robotic arms, used to move the plasma torch on the sample surface in a controlled way.

7.2.2 Physical Vapor Deposition

Quite a large number of deposition techniques are based on physical methods to produce the vapor phase from which the deposit, thin film, coating, and multilayer systems, will form (Bose 2007). The condensation of the coatings from the vapor phase requires an apparatus in which it is possible to adjust and control the working atmosphere, as concerns its pressure and composition, during the process. This in general requires a vacuum system and connections to pure gas pipelines. In this way, all needed procedures to remove unwanted gaseous impurities can be accomplished in order to control better all the reactions accompanying the deposition. In the field of coatings for gas turbine applications, electron beam physical vapor deposition (EB-PVD) is certainly the most important technique. EB-PVD derives from the electron beam melting technology (see also Sect. 4.4). The use of EB-PVD for large-scale productions began in the early 1980s, especially thanks to the improvements attained in electron guns, relevant performances, advances in the control, and vacuum systems. In the EB-PVD apparatus, focused high-energy electron beams, generated by several electron guns, are used to evaporate the ingots made of the coating material, and also for preheating the substrate, e.g., turbine blades, inside the

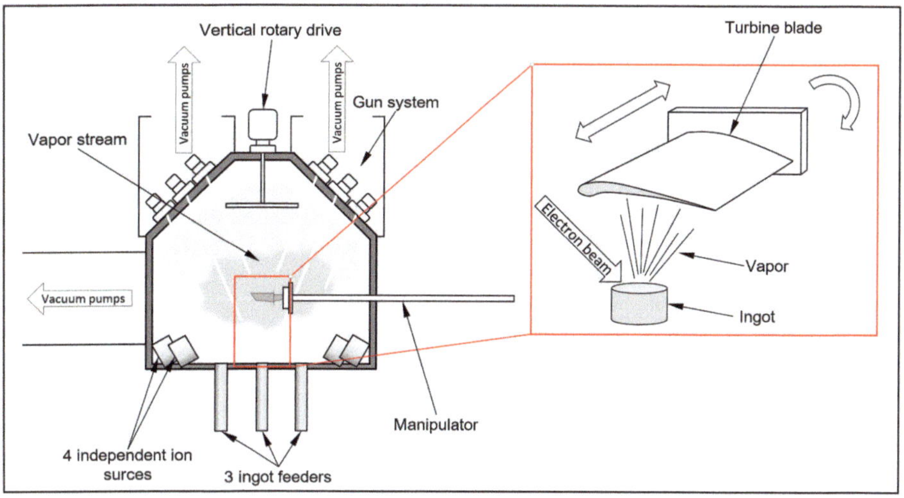

Fig. 7.7 Scheme of an EB-PVD plant with details of the vapor source and component movements during deposition. (Redrawn from Wolfe et al. 2005)

vacuum chamber. This step is important to promote the formation of the so-called thermally grown oxide (TGO, see Sect. 7.3.2) on the surface of the bond coat already present on the blades, thanks to a controlled amount of pure oxygen added to the working atmosphere. In the setup in Fig. 7.7, six electron guns are present (45 kW each). More powerful units, with gun power in excess of 100 kW, are also available, affording high deposition rates, in the range of 5–10 μm/min.

Although completely different as concerns the working principles from thermal deposition still, also in EB-PVD the coating material reaches the substrate following straight trajectories from the source. Therefore, the component needs to be rotated and moved appropriately with respect to the vapor source (Fig. 7.7), in order to obtain a homogeneous coverage, in terms of thickness, microstructure, and composition.

Early experimental works were conducted to deposit overlay coatings (Sect. 7.3.1), with a composition based on the FeCrAlY system with 10–15% aluminum and 125 μm thickness (Talboom and Grafwallner 1970). Although not particularly successful, for the occurrence of too fast interdiffusion reactions at the interface with the Ni-based superalloy substrate, the results of this research demonstrated the feasibility of using protective coatings having a composition independent from the substrate. Compositions more suitable to protect Ni- and Co-based superalloys against oxidation and hot corrosion were successively developed. Nickel, cobalt, or a combination of the two became the major constituents, together with Cr, Al, and Y, this latter as *reactive element*. Other reactive elements, added in concentrations usually not exceeding 1 wt% approx., are Si, Hf, Zr, and Ta (see Table 7.2). They improve the adhesion of the surface oxide scale, particularly under cyclic oxidation conditions (Gupta and Duvall 1984). The control of the concentration of these elements is crucial, and in this respect the limits of EB-PVD emerged. In fact, a constant quantity of reactive elements in the vapor phase during the

Table 7.2 Compositions of some overlay coatings (Nicoll et al. 1986; Reed 2006)

Overlay coating type	Ni	Co	Cr	Al	Y	Ti	Si	Hf	Others
NiCrAlY	Bal.	–	25	6	0.4	–	–	–	–
NiCrAlY	Bal.	–	22	10	1.0	–	–	–	–
NiCrAlY	Bal.	–	31	11	0.6	–	–	–	–
NiCrAlY	Bal.	–	35	6	0.5	–	–	–	–
CoNiCrAlY	32	Bal.	21	8	0.5	–	–	–	–
CoCrAlY	–	Bal.	25	14	0.5	–	–	–	–
NiCoCrAlTaY	Bal.	23	20	8.5	0.6	–	–	–	Ta = 4
NiCoCrAlYSi	Bal.	0–40	12.5–20	2–8	0–0.25	0–10	2–10	–	Fe = 0–20
									Mn = 0–5
									Mo = 0–5
									Nb = 0–4
NiCrAlTi	Bal.	–	30–40	1–10	–	1–5	–	–	–
NiCoCrAlHf	Bal.	0–40	10–40	6–25	–	–	–	0–10	–

deposition run is not easy to maintain, particularly with a sufficient degree of reproducibility, owing to the comparatively low vapor pressure of these elements. This critical aspect is still there and limits the possibility of using EB-PVD for overlay coatings (Gupta and Duvall 1984). The contemporary development of vacuum plasma spraying (VPS) technology established this technique as preferred for the deposition of overlay coatings. Considering that the compositions of overlay coatings and, thereby, the technological aspects related to their deposition are rather similar to those of bond coats in TBCs, the VPS approach became the standard choice for them too. Still concerning TBCs, EB-PVD introduced a major breakthrough in the deposition of the ceramic, zirconia-based top coat (Duvall 1981; Anderson and Sheffler 1983). Once similar reproducibility issues, like those faced for MCrAlY coatings, were settled, the columnar grain morphology of the zirconia layers and the associated, remarkable improvement in their durability established TBCs deposited using EB-PVD as the best choice for the protection of vanes and blades of the gas turbine high-pressure stages (Sect. 7.3.2) (Lammermann and Kienel 1991; Meier et al. 1990; Strangman 1987).

7.2.3 *Chemical Vapor Deposition*

Still vapor, but originated from chemical reactions, is used in the so-called chemical vapor deposition (CVD) methods. In the field of gas turbine aero-engines, most of the coatings obtained from CVD techniques are diffusion coatings, interesting for the protection of gas turbine parts against corrosion at high temperature and bond coats in TBCs. Indeed, this specific application triggered a renewed interest for diffusion coatings that, on the other hand, as anti-corrosion barriers, have largely been substituted by overlay coatings (Strangman 1996).

Two main kinds of CVD techniques are used in gas turbine applications: pack-cementation and out-of-contact processes (Goward 1998). Actually, the early use of diffusion coatings via pack cementation was not intended for gas turbines but for enhancing the oxidation resistance of heating elements made of iron wires, copper steam condenser tubes for power plants, steel furnace fixtures, and nickel combustion screens. In all cases, it was clear that the improved oxidation resistance resulted from the preferential formation of a protective alumina scale. On the basis of this know-how, the main technological aspects of the CVD processes have been extended to the gas turbine engines, including those for aerospace applications. The main components of the pack mixture used for this process are the source of the metal to be deposited, e.g., aluminum powder; the so-called activator components, e.g., NaF, NaCl, and NH_4Cl; and an inert filler, like alumina powder, whose main role is to control the kinetics of the reaction and avoid sintering of the metal source.

The main reactions, occurring during the aluminizing of base alloy components, can be summarized as follows: ammonium chloride (or one of the alternatives) acts as a reaction activator gas, and its decomposition determines the availability of hydrochloric acid that reacts with aluminum to form aluminum chloride, $AlCl_3$, according to reactions described by Eq. 7.1:

$$3HCl + Al \rightarrow AlCl_3 + \frac{3}{2}H_2 \tag{7.1}$$

Aluminum chloride vapor condensates at temperatures of 600–650 °C on the surface of the alloy components, which becomes enriched with aluminum, whereas HCl is released as exhaust, and it is eliminated from the reaction chamber (Eq. 7.2):

$$AlCl_3 + Ni \rightarrow Ni - Al + HCl \tag{7.2}$$

The phase composition and the microstructure of the coating that will form on the alloy depend on the interdiffusion kinetics and, consequently, on the aluminum activity. For high Al activities, the Ni_2Al_3 intermetallic compound would preferentially form, with a prevailing inward diffusion of aluminum. This regime is usually identified as low-temperature high activity (LTHA), to indicate also the typical temperature range for the diffusion treatment, carried out at 870 °C for 20 h or so. Alternative to this is the high-temperature low activity regime (HTLA, see also Sect. 7.3.1), during which significantly higher temperatures are employed. The diffusion treatment can be conducted at 1100 °C for 1 h, followed by an intermediate temperature step at 875 °C for 16 h (Reed 2006). In this case, the prevailing formation of NiAl is observed. In Sect. 7.3.1, the specific features of the two kinds of diffusion coatings are discussed with regard to their oxidation protection properties. In the present context, it is interesting to note that the compositions of diffusion coatings are strictly dependent on the composition of the base alloy. Specific coating systems have been developed for other than nickel-based superalloy. For instance, diffusion coatings based on the FeAl and CoAl intermetallic phases are used for the protection of Fe- and Co-based superalloys, respectively (Naji et al. 2015; Pint et al. 2001; Priest and Zhang 2015). Typical thicknesses of aluminide coatings range from 30 to 100 μm, selected on the basis of the composition, component to be protected, relevant service conditions, and expected lifetime.

The development of diffusion coatings for hot parts of gas turbine aero-engines, like inner walls of the combustors, vanes, and blades, was pursued since the early times. In fact, the first aircraft components using diffusion coatings were the inner walls of the combustion chambers made of low-alloy steel and, most probably, also the blades, made of the same steel, of the Messerschmitt ME 262 fighter aircraft that started its service in 1944 (Goward 1998). Pack-cementation aluminizing was used to protect gas turbine vanes made of cobalt-based superalloy (see Sect. 6.2) in 1957 for the first time. The process was subsequently extended to turbine blades made of nickel-based superalloys (see Sect. 6.4). Actually, in the early reported examples, no pack-cementation but hot-dip (Seelig and Steuber 1978) and slurry-fusion (Joseph 1960) processes were used. Their working principles have some affinity to pack cementation, considering that the surface metal enrichment occurs by dipping the components into hot molten metal and into a slurry containing metal powders and halides suspended in an organic binder, which is subsequently evaporated away. Diffusional thermal treatments are carried out afterward.

Starting from the early 1970s, the standard choice became aluminizing vanes and blades using pack-cementation processes (Gauje and Morbioli 1983; Parzukowski 1977; Warnes and Punola 1997). With the development of hollow blades, a limit of the technique emerged. Although aluminum is deposited onto the alloy surface through the condensation of gaseous $AlCl_3$, for the presence of the pack, this condensation and, thereby, aluminum surface enrichment occur preferentially at the contact region between the pack and the component, so that the inner parts and cooling channels of the blades remain substantially unprotected, since no aluminum surface enrichment takes place during the aluminizing treatment (Smith et al. 1999; Squillace et al. 1999). To tackle this problem, several approaches were considered: slurry aluminizing, vacuum pulse aluminizing, and finally forced flow gas phase aluminizing, which can be regarded as a precursor of the out-of-contact processes (Galmiche 1975; Restall and Wood 1986; Warnes and Punola 1997). The reactors illustrated in Fig. 7.8 rely on two slightly different working principles, as concerns the way in which aluminum is added to form the active gas phase, $AlCl_3$. In Fig. 7.8a $AlCl_3$ develops internally to the reactor, from the reaction of the gas phase with the source material. In the alternative approach illustrated by Fig. 7.8b ("above pack"), $AlCl_3$ is formed outside the reactor and can be introduced into the reaction chamber under more controlled flux rate and temperature. Surface aluminum enrichment and

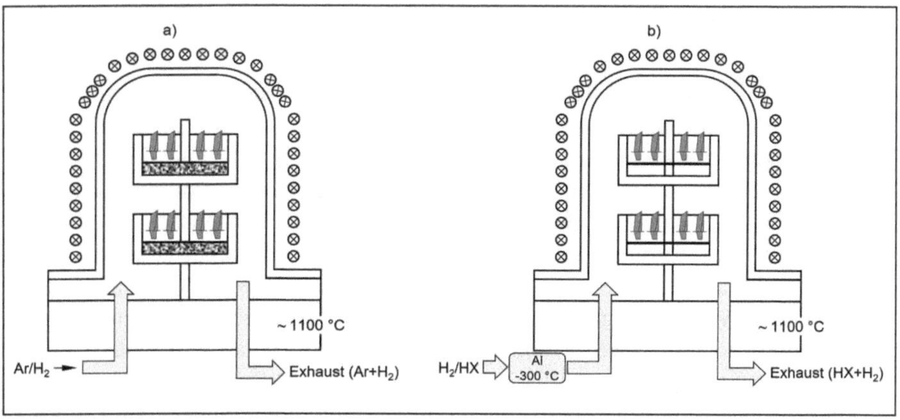

Fig. 7.8 Scheme of two alternative out-of-contact aluminizing reactors: (**a**) the vapor flux is enriched in aluminum inside the reaction chamber for the reduction of alumina powder and (**b**) the gas phase entering into the reactor chamber contains the aluminum precursor already. (Redrawn from Reed 2006)

subsequent diffusion into the alloy are the main steps to produce aluminide-based coatings capable to improve the high-temperature corrosion resistance of the underlying alloy. Several secondary additions have been considered in order to enhance the protective performances of the coating and its durability. As discussed in details in Sect. 7.3.1, these compositional changes have been pursued using different approaches. Controlled amounts of alloying elements, like silicon and chromium, have been added to the vapor source. In case of platinum-containing or platinum aluminide coatings, featuring remarkably long lifetimes, the aluminizing CVD step is preceded by the deposition of a few micrometer thick platinum layers, which take part to the diffusion reactions. Other coating systems have been developed using different diffusing elements. CVD surface treatments fully based on silicon and chromium have also been explored (Goward 1998). Although out of the scope of the present textbook, alternative applications, like coal gasification and future generation nuclear power plants, emerged as interesting for these coating systems.

7.3 Main Coating Systems in Gas Turbine Engines

After having considered the main functional aspects of coating systems in gas turbine engines, both for aircraft and terrestrial applications (Sect. 7.1), the main deposition techniques have been outlined in Sect. 7.2. In this last section of Chap. 7, the main coating systems in gas turbine aero-engines are considered.

7.3.1 Coatings Resistant to High-Temperature Corrosion

A complete discussion on the main phenomenological aspects concerning high-temperature corrosion, i.e., oxidation and hot corrosion, can be found in Chap. 8. A general approach adopted to improve the resistance of gas turbine components to these attacks relies on promoting the formation of a dense, adherent, and continuous protective layer of native oxides and other secondary reaction phases, like nitrides. These layers may form on the base alloys directly or on the protective coatings present on the alloy surface. Diffusion and overlay coatings are the main systems used for improving the high-temperature corrosion resistance of aero-engine parts, with compositions targeted at obtaining the formation of an oxide scale rich in alumina, in particular the α-Al_2O_3 polymorph. Diffusion coatings (Sect. 7.2.3) for nickel-based superalloys are based on two nickel aluminide phases, either δ-Ni_2Al_3 or β-NiAl.

Both phases are rich in aluminum, with concentrations significantly higher than those of nickel-based superalloys. Therefore, they can provide an adequate outward aluminum flux for the formation of protective layer rich in α-Al_2O_3. Moreover, the ordered crystalline structures of these intermetallic phases (see Sect. 9.4) slow down all diffusive phenomena involved with oxidation, so to reduce the relevant reaction kinetics. The Al-Ni phase diagram in Fig. 7.9 shows further important aspects, explaining the reason for the success of nickel aluminide diffusion coatings, with particular regard to the β-NiAl phase. This intermetallic phase is named "permanently ordered" since it retains an elevated degree of long-range order (see Sect. 9.4) up to the melting point, with no solid-state order-disorder transformation that in this case would be a second order one. The higher stability temperature and the wider compositional range make β-NiAl definitely preferable to δ-Ni_2Al_3. Even the larger aluminum content in the δ-Ni_2Al_3, that can be regarded as a plus in some respects, on the other hand, determines higher interdiffusion kinetics with the underlying nickel-based superalloy, having a different composition and, particularly, a significantly lower aluminum concentration than the coating itself. Together with the compositional changes resulting from the formation of the outer oxide scale, the coating-substrate interdiffusion is indeed another major lifetime limiting factor for the coating. Therefore, β-NiAl can be regarded in several respects as the preferable choice as concerns nickel-based superalloys. Similarly, for alloy systems based on iron, e.g., steels and iron-based superalloys, and cobalt, e.g., cobalt-based superalloys, overlay coating based on the FeAl and CoAl intermetallic phases have been developed. Again, the selection of iron and cobalt aluminides is suggested by their compositional affinity to the relevant substrate alloys (Lee and Kuo 2005; Pint and Zhang 2011; Ren et al. 2014; Zhang et al. 2004, 2007).

As concerns β-NiAl-based coatings, compositional changes have been introduced to enhance the coating resistance to high-temperature corrosion, also under cyclic oxidation conditions, which better represents real service requirements. Significant improvements in corrosion resistance performances have been achieved in IN100 and AISI 304 austenitic stainless steel, using mixed aluminum and chromium

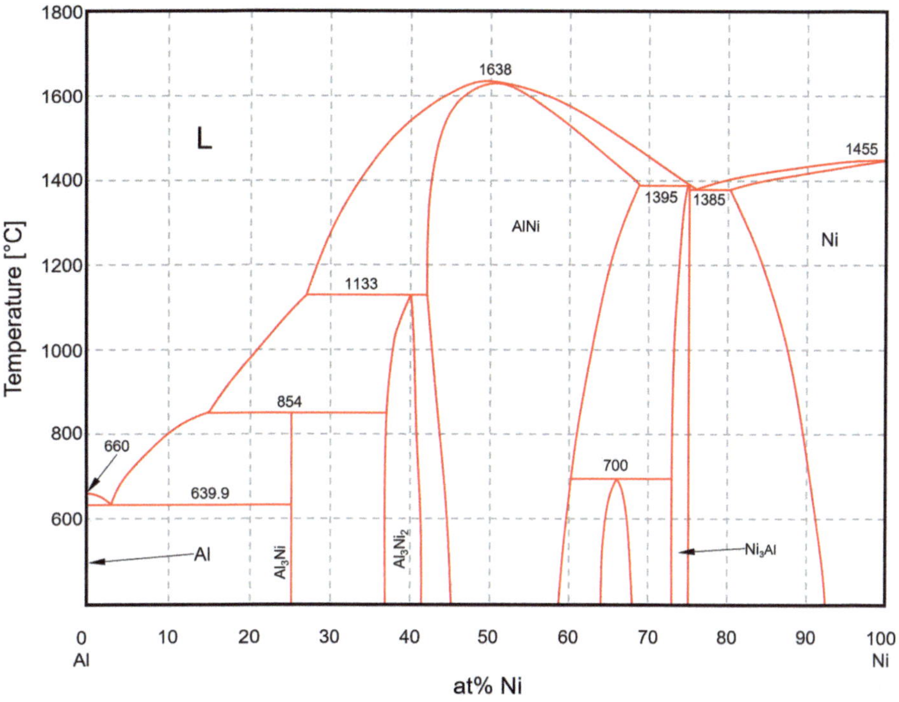

Fig. 7.9 Binary Al-Ni phase diagram

diffusion coatings (Pérez et al. 1999a, b). Chromium additions can also be attained by preparing the superalloy surface with a chromizing treatment preceding aluminizing (Luthra and LeBlanc 1987; Mévrel et al. 1986). Thus, the interdiffusion treatment, needed to promote the formation of the coating, involves both elements, resulting in a mixed composition coating. The advantage of having both aluminum and chromium as oxidation active elements is based on the reduction of the minimum concentration of aluminum necessary to the formation of a protective outer alumina layer. This same approach is exploited even more effectively in MCrAlY overlay coatings (*vide infra*). Chromium has also been considered in aluminide coatings in association with silicon (Nicholls 2003). Silicon can be co-deposited with aluminum, using a mixed slurry method, involving the application onto the component surface of a uniform layer of a slurry containing both elements. The coating forms afterward, during the diffusion heat treatment, carried out in an inert or slightly reducing atmosphere. In this regard, the Sermaloy™ coatings should be mentioned, like Sermaloy J and Sermaloy 1515. Their excellent behavior can be ascribed to the co-presence of silicon and chromium, this latter coming from the base alloy, inducing the formation of the $CrSi_2$ intermetallic. The microstructural refinement is very important too. In the Sermaloy 1515 coating, the alternate surface layers of NiAl and $CrSi_2$ are both acting as reservoirs of oxidation and corrosion active elements, i.e., Cr and Si, and thus they guarantee a continuous outward flux of these elements to form a protective surface oxide scale (Berry et al. 1995; Van Roode and Hsu 1989). Although these coatings have been successfully applied also to single-crystal superalloy blades, chromium- and silicon-alloyed coatings have primarily been investigated and are still used for improving the aluminide coating protection performances, when corrosive agents are present in the working environments (Meelu et al. 1992; Prater et al. 1981). This is the case of land-based and marine turbines, burning not very refined and pure fuels, particularly as concerns the sulfur content (Luthra and LeBlanc 1987). Silicon-alloyed coatings, featuring up to 10 wt% of this element, provide particularly good hot corrosion, both Type I and Type II, resistance (see Sect. 8.2.5) (Berry et al. 1995; Van Roode and Hsu 1989). The most successful and widely used addition to nickel aluminide coatings is platinum, a noble metal whose application to protect engine components is justified by their high added value (Allam et al. 1979; Felten 1976; Felten and Pettit 1976; Fountain et al. 1976). This element improves the oxidation resistance of nickel aluminide coatings, influencing positively some of their critical aspects. In the first place, platinum additions result in the reduction of the coating-alloy substrate interdiffusion, due to the different compositions of nickel aluminide coatings and superalloy substrates. A strong driving force is there for an inward diffusion of aluminum, which may be rather fast at elevated service temperatures. For the same reason, the outward diffusion of minority alloying elements, like titanium, niobium, tantalum, etc., is slowed down too. The consequent phase and compositional changes experienced by the coating determine eventually the loss of its protective action. Moreover, the progressive depletion of aluminum, so that precipitates of the γ'-Ni_3Al and of the martensitic polymorph of β-NiAl start to form, renders the coating particularly brittle. Concerning the mechanical stability of the coating, platinum alloying turned out to reduce also the formation of voids from vacancy coalescence at the interface with the alloy substrate. This feature turns out to be particularly important to resist better oxidation under cyclic conditions, when the interplay of mechanical stresses and stress concentrator defects may result in a premature failure, i.e., cracking and detachment, of the coating. The diffusion control operated by the platinum additions has also a positive effect on the quality of the oxide scale, that results to be made of particularly pure, compact, and adherent α-Al_2O_3, the most preferred oxide phase in this field. Platinum alloying is obtained starting with electroplating the engine components with a layer 5–10 μm thick of this element. Alternatively, the platinum layer can be deposited using EB-PVD, a choice to be made in case a particular control on composition and microstructure is required. A diffusion heat treatment, performed under inert atmosphere, e.g., argon, for several hours at about 1000 °C is conducted before the CVD aluminizing process. Depending on the overall deposition procedure, three main kinds of coatings have been reported: single platinum aluminide phase (generally $PtAl_2$) coating; two-phase coating, featuring both β-NiAl and $PtAl_2$; and single-phase coating, i.e., (Pt,Ni)Al, with platinum atoms substitutional in the β-NiAl lattice (Bose 2007; Smith and Boone 1990). The development of advanced aluminide diffusion coatings resulted in a significant increase in the component lifetime. These coatings received a particular boost to meet the requirements posed by the development of novel single-crystal nickel-based superalloys

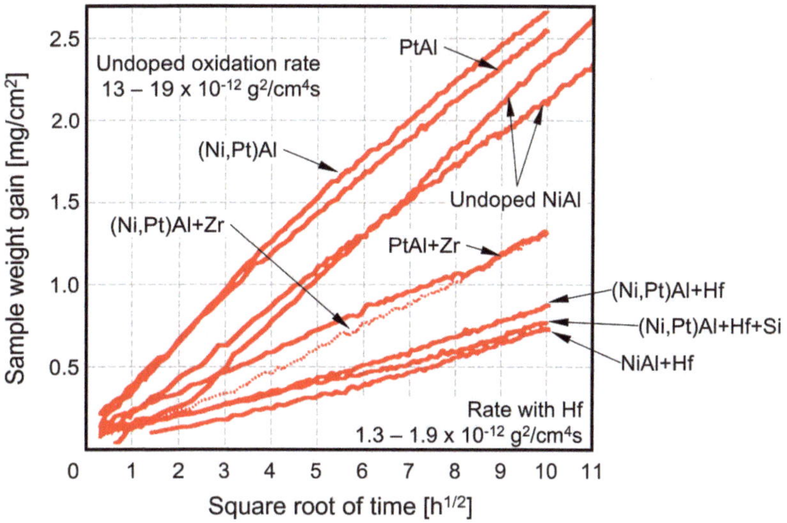

Fig. 7.10 Isothermal weight evolution of several aluminide coatings oxidized in air at 1200 °C. (Redrawn from Pint et al. 1998)

(see Sect. 6.5), oxide dispersion strengthened (ODS, see Sect. 9.3) alloys, and relevant rise of the turbine entry temperature (TET, see Sects. 2.2 and 2.5) (Giggins and Pettit 1979; Goward 1998; Nicholls 2000; Pint 1997). With particular regard to advanced superalloys, the role of protective coatings became critical, once it became clear that the novel compositions, mainly for the reduction of chromium content and overestimated beneficial effects of refractory elements, like rhenium, exhibited a lower intrinsic resistance to high-temperature corrosion (Pint et al. 2006).

Several compositions were explored to optimize all contributions to the coating durability. For instance, as concerns a better adhesion of the alumina scale, Fig. 7.10 shows the evolution of the parabolic kinetic constant (see Sect. 8.2.1, Wagner's model) for the oxidation at 1200 °C in air of several NiAl-Pt diffusion coatings with different additions of Zr, Hf, and Si (Pint et al. 1998). Incidentally, the beneficial effects on the oxidation resistance of the coatings, of reactive elements, like hafnium and zirconium, are similar to the effects that these same and other reactive elements introduce in MCrAlY overlay coatings (see Sect. 8.2.2).

An alternative approach to control the degradation kinetics of diffusion coatings is based on the reduction of their interdiffusion with the underlying alloy. The negative effect of this phenomenon depends not only on the aluminum depletion in the coating but also on the possibility that the compositional changes occurring in the so-called secondary reaction zone (SRZ), forming in the alloy subsurface regions, may result in the formation of TCP phases (see Sect. 6.4.2). In this respect, the positive influence of other alloying elements, like ruthenium, added in a similar way as platinum to the NiAl coatings, can be cited (Tyron et al. 2006, 2007; Wang et al. 2011). Major improvements of the coating lifetime have been attained thanks to the slowing down of the diffusive phenomena occurring in the SRZ: inward diffusion of aluminum from the coating and outward diffusion of refractory elements, like tungsten, molybdenum, and tantalum, from the substrate.

The research efforts made to improve the durability of the diffusion coatings, through the optimization of their critical parameters, have also been exploited for the development of high-performing bond coats in TBCs (see Sect. 7.3.2), based on these very same systems. Nonetheless, diffusion coatings have been progressively, although not completely, replaced by the MCrAlX overlay coatings, as protective coatings against high-temperature corrosion phenomena (Sect. 8.2). One of the main reasons for the success of the overlay coatings can be traced to the freedom of designing their composition independently from the composition of the base alloy. This is particularly interesting for a better control of the optimal concentration of reactive elements, like yttrium, hafnium, zirconium, etc., whose central role in enhancing the corrosion protection capabilities is discussed in the following. Differently from diffusion coatings, based on the intrinsically brittle ordered nickel aluminide phases (β-NiAl and δ-Ni_2Al_3), overlay coatings exhibit their ductile to brittle transition at temperatures (DBTT) below the operational range of the gas turbine engines. This is shown by Fig. 7.11, in which the dependence on temperature of the strain for crack initiation of several diffusion and overlay coatings is shown (Hocking et al. 1989). An alternative approach, based on the measure of the room- and high-temperature hardness of the coating, using especially designed apparatus, provides similar results (Dryepondt and Pint 2010).

The brittleness and the low fracture toughness of the aluminide coatings represent a potential factor of risk for the whole component. In fact, cracks that may form on the coating surface as a consequence of the thermal mismatch with the substrate or

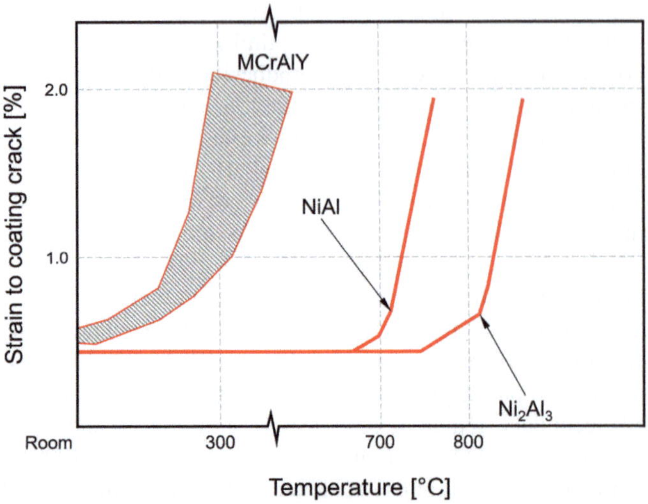

Fig. 7.11 Ductile to brittle transition temperature (DBTT) for diffusion coatings, based on the NiAl and Ni_2Al_3 intermetallics and MCrAlY overlay coatings. (Redrawn from Hocking et al. 1989)

erosion impact with particulate matter carried by the hot gas flow may propagate to the underlying superalloy. To reduce this possibility, in addition to improving the composition and microstructure of the coating to enhance its fracture toughness, another approach was to reduce the thickness and, thereby, the maximum length of the cracks that may form. In this way, the probability to have a spontaneous propagation to the substrate was automatically reduced.

An alternative approach to overcome the limits of the intrinsically brittle β-NiAl intermetallic is its substitution with tougher systems. This is precisely the route followed in the development of the MCrAlX overlay coatings, described in the following. As a sort of bridging step between diffusion and overlay coatings, it is interesting to mention an explorative study conducted on two-phase diffusion coatings based on the γ and γ' austenitic phases (Pint 2004). Although largely replaced by overlay coatings, diffusion coatings are still having an important role as bond coats for TBCs (Sect. 7.3.2). For this application, aluminide coatings are not directly exposed to the hot gases flux, since protected by the ceramic top coat. Moreover, still thanks to the presence of the top coat, lower temperatures and temperature gradients are there in the aluminide coatings, thus reducing the relevant thermomechanical stresses and the associated risk of crack formation. The situation is different for overlay coatings that, according to the Fig. 7.11, are reasonably more ductile and, therefore, tougher in-service conditions. The general composition of the overlay coatings can be expressed as MCrAlX, where "M" can be either nickel, cobalt, iron, or a combination of two of these, selected on the basis of the specific applications. "X" refers to the already introduced reactive elements, like Y, Hf, Zr, Si, and Ta, present in concentrations typically not exceeding 1 wt%. Reactive elements optimize the corrosion protection action of the coating through the improvement of the character and adhesion of the surface oxide scale. In Sect. 8.2.2, regarding the main high-temperature corrosion mechanisms, the way in which reactive elements may improve the surface protection will be illustrated. It will be also made clear why the presence of both aluminum and chromium is recommended, although alumina is still the preferred oxide in order to have a fully protective scale (See Sect. 8.2.2). In Table 7.2 the compositions of some commercial overlay coatings are listed (Nicoll et al. 1986; Reed 2006). Compositions are designed so to have austenitic γ-γ' matrix alloy, in which β-NiAl grains are dispersed and act as reservoirs for the aluminum needed to sustain the outward diffusive flux to form and subsequently mend, in case of spallation and detachment, the protective oxide scale.

The preferred deposition system for these coatings is plasma spraying (see Sect. 7.2.1), although, for standard applications, HVOF can still provide excellent results. Both thermal spraying processes are suitable for controlling better the concentrations of minority elements, i.e., the reactive elements, which have a fundamental role in determining the actual protective properties of the coating. In this regard, Fig. 7.12 shows the progressive degradation behavior, involving several damage stages, observed in overlay coatings as a function of the exposure time to a burner rig (see Sect. 8.2.6) cyclic oxidation conditions at a temperature of 1149 °C. Together with a reference composition, NiCoCrAlY, deposited using both EB-PVD and VPS,

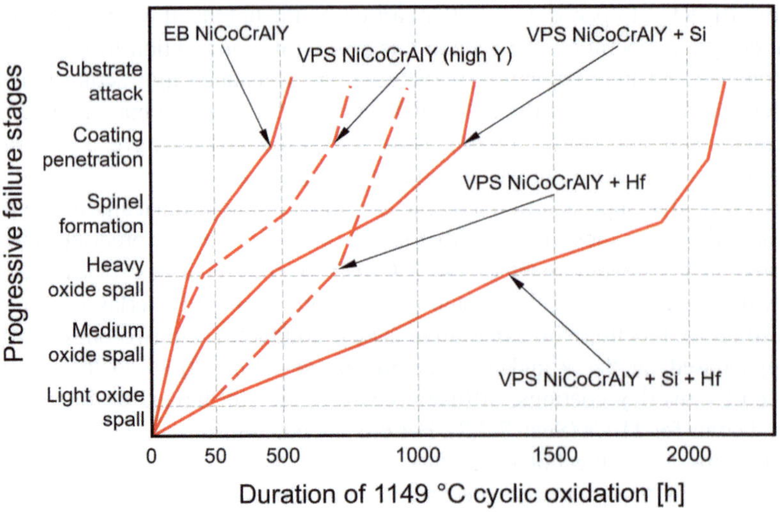

Fig. 7.12 Comparative degradation behavior of NiCoCrAlY, deposited with EB-PVD and vacuum plasma spraying (VPS in the figure). (Redrawn from Gupta and Duvall 1984)

different additions of Si and/or Hf were explored, using VPS deposition only (Gupta and Duvall 1984). As concerns the reference samples (i.e., NiCoCrAlY), the better performances of the plasma-sprayed coating are determined by its optimized microstructure with respect to EB-PVD sample. Further improvements have been obtained, thanks to the ameliorated compositions, those involving Si and/or Hf, of the plasma-sprayed coatings. This study had also demonstrated that among the different plasma spray modes, i.e., air, inert gas, and vacuum, this latter is able to produce coatings with a better microstructure and compositional homogeneity, so that the relevant trends observed in Fig. 7.12 are fully justified.

In Chap. 8 a general overview on the reaction products forming on the surface of overlay coatings will be presented (Sect. 8.2.2), depending on their compositions and testing conditions. The efficiency of overlay coatings as protective barriers against high-temperature corrosion phenomena has been optimized further through the development of functionally gradient systems (SMARTCOAT design). Layers of different compositions have been deposited, according to a design suggested by the different operating conditions, pertaining not only aircraft gas turbines but also industrial and marine land-based plants (Nicholls et al. 2002). For these latter applications, while peak temperatures are lower than in aero-engines (see Table 7.1), still harsh working conditions may be determined by lower purity fuels and critical environments, like marine and polluted industrial atmospheres. Therefore, beyond the SMARTCOAT approach, the compositions of the overlay coatings have been tuned to better respond to specific requirements, including the combined action of gas phase and salt deposit corrosion phenomena, like oxidation, Type I and Type II hot corrosion (see Sect. 8.2.5) (Goward 1998; Rajendran 2012). Thus, higher chromium concentrations, i.e., 25–40%, for situations involving Type II corrosion and lower, 12–20%, for Type I, are recommended (Mevrel 1989). At higher temperatures, in excess of 900 °C, where chromium oxide starts to be unstable, it is advisable to increase the concentration of aluminum to favor the formation of the protective α-Al_2O_3. A good choice is 18–22% Cr and 8–12%Al (Nicholls et al. 2002). The corresponding concentrations of nickel and cobalt have to be selected accordingly. Nickel guarantees a higher ductility and toughness of the coating, while cobalt controls Type I hot corrosion. As a general rule, MCrAlX overlay coatings are not suitable for long-term exposures at temperatures in excess of 1100 °C, since a too fast oxidation kinetics leads to a too rapid growth of the surface oxide layers. The stresses building up inside the oxide scale, associated with those due to thermal cycling, would result in a local detachment of the scale itself and the consequent loss of protection by the oxide layer. High-temperature conditions are normally present in new gas turbine aero-engines and have suggested using overlay and diffusion coatings as bond coats in thermal barriers coating (TBCs) systems. In TBCs, the ceramic top coat reduces the temperatures of the metallic coating and of the components down to acceptable values for a sufficient lifetime, as discussed in the next section (Sect. 7.3.2).

7.3.2 Thermal Barrier Coatings

Thermal barrier coatings have had a central role in the evolution of the turbine entry temperature (see Sects. 2.2 and 2.5). In fact, they are meant to thermally insulate the substrate, allowing higher service temperatures to be reached, thus increasing the engine efficiency (see Sect. 2.2), without increasing excessively the kinetics of thermally induced transformations that might otherwise shorten the useful lifetime of the structural components.

In Fig. 7.13 the main components of a TBC are shown (Padture et al. 2002). Although deposited onto a gas turbine blade, the structure is quite general and comprises, from the inner to the outer part, the bond coat that can be either an MCrAlX overlay coating or a diffusion coating, based on intermetallic aluminide phases; the thermally grown oxide (TGO) layer that results from the oxidation of the bond coat; and the ceramic top coat, very often made of a ZrO_2-based ceramic. Since the TGO layer should provide an important protection against further oxidation and corrosion of the bond coat, elevated standard of adhesion and compactness is crucial. This may recommend a controlled oxidation treatment prior to installation of the coated component (Wolfe et al. 2005).

An additional advantage of having an oxide layer already present on the surface of the bond coat is the reduction of the stresses in the top coat, produced by in-service oxidation. This is a central issue as concerns the TBC failure mechanisms (Sect. 7.3.3) and justifies the research efforts in developing better performing bond coats (Hazell et al. 2008). The combination of TBCs and internal cooling of superalloy blades has allowed gas turbine engines to operate even at higher than the incipient melting temperatures of the superalloy itself.

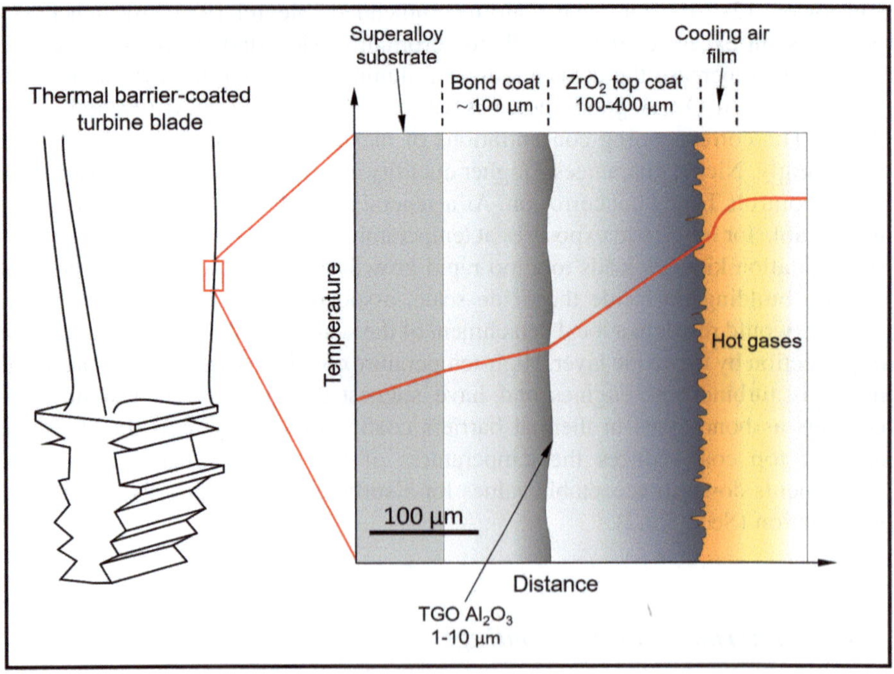

Fig. 7.13 Structure of a TBC deposited on a gas turbine blade. (Redrawn from Padture et al. 2002)

This aspect is clearly shown by Fig. 7.14, and TBCs can be regarded as one of the most important innovation factors in this field, considering the enormous impact on the performances and efficiency of the aero-engines. TBCs were first developed by NACA (National Advisory Committee for Aeronautics) and by NBS (National Bureau of Standards) (Garrett and Gyorgak 1953; Harrison 1947). The early reported aerospace applications of TBCs are probably the zirconia-calcia stabilized flame-sprayed coating deposited onto the walls of the thrust chamber of the X-15 rocket plane in the 1960s (Davies 1963; Hjelm and Bornhorst 1961). Initially, TBCs were used on static parts of the engine only, like combustors and afterburners, not subjected to excessively high mechanical stresses. For the deposition of TBCs, atmospheric plasma spray (APS) is used for both the NiCrAlY bond coat and the top coat, for which yttria-stabilized zirconia (YSZ) was identified as an excellent choice (Rigney et al. 1997). The comprehension of the main failure mechanisms and the relevant remediation approaches afforded the possibility of using these ceramic coatings also for selected regions of the high-pressure gas turbine. Starting from the 1980s, TBCs were used to protect the nozzles in the gas turbine, still using plasma spraying for the deposition. However, more demanding requirements regarding the oxidation resistance performances of the bond coat suggested the use of a VPS (or LPPS, see Sect. 7.2.1) deposition system. A major breakthrough in the coating production technology came up with the EB-PVD, which started to be used for the deposition of the top coats. The columnar microstructure of the zirconia grains,

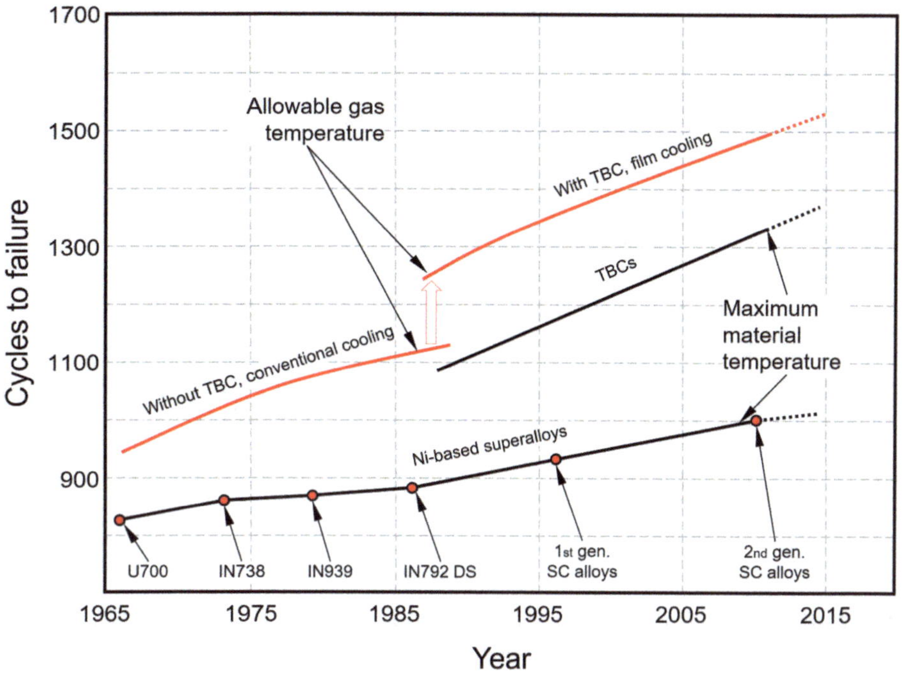

Fig. 7.14 Evolution of the materials (superalloys and TBCs) and gas temperatures in a gas turbine engine. (Redrawn from Clarke et al. 2012)

visible in Fig. 7.13, increased considerably the effective lifetime of the coating. Therefore, these coatings started to be used for the protection of turbine vanes and blades and still are, even though good progresses have been made in the plasma-sprayed counterpart. Nowadays, APS is still the main technique used to deposit the top ceramic part of TBCs on stationary components of the engine, like combustors, shrouds, vanes, and also blade baseplates. For the bond coat, either diffusion or plasma-sprayed coatings are used. In this case, vacuum plasma spraying is preferred, for a better control of the impurity concentrations. For the top coat of the rotating parts and, in general, of the hot section of the turbine, EB-PVD has become the standard choice. The particular coating deposition dynamics induce a typical columnar grain morphology and a much smoother surface finish than plasma spraying. This feature is paramount, particularly for turbine vanes and blades of the first high-pressure stage, in relation to their aerodynamic efficiency and, consequently, the efficiency of the whole engine. This aspect is shown by Fig. 7.15, which also demonstrates how the dependence of the aerodynamic efficiency of the turbine vanes and blades on their surface finish is less critical for the second and further stages, being exposed to hot gases with lower pressure and velocity. The same can be applied to the inner walls of combustors and, if present, afterburners, exposed tangentially to the gas flow. On these bases, the best combination of coatings and deposition technologies can be selected for each part, to meet both performances and

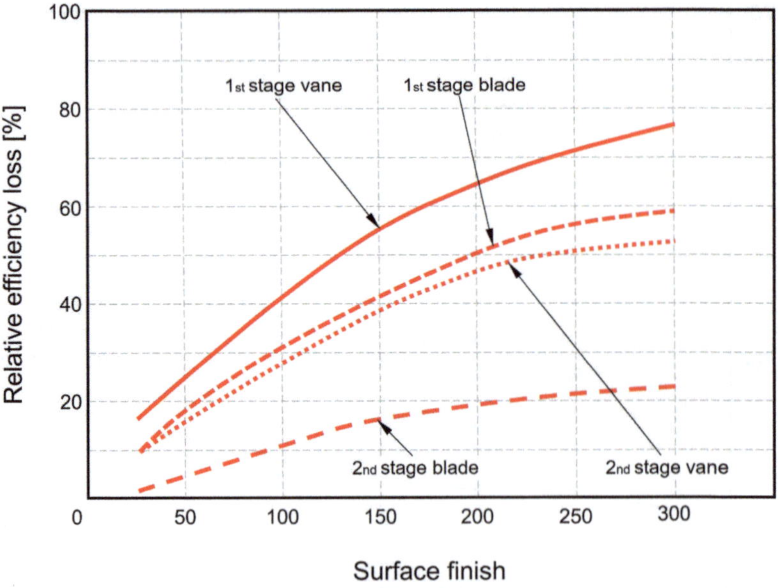

Fig. 7.15 Dependence of the aerodynamic efficiency of the first and second stage turbine blades and vanes on their surface finish. (Redrawn from Rigney et al. 1997)

cost requirements. When depositing protective coatings onto the surface of gas turbine blades and vanes, it is important to avoid that the deposited layer would obstruct the cooling holes, with obvious detrimental effects on the component lifetime and integrity. Also in this regard, EB-PVD coatings are preferable, since the parameters of the atom-by-atom physical deposition process can be controlled so to obtain a smoothly tapered coating at the hole entrance. The same cannot be achieved with plasma spraying, considering that the coating builds up through discrete single droplets, each one of which can significantly reduce the hole's opening, having diameters varying from 0.25 to 0.5 mm approximately.

Important features for the top coat of a TBC are:

- Chemical stability and corrosion resistance
- Erosion resistance
- Good adhesion to the substrate
- Resistance to high- and low-frequency fatigue
- Thermal expansion coefficient similar to that of the substrate in order to reduce the thermal mismatch stress value
- Thermal stability over all the temperature range

In addition to these properties, low thermal conductivity is obviously paramount, considering the expected main role of TBCs. In Fig. 7.16 the thermal conductivity curves, as a function of the temperature, of several ceramics, mainly oxides, are shown. ZrO_2 seems to be an excellent candidate for TBC, considering its lowest

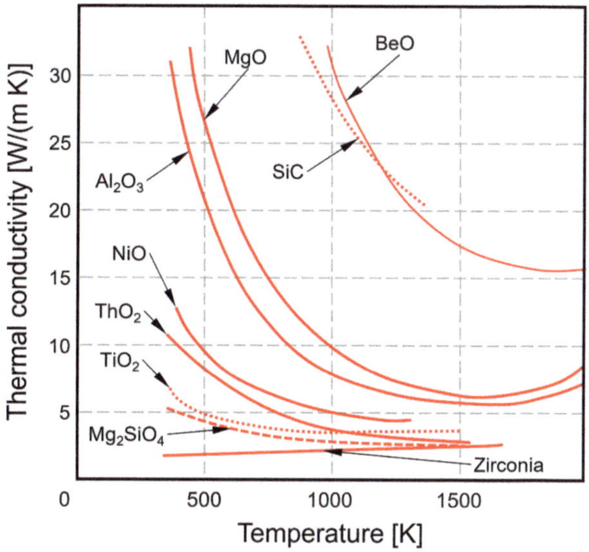

Fig. 7.16 Thermal conductivity of several ceramic compounds. (Redrawn from Clarke 2003; Kingery 1955)

value of thermal conductivity, substantially temperature-independent over a wide range, which is not the case for other candidate materials. However, the presence of three ZrO_2 polymorphs requires suitable alloying for phase stabilization. Pure zirconium oxide, at 1 atm, can indeed exist in one of the following phases, depending on the temperature:

- Monoclinic up to 1170 °C on heating
- Tetragonal from 1170 to 2370 °C on heating
- Cubic from 2370 °C up to melting (2715 °C)

All the transformations involved are martensitic and are characterized by important volume changes. For instance, the displacive tetragonal to monoclinic phase transformation is accompanied by a molar volume increase of 3–5% and up to 8% shear strain (Sato et al. 1985). The temperature changes and fluctuations under typical gas turbine servicing conditions may trigger these transformations with unavoidable consequences on the mechanical stability of the TBC. The situation is rendered even more complex because of the dependence of the transition temperatures on the microstructural parameters of the oxide samples, like defect structures and grain size that depend also on the coating deposition conditions (Garvie and Goss 1986; Tu et al. 1994). This is the reason why zirconia-based materials, with selected concentrations of phase stabilizer oxides, like Y_2O_3, MgO, CaO, are used in real coatings. The addition of the above oxides is capable to change the stability fields of the zirconia polymorphs, so to render the top coat more suitable to operate over the whole range of servicing conditions. Y_2O_3 has been identified as an excellent candidate for improving the zirconia coating performances, as concerns particularly the resistance to cyclic heating and cooling stages. This is shown by

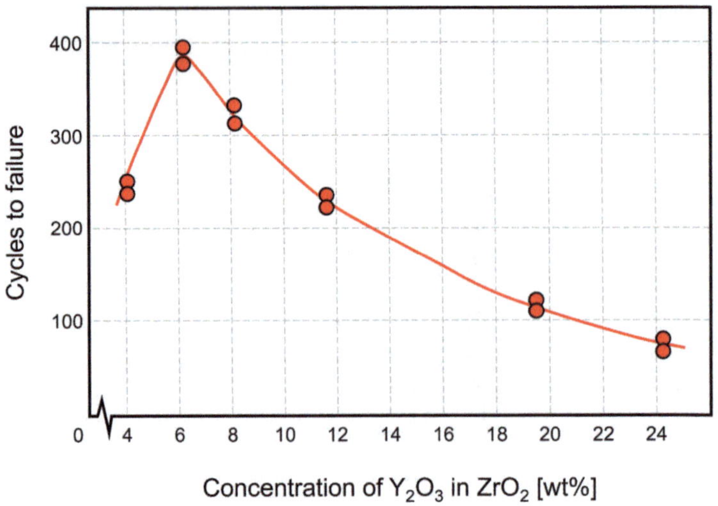

Fig. 7.17 Burner rig tests on TBC. The best performances, i.e., the largest number of cycles to failure, are achieved by the coating with composition: ZrO_2-$7Y_2O_3$. (Redrawn from Stecura 1986)

Fig. 7.17 (Stecura 1986). Data points are from burner rig tests (see Sect. 8.2.6) and illustrate the dependence of a zirconia coating lifetime on its yttria content. In particular, a concentration of Y_2O_3 equal to 7 wt% turns out to be the best choice, and, indeed, the ZrO_2-$7Y_2O_3$ composition became the standard for these applications. The reason can be inferred from the portion of the ZrO_2-Y_2O_3 phase diagram in Fig. 7.18, in which the findings of a former investigation on this same system have been applied to the definition of the phase composition of plasma-sprayed zirconia coatings. For Y_2O_3 concentrations in the 6–11 wt%, quenching from the high-temperature cubic ZrO_2 field results in the formation of the metastable "non-transformable" tetragonal ZrO_2 (Miller et al. 1981; Scott 1975). The peculiar aspect of this tetragonal polymorph of ZrO_2, as obtained by EB-PVD and identified as *t'*, is its microstructure, with two main features relevant to the observed improved coating durability. A <100> crystallographic texture means that the tetragonal c-axis is parallel to the metal-ceramic interface. Moreover, a diffuse intergranular porosity is present. Since the coating grains have a typical columnar morphology, also the pores feature an asymmetric elongated shape, aligned along a direction perpendicular to the metal-ceramic interface (Alperine and Lelait 1994; Lelait et al. 1993).

The combination of these two microstructural and crystallographic features determines the remarkable improvement in the coating lifetime for two main reasons:

- Cracks propagating parallel to the ceramic-metal interface are stopped by the intrinsic coating porosity, through the twofold action of crack blunting and crack deflection along directions perpendicular to crack propagation.

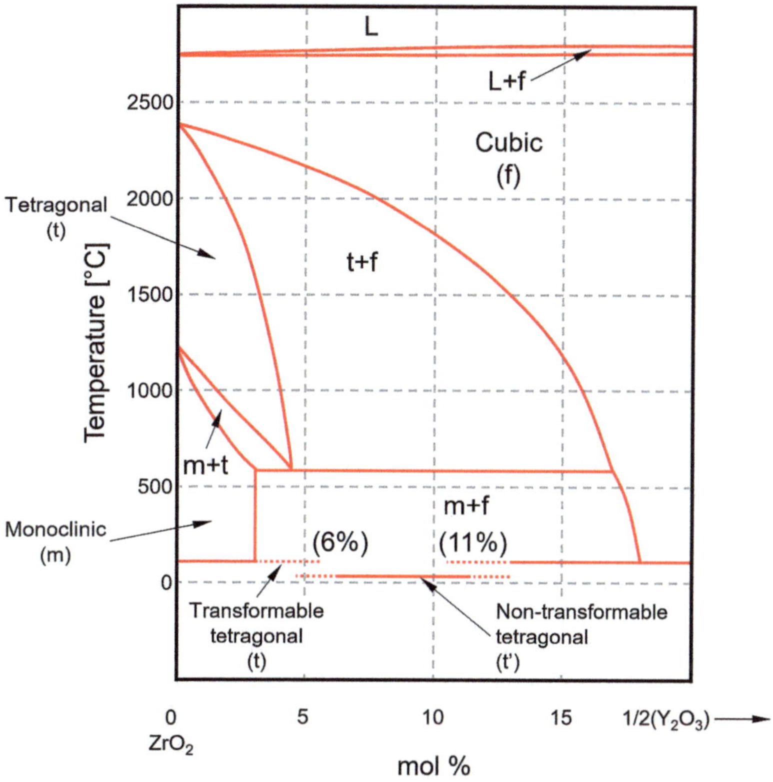

Fig. 7.18 Portion of the phase diagram showing the composition range for the formation of the *t'* tetragonal phase at room temperature. (Redrawn from Miller et al. 1981)

- Cracks propagating parallel to the ceramic-metal interface would very unlikely meet easy cleavage planes for the tetragonal structure that are perpendicular to the <100> direction and, thereby, also perpendicular to the mentioned interface. In this respect, not only crack propagation is hindered by the lack of easy cleavage planes, but an additional toughening contribution is provided by crack deflection along perpendicular directions to the top-bond coat interface.

It is evident from this description that the excellent thermal cycling resistance of the EB-PVD zirconia top coats cannot be attained with plasma-sprayed TBCs. The molten zirconia droplets are splatted either onto the substrate or on previously deposited ceramic layer, where they rapidly solidify, without any possibility of controlling microstructure and crystallographic texture.

The zirconia, in which the tetragonal t' phase is stabilized by alloying with the optimal 7 wt% concentration of yttria, is usually code-named 7YSZ. As seen, the elevated performances and reliability required in TBCs by the extreme working conditions result from the combination of different aspects, concerning the composition, microstructure, and crystallographic texture. An important factor, discussed in the next section on the failure mechanisms (Sect. 7.3.3), is the coupling of the top with the bond coat through the TGO layer. A further aspect contributes to the intrinsic fracture toughness of EB-PVD ceramic top coat; the microstructure of the zirconia layer exhibits very fine twinning structures that are extremely effective in accommodating the thermomechanical strains. This behavior, which has also been reported for zirconia coatings stabilized with other oxides, for instance, HfO_2, is based on the strain compliance through the reversible re-orientation of the tetragonal variants, with a consequent release of the local stresses that in this way are kept below critical values (Ibegazene et al. 1993; Mercer et al. 2007). This mechanism is quite similar to the alignment of variants observed in shape-memory alloys (see Sect. 9.5.1), resulting in the partial or complete de-twinning of the alloy when in the martensitic phase or in stress-induced austenite (superelastic behavior). In TBCs, the mechanism is active also at high temperatures, so that high fracture toughness is guaranteed even at the more demanding operating conditions. It is worth saying that this mechanism, even named *ferroelastic*, is different from *transformation toughening* that is active in partially stabilized zirconia. This latter mechanism involves the irreversible stress-induced transformation of zirconia from tetragonal into monoclinic phase, and the toughening effect is associated with the relevant volume expansion that hinders the propagation of the crack (Green et al. 1989), thus providing a further contribution to the fracture toughness of PSZ. In this system, a fine distribution of crystalline domains also of the other two ZrO_2 polymorphs, i.e., cubic and monoclinic, is present (Garvie et al. 1972). The irreversibility of the martensitic transformation on which this toughening mechanism is based has suggested the definition of "ceramic steels" for these zirconia-based materials (Garvie et al. 1975).

The thickness of the top coat is primarily selected considering the component to be protected. As concerns the blades, the coating thickness is a trade-off between achieving the target temperature drop in the base metal and avoiding an excessive weight increase, due to the contribution from the coating. The typical thickness of the top coat ranges from 100 to 250 μm. On the inner walls of the combustors, the shrouds, the high-pressure vanes, being all stationary components, and thicker coatings are acceptable, and indeed thicknesses ranging from 500 μm to 1 mm are typically used.

Notwithstanding the excellent performances of 7YSZ-based TBCs, new systems are being explored in order to sustain the continuous development of higher efficiency gas turbine engines, in the first place dependent on the maximum operating temperatures. Considering the primary scope of TBCs, thermal conductivity is to be regarded as a valuable selection parameter (Winter and Clarke 2007). An early development route was aimed at overcoming the limitations of the 7YSZ material, particularly when used at temperatures in excess of 1200 °C. At high temperatures, the *t'* tetragonal phase tends to separate into the equilibrium tetragonal (*t*) and cubic (*c*) phases. Subsequently, the tetragonal phase, under the effect of thermal cycling, will transform into the monoclinic low-temperature polymorph. At this stage, the integrity of the coating may be highly jeopardized. The time required to detect the early signs of the monoclinic phase is relatively short, i.e., in the 100 h range at $T = 1350$ °C (Lughi and Clarke 2005). In order to slow down this process, alternatives to yttria stabilizers have been considered. Heavier dopants have been selected, so to hinder phonon propagation and reduce the intrinsic thermal conductivity of the coating. Several rare earth oxides like Gd_2O_3, Yb_2O_3, and Nd_2O_3 have been considered in this respect. The approach was only partially successful though. As expected, the larger molecular masses of the dopants, in association with the elevated density of constitutional point defects, mostly oxygen vacancy, resulted in lower thermal conductivities. However, the novel oxide systems displayed even lower stability of the *t'* phase, as compared to the reference 7YSZ system, with the only exception of Yb_2O_3. The situation has been improved to some extent using a combination of different dopants in the same material. For instance, to the reference ZrO_2-Y_2O_3 system, with the trivalent yttrium, a given concentration of pentavalent cation oxides, like Nb_2O_5 or Ta_2O_5, were added (Raghavan et al. 2001; Raghavan et al. 2004). The complementary valence of the additions reduced the concentration of constitutional point defects, particularly oxygen vacancies. As expected, the resulting structures featured a lower thermal conductivity and a good level of stability of the *t'* phase. The trend was taken to the limit for ceramic oxide compositions in which the trivalent and pentavalent ions concentrations are substantially the same. This determines the nominal zeroing of the constitutional oxygen vacancies, resulting in a significantly reduced thermal conductivity, as compared to the reference value for 7YSZ. Moreover, the *t'* phase demonstrates an extraordinary stability (Shen et al. 2010). The combination of multiple dopants has been adopted also to improve the performances of 7YSZ, with the additions of Nd_2O_5 (or Gd_2O_5) and Yb_2O_3 or Sc_2O_3 (Jarligo et al. 2010; Song et al. 2011; Vassen et al. 2000, 2010; Wuench and Eberman 2000; Zhu and Miller 2004). Interestingly, in these systems, owing to the elevated concentrations of high molar weight dopants, nanometric oxide clusters tend to form, having the additional effects of stabilizing the microstructure of the coating against recrystallization and sintering, phenomena typically occurring during servicing and resulting in a progressive reduction of the thermal insulating capabilities of the coating, for the elimination of the initial porosity.

Molecular dynamic simulations applied to the investigation of the zirconia-based systems revealed the defect structures that appear at increasing temperatures, resembling those observed in amorphous, rather than crystalline structures, with their important effects on the thermal conductivity of the ceramic oxide. These studies indicated promising directions to be taken in order to develop more efficient TBCs (Allen et al. 1999; Pan et al. 2012). It is clear that increasingly complex crystalline structures would involve lower thermal conductivities, owing to the more difficult propagation of the acoustic phonons, a paramount aspect in heat transfer phenomena. Moreover, the higher concentrations of optical phonons encountered in low symmetry, complex crystalline structures determine an indirect reduction of thermal conductivity, owing to the reduced contribution of these phonons to heat transfer. The thermal conductivity curves, as a function of the temperatures, in Fig. 7.19a, b, illustrate the possible strategies for the development of ceramic oxides for TBC

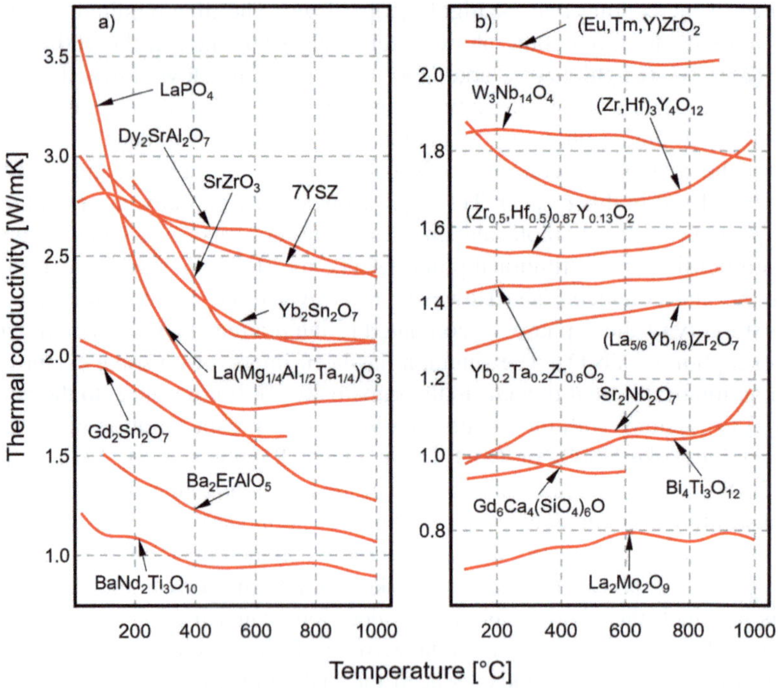

Fig. 7.19 New ceramics for TBCs: dependence of their thermal conductivities on temperature, also in comparison with the thermal conductivity of 7YSZ. (Redrawn from Pan et al. 2012)

applications. Following the direction marked by molecular dynamic investigations, several families of superlattice structures have been identified and demonstrated to have a range of similar vibrational modes, as seen with highly dense packed crystal lattice and capable of extensive solid solutioning (Cao et al. 2004; Guo et al. 2009; Qu et al. 2011; Shen et al. 2008; Sparks et al. 2010; Yang et al. 2008). This is an important prerequisite for tuning the coating composition in order to respond better to the requirements of specific applications. The main groups of ceramics that are being investigated are perovskite-structured oxides (ABO_3 type), lanthanide orthophosphates ($LnPO_4$), and rare earth silicates, like the apatite structured $Gd_8Ca_2(SiO_4)_6O$, this latter exhibiting a particularly low thermal conductivity, substantially independent from temperature.

One of the systems that is providing already very promising results, in comparison to the standard 7YSZ coating composition, is based on lanthanides (reference composition $Ln_2Zr_2O_7$, with Ln = La, Nd, Sm, Gd). Compounds, like $Gd_2Zr_2O_7$, turn out to be stable up to 1500 °C, i.e., 300 °C more than 7YSZ. A comparative investigation (see Fig. 7.20) conducted also with field tests demonstrated a definitely better durability and at the same time an excellent reliability of the novel coatings, ready to replace the classical systems (Levi et al. 2012).

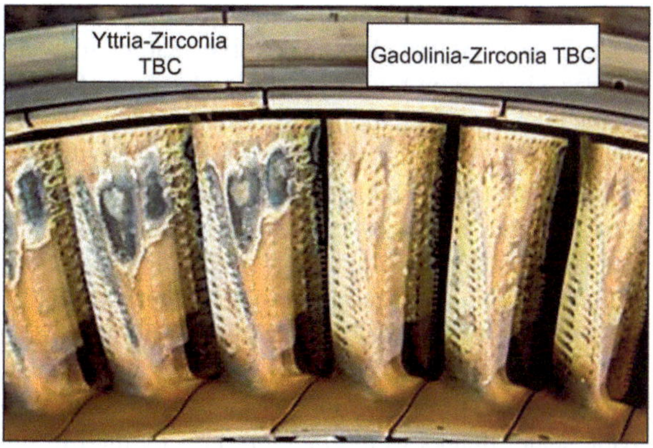

Fig. 7.20 Comparative engine test showing the different behaviors of conventional 7YSZ and novel $Gd_2Zr_{1-x}O_{2-x/2}$ ($x = 0.5$) pyrochlore structure TBCs, after simultaneous exposure to identical working conditions (Levi et al. 2012, reproduced with permission of the Publisher)

7.3.3 Failure Mechanisms of Corrosion Resistant and Thermal Barrier Coatings

Diffusion and overlay coatings, when not used as bond coats in TBCs, are directly exposed to the flux of hot gases that through chemical reactions and erosive phenomena may produce their progressive disruption. The ordinary sequence of events involves the initial formation of a protective surface oxide layer, compact, and adherent to the alloy of which either diffusion or overlay coatings are made. In the case of aluminide coatings (see Sect. 7.3.1), the surface oxide is essentially composed of α-Al_2O_3. This most stable oxide prevents the metallic substrate to get in contact with the corrosive atmosphere, so that a stationary state can be reached. In overlay, i.e., MCrAlX, coatings, the so-called synergic effect (see also Sect. 8.2.2) of chromium and aluminum is exploited. Thanks to an optimized design of the coating compositions, the formation of nickel oxide and other spinel oxide phases, still containing nickel, is significantly inhibited, except for the very early stages of oxidation. The microstructure of these phases prevents them to provide any effective protection to the underlying alloy. The occasional damage of the surface oxide layer may occur for the impact of particles carried by the high-velocity gas flux. Another reason for the local cracking and possible partial detachment of the oxide scale is the building up of internal stresses during the oxide growth process. These stresses are usually produced by the formation of the reaction products within the already existing oxide layer, a process that results from the prevailing outward diffusion of the metallic cations that can in this way react with the incoming oxygen (see Sect. 8.2.1). Thermal cycling too influences the stability of the surface oxide layer. Therefore, major improvements concerning the oxidation-resistant coatings have especially been focused on developing strategies to enhance their durability also under cyclic conditions. Reactive elements, like yttrium, zirconium, etc., present in the overlay coatings, provide several contributions for a longer lifetime (see also Sect. 8.2.2). Through their reaction with oxygen, a fine dispersion of oxide nanoparticles forms on the coating surface. These particles act as heterogeneous nucleation sites and induce a fine-grained microstructure into the growing oxide layer. This is regarded as a positive feature to resist better to thermomechanical stresses, since grain boundary sliding becomes feasible, thus reducing the incidence of cracking and spallation. As oxidation proceeds, reactive elements tend to react preferentially with the oxygen flux diffusing through the oxide layer toward the metallic substrate. This mechanism avoids an excessive internal oxidation of the alloy. Another beneficial effect of this same process is the formation of an intermediate oxide barrier, which not only limits inner oxidation but also the outward diffusion of metallic cations from the underlying alloy. This function is quite similar to the barrier effect operated by the platinum additions in aluminide coatings (Das 2013; Tawancy et al. 1991). The oxide particles of the reactive elements that form at the interface between the outer scale and the coating result in a better adhesion of the oxide scale, thanks to the *pegging effect* called in this way since it exploits also a mechanical grip component. The interface bonding is strengthened by the reduction of sulfur impurities that are captured by RE atoms to form relevant sulfide species. The high concentration of interfaces, deriving from all these secondary phases, promotes the capture of vacancies created by the Kirkendall effect as a consequence of the outer diffusion of oxide-forming elements (Khanna 2002). When this process cannot occur any longer for the saturation of the available sites, vacancy clustering becomes predominant, determined by their diffusion-assisted coalescence and leading to the formation of macroscopic voids within the bulk alloy. This phenomenon is occurring also in aluminide coatings, and it is typically observed in the final stage of the degradation process.

To summarize, the life span of oxidation-resistant coatings, i.e., diffusion and overlay coatings, can be divided into the following main phases:

- **Initial state.** The surface oxide scale starts to form and initiates its protective action of the underlying coating, thus inhibiting further oxidation. This stage may have an initial transient when the gaseous phases are in direct contact with the metallic components, so that reactions can occur directly, i.e., not assisted by diffusion mass transport. From the kinetic point of view, this corresponds to a linear dependence on time of the specific mass increase associated with the oxide growth. At this stage, the oxide scale exhibits also non-protective phases, e.g., NiO or $NiAl_2O_4$.
- **Stationary state.** A steady-state condition is reached when a protective oxide layer has formed on the metallic coating surface and the coating composition is still such, that in case of scale damage, a sufficient outgoing flux of metal is mending it with the formation of a further layer of protective oxide. This produces anyway compositional changes in the coating, leading to the formation of the so-called Al-depleted zone. At the same time, the interdiffusion between the coating and the underlying alloy results in the formation of the already mentioned Kirkendall voids and intermetallic phases. The formation of these brittle phases is promoted by the increased local concentrations of slow-diffusing atomic species, like refractory elements, present in the superalloys. At this stage, the oxidation kinetics is usually well represented by a parabolic law, according to the Wagner oxidation model (see Sect. 8.2.1).

- **Damage propagation.** When the concentration of aluminum in the overlay or diffusion coatings falls below the critical value, necessary to the formation of an outer protective scale based on α-Al_2O_3, other non-protective oxides start to form on the coating surface. This stage can be regarded as the end of the coating useful lifetime. The composition of the aluminide diffusion coatings may determine the formation of a martensite phase, exhibiting volume changes upon thermal cycling, with consequent stress generation in the coating itself (Zhang et al. 2003). This situation may be also responsible for the so-called rumpling of the coating. "Rumpling" can be regarded as a mechanical plastic instability resulting in the formation of a corrugated surface of the coating (Deb et al. 1987). In addition to the martensitic transformation occurring during thermal cycling, another drive to this phenomenon comes from the β to γ' transformation, occurring in the aluminum-depleted coating, concurrently to the martensitic transformation of the β-phase. These transformations are not the only reasons for rumpling that is observed also in overlay coatings. The phenomenon is still attracting research efforts aiming at understanding the relevant critical aspects in order to develop coatings less prone to this kind of degradation (Chen et al. 2017; Jorgensen et al. 2016; Lesnikova and Lesnikov 1986; Smialek 1971).

The phenomenology described so far with regard to the failure and degradation mechanisms of diffusion and overlay coatings is interesting also as concerns the use of these same coating systems as bond coats in TBCs (Sect. 7.3.2). In TBCs, the surface directly exposed to the hot gas atmosphere is that of the ceramic top coat. Since the ceramic layer is permeable to air, oxidation takes place (Sect. 7.3.2) with the formation of the TGO layer at the bond-top coat interface. Although the kinetics of oxidation occurring at this location is mitigated by the lower gas pressure and temperature, still, it has very important consequences on the stability and durability of the whole TBC. TBCs, particularly those protecting vanes and blades of gas turbine high-pressure stages, require an absolutely reliable estimation of their actual lifetime, since an extensive and unexpected failure would result in catastrophic effects for the structural integrity of the components, considering that gas operating temperatures can be far above the material capabilities (Fig. 7.14). High-temperature spikes may result in accelerated kinetics of diffusion-assisted phenomena and, in the most severe cases, in local melting.

The layered structure of a TBC poses the important issue of the mechanical matching and accommodation of the strain misfit and relevant residual stresses present at each one of the different interfaces: alloy-bond coat and bond-top coat. This second interface actually features the presence of the intermediate layer of the TGO. Considering the compositional affinity and the common metallic character, the alloy-bond coat interface is mainly influenced by the strains generated by interdiffusion and relevant phase and microstructural transformations. The formation of Kirkendall voids and consequent swelling, precipitation of intermetallic phases, outward diffusion of metallic atoms from the alloy substrate (Chen M et al. 2003; Tolpygo and Clarke 2004a, b), are all phenomena intervening to a different extent in the failure of TBCs, which has to be rather regarded as the result of several concurrent causes, very much dependent on the specific systems and working conditions involved.

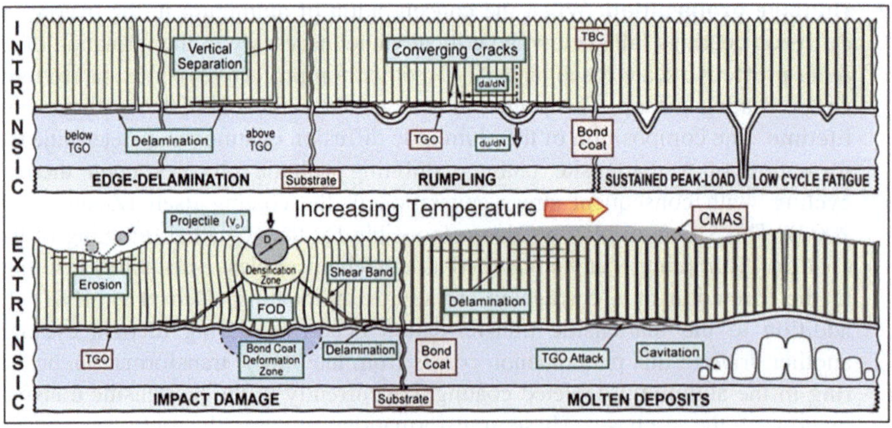

Fig. 7.21 Failure mechanisms in TBCs deposited via EB-PVD: a comprehensive overview (Levi et al. 2012, reproduced with permission of the Publisher)

Notwithstanding this complexity, still, the main mechanisms for TBC failure can be classified in two broad categories: *intrinsic* and *extrinsic* (Fig. 7.21) (Evans et al. 2008; Levi et al. 2012; Wu et al. 2017). This classification is very useful also for designing reliable lab tests, particularly as regards the *intrinsic* mechanisms, and, in general, to spot the reasons for observed failures in serviced components. The *intrinsic* failure mechanisms can be identified by analyzing the morphology of the detached coating. The detachment exposes the metallic surface of the bond coat, showing, as a general feature, the imprint of the TGO grain boundaries, scattered patches of alumina, still from the TGO layer, and some limited zirconia fragments from the top coat. The spallation has been determined by the weakening of the metal (bond coat)-TGO interface, along which crack propagates preferentially. In correspondence of heterogeneities and surface protrusions, the fracture may also cut through the TGO, and this is the reason for the limited zirconia residues observed on the detached scale (Xu et al. 2004). No metallic surface of the bond coat is exposed when the detachment of the TBC is determined by the rumpling of the TGO, with the involvement of the top coat, and producing the delamination through cracks advancing parallel to the bond-top coat interface (Davis and Evans 2006; Ruud et al. 2001; Tolpygo and Clarke 2004a, b). The detachment mechanisms of the TGO layer that have been presented herewith are fully compliant with the results achieved in a research on NiAl-Hf bond coats modified with additions of Pt and Pd (Jackson et al. 2014). The different compositions resulted in different adhesion strengths of the TGO to the bond coat. In the case of comparatively weaker adhesion, the failure of the TBC, tested under cyclic conditions, starts at the TGO-bond coat interface. If a good grip is realized by the improved bond coat composition, the TBC failure would be this time triggered by a diffuse rumpling of the bond coat, although occurring at a later stage than the former mechanism. According to the above description, the cracks initiate and propagate within the top coat and at the TGO-top coat interface. Other compositional changes of β-NiAl-based bond coats, using carbon and yttrium, together with the mentioned hafnium, have been investigated in the successful attempt to improve their durability (Tolpygo et al. 2008).

Longer times and/or higher temperature exposures may result in the formation of Kirkendall voids. Their growth can proceed, until their average size becomes so big to determine the detachment of the coating. The fracture surface, propagating within the thickness of the bond coat, would exhibit a characteristic void morphology (Mumm et al. 2001). Before achieving the final lifetime stage, a significant thickness increase, up to 10–20%, of the bond coat can be observed. It is due to the combined effect of void formation and compositional changes consequent to superalloy-bond coat interdiffusion. A net increase in nickel concentration in the coating is the consequence of the higher diffusivity of nickel, from the alloy substrate toward the bond coat, than that of aluminum in the opposite direction (Shankar and Seigle 1978). The Kirkendall voids may be further enlarged by the cavitation induced by external stress and pressure acting on the component surface. This aspect can be listed among the *extrinsic* failure mechanisms. In general, they are due to the interaction of the TBCs with damaging agents depending on the external environment, like particles impacting on the surface of the top coat, producing erosion and foreign object damage (FOD). The effects on the coating integrity depend on their size and mass and their velocity and impinging angle (Bruce 1998; Chen X et al. 2003; Nicholls et al. 1999). Erosion produces the progressive thinning of the top coat, particularly in the regions close to the leading edges of rotary blades. Detachment of the top coat fragments, micro-crack formation within the top coat, and plastic deformation of the bond coat in regions just beneath the impact spot are the main effects of impacting particles.

An important damage mechanism of TBCs is due to the penetration into the top coat of liquid deposits of calcium-magnesium-aluminosilicate (CMAS) (Kakuda et al. 2015; Jackson et al. 2015; Poerschke et al. 2016). This contaminant is originated by the ingestion into the engine of ash, dust, and sands present in the outer atmosphere (Borom et al. 1996; Krämer et al. 2008; Mercer et al. 2005). The deposition of CMAS occurs preferentially along the pressure surface of the airfoils and on the shrouds, where indeed this kind of damage is more frequently observed. For the difficulty of reproducing the phenomenology involved with CMAS attack using laboratory tests, the main studies and relevant remediation strategies have been developed mostly using data from diagnostics of really serviced components. This kind of damage turns out to be rather important in aircrafts flying at low altitudes over marine and desert regions. In the case of marine atmosphere, sulfide contamination has been often observed (Pokluda and Kianicová 2010). The molten CMAS deposit that forms at temperatures above its melting point, i.e., 1240 °C approx., penetrates into the ceramic top coat preferentially through the grain boundaries, with an important contribution by the capillary forces acting on the inner surface of the columnar pores. The solidification, upon cooling, determines a densification of the CMAS-penetrated parts of the top coat, with a consequent increase in the effective elastic modulus and reduction in fracture toughness and damage tolerance. Coherently with this picture, also the reduction in the porosity of the ceramic coating, deriving from long-term exposure to high temperatures, provides a further contribution to increasing the stiffness of the coating itself. A complex stress field is generated inside the top coat, for the mismatch of the thermal expansion coefficients induced by the top coat densification, its contamination by CMAS material, occurring inside the pores, in the intergranular regions, and on the free surface of the coating. In the long run, the stress field in the top coat may determine the nucleation of micro-cracks and its partial detachment.

The selective dissolution of the zirconia stabilizers in the top coat may be induced by the surface contamination with a layer of molten sulfate, according to the principles of the high-temperature hot corrosion in salt deposit (see Sect. 8.2.5), and may result in the destabilization of the *t'*- ZrO_2 phase and its transformation into the low-temperature monoclinic polymorph. The relevant volume change generates stresses that, if sufficiently intense, may induce the delamination of the ceramic coating.

To complete the picture, a quick glance at a terrestrial application: the land-based gas turbine plants fired by coal-derived synthesis gas (i.e., syngas) or hydrogen. In these engines, the content of water vapor is definitely higher than in conventional gas turbines, and therefore its effect on the durability of the TBC becomes an issue. The TGO observed on single-phase β-(Ni,Pt)Al and two-phase γ-γ' bond coats, oxidation tested under cyclic conditions, exhibited higher oxidation kinetics when several percentages (10, 50, 90%) of water vapor were added to the initially dry oxygen atmosphere (Unocic and Pint 2013; Wright and Gibbons 2007). Reduced lifetimes, with respect to the "dry" values, were observed, although with large fluctuations for the different test conditions. The best durability was guaranteed by the TBC with the two-phase (γ-γ') bond coat, most likely for their more stable structure, with no reported coating rumpling, as observed instead in the samples with single-phase β-(Ni,Pt)Al bond coat, after exposure to the same testing conditions. These results confirmed the findings of a former study with a similar system, tested using the same experimental protocol (Pint et al. 2011).

7.3.4 Clearance Coatings

The design of clearance coating systems stems from two opposite requirements: to avoid reciprocal rubbing between rotary and stationary components and to attain a sufficiently small level of mechanical clearance capable to ensure an adequate efficiency and thrust to the engine. For these reasons, a clearance is left beyond the tip of the blades and vanes, so that under conventional steady working conditions, they should not come into contact with the case. However, there are several reasons that may change the reciprocal rotor-stator distance, resulting in unwanted rubbing and collisions: differential thermal expansion; sudden changes in the engine working conditions; particular loading associated with specific operational requests, like during hard landing and surge; rapid shutdowns due to sudden fuel starvation; and high acceleration maneuvering. Other important aspects that may influence the correct clearance alignment are casing out-of-roundness, bearing damage, mechanical vibrations due to instabilities, and changes in the gas flow. On the other hand, having a tight sealing clearance has beneficial effects on the engine efficiency and, thereby, on fuel and other running costs. As an example, it has been demonstrated that 25 μm reduction in the clearance of the high-pressure turbine blades results in a reduction by 0.1% in the specific fuel consumption (SFC, see Sects. 2.2 and 2.6) (Lattime and Steinetz 2004). Mechanical clearance is an issue not only in the gas turbine but in the compressors too. Here, a good control of clearance results in a lower number of stages necessary to achieve a target value of the compression ratio (see Sect. 2.2), with a consequent reduction in the total weight of the engine.

The loss in engine specific power and thrust may be determined by a non-optimal management of the mechanical clearances, which leads to higher fuel consumption in order to achieve the desired engine performance. However, the lower efficiency of the thermal cycle is accompanied by a higher heat generation with immediate consequences on the exhaust gas temperatures. The kinetics of degradation and microstructural evolution of the hot structural components is enhanced, and their expected lifetime reduced. In case temperature certified limits do exist for aero-engines parts, and these are overcome, a complete, costly overhaul may be required to comply with the relevant international safety regulations. Thus, there are several important reasons for reducing mechanical clearance and for improving sealing in gas turbine engines, issues that over the years have been approached adopting several strategies. Active solutions are based on the revision of the existing part design, with the introduction of actuators and labyrinth fin seals (Dobek 1973; Scharrer and Pelletti 1995; Van Ness et al. 2006). In this way dimensional changes can be compensated either by varying in situ the part geometry or by limiting the effects of gas heating through local cooling of the components. Shrouded blades are a typical example of parts designed for the implementation of an efficient sealing. The T-shaped tips of the blades are in contact with one another for a better stiffness of the whole part, being the unshrouded blades used as a lighter alternative. As all the rest of the blade body, also the shroud is cooled, and the outer part of this ring-shaped component features one or a series of fins that cut through the seal coating deposited onto the outer rubbing surface. The tips of the blades cut directly into the abradable counterpart. In this regard, the choice of the materials for this tribological system is important to guarantee both the sealing performances and adequate blade durability. The following properties and parameters should be considered when selecting materials for abradable systems:

- Deposition technologies that guarantee repeatable results
- Good adhesion to the substrate, in order to resist adequately against the shear stresses by the rubbing tips of the blades or by the fins of shrouded components
- Production of small-sized wear debris that should not provoke any major erosion damage to the parts found along the path out of the engine
- Resistance to thermal shock and thermal cycling
- Surface stability under the normal servicing conditions, as concerns both hot corrosion and erosion resistance, in order to retain the above properties during the whole servicing lifetime

These requirements are rather different in case of land-based with respect to aero turbomachinery, just considering the different environments and standard operational cycles that these latter would rather face. For instance, large land-based gas turbines on average will have a stop and start once every 3–5 years, whereas commercial aircraft gas turbine engines have on average 1500 stop and start cycles per year (Markou et al. 2011). Moreover, an important aspect of thermal cycling is the possibility for liquid and solid layers of contaminants to condensate on the exposed surface, like sulfates and silicates, i.e., CMAS (Sect. 7.3.3), inducing hot corrosion phenomena (Sect. 8.2.5), involving important changes in the materials properties.

As a general rule, the abradable coating materials should have a lower hardness than the material of the rubbing counterpart. As seen in Chaps. 3, 4, 5 and 6, the main classes of materials for structural components in an aero-gas turbine engine are metal laminate composites (see latest fan material), titanium alloys, steels, and superalloys. The main wear phenomena that may intervene and whose incidence should be kept under control are essentially related to the wearing out of the rotating components (e.g., the blade tip or the fins on the outer rim of shrouded blades) due to abrasive or oxidation wear. In the case of titanium alloys, overheating, in association with high gas pressure conditions, may trigger ignition (Sect. 4.3.1.5). The frictional overheating in association with the enthalpy output coming from the rapid oxidation of the titanium alloy components may determine the complete loss of an acceptable aerodynamic geometry. Wear mechanisms may also involve adhesive contributions. In this case, material transfer from the abradable coating to the rubbing component can be observed, the phenomenon being known as adhesive transfer or "galling." An excessive material transfer to the tip of the blades or to the fins can result in deep grooving on the inner shroud wall accompanied by a temperature rise due to frictional heating. To avoid these drawbacks, a composite material approach is mandatory, and indeed all coating systems especially developed for improving mechanical clearances in different parts of a gas turbine aero-engine are based on this principle. The thermal stability requirements have been fulfilled by choosing suitable combinations of matrix and strengthening secondary phases. Compressor coatings are based on a polymeric matrix, usually a polyester, reinforced with Al-Si or Al-bronze additions. The strength of the shearable polymer matrix is tuned to the specific working conditions by the inorganic additions, in order to reduce the frictional heating and at the same time to prevent excessive adhesive mass transfer. Higher working temperatures render the polymeric matrix inadequate as concerns its mechanical stability, worsened by the faster kinetics of the oxidation phenomena. Using oxygen-acetylene flame spraying deposition (Sect. 7.2.1), nickel and nickel alloy-based systems are adopted for higher temperature use and relevant more demanding stability requirements. On the other hand, to avoid that these mechanically stronger than polymer-based materials may damage the rotating counterparts, the deposition parameters, the powder grain morphology, and size are selected so to obtain relatively porous coatings. Powders used for these coatings are:

- **Nickel-graphite.** In this case graphite is encapsulated inside a nickel shell using an autoclave-based process. These coatings are typically coupled with steels and nickel superalloy blades or fins of shrouded parts. However, selecting a suitable hardness value and playing with the coating porosity, they can also be used with titanium alloy components.

- **NiCrAl-bentonite.** A similar microstructure as in the former system is present in these powders. Bentonite grains are embedded into pre-alloyed Ni-Cr-Al particles. The coatings, with suitable degree of porosity and, thereby, hardness, are used in association with steel and superalloys rubbing components. A graded structure can be attained by changing the deposition parameters and thereby the hardness and porosity across the deposit thickness. The aluminosilicate component renders these coatings too abrasive for titanium alloys.
- **NiCrFeAl-hBN (hexagonal boron nitride).** CERMET composite coatings are obtained from powders made of pre-alloyed NiCrFe mixed with aluminum and boron nitride. The reaction of aluminum with the pre-alloyed Ni-Cr-Fe solid solution produces an enthalpy output, which sustains the melting of the metallic powders. Boron nitride (hexagonal polymorph) reduces grain boundary diffusivity, thus enhancing the oxidation resistance and promoting the incursion by the rotating components, since it helps the cutting through of the abradable matrix acting as a solid lubricant, with the additional beneficial effect of keeping the temperature rise within acceptable values. This functional aspect and the overall performances of these coatings can be optimized for the different blading materials: titanium alloys, steel, and nickel-based superalloys.

An important boost to the technology of abradable coatings came from the introduction of APS (Sect. 7.2.1) and its flexibility in modulating the deposition conditions and to adapt them to each specific material system. According to the scheme proposed in Wilson (2012), the following main coating systems can be deposited by APS. The first is obtained from powder mixtures containing one of the following combinations of materials: aluminum-polymer, aluminum-hexagonal boron nitride, and aluminum-bronze-polymer. Polymers are typically heat-resistant polyesters, and the resulting coatings are by all means similar to those described above (Al-Si in a polymer matrix). They are suitable for fan and low-pressure compressor (LPC) stages, operating at temperatures up to 325 °C. Higher temperatures, up to 450 °C, require more stable matrix and solid lubricants, like hBN. At this stage the shearing of the coating matrix is ruled by the lubrication action of boron nitride, having a similar role as graphite in the Al-Si-graphite and Ni-graphite flame-sprayed coatings. A further group of APS coatings are those based on the MCrAlY-hBN-polymer system, designed for higher temperatures, up to 750 °C. The composition of the metallic components is similar to the composition of the overlay coatings (see Sect. 7.3.1). In these coatings the role of the polymer is paramount in determining the porosity, which becomes the ruling factor of the abrasion behavior. In fact, a full density MCrAlY material would have a comparable or higher hardness than the materials of the rubbing counterparts, and a 35–60% porosity is necessary to render these coatings actually abradable. Therefore, a controlled amount of polymer is deliberately entrapped into the coating material and afterward eliminated with a post-deposition heat treatment. This process leaves behind the wanted percentage of porosity that can even be distributed in a graded way within the coating thickness, from the top, higher porosity values, down to the interface with the substrate (lower porosity and therefore higher density). In these higher-temperature abradable coatings, the polymer component has a role in the manufacturing stage only, since it is exploited to attain the right degree of porosity, whereas in the abradable coatings for lower-temperature applications, a significant polymer fraction is retained also in the coating material. This induces two different wear mechanisms which are paramount to ensure an effective tolerance and sealing effect. In the low-temperature coatings, the comparatively softer polymer matrices are shear cut by the intruding components. The wearing out of the high-temperature abradable systems is guaranteed by a sort of friable cutting of the highly porous surface layer favored by the presence of a solid lubricant, like hBN. This same approach is adopted for the refractory ceramic abradable coatings used for the hot parts of the gas turbine engine, like the high-pressure stage of the turbine, where temperatures as high as 1200 °C can be reached. Ceramic coatings are necessary, and the obvious candidate is ZrO_2, already providing excellent performances in TBCs (see Sect. 7.3.2). In fact, yttria partially stabilized zirconia (PSZ) is used in abradable coatings, adopting the microstructural features recommended for these applications (Zhao et al. 2012). However, specific compositions have also been developed, since the PSZ materials are not fully compliant with the specific requirements recommended for these applications. A significant porosity is still interesting, to have reasonably abradable coatings that should be stable against sintering under the extreme temperature conditions, up to 1200 °C that they may face in service. In addition to that, ceramic abradable coatings require thicknesses up to 2 mm, which are definitely higher than the typical thicknesses of TBCs, usually 50–200 μm. This enhances the spallation probability, particularly under cyclic thermal conditions, so that adequate fracture toughness and thermal shock resistance are paramount (Vassen and Stöver 2007). To tackle these critical aspects, alternative stabilizer oxides have been considered, and important improvements have been achieved with dysprosia (Dy_2O_3) and, although to a lower extent, also with ytterbia (Yb_2O_3) (Dorfman et al. 2008; Oerlikon Metco 2014; Schmid 1997; Sporer et al. 2011). Both of them have introduced significant improvements in the high-temperature durability, with respect to the YSZ coatings. Similarly to TBCs, zirconia-based abradable coatings are deposited not on the bare alloy surface but onto a bond coat, necessary for protecting the underlying components by an excessive oxidation and corrosion. Nickel aluminide coatings or overlay coatings are used in this regard. In Table 7.3 the main abradable coating systems are listed with the indication of the relevant temperature limits and corresponding engine part in which they can be used. The abrasive counterpart material systems are also indicated together with the dominant wear mechanisms.

7.3.5 *Erosion- and Fretting-Resistant Coatings*

As concerns the coating systems for gas turbines involving metallic components, two more surface degradation mechanisms have to be considered. The first is erosion, and the other one is fretting. Erosion can be defined as the progressive removal of material from the surface of a structural component, due to the

Table 7.3 Overview on the main abradable coating systems

Materials systems	Temperature limits [°C]	GT part	Intruding blading material compatibility	Cutting mechanisms
Polymers	200	Fan	Composites	Softer matrices: shear cutting
Al-Si + polymer	325	LPC	Titanium alloys	
			Steels	
			Ni-based Superalloys	
Al-Si +graphite	450	LPC	Titanium alloys	
Al-Si+hBN			Steels	
Ni+graphite			Ni-based Superalloys	
Al-bronze+polymer	550	LPC	Titanium alloys	Harder matrices: friable cutting
CoNiCrAlY+hBN+polymer		HPC	Steels	
NiCrFeAl + hBN			Ni-based Superalloys	
NiCrAl + bentonite	650	HPC	Ni-based Superalloys	
NiCrFeAl + hBN	750	HPC	Ni-based Superalloys	
CoNiCrAlY+hBN+polymer				
YSZ + hBN + polymer	1100	HPT	Ni-based Superalloys	
DySZ + hBN + polymer	1200		Recommended h(c)BN on blade tip or coating porosity >30% with bare superalloys	

h(c)BN hexagonal (cubic) boron nitride, *Y(Dy)SZ* yttria (dysprosia) stabilized zirconia, *LPC* and *HPC* low- and high-pressure compressor, respectively, *HPT* high-pressure turbine (Wilson 2012)

interactions with solid and even liquid particles/droplets contained in the gas flux. This kind of attack is rather important for compressors, since they are the parts of the engines directly exposed to the incoming air flux, which may contain sand, fly and volcanic ashes, salt, and ice crystals (Helm and Olaf Roder 2007). The wearing out of the compressor blades due to erosion may lead to an increase in the aerodynamic losses with a consequent decrease in the compression ratio and thereby a reduction in the overall engine efficiency. Similar issues are there for land-based gas turbine engines (Alman et al. 2013; Muboyadzhyan 2009). Considering that erosion is by far a critical aspect for compressors, coatings for titanium alloys and steels have been developed the most. The successful approach is based on the combination of very hard, brittle layers alternate to tough metallic ones. These combined brittle-tough systems in general guarantee an optimal protection against erosion and its dependence on the impinging angles of the eroding agents. Material removal by the impinging particles is reduced by the hard, brittle components of the coating, like nitrides (e.g., TiN) or carbides (e.g., Cr_3C_2). If cracks nucleate in this brittle layer, their propagation is stopped by the tough metallic layer, like Ti, Cr, etc., that is also present in these coating systems. The actual coating composition to be selected will depend, in the first place, on compatibility criteria with the substrate and, secondly, on the required servicing conditions. These may be rather demanding in case of inner, high-pressure stages of the compressor. Therefore, the erosion protective coatings are also required to have a good thermal stability and to provide adequate protection against oxidation. Several categories of these coatings can be found in the literature. One possibility, are just two-layer coatings, like Ti-TiN, Ti-TiC, Zr-ZrN, Cr-Cr_3C_2, and V-VC, with a total thickness of 70–100 μm. The alternative approach is based on the same material systems but with coatings that are now multilayered, with a typical thickness of each single layer equal to 1–2 μm, up to an overall thickness of 25–30 μm (Muboyadzhyan 2009; Rajendran 2012). More systems, like TiN-Zr and TiN-Nb, have been considered, and from the characterization tests, it turns out the paramount role of the ductile phase in maximizing the erosion resistance (Brendel et al. 2008; Borawski et al. 2011). The latest evolution, influenced by the positive results obtained in the tribological field, is still based on multilayered coatings but with the thickness of each layer far below 1 μm. The finer structures exhibit intrinsic toughness even when composed of alternate layers of nominally brittle phases, like TiN-CrN and TiC-CrC (Kablow and Muboyadzhyan 2017).

The second surface degradation mechanism considered herewith, fretting, is associated with the standard operating conditions and is due to the unavoidable contact between surfaces under cyclic loading with typical oscillation amplitudes of a few tenths of millimeter (Cortez et al. 1999; Kermapur et al. 2015; Park et al. 2006). When fretting occurs in a corrosive environment, it would be classified as a mechanochemical corrosion phenomenon (see Sect. 8.1.6). Critical parts for fretting are the dovetail joints in compressors, where the blades are connected to the discs. The sequence of events that may lead to damaging the parts, even macroscopically, can be summarized as follows. The very early stage of fretting features the progressive removal of the thin oxide layers that are initially present on the surface of the metallic components, when they start rubbing each other. When these oxide layers are interrupted, metal-metal adhesive junctions start to form. Intense plastic deformation can occur, accompanied by material transfer when the junctions get fractured. The process continues and protrusions start to form on the mating surfaces. The increasing roughness of the mating surfaces enhances the formation of more junctions, hillocks interlocking, and a reciprocal plowing. Eventually, these phenomena may lead to the formation of macroscopic cracks and to the detachment of wear fragments. Wear debris may either be released or remain trapped in between the contacting surfaces and get oxidized by the reactive atmosphere. The hard oxide particles may both abrade the metallic surface and initiate further fatigue cracks, so that the damage would progressively propagate to subsurface regions of the component. To prevent or to reduce the incidence of fretting, *triballoy* coatings are deposited onto the blade dovetail root, via thermal spraying deposition routes, like APS, flame, and arc spraying (Sect. 7.2.1). Common coating compositions are based on the Cu-Ni-In system, Cu-38Ni or Cu-36Ni-5In, deposited with thickness ranging from 10 to 100 μm (Oerlikon Metco 2016; Praxair Surface Technologies 2000). To optimize the mechanical coupling between the inner surface of the dovetail on the disc, the surface of the coating is usually covered with a layer of a composite paste, made of a polymeric matrix with a suspension of $MoSi_2$ particles (Molydag) (Zhou 1999).

References

Allam I M et al (1979) Influence of small Pt additions on Al_2O_3 Scale Adherence. Oxidation of Metals 14: 517

Allen P B et al (1999) Diffusion, Locons and Propagons; Character of atomic Vibrations in amorphous Si. Philosophical Magazine B 79 (11–12): 1715–1731

Alman et al (2013) Erosion-resistant Nanocoatings for improved Energy Efficiency in Gas Turbines. U.S. Department of Energy

Alperine S, Lelait L (1994) Microstructural Investigations of Plasma-sprayed Yttria partially stabilised Zirconia TBC. Transactions ASME: Journal of Engineering for Gas Turbines and Flows 116: 258–265

Anderson P N, Sheffler K D (1983) Development of Strain Tolerant Thermal Barrier Coating Systems. Available via DIALOG. https://ntrs.nasa.gov/search.jsp?R=19840004244. Accessed 25 Jul 2017

Berry D et al (1995) Enhancing Performance of Silicon-modified Slurry Aluminides on Turbine Components operating in Marine Environments. In: International Gas Turbine and Aeroengine Congress and Exposition, Huston, Texas, 5–8 June 1995

Borawski B et al (2011) The Influence of ductile interlayer Material on the Particle Erosion Resistance of Multilayer TiN based Coatings. Wear 271 (11–12): 2890–2898

Borom M P et al (1996) Role of Environmental Deposits and operating Surface Temperature in Spallation of Air Plasma sprayed Thermal Barrier Coatings. Surface and Coatings Technology 86–87 Part 1: 116–126

Bose S (2007) High Temperature Coatings. Butterworth Heinemann

Brendel et al (2008) MTU Solutions against Erosive Attack and Loss of EGT Margin in Turbo Engines-ER coat. MTU Technical Publication

Bruce R W (1998) Development of 1232 °C (2250 F) Erosion and Impact Tests for Thermal Barrier Coatings. Tribology Transactions 41 (4): 399–410

Cao X Q et al. (2004) Ceramic Materials for Thermal Barrier Coatings. Journal of the European Ceramic Society 24 (1): 1–10

Chen M W et al (2003) Characterization and Modeling of a Martensitic Transformation in a Platinum modified Diffusion Aluminide bond Coat for Thermal Barrier Coatings. Acta Materialia 51 (14): 4279–4294

Chen X et al (2003) Foreign Object Damage in a Thermal Barrier System: Mechanisms and Simulations. Materials Science and Engineering A 352 (1–2): 221–231

Chen Y et al (2017) A Mechanistic Understanding on Rumpling of NiCoCrAlY bond Coat for Thermal Barrier Coating Applications. Acta Materialia 128: 31–42

Clarke D R (2003) Materials Selection Guidelines for low thermal Conductivity Thermal Barrier Coatings. Surface and Coatings Technology 163–164: 67–74

Clarke D R et al (2012) Thermal-barrier Coatings for more efficient Gas-turbine Engines. MRS Bulletin 37 (10):891–898

Cortez R et al (1999) Investigation of variable Amplitude Loading on Fretting Fatigue Behavior of Ti-6Al-4V. International Journal of Fatigue 21 (7): 709–717

Das D K (2013) Microstructure and High Temperature Oxidation Behavior of Pt-modified Aluminide bond Coats on Ni-base Superalloys. Progress in Materials Science 58 (2): 151–182

Davies H (1963) The Design and Development of the Thiocol XRL99 Rocket Engine for the X-15 Aircraft. Journal of the Royal Aeronautical Society 167: 79–91

Davis A W, Evans A G (2006) Effects of bond Coat Misfit Strains on the Rumpling of thermally grown Oxides. Metallurgical and Materials Transactions A 37 (7): 2085–2095

Deb P et al (1987) Surface Stability of Platinum Modified Aluminide Coatings during 1100 °C Cyclic Test. Journal of Vacuum Science and Technology A 5 (6): 3366–3372

Dobek L J (1973) Labyrynth Seal Testing for Lift Fan Engines. Available via DIALOG. https://ntrs.nasa.gov/archive/nasa/casi.ntrs.nasa.gov/19730007797.pdf. Accessed 4 September 2017

Dorfman M R et al (2008) Recent Development of High Temperature Thermal Spray Abradable Coatings. Ceramics Japan, The Ceramic Society of Jaan 43 (5): 389–395

Driver M (2012) Coatings for Biomedical Applications. Woodhead Publishing Ltd, Cambridge (UK)

Dryepondt S, Pint B A (2010) Determination of the Ductile to Brittle Temperature Transition of Aluminide Coatings and its Influence on the mechanical Behavior of coated Specimens. Surface Coatings and Technology 205: 1195–1199

Duvall D S (1981) Processing Technology for Advanced Metallic and Ceramic Turbine Airfoil Coatings. In: Proceedings of 2nd Conference of Advanced Materials for Alternative-Fuel-Capable Heat Engines, Electric Power Research Institute, 24–28 August 1981

Evans et al (2008) The Influence of Oxides on the Performance of Advanced Gas Turbines. Journal of the European Ceramic Society 28 (7): 1405–1419

Fauchais P L et al (2014) Thermal spray Fundamentals: From Powder to Part. Srpinger

Felten E J (1976) Use of Platinum and Rhodium to improve Oxide Adherence on Ni-8Cr-6Al Alloys. Oxidation of Metals 10 (1): 23–28

Felten E J, Pettit F S (1976) Development, Growth, and Adhesion of Al_2O_3 on Platinum-Aluminum Alloys. Oxidation of Metals 10 (3): 189–223

Firdaouss M et al (2017) Tungsten Coating by ATC Plasma Spraying on CFC for WEST Tokamak. Physica Scripta T170

Fountain J G et al (1976) The Influence of Platinum on the Maintenance of α-Al_2O_3 as a protective Scale. Oxidation of Metals 10 (5): 341–345

Galmiche P M (1975) US Patent 3,900,613

Garrett E G, Gyorgak C A (1953) Adhesive and Protective Characteristics of Ceramic Coating A-417 and its Effects on Engine Life of Forged Refractory-26 (AMS 5760) and Cast Stellite (AMS 5385) Turbine Blades. Available via DIALOG. https://ntrs.nasa.gov/search.jsp?R=19930082087. Accessed 20 July 2017

Garvie R C et al (1972) Structure and Thermomechanical Properties of Partially Stabilized Zirconia in the CaO-ZrO_2 System. Journal of the American Ceramic Society 55 (3) 152–157

Garvie R C et al (1975) Ceramic Steel? Nature 258: 703–704

Garvie RC, Goss M F (1986) Intrinsic Size Dependence of the Phase Transformation in Zirconia Microcrystals. Journal of Materials Science 21 (4): 1253–1257

Gauje G, Morbioli R (1983) Vapor Phase Aluminizing to Protect Turbine Airfoils. In: Singhal S C (ed) High Temperature Protective Coatings. The Metallurgical Society of AIME, Atlanta, GA, p 13–26

Giggins C S, Pettit F S (1979) Hot Corrosion Degradation of Metals and Alloys – A Unified Theory. Pratt & Whitney

Goward G W (1998) Progress in Coatings for Gas Turbine Airfoils. Surface and Coatings Technology 108-109: 73–79

Green D J et al (1989) Transformation Toughening of Ceramics. CRC Press, Boca Raton, Florida, USA

Guo H et al (2009) Thermo-physical and Thermal Cycling Properties of Plasma-sprayed $BaLa_2Ti_3O_{10}$ Coating as Potential Thermal Barrier Materials. Surface and Coatings Technology 204 (5): 691–696

Gupta D K, Duvall D S (1984) A Silicon and Hafnium modified Plasma Sprayed MCrAlY Coating. In: Bricknell R H et al (eds) Superlloys. The Metallurgical Society of AIMR, Warrendale, PA, p 711–720

Harrison W N (1947) Review of an Investigation of Ceramic Coatings for Metallic Turbine Parts and Other High Temperature Applications. Available via DIALOG. https://ntrs.nasa.gov/archive/nasa/casi.ntrs.nasa.gov/19930087630.pdf. Accessed 20 July 2017

Hazell B et al (2008) Development of improved bond Coat for enhanced Turbine Durability. Superalloys 2008: 753–760

Heimann R B (2006) Plasma Spray Coating: Principles and Applications 2nd edn. Wiley VHC

Helm D, Olaf Roder O (2007) Recent Titanium Research and Development in Germani. In: Ninomi M et al (ed) Ti-2007 Science and Technology. The Japan Institute of Metals, p 25–32

Hermanek F J (2001) Thermal Spray Terminology and Origins. ASM International, Materials Park, OH

Hjelm L N, Bornhorst B R (1961) Research-Airplane-Committee on Conference on the Progress of the X-15 Project. Available via DIALOG. https://ntrs.nasa.gov/search.jsp?R=19710070129. Accessed 20 July 2017

Hocking M G et al (1989) Metallic and Ceramic Coatings. Longman Scientific & Technical

Ibegazene H et al (1993) Microstructure of Yttria stabilized Zirconia-Hafnia Plasma sprayed Thermal Barrier Coatings. Journal De Phisique IV 3: 1013–1016

Jackson R W et al (2014) Thermal Barrier Coatings Adherence to Hf-modified B2 NiAl bond Coatings. Acta Materialia 80: 39–47

Jackson R W et al (2015) Interaction of molten Silicates with Thermal Barrier Coatings under Temperature Gradients. Acta Materialia 89: 396–407

Jarligo M O et al (2010) Atmospheric Plasma Spraying of High Melting Temperature Complex Perovskites for TBC Application. Journal of Thermal Spray Technology 19 (1–2): 303–310

Jorgensen D J et al (2016) Bond Coatings with high Rumpling Resistance: Design and Characterization. Surface and Coating Technology 300: 25–34

Joseph A D (1960) US Patent 3,102,044

Kablow E N, Muboyadzhyan S A (2017) Erosion-Resistant Coatings for Gas Turbine Engine Compressor Blades. Russian Metallurgy 2017 (6): 494–504

Kakuda T R et al (2015) The Thermal Behavior of CMAS-infiltrated Thermal Barrier Coatings. Surface and Coatings Technology 272: 350–356

Kermapur A et al (2015) Failure Analysis of Ti6Al4V Gas Turbine Compressor Blades. Engineering Failure Analysis 15 (8): 1052–1064

Khanna A S (2002) Introduction to High Temperature Oxidation. ASM International

Kingery W D (1955) Thermal Conductivity: Temperature Dependence of Conductivity for Single-phase Ceramics. Journal of the American Ceramic Society 38: 251–255

Krämer S et al (2008) Mechanism of Cracking and Delamination within Thermal Barrier Systems in Aeroengines subject to Calcium-Magnesium-Alumino-Silicate (CMAS) Penetration. Materials Science and Engineering A 490 (1–2): 26–35

Lammermann H, Kienel G (1991) PVD Coatings for Aircraft Turbine Blades. Advanced Materials and Processes 140 (6): 18–23

Lattime S B, Steinetz B M (2004) Turbine Engine Clearance Control Systems: Current Practices and Future Directions. Journal of Propulsion and Power 20 (2): 302–311

Lee J W, Kuo Y C (2005) Cyclic Oxidation Behavior of a Cobalt Aluminide Coating on Co-base Superalloy AMS 5608. Surface and Coatings Technology 200 (5–6): 1225–1230

Lelait L et al (1993) Microstructural Investigations of EBPVD Thermal Barrier Coatings. Journal De Physique IV 3: 645–654

Lesnikova E G, Lesnikov V P (1986) Influence of the β-Phase of the Aluminide Coating on the Condition and Scale Resitance of the Surface Layer of NBi-Al alloys. Metal Science and Heat Treatment 28 (5): 372–376

Levi C G et al (2012) Environmental Degradation of Thermal Barrier Coatings by molten Deposits. MRS Bulletin 37: 932–941

Lughi V, Clarke D R (2005) High Temperature Aging of YSZ Coatings and subsequent Transformation at low Temperature. Surface and Coatings Technology 200 (5–6): 1287–1291

Luthra K L, Leblanc O H (1987) Low Temperature Hot Corrosion of CoCrAl Alloys. Materials Science and Engineering 87: 329–335

Markou C et al (2011) IATA Maintenance Cost Executive Commentary. Available via DIALOG. https://www.iata.org/whatwedo/workgroups/Documents/MCTF/AMC-Exec-Comment-FY14.pdf. Accessed 12 November 2017

Meelu M C et al (1992) Sermaloy J (Silicon modified Aluminide) Coating mechanical Improvements. Processing, Properties and Applications of Metallic and Ceramic Materials 2: 1247–1253

Meier S M et al (1990) Status of Ceramic Thermal Barrier Coatings – Gas Turbine Applications and Life Prediction Methods. In: Proceedings of Coatings for Advanced Heat Engines Workshop, US Department of Energy, 6–9 August 1990

Mercer C et al (2005) A Delamination Mechanism for Thermal Barrier Coatings subject to Calcium-Magnesium-Alumino-Silicate (CMAS) Infiltration. Acta Materialia 53 (4): 1029–1039

Mercer C et al (2007) On a ferroelastic Mechanism governing the Toughness of Metastable tetragonal-prime (t') Yttria-stabilized Zirconia. Proceedings of the Royal Society A: Mathematica, Physical and Engineering Sciences 463 (2081): 1393–1408

Mévrel R (1989) State of Art on High Temperature Corrosion resistant Coatings. Materials Science and Engineering A 120–121 Part 1: 13–24

Mévrel R et al (1986) Pack Cementation Process. Materials Science and Technology 2: 201–206

Miller R A et al. Phase Stability in Plasma-sprayed partially stabilised Zirconia-Yttria. In: Heuer A H, Hobbs L W, editors. Advances in Ceramics Volume 3 – Science and Technology of Zirconia. The American Ceramic Society, Columbus, Ohio; 1981. p. 241–253

Muboyadzhyan S A (2009) Erosion Resistant Coatings for Gas Turbine Compressor Blades. Russian Metallurgy 2009 (3): 183–186

Mumm D R et al (2001) Characterization of a Cyclic Displacement Instability for a thermally grown Oxide in a Thermal Barrier System. Acta Materialia 49 (12): 2329–2340

Naji A et al (2015) Improvements in the Thermodynamic and Kinetic Considerations on the Coating Design for Diffusion Coatings formed via Pack Cementation. Materials and Corrosion 66 (9): 863–868

Nicholls et al (1999) A Comparison between the Erosion Behavior of Thermal Spray and Electron-beam Physical Vapour Deposition Thermal Barrier Coatings. Wear 233: 352–361

Nicholls J R (2000) Designing Oxidation-resistant Coatings. JOM 52 (1): 28–35

Nicholls J R (2003) Advances in Coating Design for High-Performance Gas Turbines. MRS Bulletin 28 (9): 659–670

Nicholls J R et al (2002) Smart Overlay Coatings – Concept and Practice. Surface and Coatings Technology 149: 236–244

Nicoll A R et al (1986) Future Developments in Plasma Spray Coating. Materials Science and Technology 2 (3): 214–219

Oerlikon Metco (2014) Solutions Flash: Improve Efficiency and Reduce Emissions with High Pressure Turbine Abradable Coatings for Industrial Gas Turbines. Available via DIALOG. https://www.oerlikon.com/ecomaXL/files/metco/oerlikon_SF-0015.1_IGT_HPT_Abradables_EN.pdf&download=1. Accessed 12 November 2017

Oerlikon Metco (2016) Material Product Data Sheet – Copper Nickel and Copper Nickel Indium Thermal Spray Powders. Available via DIALOG. https://www.oerlikon.com/ecomaXL/files/metco/oerlikon_DSMTS-0061.5_CuNi_CuNiIn_Powders.pdf&download=1. Accessed 12 November 2017

Padture N P et al (2002) Thermal Barrier Coatings or Gas-Turbine Engine Applications. Science 296: 280–284

Pan W et al (2012) Low Thermal Conductivity Oxides. MRS Bulletin 37: 917–922

Park Y W et al (2006) Effect of Fretting Amplitude and Frequency on the Fretting Corrosion Behavior of Tin plated Contacts. Surface and Coatings Technology 201: 2181–2192

Parzukowski R S (1977) Gas-Phase Deposition of Aluminum on Nickel Alloys. Thin Solid Films 45: 349–355

Pérez F J et al (1999a) Aluminizing and chromizing Bed Treatment by CVD in a fluidized Bed Reactor on Austenitic Stainless Steels. Surface and Coatings Technology 120–121: 151–157

Pérez F J et al (1999b) Kinetic Studies of Cr and Al Deposition using CVD-FBR on different metallic Substrates. Surface and Coatings Technology 122: 281–289
Pint B A (1997) Study of the Reactive Element Effect in ODS Iron-Base Alumina Formers. Materials Science Forum 251–254: 397–404
Pint B A (2004) The Role of Chemical Composition on the Oxidation Performance of Aluminide Coatings. Surface and Coatings Technology 188–189: 71–78
Pint B A et al (1998) Substrate and Bond Coat Compositions: Factors affecting Alumina Scale Adhesion. Materials Science and Engineering A 245 (2): 201–211
Pint B A et al (2001) Evaluation of Iron-Aluminide CVD Coatings for High Temperature Corrosion Protection. Materials at High Temperatures 18 (3): 185–192
Pint B A et al (2006) Oxidation Resistance: One Barrier to moving beyond Ni-base Superalloys. Materials Science and Engineering A 415 (1–2):255–263
Pint B A et al (2011) Effect of increased Water Vapor Levels on TBC Lifetime with Pt-containing bond Coatings. Surface and Coatings Technology 206: 1566–1570
Pint B A, Zhang Y (2011) Performance of Al-rich Oxidation-resistant Coatings for Fe-base Alloys. Materials and Corrosion 62 (6): 549–560
Poerschke D L et al (2016) Equilibrium Relationships between Thermal Barrier Oxides and Silicate Melts. Acta Materialia 120: 302–314
Pokluda J, Kianicová M (2010) Damage and Performance Assessment of Protective Coatings on Turbine Blades. Available via DIALOG. https://www.intechopen.com/download/pdf/12092. Accessed 28 October 2017
Prater J T et al (1981) Proceedings of the second Conference on Advanced Materials for alternate Fuels capable Heat Engines. Palo Alto, CA 1981
Praxair Surface Technologies (2000) Praxair and TAFA Arc Spray Copper-Nickel-Indium Wire-58T. Available via DIALOG. http://tatiscia.com/wp-content/pdf/tafa/1.9.1.2-58T%20-%20Copper%20Nickel%20Indium%20Wire.pdf. Accessed 13 November 2017
Priest M S, Zhang Y (2015) Synthesis of Clean Aluminide Coatings on Ni-based Superalloys via modified Pack Cementation Process. Materials and Corrosion 66 (10): 1111–1119
Qu Z et al (2011) Thermal Conductivity of the Gadolinium Calcium Silicate Apatites: Effect of Different Point Defect Types. Acta Materialia 59 (10): 3941–3850
Raghavan S et al (2001) Thermal Properties of Zirconia co-doped with trivalent and pentavalent Oxides. Acta Materialia 49 (1): 169–179
Raghavan S et al (2004) Ta_2O_5/Nb_2O_5 and Y_2O_3 Co-doped Zirconias for Thermal Barrier Coatings. Journal of the American Ceramic Society 87 (3): 431–437
Rajendran R (2012) Gas Turbine Coatings – An Overview. Engineering Failure Analysis 26: 355–369
Reed R C (2006) The Superalloys – Fundamentals and Applications. Cambridge University Press, Cambridge (UK)
Ren W et al (2014) Oxidation and Microstructure Evolution of Cobalt Aluminide Coatings on directionally-solidified Superalloys during long term Exposure at 1000 °C. Materials Research Innovations 18 (S4):945–951
Restall J E, Wood M I (1986) Alternative Processes and Treatments. Materials Science and Technology 2: 225–231
Rigney D V et al (1997) PVD Thermal Barrier Coating Applications and Process Development for Aircraft Engines. Journal of Thermal Spray Technology 6 (2): 167–175
Rolls-Royce (1996) The Jet Engine 5th edition. Rolls-Royce plc, Derby
Ruud J A et al (2001) Strength Degradation and Failure Mechanisms of Electron-beam Physical-vapor-deposited Thermal Barrier Coatings. Journal of the American Ceramic Society 84 (7): 1545–1552
Sato T et al (1985) Transformation of Yttria partially stabilized Zirconia by low Temperature Annealing in Air. Journal of Materials Science 20 (4): 1466–1470

Scharrer and Pelletti (1995) Leakage and Rotordynamic Effects of Compressor annular Seals. In: Proceedings of 24th Turbomachinery Symposium, Huston, Texas, 26–28 September 1995, p 175

Schmid R (1997) New High Temperature Abradable Coatings for Gas Turbines. PhD Thesis, Swiss Federal Institute of Technology

Scott H G (1975) Phase Relationship in the Zirconia-Yttria System. Journal of Materials Science 10 (9): 1527–1535

Seelig R P, Steuber R J (1978) High-Temperture-resistant Coatings for Superalloys. High Temperatures – High Pressures 10 (2): 207–213

Shankar S, Seigle L L (1978) Interdiffusion and intrinsic Diffusion in the NiAl (δ) Phase of the Al-Ni System. Metallurgical Transactions A 9 (10): 1467–1476

Shen et al (2008) Anisotropic thermal Conductivity of the Aurivillus Phase, Bismuth Titanate ($Bi_4Ti_3O_{12}$): A natural nanostructured Superlattice. Applied Physics Letters 93 (10) 102907

Shen Y et al (2010) Low thermal Conductivity without Oxygen Vacancies in equimolar $YO_{1.5}$+$TaO_{2.5}$- and $YbO_{1.5}$+$TaO_{2.5}$- stabilized Zirconia Ceramics. Acta Materialia 58 (13): 4424–4431

Smialek J L (1971) Marteniste in NiAl Oxidation Resistant Coatings. Metallurgical Transactions 2 (3): 913–915

Smith A B et al (1999) Vapour Aluminide Coating of Internal Cooling Channels in Turbine Blades and Vanes. Surface Coatings Technology 120–121: 112–117

Smith J S, Boone D H (1990) Platinum Modified Aluminides-Present Status. In: Gas Turbine and Aeroengine Congress and Exposition, Brussels, Belgium, 1980

Song X W et al (2011) Influence of the partial Substitution of Y_2O_3 with Ln_2O_3 (Ln=Nd, Sm, Gd) on the Phase Structure and thermophysical Properties of ZrO_2-Nb_2O_3-Y_2O_3 Ceramics. Acta Materialia 59 (10): 3895–3902

Sparks T D et al (2010) Anisotropic Thermal Diffusivity and Conductivity of La-doped Strontium Niobate $Sr_2Nb_2O_7$. Journal of the American Ceramic Society 93 (4): 1136–1141

Sporer D R et al (2011) US Patent 7981530 B2

Squillace A et al (1999) The Control of the Composition and Structure of Aluminide Layers formed by Vapour Aluminising. Surface and Coatings Technology 120–121: 118–123

Stecura S (1986) Advanced Thermal Barrier System bond Coatings for Use on Nickel-, Cobalt- and Iron-base Alloy Substrate. Thin Solid Films 136 (2): 241–256

Strangman T E (1987) Development and Performance of Physical Vapor Deposition Thermal Barrier Coating Systems. In: Proceedings of Coatings for Advanced Heat Engines Workshop, US Department of Energy, 27–30 July 1987

Strangman T E (1996) US Patent 5,514,482

Talboom F P, Grafwallner J (1970) US Patent 3,542,5330

Tawancy H M et al (1991) Role of Platinum in Aluminide Coatings. Surface and Coatings Technology 49 (1-3):1–7

Thorpe M L, Richter J (1992) A pragmatic Analysis and Comparison of HVOF Processes. Journal of Thermal Spray Technology 1 (2): 161–170

Tolpygo V K et al (2008) Effect of Hf, Y and C in the underlying Superalloy on the Rumpling of diffusion Aluminide Coatings. Acta Materialia 56 (3): 489–499

Tolpygo V K, Clarke D R (2004a) On the Rumpling Mechanism in Nickel-Aluminide Coatings: Part I: An experimental Assesment. Acta Materialia 52 (17): 5115–5127

Tolpygo V K, Clarke D R (2004b) On the Rumpling Mechanism in Nickel-Aluminide Coatings: Part II: Characterization of Surface Undulations and bond Coat Swelling. Acta Materialia 52 (17): 5129–5141

Tu J et al (1994) The Size Effect of the martensitic Transformation in ZrO_2-containing Ceramics. Journal of Materials Science 29 (6): 1662–1665

Tyron B et al (2006) Multilayered Ruthenium-modified bond Coats for Thermal Barrier Coatings. Metallurgical and Materials Transactions A 37 (11): 3347–3358

Tyron B et al (2007) Hybrid Intermetallic Ru/Pt modified bond Coatings for Thermal Barrier Systems. Surface and Coatings Technology 2020 (2): 349–361

Unocic K A, Pint B A (2013) Effect of Water Vapor on thermally grown Alumina Scales on bond Coating. Surface and Coatings Technology 215: 30–38

Van Ness D et al (2006) Turbine Tip Clearance Flow Control Using Plasma Actuators. Paper presented at the 44th AIAA Aerospace Sciences Meeting and Exhibit, Reno, Nevada, 9–-12 January 2006

Van Roode M, Hsu L (1989) Evaluation of the Hot Corrosion Protection of Coatings for Turbine hot Section Components. Surface and Coatings Technology 37 (4): 461–481

Vassen R et al (2000) Zirconates as New Materials for Thermal Barrier Coatings. Journal of the American Ceramic Society 83 (8): 2023–2028

Vassen R et al (2010) Overview on advanced thermal Barrier Coatings. Surface and Coatings Technology 205 (4): 938–942

Vassen R, Stöver D (2007) Influence of Microstructure on the Thermal Cycling Performance of Thermal Barrier Coatings. In: Marple B R et al (ed) Thermal Spray 2007: Global Coating Solutions: Proceedings of the 2007 International Thermal Spray Conference. ASM International, Materials Park, Ohio, 2007

Wang et al (2011) Diffusion Barrier behaviors of (Ru, Ni)Al/NiAl Coatings on Ni-based Superalloy Substrate. Intermetallics 19 (2): 191–195

Warnes B M, Punola D C (1997) Clean Diffusion Coatings by Chemical Vapor Deposition. Surface and Coatings Technology 94–95: 1–6

Wilden J et al (2006) Plasma transferred Arc Welding – Modeling and experimental Optimization. Journal of Spray Technology 15 (4): 779–784

Wilson S (2012) Thermally sprayed abradable Coating Technology for Sealing in Gas Turbines. Oerlikon Metco White Paper – Thermally Sprayed Abradable Coatins 10: 1–9

Winter M R, Clarke D R (2007) Oxide Materials with Low Thermal Conductivity. Journal of the American Ceramic Society 90 (2): 533–540

Wolfe D E et al (2005) Tailored Microstructure of EB-PVD 8YSZ Thermal Barrier Coatings with low thermal Conductivity and high thermal Reflectivity for Turbine Applications. Surface and Coatings Technology 190: 132–149

Wright I G, Gibbons T B (2007) Recent Developments in Gas Turbine Materials and Technology and their Implications for Syngas Firing. International Journal of Hydrogen Energy 32 (16): 3610–3621

Wu L T et al (2017) A prominent Driving Force for the Spallation of Thermal Barrier Coatings: Chemistry dependent Phase Transformation of the bond Coat. Acta Materialia 137: 22–35

Wuench B J, Eberman K W (2000) Order-Disorder Phenomena in $A_2B_2O_7$ Pyrochlore Oxides. JOM 52 (7): 19–21

Xu T et al (2004) Observations and Analyses of Failure Mechanisms in Thermal Barrier Systems with two Phase bond Coats based on NiCoCrAlY. Acta Materialia 52 (6): 1439–1450

Yang S et al (2008) Anisotropic Thermal Conductivity of the Aurivillus Phase, Bismuth Titanate ($Bi_4Ti_3O_{12}$): A Natural Nanostructured Superlattice. Applied Physics Letters 93 (10): 102907–102907-3

Zhang Y et al (2003) Martensitic Transformation in CVD NiAl and (Ni, Pt)Al bond Coatings. Surface and Coatings Technology 163–164: 19–24

Zhang Y et al (2004) Effect of Cycle Length on the Oxidation Performance of Iron Aluminide Coatings. Surface and Coatings Technology 188–189: 35–40

Zhang Y et al (2007) Interdiffusional Degradation of Oxidation-resistant Aluminide Coatings on Fe-base alloys. Materials and Corrosion 58 (10): 751–761

Zhao M et al (2012) Properties of Yttria-Stabilized-Zirconia Based Ceramic Composite Abradable Coatings. Key Engineering Materials 512–515: 1551–1554

Zhou Z R (1999) Lubrication in Fretting – A Review. Wear 225–229: 962–967

Zhu D M, Miller R A (2004) Development of Advanced Low Conductivity Thermal Barrier Coatings. International Journal of Applied Ceramic Technology 1 (1): 86–94

Further Reading

Davis J R (2004) Handbook of Thermal Spray Technology. ASM International
Dorfman M R et al (2013), Thermal Spray Technology Growth in Gas Turbine Applications. In: Tucker R C (ed) Thermal Spray Technology vol. 5A, ASM International, pp 280–286
Xu H and Guo H (2011) Thermal Barrier Coatings. Woodhead publishing
Kanna A S (2016) High Temperature Corrosion. World Scientific
Tamarin Y (2002) Protective Coatings for Turbine Blades. ASM International

Chapter 8
Corrosion

8.1 Introduction

Corrosion can be defined, in general terms, as a deterioration process that metallic materials undergo in specific environmental conditions, involving chemical reactions that may significantly change the original properties of the base alloys, including those used in aerospace applications. The concept of corrosion can be applied beyond the limits of metallic materials, and it is nowadays used to describe deterioration of a wider range of nonmetallic materials too, like rubbers, polymers, and ceramics (ASM International 2003). Liquid metal corrosion and deterioration of paints exposed to sunlight are just a few examples of the relevant phenomenology (Fontana 1987). In some respects, corrosion of metals can be regarded as the opposite transformation to the metallurgical extraction process. Apart from the so-called noble metals, the other metallic elements are usually found in nature as compounds, like oxides, sulfides, sulfates, silicates, and carbonates, just to mention the main ore components. These same compounds are often present as products of the corrosion reactions.

Corrosion has a large impact in several technological fields, like transportation, food industry, energy production, petrochemical, and aerospace. Effects of corrosion phenomena can be observed on a great variety of products, objects, and piece of equipment, like metallic structures, pipelines, vehicles, biomedical implants, chemical and nuclear reactors, structural parts of aircrafts, and parts of engines (Pedeferri 2010). An important step in the development of products that need to resist corrosion is their accurate structural and functional design, taking into account the environmental working conditions, which will thus complete the set of parameters to be accounted for in view of a reliable life cycle and life cycle cost assessment. However, it is not always that simple to predict and anticipate all service conditions and relevant protection measures against corrosion phenomena that may even lead to

S. Gialanella, A. Malandruccolo, *Aerospace Alloys*, Topics in Mining, Metallurgy and Materials Engineering, https://doi.org/10.1007/978-3-030-24440-8_8

Table 8.1 Frequency of failure mechanisms for engineering and aircraft components (Findlay and Harrison 2002)

	Percentage of component failures [%]	
Mechanism	Aircraft	General Engineering
Fatigue	55	25
Corrosion	16	29
Overload	14	11
Mechanochemical corrosion and hydrogen embrittlement	7	6
Abrasion/erosion/wear	6	3
High-temperature corrosion	2	7
Brittle fatigue	–	16
Creep	–	3

part failure, when the component has lost its load-bearing capabilities. Failures in aircraft structures are commonly associated with stress concentration (see Sect. 1.2) due to:

- Design mistakes, e.g., holes in the wrong position, notches, and too tight fillet radii
- Corrosion attack
- Microstructural features of the material, e.g., segregations and voids and unfitted grain size

Records and case histories data show that for aircraft components the dominant failure mode is fatigue (Findlay and Harrison 2002). Table 8.1 shows a comparison between the failure modes for general engineering and aircraft components with the relative incidence of each one of them. After fatigue (55%), corrosion-related mechanisms turn out to be the second reason for failure (25% in total). Moreover, the efforts for detection and repair of corrosion damages are prevailing over other similar activities, like those involved with fatigue.

Different from fatigue damage, corrosive phenomena are generally slower and their effects easier to be detected and, if possible, mended with routine maintenance activities (Findlay and Harrison 2002). Corrosion protection involves both thermodynamic and kinetic issues. The strategy aims not much at avoiding corrosion at all but rather slowing it down, so to guarantee a reasonable and well-predictable service lifetime. Thermodynamics provides a quantitative indication of the driving force for the corrosion process to occur. The rate of reactions can be determined by means of kinetic laws that help distinguishing among different corrosion processes, since they mainly depend on the environment in which corrosion takes place (ASM International 2003).

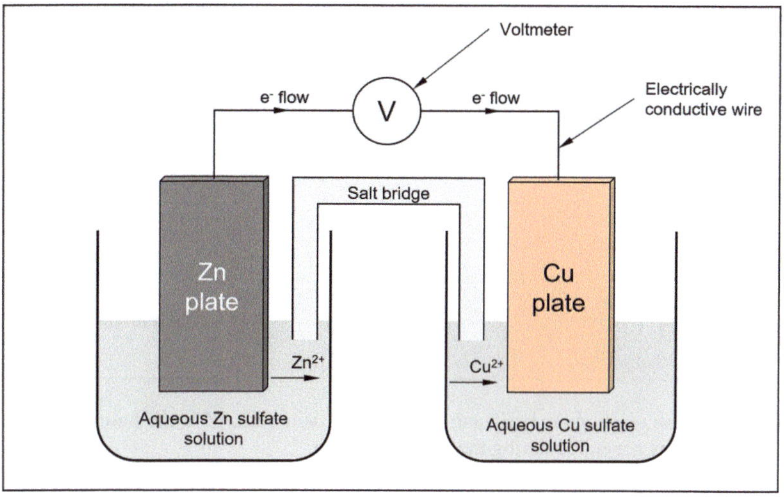

Fig. 8.1 The Daniell battery: a reference electrochemical system, useful to discuss the main reactions involved with electrochemical (aqueous) corrosion. The zinc plate is the anode, while the copper plate is the cathode

Two main groups of corrosion phenomena can be identified: electrochemical corrosion and high-temperature corrosion. Electrochemical corrosion is quite a common process, which depends on reactions involving transfer of electrons from and to metallic alloys. In this regard, batteries can be considered as prototypes of electrochemical systems. Referring to the copper-zinc cell developed by John Frederic Daniell, Cu and Zn electrically connected plates are immersed in solutions containing their salts (copper and zinc sulfate in the present example) (see Fig. 8.1). A salt bridge between the two solutions allows the electrochemical cell circuit to be completed. The two metals exhibit a different tendency to be corroded. In particular, zinc is oxidizing, according to the reaction: $Zn^{2+} + 2e^{-}$ resulting in the dissolution of zinc in the zinc sulfate solution. Copper is reducing, i.e., copper ions (Cu^{2+}) contained in the copper sulfate solution turn into metallic Cu by capturing the electrons produced by zinc oxidation and reaching the copper plate. Common features to all electrochemical corrosion processes are:

- The overall corrosion reaction can be split into two "half-reactions": anodic and cathodic.
- The two half reactions occur either in the same region at the interface between the metal and the environment or in separate spots, depending on the local electrochemical potential.
- Since the half reactions are sustained by electron release and trapping, they proceed at the same rate.
- Liquid water or alternative electrolytes are necessary.

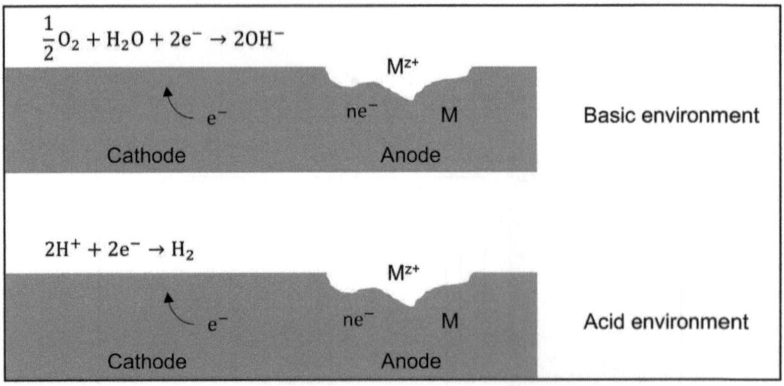

Fig. 8.2 Corrosion reactions in basic and acid aqueous solutions. M is the base metal; ne^- is the number of electrons generated during the process

For aqueous solutions, it is important to consider also the water electrolysis reactions, since they have an active role in the corrosion process. With reference to the scheme in Fig. 8.2, depicting a metal in contact with pure water, the oxidation reaction involving the metal is:

$$M \rightarrow M^{z+} + ze^- \tag{8.1}$$

The reduction reactions involving water are:

$$\frac{1}{2}O_2 + H_2O + 2e^- \rightarrow 2OH^- \tag{8.2}$$

$$2H^+ + 2e^- \rightarrow H_2 \tag{8.3}$$

Equations 8.2 and 8.3 describe the reduction reactions that take place in a basic and acid environment, respectively.

Considering the oxidation reaction, z indicates the oxidation number, i.e., the valence of the metal involved in the process, influencing the density of charge carriers. The electron flow is generated by a difference in electrical potential between cathodic and anodic regions.

The chemical equilibrium reached by a metal in contact with an electrolyte can be described in terms of an electrical potential of the relevant half reaction. The electrical potential of each half reaction, E^0, is usually measured with respect to a standard electrode potential, taken equal to zero. This is the standard hydrogen electrode (SHE) potential. The E^0 series in Table 8.2 is obtained in standard conditions, i.e., considering each single chemical species in a 1 mole concentration solution of the relevant ion. The series is obtained under equilibrium conditions and cannot account for the kinetic aspects of the corrosive process. The kinetics of the corrosive reaction strongly depends on the passivation phenomena. Passivation

Table 8.2 Selected standard reduction potentials, E^0, referred to the standard hydrogen electrode, SHE at 298 K. The concentration of each aqueous solution is 1.0 mol/dm^3. The pressure of a gaseous component is 10^5 Pa (1 atm) (Housecroft and Constable 2006)

Reduction half-reaction	E^0 [V]
$Li^+(aq) + e^- \leftrightarrow Li(s)$	−3.04
$K^+(aq) + e^- \leftrightarrow K(s)$	−2.93
$Ca^{2+}(aq) + 2e^- \leftrightarrow Ca(s)$	−2.87
$Na^+(aq) + e^- \leftrightarrow Na(s)$	−2.71
$Mg^{2+}(aq) + 2e^- \leftrightarrow Mg(s)$	−2.37
$Al^{3+}(aq) + 3e^- \leftrightarrow Al(s)$	−1.66
$Zn^{2+}(aq) + 2e^- \leftrightarrow Zn(s)$	−0.76
$S (s) + 2e^- \leftrightarrow S^{2-} (aq)$	−0.48
$Fe^{2+}(aq) + 2e^- \leftrightarrow Fe(s)$	−0.44
$Cr^{3+}(aq) + e^- \leftrightarrow Cr^{2+}(s)$	−0.41
$Co^{2+}(aq) + 2e^- \leftrightarrow Co(s)$	−0.28
$Pb^{2+}(aq) + 2e^- \leftrightarrow Pb(s)$	−0.13
$Fe^{3+}(aq) + 3e^- \leftrightarrow Fe(s)$	−0.04
$2H^+(aq) + 2e^- \leftrightarrow H_2$(gaseous at 1 bar)	0.00
$2H^+(aq) + S(s) + 2e^- \leftrightarrow H_2S(s)$	+0.14
$Cu^{2+}(aq) + e^- \leftrightarrow Cu^+(s)$	+0.15
$Cu^{2+}(aq) + 2e^- \leftrightarrow Cu(s)$	+0.34
$O_2(g) + 2H_2O(l) + 4e^- \leftrightarrow 4[OH]^- (aq)$	+0.40
$Cu^+(aq) + e^- \leftrightarrow Cu(s)$	+0.52
$I_2(aq) + 2e^- \leftrightarrow 2I^- (aq)$	+0.54
$Fe^{3+}(aq) + e^- \leftrightarrow Fe^{2+}(aq)$	+0.77
$Ag^+(aq) + e^- \leftrightarrow Ag(s)$	+0.80
$Br_2(aq) + 2e^- \leftrightarrow 2Br^-(s)$	+1.09
$O_2(g) + 4H^+(g) + 4e^- \leftrightarrow 2H_2O (l)$	+1.23
$[Cr_2O_7]^{2-} (aq) + 14H^+ (aq) + 6e^- \leftrightarrow 2Cr^{3+}(aq) + 7H_2O(l)$	+1.33
$Cl_2(aq) + 2e^- \leftrightarrow 2Cl^- (aq)$	+1.36
$[MnO_4]^- (aq) + 8H^+ (aq) + 5e^- \leftrightarrow Mn^{2+}(aq) + 4H_2O(l)$	+1.51
$Ce^{4+}(aq) + e^- \leftrightarrow Ce^{3+}(aq)$	+1.72
$F_2(aq) + 2e^- \leftrightarrow 2F^-(aq)$	+2.87

aq aqueous, *g* gaseous, l liquid, *s* solid

involves the formation of surface layer of metal native oxides and hydroxides. These reaction products interfere with the electrochemical degradation process, creating an insulating barrier on the metal surface. This leads to a protective condition, generated by the presence of corrosion products on the corroding metal surface. The actual protection of the surface corrosion layer strongly depends on the characteristics of the environment in which the corrosion process is taking place. For instance, iron undergoes passivation only in alkaline environments (pH > 7). The passivation capability of metallic materials can be accounted for in the definition of an electrochemical classification. Namely, those metals that exhibit passivation would occupy places in the upper part of this ranking. Real alloys do not have a single precise but rather a range of electrochemical potentials. The potential range is determined by the presence of alloying elements and impurities. Moreover, the way the material undergoes corrosion is a function of both the potential and of the environment in which the corrosive phenomena take place (McCafferty 2010). A specific combination of environment and stress field, both applied and internal, will determine the actual corrosion behavior of any material. The surface conditions, concerning finishing, defect structures, actual composition, and relevant evolution with time, would strongly influence the kinetics and thermodynamics of corrosion processes. The combination of these factors determines diverse corrosion types that can be summarized as follows:

- General corrosion (uniform or nonuniform)
- Pitting
- Crevice
- Selective corrosion
- Galvanic corrosion (two metal corrosion)
- Mechanochemical corrosion, including fretting and stress corrosion cracking (SCC)
- Hydrogen embrittlement (HE)

The most important corrosion types, fully discussed in (McCafferty 2010), will be herewith illustrated with reference to specific cases, relevant to the aerospace field.

8.1.1 General Corrosion

General corrosion occurs either on the entire surface of a metal component or on a very wide part of it, on which the general attack causes the metal to be corroded homogeneously. General corrosion occurs in aircraft structures under normal service conditions, in particular, in those parts of the airplane where condensation of water may intervene. Cathodic and anodic reactions change dynamically over the time, with reaction products that determine a uniform weight change all over the involved surface area. A weight loss is observed as a consequence of non-adherent corrosion products that may spall off after some time. Nonuniform general corrosion is characterized by an irregular spatial distribution of the corrosive attack and consequent weight change. Uniform general corrosion can be tolerated if it proceeds at a sufficiently low rate to guarantee an adequate lifetime to the component. Actually, general corrosion is not necessarily to be regarded as detrimental, since it might also have positive effects; therefore it can be deliberately induced, like in the following application cases:

- Increasing the roughness of the surface for specific applications, e.g., osteogenesis improvement in orthopedic implants.
- Metallographic attack for enhancing the surface contrast of the microstructure of crystalline materials.

- Pickling of the surface as preparation treatment for further surface operations, like painting.

However, if prolonged in time, general corrosion may lead to a weakening of the base metal, with a load redistribution causing local stress concentration and, possibly, secondary bending (Benavides 2014).

If the areas subjected to corrosion damage are accessible to visual inspection, once the attack has been detected, protection and remediation activities can be initiated. On the other hand, when the corrosive attack involves inner parts, where direct visual inspection may be difficult or impossible at all, the corrosion damage may become serious. Therefore, inspection methods, such as radiographic- and ultrasonic-based techniques (see Sect. 4.5), are recommended for an early detection and diagnostics, followed by a proper maintenance intervention, if needed. Aluminum and magnesium alloys are prone to general corrosion. Concerning Al-alloys, general corrosion can affect stringers when the chromate pre-treatment and the primer coat fail, due to aging and exposure to moisture. General corrosion of Mg-alloys can occur on external parts, such as the skin, again for the localized failure of the chromate pre-treatment (Wallace et al. 1985). As concerns general corrosion, relative humidity, environmental, and relevant pH conditions are all extremely important parameters. Their effects can be discussed considering the relevant Pourbaix diagrams (see Fig. 3.4, Fig. 3.31, and Fig. 4.18 for the corrosion behavior of Al, Mg, and Ti).

8.1.2 *Pitting Corrosion*

Pitting corrosion is a localized degradation process, regarding those alloys usually showing a self-passivation behavior, like stainless steels, aluminum, and titanium alloys. Reactive metals, e.g., magnesium, may suffer from this kind of corrosive attack too, as a result of local wetting of their surface. The possible reasons for pitting corrosion are defects and cracks in the protective self-passivation surface layer, which allow a very localized exposure of the alloy surface to the outer environment. Halide ions, like Cl^-, highly soluble and present in diverse environments, like seawater and ground, render the attack more aggressive. Indeed, the formation of metal chlorides slows down or inhibits at all the re-formation of the protective oxide layer on the alloy surface, which progressively loses its self-passivating attitude. The occurrence of pitting corrosion depends on the characteristics of the passivation layer: if its thickness is not uniform on the entire surface, and microstructural and lattice defects are present, this renders specific areas preferential sites for a localized corrosion. Figure 8.3 illustrates the main mechanisms involved with the pitting corrosion process of a metal in contact with an electrolyte containing chlorine. If the concentration of oxygen inside the surface cavity is comparatively low, this may locally inhibit the passivation of the metal. Consequently, two types of reactions occur inside the defect: an anodic reaction involving the metal and the reduction of hydrogen, this latter

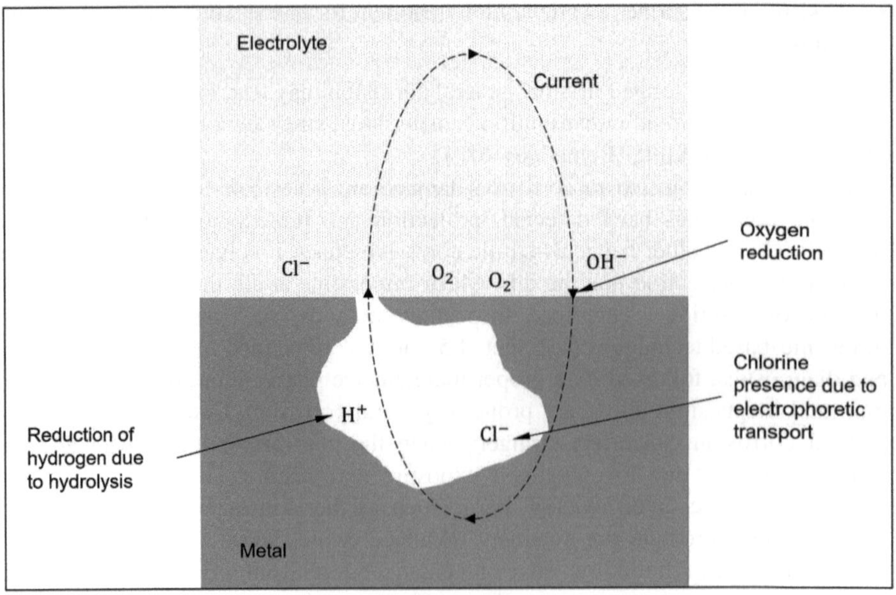

Fig. 8.3 Schematic of pitting corrosion

being favored by the increasing concentration of metallic ions (M^+). The metal ions diffuse out of the defect, whereas a counter-flux of chlorine ions tends to increase their concentration inside the defect itself, thanks to a mass transport mechanism sustained by the electric field (electrophoresis) associated with the ionic charges. This situation enhances metal dissolution and hinders its passivation further. The anodic reaction brings metallic ions into solution, sustaining a localized corrosion and subsequent extension of the pit. The alloy resistance to pitting corrosion depends on its chemical capability to re-form a protective layer in case it gets damaged. The AISI 316 stainless steel, thanks to its molybdenum content, exhibits a better pitting resistance than AISI 304. In fact, molybdenum acts as a stabilizer for the passivation layer and slows down the outward metal diffusion.

There are several spots in aircraft structures that may be subjected to pitting corrosion, in particular those exposed to local contamination by highly corrosive substances. Some examples are battery compartments, galley areas, and toilets. Aircrafts operating in marine environments are particularly prone to pitting corrosion phenomena because of the pervasive presence of chlorine ions. It has been reported that the failure of an AISI 422 stainless steel (see Sect. 5.4.3.2) component caused the loss of an engine's thrust reverser door from a large commercial airliner during cruise flight. The door driver link, made of AISI 422, broke down, allowing the door to lower in the 1000 km/h airstream. More detailed inspections showed that small fatigue cracks were growing from each of the many pits on the part surface. The root cause for pitting corrosion was then identified in the entrapment of moisture and fuel combustion by-products collected in fissures in the coating in direct contact with the stainless steel surface (Wallace et al. 1985).

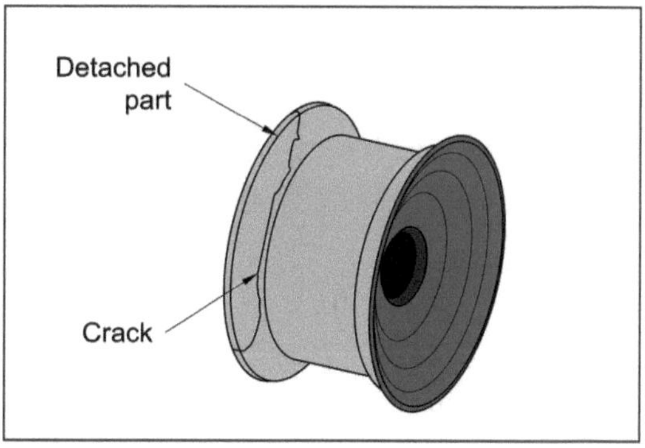

Fig. 8.4 Schematic of the P-180 aircraft main landing gear wheel flange with indication of the crack path. (Redrawn from Bagnoli and Bernabei 2008)

Aluminum alloys too can be subjected to pitting corrosion. The high-strength 7075-T6XXX aluminum alloy (see Chap. 3), extensively used in the fabrication of fuselage skin and airframes (see Tables 3.14 and 3.15), is susceptible to both pitting and stress corrosion cracking (see Sect. 8.1.4). Pitting corrosion of 7075 alloy may have serious consequences on the fatigue life of the component. Experimental observations reported by DuQuesnay et al. (2002) have shown that, owing to the defects introduced by pit corrosion, fatigue life of the material, in simulated flying hours (SFH), can decrease from 25,000 SFH with a defect depth of 0.1 mm down to 5000 SFH, when the defect has a depth of 0.75 mm. Furthermore, it has been shown that the average depth of the pit can be regarded as the critical factor influencing the phenomenon (DuQuesnay et al. 2002). Still concerning the effect of pitting on the failure of Al alloy component, Bagnoli and Bernabei (2008) reported on the crack of the inner wheel flange of the main landing gear of a Piaggio P-180 Avanti aircraft after landing, during taxing on the runway. At that very moment, the crew heard a sudden loud bang from the right side of the aircraft. An inspection carried out afterwards showed that a part of the inner flange wheel fillet had detached from the rest of the landing gear (Fig. 8.4). The failed wheel was made of a 2014-T6 Al alloy forging and had been in service for 22 years before the accident.

The failure analysis revealed the presence of three different regions on the fracture surface: bright and smooth surface, containing the typical "beach marks" (fatigue crack growth region); rough and dull surface alternated to a bright one (unstable crack propagation); and a dull, rough with a coarse-grained region (final unstable fracture). The crack initiation area was identified to be along the inner surface of the flange, near the fillet. In this area, corrosion pits were found, suggesting multiple initiation sites. Moreover, a little amount of corrosion products were found on the fracture surface (Bagnoli and Bernabei 2008). It has been reported over the years, on several occasions, that aluminum and magnesium alloys, used in aircraft wheels, had similar cracks. The wheels had failed similarly to what is observed in this specific case history, and many of these failures were clearly initiated by cracks due to pit corrosion that damaged the protective coatings or passivation layer on the wheel part (ASM International 1992).

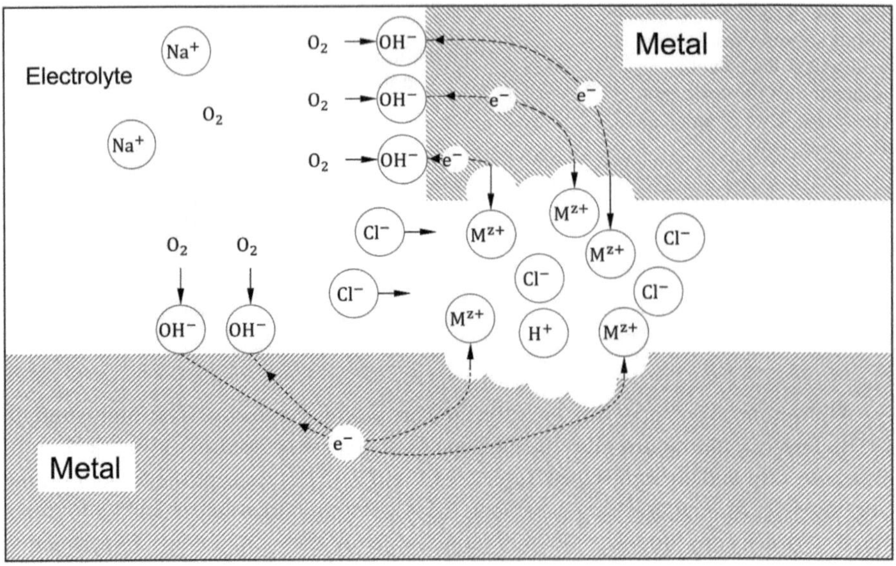

Fig. 8.5 Schematic of the crevice corrosion mechanism

8.1.3 Crevice Corrosion

Crevice corrosion is another type of localized corrosion, somehow similar to pitting corrosion. The mechanism is indeed triggered by a surface defect, typically a tiny gap between two alloy surfaces (crevice, Fig. 8.5). Any alloy may suffer from this corrosion attack, irrespective of the composition of the electrolyte. It is also known as "concentration cell corrosion," and it may take place in interstices of mechanical couplings, where water stagnates. Concerning aircrafts, crevice corrosion is usually associated with small volumes of stagnant aqueous solutions trapped in between gasket surfaces, lap joints, and holes; under rivets and loose detached paint; and in crevices under fasteners heads, within delaminated bonds or unsealed joints. It is the most common type of corrosion phenomenon that can be found in old aircrafts, when maintenance had not been performed correctly (Mouritz 2012). Crevice corrosion is triggered by variations in the composition of the electrolyte between the inner and the outer part of the interstice (similarly to pitting corrosion), due to either corrosion reactions, or obstacles to ionic transport phenomena, which would otherwise homogenize the composition of the electrolyte, thus reducing the crevice corrosion incidence. The way the corrosive process happens is very similar to pitting corrosion. However, one important difference is that crevice corrosion can be prevented with an accurate component design that would minimize the presence of interstices, and using appropriate rubber gaskets, if necessary.

Figure 8.5 shows a schematic of the crevice corrosion mechanism, involving two metal parts in contact with an aqueous electrolyte. At the initial stage, there is the oxidation of part of the metal, leading to the formation of M^{z+} ions. The resulting electron flow allows OH^- to form at the meta-electrolyte interface. After an initial stage, the oxygen within the interstice is consumed for the restrictions to mass transport from and to the crevice. Therefore, the electrolyte compositions inside and outside the interstice become progressively different. In particular, inside the crevice, with a very low oxygen concentration, the cathodic semi-reaction cannot occur any longer. Hence, the anodic semi-reaction is localized into the crevice, while the cathodic can still occur outside. OH^- ions balance the production of M^{z+} and H^+ ions so that the corrosive process is sustained. The volume and reaction surface in the interstice, available to the anodic semi-reaction, are relatively smaller than for the cathodic one. Therefore, the corrosive phenomenon is largely localized in between the two metal parts and is characterized by the severity of the anodic process. Due to its particular phenomenology, crevice corrosion is also known as *differential aeration* corrosion. Depending on the alloy composition and microstructure, and, obviously, on the environmental conditions, crevice corrosion can quickly evolve into exfoliation corrosion (see Sect. 8.1.4) or pitting, if it is not adequately counteracted (Mouritz 2012). This corrosion type may be particularly important in Mg-alloys, since they are the most anodic among the structural alloys (see Table 8.2), although this corrosive attack is reported also in Al-alloys. In fact, crevice corrosion was observed in the outer wing coupling flange of old small transport aircrafts that had been operated in the Mediterranean region. The structure was predominantly made of Al-alloy sheet and flanges with steel fasteners. Moreover, a protective TiO_2-based painting was applied onto the structure for protective purposes. Crevice corrosion was evident at the aluminum lap joints and between the aluminum flanges and sheet. Furthermore, corrosion cavities were found under the nuts and bolts, and the corrosive attack extended beyond the area of contact of the fasteners (Wallace et al. 1985).

8.1.4 Selective Corrosion

Selective corrosion of metallic alloys implies the preferential dissolution of the alloying elements, or alloy phases, with lower electrochemical potentials. Selective corrosion occurs in environments that are not particularly aggressive and typically results in the formation of a porous surface layer. An interesting example of selective corrosion is intergranular corrosion, i.e., a corrosion process that occurs at the alloy grain boundaries. The driving force for intergranular corrosion is the difference in electrochemical potential between the inner and boundary regions of the same grain. In the corrosive process, the inner region of the grain becomes cathodic, and the grain boundary becomes anodic. The anodic character of the grain boundary can mainly be explained through some of its features, which imply an excess of energy and, therefore, a higher reactivity. In particular, the chemical composition at the grain boundary regions is different from the grain core because of segregation of impurities and precipitation of secondary phases. Grain boundary precipitation is often accompanied by the formation of a precipitation-free zone (PFZ, see Fig. 3.21) due to the depletion of alloying elements along the grain boundaries (Mouritz 2012). A common intergranular corrosion process is the *sensitization* occurring in austenitic stainless steels (see Sect. 5.4.3.1), often as a consequence of welding processes. The increase in temperature near the welded zone causes the precipitation of Cr-rich carbides ($M_{23}C_6$) at the grain boundary with a consequent depletion of chromium in the surrounding region, where a lowering of the corrosion resistance is observed. This leads, in the worst condition, to the absence of the passivation layer on the steel surface and to the localization of corrosive phenomena in this area. There are different ways to avoid sensitization of stainless steels:

- Addition of alloying elements, like titanium, that have higher than chromium affinity to carbon, thus avoiding chromium carbides formation and consequent chromium depletion from the bulk.
- Fast cooling after the welding in order to avoid crossing the carbide transformation curves.
- Low-carbon content. This reduces the possibility for chromium to combine with carbon during cooling after the welding.
- Welding process involving a reduced heating of the part, using, for instance, laser welding.

Another particular type of intergranular corrosion is exfoliation corrosion. The corrosive attack occurs at the grain boundaries, and the resulting corrosion products exert a pressure on the surface grains. This pressure may lift off surface alloy grains that then are removed by abrasion or other mechanical actions, which in this way may sustain the process (Mouritz 2012). In the mid-1960s, exfoliation corrosion was one of the main corrosive problems faced by the US Navy with the 7075-T651 aluminum alloy extrusions, used for the wing spars of air-sea rescue aircrafts. Metallographic analyses of failed specimens highlighted that the deteriorated laminar paths were preferential sites for fatigue crack nucleation. For this reason, the fatigue life of the exfoliated material was strongly reduced, even down to the 70%. Exfoliation corrosion originated deep inside rivet holes under crevice corrosion conditions (see Sect. 8.1.3). The resulting deterioration was not detectable by standard visual inspection, and in all cases, the intergranular corrosion propagated along corrosion pits preferentially (Mills et al. 2014). Intergranular corrosion was also detected in an aluminum alloy, used for wing attachment fitting of a fighter aircraft during periodic inspections. The component was made of an Al-Zn-Mg-Cu alloy with a chemical composition similar to that of 7075 alloy but with a little amount of Ag. The mechanical properties were those induced by a solution treatment followed by aging (T6 temper condition, see Table 3.9) and shot peening on all external surfaces. Chemical conversion coatings, using Alodine 1200 and Zn chromate primer, were also employed as surface finishing. Intergranular corrosion occurred at the wing attachment fitting due to a combination of several adverse conditions that were not inhibited by the surface protective treatment and coating (Wallace et al. 1985).

8.1.5 Galvanic Corrosion (Two-Metal Corrosion)

Galvanic corrosion takes place when metals and alloys with different electrochemical potentials (see Table 8.2) are in contact in a corrosive medium. The resulting corrosive attack is the consequence of the galvanic coupling among the involved metals and alloys. However, galvanic corrosion does not necessarily require the electrical contact between different alloys, since a galvanic coupling can also be established within the same material between regions with different electrochemical potentials. Quite common examples are the inner parts and the boundaries of crystalline grains; precipitates in a nobler parent phase matrix; and regions with different residual stress values, resulting from heat treatments and mechanical working. In fact, the only strict requirement for galvanic corrosion to occur is a difference in electrochemical potential and the presence of a suitable electrolyte (see Daniell battery in Fig. 8.1). The less noble metal will be subjected to the anodic semi-reaction, thus becoming the anode of the equivalent cell. The nobler metal is instead subjected to the cathodic semi-reaction and acts as the cathode of the electrolytic cell equivalent. The metal, or part of it, subjected to the anodic process will show a progressive corrosive damage.

Systems in which galvanic corrosion is observed are, for instance, bronze propeller in contact with steel hull, copper water pipes coupled with iron pipes, and cold-worked alloy in contact with the same alloy but in a fully annealed condition.

One of the most common examples in aircraft structures is the corrosion occurring at fasteners holes of Al- or Mg-alloy skin coupled with steel bolts and rivets or in contact with a conductive cement. The use of graphite-based products or composite materials, if not properly designed, can lead to galvanic corrosion too. Galvanic corrosion has been reported for an Al-alloy wing skin, as a consequence of the use of a graphite-based marker, i.e., a pencil, used for identifying a small nick in the lower wing skin of a fighter aircraft (Wallace et al. 1985).

8.1.6 Mechanochemical Corrosion

Mechanochemical corrosion, also known as mechanically-assisted corrosion, is the degradation process that takes place when mechanical stresses are combined with a corrosive environment (McCafferty 2010). The intensity of the overall corrosive attack would strongly depend on the balance between mechanical stress and the main characteristics of the corrosive environment. Fretting is a typical example of a mechanochemical corrosion process, resulting from the combined action of the corrosive environment and a relative oscillatory movement between two mating contacting surfaces (Fig. 8.6). Fretting may even occur between ball bearings and their cases and in flexible mechanical couplings (Wallace et al. 1985). Although the phenomenology of fretting is rather variate, two important mechanisms are intervening in most systems:

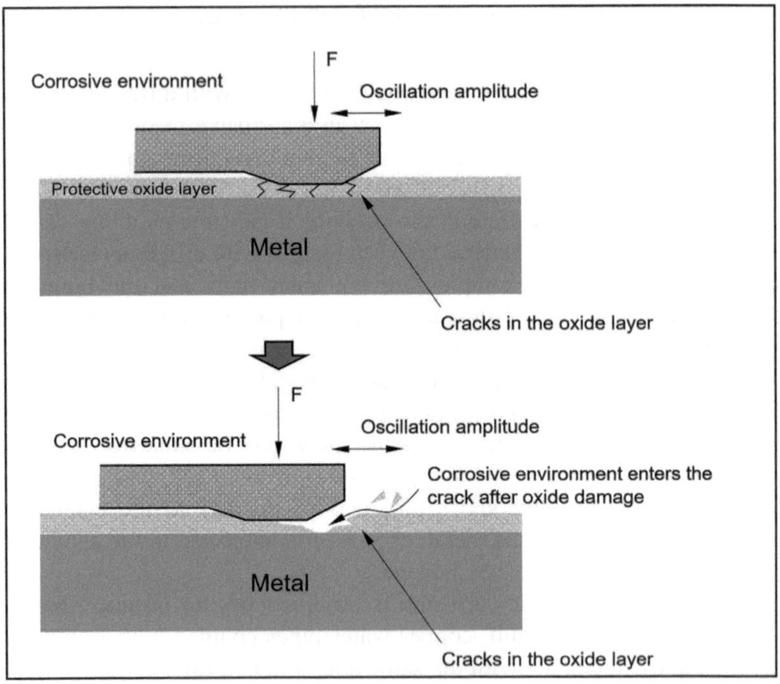

Fig. 8.6 Schematic of fretting corrosion initiated by the damage of the passivation layer

- Damage of passivation layer on the alloy surface. Mechanical damage of the passivation layer may be induced by the relative in contact movement of two surfaces, with the partial removal of the protective oxide layer and consequent localization of the corrosive phenomenon at the freshly exposed metallic surfaces.
- Oxidation wear. A mechanical damage of the surface is observed as a consequence of the removal of the products of the oxidation induced by the localized temperature rise.

Fretting damage can be particularly severe, since it may evolve into a sudden and unexpected fatigue failure of the components, notwithstanding a correct design to sustain normal service conditions.

Fretting corrosion has been identified as the main reason for the in-flight engine failure of a piston engine helicopter. The failure was caused by fretting-induced fatigue of an engine connecting rod (in this case, it is rather more appropriate to talk of fretting fatigue). Fretting was caused by the rotational movement of the shell bearing within the connecting rod big end bore (Wallace et al. 1985). One of the most important reported cases of fretting corrosion in the aerospace field was the corrosion of electrical contacts in the fuel control system of the General Dynamics/ Lockheed Martin F-16 *Flying Falcon*. The fretting damage caused an automatic shut off of the valves of the main fuel supply without any preliminary warning. Before the failure mechanism could be identified and the problem fixed, this anomaly was the

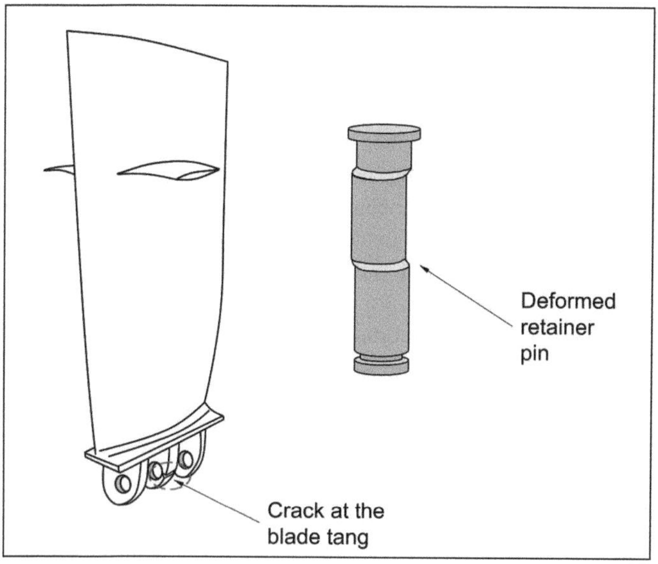

Fig. 8.7 Schematic of the compressor blade with the detail of the crack at the tang and deformed retainer pin. (Redrawn from Lee et al. 2011)

main reason for a high number, at least six, of F-16 crashes (Mouritz 2012). Fretting is also an important issue in gas turbine engine compressor and turbine discs, in particular at the dovetail joint between discs and blades. In those regions, contact stresses are very high during flight operations, and, even in view of the relatively low damage tolerance of the parts, the failure consequences are potentially serious (Mills et al. 2014). In this regard, Lee et al. (2011) reported on the fracture of a Ti-6Al-4 V (see Chap. 4) compressor blade installed on a J85 engine and connected to the rotating disc through retainer pins made of Alloy 718 (see Sect. 6.3.2), inserted through the tang-type root of the blade. The failure occurred at the blade tang, as evidenced in Fig. 8.7.

The blade-disc connection is subjected to particularly severe service conditions, involving both high rotational frequencies, vibrations, and centrifugal loads. The highest centrifugal loads are localized in the first stages of the compressor at the blade-disc connection regions, owing to the high rotational speed, the large radius of the rotating disc, and the length of the blades. The retainer pins of the failed blade resulted in severe deformation, and the detected diameter was smaller than that of new parts (see Fig. 8.7), probably for the prolonged oscillating motion during the engine operation. The damage of the retainer pin produced limited oscillatory movements of the pin in contact with the inner surface of the pinhole in the blade tang. The diagnostic investigation that was conducted on the damaged blade and pin confirmed fretting corrosion to be the main cause for the observed failure. Debris loss and mud cracks were observed at the crack initiation spot. The mud crack was an evidence of dried liquid lubricant, probably leaked out once the crack had nucleated.

The damaged retainer pin accelerated the propagation of the fretting damage because of the increased clearance between the two contacting surfaces and consequent increase of mechanical tolerance (Lee et al. 2011).

Another type of damage depending on mechanochemical corrosion is the stress corrosion cracking (SCC). SCC implies the failure induced by a slow, environmentally driven crack propagation, occurring as a consequence of the synergistic interaction between corrosion reaction and mechanical stress. Concerning the aerospace field, routine maintenance and airframe inspection activities often include the search for SCC signs. Although SCC is not the most common form of corrosive attack in aerospace components, it accounts for 5–10% of all failures of aircraft components (Mouritz 2012). Therefore, it is important to map the materials that can be subjected to it, the type of stresses involved, and the environments and relevant interactions that lead to SCC. Here is the summary list of the main representatives of these SCC factors:

- Materials: α-brass; several steels, like AerMet steels, austenitic stainless steels, carbon and high-strength steels in general, maraging steels, and martensitic and PH stainless steels; magnesium, aluminum, and titanium alloys; some nickel-based superalloys; zirconium alloys.
- Stress conditions: tensile loads; external applied stresses, such as structural or aerodynamic loads; overtightening of fasteners heads causing stresses in the material underneath. Internal stresses, resulting from processing like bending or rolling, nonuniform cooling after heat treatment (Mouritz 2012).
- Environmental agents: aqueous electrolytes, especially when containing H_2S (high-strength steels). Ammoniacal solutions (α-brass), solutions containing chlorine compounds (austenitic stainless steels, Mg- and Zr-alloys), high purity steam (Ni-alloys), chloride, bromides, and iodide solutions (Al- and Ti-alloys). These contaminants may be present in the outer atmosphere, in lubricants and products deriving from exhaust condensation, just to mention a few main sources.
- Interactions: it is possible to distinguish between corrosion-dominated failure and stress-dominated failure. In the first case, either intergranular or transgranular fracture occurs. Solution requirements are alloy specific: NO_3^- solutions for carbon steels; chloride solutions for Al-Zn-Mg alloys and for Fe-Cr-Ni alloys; NH_3 solutions for Cu-Zn alloys. In case of stress-dominated failure, brittle fracture would generally occur, and the influence of the electrolyte is not that critical, if not just as a precursor that promotes the formation of reaction products. They may act as nucleation sites for fracture and also favor its subsequent propagation, although under the prevailing action of the mechanical drive (Jones 1992).

SCC presents an early stage, usually propagating slowly and corresponding to crack initiation. Afterward, a steady-state crack propagation follows, rapidly leading to the third and final stage, featuring a sudden complete failure. The unpredictability of the whole time evolution of SCC, owing to the extended hidden period of the process, renders this phenomenon rather serious as concerns the assessment of a component lifetime. For this reason, prevention is very much recommended. Several strategies and relevant practical approaches are nowadays available (Table 8.3) and are systematically implemented.

Table 8.3 Prevention strategies against stress corrosion cracking

Mechanical approach	Metallurgical approach	Environmental approach
Compression stresses on the surface	Changes in the microstructure	Anodic or cathodic protection
Limitation of the stress concentration (i.e., optimization of the geometry)	Optimization of the composition by choosing a proper alloy (e.g., duplex stainless steels instead of AISI 316)	Environmental control (e.g., limitation of chlorides presence in the environment)
Maintenance and control-oriented design	Use of surface coatings	Temperature control (the lower the temperature, the slower the corrosive phenomenon)
Reduction of the applied loads (e.g., increase in dimensions of the component)	–	Use of inhibitors of corrosive phenomena
Relaxation of residual stresses	–	–

Regarding Al-alloys used in aerospace applications, SCC has been reported for the front and rear spars of a fighter aircraft vertical fin. The spars were made of 7079-T6 Al-alloy forgings, age tempered to the peak strength condition. Stress corrosion cracking was detected during standard inspections, carried out with the eddy current NDT method. The spars were actually coated. Nonetheless, the coating was damaged during the part assembly, using bolts. This provoked the exposure of the bare metal to the outer environment and SCC developed in the material as a consequence of the humidity combined with high tensile stresses of the installed fasteners (Wallace et al. 1985). SCC has also been reported to occur in a 7075-T6 fuselage frame flange of a fighter aircraft. Cracks were detected during routine inspection and were identified along a row of rivets in the upper flange, linking together adjacent rivet holes. The microstructural characterization of the damaged part evidenced a heavily wrought microstructure characterized by pancake-shaped grains. The crack plane turned out to be coincident with the plane containing the longitudinal-long transverse load direction, whereas the transverse-short direction was normal to the crack plane. Further analyses confirmed the absence of sealants or protective substances, so that the damaged alloy was actually free from any corrosion protection. SCC turned out to have initiated at corrosion pits in the bore of the fastener holes. Stresses were originated both by the part assembling and by the in-service operating loads, especially in the short-transverse direction close to the fastener holes. In this latter area, stresses were higher than the threshold value for SCC (Wallace et al. 1985).

Concerning Mg-Alloys, stress corrosion cracking was identified as the main reason for the failure of an AZ91C-T6 brake housing casting of a popular light aircraft. The failure was induced by cracks and leaks at the brake-liner union. SCC may occur also in Mg-alloys, containing more than 2.5% Al (see Table 3.23), and when the applied loads are higher than 90% of the yield strength.

Regarding SCC in aircraft steels, an interesting case history concerns the failure of a Nitronic 60 (see Table 5.13) pin in a rear compressor variable vane lever arm assembly of an F-16 fighter, reported to be occurred in February 1992. The engine failure caused the aircraft crash in the Dutch city of Hengelo. At the time, the susceptibility of Nitronic 60 to SCC was unknown, and, therefore, the failure was absolutely unpredictable and unexpected. Subsequent analyses showed that the failure began with crevice corrosion, and the fracture was caused by SSC. At the same time, many more pins were subjected to investigations, and it was found that a large amount of them contained cracks, caused by a combination of salt solutions in crevices between the lever arms and the pins and by the presence of residual stresses in the pins caused by the production process. In order to prevent further failures, the engine manufacturer adopted two remediation measures: change of the pin material and progressive worldwide replacement of all old lever arm assemblies with new ones. The Ni-based Inconel 625 superalloy, immune to SCC in salt solutions, replaced the austenitic stainless steel Nitronic 60 (Wanhill et al. 2011). Still regarding high-strength steels, the failure of a cylindrical pressure vessel made of a 18Ni maraging steel (see Sect. 5.4.1) has been reported. The cylinder, containing He-gas, failed and fragmented while having an internal pressure of 55 MPa. The component was the auxiliary power unit of a missile and was supposed to leak and not to crack down if and when subjected to overpressure conditions. Crack initiation occurred on the inner surface of the cylinder and propagated through about 80% of the wall thickness before the final overload rupture was achieved. Before this final stage, the crack propagation occurred in a slow growth mode, typical of SCC. The slow crack growth region was characterized by the presence of corrosion products and intergranular fracture. These latter factors suggested an environmentally induced failure. In fact, the inner surface of the cylinder showed the presence of a thin layer of hydraulic fluid that provoked the corrosive attack to the vessel wall (Wallace et al. 1985).

8.1.7 Hydrogen Embrittlement

Hydrogen embrittlement is a particular type of degenerative phenomenon, occurring in metals as a consequence of hydrogen permeation, resulting in a catastrophic failure. There are actually several types of hydrogen-related degradation processes occurring in metals, in addition to hydrogen embrittlement, like hydrogen attack and hydrogen blistering (ASM International 2003). Hydrogen attack consists in the solutioning of this element in the lattice of the host metal, up to the solubility limit. At this stage, hydride precipitation may occur and a general embrittlement is also observed. Blistering refers to the formation of gaseous hydrogen, generally localized on the material surface, resulting from metal catalyzed hydrolysis, as observed in aluminum alloys (see Fig. 3.22). Since hydrogen is present in the environment of several processes, e.g., hydrocarbon processing plants, melting and welding processes without controlled atmosphere, etc., there are many possibilities for metals to be subjected to hydrogen inward diffusion. Hydrogen embrittlement implies a sort of ductile to brittle transition (DBT), caused by the interaction of hydrogen with metal, changing the chemical bond character from exclusively metallic into partially covalent.

Although the process of hydrogen embrittlement has been extensively analyzed in the literature, the nature of the atomistic mechanism behind the phenomenon is still controversial, and three main models can be considered:

- Hydrogen-enhanced decohesion (HEDE). This process results from the reduction in the cohesive strength of the atomic bonding, produced by hydrogen accumulation in the crystalline lattice. This mechanism is coherent with the fact that the relevant damage occurs at the crack tip fracture preferentially. In this region, the local tensile load exceeds the maximum value of the local cohesion strength, which was lowered by the presence of hydrogen. HEDE is emerging as the dominant mechanism for hydrogen environmental assisted cracking (HEAC) and for internal hydrogen assisted cracking (IHAC) in high-strength alloys that are not prone to the internal formation of metal hydrides (Popov 2015).
- Hydrogen-enhanced localized plasticity. Different from hydrogen embrittlement, HELP (hydrogen-enhanced localized plasticity) is a phenomenon in which hydrogen may act as an enhancer of the dislocation mobility, resulting in a localized plasticity and a consequent subcritical crack growth, still exhibiting, though, brittle features on the macroscopic scale. HELP model is based on the effect of hydrogen either in reducing the stacking fault energy or preventing the dislocations from interacting with other obstacles, allowing them to move under lower applied stresses. The evidence for HELP has been obtained through in situ ETEM (environmental transmission electron microscopy) of specimens subjected to plastic deformation either in vacuum or in hydrogen atmosphere. These observations revealed at a microscopic level the dislocation movement and their increased number in pile-ups, when hydrogen was introduced in the microscope sample region, coherently with a reduction in the stacking fault energy. A plastic deformation similar to the one described above, accompanied crack growth observed in the TEM in situ tests (Popov 2015).
- Hydride-induced embrittlement. Stress-induced hydride formation and cleavage are established hydrogen embrittlement mechanisms, as confirmed by a large theoretical and experimental support. Hydrides nucleate and grow in the volume ahead of a propagating crack, according to a sort of autocatalytic process. The brittle fracture of carbides, if present, may contribute to the process (ASM International 2003).

Hydrogen embrittlement was analyzed by Lee and Krakkainen (2018), in connection with the discovery of a 1 m long crack in the inboard trailing edge of the horizontal stabilizer of a jet aircraft during the post flight routine inspection (see Fig. 8.8). The stabilizer is a control surface (see also Fig. 3.1), and its failure can lead to serious consequences for the loss of aircraft maneuverability. The jet fighter, whose stabilizer failed, belongs to a group of aircrafts that had been operated for more than 40 years. The failed stabilizer structure consisted of an aluminum alloy square honeycomb core with high-strength steel skin that, before failing, accounted

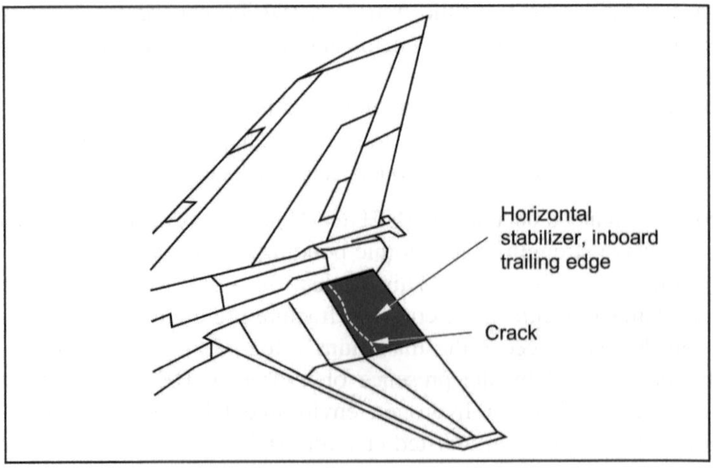

Fig. 8.8 Schematic of the stabilizer and of the location of the detected crack. (Redrawn from Lee and Karkkanien 2018)

for more than 700 h of time-flight. Before this failure, there were only two reported cases of a complete loss of the inboard stabilizer trailing edge. The cracking pattern usually associated with this type of structures starts from a bolt hole edge and extends along adjacent holes, due to fatigue. In this case, fastener holes were not characterized by any crack propagation, but the crack was localized at a certain distance from the fasteners holes line with a propagation path that was almost linear across the entire panel. The analyses of the fracture surface were the key point in determining the real cause of the failure. SEM observations revealed various morphologies, although the dominant fracture mode, accounting for more than 75% of the total fracture surface, was an intergranular fracture. The crack initiation region was characterized by a fatigue morphology, while the transition region, between fatigue cracking and intergranular fracture, showed a quasi-cleavage morphology. The fracture surface close to the overload area exhibited typical ductile micro-void coalescence. This fracture morphology is typical of hydrogen-induced cracking in high-strength steels.

Samples were taken from the fracture surface and from the opposite side of the skin panel, in order to measure the concentration of hydrogen. The results indicated a concentration of 154 ppm at the fracture surface, while on the opposite side, the concentration was 9.5 ppm only. It is worth mentioning that a concentration of hydrogen in the 100–1000 ppm range is sufficient for degrading mechanical properties of several steels significantly. The part failure is induced by the piling up of residual internal stresses, resulting from maintenance activities, missing shims, and aerodynamic loads, generated during the flight. Moreover, the notorious sensitivity of honeycomb structures to moisture intake, like that coming from fastener holes and trapped into the core cells, may have worsened further the situation (Lee and Karkkainen 2018).

8.2 High-Temperature Corrosion

High-temperature corrosion occurs in several technological fields. Similar to electrochemical corrosion, the relevant phenomenology influences the in-service performances, particularly of the engine components, and represents an important lifetime limiting factor. In aircraft and aerospace structures, high-temperature corrosion is particularly interesting for the metallic alloys used in the engines, although it is an issue also for the skin and fuselage of special structures, like the thermal protection system of the Space Shuttle Orbiter (Buckley et al. 1981; Korb et al. 1981). In this section, the main principles of oxidation will be illustrated with reference to simple models, based on the Ellingham diagrams, as concerns the thermodynamics, and on the Wagner theory for the main kinetic aspects. Although high-temperature oxidation will be the leading topic, a few aspects of mixed gas phase corrosion, with particular regard to sulfidation, will be also recalled. These principles are useful to investigate the reactions occurring in real systems and laboratory testing rigs. Moreover, they provide indications and first principles for the development of high-temperature corrosion protection systems, like the coatings, discussed in Sect. 7.3, and relevant strategies. An alternative protection approach will be presented, based on the nanocrystallization of the alloy surface. Examples of the systems for which this protection approach has provided good results will be illustrated. The last part of the section is devoted to the important phenomenon of hot corrosion. This complex process involves not only a gaseous environment but also sulfate deposits that condensate on the alloy surface. This introduces a further attack mechanism, based on the formation of non-protective reaction products and the dissolution of potentially protective oxide layers, present on the alloy surface before salt deposition or even forming in the presence of the condensate deposit. The complexity of these systems renders effective characterization protocols, using laboratory tests, very important to interpret real-system phenomena, which in fact are presented in the final part of the section.

8.2.1 High-Temperature Oxidation: General Thermodynamic and Kinetic Aspects

From a chemical point of view, an oxidation is a reaction in which a metal forms more stable chemical compounds, like oxides, nitrides, carbides, etc. A general high-temperature oxidation reaction between a metal, M, and oxygen, O_2, can be written as:

$$M(s) + O_2(g) = MO_2(s) \tag{8.4}$$

The condition for this reaction to occur depends on the higher stability of the oxide phase with respect to the reagents in the initial state. This condition can be expressed in terms of the standard free energy of the oxidation reaction:

$$\Delta G^0 = G^0_{MO_2}(s) - G^0_{O_2}(g) - G^0_M(s) = RTln(PO_2) \tag{8.5}$$

where:

- ΔG^0 is the standard Gibbs free energy variation of the oxidation reaction.
- $G^0_{MO_2}(s)$ is the standard Gibbs free energy of the oxide.
- $G^0_{O_2}(g)$ is the standard Gibbs for oxygen.
- $G^0_M(s)$ is the standard Gibbs free energy of the reacting metal.
- R is the gas constant.
- T is the absolute temperature.
- P_{O_2} is the partial pressure of oxygen.

Equation 8.5 can be made simpler by the assumption that the activities of the condensed phases, i.e., the oxide product and the metal, both considered to be solid in the present example, can be taken equal to unity. The first outcome of Eq. 8.5 is the value of the oxygen pressure in equilibrium with the oxide, i.e., the minimum oxygen partial pressure needed to form a stable oxide.

The British chemist, Harold Johann Thomas Ellingham, created a graphical representation of experimentally determined variations of the standard Gibbs free energy (ΔG^0) versus temperature for the oxidation and sulfidation of several metals as a function of the temperature. In Fig. 8.9, the Ellingham diagram for the oxidation reactions is shown. Similar diagrams are available also for other gas phase reactions and are extremely useful to predict the system behavior as a function of the temperature and relevant gas phase pressures. Starting from a general simplified expression for ΔG^0 in terms of the standard enthalpy, ΔH^0, and entropy, ΔS^0, of the reaction, the Eq. 8.6 can be written:

$$\Delta G = \Delta H^0 - T\Delta S^0 \tag{8.6}$$

This expression is coherent with the linear trend of the curves in Fig. 8.9. The intercepts of these lines are the standard enthalpies of the oxidation reactions; the slopes are the standard entropies. The slope changes observed in some of the reaction lines correspond to the melting (M points) and boiling (B points) points. The positive slope usually displayed by the free energy lines indicates that a decrease in entropy would accompany the oxidation reactions. This is so when the oxide is a condensate phase, since the corresponding configurations are definitely less than the configurations of the reagents, including also a gas phase. Of course, the carbon dioxide formation occurs at constant entropy, since 1 mole of gaseous reagent (oxygen) is replaced by 1 mole of another gas phase, i.e., CO_2. A negative slope is there for the CO formation reaction, corresponding to an entropy increase, since 1 mole of oxygen is replaced by 2 moles of carbon monoxide, both phases being gaseous.

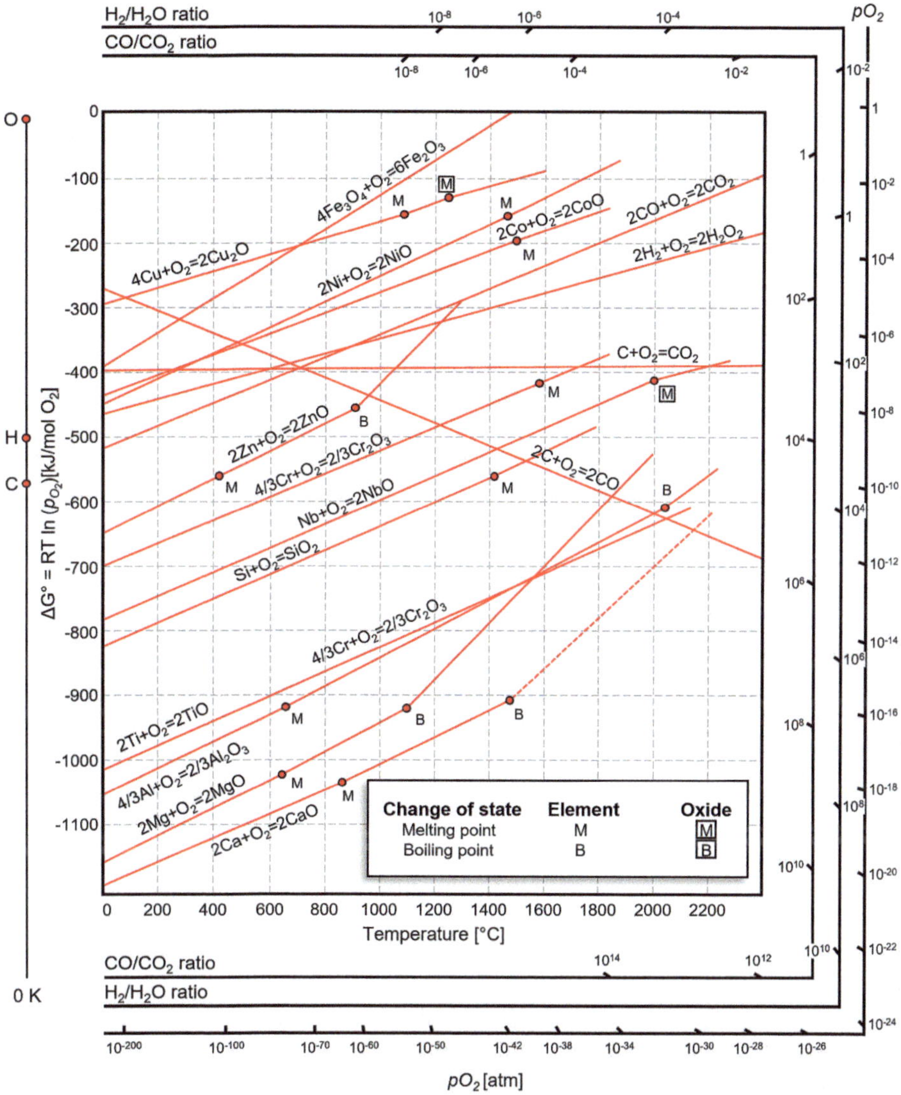

Fig. 8.9 Ellingham diagram for the formation of oxides. The point "O" is used as reference to draw constant oxygen pressure lines, useful for graphical thermodynamic calculations, like the equilibrium partial pressure of any metal-metal oxide couple. Alternatively, the pressure of oxygen, as a function of the temperature, may be established by considering the H_2/H_2O and CO/CO_2 activity ratios in water vapor and carbon oxides mixtures. The points "H" and "C", instead of point "O", are used for the relevant graphical constructions (Kofstad 1988)

The position of the lines in the Ellingham diagram indicates the stability of the relevant oxides, as a function of the temperature. The higher stability oxides correspond to lower laying curves in the diagram. CaO turns out to be the most stable oxide among those in Fig. 8.9; whereas the curves for copper and iron oxides are located in the upper end of the diagram, indicating that they are the less stable oxides in this set. This information provides an important indication on the oxide species to be preferred for alloy protection. Indeed, a common approach is to promote the formation of a protective scale of native oxides either on the alloy itself or on the oxidation-resistant coating. The oxide phases of nickel, cobalt, and iron, the elements on which the main families of superalloys are based (see Chap. 6), are by far less stable than aluminum and chromium oxides. Suitable alloy additions (see Sect. 6.4) or protective coatings (Sect. 7.3.1) are designed so to induce the preferential formation of the mentioned, more stable oxide phases. For the same reason, the position of SiO_2 turns out to be a good choice to protect niobium-based alloys, on which the formation of a silica-rich surface layer, can be obtained with specific alloying strategies or coating systems with the appropriate composition (see Sect. 9.2.2). In addition to the thermal stability, other features are fundamental for an effective protection of the alloy, like adhesion to the substrate, compact microstructure, and good thermomechanical compatibility. These aspects are also influenced by the kinetics of the oxidation reactions, which are paramount in most of real applications, including those involving high-temperature structural alloys. The actual kinetics of the process is fundamental in the evaluation of the real life of the component or to select the right coating system, capable to achieve the target performances. Although rather simple and, in many respect, basic, the Wagner's theory of oxidation is considered herewith. Wagner's theory is based on a purely diffusive model of the reacting species through a comparatively thick oxide, resulting in a parabolic kinetics. Its validity depends on some important assumptions:

- Electrostatic interaction range is little if compared to the thickness of the oxide layer.
- Negligible solubility of oxygen in the base metal.
- The oxidation process is assisted by the diffusion of ions and electrons through the oxide layer.
- The oxide is perfectly adherent, i.e., there are no gaps at the interface with the base metal.
- The reactions produce exclusively stoichiometric oxides.
- Thermodynamic equilibrium at the oxide-atmosphere interface and within the oxide.

Wagner's model was found to be qualitatively and quantitatively accurate for many real cases, like sulfidation of silver and iron and oxidation of nickel, cobalt, and copper. However, also for those alloys that are far too complex for an accurate description in terms of Wagner's model, still, the predictions of this approach can profitably be used as reference for the interpretation of the real results. A condition for diffusion to be the rate-controlling process is that a local equilibrium must be achieved at the scale boundaries. This latter condition requires that the reactions occurring at the metal-oxide and oxide-atmosphere interfaces are fast enough, and are not the rate control processes (Young 2015).

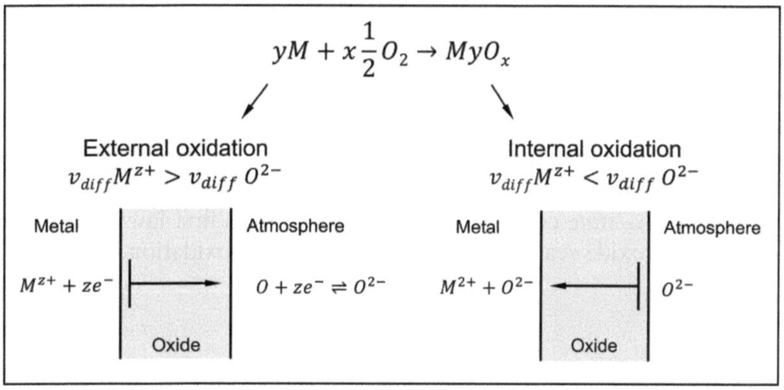

Fig. 8.10 Oxidation reactions according to the Wagner's model. External oxidation occurs when the diffusion rate (ν_{diff}) of the metal cations through the oxide layer is higher than that of the oxygen from the outer atmosphere. On the other hand, internal oxidation occurs when the diffusion rate of oxygen through the oxide layer is larger than that of the metal cations

Wagner's model is illustrated with reference to Fig. 8.10, which describes a general oxidation process: $y\text{M} + x\frac{1}{2}\text{O}_2 = \text{M}_y\text{O}_x$. Depending on material properties and environmental conditions, metal oxidation generally involves two main mechanisms: external and internal oxidation (Fig. 8.10).

External oxidation is driven by the outward diffusion of metal cations through the surface oxide layer; the alternative process, internal oxidation, occurs when the inward oxygen diffusion prevails on the metal outward flux. External and internal oxidation should be regarded as limit situations that in real systems would generally coexist, to different extent depending on the oxidation regime and degree of advancement of the process. In fact, three main stages can be identified for the high-temperature corrosion phenomena, including oxidation. The initial stage, featuring the metal or alloy surface in direct contact with the outer atmosphere, does not require mass transport, and the products that start to form are those involving the more reactive metals present in the alloy. At this stage, the rate of reaction is directly proportional to the time. After this *initiation* stage, if the reaction products are thermodynamically stable, adherent to the substrate, and reasonably compact, a surface reaction layer starts to form on the alloy surface, and mass transport mechanisms are needed to sustain the reactions. The kinetics, initially linear, starts to level up, and a logarithmic law (direct or inverse) would typically describe this further regime, and the scale thickness (or specific mass change), x, as a function of time, t, can be expressed as:

$$x = K\log(t) + A \quad \text{and} \quad x^{-1} = B + K'\log(t) \tag{8.7}$$

where:

- K and K' are the rate constants of the reactions.

- A and B are integration constants.

The logarithmic kinetics is compatible with the presence of a thin oxide scale on the alloy surface, through which the transport of ions and electrons takes place. The transport is not purely diffusive and includes also the electrostatic attraction among opposite charges. As the oxide thickness increases, the system may get the features described with reference to the Wagner's model, which can therefore be used to describe the *steady- state* condition. According to Fick's first law of diffusion, the kinetic law of the oxide scale growth under *steady-state* oxidation conditions is:

$$\frac{\mathrm{d}x}{\mathrm{d}t} = D\frac{\mathrm{d}C}{\mathrm{d}x} \tag{8.8}$$

where:

- D is an equivalent diffusion coefficient, taking into account the contributions from all reagents.
- $\frac{\mathrm{d}C}{\mathrm{d}x}$ is the concentration gradient across the oxide layer thickness.

From the integration of Eq. 8.8, the following parabolic expression is obtained:

$$x^2(t) - x^2(t_0) = K'_p(t - t_0) \tag{8.9}$$

Assuming that x represents the oxide scale thickness increase starting from the initial time, $t = 0$, when the diffusive model starts to be applicable, the above equation becomes:

$$x^2(t) = K'_p t \tag{8.10}$$

This is a very practical form for the interpretation and modeling of the experimental data. Similar expressions are used for the other oxidation kinetics, i.e., linear and logarithmic kinetics (Fig. 8.11).

It is common practice to consider the specific weight, i.e., the weight of reaction products per unit surface area of the alloy, in order to obtain kinetics parameters easily comparable. This can be done by multiplying Eq. 8.10 by the oxide scale density (ρ), so to obtain an expression for the parabolic oxidation law that can be written as:

$$(x\rho)^2 = \rho K'_p t \rightarrow \left(\frac{\Delta M}{A}\right)^2 = K_p t \tag{8.11}$$

where K_p is the parabolic constant of the oxidation reaction, ΔM is the weight change due to oxidation, and A is the alloy sample free surface. In case of laboratory tests, the alloy free surface area is directly measured on the test sample before

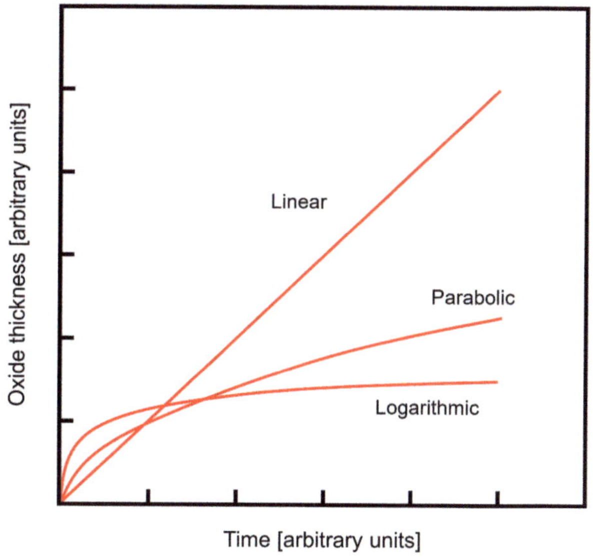

Fig. 8.11 The linear, logarithmic, and parabolic oxidation kinetics. (Redrawn from McCafferty 2010)

starting the oxidation experiment. Since K_p includes a diffusion coefficient (see Eq. 8.8), it may be expressed with an Arrhenius-like relation:

$$K_p = K_0^{-\frac{E_{act}}{RT}} \tag{8.12}$$

from which the equivalent activation energy, E_{act}, of the process can be evaluated. Although the solution of Wagner's model has been obtained starting from very simple assumptions, still, as it was the case for Ellingham diagram, also Wagner approach provides interesting information concerning the oxidation kinetics of real systems. The parabolic constants for the formation reactions of several oxides are shown by Fig. 8.12. Interestingly, the Al_2O_3 and Cr_2O_3, mentioned already as thermodynamically more stable than nickel, cobalt, and iron oxides, would also feature slower formation kinetics, which rationalize further the important role of these phases in the oxidation protection strategies.

The *steady-state* condition would last as long as the outward metallic ion flux is sufficient to guarantee the formation of a protective oxide scale also in case of damages and spallation phenomena. The thermogravimetric tests conducted with thermo-analytical instrumentation under conditions simulating those faced under real service can be used to predict the real alloy behavior. For instance, in Fig. 8.13, the ideal "theoretical" parabolic behavior (dotted curve) is compared with a "real" system oxidation curve, in which two different oxidation stages can be identified, each one featuring a specific parabolic constant. This behavior can be explained with the compositional changes in the base alloy during the process. Another possible phenomenon is illustrated by the other "real" curve, describing a kinetics involving cracking and partial spallation of the surface oxide layer.

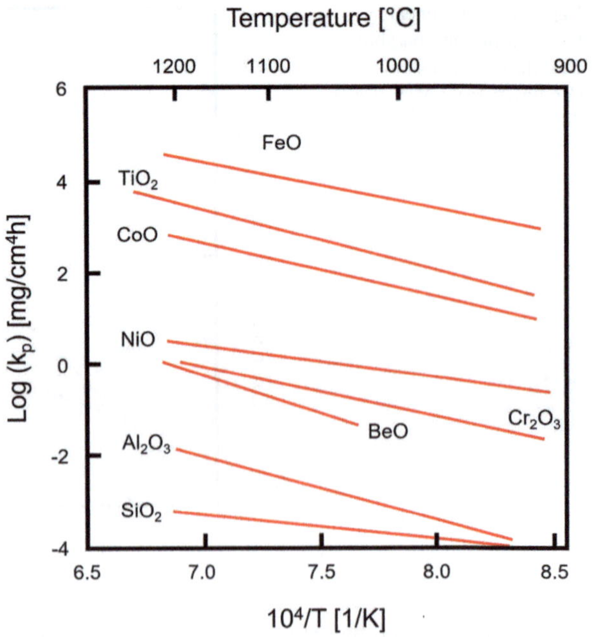

Fig. 8.12 Parabolic constants of several oxides as a function of the temperature. (Redrawn from Okafor and Reddy 1999)

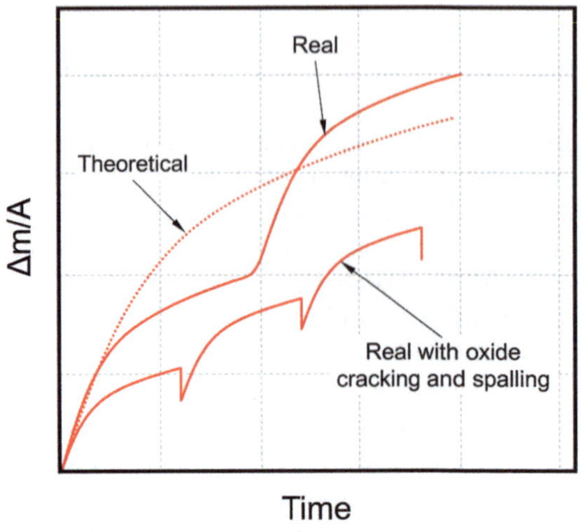

Fig. 8.13 Schematic of possible real kinetic curves for the oxidation of alloys under different conditions: specific weight change (mass change over sample area) versus time

The partial detachment of the oxide scale, occurring in correspondence of the sharp drops visible in the relevant thermogravimetric curve, may be originated by mechanical stresses associated with thermal cycling. In the long run, the alloy, or the protective coating, is no longer capable to provide a sufficient amount of reagents to mend the damaged scale. The damage can therefore propagate to the alloy and this would correspond to the end of life of the component. This latter phenomenon is known as "propagation" or final breakdown stage. When partial detachment of the surface oxide scale occurs, the kinetics may resume the linear trend, if the metal gets back in direct contact with the gaseous atmosphere.

Oxidation-resistant alloys should exhibit prolonged *steady-state* regimes, characterized by a slow progression of the corrosion damage. Although high-temperature oxidation phenomena are complex, experimental approaches and modelling tools are available, affording important inputs to design more effective protection strategies, as illustrated in the following (Young 2015). Wagner's model discussed so far would strictly refer to the gas phase oxidation of pure metals, which is actually quite a peculiar situation, certainly not interesting for real high-temperature applications, including aerospace engines. Actually, as discussed with reference to titanium alloys (see Chap. 4), steels (see Chap. 5), superalloys (see Chap. 6), and refractory alloys (see Sect. 9.2), a common requirement for these materials, in order to extend at higher temperatures their usability range, has always been to have adequate high-temperature corrosion resistance capabilities. The common protection strategy, as emerging also from the above discussion, relies on the design of alloy bulk composition or surface coating capable to form on their own surface a layer of reaction products (initiation stage) suitable to protect the underlying alloy for a conveniently long time, corresponding to the *steady-state* oxidation regime. These aspects will be considered next with reference to important examples in the field of aerospace alloys.

8.2.2 Alloy Oxidation

In the previous section, Wagner's model has been introduced and discussed with reference to gas phase oxidation of pure metals, which is not the situation usually faced in real applications, including the aerospace field, where alloys are typically employed. Therefore, the study and control of high-temperature oxidation of real systems require accounting for the more complex reactions that multi-elemental alloys necessarily involve. However, the main principles introduced as regards pure metals are still valid, and also the protection strategy aiming at reducing the effects of high-temperature gas phase reactions is still the same as those discussed in the previous section. In particular, obtaining on the alloy, or coating, surface, a protective scale made of reaction products of the base alloy elements with the outer environment is still the standard protection approach, whose effectiveness depends upon thermodynamic stability, slow growth rate, low diffusivity, compactness, good adhesion strength, and capacity of self-repairing, in case of damage. With the help of fundamental parameters, like energy of formation, reaction constants, and relevant activation energies, which can be obtained from laboratory tests, the design of alloys and protective coatings has been possible, resulting in the extension of the lifetimes of real components. A typical example in this regard is the development of oxidation-resistant steel alloys, in which the interplay of chemical, i.e., chromium alloying, and structural, i.e., austenite stabilization with nickel addition, factors have been profitably exploited. In Fig. 8.14, the progressive reduction in the oxidation rate

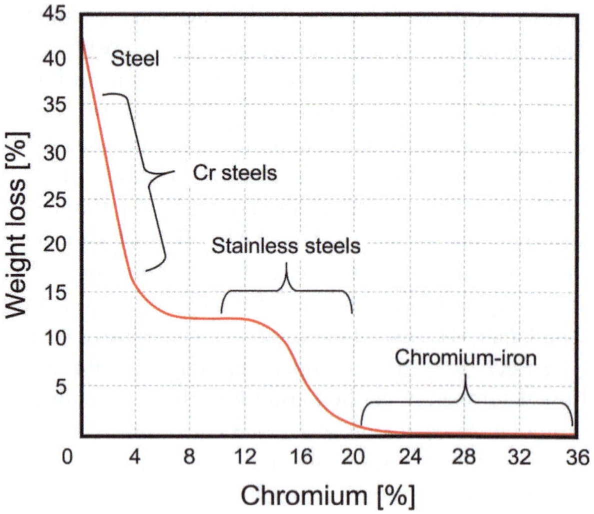

Fig. 8.14 Effect of the Cr content on the oxidation resistance of steel alloys at 1000 °C. (Redrawn from ASM International 2003)

at 1000 °C of diverse steel alloys, featuring increasing chromium concentrations, is shown. Higher concentrations of chromium modify the character of the surface oxide scale. The formation of spinel $Fe(FeCr)_2O_4$ reduces the diffusivity of the iron oxide scale, which is typically composed of all three polymorphs: wüstite (FeO), magnetite (Fe_3O_4), and hematite (Fe_2O_3).

A further improvement is introduced by alloying with nickel and relevant stabilization of the γ phase (see Appendix 4 and Chap. 5), whose closest packed structure has an intrinsically lower diffusivity than the "open" *bcc* structure of ferrite. This effect is still there if nickel is either the majority, like in nickel-based superalloys (see Sect. 6.4), or is added in significant concentrations, above 20 wt% (Fig. 8.14), like in iron-based superalloys (see Sect. 6.3).

Dealing with the high-temperature oxidation behavior of superalloys (see Sect. 6.4.2) and oxidation-resistant coatings, like diffusion and overlay coatings (see Sect. 7.3.1), the importance of Al_2O_3 and Cr_2O_3 has been discussed already. The thermodynamic aspects emerging from the Ellingham diagram (Fig. 8.9) confirm the elevated stability of these oxides, as compared to the oxide of the majority elements present in the different groups of superalloys, i.e., nickel, cobalt, and iron. Similarly, the refractory alloys (see Sect. 9.2) can rely on the formation of a layer of native SiO_2, amorphous at sufficiently high temperatures, for resisting better to high-temperature aggressive environments. For the lack of grain boundaries and associated diffusion paths, and for the adaptive capacity to the changes of the underlying substrate alloy, the silica layer represents a particularly effective protection barrier. However, the aluminum, chromium, and silicon oxides are also those featuring the comparatively lower kinetic constants (Fig. 8.12). Therefore, their preferential formation requires to overcome kinetics constraints, through a *selective oxidation*, in order to overtake faster growing, not protective, competitor oxides. This process can be enhanced through the design of suitable alloy and protective coating

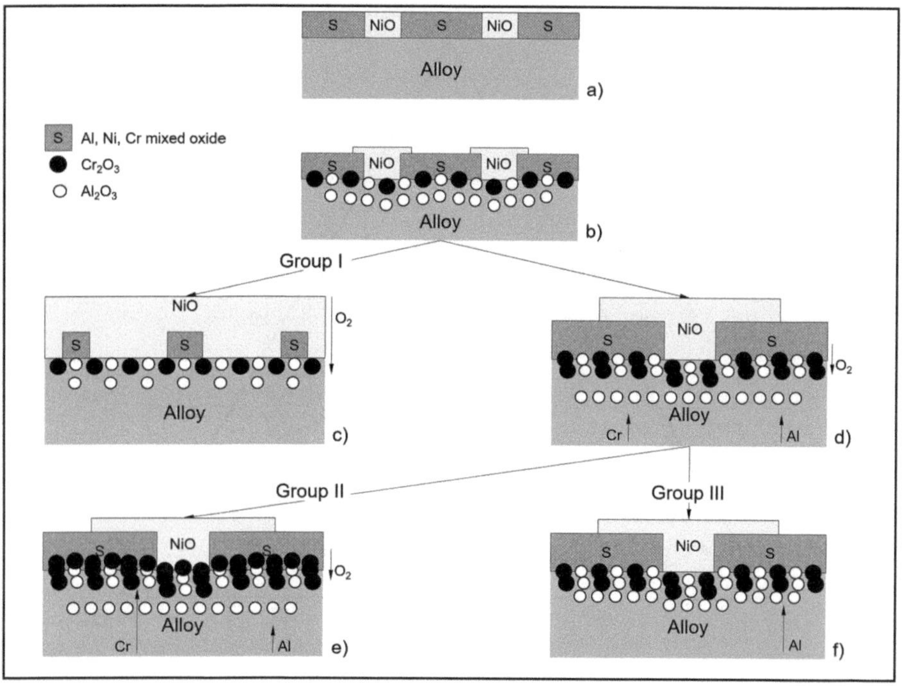

Fig. 8.15 Schematic representation of the evolution of the surface oxide layer of a Ni-Cr-Al alloy in subsequent time steps from (**a–f**). Detailed description of each phase is given in main text. (Redrawn from Giggins and Pettit 1971)

compositions, taking into account the compositional changes in the outer region of the base alloy or coating, in the oxide scale and in the local atmosphere. The case of alumina- and chromia-forming nickel-based alloys has been thoroughly considered in Giggins and Pettit (1971), a study that provides an excellent insight into the main oxidation reactions occurring in Ni-Al-Cr systems. The outcome is interesting for interpreting the oxidation behavior of both nickel-based superalloys and overlay coatings. With reference to Fig. 8.15, it turns out that, irrespective of the actual alloy composition, the early stages of oxidation occur by a direct reaction of the alloy with the oxygen present in the working atmosphere, with a very limited contribution from diffusive phenomena. The main oxide phases that form under these conditions are Ni $(Cr,Al)_2O_4$ and NiO (Fig. 8.15a). Depending on the actual composition of the base alloys, these two phases may be "doped" to different extent by outward diffusing chromium and aluminum atoms. Moreover, since nickel oxide and the spinel are not preventing oxygen inward diffusion, the formation of a mixed layer, made of chromia and alumina precipitates, can be observed (Fig. 8.15b), favored by the oxygen pressure reduction and relative increase in the concentrations of these two elements, determined by the outward diffusion of nickel. After this initial transient, the oxidation kinetics enters a steady-state condition featuring a parabolic oxidation, with products having different characteristics, depending on the alloy composition.

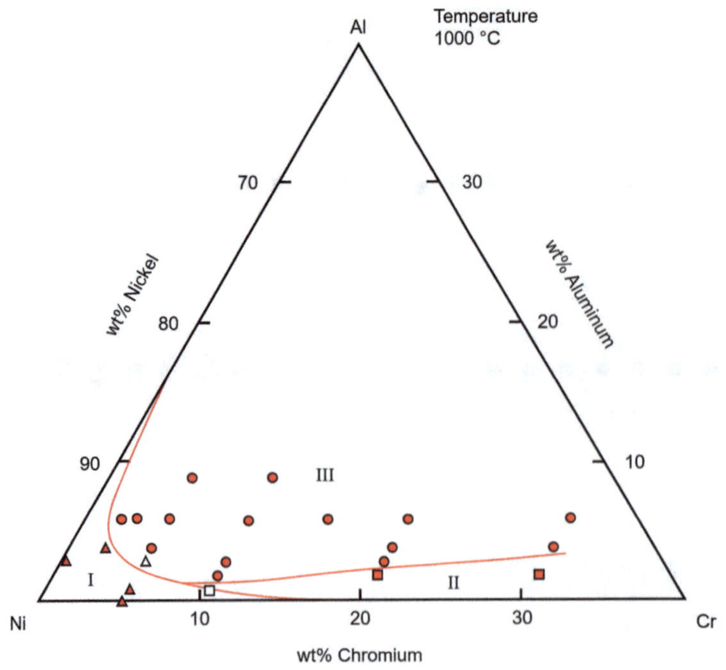

Fig. 8.16 Isothermal diagram at 1000 °C at 0.1 atm O_2 from showing the classification of Ni-Cr-Al alloys as concerns their oxidation behavior. Group I: external scale of NiO; subscale made of a mixture of Cr_2O_3 and Al_2O_3. Group II: the initial oxidation transient would feature the formation of a non-protective scale made of a mixture NiO and the spinel $Ni(Cr,Al)_2O_4$. Group III: Cr_2O_3 and Al_2O_3 are the prevailing phases forming on the alloy surface. (Redrawn from Giggins and Pettit 1971)

In case of nickel alloys with low concentrations of chromium and aluminum (Group I), the external oxide layer is made of NiO, whereas the internal oxidation leads to the formation of a discontinuous mixed Cr_2O_3-Al_2O_3 subscale (Fig. 8.15c). In Group II, the subscale Cr_2O_3 becomes continuous, although, it would not prevent internal oxidation of aluminum, since the scale contamination by the alloy elements would still allow the inward diffusion of oxygen (Fig. 8.15e). Eventually, if aluminum concentration is sufficiently high (Group III), the Al_2O_3 subscale becomes continuous, and the oxidation kinetic is generally quite slow (Fig. 8.15f). This alloy classification is summarized in Fig. 8.16, showing an isothermal (1000 °C) section of the ternary Ni-Cr-Al phase diagram, in which data points referring to the alloys of the three groups are indicated (Giggins and Pettit 1971).

Although the results discussed so far refer to oxidation experiments carried out in a low pure oxygen pressure, still, they provide reliable indications for the development of coating systems, alloy materials, and relevant oxidation protection strategies. The coexistence of aluminum and chromium reduces, with respect to binary Ni-Al or Ni-Cr alloys, the concentration that each one of these elements should have in order to produce a continuous oxide scale. This aspect has been exploited in the

Table 8.4 Standard free energy of formation of some sulfides and relevant oxides at 1000 °C (Reed 2006)

Sulfides	ΔG^0 [kJ/mole]	Oxides	ΔG^0 [kJ/mole]
$1/3Al_2S_3$	−219	$1/3Al_2O_3$	−429
$1/3Cr_2S_3$	−135	$1/3Cr_2O_3$	−269
CoS	−80	CoO	−136
FeS	−86	FeO	−176
MnS	−190	MnO	−294
NiS	−88	NiO	−127

development of MCrAlY overlay coatings (see Sects. 7.3.1 and 7.3.2) and has important consequences in keeping the concentration of chromium at sufficiently low levels, thus reducing the risk of forming the TCP σ phase (see Sect. 6.4.2).

An optimum oxidation protection is achieved when the external selective oxidation produces a stable, compact and adherent scale. In real working atmospheres, containing other gaseous species than oxygen, the attempt is still to obtain a protective surface layer mainly made of stable oxides. In this respect, the thermodynamic drive may help, considering the higher stability of most common oxides than the relevant compounds from which they may form. As an example, in Table 8.4, the standard free energies of some sulfides and of the corresponding oxides are listed.

The classification of superalloys and oxidation protective coatings in terms of alumina and chromia formers is coherent with the compatibility of their parabolic oxidation constants with those of the two oxides. However, two aspects should be considered. The equivalence of the kinetic constants refers to steady-state conditions, which are not always the same all over the lifetime of a real component. In fact, several high-temperature oxidation regimes can be rather envisaged for an alloy exposed to real operating conditions (see also Fig. 8.13). Moreover, reactions involving multicomponent alloys result in the formation of surface scales containing also other compounds, e.g., nitrides, than just the oxide phases described so far (Liu et al. 2010; Pei et al. 2017). Therefore, chromia and alumina should rather be regarded as the majority phases in real surface scales, whose real composition can be better predicted on the basis of more accurate models, including the effect also of the constitutional defect structures present in the growing scale (Sato et al. 2011).

An effective protection by the surface oxide scale depends also on its degree of adhesion to the substrate. The surface scale has to resist to the shear stresses determined by the volume changes induced by internal, to the scale, oxidation reactions. Another shear stress contribution may come from the mechanical mismatch associated with temperature gradients across the alloy-scale interface. Additions of reactive elements, like yttrium, zirconium, just to mention the most used ones, have introduced remarkable improvements in the adhesion strength of the oxide scale to the alloy, as discussed already with reference to Ni-based superalloy composition selection (see Sect. 6.4.2) and overlay coatings (see Sect. 7.3.1). These additions are either pure elements or oxide particles, e.g., Y_2O_3. This latter condition is rapidly reached anyway, owing to the elevated affinity to oxygen of these elements, which are thereby easily oxidized by both the oxygen present in the working atmosphere or scavenged from the alloy matrix. In case of elemental additions, concentrations not exceeding 1–2% are recommended, in order to limit any excessive contribution to the oxidation process. Several mechanisms have been proposed to explain the beneficial effects of reactive elements on the adhesion strength of the oxide scale (Hou and Stringer 1995; Huntz 1987; Stott and Wood 1987). The fine oxide particles act as heterogeneous nucleation sites for the scale forming on the alloy surface. This results in a finer grain size of the surface oxide layer that will exhibit a higher intrinsic plasticity and fracture toughness. Consequently, internal and interfacial thermomechanical stresses will be better accommodated. Owing to their elevated reactivity with oxygen, but also with sulfur, these elements tend to act as purifying getters of the alloy. Since the active elements are usually much slower diffusers than the other scale-forming elements, like aluminum and chromium, the particles of oxides, mostly, and sulfides are localized at the scale-alloy interface. This has a twofold beneficial effect. The first regards the mechanical gripping action that these particles would deploy, behaving like a sort of roots of the outer scale (pegging effect). Moreover, the interfaces of the oxide particles with the alloy matrix act as vacancy sinks (see a similar effect also in ODS alloys in nuclear plant applications in Sect. 9.3), thus keeping the Kirkendall voids, produced by vacancy coalescence during the oxidation process, smaller. These mechanisms contribute to different extent to enhance the oxidation resistance of superalloys, ODS superalloys, containing in their matrix a fine oxide particle dispersion (see Sect. 9.3), and overlay coatings (see Sect. 7.3.1).

8.2.3 Surface Nanocrystallization

In Sect. 7.3.1, the main coating systems used for the protection of superalloys against high-temperature oxidation have been introduced. The base idea is to have a reservoir of those metallic elements promoting the formation of a protective oxide scale and, at the same time, to provide effective adhesion mechanisms of the oxide scale growing on the coating surface. Diffusion coatings, based on aluminide phases, e.g., NiAl, FeAl, and CoAl, have been for quite sometimes the best choice. In several respects, overlay coatings have introduced important improvements in the oxidation protection, thanks to a better design of their composition. Indeed, the coexistence of chromium and aluminum affords a significant reduction of the minimum concentration, particularly of aluminum, necessary to the formation of a protective outer oxide scale. An indirect consequence of this situation is that lower concentration gradients are present across the coating-alloy interface, with obvious advantages for the durability of the coating, whose degradation mechanisms are also dependent on the interdiffusion into the base alloy. A further reduction in the interdiffusion kinetics may come from the modification of the microstructure of the alloy surface without significantly changing its composition. This can be achieved through surface nanocrystallization (SNC), an approach that is attracting increasing interests, within the broader field of the nanotechnological research (Lu and Lu 1999).

Notwithstanding the elevated concentration of grain boundaries and, consequently, the higher diffusivities, several studies have demonstrated the possibility of improving the oxidation resistance of fine-grained alloys, with respect to their macro-grained counterparts. The earlier studies on this matter regarded Ni-Cr alloys and AISI 304 stainless steels (Baer and Merz 1980; Liu et al. 2000; Merz 1979). The better reported oxidation resistance was ascribed to a more efficient oxidation of chromium, to form a chromia-rich surface scale, due to an increased outward flux of this element determined by the increased contribution of grain boundary diffusion. Similar results have been reported as concerns the oxidation of nickel-based superalloys, both chromia and alumina formers, in experiments carried out in air up to 1000 °C (Lou et al. 1996). The K38G nickel-based superalloy (Ni-16.3Cr-8.4Co-3.81Ti-4.01Al-2.7W-1.8Mo-1.75Ta-0.76Nb-0.16C-0.01B wt%) was studied in the as-cast bulk condition and as a sputter coating, with a thickness of 40 μm and an average grain size of 0.1 μm (Lou et al. 1995, 1996). In this latter, lower oxidation rates, than in bulk microcrystalline specimens, were observed under isothermal conditions at temperatures of 800 °C, 900 °C, and 1000 °C. The K38G superalloy, owing to the comparatively high Cr concentration, is regarded as a chromia-former, as indeed confirmed by high-temperature oxidation experiments on bulk samples. However, particularly at higher temperatures, the scale exhibited a tendency to spall-off. A dense, well-adherent alumina scale was observed instead on the nanograined K38G coatings, coherently with the reported slower oxidation kinetics. The transition from chromia to alumina former, induced by the reduction in the average grain size of the alloy specimens, is determined by the relevant reduction in the critical concentration of aluminum necessary to the formation of a protective alumina scale. As predicted by the theory of Wagner for the isothermal oxidation of binary alloys (Kofstad 1988), the outward flux of solute atoms capable of sustaining external oxidation leading to the formation of an external scale depends on the critical solute concentration, N_B (B = Al in the present alloy system), that can be expressed as:

$$N_{\mathrm{B}} = \sqrt{\frac{\pi g^* P_{\mathrm{O}} V_{\mathrm{mM}}}{2 D_{\mathrm{B}} V_{\mathrm{mOX}}}} \tag{8.13}$$

where:

- P_{O} is the oxygen permeability in the alloy matrix, i.e., the product of oxygen concentration times its diffusivity.
- D_{B} is the solute diffusivity.
- g^* is the critical volume fraction of oxide required for the transition from internal to external oxidation. The value is often adopted as about 0.3.
- V_{mM} and V_{mOX} are the molar volumes of the alloy and oxide, respectively.

The diffusivity of the solute atoms increases upon reducing the average grain size of the alloy since new diffusion paths, i.e., the grain boundaries in the nanostructured alloy, become available. Therefore, an alumina scale can even form on a substrate alloy having a sufficiently refined grain structure, although the same alloy would rather behave like a chromia-former in a coarse-grained condition (Wang 1997). This aspect will be considered further on, in connection to the oxidation of γ-TiAl alloys.

It is now interesting to analyze the factors that are responsible for the improved oxidation resistance of nanostructured alloy systems. The oxide scale growing on a nanostructured surface would feature a nanometric microstructure too. This is so, considering the heterogeneous nucleation action induced by the alloy grain boundaries. The refined microstructure of the oxide scale will better relieve internal stresses and those induced by thermal cycling. Stress relief may more easily occur thanks to the plastic flow of the oxide scale, assisted by the sliding along grain boundaries. In this way, the spallation limit stress of the metal-scale interface is not reached, and the surface oxide continues to protect the underlying alloy. Thermal stresses in the oxide scale are proportional to the mismatch in thermal expansion coefficients of the scale and the substrate material through the elastic modulus of the oxide. Dealing with thermodynamic and mechanical properties of nanocrystalline materials, it has been shown that they typically display a higher coefficient of thermal expansion and a lower elastic modulus (Fecht 1990; Klam et al. 1987). These two conditions contribute to the reduction of the stress in the surface oxide layer. The oxide scales observed on oxidized nanostructured alloys or coatings are generally characterized by a compact microstructure with limited porosity, also at the interface with the alloy. This is another consequence of a substantial levelling up of the diffusion coefficients of the involved atomic species. The important fraction of diffusive flux occurring along grain boundaries, of the alloy only initially, and, afterwards, also of the newly formed oxide scale, guarantees a fast diffusion to occupy the lattice sites, thus keeping the vacancy concentration comparatively low and reducing the clustering tendency to form larger pores.

The formation of a nanocrystalline layer on metallic alloys can be obtained with different approaches that can be classified into three main categories: hybrid surface nanocrystallization, deposition of surface coatings, and surface self-nanocrystallization (Gupta et al. 2012; Lu and Lu 2004).

Hybrid surface nanocrystallization is achieved in two steps, conducted in sequence or simultaneously. A possible approach might involve a surface mechanical attrition treatment (SMAT) for the production of a nanostructured layer that can subsequently be transformed through a metallurgical, chemical, or thermal process (Lu and Lu 2004).

For the deposition of nanocrystalline coatings, several PVD techniques have been used, also with nickel-based superalloys (Wang et al. 2014, 2015). By tuning the deposition parameters, selected features of the nano-grained coatings can be attained, as concerns average grain size, grain size distribution, adhesion strength to the alloy substrate, residual stresses, and possible composition fluctuations. To overcome the reported limited hot corrosion resistance in sulfate environment of nickel-based superalloys (see Sect. 8.2.5), a two-layer structure was developed (Yang et al. 2016). On top of the sputter-deposited nanocrystalline coating, having the same composition as the base superalloy N5 (Ni-7.5Co-7.0Cr-1.5Mo-5.0W-6.5ta-6.2Al-3.0Re), a second NiCrAlY coating was deposited, using an arc-ion

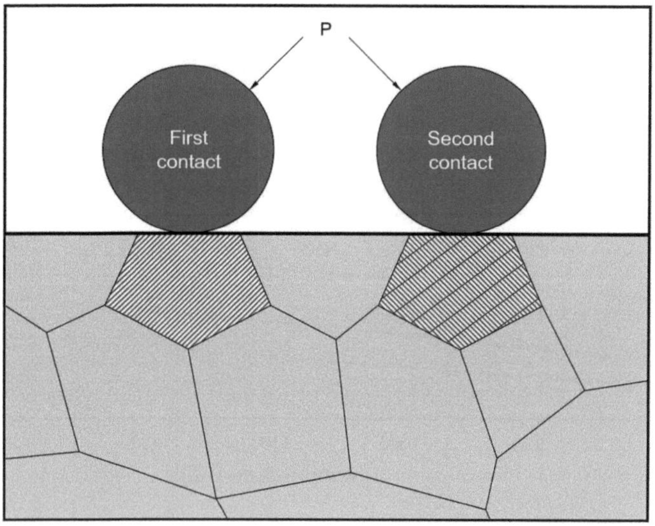

Fig. 8.17 Shot peening. Repeated multidirectional plastic deformation and consequent formation of shear bands with a high density of dislocation arrays within the grain results in the formation of a nanostructured surface layer, whose composition is substantially the same as the base alloy. (Redrawn from Lu and Lu 1999)

plating PVD technique. The combined protective action of the two layers turned out to be effective over a broad range of oxidation conditions, and, as expected, improved also the sulfate hot corrosion behavior.

The main idea behind the self-nanocrystallization techniques is to produce nanocrystalline surface layers by direct modification of the alloy microstructure, without significant compositional changes. There are two main types of surface self-nanocrystallization processes: mechanically induced surface self-nanocrystallization (MISS-NC) and thermally induced surface self-nanocrystallization (TISS-NC). An interesting example of the first process is *shot peening*, a SMAT widely used to improve the mechanical properties of an alloy through its surface hardening (Sheng 2011). Nanocrystallization can be induced by shot peening and other SMATs if the applied loads are sufficient to determine a severe plastic deformation field (Fig. 8.17), accompanied by the formation of arrays of dislocations. Following complex reactions of dislocation annihilation and recombination, the formation of a domain structure is obtained, featuring small-angle grain boundaries separating individual nanometric grains (Xia et al. 2015).

Mechanical surface treatments are not only meant to modify the properties of the alloy surface layer but also its surface finish, like can be done using sand blasting. Still using the superalloy N5, the effects of different roughness values, obtained by sand blasting (using 100-mesh Al_2O_3 particles at a pressure of 0.2 MPa for 2 min), mechanical polishing (1 μm diamond paste), and grinding (using a 2000 grit SiC paper), have been investigated (Yang et al. 2016).

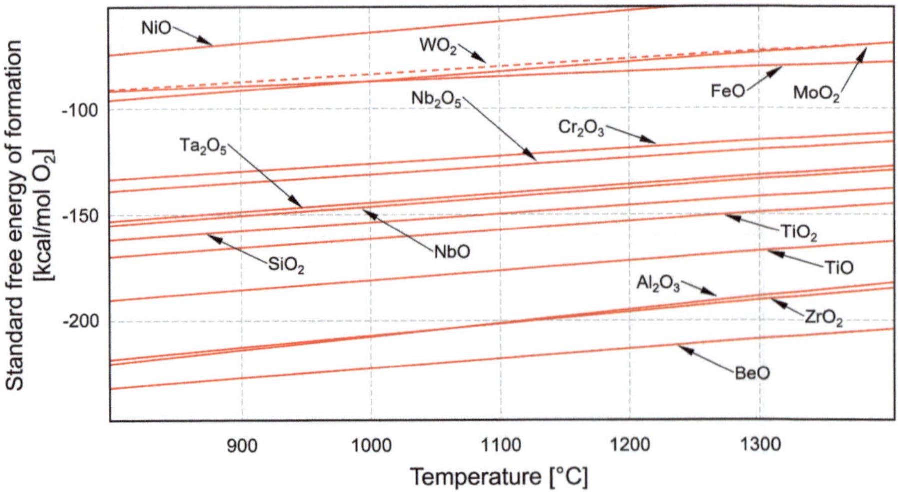

Fig. 8.18 Simplified Ellingham diagram, showing the position of Ti-O oxides with respect to Al_2O_3. (Redrawn from Okafor and Reddy 1999)

The topographies obtained on the alloy surface and transferred also to the coating deposited onto it influence the morphology and the spalling mode of the oxide scale. The higher surface roughness of the sand-blasted alloy determines a layer-by-layer spalling off of the scale, which would otherwise display a bulk fracture mode, in case of smoother surfaces, like those obtained with polishing and grinding.

The TISS-NC processes are considered also in connection to the oxidation protection of γ-TiAl alloys (see Sect. 4.3.2). As shown by the simplified Ellingham diagram in Fig. 8.18, the formation of an alumina-rich scale would be favored in view of their more negative energy of formation as compared to titanium oxides, in particular the stable polymorph, TiO_2 (Okafor and Reddy 1999). However, also in TiAl alloys, notwithstanding the thermodynamic drive and the elevated concentration of aluminum, the kinetic constraint prevails, and titanium oxide tends to form preferentially, as it can actually be inferred from the values of the parabolic constants for oxidation shown in Fig. 8.12. Owing to their columnar morphology, the titania grains cannot provide an effective protection, and any increase in the alumina concentration in the surface oxide scale is very much appreciated. A proposed strategy for obtaining this result is based on a TISS-NC approach. The surface of a TiAl has been modified using rapid solidification, following the flash melting of the alloy surface obtained with a plasma torch, generally used for the deposition of TBCs (see Sect. 7.2) (Tomasi et al. 2002). This surface treatment results in the formation of a nanostructured surface layer, with a thickness of a few micrometers (Fig. 8.19a), which introduces remarkable improvements in the quality of the oxide scale that forms on the alloy surface during isothermal oxidation treatments carried out in air at 800 °C and 900 °C. The relevant thermogravimetric curves (Fig. 8.19b) show the weight increase of alloy samples, having the same composition, but two

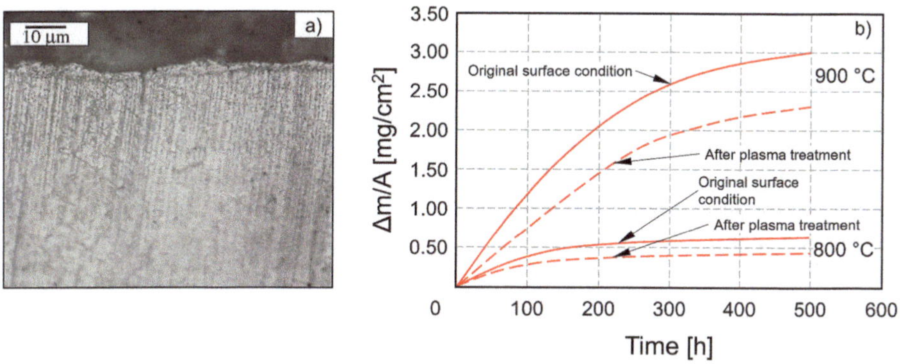

Fig. 8.19 (**a**) Cross section of the plasma-treated TiAl, showing a surface layer with a nanometric grain size; (**b**) thermogravimetric curves showing the oxidation kinetics in air at 800 °C and 900 °C of specimens of the TiAl alloy, tested in the original surface condition and after plasma surface treatment. (Images property of the Authors, unpublished data)

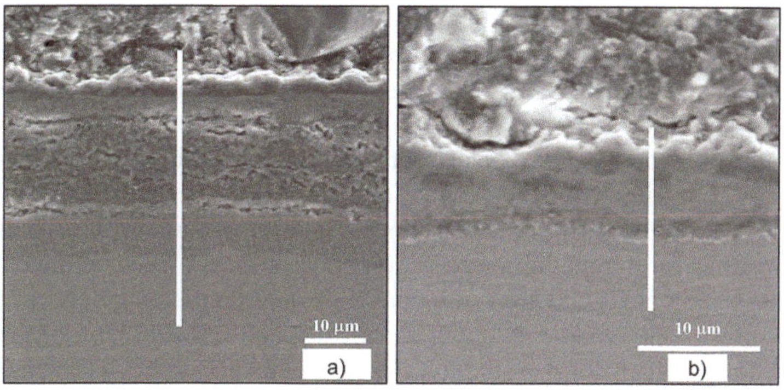

Fig. 8.20 Cross-sectional micrographs showing the oxide scale grown in a TiAl-based alloy. (**a**) Oxidized surface in its original condition; (**b**) surface after a plasma TISS-NC treatment. (Images property of the Authors, unpublished data)

different surface conditions. The oxidation kinetics of the alloy in the pristine condition turns out to be faster than the kinetics of the alloy with a nanocrystalline surface layer. Similar results have been obtained using an alternative TISS-NC approach based on laser ablation to produce the nanostructured surface layer (Mandelli et al. 2007).

The composition of the nanostructured layer obtained on the surface of the TiAl alloy is substantially unchanged with respect to the base alloy. The protective action of the surface nanocrystallization treatment is based on the balance of the diffusivities of aluminum and titanium, so to favor from the beginning the formation of an alumina-rich oxide scale, compact and well adherent. This is actually the case, when comparing the oxide scale grown on the untreated (Fig. 8.20a) and plasma-treated (Fig. 8.20b) TiAl-based alloy.

The approach discussed herewith is one of the several developed so far for the protection against high- temperature corrosion attack of TiAl-based alloys. As emerges from review publications on this theme, research efforts have developed different protection strategies, which can be selected in view of the specific applications and oxidation mechanisms (Leyens et al. 2006, Pflumm et al. 2015). Considering the growing interest in these alloys for their established position as strategic materials for gas turbines applications, it is likely that novel techniques will be developed soon.

8.2.4 Gas Phase Corrosion: Sulfur-Induced Corrosive Processes

Combustion in gas turbine jet engines produces exhaust gases containing substances, like SO_x, NO_x, CO_x, HO_x, H_xSO_y, and HNO_x (see also Fig. 2.21), that activate corrosion mechanisms that are definitely more complex than a simple oxidation. This, for instance, is the case of sulfidation and other reactions involving sulfur and its compounds. Sulfur is usually present in fossil fuels, also refined and purified, like those used to power aircraft engines, in which this element is usually below the 0.3 wt%. Sulfidation reactions, relevant to gas turbine engines, can either occur in mixed gaseous environments, with sufficiently high activities of oxygen, to keep the oxides present on the alloy surface stable, or under a salt deposit (i.e., salt-induced attack) that may form from the condensation of the sulfates present in the exhaust gas stream. Reducing environments, with partial oxygen pressures below the dissociation pressures of the oxide containing chemical species, would be a further option, although not that interesting for the oxygen-rich atmospheres of gas turbine engines.

Thermodynamic aspects can be dealt with relevant Ellingham diagrams (see Fig. 8.9). However, considering that very often sulfidation environments would feature both oxygen and sulfur partial pressures, metal-sulfur-oxygen stability diagrams are very useful in predicting the stable phases that actually form on the alloy surface, depending on the local relative pressures. The formation of metal sulfides may occur as a result of a solid-state diffusion through the oxide scale, sustained by the gradient of sulfur existing between metal/oxide and oxide/atmosphere interfaces and determined by the reduction in the partial pressure of sulfur due to the presence of the oxide scale. The resulting parabolic kinetic law of sulfidation features rate constants definitely higher than the corresponding oxides. Moreover, the reaction products are generally not protective. Therefore, this corrosive process is generally regarded as more aggressive and severe to the metal alloys than pure oxidation. Sulfidation rate constants for various metals under specific temperature and partial pressure of S_2 are listed in Table 8.5. In this table, the ratios between the sulfidation and the oxidation kinetic constants are also indicated. They provide a direct evidence of how faster sulfidation typically is with respect to oxidation, owing to the elevated reactivity of metals versus sulfur. For instance, the sulfidation rate of chromium is

Table 8.5 Sulfidation rate data for some pure metals. $K_{(S)}$ is the sulfidation rate constant, $K_{(O)}$ is the oxidation rate constant, and E_A is the activation energy for the growth of the sulfides (Young 2015)

Metal	T (°C)	pS_2 (atm)	E_A (kJ/mol)	$K_{(S)}$ (g^2/cm^4s)[a]	$K_{(S)}/K_{(O)}$[b]
Co	650–800	1	142	3×10^{-6}	450
	500–700	0.013	78	–	–
	800–1000	2×10^{-3}	210	–	–
Cr	800–1000	1	192	3×10^{-8}	>10^4
	700–900	1	72	1×10^{-7}	–
Fe	650–750	1	92	–	10^2
	700–850	0.26	84	1×10^{-5}	–
	650–900	0.13	70	–	–
	600–1000	1×10^{-4}	61	–	–
		1×10^{-8}	12	–	–
	527–697	2×10^{-6}	26	–	–
Mn	800–1000	10^{-8}–0^{-4}	82–96	–	≈ 1
	700–1000	1	130	4×10^{-9}	–
	550–711	5×10^{-9}– 4×10^{-3}	72	–	–
Mo	850–1050	50% H_2S	199	8×10^{-11} (900 °C)	–
	700–950	1.3×10^{-3}	147	7×10^{-12}	–
Nb	800–1200	1	58	1×10^{-13}	–
Ni	420–640	1	92	8×10^{-7} (600 °C)	1×10^5 (600 °C)
Ta	800–1000	1	64	7×10^{-10}	–
W	380,980	50% H_2S	135	3×10^{-10}	–

[a]Rate is obtained at 800 °C, $pS_2 = 1$ atm unless differently specified
[b]Ratio between sulfidation and oxidation rate constants at 800 °C, unless differently specified

about four orders of magnitude larger than the corresponding oxidation rate. Since chromium is a key constituent of several alloys that have to withstand high-temperature working conditions, this is certainly an issue to be considered as concerns the corrosion resistance of structural components and protective coatings (Young 2015). The presence of cracks in the oxide scale or even detached parts may produce a local change in the relative pressures of sulfur and oxygen. The oxidation occurring at the metal/oxide interface, determines an increase in local pressure of sulfur, with a consequent rise in the sulfidation rate. Since sulfide formation involves also elements like chromium and aluminum, alloy surface depletion takes place, thus reducing the possibility for the scale self-healing. This is a critical condition, since it may lead to a fast sulfidation kinetics and propagation of the attack to the alloy.

8.2.5 *Hot Corrosion*

The condensation of salt deposits, in particular, sodium sulfate, on hot parts of gas turbine engines, triggers important corrosion phenomena, usually named as *hot corrosion*, to distinguish them from electrochemical, low-temperature corrosion (see Sect. 8.1). The main components of the salt deposit come from combustion exhaust gases, mainly, but not exclusively, sulfur compounds, and the outer atmosphere. In marine environment, the air entering the engine contains sodium chloride, so that reactions like the one given by Eq. 8.14 may very well occur, considering the high air to fuel ratio and the elevated stability temperature of sodium sulfate.

$$SO_2 + H_2O + 2NaCl + \frac{1}{2}O_2 \rightarrow Na_2SO_4 + 2HCl \tag{8.14}$$

High-pressure turbine vanes and blades are the engine components more liable to salt condensation, when their temperature falls below the dew point of sodium sulfate, whose melting point is 884 °C. Sodium sulfate is not the only substance interesting for this phenomenology, and other sources of contaminants can be identified. Seawater, from which the aerosol carrying sodium chloride into the engine comes, contains also other impurity elements, like Mg, S, and K, just to mention those more interesting for corrosive phenomena. As concerns the fuel, sulfur is not the only impurity that may enter the hot corrosion reactions. Depending on the provenance of the raw material (crude oil) and the refining process, concentrations of minor elements can be found in the fuel, and some of them are particularly important in the present context (Molière and Sire 1993). Although the main contribution to the formation of sulfate comes from the outer atmosphere, sodium is also one of the impurity minority elements present in oil. Comparatively low concentrations of sulfur are tolerated in aviation fuels, typically below 0.3 wt%, whereas up to 4 wt% can be found in heavy oils. Elemental sulfur reacts with oxygen to form the metastable SO compound, which is oxidized to form SO_2 and, to partial extent, SO_3. Thermodynamic (especially over the temperature range 1300–2300 °C) and kinetic (below 1000 °C) constraints limit to a few percent the conversion of SO_2 into SO_3. This is an important aspect, considering the strongly acidic character of SO_3 that may induce aggressive corrosive attacks, particularly in the 650–750 °C range (*vide infra*).

Another alkaline metal, i.e., potassium, may contribute to the formation of the sulfate deposit (K_2SO_4). Its origin is still seawater, oil, and also water contamination coming from the oil extraction process and ballast operation in the tankers, considering that very low limits for alkali metals altogether are admissible for aviation fuels, i.e., below 1 ppm (Chevron Product Company 2007). Alkali metal, and also earth-metal e.g., Mg, sulfates reduce the melting temperature of pure Na_2SO_4, with a consequent speeding up of the corrosive reactions also at temperatures lower than 884 °C.

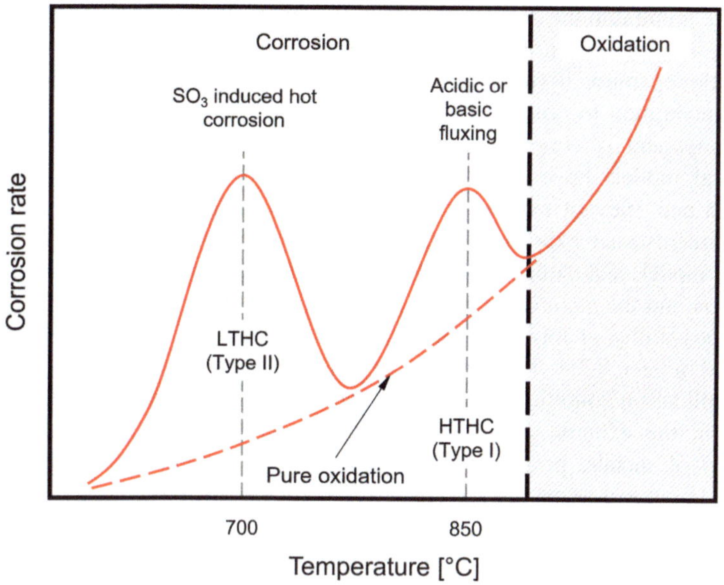

Fig. 8.21 Qualitative plot of corrosion rate vs temperature featuring the two maxima, corresponding to Type I and Type II hot corrosion. In the graphs, the main mechanisms and relevant agents are indicated and discussed in the main text

Other hot corrosion-relevant elements are vanadium, nickel, and iron, structurally embedded in porphyrin molecules, present in crude oil, with concentration and character depending on the provenance regions (Ali et al. 1993). Vanadium may contribute to the formation of corrosive deposits and ashes, with an impact that depends on its concentration, ranging from a few ppm up to 1200 ppm. V_2O_5, whose melting temperature is 690 °C, is produced during combustion and may contribute to the formation of vanadate by reacting with the sulfate components (see Eq. 8.20).

Hot corrosion phenomena can be classified as type I and type II (see Fig. 8.21). Type I high-temperature hot corrosion (HTHC) has a maximum around 850 °C, a temperature that is compatible with the presence of a molten salt deposit on the corroding alloy surface. Type II corrosion is observed at lower temperatures, i.e., over the 650–750 °C range, where another maximum in the corrosion rate curve (Fig. 8.21) is present. For this reason, type II corrosion is also called low-temperature hot corrosion (LTHC). Temperatures at which type II process occurs are well below the melting point of pure sodium sulfate, and this suggests that in this corrosion regime, the deposit would be generally solid, unless the presence of "contaminants" is reducing its melting temperature, so to have localized liquid regions in the deposit. Type II corrosion, as discussed below, is ruled in several respects by the partial pressure of SO_3, which depends on the composition of the gaseous atmosphere and the contribution from reactions involving the oxides of the refractory metals contained in the alloy.

Before going into the details of the models developed for the interpretation of the hot corrosion phenomenology, it is important to note that, owing to the comparatively little separation between the two temperature ranges of HTHC and LTHC, it is not so uncommon to observe both hot corrosion types in different regions of the same component, of course featuring different local conditions (Nicholls 2003).

Several models have been proposed to interpret the phenomenology of hot corrosion and relevant experimental observations, obtained both from diagnostics of components after exposure to real service conditions and from laboratory tests, used to model real situations (see Sect. 8.2.6). The approaches based on the *sulfidation* and the *fluxing* model are presented herewith.

The possibility of forming sulfides out of gas phase reaction has already been discussed in Sect. 8.2.4. The presence of a Na_2SO_4 deposit on the alloy surface may create sulfidation conditions, essentially through the oxidation of pyrolytic carbon, present in the exhaust gases and due to an incomplete fuel combustion. The oxidation of metals, present in the alloy, may also contribute to the oxidation process. In general terms, the two reactions can be written as:

$$\begin{aligned} &NaSO_4 + R_1 \rightarrow Na_2O + 3CO + S \\ &\text{and} \\ &NaSO_4 + R_2 \rightarrow Na_2O + 3MO + S \end{aligned} \tag{8.15}$$

where:

- R_1 and R_2 are representing the reducing agents, i.e., pyrolytic carbon and an alloy metal.
- M is a metal originally present in the alloy, particularly in the surface region involved with the corrosive process.

Sulfur liberated by the reactions (Eq. 8.15) may contribute to the formation of sulfides. In this regard, the salt deposit behaves like the oxide scale described in Sect. 8.2.4, since it reduces the partial pressure of oxygen and favors the formation of the sulfide phases. This oxidation stage of the process can be regarded as the *initiation* of the corrosive attack. It is followed by the *propagation*, featuring the oxidation of the sulfides that start to form at the alloy/deposit interface:

$$NaSO_4 + MS \rightarrow MO + Na_2O + 2S \tag{8.16}$$

Since this oxidation produces more sulfur that may then react to form more sulfides, the reaction can be regarded as *autocatalytic*. Table 8.6 gives the standard free energy of formation of some metal sulfides with the relative formation reactions, showing that, according to the free energies of metal sulfide formation, sulfur is a strong oxidizing agent.

The reactions Eqs. 8.15 and 8.16 describe one of the key aspects of the hot corrosion process, i.e., sulfidation (see Sect. 8.2.4). However, the proposed model is not so flexible to describe other reactions that may take place during a corrosive

Table 8.6 Standard free energy (ΔG^0) values normalized to ½ $S_2(g)$ of metal sulfide formation reactions (Young 2015)

		$\Delta G^0 = A + BT$ [J/mol]	
Reaction	T [K]	A	B
$Fe + \frac{1}{2}S_2 = FeS$	411.15–1190	−148.53	52.80
$\frac{9}{8}Co + \frac{1}{2}S_2 = \frac{1}{8}Co_9S_8$	<1053.15	−165.77	83.28
$\frac{4}{3}Co + \frac{1}{2}S_2 = \frac{1}{3}Co_4S_3$	1053.15–1153.15	−129.49	50.04
$Co_9S_8 + \frac{1}{2}S_2 = 9CoS$	733.15–1108.15	Not linear with temperature	
$\frac{3}{2}Ni + \frac{1}{2}S_2 = \frac{1}{2}Ni_3S_2$	733.15–808.15	−165.56	78.03
$\frac{3}{2}Ni + \frac{1}{2}S_2 = \frac{1}{2}Ni_3S_{2\text{-}x}$	808.15–923.15	−140.08	34.94
$Ni_3S_2 + \frac{1}{2}S_2 = Ni_6S_5$	673.15–798.15	−102.84	40.71
$Ni_3S_{2+x} + \frac{1}{2}S_2 = 3NiS$	833.15–108.15	−159.54	118.74
$Ni_6S_5 + \frac{1}{2}S_2 = 6NiS$	673.15–833.15	−111.92	62.93
$Cr + \frac{1}{2}S_2 = CrS$	1173.15–1313.15	−204.06	56.80
$3CrS + \frac{1}{2}S_2 = Cr_3S_4$	973.15–1173.15	−226.17	169.00
$2Cr_3S_4 + \frac{1}{2}S_2 = 3Cr_2S_3$	873.15–1173.15	−42.719	52.00
$Mn + \frac{1}{2}S_2 = MnS$	<1517.15	−263.38	60.00
$\frac{2}{3}Mo + \frac{1}{2}S_2 = \frac{1}{3}Mo_2S_3$	973.15–1373.15	−179.43	75.90
$Mo_2S_3 + \frac{1}{2}S_2 = MoS_2$	973.15–1373.15	−179.72	82.20

attack. For instance, the sulfate deposit may form not on the alloy surface directly but on the already existing scale of native oxides (see Fig. 8.22).

From a thermodynamic point of view, the stability field of various substances in a molten salt deposit can be described by means of relevant phase stability diagrams, calculated starting from the standard free energy of the involved phases. For the sodium-sulfur-oxygen system, the stability diagram is given in Fig. 8.23, with different stability fields depending on the relative concentration of the basic (Na_2O) and acidic (SO_3) components in which sodium sulfate may decompose according to Eq. 8.17:

$$Na_2SO_4 \rightarrow Na_2O + SO_3(g) \tag{8.17}$$

Phase stability diagrams are evaluated at a fixed temperature, and the boundaries between different fields are obtained by considering unit activity for the corresponding condensed phases. The thermodynamic state of the molten salt at a fixed temperature is therefore determined.

With a similar approach, the phase stability diagram including also the metallic species undergoing a corrosive attack can be evaluated, as shown by Fig. 8.24 for the Ni-O-S system at 900 °C. Depending on the basic or acidic character of the salt deposit, different reactions may occur. Particularly interesting for the hot corrosion phenomenology is to consider the dissolution reactions (fluxing) that may involve oxides, either originally present on the alloy surface before the deposition of the salt

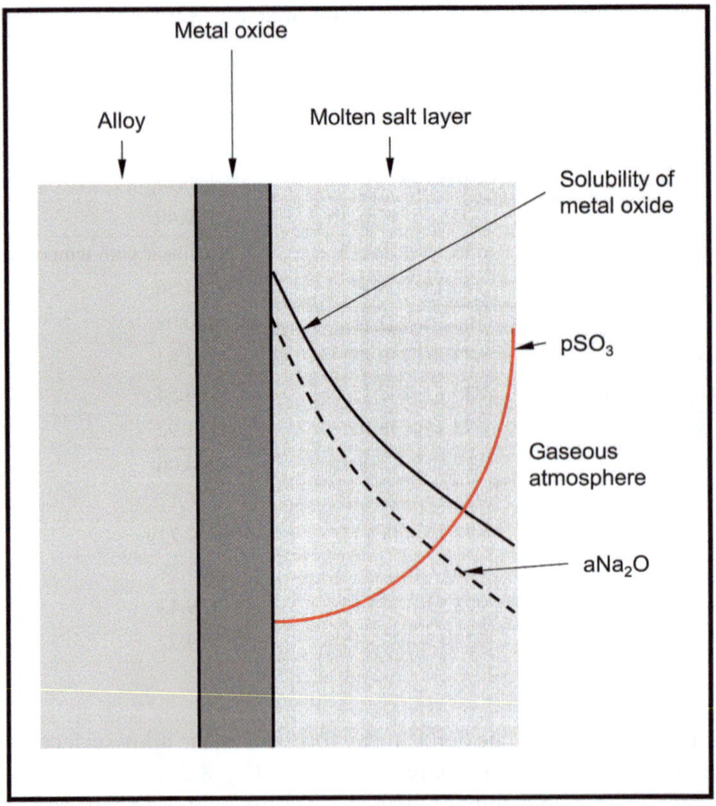

Fig. 8.22 Hot corrosion by dissolution and reprecipitation of the oxide scale under a molten salt layer based on the fluxing model, developed by Otsuka and Rapp. In this example, the system exhibits a decreasing Na_2O activity moving from the oxide/deposit interface outward. The pressure of SO_3 is decreasing in the opposite directions. This condition is typical of basic dissolution of the metal oxide (basic fluxing). (Redrawn from Young 2015)

or resulting from the ongoing oxidation taking place under or within the deposit. With reference to the diagram in Fig. 8.24, it turns out that in a Na_2O-rich deposit, i.e., a basic salt, the fluxing reaction that may take place is:

$$NiO + Na_2 \rightarrow Na_2NiO_2 \tag{8.18}$$

whereas, in an acidic deposit (relatively higher SO_3 pressure), the favored reaction is:

$$NiO + SO_3 \rightarrow NiSO_4 \tag{8.19}$$

The two reactions (Eqs. 8.18 and 8.19) can be taken as reference for *basic* and *acidic* fluxing phenomena, respectively. Similar diagrams as the one in Fig. 8.24 are

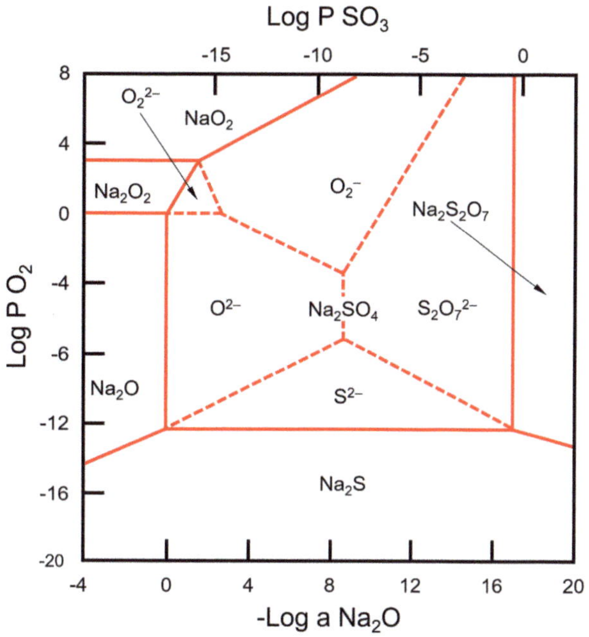

Fig. 8.23 The Na-S-O phase stability diagram at 900 °C. (Redrawn from Young 2015)

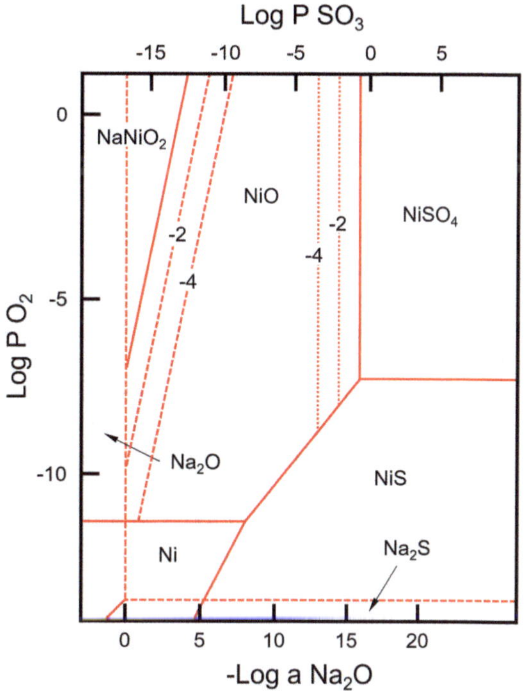

Fig. 8.24 Phase stability diagram for the Ni-O-S system at 900 °C in the presence of Na_2SO_4. (Redrawn from Young 2015)

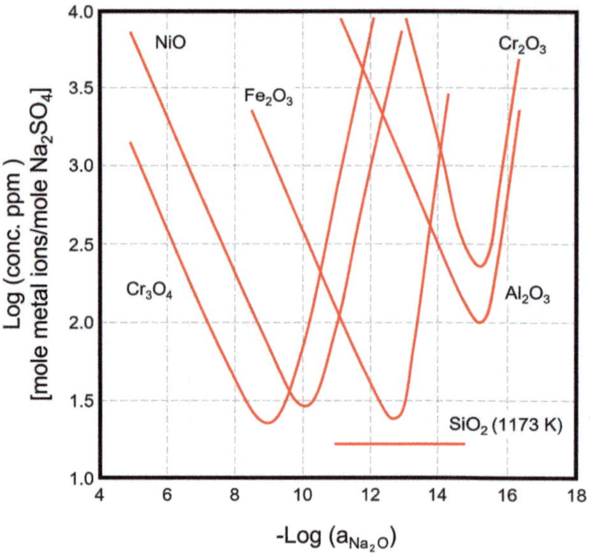

Fig. 8.25 Oxide solubility curve in molten sodium sulfate (Na_2SO_4) at 1200 K (927 °C) and pO_2 of 1 atm. (Redrawn from Rapp and Zhang 1994)

available for single metals or, if necessary, for a combination of several of them, in order to take into account also their relative interactions. Since these diagrams are equilibrium ones, they cannot predict the effects of kinetic constraints that may limit and modify the actual stability fields. Moreover, the composition of the deposit is generally different at different depths in its thickness and may change for the release of the reaction products and variations in the working atmosphere.

Considering the reactions for the fluxing model (Eqs. 8.18 and 8.19), it turns out that the basic fluxing tends to increase the oxide ion concentration in the deposit.

A concentration gradient of the reacting species, like shown in Fig. 8.22, may result in the solutioning of the oxide at the scale/salt interface and its re-precipitation inside the salt and at the salt/atmosphere interface. This picture suggests that the hot corrosion process is sustained by the presence of a negative solubility gradient, as proposed in Rapp and Goto (1981). Referring to the curves in Fig. 8.25, it can be noticed that the solubility of several oxides in molten Na_2SO_4 exhibits relatively larger values in the two lateral branches, corresponding to acidic and basic salt conditions. In between, a solubility minimum is present. Referring to these curves, it is possible to predict if hot corrosion can proceed or if a sort of buffer condition is reached at some stage and the corrosive attack is consequently interrupted.

As suggested by Rapp (2002), and shown by the general examples in Fig. 8.26, the solubility values for the fluxing oxide species at the oxide scale/salt (point I in Fig. 8.26) and salt deposit/atmosphere (point II in Fig. 8.26) interfaces can be marked on the relevant solubility curve. In cases A and C, a basic and acidic fluxing may occur, with the dissolution of the oxide scale present on the alloy surface and its re-precipitation in the salt deposit and particularly at the free salt surface, which is a preferential nucleation site. In all cases, the re-precipitated oxide is non-protective, being discontinuous and porous. Cases B represent in principle situations in which

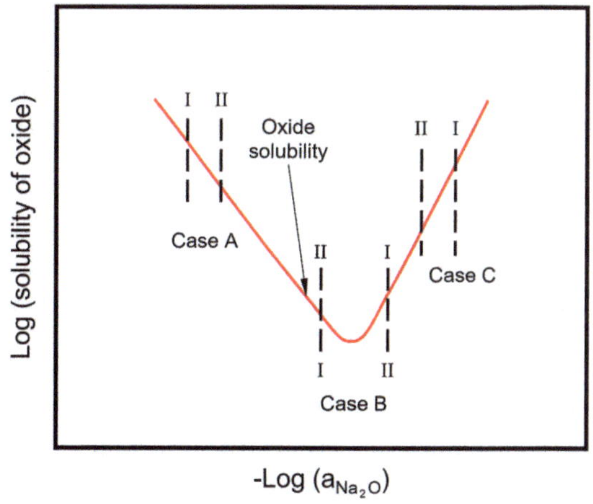

Fig. 8.26 Hot corrosion general cases of a pure metal. I and II are the solubility at the oxide/salt and salt/atmosphere interfaces, respectively. Case A and C basic and acidic fluxing are sustained by the negative solubility gradients. In cases B, in principle, hot corrosion cannot occur, since no solubility gradient is present. However, any minimum change toward the solubility minimum would activate corrosion. (Redrawn from Rapp 2002)

no solubility gradient is present within the salt deposit. From a practical point of view, this may correspond to a compositional homogeneity of the molten salt determined by thermal convective remixing. However, the equilibrium is very unstable indeed, and as soon as a local minimum in the solubility is reached, fluxing will start. Incidentally, it is worth noting that both chromia and alumina are close to the minima in their solubility curves under the operating conditions of a gas turbine engine. Therefore, they are not so prone to hot corrosion fluxing, a further positive aspect of these oxides that, as discussed in the previous sections (Sects. 8.2.1 and 7.3.1), justifies their role in the protection strategy of gas turbine components.

The solubility curves can also be used to follow the evolution of the salt composition as the oxide dissolution proceeds. Basic fluxing consumes the sodium sulfate deposit and replaces it with sodium-metal oxide (see, for instance, Eq. 8.18 for the formation of Na_2NiO_2), until the reaction subsides. From this point of view, acidic fluxing may be more deleterious for the integrity of the component. The intermediate stage that is implicit in the reaction Eq. 8.19 is the decomposition of the NiO in its component ions: Ni^{2+} and O^{2-}. The dissociation of the oxide may continue as long as the activity of the acidic agent, i.e., SO_3, renders the deposit aggressive to the alloy. Also in view of developing protective strategies, it is important to identify the main possible sources of SO_3 that may determine the acidic fluxing. In the first place, part of SO_3 may come from the exhaust, in which the oxidation of sulfur present as an impurity in the fuel may take place, as described earlier on. The same may occur to vanadium, to form V_2O_5. The reaction of vanadium oxide with the salt deposit determines an increase of its acidity, owing to the increase in the partial pressure of SO_3:

$$V_2O_5 + Na_2SO_4 \rightarrow 2NaVO_3 + SO_3 \tag{8.20}$$

More critical, since it may result in the catastrophic corrosion of the alloy, is the acidic fluxing sustained by the reaction of sodium sulfate with the oxide of refractory metals present in the alloy, like molybdenum and tungsten. The formation of sodium molybdate is induced by the following reaction, involving MoO_3:

$$Na_2SO_4 + MoO_3 \rightarrow Na_2MoO_4 + SO_3 \tag{8.21}$$

A similar reaction, involving WO_3, would result in the formation of sodium tungstate, Na_2WO_4. This reaction (Eq. 8.21) is sustained by an alloying element, and therefore it can continue as long as this element is available. Therefore, the acidic corrosive attack may propagate to significant regions of the alloy, even considering that the acidic character of the salt would determine the dissociation of possible oxide scale originally present on the alloy surface, including stable oxides, as alumina, according to dissolution reactions, like:

$$Al_2O_3 + 3MoO_3 \rightarrow 2Al^{3+} + 3MoO_4^{2-} \tag{8.22}$$

The particularly aggressive character of this type of acidic fluxing, called alloy-induced acidic fluxing and normally observed over a temperature range of 900–1000 °C (i.e., Type I hot corrosion), is due to the combination of the two acidic agents: SO_3 and MoO_3. They are both present in the salt deposit, and, since the acidic reaction involves an alloy component, e.g., molybdenum, the negative solubility gradient criterion (see Fig. 8.26) is generally satisfied. Interestingly, it has been demonstrated that by increasing the SO_3 pressure in the gas atmosphere, the rate of alloy-induced corrosion is reduced, owing to a lower equilibrium activity of MoO_3 in the deposit (Mishra 1986).

The fluxing mechanisms, both basic and acidic, involve the solutioning or dissociation of native oxides present already on the alloy surface before the salt deposit would form, or growing during the hot corrosion process. There is a third category of oxides that may be destabilized by hot corrosion, that is to say thermal barrier coatings (TBCs, see Sect. 7.3.2). The zirconia-based coatings, used to protect the internal wall of combustors and the high-pressure stage vanes and blades of the turbine from an excessive temperature rise, may suffer from fluxing effects due to sulfate deposits. This may damage the TBCs reducing their protection efficiency in case their thickness is reduced. The critical aspect is the heterogeneity that typically characterizes the corrosive attack and consequent thermal gradients induced in the alloy component (Chen et al. 2003). Moreover, it should be considered that a selective fluxing of the zirconia stabilizing oxides, e.g., Y_2O_3, may take place, for the different solubility of the diverse oxides (Zhang and Rapp 1987). The result is the destabilization of the zirconia matrix, with consequent phase change and volume expansion. The stresses building up within the coating may determine the local spalling off of the TBC and the direct exposure of the alloy to the flux of hot gases. The critical consequences that this failure mechanism may have on the stability of the component require a careful consideration.

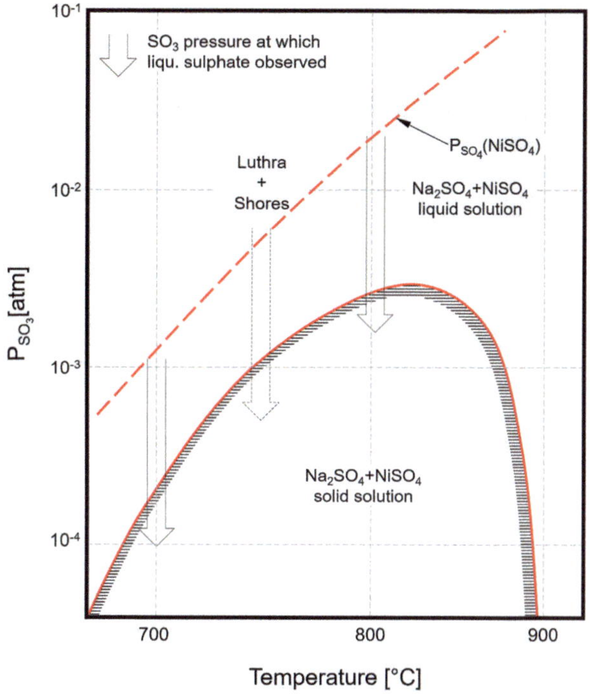

Fig. 8.27 Stability diagram for the Na_2SO_4-$NiSO_4$ system. Below the shadowed curve, a Na_2SO_4-$NiSO_4$ solid solution is stable. Above it, a liquid solution Na_2SO_4-$NiSO_4$ is stable, still containing some solid fractions both of the solid solution (Na_2SO_4-$NiSO_4$) and of NiO. The dotted line shows the pressure values of SO_3 above which $NiSO_4$ is more stable than NiO, which would therefore tend to disappear. (Redrawn from Luthra and Shores 1980)

The formation reaction of nickel sulfate (Eq. 8.19) is interesting also for introducing Type II hot corrosion (LTHC), occurring at lower temperatures than Type I (HTHC), below the melting temperature of pure Na_2SO_4. In fact, the formation of metal sulfates contributes to the reduction of the incipient melting temperature of the salt deposit. The crucial aspect is that in this context the formation of these compounds is mainly due to a sufficiently high pressure of SO_3, which is the main agent of this corrosion attack. In the stability diagram for the Na_2SO_4-$NiSO_4$ system (Fig. 8.27), the existence of molten phases for sufficiently high pressures of SO_3, depending on the temperature, is displayed. Still referring to the SO_3 pressure, the (dotted) line above which $NiSO_4$ is more stable than NiO is also shown.

Similar reactions as Eq. 8.19, involving cobalt oxides, may lead to the formation of cobalt sulfate, starting from its main oxide phases (Luthra 1982):

$$\begin{gathered} CoO + SO_3 \rightarrow CoSO_4 \\ \text{and} \\ Co_3O_4 + 3SO_3 \rightarrow 3CoSO_4 + \frac{1}{2}O_2 \end{gathered} \tag{8.23}$$

The reduction of the liquidus temperature of the sulfate solution is fully compatible with the LTHC regime, considering that the eutectics for the binary systems are Na_2SO_4 – $CoSO_4$ (T_E = 671 °C) and Na_2SO_4 – $NiSO_4$ (T_E = 565 °C). Other substances may contribute to the reduction of the liquidus temperature of the deposit,

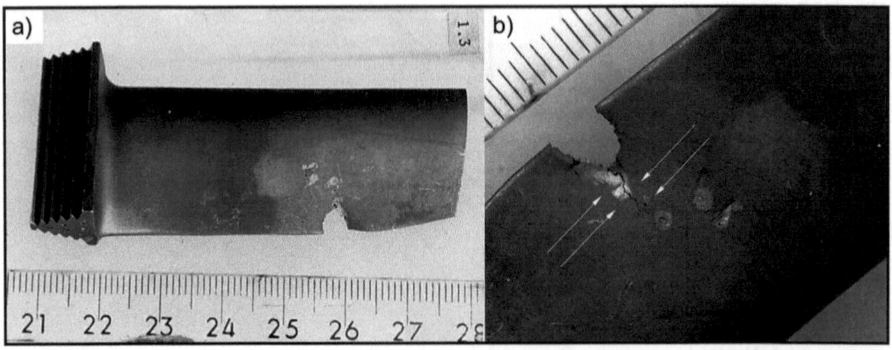

Fig. 8.28 (**a**) Macroscopic view of the damaged turbine blade (concave surface); (**b**) macroscopic view of the damaged turbine blade (convex surface) with arrows indicating a crack propagating along the blade section. (Eliaz et al. 2002, reproduced with permission of the Publisher)

like $MgSO_4$ and other contaminant either of the fuel or from the environment, including the unwanted dust that can be sucked into the engine from the runaways, during taxing and take-off operations. As concerns fuel contaminants, the formation of sodium vanadate (Eq. 8.20) may contribute to reducing the incipient melting temperature of the $Na_2SO_4 - NaVO_3$ solution, in principle down to the minimum of 500 °C at the eutectics (Singh et al. 2007).

Type II corrosion is typically a nonuniform corrosive attack with a characteristic pitting morphology produced by the mentioned low-melting sulfate solutions that form in those regions of the deposit where particular compositional conditions are realized. Hot corrosion attacks determine different kinds of damages, which may lead to part failure. Two case studies are discussed in Eliaz et al. (2002) and illustrate common ways in which hot corrosion may affect the integrity of structural components. The first, refers to MAR-M200 (see Sect. 6.4) blades from the first stage of a PT6 gas turbine engine, presenting diffuse material loss and localized deformation. Cross-sectional observations revealed the presence of a porous oxide surface scale, showing in the subscale region the typical product of corrosion reactions, e.g., metal sulfides. In this case, the reduction of the load-bearing section caused the detected plastic deformation of the component, which was thereafter replaced. Another case history is that of a low-cycle low-stress fatigue crack (arrowed in Fig. 8.28b) observed in a IN713 C (see Sect. 6.4) blade (Fig. 8.28a) that suffered from the detachment of a large fragment from its trailing edge. As revealed by microanalytical investigations, traces of chlorine, sulfur, and potassium were present on the exposed surface of the crack, confirming that the initial macroscopic defects had been created by localized hot corrosion. If not replaced on time, the crack initiated by stress concentration after the spalling off of the initial defect, might have propagated to the whole section of the blade, thus resulting in a catastrophic failure.

These examples and the mechanisms underlying Type I and II hot corrosion suggest possible protection and remediation strategies that have been actually adopted to counteract the very detrimental effects of this process (Eliaz et al. 2002).

Nickel and cobalt base superalloys are the first choice for the hottest and more demanding, in terms of in- service conditions, components of gas turbine engines. For these alloys, hot corrosion is definitely an important aspect to consider. However, as concerns the improvement of the corrosion resistance of the superalloys, the requirements are often conflicting with those for better mechanical properties, both static and time-dependent, and high-temperature structural stability. Refractory elements, like tungsten and molybdenum, are excellent solid solution strengtheners and, through the reduction of the alloy diffusivity (see Fig. 6.24), have also beneficial effects on creep resistance. Nonetheless, under particular environmental conditions, they may determine an important hot corrosion phenomenon (alloy-induced hot corrosion) that is particularly critical since sustained by alloy elements, in the first place, and therefore potentially severely destructive for the component. Chromium and aluminum provide the best protection against high-temperature corrosion, in view of their elevated stability and capacity to form a protective scale on the alloy surface. These and similar principles have been followed to rank different alloys as concerns their corrosion resistance, so to provide helpful material selection guidelines (Meier 1989; Pettit 2011; Sidhu et al. 2006).

Better compositions to resist hot corrosion have been developed to comply more effectively with specific application requirements. Cobalt-based superalloys have demonstrated an excellent corrosion resistance, which has been ascribed to some extent to the higher eutectic temperature of the Co-Co_3S_4 mixture, higher than that of the corresponding mixture forming in nickel-based superalloys (i.e., Ni-Ni_xS_y). Therefore, the novel generation of the low-density cobalt-based superalloys (see Sect. 6.2.2) has this further specific characteristic that justifies the growing interests in these materials. Also in the development of single-crystal nickel-based superalloys, the possibility of improving hot corrosion resistance has been considered as an important design parameter, to improve a strategic material property (Zhang et al. 1993a, b).

Hot corrosion protection has also become an issue for the TiAl-based materials, considering their introduction as material for gas turbine components (see Sects. 4.3.2 and 4.6), and similar protection strategies are being considered, with the necessary adaptations to specific compositional systems (Tang et al. 1999; Zhao et al. 2010). The protective coatings developed for enhancing the oxidation resistance of high- temperature aerospace alloys (see Sect. 7.3.1) would indirectly improve also their resistance against hot corrosion attack. In fact, the oxide scales that form on the coating surface, owing to the optimized composition and microstructure, are relatively more stable and better adherent to the substrate, so to be more resistant also in the salt deposit aggressive environment. In case the corrosive damage would produce the partial removal of the oxide scale, this can reform, owing to the larger reservoir of scale-forming elements in the coatings. In this respect, MCrAlX overlay coatings (see Sect. 7.3.1) have proved to be particularly successful, since their composition can be tuned to better respond specific external conditions. For prevailing LTHC, high chromium and low aluminum are preferable compositions, whereas HTHC can be better tackled with lower chromium and higher aluminum coatings (Sudiro et al. 2011). As concerns the aluminide diffusion coatings, platinum additions, beneficial to the oxidation resistance, turn out to be important also for better hot corrosion performances (Yang et al. 2017).

Optimal operating conditions, in terms of air and fuel purity, and engine maintenance procedures are additional effective tools for containing the negative effects of hot corrosion on the engine lifetime. Air filters, quite common in land-based large gas turbine plants, have been developed also for aerospace applications with particular regard for helicopter engines (Scimone 2003; Wilcox et al. 2012).

Sulfur, vanadium, and alkali metals, i.e., Na and K, are the main fuel impurities that may initiate hot corrosion once they react to form phases that may condensate and form a liquid or a solid deposit. Therefore, there are maximum acceptable values for the above impurities. Typical limiting concentrations are $(Na + K) \leq 0.2$–0.6 ppm, $V \leq 0.5$ ppm, and $S \leq 1\%$. These limiting concentration may be overcome if the more exposed components are adequately protected, using appropriate coatings or suitable additives. These additives are meant to reduce the corrosion rate by promoting competitive reactions. For instance, an excess of vanadium may be tolerated if a suitable concentration of MgO is added, which reacts preferentially with V_2O_5 to form $Mg_3V_2O_8$. This vanadate remains solid even above the HTHC temperature interval (Stringer 1987).

To complete the picture on the diverse approaches used to protect the structural components from hot corrosion attack, one possibility is based on the water wash of the engine to remove all those contaminating substances that may initiate the hot corrosion attack. Standard washing protocols have been established and delivered in specific manuals (Need et al. 2015). The frequency and other parameters of the procedure are determined on the basis of the specific risk of corrosive attack both during operation, or for engines exposed to aggressive atmospheric conditions, like marine environments. In this case, washing is very effective in reducing the corrosion load also of the compressor, particularly as concerns the parts exposed to the air intake flux, since in these cases the prevailing corrosion agents are ingested atmospheric pollutants. These are removed by introducing water and detergent substances into the intake air flux of an operating engine. The application of protective plugs to the open ends of an engine, i.e., air intake and exhaust duct, is a common practice to reduce the access of contaminants into the engine during prolonged idle periods, particularly for aircrafts exposed to critical and harsh environmental conditions (Carter 2005; Hoyland and Antoine 2005).

8.2.6 Hot Corrosion Testing

The experimental investigation on the hot corrosion behavior of alloys is not a simple matter, mainly for the complexity of the involved sub-systems, alloy, deposit, atmosphere, and relevant compositional changes, as the corrosion reactions proceed.

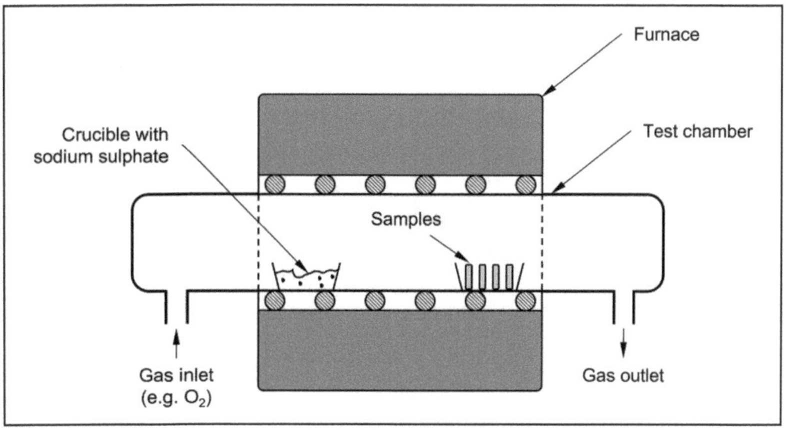

Fig. 8.29 Schematic of the exposure test

However, there are established methods for hot corrosion testing that provide extremely useful indications for predicting material behavior under service conditions. The results of real part inspections are very important for the definition of testing conditions that may provide response as concerns maintenance procedures, safe lifetime estimation, and part replacement. The three most commonly used testing methods are:

- Exposure test
- Crucible test
- Burner rig test

The exposure test is usually performed in an electrically heated furnace. The test implies the contact between the specimens and a sulfur-containing atmosphere. Exposure test is also known as bead test. The molten salt deposition is reproduced by evaporation and condensation of sodium sulfate or other sulfur-containing salts. Condensation is obtained through a suitable temperature profile inside the furnace chamber and using a gas carrier, to simulate also the effect of the different atmospheres (see Fig. 8.29).

A sample holder may contain up to 10–20 specimens, whose initial weight and exposed surface are measured before starting the test. In isothermal conditions, the samples are extracted at regular intervals, and their weight is measured, thus giving information about the degradation process they have been subjected to. The *crucible test* (Fig. 8.30) requires the immersion for fixed time intervals (exposure time) of specimens with defined cross-section geometry (circular for cylindrical samples, rectangular for sheet coupons, etc.) into a crucible filled with a molten Na_2SO_4-based salt mixture, kept at constant temperature. The aim of the *crucible test* is both to measure the metal loss after the corrosive attack and to evaluate the depth reached by the corrosive attack. For this aspect, suitable material characterization methods are

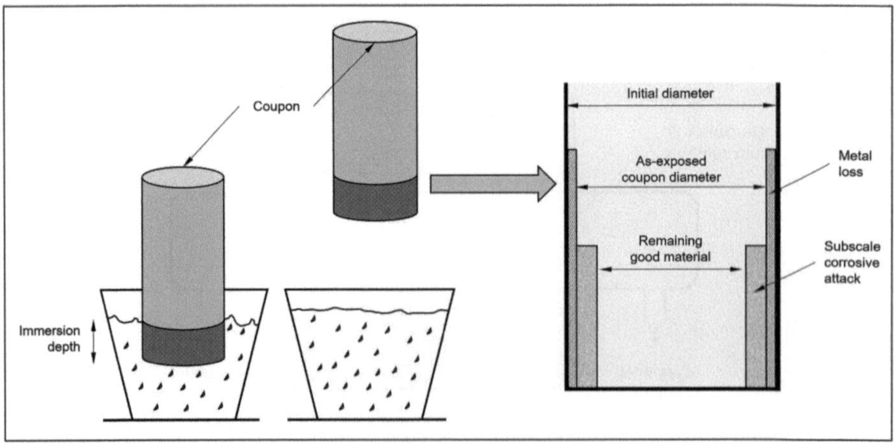

Fig. 8.30 Schematic of the crucible test. (Redrawn from Baboian et al. 2005)

used. The test is quite simple and definitely useful for comparative studies, although it exhibits evident limitations for an effective study of the hot corrosion mechanism. This is mainly due to the following aspects:

- The composition of the salt bath changes during the test, owing to the reactions with the sample.
- The effect of gas atmosphere cannot be accounted for, since the partial pressures of the gaseous components are different at different depths in the bath.

In this regard, the *burner rig* test is definitely much more effective and reliable for the assessment of the hot corrosion resistance of gas turbine alloys (Sumner et al. 2013).

A salt solution is injected into a flame produced in a burner chamber. The exhaust flux, containing the salt droplets, impinges onto the surface of the samples, placed on a rotating carousel (sample stage in Fig. 8.31). The samples can also be moved in and out of the hot gas flux, so to test thermal cycling effects. For a full understanding of the reactions occurring during any of the tests, to evaluate the volume of corroded material and the depth beneath the alloy surface reached by the corrosion attack, a set of microscopy and crystallo-chemical investigations are absolutely necessary for a complete characterization of the corrosion products. Surface analyses, like X-ray diffraction (XRD) and X-ray fluorescence (XRF), are important to characterize the phase composition and crystallographic phases present in the scale and in the subscale interdiffusion layer, if the thickness of the surface layer of salt deposit and reaction products is not too large. The surface topography can be surveyed with scanning electron microscopy (SEM) observations. Cross-sectional analyses are fundamental to understand the extension of the corrosion damage inside the alloy, the internal corrosion products, and their composition. Metallographic sample preparation is necessary, and it is advisable to protect the surface layer to avoid damage during handling and polishing operations. SEM observations can be accompanied by

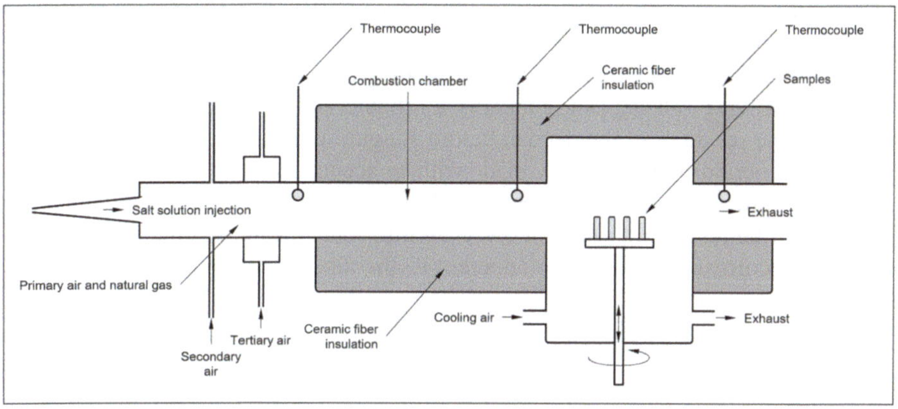

Fig. 8.31 Schematic of a burner rig

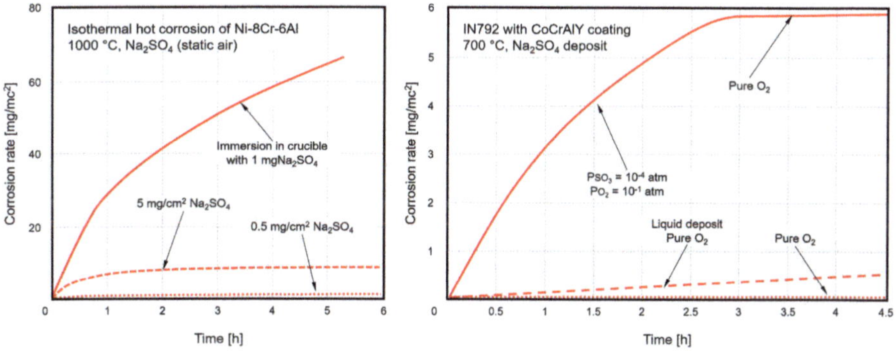

Fig. 8.32 Examples of thermogravimetric curves obtained from hot corrosion tests. (**a**): Weight gain of the Ni-8Cr-6Al-3Mo alloy hot corroded under different thicknesses of molten Na_2SO_4 deposit; (**b**) hot corrosion tests of the IN792 superalloy protected by a CoCrAlY overlay coating. The effect of different compositions of the atmosphere, as indicated in the graph, is considered. The tests are conducted under LTHC regime, with solid salt deposit, with the exception of the dotted curve referring to a liquid deposit obtained by mixing Na_2SO_4 win $MgSO_4$. (Redrawn from Pettit 2011)

energy dispersive X-ray spectroscopy (EDXS) analyses, which can be used for evaluating the chemical composition of the different regions in the corrosion altered alloy (X-ray mapping). To complete the picture, microhardness profile can be acquired across this same altered layer as a further indication of the changes brought about by the corrosion attack. Scientific journals and monographs on material characterization techniques provide a complete information on the possibilities of the different approaches, as concerns also the sample preparation procedures of each technique (Brandon and Kaplan 2008). Thermogravimetric curves, like those presented in the previous section, can be used also for hot corrosion tests data, by plotting the specific weight change as a function of the time of exposure to the hot corrosion conditions (Fig. 8.32).

An important aspect to consider in thermogravimetric tests in hot corrosion studies, involving salt deposit, concerns a reliable and reproducible procedure of removing the residuals of the deposit before weighting the tested samples at the end of each test period. This is usually made with an accurate washing of the specimens. A possible procedure requires washing in boiling water for 30 min, mostly for the removal of the remaining salts, followed by ultrasonic cleaning, effective for eliminating the corrosion product not adherent to the alloy substrate (Yuan and Wang 2010).

References

Ali MF et al (1993) Nickel and Vanadyl Porphyrins in Saudi Arabian Crude Oils. Energy & Fuels 7: 179–184

ASM International (1992) Handbook of Case Histories in Failure Analysis – Vol. 1. ASM International, Materials Park, Ohio

ASM International (2003) ASM Handbook Vol. 13A – Corrosion: Fundamentals, Testing and Protection. ASM International

Baboian et al. (2005) Corrosion Tests and Standards: Application and Interpretation. ASTM International

Baer D R, Merz M D (1980) Differences in Oxides on Large- and Small-Grained 304 Stainless Steel. Metallurgical Transactions A 11 (12): 1973–1980

Bagnoli F, Bernabei M (2008) Fatigue analysis of a P180 Aircraft Main Landing Gear Wheel Flange. Engineering Failure Analysis 15: 654–665

Benavides S (2014) Corrosion Control in the Aerospace Industry. CRC Press

Brandon D, Kaplan W (2008) Microstructural Characterization of Materials. John Wiley and Sons Ltd

Buckley et al (1981) Early Development of Ceramic Fiber Insulation for Space Shuttle. Ceramic Bulletin 60: 1196–1200

Carter T J (2005) Common Failures in Gas Turbine Blades. Engineering Failure Analysis 12 (2): 237–247

Chen Z et al (2003) Effect of Al_2O_3 Overlay on Hot-Corrosion Behavior of Yttria-stabilized Zirconia Coating in Molten Sulfate-vanadate Salt. Thin Solid Films 443 (1): 46–52

Chevron Product Company (2007) Aviation Fuels – Technical Review. Chevron Product Company, CA, USA

DuQuesnay D L et al (2002) Fatigue Crack Growth from Corrosion Damage in 7075-T6511 Aluminum Alloy under Aircraft Loading. International Journal of Fatigue 25 (5): 371–377

Eliaz N et al (2002) Hot Corrosion in Gas Turbine Components. Engineering Failure Analysis 9 (1): 31–43

Fecht H J (1990) Thermodynamic Properties and Stability of Grain Boundaries in Metals based on the Universal Equation of State at Negative Pressure. Acta Metallurgica et Materialia 38 (10): 1927–1932

Findlay J S, Harrison N D (2002) Why Aircraft Fail. Materials Today 5 (11): 18–25

Fontana M G (1987) Corrosion Engineering. McGraw-Hill, Singapore

Giggins C S, Pettit F S (1971) Oxidation of Ni-Cr-Al Alloys Between 1000 °C and 1200 °C. Journal of the Electrochemical Society 118 (11): 1782–1790

Gupta R K et al (2012) Oxidation Resistance of Nanocrystalline Alloys. Available via DIALOG https://research.monash.edu/en/publications/oxidation-resistance-of-nanocrystalline-alloys. Accessed 23 Dec 2018
Hou P Y, Stringer J (1995) The Effect of Reactive Element Additions on the Selective Oxidation, Growth and Adhesion of Chromia Scales. Materials Science and Engineering A 202 (1–2): 1–10
Housecroft C E, Constable E C (2006) Chemistry. Pearson
Hovland G, Antoine M (2005) Scheduling of Gas Turbine Compressor Washing. Intelligent Automation and Soft Computing 11: 1–12
Huntz A M (1987) Influence of Active Elements on the Oxidation Mechanism of M-Cr-Al Alloys. Materials Science and Engineering 87: 251–260
Jones R H (1992) Stress-Corrosion Cracking. ASM International
Klam H J et al (1987) The Thermal Expansion of Grain Boundaries. Acta Metallurgica 35 (8): 2101–2104
Kofstad P (1988) High Temperature Corrosion. Elsevier Applied Science, London
Korb et al (1981) The Shuttle Orbiter Thermal Protection System. Ceramic Bulletin 60: 1188–1193
Lee B et al (2011) Investigation on Fretting Fatigue in Aircraft Engine Compressor Blade. Engineering Failure Analysis 18: 1900–1908
Leyens C et al (2006) Recent Progress in the Coating Protection of Gamma Titanium Aluminides. JOM 58 (1): 17–21
Liu C T et al (2010) Oxidation Behavior of a Single-Crystal Ni-base superalloy between 900 and 1000 °C in Air. Journal of Alloys and Compounds 491 (1–2): 522–526
Liu et al (2000) Surface Nanocrystallization of 316L Stainless Steel Induced by Ultrasonic Shot Peening. Materials Science and Engineering A 11 (4): 91–95
Lou H et al (1995) Rehealing Ability of Oxide Scales Formed on Microcrystalline K38G Coatings. Oxidation of Metals 43 (3–4): 317–328
Lou H et al (1996) Oxidation Behavior of Sputtered Microcrystalline Coating of Superalloy K17F at High Temperature. Materials Science and Engineering A 207 (1): 121–128
Lu K, Lu J (1999) Surface Nanocrystallization (SNC) of Metallic Materials-Presentation of the Concept behind a new Approach. Journal of Materials Science and Technology 15 (3): 193–197
Lu K, Lu J (2004) Nanostructured Surface Layer in Metallic Materials induced by Surface Mechanical Attrition Treatment. Materials Science and Engineering A 375–377: 38–45
Luthra K L (1982) Low Temperature Hot Corrosion of Cobalt-Base Alloys: Part I. Morphology of the Reaction Product. Metallurgical Transactions A 13 (10): 1853–1864
Luthra K L (1982) Low Temperature Hot Corrosion of Cobalt-Base Alloys: Part II. Reaction Mechanism. Metallurgical Transactions A 13 (10): 1853–1864
Luthra K L, Shores D A (1980) Mechanism of Na_2SO_4 Induced Corrosion at 600–900°C. Journal of Electrochemical Society 127 (10): 2202–2210
Mandelli M et al (2007) Surface Treatments and Oxidation Resistance of a γ-α_2 TiAl Alloy (Ti-30.5 Al-9W-0.4Si). La Metallurgia Italiana March issue: 21–28
McCafferty E (2010) Introduction to Corrosion Science. Springer Verlag, New York
Meier G H (1989) A review of Advances in High-Temperature Corrosion. Materials Science and Engineering A 120–121: 1–11
Merz M D (1979) The Oxidation Resistance of Fine-Grained Sputter-Deposited 304 Stainless Steel. Metallurgical Transactions A 10 (1): 71–77
Mills T et al (2014) Corrosion and the Threat to Aircraft Structural Integrity. In: Benavides S (ed) Corrosion Control in the Aerospace Industry. Woodhead Publishing Limited, p 35–66
Mishra A K (1986) Mechanism of Na_2SO_4-1nduced Corrosion of Molybdenum Containing Nickel-Base Superalloys at High Temperatures II. Journal of the Electrochemical Society 133: 1038–1042
Molière M, Sire J (1993) Heavy Duty Gas Turbine Experience with Ash-forming Fuels. Journal de Physique IV Colloque 03 (C9): 719–730
Mouritz A (2012) Introduction to Aerospace Materials. Woodhead Publishing Limited

Need R et al (2015) Reliability Evaluation of a Gas Turbine Water Wash System – A Case Study. Journal of Natural Gas Science and Engineering 24: 457–463
Nicholls J R (2003) Advances in Coating Design for High-Performance Gas Turbines. MRS Bulletin 28 (9): 659–670
Okafor I C I, Reddy R G (1999) The Oxidation Behavior of High-Temperature Aluminides. JOM 51 (6): 35–40
Pedeferri P (2010) Corrosione e Protezione dei Materiali Metallici Vol. 1 and 2. Polipress
Pei H et al (2017) Long-Term Oxidation Behavior and Mechanism of DD6 Ni-based Single Crystal Superalloy at 1050 °C and 1100 °C in Air. Journal of Alloys and Compounds 704: 218–226
Pettit F (2011) Hot Corrosion of Metals and Alloys. Oxidation of Metals 76 (1–2): 1–21
Pflumm R et al (2015) Oxidation Protection of γ-TiAl-based Alloys – A Review. Intermetallics 56: 1–14
Popov B (2015) Corrosion Engineering: Principles and Solved Problems. Elsevier
Rapp A (2002) Hot Corrosion of Materials: A Fluxing Mechanism? Corrosion Science 44 (2): 209–221
Rapp R A, Goto K S (1981) The Hot Corrosion of Metals by Molten Salts. Proceedings of the Electrochemical Society 1980: 159–177
Rapp R, Zhang Y (1994) Hot Corrosion of Materials: Fundamentals Studies. Journal of the Minerals 46 (12): 47–55
Reed R C (2006) The Superalloys – Fundamentals and Applications. Cambridge University Press, Cambridge (UK)
Sato A et al (2011) Oxidation of Nickel-based Single-Crustal Superalloy for Industrial Gas Turbine Applications. Acta Materialia 59 (1): 225–240
Scimone M J (2003) Aircraft Engine Air Filter and Method, US Patent 6.595.743, B2
Sheng G (2011) Investigation of Surface Self-Nanocrystallization in 0Cr18Ni9Ti Induced by Surface Mechanical Attrition Treatment. International Scholarly Research Network Nanotechnology. Available via DIALOG https://www.hindawi.com/journals/isrn/2011/583215/. Accessed 12 Nov 2018
Sidhu T S et al (2006) Hot Corrosion and Performance of Nickel-based Coatings. Current Science 90 (1): 41–47
Singh H et al (2007) An overview of Na_2SO_4 and /or V_2O_5 induced Hot Corrosion of Fe- and Ni-based Superalloys. Reviews on Advanced Materials Science 16 (1): 27–50
Stott F H, Wood G C (1987) Growth and Adhesion of Oxide Scales on Al_2O_3 forming Alloys and Coatings. Materials Science and Engineering 87: 267–274
Stringer J (1987) High-Temperature Corrosion of Superalloys. Materials Science and Technology 3 (7): 482–493
Sudiro T et al (2011) High Temperature Corrosion of CoNiCrAlY-Si Alloys in an Air-Na_2SO_4-NaCl Gas Atmosphere. Materials Transactions 52 (3): 433–438
Sumner J et al (2013) High Temperature Oxidation and Corrosion of Gas Turbine Materials in Burner Rig Exposure. Materials Science and Technology 29 (7): 813–821
Tang Z et al (1999) Hot-Corrosion Behavior of TiAl-Base Intermetallics in Molten Salts. Oxidation of Metals 51 (3–4): 235–250
Tomasi A et al (2002) Oxidation Protection of γ-TiAl Alloys by Surface Heat Treatments. In: Proceedings of the International Symposium on High Temperature Corrosion in Energy Related Systems, Angra dos Reis, RJ, September 2002
Wallace W et al (1985) AGARD Corrosion Handbook Volume 1 – Aircraft Corrosion: Causes and Case Histories. AGARD-Advisory Group for Aerospace Research and Development
Wang F (1997) The Effect of Nanocrystallization on the Selective Oxidation and Adhesion of Al_2O_3 Scales. Oxidation of Metals 48 (3–4): 215–224
Wang M J et al (2015) Comparative Study of Oxidation and Interdiffusion Behavior of AIP NiCrALY and Sputtered Nanocrystalline Coatings on a Nickel-based Single-Crystal Superalloy. Corrosion Science 98: 530–540

Wang X et al (2014) Influence of Sputtered Nanocrystalline Coating on Oxidation and Hot Corrosion of a Nickel-based Superalloy M951. Journal of Materials Science & Technology 30 (9): 867–877
Wanhill R et al (2011) Stress Corrosion Cracking (SCC) in Aerospace Vehicles. In: Raja V S, Shoji T (eds) Stress Corrosion Cracking: Theory and Practice, Whoodhead Publishing Limited, p 608–650
Wilcox M et al (2012) Technology Review of Modern Gas Turbine Inlet Filtration Systems. International Journal of Rotating Machinery 2012 (12)
Xia Z X et al (2015) Improve Oxidation Resistance at High Temperature by Nanocrystalline Surface Layer. Scientific Reports 5. Available via DIALOG https://www.nature.com/articles/srep13027. Accessed 11 Dec 2018
Yang L et al (2016) A Duplex Coating for High Temperature Applications on Single-Crystal Superalloys. Corrosion Science 102: 72–83
Yang L et al (2016) Effects of surface Finish of Single Crystal Superalloy Substrate on Cyclic Thermal Oxidation of Its Nanocrystalline Coating. Corrosion Science 111: 313–324
Yang Y F et al (2017) Hot Corrosion Behaviour of Single-Phase Platinum-modified Aluminide Coatings: Effect of Pt Content and Pre-oxidation. Corrosion Sience 127: 82–90
Young D J (2015) High Temperature Oxidation and Corrosion of Metals. Elsevier
Yuan L, Wang H M (2010) Hot Corrosion Behaviors of a $Cr_{13}Ni_5Si_2$-based Metal Silicide Alloy in Na_2SO_4+ 25 wt.% K_2SO_4 and Na_2SO_4+25 wt.% NaCl molten Salts. Intermetallics 18 (3): 324–329
Zhang J S et al (1993a) Design and Development of Hot Corrosion-Resistant Nickel-Base Single-Crystal Superalloys by the d-Electrons Alloy Design Theory: Part I – Characterization of the Phase Stability. Metallurgical Transactions A 24: 2443–2450
Zhang J S et al (1993b) Design and Development of Hot Corrosion-Resistant Nickel-Base Single-Crystal Superalloys by the d-Electrons Alloy Design Theory: Part II – Effects of Refractory Metals Ti, Ta and Nb on Microstructures and Properties. Metallurgical Transactions A 24: 2451–2464
Zhang Y S, Rapp R A (1987) Solubilities of CeO_2, HfO_2 and Y_2O_3 in Fused Na_2SO_4 -30 mol% $NaVO_3$ and CeO_2 in Pure Na_2SO_4 at 900 °C. Corrosion 43 (7): 348–352
Zhao W et al (2010) The Effect of High Temperature Corrosion on Mechanical Behavior of a Gamma TiAl Alloy. International Journal of Modern Physics B 24 (15–16): 2970–2975

Further Reading

Craig B, Anderson D (1995) Handbook of Corrosion Data. ASM International
De Luccia J et al (1987) AGARD Corrosion Handbook Volume 2 – Aircraft Corrosion Control Documents: A Descriptive Catalogue. AGARD-Advisory Group for Aerospace Research and Development
DeHoff R T (2006) Thermodynamics in Materials Science 2nd ed. CRC Press
Embury J (1990) High-Temperature Oxidation and Sulphidation Processes. Pergamon Press
Khanna A S (2002) Introduction to High Temperature Oxidation and Corrosion. ASM International Materials Park, OH (USA)
Lai G Y (2007) High-Temperature Corrosion and Materials Applications. ASM International
Lee B, Karkkainen R L (2018) Failure investigation of hydrogen induced cracking of stabilator skin in jet aircraft. Engineering Failure Analysis 92:182-194
Rangone D V (1995) Thermodynamics of Materials – Volumes 1 and 2. Wiley & Sons
Roberge P (2008) Corrosion Engineering: Principles and Practice. McGraw-Hill
Wanhill R, Windisch M (2017) Corrosion and Stress Corrosion Testing of Aerospace Vehicles Structural Alloys. Springer

Chapter 9
Other Interesting Alloys for Aerospace and Related Applications

9.1 Introduction

This last chapter is dedicated to four classes of materials that have been considered and are still investigated in view of their potential applications in the aerospace field, as concerns in particular hot parts of aero-engines and functional applications, like actuators and control systems.

New materials for aircraft structures, and relevant production techniques, have continuously been developed over more than one hundred years now. This trend seems to be substantially unaltered in the future perspective, considering that innovation has always been a pillar in aerospace field. Initially, in view of the limited power of the engines, weight reduction of all parts of an aircraft was the most effective approach to enhance its aerodynamic efficiency. For achieving this, since the early attempts by Orville and Wilbur Wright (Chap. 1), wood, varnished fabric, and steel, this latter for the load-bearing structures, were the main materials used for the Flyer I and successive aircraft structures. Following the same principle, for the engine, the Wright brothers used a cast aluminum alloy (King et al. 2009). Mass reduction and better specific structural properties of materials continued to be fundamental selection parameters even after the pioneering times of aviation. However, more issues came up later, playing a decisive role not only in the development of new materials but also in the optimization of the properties of the already existing ones and in the implementation and validation of novel process technologies. In recent times, considering the planetary dimension achieved by airborne transportation, the environmental impact has become a crucial aspect (see Sect. 2.6). The same can be said for fuel saving and, thereby, higher propulsion efficiency, just to mention a couple of issues that are nowadays central in the development of the field, whereas they were out of the agenda until comparatively recent times (see Fig. 9.1). The ongoing related research efforts are proceeding in quite many directions. In this final chapter, some of the systems that are under investigation for answering novel and even more demanding technological requirements are presented. The first three

S. Gialanella, A. Malandruccolo, *Aerospace Alloys*, Topics in Mining, Metallurgy and Materials Engineering, https://doi.org/10.1007/978-3-030-24440-8_9

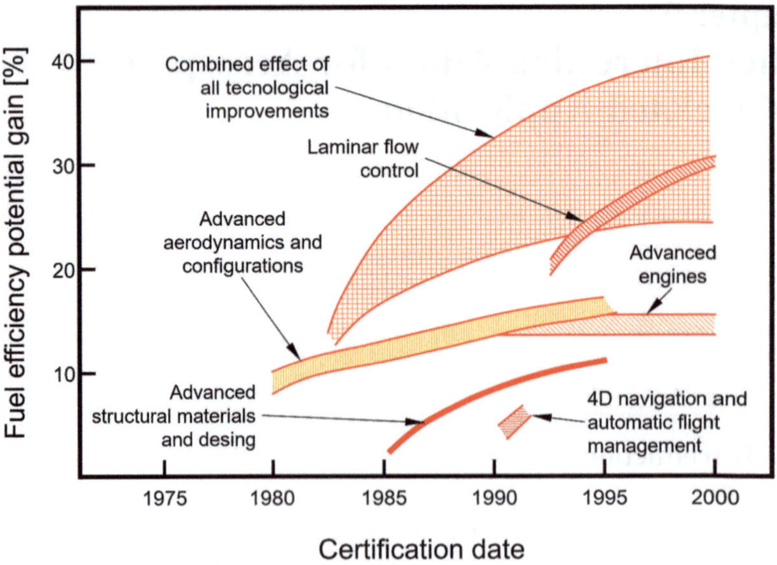

Fig. 9.1 Effect of different technological approaches on the improvement in aircraft fuel consumption. (Redrawn from Cantor et al. 2001)

examples refer to alloys having as common ground the search for higher engine operating temperatures: refractory element, oxide dispersion strengthened (ODS) and intermetallic alloys (Sects. 9.2, 9.3, and 9.4). Actually, intermetallic compounds would comprise also Nitinol alloys, based on the NiTi-ordered phase, a fundamental group of shape memory alloys (SMAs, Sect. 9.5). These alloys are rather interesting for their functional properties and relevant possibility to manufacture *smart* components. They might allow important improvement in the overall aircraft performances, coherently with the mentioned guidelines for the present development strategies of aerospace structures.

9.2 Refractory Alloys

9.2.1 General Aspects

Increasing the operating temperatures of gas turbine engines has been a constant challenge for machine designers and materials scientists over the years. Together with the improvement of engine thrust performances, other aspects are nowadays regarded as crucial, like the need for cleaner emissions and a general saving of fossil fuels, in order to reduce the associated greenhouse effects (see Sect. 2.6). The increase of the operating temperatures, particularly the raise of the turbine entrance temperature (TET, see Sect. 2.2 and Fig. 2.20), is an obvious approach to achieve

Table 9.1 Melting temperature and density of *bcc* refractory metals

Symbol	Element	Melting temperature [°C]	Density [g/cm^3]
Cr	Chromium	1863	7.2
Nb	Niobium	2469	8.6
Mo	Molybdenum	2623	10.2
Ta	Tantalum	2996	16.6
W	Tungsten	3410	19.3

this goal. Structural materials featuring adequate thermal stability are necessary. In this regard, nickel-based superalloys, once the metal temperature overcomes 1100 °C, start creeping at intolerable rates. Nonetheless, this temperature limit can definitely be regarded as absolutely remarkable, considering that it is just 250 °C below the incipient melting point of state-of-the-art superalloys. Moreover, the gas temperature in the hottest parts of the engine, i.e., the combustion chamber, can reach peaks as high as 2000 °C (see Fig. 2.12). It is thanks to a combination of efficient design and operation of the cooling system present in the hollow vanes and blades, associated with the protective action by TBCs (see Sect. 7.3.2), that the components can resist these extreme operating conditions. Refractory metals are characterized by melting temperatures that make them potentially interesting as base materials for components capable to break the limits of superalloys, as concerns structural applications at high temperatures (Perepezko 2009). The melting points of some interesting metals are given in Table 9.1, all featuring values of their melting temperature above that of pure nickel, i.e., 1455 °C. Although this is coherent with their excellent intrinsic creep resistance, the early investigations on the potentials of refractory metals and their alloys highlighted several restrictions and critical aspects (Wadsworth et al. 1988). Some descend from the *bcc* structure that all of the metals listed in Table 9.1 have, with a ductile to brittle transition (DBTT) above room temperature, and consequent limited room-temperature ductility. This renders their processing and forming rather critical even in view of the elevated strength that these metals would usually exhibit. Joining and welding are also rather challenging, since the resulting embrittlement has consequences not only on the production but also on the in-service alloy performances. Eventually, the exposure to high-temperature oxidation conditions has demonstrated the insufficient intrinsic resistance of these alloys.

Notwithstanding this complex picture, mixing together positive and critical aspects, several applications have importantly benefitted from using refractory metal alloys. These materials are today the top priority choice for several parts working under particularly demanding conditions, like dies for precision metal forging, used also in the production of aircraft fan blades and rotors (Starck Fabricated Products 2013). The alloy systems that have actually been employed in several fields, including aerospace, are those based on niobium, molybdenum, tungsten, and tantalum. More recently, feasibility studies have been conducted to evaluate alloys based on chromium, listed among other refractory metals in Table 9.1. An overview

Table 9.2 Code name and compositions of the commercial refractory metal alloys, subdivided according to the main element on which they are based (Lipetzky 2002)

Base element	Alloy name	Composition (wt%)
Mo	TZC	Mo-1.2Ti-0.25Zr-0.15C
	TZM	Mo-0.5Ti-0.1Zr-0.02C
	Mo-Re	Mo-50Re
	Mo-13Re	Mo-13Re
Nb	Nb-1Zr	Nb-1.0Zr-0.005C
	Nb1	Nb-30W-1Zr-0.06C
	B88	Nb-28W-2Hf-0.067C
	FS85	Nb-28Ta-10W-1Zr-0.004C
	C129Y	Nb-l0W-10Hf-0.2Y-0.015C
	C-103	Nb-10Hf-1.0Ti-0.7Zr-0.015C
Ta	T-111	Ta-8W-2Hf
	T-222	Ta-10W-2.5Hf-0.01C
	ASTAR 811C	Ta-8W—1.0Hf-1.0Re-0.025C
W	W-Re	W-25Re
	3D	W-3Re

on the present knowledge on chromium-based alloys will be presented in Sect. 9.2.3, together with other alloy systems, including the alloys of the platinum group metals (PGMs).

9.2.2 *Refractory Metal Alloys*

The refractory alloy systems that have attracted the main interests so far are based on Mo, Nb, Ta, and W. Table 9.2 gives a list of alloys based on these refractory elements, developed over the years and successfully employed in several high-temperature applications.

As concerns niobium, in the 1960s, several high-strength alloys were developed (Wojcik 1991). The typical production cycle for Nb-based alloys starts with an electron beam furnace for melting the pure metal, followed by vacuum arc remelting (VAR, see Sect. 5.3.6), during which alloying elements are added. For the final refinement, hot extrusion or forging processes close to the recrystallization temperature range, i.e., 1100–1380 °C, are employed. Due to an inadequate fracture toughness observed in some Nb-based alloys, medium strength, higher ductility, and better weldable materials, like nickel-based superalloys, are still preferred. However, the C103 alloy, which is one of the most used niobium-based alloys ever, is still an excellent choice for aerospace applications requiring exposure to temperatures ranging from 1200 to 1550 °C. This alloy exhibits an excellent weldability in addition to a good workability, so that also complex forms, like rocket thrust cones, high-temperature valves, and connections, can be fabricated. In particular, the C103 alloy was used for the descent nozzle of the Apollo lunar module and thrusters for extra-atmospheric spacecraft positioning. However, the C103 alloy, as other niobium alloys and like, in general, all alloys based on refractory metals, exhibits a very limited oxidation resistance that hinders their use in oxidizing environments. In case of Nb-alloys, the oxide phase that tends to form on the surface of these materials, Nb_2O_5, is not protective, since it does not form a continuous, adherent, and compact layer. Protective coatings, based on silicide phases, have been developed to provide a sufficient level of protection to the refractory metal alloys (Bewlay et al. 2003a, b). These coatings are typically obtained using a halide-activated pack-cementation process (Kung and Rapp 1989), similar to that considered for nickel aluminide diffusion coatings (see Sect. 7.2.3). The growth of the coating requires two main steps: first, deposition of silicon on the alloy surface, and second, diffusion to form the silicide phases. Actually, not only silicon is deposited onto the surface to be protected but also a series of other alloying elements, like titanium, chromium, iron, etc., that are introduced to improve the coating performances and durability. For this reason, setups for co-deposition are commonly used, selecting the halide species in order to achieve different values of partial pressures of the metal precursors. A number of compositions and phases have been investigated, guided by the knowledge of the relevant phase diagrams (Harper and Rapp 1994; Vilasi et al. 1998a, b). Coatings with general compositions like M_7Si_6 and M_6Si_5 (M being a combination of several elements like Co, Cr, Fe, Nb, Ni, and Ti) have been developed (Knittel et al. 2013). These coatings provide an adequate protection to Nb-based alloys and composites, even because their composition can be tuned and modified to better match the substrate features and comply with the requirements of the real servicing conditions (Vilasi et al. 1993; Vilasi et al. 1998a, b). Boron is often present in the coating formulations, having a general stabilization role. Boron additions result in lower diffusivities, thus rendering the coating less permeable to gas infiltration but also less prone to interdiffusion with the substrate, a positive factor for extending the coating lifetime. In case of surface cracking, the borides present in the outer surface layer act as a reservoir of this element, which contributes to the formation of a low-temperature melting borosilicate glass. At the operating temperatures, this glass can flow and mend the damaged regions of the coating, with a self-healing action (Portebois et al. 2014). The excellent surface stability and oxidation resistance achieved thanks to these different approaches, coupled with a relative ease of deformation, render niobium-based alloys promising candidates for hot components of future gas turbine engines, as indirectly demonstrated by the industrial development projects that have been carried out on this matter (Bewlay et al. 2003a, b; Jehanno et al. 2005; Sankar et al. 2015).

Similar issues, as concerns oxidation protection, apply also to molybdenum-based alloys. Molybdenum has a higher density than niobium (Table 9.1), and its alloys are used for ultrahigh-temperature applications, under protective atmosphere, so that gas-phase corrosion cannot take place (Dimiduk and Perepezko 2003). In these conditions, the structural stability and the good mechanical and creep resistance of molybdenum alloys can be fully exploited up to temperatures in excess of 1200 °C. The oxidation behavior of molybdenum-based alloys is somehow similar to that of the chromia-forming ones. In fact, the main oxide phase that forms on the exposed surfaces of molybdenum alloys, i.e., MoO_3, is volatile above about 700 °C. Therefore, this phase cannot provide any protection to the underlying alloy above this temperature but can rather promote gas-phase hot corrosion (see Sect. 8.2.5; Jones et al. 1958). The similarity with chromia-forming alloys is that also Cr_2O_3 transforms into the CrO_3 volatile phase above 950 °C approximately (see Eq. 6.13). Successful strategies to overcome this limit are either based on bulk compositional changes or on the alloy protection using suitable coatings. As for niobium alloys, silicon additions provide an essential constituent for the formation of a protective silicon oxide-based surface scale. Boron is also added in order to enhance the silicon storage capabilities of the alloy, beyond the solid solubility limit of 3 at% for silicon and far below 1 at% for boron itself (Heilmeier et al. 2009), through the formation of molybdenum and boron-silicide precipitates. After an initial oxidation transient, corresponding to a weight reduction for the formation of the volatile molybdenum oxide phase, MoO_3, a steady-state oxidation regime is entered, during which a borosilicate surface layer starts to form. This barrier protects the alloy and slows down the overall oxidation kinetics (Parthasarathy et al. 2002). To reach this condition, at least temperatures as high as 1000 °C are necessary. In fact, at intermediate temperatures, in between 700 and 1000 °C, the formation of a porous scale is still observed, featuring the coexistence of MoO_3 with the borosilicate component. At intermediate temperatures, below 1000 °C though, a continuous layer of SiO_2-B_2O_3 glassy oxide forms on the alloy surface with an underlying MoO_2-rich layer. This stable molybdenum oxide phase may form thanks to the reduction in the oxygen residual partial pressure attained beneath the outer borosilicate scale. The residual formation of a limited amount of MoO_3, if any, can be inferred from the presence of gas bubbles, trapped into the outer glassy scale. A truly parabolic oxidation kinetics can be observed at temperatures of 1300 °C or so (Azim et al. 2017; Burk et al. 2010). The viscosity of the protective outer scale can be modulated by changing the concentration of boron. Moreover, as reported for Nb-Si-B alloys, the borosilicate glassy phase may contribute to heal cracks forming in the oxide surface layer (Akinc et al. 1999; Sakidja et al. 2005). From the structural point of view, the silicate and borosilicate precipitates introduce also an important strengthening effect, which, in association with the intrinsic creep resistance, renders molybdenum alloys of potential interests for structural applications, particularly under important mechanical loads. Nonetheless, the limited room-temperature ductility is still a main issue to consider when dealing with molybdenum and other refractory metal alloys. In this respect, important steps forward have been made both in terms of composition design and alloy processing (El-Genk and Tournier 2005; Wadsworth et al. 1988). The challenge is to render the alloys sufficiently deformable, without reducing their mechanical strength. Alloying molybdenum with rhenium, which enhances the dislocation mobility, is a strategy that has been abandoned, since, in addition to an excessive softening of the alloy, the material cost was intolerably high, owing to the high price of the raw materials. An alternative route was initiated starting from the experimental evidence that the reason for the limited ductility of Mo alloys strongly depends on their sensitivity to the impurities, e.g., oxygen and nitrogen, segregating along the alloy grain boundaries

(Brosse et al. 1981; Medvedeva et al. 2007; Trinkle and Woodward 2005; Wadsworth et al. 1988). Several strategies can be implemented to modify the size distribution of grain boundaries and solute segregation. One possibility is promoting the precipitation of secondary phases, in the form of carbides and oxides (Cockeram 2009; Majumdar et el. 2009; Miller et al. 2003; Mueller et al. 2000). They can act as nucleation sites for new alloy grains and hinder the recrystallization and coarsening of the already existing ones. A possible route to obtain these structures is mechanical alloying, a powder metallurgy approach described in Sect. 9.3, concerning oxide dispersion strengthened alloys. An innovative liquid-based process is reported herewith, most promising as concerns the production of Mo alloys featuring extremely large deformation to rupture (Liu et al. 2013). The key factor that determines the exceptional mechanical behavior of these alloys is their extremely refined microstructure, with molybdenum alloy grains below 1 μm and a uniform nanometric oxide dispersion. The process is based on a liquid-liquid mixture of the precursors of the two components, i.e., the molybdenum matrix and the lanthanum oxide dispersion. Lanthanum oxide (La_2O_3) is dissolved into an aqueous solution of $La(NO_3)_3$ (lanthanum nitrate), subsequently mixed with another solution, containing ammonium bi-molybdate, which promotes the precipitation of a $NH_4La(Mo_2O_7)_2$ powder. The grains of this powder act as heterogeneous nucleation sites for the $(NH_4)_2Mo_2O_7$, so that a core-shell structure is obtained. Subsequent thermal treatments and hydrogen reduction result in the formation of molybdenum grains, still core-shelled, with an inner part made of La_2O_3. The compaction and densification of this powder results in a molybdenum matrix reinforced by a dispersion of finely distributed lanthanum oxide particles. The mechanical properties of this nanostructured (NS) alloy have been measured and compared with those of another molybdenum alloy, having the same composition but obtained using a conventional powder metallurgy route, typical of the oxide dispersion strengthened (ODS) alloys (Cockeram 2006; Sturm et al. 2007). Some of these results are summarized in Table 9.3, where the data for a commercially pure (CP) Mo alloy are shown for a further comparison and better illustration of the excellent properties of the newly developed alloy.

The excellent values of the tensile ductility and mechanical strength can mainly be ascribed to the refined microstructure of this alloy and to its effect on the dislocation dynamics, as demonstrated by the thorough electron microscopy

Table 9.3 Comparison between the mechanical properties of two ODS Mo alloys, one of which having a nanostructured (NS) grain. Commercially pure (CP) molybdenum is quoted as reference (Liu et al. 2013)

Alloy	Average grain size [μm]	Yield strength [MPa]	UTS [MPa]	E%	Fracture toughness [MPa√m]	DBTT [°C]
CP-Mo	97 ± 8	385	534	13.0	24.2 ± 2.3	25
ODS Mo	1.4 ± 0.7	710 ± 37	746 ± 49	14.1 ± 6.5	52.8	−60
NS-Mo	0.5 ± 0.3	813 ± 3	865 ± 5	37.5 ± 3.0	121.5 ± 3.5	−100

characterization carried out on the mechanically tested samples. Interestingly, this approach may in principle be applied to other refractory alloys, like tantalum- and tungsten-based ones, that also suffer from a limited ductility. These alloys (see Table 9.2) have been developed with a particular focus related to aerospace applications, for high-density components (*vide infra*), for which these systems are particularly suited. Tantalum, different from other refractory metals, has a comparatively good oxidation resistance and, like niobium, exhibits an acceptable deformability at room temperature (Schmidt and Ogden 1963). In fact, it can be processed using liquid nitrogen, as protective environment, so to avoid oxidation losses and, at the same time, not induce unwanted embrittlement, since its absolutely cryogenic DBTT is below −269 °C (Buckman 2000). As concerns tungsten, the main aspects of its metallurgy, including extraction, processing, and corrosion resistance can be found in Lassner and Schubert (1999).

Although the main feature of the refractory element alloys discussed so far is their high-temperature stability, tantalum and tungsten alloys have found their main aerospace applications for another specific property, i.e., their elevated density. This is the case of the so-called balancing weights, introduced into aircraft or helicopter structures to reduce their vibrations, thus improving the comfort of the passengers and the crew. These weights are produced with different shapes and geometries, so that they can be mounted on the control surfaces of the ailerons, elevators, and rudder sections of aircraft, rotor blades of helicopters, and cockpit instrumentation as well. Since for these components a limited volume is paramount, heavier alloys, i.e., tungsten-based, are the first choice (Starck Fabricated Products 2013). In addition to aircraft *balancing weights*, other applications of tungsten- and tantalum-based alloys in aerospace are rim components of flywheels or rotating members in governors and airborne antenna bases. As concerns the high-temperature applications, the explorative use of tungsten wires to reinforce nickel-based superalloy turbine blades can be mentioned as one of the few involving this class of alloys (McLean 1983). Niobium and molybdenum alloys have had instead rather important high-temperature application in aerospace. Niobium alloys are the standard choice for ablative systems of the exhaust cones of missiles and rockets, including the recommended protection by suitable silicide coatings (Wojcik 1991). A creep resistance better than the benchmark CMSX4 nickel-based single-crystal superalloy has been recorded for the Mo-9Si-8B alloy that in this respect can be regarded as an excellent candidate for sustaining the search for more performing engines, as other alloys of the Mo-Si-B system mentioned in this section (Heilmeier et al. 2009; Muktinutalapati 2011). All refractory metals, together with rhenium that differs from the other *bcc* metals for its hexagonal close-packed (*hcp*) structure, are used as alloying elements in nickel-based superalloys (see Sect. 6.4), as solid solution strengtheners and creep resistance enhancers, especially for their influence on the alloy diffusivities. Indeed, as emerges from the above overview, the interest for refractory metals and relevant alloys has been initially driven by their potential at temperatures exceeding those accessible to advanced nickel-based superalloys. Notwithstanding the research efforts, successful in many respects to overcome the main critical aspects of refractory element alloys, the feeling is that other materials systems in the end will take the lead. The reason is that the search for more efficient and better performing engines and, in general, aerospace structures cannot override the environmental aspects, for which weight saving is a central prerequisite. Tougher ceramics and carbon-based composites are nowadays regarded as most attractive systems in this new development strategy (Lu et al. 2013; Richerson 2004). They might be preferred to metallic systems considering the remarkable weight saving and the temperature raise that they would afford. From this point of view, the evolution of structural materials for aero-engines, to be used in the hottest spots, shows some analogy and reminds some aspects of the development research dedicated to the space shuttle thermal protection system (TPS), for which the possibility of using niobium-based shield was initially considered (Buckley et al. 1981). It is well known that the final choice was completely different, i.e., based on the lighter, cheaper, and more performing TPS, made of ceramic felt tiles and carbon-carbon composites for the hottest parts (Korb et al. 1981). However, the research invested in the development of refractory alloys for aerospace has demonstrated the feasibility of using these alloys also for other, different applications (Briant 2000). Molybdenum was employed for microelectronic device, in flat-panel displays, using field emission tips, rather than liquid crystals. The same occurred for tantalum-based alloys, employed for the manufacture of electronic components, like capacitors. Moreover, the high density of Ta alloys suggested a number of military applications, like penetrators or gun barrel liner, capable to resist to extremely high flame temperatures, in excess of 3000 °C, reached during the shot firing. To complete the picture, regarding the field of space exploration, tantalum alloys listed in Table 9.2 were all developed for space nuclear power systems (Buckman 2000). This is an issue that will be considered further in Sect. 9.3 dealing with ODS alloys. Tantalum alloys were selected for the best combination of mechanical strength, in particular creep, also under the effect of radiation-induced deformation, and corrosion resistance, and for their compatibility with liquid alkali metals and molten plutonium. Moreover, tantalum alloys exhibit good fabricability and weldability, so that they are the best candidates for nuclear applications, notwithstanding the comparatively high density.

9.2.3 Other Refractory Alloys: Chromium and the Platinum Group Metals

Two more groups of refractory alloys are considered in this section. The first one is based on chromium, another *bcc* metal (see Table 9.1). This element can be regarded as “refractory” in the present context since its melting temperature falls above that of nickel and, consequently, above the incipient melting temperature of nickel-based superalloys, of which chromium is a most important alloying element (see Sect. 6.4) (Gu et al. 2004). Chromium exhibits specific properties interesting for structural applications at high temperatures. First of all, a low density 20% lower than nickel and significantly lower than the density of tantalum and tungsten (see Table 9.1).

Moreover, chromium features an elevated solubility for other elements, so that an effective solid solution strengthening is potentially achievable. This latter feature may also be accompanied by precipitation hardening, thanks to the intervention of strengthening intermetallic precipitates, like Cr silicides. Alloying with silicon is interesting also as concerns the oxidation behavior of these alloys, in analogy with the observations reported for niobium and molybdenum alloys (Sect. 9.2.2). In particular, chromium alloys can rely on the formation of a surface protective oxide scale (Cr_2O_3), stable up to 950 °C approximately. At higher temperatures, the presence of a sufficient amount of silicon guarantees an outward diffusion flux of this element, resulting in the formation of a protective, extremely stable layer of amorphous silica (SiO_2) (Soleimani-Dorcheh et al. 2014). In this way, the alloys can be operated even up to very high temperatures without the need of applying any protective coating, unless particularly severe conditions are faced (Gu et al. 2004). The thermal conductivity of chromium alloys is higher than that of the majority of nickel-based superalloys, so that thermal gradients and associated stresses remain below critical values, with obvious advantages for the integrity and lifetime of the structural components. Eventually, chromium is also less expensive than all the other refractory elements in Table 9.1, and its availability on the international market is not a critical issue. However, two main aspects have still to be improved for a wide and reliable use of this metal and its alloys. First of all, chromium alloys show a tendency to become brittle when exposed to air for the combined effect of the contamination with nitrogen and oxygen, similarly to what is reported for molybdenum and tungsten. In the second place, coherently with the previous datum, chromium alloys show a low-impact ductility, especially at room temperature and even with a particular sensitivity of the DBTT on the level of impurities (Gu et al. 2005). This aspect can be remediated by introducing suitable alloying elements, like cobalt, iridium, iron, and manganese, which scavenge and clean up the alloy matrix from the undesired contaminants. Alloy design developments have introduced remarkable improvements in the mechanical properties, so that chromium alloys can now be regarded as reliable replacements for conventional superalloys (Gao et al. 2008).

From the discussion on refractory metal alloys, it turns out that a critical aspect, common to all these materials, although to different extents, is their limited intrinsic resistance to high-temperature corrosion. This aspect has been tackled with valuable solutions, which have rendered these alloys useable also in environmentally aggressive conditions. In Sect. 7.3.1, the importance of platinum additions to nickel aluminide diffusion coatings (see Sect. 7.2.3) for improving their oxidation resistance has been considered. Platinum and other elements of the same group, like rhodium and iridium, indeed named platinum group metals (PGMs), do show high melting temperatures, elevated oxidation resistance, and an adequate corrosion resistance also when exposed to molten salts (Alven 2004; Fischer et al. 1999). The critical issue is that the global availability of these metals is extremely limited and consequently their prices are very high. Just as a reference value, the worldwide extraction of iridium is estimated in 22 metric tons per year. Nonetheless, there have been ongoing researches aiming at assessing the mechanical properties and corrosion resistance of PGM alloys (Potgieter and Maledi 2014; Yamabe-Mitarai and Murakami 2014). The focus has mainly been the high-temperature behavior, considering the fields of interest for future applications of these materials. In addition to the excellent mechanical properties and high-temperature stability, an aspect that contributes to render these refractory alloys particularly attractive is the similarity encountered in the microstructural features of two-phase platinum and iridium alloys with those of γ-γ' nickel-based superalloys, including the dislocation structures and dynamics (Liebscher and Glatzel 2014; Vorberg et al. 2005). This aspect is definitely useful for the validation of material properties important for any specific application and related thermal treatment parameters, in view of their structural applications (Cornish et al. 2006; Vorberg et al. 2004; Wenderoth et al. 2005; Yamabe-Mitarai et al. 2004).

9.3 ODS: Oxide Dispersion Strengthened Alloys

The leading idea that was followed in the early stages of the development of oxide dispersion strengthened (ODS) superalloys was to extend the temperature range of the reinforcement beyond the stability limit of the γ' precipitates, i.e., the γ' solvus. Closely related to this aspect was the search for a strengthening mechanism acting on the dislocations, once their prevailing interaction with γ' precipitates is shifting from cutting to climbing, as a consequence of the coarsening of the precipitates due to the exposure to servicing conditions (Ostwald ripening, see Sect. 6.5.2). Nanometric oxide particles were identified soon as potentially effective strengthening agents, complying with the above requirements. In the earliest unsuccessful attempts to obtain a fine dispersion of the particles, they were insufflated into the molten alloy, resulting, though, in uncontrolled segregations. Therefore, this approach was abandoned soon. Mechanical alloying (MA) instead turned out to be the right choice. MA, that is indeed often in the relevant alloy code name (see Table 9.4), was an already established powder metallurgy approach, developed earlier on, originally intended to mix up hard metal powders, like WC-Co, and to reinforce aluminum alloys, the so-called SAP alloys (Benjamin 1990), from *sinter-aluminum-pulver* (German), after the definition introduced in 1946 by the Austrian chemist Roland Irmann. In collaboration with his co-workers, he did use high-energy ball-milling and hot-pressing sintering to prepare aluminum samples for spectrographic measurements (Pickens 1991). The unexpectedly high mechanical strength obtained in the powder compacts was ascribed to the interplay between the oxide strengthening and the effective sintering induced by mechanical attrition. The oxide dispersion came, mainly, from the fragments of the aluminum oxide layer initially present on the surface of the powder particles. The intense deformation imparted during milling to the grains of powder determines also the formation of fresh metal-metal contact surfaces, promoting intense intergranular bonding and diffusional phenomena during sintering. In this regard, these experiments indicated the main effects that a mechanical milling process, using high-energy tools, may determine in a ductile metallic powder. In Table 9.4, the compositions of nickel-based superalloys and aluminum alloys reinforced by a dispersion of oxide particles are listed (see Table 9.5 for ferritic matrix alloys). The oxides used in these materials are selected

Table 9.4 Compositions of some mechanically alloyed ODS nickel-based superalloys and aluminum alloys. Please note that oxygen concentrations quoted for these latter are those referring to Al_2O_3 reinforcing particles. The composition of the main ferritic ODS alloys can be found in Table 9.5 (Elzey and Artz 1988)

Alloy	Ni	Fe	Al	Cr	Mg	Li	C	Ti	Y_2O_3	Others
Alloy 51	Bal.	–	8.5	9.5	–	–	0.05	–	1.1	B = 0.01 Mo = 3.4 W = 6.6 Zr = 0.15
Alloy 92	Bal.	–	6.5	8.0	–	–	0.05	1.0	1.0	B = 0.01 Mo = 1.5 Ta = 3.0 W = 6.0 Zr = 0.15
IncoMAP AL-9052	–	–	Bal.	–	4.0	–	1.15	–	–	O = 0.5
IncoMAP AL-905XL	–	–	Bal.	–	4.0	1.3	1.15	–	–	O = 0.5
Inconel MA 6000	Bal.	–	4.5	15.0	–	–	0.05	2.5	1.1	B = 0.01 Mo = 2.0 Ta = 2.0 W = 4.0 Zr = 0.15
Inconel MA 760	Bal.	–	6.0	20.0	–	–	0.05	–	0.95	B = 0.01 Mo = 2.0 W = 3.5 Zr = 0.15
Inconel MA754	Bal.	1.0	0.3	20.0	–	–	0.05	0.5	0.6	–
MA 758	Bal.	–	0.3	30.0	–	–	0.05	–	0.6	W = 0.5
PM1000	Bal.	3.0	0.3	20.0	–	–	0.05	0.5	0.6	–
PM3030	Bal.	–	6.0	17.0	–	–	0.05	–	0.9	Mo = 2.0 Ta = 2.0 W = 3.5 Zr = 0.15

Table 9.5 Compositions of some mechanically alloyed ODS ferritic/martensitic steels (Donachie and Donachie 2002)

Alloy	Fe	Cr	C	Ti	Mo	Y_2O_3	TiO_2
DT2906	Bal.	13.0	0.05	2.9	1.5	–	1.8
DT2203Y05	Bal.	13.0	0.05	2.2	1.5	0.5	–
DT	Bal.	13.0	0.05	2.9	1.5	–	1.8
DY	Bal.	13.0	0.05	2.2	–	0.5	0.9
MA957	Bal.	14.0	0.01	1.0	0.3	0.25	–

among the phases not soluble in the metallic matrix, with incoherent interfaces, and that at the same time may guarantee a good thermomechanical compatibility over the whole range of working conditions. In Al-alloys, in addition to oxide particles, aluminum carbides, e.g., Al_4C_3, that form in situ from the reaction between graphite and aluminum powders guarantee a further strengthening action (Gilman and Nix 1981).The main features of the strengthening dispersoid particles, i.e., incoherence, no solubility in the metallic matrix, and nanometric grain size, are paramount as concerns the resulting microstructure and relevant mechanical properties of the alloys. The application of MA process to ODS superalloys dates back to 1966, with the pioneering works conducted at INCO by John Benjamin, using high-energy ball-milling to mix and refine the microstructure of the different ingredients before proceeding with the compaction and sintering steps. Nickel- and iron-based superalloys were the first investigated systems within a research activity that lead to the use of the MA754 ODS nickel-based superalloy for gas turbine vanes in the F404 engine in 1977 (Froes 1990; Muktinutalapati 2011).

The production of ODS alloys starts with the preparation of a master batch mix, comprising elemental metal, pre-alloyed alloy powders and the oxide particles. Several oxide phases have been considered, including, in the early times, ThO_2, subsequently abandoned for its radioactivity, intolerable particularly during the manipulation involved with processing and part manufacturing. For high-temperature alloys, with a nickel- and iron-based matrix, the standard choice is Y_2O_3, an oxide that guarantees excellent stability particularly at the highest temperatures, good thermomechanical compatibility, and incoherency with the metallic matrix. For MA superalloys, in general no process control agents (PCAs), like graphite, stearic acid, or other surfactants, are used, in order to avoid the introduction of possible contaminations. In other systems, like MA aluminum alloys, PCAs, including ethanol, are used instead, for reducing cold welding among particles, and the sticking the particles to the milling media, both effects that reduce the efficiency of the process drastically. Suitable PCAs, like graphite, are introduced also for inducing mechanochemical reactions with the metallic powder particles, thus promoting the in situ formation of carbide strengthening dispersions (Benjamin and Bomford 1977; Cao et al. 2013). Eventually, if not removed, before proceeding with the further processing steps, PCAs can be left in the milled powder on purpose, helping the subsequent compaction and sintering stages, by reducing the inner friction among the grains of powder and between the powder and the walls of the compaction die.

After milling the following main processing steps are necessary to obtain the final products:

- Preliminary consolidation of the powder up to approximately 80% of the full density
- Hot isostatic pressing (HIP), usually in austenitic steel vacuum sealed cans to full density
- Hot forging or extrusion for pre-forming the final product and inducing a crystallographic texture in the alloy grain
- Directional recrystallization (DRX), used for ODS superalloys only

After HIPing, the alloy still presents a metallic component with a micrometric grain size, as attained by the mechanical grinding, with a nanometric oxide dispersion, having a stabilizing effect against recrystallization of the metal grains along transverse to the texture, due to forging, direction. DRX induces instead a preferential grain growth precisely in the texture direction. This process is favored by the direction of movement of the ODS product with respect to a localized thermal gradient, realized with an induction coil heat source. Before entering the hot zone, the alloy can be kept in a cooling bath or in an insulated region of the furnace, sufficiently far away from the heat source. In this way, any spontaneous recrystallization is limited, and the dominant grain growth occurs when the alloy piece goes through the hot zone. An optimized microstructure is obtained when the advancement speed of the alloy in the thermal gradient does not exceed the natural grain growth velocity. This parameter depends on the alloy composition but also on the degree of deformation texture induced by the thermomechanical treatments that precede DRX. The microstructure induced by the DRX can be qualified and ranked according to the grain aspect ratio (GAR), defined by the ratio between the longitudinal and transverse average grain size. In general, the larger GAR, the better are the resulting mechanical properties of the alloy, especially as concerns fatigue and creep resistance with load applied along the longitudinal direction.

As concerns creep, the ODS alloys exhibit quite a peculiar trend of the creep rate ($\dot{\varepsilon}$) vs applied stress (σ) curves, over the temperature and stress range where dislocation creep is the prevailing mechanism, and a power-law creep can be assumed to model the experimental data. The creep rate of an ODS material can be expressed as:

$$\dot{\varepsilon} = D\left(\frac{\sigma - \sigma_{\text{th}}}{E}\right)^n \tag{9.1}$$

where

- D is a diffusion factor.
- E is the average Young's modulus of the alloy.
- n is the creep exponent.
- σ_{th} is the so-called threshold stress.
- σ is the applied stress.

The elevated values of n, as high as 20–30, and the presence of a stress below which no detectable creep is observed can be explained in terms of an attraction exerted by the oxide particles on the climbing dislocations.

The relaxation of the dislocation stress fields and, thereby, of their energy reduction explains the tendency for a dislocation line to remain stuck onto the surface of the incoherent oxide particles. Once the applied stress overcomes the *threshold stress*, a fast rise of creep rate is observed, since at this stage the applied load is by far larger than any value the alloy can possibly stand. The *threshold stress*, being influenced by the whole alloy response to diffusion-assisted phenomena, is also dependent on the GAR of the alloy, increasing linearly with this microstructural parameter (Howson et al. 1980). Thanks to the available information, it can be

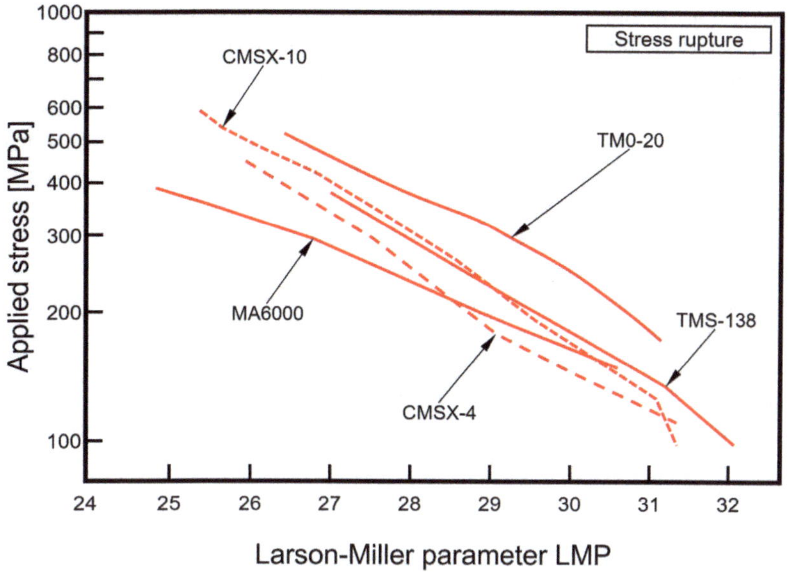

Fig. 9.2 Larson-Miller creep curves for different ODS alloys, compared to single-crystal superalloys (for alloy compositions see Tables 6.14, 6.16, and 9.4). (Redrawn from Harada 2003)

concluded that the *threshold stress* marks an important stress limit, to be considered in the part design, to extend its lifetime. Still concerning the creep behavior, ODS superalloys can actually reach comparable performances as single-crystal nickel-based superalloys (Fig. 9.2), although these latter have always been, and still are, the priority choice for blades and vanes in state-of-the-art GTEs. Several reasons can explain this situation. The early steps of development of the ODS superalloys occurred at the same time as the investment cast technology, applied to the production of directionally solidified and single-crystal components. The better process control and consequently better and more reproducible properties of the products obtained from investment casting turned out to be particularly important in determining the success of DS and SC components with respect to the mechanically alloyed ones. Indeed, the microstructural defects typically present in ODS superalloys, like recrystallization defects, i.e., fine post-milling grains surviving the DRX process or clusters of oxide particles, are responsible for crack initiation, particularly under low-cycle fatigue (LCF) and creep-fatigue conditions (Elzey and Artz 1988). Therefore, ODS superalloys were used as a possible alternative to DS and SC superalloys for high-pressure gas turbine vanes only, components that may definitely benefit from the stability and creep resistance of ODS alloys, although less critical than turbine blades, for the lack of applied centrifugal forces. However, also this application of ODS is now largely abandoned, in view of the weight reduction afforded by the investment cast hollow vanes, associated with the high-reliability standards achieved by the EB-PVD thermal barrier coatings and consequent reduction of the peak metal temperatures.

Although a substantial loss of interest for ODS superalloys is there, as concerns aircraft gas turbine applications, in more recent years, ODS alloys have attracted a renewed interest in view of their use for nuclear power plant applications, with regard to both space and terrestrial reactors. Nuclear energy has been used in the past for several space applications, before solar energy took over, whenever feasible. USA and USSR used nuclear reactors for satellites, capable to provide the energy required for the relevant operations. The SNAP 10A was actually the only nuclear reactor installed on a US satellite, back in 1965, i.e., the very early years of the "space rush." More numerous are the reactors launched by the USSR: 33 altogether, the majority being the BES-5 and just 2 TOPAZ reactors. Incidentally, the relicts of these satellites continue to orbit around the Earth, if not on some occasion moved to more delocalized trajectories, the so-called disposal orbits, not to interfere and create potential risks and damages to newer satellite deployment. Notwithstanding these cares, a number of disposed reactors have released radioactive matter, especially NaK used as coolant liquid, which is now contaminating Earth orbits with droplets and fragments with estimated diameters ranging from tenths of millimeters up to several centimeters and an average velocity of 7–13 km/s (Reese and Vick 1983; Wiedemann et al. 2005). The total amount of released nuclear waste is unknown and has raised the concern of the scientific community, and the UN has issued two recommendations on the peaceful use of nuclear power in the outer space (United Nations General Assembly 1992). Indirectly, this situation promoted the use of solar panels, having the advantage of being not polluting and to be deployable using comparatively simple devices, like smart actuators relying on shape memory alloys (see Sect. 9.5). Although the use of solar panels has been a strong push to abandon the nuclear technology for space missions, on the other hand, there are still some that cannot rely on energy harvesting via photovoltaic solar panels only. A few important examples of these actual and future space mission are: outer solar system missions, like the Rosetta ESA mission, and, considering closer to Earth scenarios, the Mars and Moon lander missions and in the future, as concerns Mars, also manned missions. These latter will require the transportation to the Red Planet of several tons, probably tens of tons of equipment and components. The duration of the tests requires also overnight measurements, and test might be conducted in no ideal conditions, like Martian nights and dust storms. From this point of view, nuclear power plants are essential energy providers, also considering their undisputable additional advantages, like the possibility of generating energy independently from the orbit and distance from the sun. In principle, high power density can be reached; different forms of energy and not just electrical can be obtained selecting the more appropriate reactor design and heat conversion systems (El-Genk 2009). Other missions that would be possible thanks to the deployment of nuclear reactors for spacecraft energy production are related to robotic exploration, like removal of abandoned spacecrafts and orbital stations, but also to the active mission of deflecting Near-Earth objects (NEO), to protect the Earth from meteorite deep impacts.

Nuclear reactors may also provide power for electrical or combined thermal and electrical thrusters on interplanetary spacecrafts, as demonstrated by the NASA

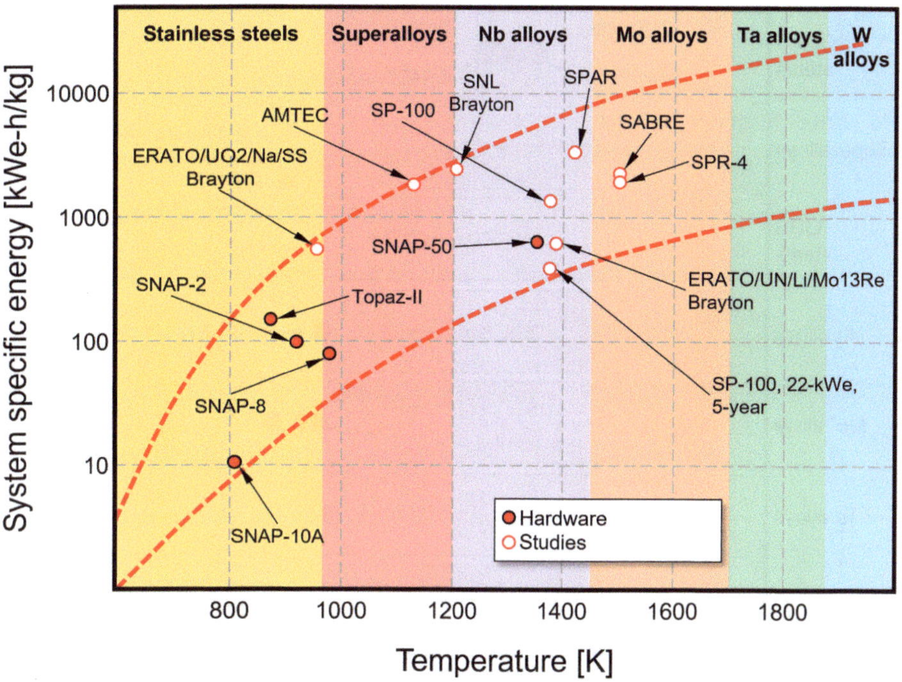

Fig. 9.3 Specific energy of space nuclear power systems versus reactor exit temperature. (Redrawn from El-Genk and Tournier 2005)

concept studies on the nuclear thermal propulsion (NTP) and nuclear electric propulsion (NEP) technologies. Irrespective of the technology used to convert energy into electricity, some heat must be released at lower temperatures for obvious thermodynamic reasons. Differently from terrestrial applications, heat rejection in space environment is a rather critical operation and can only be achieved by irradiation from suitable surfaces or by ejecting heated matter directly from the spacecraft. The radiator solution has the advantage not to contribute to space pollution, affords longer operational times if needed, and, moreover, provides solutions that might be transferred to terrestrial plants.

Although lower working temperatures would be less critical for the materials, which might be chosen among established alloys, still, to improve the thermodynamic efficiency, both peak and radiation temperatures have to be increased (El-Genk and Tournier 2005). As shown by Fig. 9.3, different classes of materials are potentially suitable for the different reactor types. In Fig. 9.4, the temperature ranges faced by different materials, with relevant allowable domains, considering real working conditions and material limitations, are indicated. They depend on the DBTT, the irradiation embrittlement, and the embrittlement determined by unwanted contamination during processing or installation procedures.

The elevated standards in terms of performances, safety, and reliability are central for a successful development and deployment of nuclear space reactors. Although

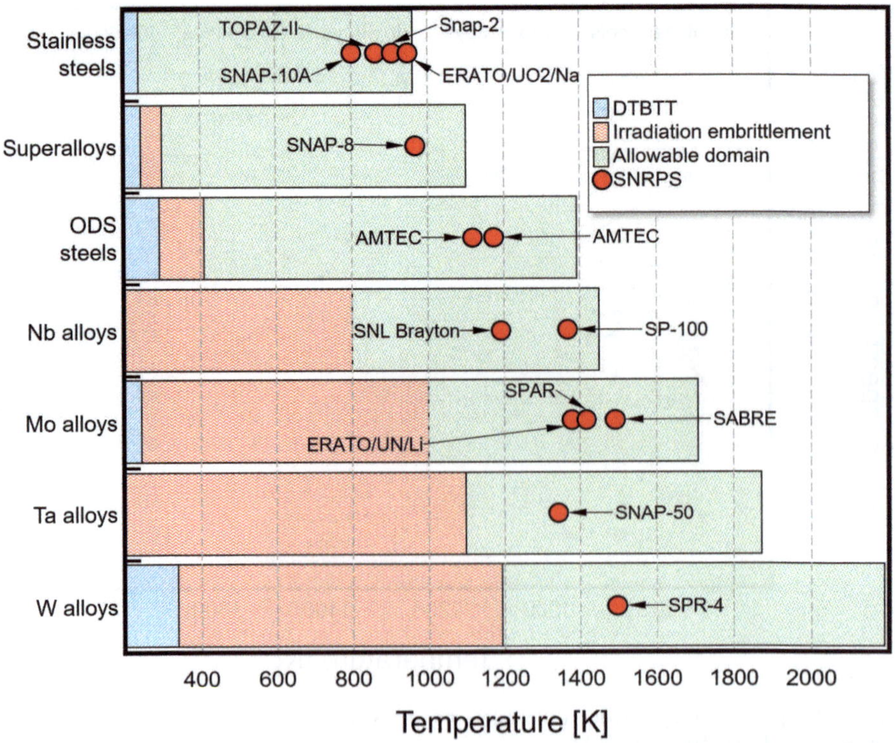

Fig. 9.4 Recommended operation temperature range for structural materials in space nuclear power systems. (Redrawn from El-Genk and Tournier 2005)

fully different in size and layout, still similar material requirements are there for generation-IV terrestrial nuclear reactors. These new reactors represent the evolution of the present technology and should introduce significant improvements as concerns safety, operational reliability, sustainability, and useful life, as specified in the initiative statements launched in year 2000 by the US Department of Energy (Mansur et al. 2004). The structural materials for the space nuclear reactors (Figs. 9.3 and 9.4) are being considered also for the novel terrestrial plants. Two broad categories can be identified. The first one includes materials already available and with consolidated and widespread applications in the nuclear field, whose performances have to be reconsidered and verified for the more demanding operating conditions foreseen in the new plants and depending on higher operating temperatures and neutron doses, coupled with corrosive phenomena, determined by the interaction of the structural materials with the coolants and fuel. In addition to these, there is a second category, including other materials that would potentially provide adequate responses to the Gen-IV reactor technology, although still requiring research and investigations to assess and qualify their performances under the operating conditions they would face. Ferritic martensitic (F/M) steels, initially developed for fossil-fuel power plants, are still important structural alloys for Gen-IV reactors (Klueh 1993; Klueh and Nelson 2007). The compositions that should provide the better results, having tuned the right temper conditions, are those with 9–12 wt% Cr (see also Table 5.16). The relatively high chromium concentration is interesting since it guarantees excellent corrosion and oxidation resistance. Moreover, these steels have demonstrated a reduced activation (RA) tendency, i.e., a rapid decay of the radioactivity induced by the exposure to a neutron flux. This aspect is crucial for the decommissioning stage and disposal or storage of the reactor parts, with relevant effects on the life cycle and life cycle cost assessments, essential parameters to take the right and more convenient decision in economic terms. Another aspect that renders the F/M steels particularly interesting is their dimensional stability with respect of swelling and void formation, much better than in other potential candidates, like austenitic steels and superalloys. Irradiation swelling is a further deformation mechanism, active when alloys are used under intense radiation and nuclear particles, especially neutron, fluxes (Hoffelner 2011). It can be implemented in material-specific Ashby deformation maps, like those shown in Fig. 9.5 for the AISI 316 stainless steel, with the occurrence of an irradiation deformation field (Mutry and Charit 2008).

In addition to the mechanical instability and dimensional changes due to irradiation swelling, the austenitic steels also suffer from microstructural evolution, involving recrystallization and precipitate ripening, which may deeply change the overall mechanical strength both under stationary and time-dependent stress and irradiation conditions. Moreover, the corrosion resistance for the exposure to typical coolant liquids used in nuclear plants, like lead and lead alloys, is insufficient to comply with the required useful lifetime and safety requirements. These results have confirmed a preferential choice for F/M steels and indicate a promising strategy to enhance their properties through the development of ODS F/M steels (Byun et al. 2013; Chen 2013). The dispersion of fine oxide particles introduces vacancy sinks at their interfaces with the metallic matrix, in which diffusional flux determines an accumulation of these defects (vacancies). This vacancy sequestration mechanism reduces their concentration in the alloy, with a consequent slowing down of the diffusive phenomena. Moreover, as expected, the oxide particles hinder the dislocation movements in the alloy, particularly in the higher-temperature range, where dislocation creep, determined by the diffusive climb of dislocations, is prevailing (Yvon and Carré 2009). The combination of these mechanisms makes F/M ODS steels extremely stable over a wide range of operational conditions, as demonstrated in specific studies that have been conducted to assess the performances of the alloys listed in Table 9.5 (Chen and Hoffelner 2009; Miller et al. 2005, 2006).

The ODS approach has also been used for refractory metals, to provide additional strengthening mechanisms to these alloys (Patra et al. 2015). However, the limited intrinsic ductility of refractory metals and their alloys (Sect. 9.2) renders the control of the defects introduced by the mechanical alloying technology even more critical than in standard alloys.

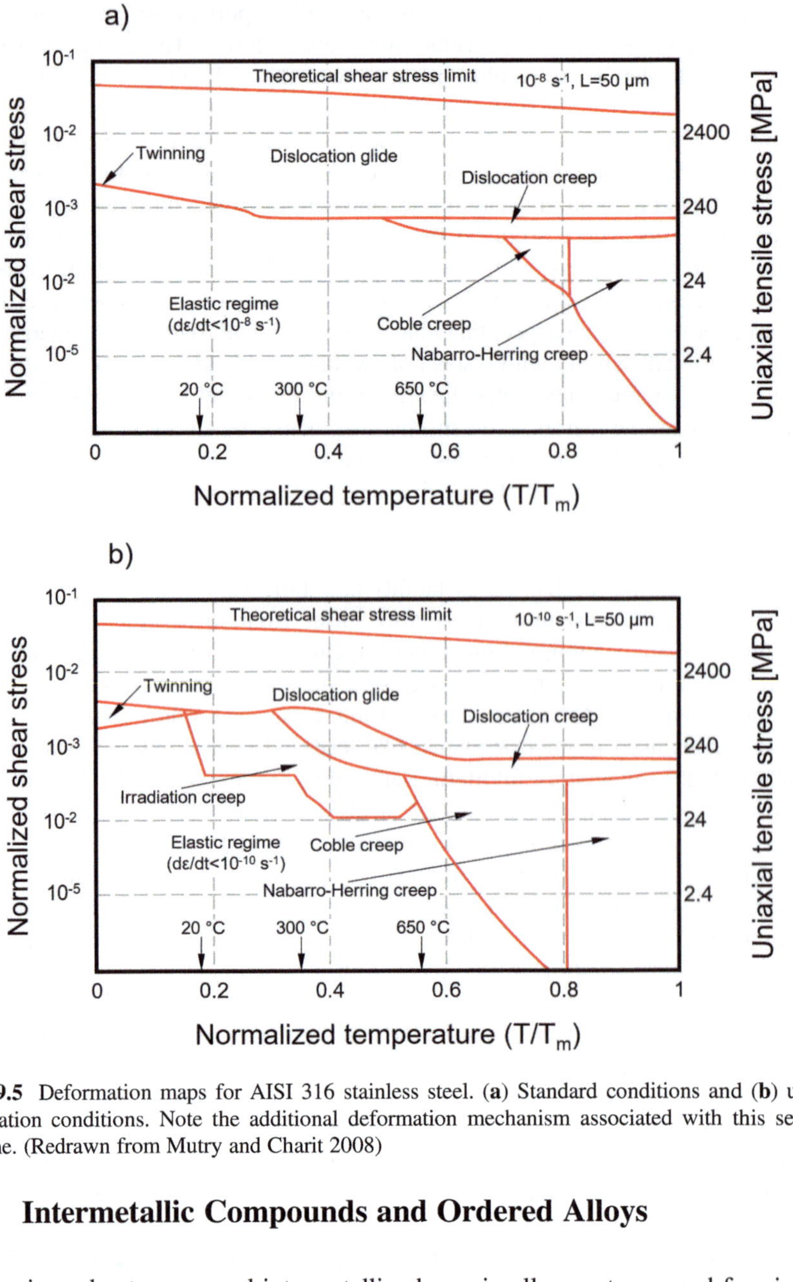

Fig. 9.5 Deformation maps for AISI 316 stainless steel. (**a**) Standard conditions and (**b**) under irradiation conditions. Note the additional deformation mechanism associated with this second regime. (Redrawn from Mutry and Charit 2008)

9.4 Intermetallic Compounds and Ordered Alloys

In previous chapters, several intermetallic phases in alloy systems used for aircraft structures and engine parts have been cited. In hardenable Al-alloys, equilibrium intermetallic phases, like Al_2Cu, in 2xxx or Mg_2Si in 6xxx alloys, would form in the overaged conditions (see Sect. 3.2). These phases should be avoided, since their presence is in general indicative of an alloy temper featuring poor mechanical

Table 9.6 Main advanced aluminide alloys, based on the Ni_3Al phase (Sikka et al. 2000)

Alloy	Al	Cr	Fe	Zr	Mo	B	Ni
IC50	21.73	–	–	1.15	–	–	77.12
IC-74M	23.51	–	–	–	–	0.04	76.45
IC-218	16.66	15.29	–	1.57	–	0.10	66.39
IC2181Zr	17.00	15.82	–	0.39	–	0.04	66.75
IC-221	16.73	15.35	–	3.35	–	0.04	64.54
IC357	18.36	13.53	21.64	0.77	2.51	0.04	43.15
IC-396M	15.96	15.36	–	1.60	5.98	0.02	61.08

properties. The γ' ordered phase, based on the $L1_2$-ordered Ni_3Al compound, is fundamental as concerns the mechanical properties, both static and time-dependent, of iron- and nickel-based superalloys (see Sects. 6.3 and 6.4). The dependence of the yield strength of Ni_3Al on the temperature (see Fig. 6.21), a consequence of the dislocation structure and dynamics in the ordered crystallographic structure of this alloy, has driven the research for single-phase intermetallic alloys, based on the abovementioned binary compound. An alloy fully made of the strengthening phase of the nickel-based superalloys was expected to have remarkable mechanical performances. Alloy design studies provided indications on how to improve mechanical strength of the binary intermetallic and, at the same time, increasing its ductility and fracture toughness up to values suitable for both processing and structural applications (Jiao 2016; Pettifor 1988a, b). Boron "doping" (concentration in the 0.1 wt% range) was employed and turned out to be an effective approach to strengthen grain boundary bonding, with a consequent beneficial effect on room-temperature ductility. Advanced aluminides (Table 9.6) were the alloys developed in this context (Aoki and Izumi 1979; Liu et al. 1985; Sikka et al. 2000). However, they never found their way to aerospace applications, since the overall mechanical behavior was definitely worse than state-of-the-art nickel-based superalloys: the lack of the γ-γ' interfaces and the optimized microstructure that can be achieved in multiphase alloys (see Sect. 6.4), using appropriate thermal treatments, are the main reasons for this outcome.

In view of the high oxidation and carburization resistance, as it is for wear and cavitation-erosion resistance of these alloys, several terrestrial applications have been considered instead: furnace rolls, radiant burner tubes for steel industry, forging dies, and corrosion resistant fixtures for chemical industries, center posts for pit-carburization furnaces (Stoloff 2000). A similar development has regarded also the iron aluminide alloys, based on the FeAl and Fe_3Al compounds. Their excessive brittleness was mitigated using ternary additions, like Mo, Nb, Ti, and Zr also in association with second phase particles, like TiB_2 (Liu and Kumar 1993; McKamey et al. 1991). Boron doping turned out to be detrimental for Fe_3Al, whereas it provided strengthening and toughening effects in FeAl, depending on the actual composition of the alloy (Baker et al. 1998). Several alloys have been developed and used as steel replacement in components working in aggressive environments, like coal gasification plants (Saunders et al. 1997). With reference to nickel aluminides, another intermetallic has been extensively investigated: NiAl. NiAl, together with

CoAl and FeAl, is the main phase in diffusion coatings (see Sect. 7.3.1), the oxidation protective coatings used for protecting nickel- and cobalt-based superalloys and steels, respectively, against oxidation and other high-temperature corrosion phenomena. NiAl is also investigated as possible replacement for nickel-based superalloys, considering the interesting properties that this intermetallic phase exhibits, including a low density, i.e., 5.9–6.3 g/cm^3, depending on the actual alloy composition, that is, approximately two thirds the density of nickel-based superalloys, typically in the 8.6 g/cm^3 range (Darolia 1991; Yu et al. 1993). The thermal conductivity of NiAl, in the 100 W/mK range, is from four to eight times higher than the thermal conductivity of nickel-based superalloys, an important prerequisite to keep the internal thermal stresses below critical values. Coherently with its mentioned use in diffusion coatings, NiAl has an excellent oxidation resistance. Although based on the simple cubic structure, featuring a lower number of active slip systems than the austenitic structure of the superalloys, still this crystallographic structure can guarantee a sufficient level of plastic deformation, far better than the complex low-symmetry structures of silicides and other exotic intermetallics, discussed further on in this section (Miracle 1993). Eventually, the significantly higher incipient melting temperature (1638 °C for the stoichiometric NiAl composition) corresponds to a higher material stability, particularly as concerns creep deformation and other diffusion-assisted phenomena. Several processing routes have been adopted over the years to produce NiAl-based alloys and to manufacture components, including prototypes of jet engine parts. Following the approach used for single-crystal nickel-based superalloys, single-crystal and directionally solidified NiAl materials have been developed (Cui et al. 2005). Directional solidification is particularly interesting for eutectic systems, like NiAl-Mo, NiAl-Cr, NiAl-Cr(Mo), and NiAl-Cr(Fe), although the fracture toughness of these alloys turned out to be not always adequate for structural applications (Bei and George 2005; Wang et al. 2016, 2017; Yang et al. 1997). The microstructure of these alloys resembles that of composite materials, with a matrix phase, i.e., NiAl, reinforced and toughened by "long fibers" or lamellae of chromium and/or molybdenum. For this reason, eutectic alloys are often named in situ composites, to highlight the compositional and microstructural features of these materials, obtained directly from the melt. Eutectic alloys considered for high-temperature applications involve also other intermetallic compounds, like the Nb-Nb_3Al, Cr-Nb, NiAl-NiAlNb, Nb-Nb_5Si_3, Nb-Nb_5Si_3-Nb_3Si, $MoSi_2$-Mo_5Si_3 (Bewlay et al. 1995; Cao et al. 1994; Mason and Van Atkin 1993; Mendiratta and Dimiduk 1993; Reuss and Vehoff 1990; Reviere et al. 1992; Whittenberger et al. 1992a; Yu et al. 2017). Another group of these materials comprises composites in which a metallic matrix is reinforced by refractory metal carbides, in the form of long fibers: Co-TaC (COTAC), NiTaC (NITAC), and Cr-Cr_2C_3 (McLean 1983). Because of the exceptional mechanical properties, these alloys had attracted a great deal of interest in the early days of their development. However, although part of this interest still survives, the critical aspects associated with material processing, like an intrinsic low growth rate, still pose serious limitations to a widespread application of these materials in structural applications at high temperatures.

A further development of NiAl alloys concerns alloys strengthened by the precipitation of secondary phases. The early investigations on these systems date back to the study of the two phase, β-NiAl-β′-$NiAl_2Ti$, alloy. The β′-$NiAl_2Ti$ Heusler phase displays a particularly good creep resistance that can be improved further when combining the two intermetallic phases in a single alloy and optimizing the relative fraction of the two (Polvani et al. 1976; Strutt et al. 1976). Microscopy observations confirmed that the increased mechanical resistance of the two-phase alloys, reaching the level of a nickel-based superalloy, i.e., MAR-M200, was strongly dependent on the interaction of the dislocations with the semi-coherent interfaces between β and β′ phases. Being based on cubic structure with *bcc*-like symmetries, these alloys were regarded as the *bcc* counterparts of the conventional nickel-based austenitic superalloys so that they were named the "*bcc* nickel-based superalloys." The concept of two-phase intermetallic alloy was extended also to the FeAl-Fe_3Al alloys, which coherently with the previous denomination can be defined as "*bcc* iron-based superalloys" (Morris and Gunther 1997). In the above studies, the two-phase alloys were polycrystalline and obtained by conventional casting, featuring quite a low room-temperature ductility to fracture, below 2%. From this point of view, a significant improvement was obtained by setting the composition in the three-phase field, in which the β and β′ phases would coexist with the γ′ phase (Yang et al. 1991, 1992). The three-phase materials, developed with different compositions, introducing additional alloying elements, turned out to have very interesting mechanical properties, including better deformability, also in this case comparable to a standard nickel-based superalloy, i.e., Udimet 720 (Yang et al. 1992). An important step forward was the development of multiphase intermetallics, based on NiAl precipitation hardened by another Heusler phase: Ni_2HfAl. The formation of Ni_2HfAl is promoted by 1 at% of hafnium approximately in the formulation, on some occasion added together with zirconium (Locci et al. 1996). Silicon was also introduced, with generally positive effects on the resulting properties of the alloy, thanks to the formation of hardening precipitates (G-$Ni_{16}Hf_6Si_7$) (Chen et al. 2000; Whittenberger 1999). These studies highlighted an important processing issue of these alloys, having either a single-crystal or directionally solidified microstructure. Indeed, the presence of silicon oxide coming from the erosion of the mold was observed. The prolonged contact between the molten metal and the mold inner wall and the very high temperatures involved may easily result in the contamination of the alloy during the solidification process. To improve this aspect, specific casting technologies have been developed for a better control of the alloy final composition and microstructure (Scheppe et al. 2002). Alternatively, a powder metallurgy approach, using plasma-spheroidized and subsequently HIPed powders, has been developed, with excellent performances of the consolidated products (Kaplanskii 2018). Another approach that was successfully applied to the strengthening of the NiAl intermetallic was through the dispersion of secondary ceramic phases using the mechanical alloying process, discussed already with regard to ODS alloys (Sect. 9.3). Two main powder metallurgy techniques have been used for processing NiAl-based alloys: a standard mechanical alloying process, similar to that used for producing an ODS NiAl alloy, and cryomilling, i.e., high-energy ball-milling of a

NiAl alloy in liquid nitrogen (Whittenberger et al. 1990a, b). In the first approach, as for ODS superalloys, a limited concentration, usually not exceeding 5 wt%, of a fine-grained thermodynamically stable phase is distributed throughout the intermetallic matrix, so to provide an obstacle to the dislocation movement, particularly at high temperatures. The very first attempts to obtain the oxide dispersion strengthening of the NiAl alloy involved several different oxides, like Al_2O_3, ThO_2, and Y_2O_3, although the thoriated material only resulted to be present as a nanometric dispersion within the intermetallic alloy, whereas the particles of the other oxides clustered together in rather coarse agglomerates (Koch and Whittenberger 1996). The alternative approach, based on cryomilling, was quite effective in attaining a good dispersion and grain refinement, also as concerns the metallic component, by avoiding cold welding. An important output of the mentioned research was not only an effective metallic grain comminution and fine dispersion of the ceramic phase, Y_2O_3 in this case, but also the formation of AlN particles (Whittenberger et al. 1990a, b; Witkin and Lavernia 2006). This occurred for the reaction of liquid nitrogen with the metallic powder, sustained by the mechanical grinding. Cryomilling of NiAl and the formation of AlN were investigated further in order to optimize the properties of the products, particularly in their consolidated condition. Several processing aspects were considered, like the interplay between in situ-formed AlN and the ceramic dispersion added to the NiAl powder on purpose and the possibility of obtaining more stable nitride phases, with the introduction of suitable reacting metals, e.g., zirconium (Whittenberger et al. 1999). In view of the positive strengthening effects of the AlN dispersion, alternative routes were attempted, with the idea of bypassing cryomilling. Roasting treatments of NiAl powders in a nitrogen atmosphere resulted in the formation of the nitride phase, although the mechanical properties of the consolidated samples did not reach the level of cryomilled materials. However, the mentioned treatment produced a general improvement in the powder quality, as regards sinterability, most probably for the elimination of the surface contamination present in the grains of metallic powder (Whittenberger et al. 2000). Nowadays, it is acknowledged that the powder metallurgy route is to be regarded as most promising for the technological exploitation of NiAl-based alloys. Interestingly, also for NiAl alloys, as it happened in the early stages of the development of ODS superalloys (Sect. 9.3), a molten metal approach was adopted. Rapid solidification was in fact used to incorporate HfC, TiB_2, and TiC in NiAl (Jha et al. 1989). The HfC-hardened material provided mechanical properties, also at high temperatures, definitely better than the un-reinforced matrix (Ray et al. 1989; Whittenberger et al. 1990b, 1992b). Creep curves exhibited a very similar trend as for ODS alloys, i.e., an increase of the stress exponent, up to values as high as 20, and the presence of a threshold stress, below which no detectable creep rate is recorded. Nonetheless, the excessive costs of the rapid solidification plants, the comparatively limited productivity, and the difficulty of scaling up make this technology not so attractive, particularly if compared to mechanical alloying.

Beyond nickel aluminides, intermetallic compounds continue to attract the interest in the search for materials capable to be used at temperatures above the present limits of nickel-based superalloys, i.e., a metal temperature above 1100 °C. The melting temperature might definitely be an effective selection parameter, considering that higher-melting temperatures correspond to lower diffusivities and therefore better creep performances (Fairbank et al. 2000). Moreover, the ordered structure of intermetallics enhances this aspect further, in view of the lower diffusivities in the ordered rather than disordered state (Schoijet and Girifalco 1967; Schoijet and Girifalco 1968; Wang and Akbar 1993). These factors have addressed the research for high-temperature ordered intermetallics toward PGM and silicide alloys, both systems being considered already in Sect. 9.1 (Zhao and Westbrook 2003). The counterpart to the ordered structure of intermetallic compounds is their extremely limited ductility, particularly at room temperature, and fracture toughness. In fact, the strong covalent character of the chemical bonding and the limited number of slip systems together with the complex, low-symmetry crystallographic structures render these alloys in many respects similar to ceramics. The approach successfully applied to nickel aluminide intermetallic alloys, based on the development of multiphase systems, is being followed also for these higher-temperature systems, with excellent results (Mitra 2006; Petrovic and Vasudevan 1999; Perepezko et al. 2014). This concerns not only the structural properties but also the oxidation and high-temperature corrosion resistance (see Chap. 8).

Still referring to multiphase systems, intermetallic matrix composites have been investigated (Ward-Close et al. 1996). These materials feature a matrix made of one or more intermetallic phases, like nickel, iron, titanium aluminides, niobium, and molybdenum silicides, to be selected on the basis of the target applications. Since the main task for intermetallic matrix composites is increasing the fracture toughness of the matrix, ductile metallic reinforcements are used, like in case of NiAl-Mo and NiAl-Cr eutectic alloys. Using directional solidification, lamellar NiAl-Mo structures with excellent fracture toughness have been produced. Alternatively, the intermetallic matrix can be reinforced with high-strength ceramic particles or fibers, which may have a beneficial effect not only on static strength but also on creep resistance at higher temperatures. High-strength NiAl particle composites have been obtained using a combustion synthesis, a processing technology particularly suitable for intermetallics that usually have relatively large negative enthalpies of mixing that can be exploited to sustain the composite formation reaction. In this example the reinforcement is provided by NbB_2 and Nb_xC, although a very broad range of matrix reinforcement can be found in the literature, involving NiAl and several other intermetallic systems (Bochenek and Batista 2015; Mitra 2018). The specific features of the intermetallic phases have required design strategies and processing techniques not necessarily coincident with those applied to metal matrix and ceramic matrix composites. In the first ones, the ductile and tough metal matrix is reinforced by strong and stiff ceramics, like Al_2O_3, SiC, TiB_2, etc., introduced as fibers or particles. In ceramic matrix composites, the brittle matrix is toughened by the presence of weak interfaces, which become preferential pathways for crack propagation, whose energy is thus absorbed by the fracture surface increase.

Nonetheless, the "multiphase" approach will help overcoming the present obstacles to the aerospace applications of potentially interesting intermetallic compounds, thus extending the list that comprises already very important examples, like nickel, cobalt, and iron aluminides in overlay coatings (see Sect. 7.3.1) and titanium aluminides, especially γ-TiAl-based alloys, in low-pressure stage gas turbine blades (see Sect. 4.6) (Lasalmonie 2006).

9.5 Shape Memory Alloys

Another family of intermetallic compounds, interesting for the aerospace research, is that of shape memory alloys (SMAs), of which the B2 compound NiTi is the main and most notorious representative. The potential of shape memory alloys in aerospace applications stems from the phenomenology of these materials. SMAs are capable to react directly, in terms of strain and shape changes, to temperature, magnetic field, and stress variations. The shape changes and the relevant developed stresses can be used to operate switches, actuate devices, produce displacements, etc. This aspect allows to reduce the number of components to be used and also the design complexity of thermomechanical devices. An immediate consequence of these features is the weight reduction of the systems involving SMAs parts, an aspect that is always interesting in aerospace components and structures. The possibility of tuning the operating temperature range, the recoverable deformation and stress, adjustable through the selection of a suitable alloy, and relevant sequence of thermomechanical treatments are all additional qualifying properties of SMAs. In this section, a few important aspects of the SMAs phenomenology interesting for the aerospace field are recalled. Then, the main applications of SMAs to aerospace structures and devices are illustrated with the indication of the main and most promising developments routes.

9.5.1 Main Phenomenological Aspects and Relevant Mechanisms

The main aspects of the shape memory phenomenology derive from a few structural and microstructural aspects, which are related to the martensitic transformation that occurs in these alloys. Different from the martensitic transformation in steels that is notoriously irreversible and as such should not be treated as a genuine phase transition, the martensitic transformation occurring in SMAs is reversible. On the other hand, the steel-related naming is adopted for the involved SMA phases. The low-temperature phase is indeed called *martensite*, the high temperature one *austenite*. The denominations for the transition temperatures are coherent with this convention: upon cooling, *austenite* will transform into *martensite* over a temperature range going from *martensite start* (M_s) down to *martensite finish* (M_f). The reverse transformation occurs from A_s to A_f, *austenite start* and *finish*, respectively,

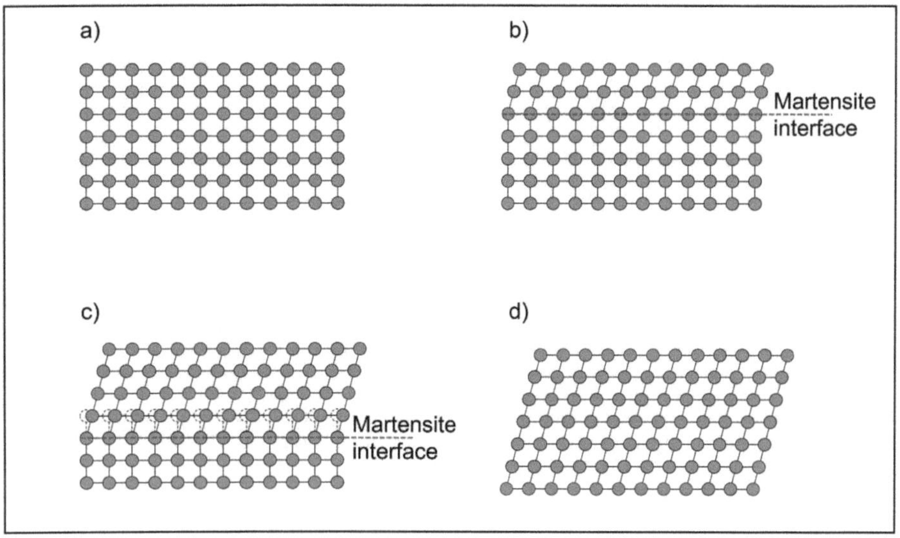

Fig. 9.6 Schematic of the displacive martensitic transformation occurring in SMAs. Upon cooling ($T< M_s$) the initially fully austenitic alloy (**a**) starts to transform into martensite (**b**). The transformation proceeds for the coherent displacement of large number of atoms for small distances. During the transformation the two phases coexist across a common interface plane (**b–c**). At $T< M_f$, all the austenite is transformed into martensite (**d**). The reverse path from (**a–d**) may occur, because the transformation is reversible. (Redrawn from Duerig et al. 1990)

on heating *martensite* until it fully transforms back to *austenite*. In the NiTi alloys that are by far the most widely used SMAs, austenite has a cubic CsCl B2 structure, isomorphous to the β-NiAl phase (see Sect. 7.3.1). Martensite in binary NiTi alloys is generally monoclinic. The presence of dislocations and other defects introduced by plastic deformation may result in the formation of an intermediate rhombohedral phase (R-transformation). Eventually, ternary additions, e.g., copper in concentrations exceeding 10 wt%, determine the formation of a martensitic phase based on the orthorhombic structure (Duerig et al. 1990). All the involved transformations are displacive. The displacive character depends on the fact that the transformation from one phase to the other does not involve any compositional change and, therefore, no diffusive mass transport is needed. Moreover, in analogy with the martensitic transformation in steels, the crystalline lattice of the austenite can be obtained by a distortion of martensite and vice versa (Bain strain). Therefore, the transformation occurs through a coordinated displacement of large number of atoms (*military* transformation), moving for very short distances, smaller than the average interatomic distance. The transition is a first-order one, i.e., a latent heat exchange is involved, and the coexistence of the two phases is observed as the transformation proceeds. The two-dimensional scheme in Fig. 9.6 depicts the essential aspects described so far.

In real alloys, the transformation takes place over a finite temperature range and with a hysteresis on cooling with respect to heating. This is due to the mechanical

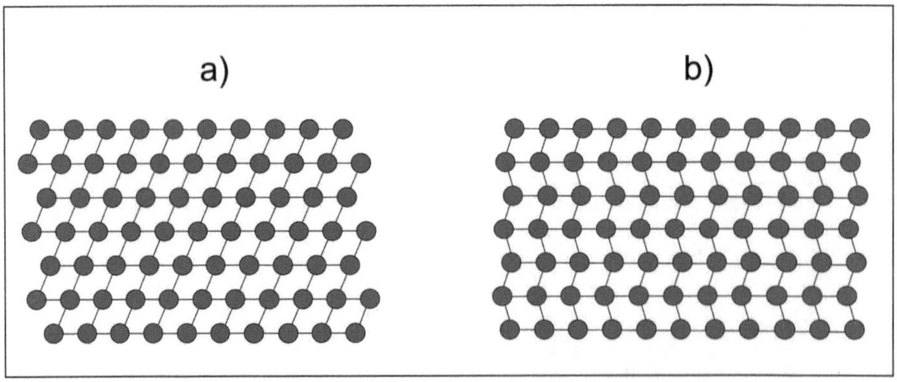

Fig. 9.7 The two mechanisms intervening for accommodating the shape change due to a martensitic transformation: (**a**) slip and (**b**) twinning. (Redrawn from Duerig et al. 1990)

constraint faced when the new phase is forming from the parent one. Therefore, both austenite and martensite are stabilized below and above the equilibrium transition temperature, respectively.

The crystallographic structure of austenite can be obtained through the lattice strain of martensite. The transformed volume of the alloy would generally have a different shape from the initial one (see Fig. 9.6a, d). This requires the accommodation of the new shape within the volume initially occupied by the parent phase. Slip and twining are the two mechanisms intervening in this process. In general, slip is observed in those systems exhibiting differences in the molar volume between austenite and martensite larger than 1%, so that shape accommodation involves comparatively high stresses. When martensite and austenite have similar molar volumes, twinning is the main shape accommodating mechanism. The fundamental difference between slip and twinning is that slip involves breaking of atomic bonds, so that the structure is irreversibly deformed. In twinned structures, the shape accommodation is attained through the orientation of the so-called twin variants, just two in the simple bidimensional model in Fig. 9.7b, more numerous and with complex orientation relationships in case of real 3D systems. The twin variants have highly mobile interfaces (twin boundaries) as a consequence of their low surface energy. In case an external stress is applied, all variants tend to align parallel to the stress direction, a process that is assisted by the movement of the twin boundaries. At the same time, all variants that are not favorably oriented tend to disappear. Martensite de-twinning is a particularly important deformation mechanism for shape memory alloys, occurring at nearly constant stress, in agreement with the claimed mobility of the twin boundaries, and reaching strain values up to 8% approximately. Once de-twinning is complete, the elastic deformation of de-twinned martensite will start, up to the yield stress of the alloy. These structural and microstructural aspects are the basis for two important phenomena exhibited by SMAs: the shape memory effect (SME) and the superelastic behavior.

As concerns the SME, the main idea is that the initial shape that the SMA part is having in its austenitic phase will be recovered once the part is deformed, so that the

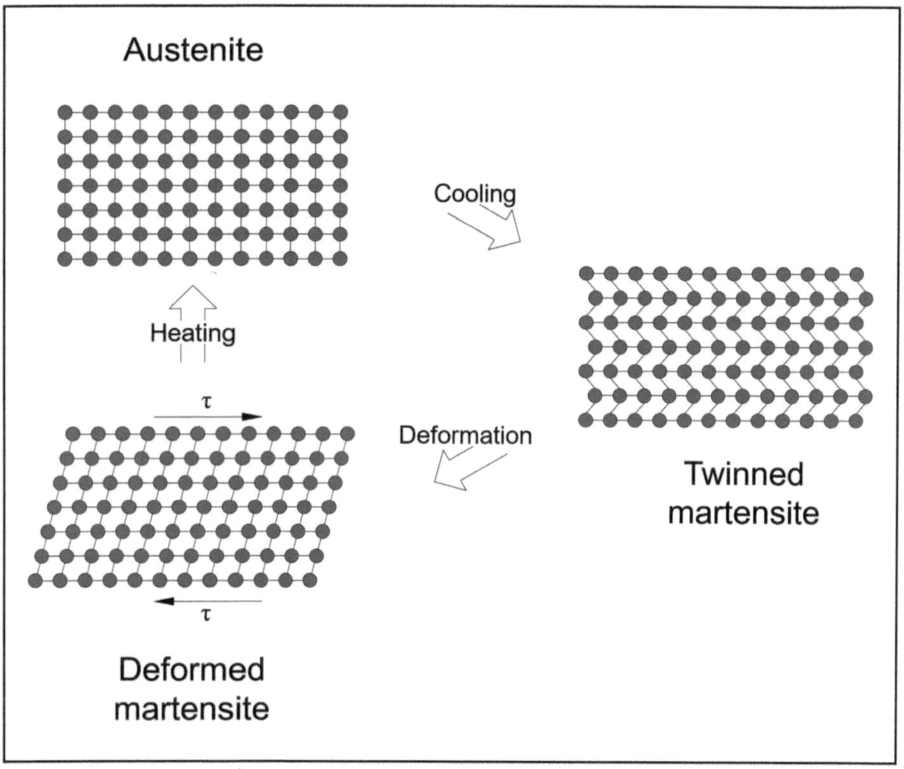

Fig. 9.8 Schematic of the base principles of the shape memory effect (SME). The austenitic alloy is cooled below M_f, so that it turns to martensite, with the usual twinned structure. The deformation of martensite involves the rearrangement of the twin variants. Upon heating up the deformed martensite, the reverse transition to austenite is accompanied by the recovery on the original form. (Redrawn from Duerig et al. 1990)

original shape is changed, in the martensitic field. With reference to the sequence of events depicted in Fig. 9.8, the austenitic alloy is cooled down in the martensitic field, and its shape is changed with a deformation based on the reorientation of the twin variants. Once the alloy is heated up back to austenite, the transition to the high-temperature phase is accompanied by the return to the initial shape of the alloy.

Two aspects are essential to the shape memory effect. First, the shape change imparted to the alloy in the martensitic phase that can be actually recovered is only due to the rearrangement of the twin structure. Any fraction of deformation involving dislocation movement and slip will not be recovered, and it is thereby called alloy "amnesia." Second, the springback effect that allows the return to the initial austenite shape, when the reverse martensite-austenite transition occurs, is an implicit consequence of the lower crystal symmetry of the martensitic, with respect to the austenitic phase. This means that, while there are several ways of obtaining martensite from austenite, there is just one to obtain austenite starting from

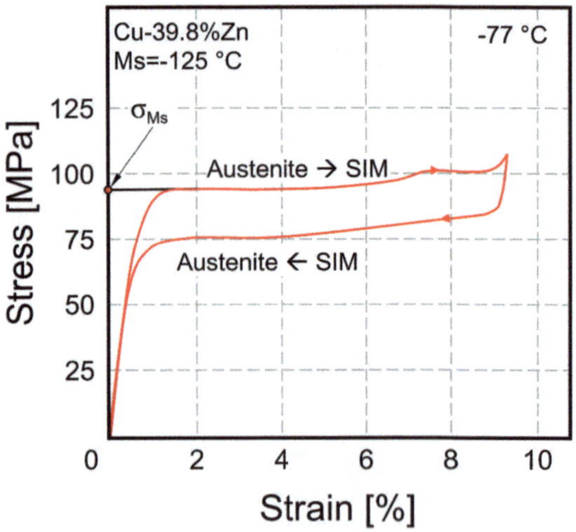

Fig. 9.9 Stress-strain curve of a superelastic SMA (Cu-39.8%Zn) starting from the austenite field. (Redrawn from Duerig et al. 1990)

martensite. In the simple bidimensional scheme in Fig. 9.8, this corresponds to the fact that from the square symmetry of austenite, two martensite variants can be obtained. On the contrary, starting from these latter, just one austenite variant is obtained.

The Bain strain relationships existing between martensite and austenite can be exploited to induce the transformation by deforming the austenite structure to martensite also above M_s. This is the so-called stress-induced martensite (SIM) and on this mechanism superelasticity (SE) is based. SE is the capability of SMAs to fully recover even relatively large deformations, i.e., 8% or so. Actually, the applied stress inducing the austenite to martensite transformation is by all means acting as a temperature reduction. Therefore, SE is regarded as a thermoelastic process, since there is a complete equivalence in the temperature and stress response of the alloy. Once the stress overcomes the onset value for starting the transformation (σ_{Ms}), the first martensitic nuclei start to form, aligned along favorable variant directions. The transformation is completed once all austenite is transformed under the effect of the applied stress to martensite. At still higher stresses, the elastic deformation of the SIM is observed up to its yield stress ($\sigma_y{}^M$). At any stress below $\sigma_y{}^M$, upon unloading the alloy, a complete recovery of the deformation is attained for the reverse martensite to austenite transformation (Fig. 9.9).

The stress needed to initiate the martensitic transformation (σ_{Ms}) depends on the temperature, as it is to be expected for the equivalence of stress and temperature involved with thermoelasticity. In particular, the higher is the temperature of the alloy above M_s, the larger is the stress required for its transformation from austenite to martensite. At temperatures above M_d (see Sect. 5.4.3.1), the stress necessary to induce the martensitic transformation would be above the yield stress of austenite. Therefore, no transformation to martensite is obtained but just the plastic deformation of austenite. An interesting approach to the combined exploitation of both SME and SE has been proposed. The leading idea is not to use a thermally but a mechanically, via SE, induced martensite in SME applications (Casati et al. 2014). The relevant thermomechanical cycle for devices using the SME is illustrated in Fig. 9.10a. In practice, the stress and strain available to any potential applications based on the SME are those provided by the de-twinning of martensite and relevant values of stresses, particularly interesting in case of constrained recovery of the shape. Figure 9.10b shows how the stress field that may be provided by the SMA active part can be enhanced if the working stress field is raised to the value of the upper stress of the superelastic cycle. This renders particularly attractive SMA actuators based on high-performance SME, for the increased operating performances.

SME and SE are intrinsic properties of the SMAs. It is clear from the description of SME that after the first transformation loop, illustrated in Fig. 9.8, once the alloy has returned to the initial austenite shape, a further loop requires necessarily an external stress to deform the alloy in the martensite field. It is possible to "train" the alloy and induce another effect, the so-called two-way SME (TWSME). It consists in the repeated change of the alloy shape by just varying the temperature across the martensitic transition interval. Thanks to the TWSME, in the same alloy, two different shapes can be recovered by just changing the temperature above A_f and below M_f. The "training" of the alloy, which can be conducted in different ways (Auricchio et al. 2003), generates a matrix of structural defects and associated stress fields, introducing into the alloy an anisotropic texture condition. Therefore, also the twinned martensite that forms upon cooling austenite will produce a macroscopic shape change for the arrangement of the variants induced by the textured microstructure. As compared to SME that can also be called one-way SME (OWSME) for obvious reasons, the TWSME, although very interesting in several respects, exhibits some important limitations, to be taken into the right consideration for application purposes. In the first place, the strain excursion required to oscillate between the two shapes tends to degrade very rapidly, so that TWSME is not recommended in case a large number (more than 100) of cycles are required. Secondly, the stress levels achievable via TWSME are by far lower than those involved with OWSME. Nonetheless, there are several applications, particularly those involving controlled displacements that may benefit from TWSME.

9.5.2 Aerospace Applications of Shape Memory Alloys

The early observation of the SME was made in Au-Cd alloys, although the potential and consequent interest for the shape memory effect roused after the research work on NiTi alloys (Buehler et al. 1963; Kaufman and Mayo 1997; Ölander 1932). The unexpected results were obtained within a study on alloys for nuclear applications, carried out at the Naval Ordnance Laboratory, after which the new family of alloys was named (Nitinol). The first Nitinol alloy that was commercialized was registered as Tinel®. Its first application was on the US Navy's Grumman F-14 "Tomcat" jet fighter in 1971, as a coupling system for hydraulic pipes, code-named Cryofit, after

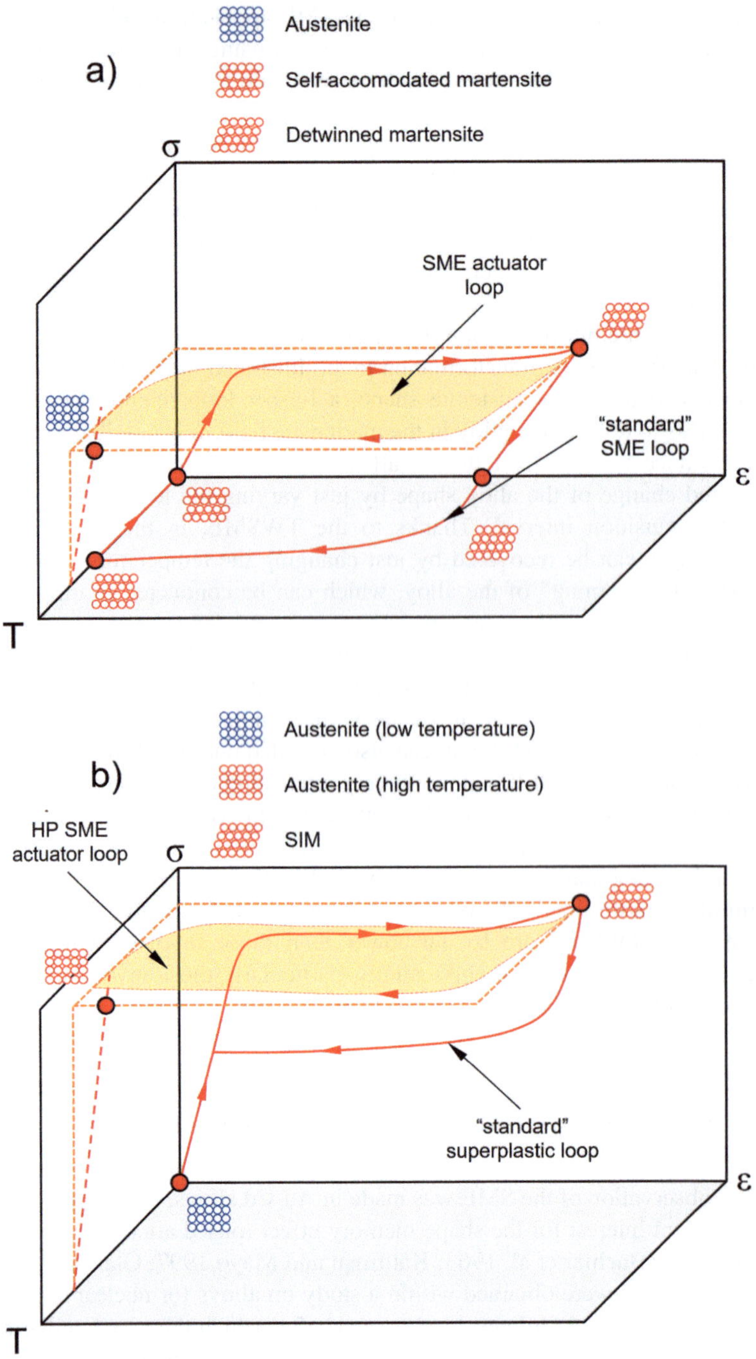

Fig. 9.10 (**a**) Thermomechanical loop for the standard SME and for (**b**) the high-performance SME. (Redrawn from Casati et al. 2014)

the cryogenic temperatures involved in its installation (Melton 1999). This cylindrical coupling, machined from a bar of Nitinol, had in the austenitic phase a diameter smaller than the external diameter of the two pipes to be connected. The working principle is the OWSME, and therefore an important parameter is the temperature at which the joint is meant to operate. At this temperature the alloy has to be fully austenitic. The positioning of the Cryofit across the joint between the two pipes requires its deformation in the martensite field. This is obtained using liquid nitrogen and an appropriate tool, like a tapered mandrel. The installation is accomplished by heating up the SMA coupling, back to the austenitic field. This reverse transition may also occur spontaneously, just leaving the Cryofit warming up at ambient temperature. To enhance the contact stress and to avoid unwanted slippages, radial teeth are present on the inner surface of the Cryofit. Therefore, the installation of the Cryofit can be completed quite quickly and using very basic tools only, mainly for handling the cryogenically cooled part and to deform it in a controlled way, i.e., without introducing unwanted plastic deformations that might depress the recovery capabilities of the SMA. The space required to conduct the whole procedure is relatively limited, which renders the approach very interesting in case of a high density of piping connections. A connection based on SMAs exhibits also the so-called "live joint" behavior, i.e., the capability to accommodate even large deformations without changing the tightening stress. This behavior descends from the superelastic effect of the austenite phase, i.e., the constant stress plateau above σ_{Ms} (Fig. 9.9). After the early applications of Cryofit in the military field, for which it was certified soon after the successful tests, the use was subsequently extended to commercial aircrafts too, and since then Cryofit has become a standard for the aerospace pipe connections. These fittings are applicable to connect pipelines with maximum internal pressures of 40 MPa and represent an interesting solution not only for new installation but also for repair operations.

The potentials of SMAs disclosed more aerospace applications, particularly once the main issues concerning the reproducibility of the mechanical behavior of these alloys and their critical dependence on the actual composition have been better understood and managed (Lewis 1997). The weight saving and, most importantly, the performance reliability introduced by SMA actuators are particularly attractive for reducing the weight of satellites and extraterrestrial mission spacecrafts. These reasons have guided the development of several devices that have become an absolute must for the release and deployment of payloads from space carrier vehicles. The Frangibolt, developed and commercialized by the TiNi Aerospace Inc., is a certified deployment system, which has replaced, in most space missions, explosive devices (Johnson 1992). Its working principle is still based on the OWSME. A NiTi cylinder is coaxially mounted to a notched bolt, which is holding the payload in place during the take-off and preliminary steps of the flight. In this condition the SMA is martensitic. A resistive heating system is integrated into the cylinder, and it is switched on to activate the payload release. The SMA cylinder becomes austenitic and tries to recover the relevant extended shape that, for this application, is the shape to be "remembered." The stress developed in the process breaks the notched bolt and thus releases the payload. The typical weight of each active element is 100 g approximately, and the force developed is 45–50 kN, to be compared with a force of 20–25 kN of the load-holding notched bolt. The main advantage of the Frangibolt, with respect to an explosive deployment device, is the possibility of pretesting each component, whose correct operation is paramount and absolutely critical for the full success of the space mission. The deployment dynamics of the Frangibolt involves a much more limited shock energy than the explosive approach. Moreover, being essentially based on an electrical current regulation, it is definitely more controllable than an explosion, and this assures a better control of the correct orbital positioning of the released object, without the need of subsequent adjustments.

The Pinpuller is another release device using a Nitinol wire, in which a pin is unlocked and actuates the deployment. This device was used for the first time in a space mission on the Mars Global Surveyor in 1996 (Bokaie et al. 1998). In the rest condition, the pin, on which acts a compressed drive spring, is kept in the extended outward position by a retractable detent system. The remaining blockage of the pin is guaranteed by a set of balls kept in outward position by a suitable ball keeper and a latch, connected to the NiTi wire, martensitic in the rest position of the Pinpuller. When the wire is resistively heated and goes back to the austenitic phase, it contracts, for the SME, and, through the latch, drives the balls inward, thus unlocking the pin that is retracted under the effect of the drive spring. Typical operating parameters for NiTi wire that is the active part in Pinpullers are pull force ranging from 20 to 450 N, with a corresponding stroke range from 6.3 to 16.0 mm. Another important and interesting feature of the Pinpuller is its extremely short actuation time, estimated in a few milliseconds.

Another operation very important for space orbital stations and satellites is the deployment and unfolding of solar panels and antennas that are launched stowed and fully folded for obvious space restriction in the hold. The conventional approach, before the introduction of SMA systems, was again based on nuts explosively fractures and hinges connected to spring dampers. All these techniques have been now replaced by a deployment system made of a NiTi strips (Carpenter and Lyons 2002). The strips, folded in the rest condition (martensite), are heated up to the austenite unfolded condition by a nickel-chrome resistive heater strip, stuck onto the inner surface of the SMA element. After deployment the SMA part goes back to martensite, but the deployed shape is of course retained. In principle, as seen when introducing the OWSME, the "smart" strip might be folded back again by an external force and the deployment repeated. In real applications, this is not really the case, since, once deployed, the solar panels and satellite antennae remain in position for the rest of their lifetime.

An active device, comprising also a reloading mechanism for the SMA wire, was developed to move a glass screen back and forth over an element of the solar panel installed on the Mars Pathfinder micro-rover (Stone 1996). This device was meant to monitor and evaluate the effects of the accumulation of Martian dust on the solar panel efficiency under different coverage conditions (Jenkins and Landis 1995). This is a paramount aspect to be considered in view of future expeditions and possible human settlements on the Red Planet.

In addition to the above applications, since the early 1990s, several studies aimed at applying SMAs to active aircraft structures. The DARPA "Smart Wing" project considered the feasibility of using SMA actuators, located within the wings, to modify their profile (Jardine et al. 1997; Kudva 2004). A similar approach, relying on actuators inside the components, had been proposed earlier on for changing the geometry of the aircraft wings (Strelec et al. 2003). A similar approach has been adopted for the intake engine and exhaust ducts, demonstrated in a realistic environment on a full-scale F-15 smart structure within the DARPA-funded SAMPSON project (Pitt et al. 2001, 2002). This idea was developed further, with an approach still based on the OWSME, developed to modulate the gap of the nozzle of a bypass fan gas turbine engine, in order to reduce the jet noise during take-off and the shock-cell noise under cruising flight conditions. At the same time, the control of the fluxes emerging from the rear part of the engine should also optimize specific fuel consumption, particularly during altitude cruise flight (Calkins et al. 2006). SMA-functionalized chevrons have been developed and dislocated along the trailing edge of the thrust reverser. Three Nitinol elements, including resistive heaters, are embedded into the chevron body, made of composite material (typically carbon fiber-reinforced polymer matrix). The correct selection of the composite stiffness is paramount, since a springback action is required to push back and reload the active SMA elements, once they have completed their deployment action. These variable geometry chevrons (VGCs) can be operated in two different ways. In the *powered* mode, the tip of each chevron is actively modulated by the heaters connected to each one of the SMA elements, in order to optimize noise reduction, without affecting too much the propulsion efficiency of the engine (Mabe et al. 2007). Alternatively, the profile modifications of the chevrons are induced autonomously (*autonomous*, unpowered mode) by the temperature changes induced by the take-off and altitude cruise flight conditions. Specific NiTi alloy compositions, also with ternary additions, were investigated and characterized to optimize the part performances, over the complete temperature range (Hart et al. 2010; Noebe et al. 2009; Van Humbeeck 2012). Noise reduction is attained, through the chevrons, by promoting an optimal mixing of the free gas stream with the fractions coming from the engine fan and from the hot engine core, i.e., combustor and turbine (Mabe et al. 2005; Saiyed et al. 2000). The VGCs have been successfully tested for the first time on a Boeing 777 aircraft using GE-115B engines, showing a significant noise reduction under different flight conditions, in association with the expected optimization of the engine performances as concerns fuel consumption. These results were very much appreciated, also in view of the growing concern on the impact of the noise from aircraft jet engines and relevant legislation (Brueckner and Girvin 2008). The VGC program has demonstrated the potentials of SMAs to support the development of aircraft structures and systems capable of self-adaptation and optimization to all flight conditions (Calkins et al. 2008). These aspects are fundamental for morphing (from metamorphose), i.e., changing the shape of aero-structures (Barbarino et al. 2011). In fact, conventional morphing technologies usually involve complex systems for actuating the relevant shape changes. Hydraulic pumps, compressors, and electric motors operating mechanical actuators are not only complex but also rather expensive and heavy. Moreover, the route to a complete acceptance of these solutions, notwithstanding the undisputable benefits they bring, is rather complex, and delays are regularly faced until a full certification is issued. SMAs but also other functional materials, like piezoelectrics, electro- and magnetostrictive materials, and electroactive polymers, may enable unique morphing capabilities (Loewy 1997). In fact, different from traditional approaches, "smart" components do not require sensors and servomechanisms to be actuated, in the first place, since they are sensitive and directly responsive to a suitable external input, like a temperature and/or stress change. Moreover, these materials are not only self-activated but, owing to their peculiar thermomechanical properties, can directly perform the required actuation (Calkins and Habe 2010). In this regard, SMAs may represent a major breakthrough in the field of morphing of aerospace structures, as demonstrated by several research studies. The reconfigurable rotor blade (RRB) program was aimed at improving the performances of a rotorcraft, a particularly sensitive aspect for VSTOL (vertical and/or short take-off and landing) aircrafts. The attempted approach was to develop an SMA actuator, embedded in the blades, capable of changing their twist angle during cruise and hover flight conditions (Caldwell et al. 2007; Jacot et al. 2002; Mabe et al. 2006; Prahlad and Chopra 2001; Ruggieri et al. 2008). A prototype was wind tunnel-tested and provided extremely positive results, with no recorded losses in performances all through the duration of the test that, for its extreme conditions, is now regarded as a new benchmark for SMA actuators and active rotor blades (Calkins and Habe 2010). An outcome of the RRB program was the development of a deployable rotor blade aerodynamic device to reduce the noise due to the blade-vortex interaction (BVI). This is quite a well-known issue, and whereas noise reduction is strictly required, a successful solution is based on applying small tabs on the rotor blades that reduce the noise associated with the BVI. The counterpart of this solution is the efficiency reduction due to the drag introduced by the tabs. The improvement, whose feasibility has been demonstrated by wind tunnel tests, is the use of deployable tabs, operated by SMA actuators, which can fit inside the blades. The tabs are extended when noise reduction and quiet operations are desirable or, like in some urban areas, mandatory. In normal conditions, they can be retracted and stowed within the blade profile, thus not influencing negatively the aerodynamics (Calkins et al. 2006; Kennedy et al. 2000; Singh and Chopra 2002). Still concerning rotorcraft morphing to reduce noise emission, SMA may provide interesting solutions for its control, also acting on the blade tips (Loewy 1997; Testa et al. 2005). This latter study is based on a theoretical model and numerical investigation. The results show the beneficial effect of a SMA actuator, used to maximize the separation between the tip vortex and the rotor plane, a condition to reduce the noise produced by VBI. The effect can be enhanced if used in association with a magneto-rheological fluid that selectively reduces the bending stiffness spanwise in order to increase the actuation efficiency. Noise reduction, and also an improvement in the specific fuel consumption, can be obtained by varying the area of the fan entrance nozzle. A larger diameter at take-off and approach reduces the jet velocity and thereby the noise produced by the engine. Variable area fan nozzles (VAFNs) employ SMA flexure actuators. Test on full-scale prototypes has demonstrated also for this application the successful design that can be implemented thanks to SMAs (Mabe et al. 2008).

The perspectives of using SMA actuators and parts in adaptive aircraft components appear promising and, in some cases, the only feasible way for implementing novel solutions. Design and modeling studies have demonstrated already the potential of SMA in wing morphing (Icardi and Ferrero 2009; Sofla et al. 2010). Considering the increasing interest for more efficient, less polluting, and noisy aircrafts, it is likely that SMA-based applications for which the feasibility has been assessed already will be certified soon and applied both on the existing fleet, whenever possible in replacement of conventional components, and on the newly designed aircrafts (Hartl and Lagoudas 2007).

References

Akinc M et al (1999) Boron-doped Molybdenum Silicides for Structural Applications. Materials Science and Engineering A 261 (1–2): 16–23

Alven D A (2004) Refractory and Precious Metal-Based Superalloys. JOM 56 (9): 27

Aoki K, Izumi O (1979) Improvement in Room Temperature Ductility of the $L1_2$ Type Intermetallic Compound Ni_3Al by Boron addition. Journal of the Japan Institute of Metals 43 (12): 119–1196

Auricchio F et al (2003) Modelling of SMA Materials: Training and Two Way Memory Effects. Computers and Structures 81: 2301–2317

Azim M A et al (2017) Characterization of Oxidation Kinetics of Mo-Si-B Based Materials. Oxidation of Metals 87 (1–2): 89–108

Baker I et al (1998) The Room Temperature Strengthening Effect of Boron as a Function of Aluminum Concentration in FeAl. Intermetallics 6 (3): 177–183

Barbarino S et al (2011) A Review of Morphing Aircraft. Journal of Intelligent Material Systems and Structures 22: 823–877

Bei H, George E P (2005) Microstructure and Mechanical Properties of a Directionally Solidified NiAl-Mo Eutectic Alloy. Acta Materialia 53 (1): 69–77

Benjamin J S (1990) Mechanical Alloying – A Perspective. Metal Powder Report 45 (2): 122–127

Benjamin J S, Bomford MJ (1977) Dispersion Strengthened Aluminum Made by Mechanical Alloying. Metallurgical Transactions A 8 (8): 1301–1305

Bewlay B P et al (1995) Solidification Processing of High Temperature Intermetallic Eutectic-Based Alloys. Materials Science and Engineering A 192: 534–543

Bewlay B P et al (2003a) A Review of Very-High-Temperature Nb-Silicide-Based Composites. Metallurgical and Materials Transactions A 34 (10): 2043–2052

Bewlay B P et al (2003b) Ultrahigh-Temperature Nb-Silicide-Based Composites. MRS Bulletin 28 (09): 646–653

Bochenek K, Batista M (2015) Advances in Processing of NiAl Intermetallic Alloys and Composites for High Temperature Aerospace Applications. Progress in Aerospace Sciences 79: 136–146.

Bokaie M D et al (1998) Release Device for Retaining Pins. US Patent 5,771,742

Briant C L (2000) New Applications for Refractory Metals. JOM 51 (3): 36

Brosse J B et al (1981) Intrinsic Intergranular Brittleness of Molybdenum. Scripta Metallurgica 15 (6): 619–623

Brueckner J, Girvin R (2008) Airport Noise Regulation, Airline Service Quality, an Social Welfare. Transport Research Part B: Methodological 42 (1): 19–37

Buckley J D et al (1981) Early Development of Ceramic Fiber Insulation for Space Shuttle. Ceramic Bulletin 60: 1196–1200
Buckman R W (2000) New Applications for Tantalum and Tantalum Alloys. JOM 52 (3): 40–41
Buehler W J et al (1963) Effect of Low-Temperature Phase Changes on the Mechanical Properties of Alloys near Composition TiNi. Journal of Applied Physics 34 (5): 1475–1477
Burk S et al (2010) Effect of Zr Addition on the High-Temperature Oxidation Behaviour of Mo-Si-B Alloys. Oxidation of Metals 73 (1): 163–181
Byun T S et al (2013) Irradiation Dose Temperature Dependence of Fracture Toughness in High Dose HT9 Steel from the Fuel Duct of FFTF. Journal of Nuclear Materials 432 (1–3): 1–8
Caldwell N et al (2007) Heat Transfer Model for Blade Twist Actuator System. Journal of Thermophysics and Heat Transfer 21 (2): 350–360
Calkins F T et al (2006) Variable Geometry Chevrons for Jet Noise Reduction. Paper presented at the 12th AIAA/CEAS Aeroacoustics Conference (27th AIAA Aeroacoustics Conference), Cambridge, Massachusetts, 8–10 May 2006
Calkins F T et al (2008) Overview of Boeing's Shape Memory Alloy Based Morphing Aerostructures. Paper presented at the ASME 2008 Conference on Smart Materials, Adaptive Structures and Intelligent Systems, Ellicott City, Maryland, USA, 28–30 October
Calkins F T, Habe J H (2010) Shape Memory Alloy Based Morphing Aerostructures. Journal of Mechanical Design 132 (11): 111012
Cantor B et al (2001) Aerospace Materials. CRC Press
Cao H et al (1994) Mechanical Properties of an In Situ Synthesized Nb/Nb_3Al Layered Composite. Materials Science and Engineering A 185 (1–2): 87–95
Cao H et al (2013) Preparation of Dispersion Strengthened Aluminum Alloy by High Energy Ball Milling. Advanced Materials Research 602–604: 598–601
Carpenter B, Lyons J (2002) Lightweight Flexible Solar Array Validation Report. Available via DIALOG. https://eo1.gsfc.nasa.gov/new/validationReport/Technology/Documents/Reports/LFSA.pdf. Accessed 17 Sept 2018
Casati R et al (2014) Thermal Cycling of Stress-Induced Martensite for High-Performance Shape Memory Effect. Scripta Materialia 80 (13–16)
Chen J, Hoffelner W (2009) Irradiation Creep of Oxide Dispersion Strengthened (ODS) Steels for Advanced Nuclear Applications. Journal of Nuclear Materials 392 (2): 360–363
Chen Y (2013) Irradiation Effects of HT-9 Martensitic Steel. Nuclear Engineering and Technology 45 (3): 311–322
Chen Y X et al (2000) Microstructure and Phase Stability Studies on Heusler Phase Ni_2AlHf and G-Phase $Ni_{16}Hf_6Si_7$ in Directionally Solidified NiAl-Cr(Mo) Eutectic Alloyed with Hf. Journal of Materials Research 15 (6): 1261–1270
Cockeram B V (2006) The Mechanical Properties and Fracture Mechanisms of Wrought Low Carbon Arc Cast (LCAC), Molybdenum-0.5pct Titanium-0.1pct Zirconium (TZM), and Oxide Dispersion Strengthened (ODS) Molybdenum Flat Products. Materials Science and Engineering A 418 (1–2): 120–136
Cockeram B V (2009) The Fracture Toughness and Toughening Mechanism of Commercially Available Unalloyed Molybdenum and Oxide Dispersion Strengthened Molybdenum with an Equiaxed, Large Grain Structure. Metallurgical and Materials Transactions A 40: 2843–2860
Cornish L A et al (2006) New Pt-based Alloys for High Temperature Application in Aggressive Environments: The Next Stage. International Platinum Conference "Platinum Surges Ahead", The South African Institute of Mining and Metallurgy: 57–66
Cui C Y et al (2005) High Tensile Elongation of a Directionally Solidified NiAl Multiphase Alloy at High Temperatures. Materials Science and Engineering A 396 (1–2): 194–201
Darolia R (1991) NiAl Alloys for High-Temperature Structural Applications. JOM 43 (3): 44–49
Dimiduk D M, Perepezko J H (2003) Mo-Si-B Alloys: Developing a Revolutionary Turbine-Engine Material. MRS Bulletin 28 (9): 639–645
Donachie M J, Donachie S J (2002) Superalloys – A Technical Guide. ASM International
Duerig T W et al (1990) Engineering Aspects of Shape Memory Alloys. Butterworth-Heinemann

El-Genk M S (2009) Deployment History and Design Considerations for Space Reactor Power Systems. Acta Astronautica 64 (9–10): 833.849

El-Genk M S, Tournier J M (2005) A Review of Refractory Metal Alloys and Mechanically Alloyed-Oxide Dispersion Strengthened Steels for Space Nuclear Power Systems. Journal of Nuclear Materials 340 (1): 93–112

Elzey D M, Artz E (1988) Oxide Dispersion Strengthened Superalloys: The Role of Grain Structure and Dispersion during High Temperature Low Cycle Fatigue. In: Reichman S et al (ed) Superalloys, The Metallurgical Society, p 595–604

Fairbank G B et al (2000) Ultra-High Temperature Intermetallics for The Third Millennium. Intermetallics 8 (9–11): 1091–1100

Fischer B et al (1999) High Temperature Mechanical Properties of the Platinum Group Metals. Platinum Metals Review 43 (l): 18–28

Froes F H (1990) The Structural Applications of Mechanical Alloying. In: Froes F H, deBarbdillo J J (eds) Proceedings of an ASM International Conference, Myrtle Beach, South Carolina, 27–29 March 1990. ASM International, Materials Park, Ohio

Gao M C et al (2008) The First-Principles Design of Ductile Refractory Alloys. JOM 60 (7): 61–65

Gilman P S, Nix W D (1981) The Structure and Properties of Aluminum Alloys Produced by Mechanical Alloying: Powder Processing and Resultant Powder Structures. Metallurgical and Materials Transactions A 12 (5): 813–824

Gu Y F et al (2004) Chromium and Chromium-Based Alloys: Problems and possibilities for High Temperature Service. JOM 56 (9): 28–33

Gu Y F et al (2005) Microstructural Evolution and Mechanical Properties of Cr-Ru Alloys. Metallurgical and Materials Transactions A 36 (3): 577–582

H C Starck Fabricated Products (2013) Defense and Aerospace Applications of Molybdenum, Tungsten and Tantalum Products. Available via DIALOG. https://www.azom.com/article.aspx?ArticleID=9073. Accessed 15 Nov 2018

Harada H (2003) High Temperature Materials for Gas Turbines. The Present and Future. In: Proceedings of the International Gas Turbine Congress, Tokyo, 2–7 November 2003

Harper M A, Rapp R A (1994) Codeposited Chromium and Silicon Diffusion Coatings for Fe-Base Alloys via Pack Cementation. Oxidation of Metals 42 (3–4): 303–333

Hart D J et al (2010) Jet Engine Chevron Application: II Experimentally Validated Numerical Analysis. Smart Materials and Structures 19 (1): 015021

Hart D J et al (2010) Use of a Ni60Ti Shape Memory Alloy for Active Jet Engine Chevron Application: I Thermomechanical Characterization. Smart Materials and Structures 19 (1): 015020

Hartl D J, Lagoudas D C (2007) Aerospace Applications of Shape Memory Alloys. Proceedings of the Institution of Mechanical Engineers, Part G: Journal of Aerospace Engineering 221, p 535–552

Heilmeier M et al (2009) Metallic Materials for Structural Applications Beyond Nickel-based Superalloys. JOM 61 (7): 61–67

Hoffelner W (2011) Design Related Aspects in Advanced Nuclear Fission Plants. Journal of Nuclear Materials 409 (2): 112–116

Howson T E et al (1980) Creep and Stress Rupture of Oxide Dispersion Strengthened Mechanically Alloyed Inconel Alloy MA754. Metallurgical Transactions A 11 (9): 1599–10607

Icardi U, Ferrero L (2009) Preliminary Study of an Adaptive Wing with Shape Memory Alloy Torsion Actuators. Materials & Design 30 (10): 4200–4210

Jacot A D et al (2002) Shape Memory Alloy Device and Control Method. US Patent 6,499,952 B1

Jardine A P et al (1997) Smart Wing Shape Memory Alloy Actuator Design and Performance. In: Goldstein D H, Chipman R A (eds) Proceedings of SPIE – The International Society for Optical Engineering. Polarization: Measurement, Analysis, and Remote Sensing, San Diego, CA, United States, 30 July – 1 August 1997

Jehanno P et al (2005) Assessment of a Powder Metallurgical Processing Route for Refractory Metal Silicide Alloys. Metallurgical and Materials Transactions A 36 (3): 515–523

Jenkins P P, Landis G A (1995) A Rotating Arm using Shape Memory Alloy. In NASA (ed) The 29th Aerospace Mechanisms Symposium, p 167–171
Jha SC et al (1989) Dispersoid in Rapidly Solidified B2 Nickel Aluminides. Scripta Metallurgica 23: 805–810
Jiao Z B et al (2016) Strategies for Improving Ductility of Ordered Intermetallics. Progress in Natural Science: Materials International 26 (1): 1–12
Johnson A D (1992) Non-Explosive Separation Device. US Patent 5,119,555
Jones E S et al (1958) The Oxidation of Molybdenum. Corrosion 14 (1): 20–26
Kaplanskii Y Y (2018) Microstructure and Thermomechanical Behavior of Heusler Phase Ni_2AlHf-strengthened NiAl-Cr(Co) Alloy produced by HIP of Plasma-spheroidized Powder. Materials Science and Engineering A 729: 398–410
Kaufman G, Mayo I (1997) The Story of Nitinol: The Serendipitous Discovery of the Memory Metal and its Applications. The Chemical Educator 2 (2): 1–21
Kennedy D K et al (2000) Development of an SMA Actuator for In-Flight Rotor Blade Tracking. Journal of Intelligent Material Systems and Structures 15 (4): 235–248
King D et al (2009) Advanced Aerospace Materials: Past, Present and Future. Aviation and the Environment 3: 22–27
Klueh R L (1993) Tensile Behavior of Neutron-Irradiated Martensitic Steels: A Review. Nuclear Technology 102 (3): 376–385
Klueh RL, Nelson A T (2007) Ferritic/martensitic steels for next generation reactors. Journal of Nuclear Materials 371:37–52
Knittel S et al (2013) Development of Silicide Coatings to Ensure the Protection of Nb and Silicide Composites against High Temperature Oxidation. Surface and Coatings Technology 235: 401–406
Koch C C, Whittenberger J D (1996) Mechanical Milling/alloying of intermetallics. Intermetallics 4 (5): 339–355
Korb L J et al (1981) The Shuttle Orbiter Thermal Protection System. Ceramic Bulletin 60: 1188–1193
Kudva J (2004) Overview of the DARPA Smart Wing Project. Journal of Intelligent Material Systems and Structures 15: 261–267
Kung S C, Rapp R A (1989) Analyses of the Gaseous Species in Halide-Activated Cementation Coating Packs. Oxidation of Metals 32 (1–2): 89–109
Lasalmonie A (2006) Intermetallics: Why is it so Difficult to Introduce Them in Gas Turbine Engines? Intermetallics 14 (10–11): 1123–1129
Lassner E, Schubert W D (1999) Tungsten – Properties, Chemistry, Technology of the Element, Alloys, and Chemical Compounds. Kluwer Academic/Plenum Publisher, New York
Lewis P (1997) Aircraft Industry Accepts Shape Memory Alloy Technology. Aircraft Engineering and Aerospace Technology 69 (1): 31–34
Liebscher C H, Glatzel U (2014) Configuration of Superdislocations in the γ'-Pt_3Al Phase of a Pt-based Superalloy. Intermetallics 48: 71–78
Lipetzky P (2002) Refractory Metals: A Primer. JOM 54 (3): 47–49
Liu C T et al (1985) Effect of Boron on Grain-Boundaries in Ni_3Al. Acta Metallurgica 33 (2): 213–229
Liu C T, Kumar K S (1993) Ordered Intermetallic Alloys, Part I: Nickel and Iron Aluminides. JOM 45 (3): 38–44
Liu G et al (2013) Nanostructured High-Strength Molybdenum Alloys with Unprecedented Tensile Ductility. Nature Materials 12 (4): 344–350
Locci I E et al (1996) Microstructure and Phase Stability of Single Crystal NiAl alloyed with Hf and Zr. Journal of Materials Research 11 (12): 3024–3038
Loewy R G (1997) Recent Developments in Smart Structures with Aeronautical Applications. Smart Materials and Structures 6 (5): R11–R42
Lu Z L et al (2013) Fabricating Hollow Turbine Blades using Short Carbon Fiber-reinforced SiC Composite. International Journal of Advanced Manufacturing Technology 69 (1–4): 417–425

Mabe J H et al (2005) Design and Control of a Morphing Chevron for Takeoff and Cruise Noise Reduction. In: Proceedings of the 26th Annual AIAA Aeroacoustics Conference, Monterey, CA, 23–25 May 2005
Mabe J H et al (2006) Boeing's Variable Geometry Chevron, Morphing Aerostructure for Jet Noise Reduction. Paper presented at the 47th AIAA/ASME/AHS/ASC Structures, Structural Dynamics, and Materials Conference, Newport, Rhode Island, 1–4 May 2006
Mabe J H et al (2007) Full-Scale Flight Tests of Aircraft Morphing Structures using SMA Actuators. In: Proceedings Volume 6525, Active and Passive Smart Structures and Integrated Systems 2007. SPIE Smart Structures and Materials + Nondestructive Evaluation and Health Monitoring, San Diego, CA, United States, 2007
Mabe J H et al (2008) Variable Area Jet Nozzle Using Shape Memory Alloy Actuators in an Antagonistic Design. In: Proceedings Volume 6930, Industrial and Commercial Applications of Smart Structures Technologies 2008. SPIE Smart Structures and Materials + Nondestructive Evaluation and Health Monitoring, San Diego, CA, United States, 2008
Majumdar S et al (2009) Densification and Grain Growth during Isothermal Sintering of Mo and Mechanically Alloyed Mo-TZM. Acta Materialia 57 (14): 415–4168
Mansur L F et al (2004) Materials Needs for Fusion, Generation-IV Fission Reactors and Spallation Neutron Sources – Similarities and Differences. Journal of Nuclear Materials 329–333 Part A: 166–172
Mason D P, Van Atkin D C (1993) The Effect of Microstructural Scale on Hardness of $MoSi_2$-$Mo5Si_3$ Eutectics. Scripta Metallurgica et Materialia 28 (2): 185–189
McKamey C G et al (1991) A Review of Recent Developments in Fe_3Al-based Alloys. Journal of Materials Research 6 (8): 1779–1805
McLean M (1983) Directionally Solidified Materials for High Temperature Service. The Metal Society, London, UK
Medvedeva N I et al (2007) Solid Solution Softening and Hardening in the Group-V and Group-IV bcc Transition Metals Alloys: First Principles Calculations and Atomistic Modeling. Physical Review B 76 (21): 212104
Melton K N (1999) General Applications of Shape Memory Alloys and Smart Materials. In: Otsuka K, Wayman C M (eds) Shape Memory Materials. Cambridge University Press, p 220–239
Mendiratta M G, Dimiduk D M (1993) Strength and Toughness of a Nb/Nb_5Si_3 Composite. Metallurgical and Materials Transactions A 24 (2): 501–504
Miller M K et al (2003) Improvement in the Ductility of Molybdenum Alloys due to Grain Boundary Segregation. Scripta Metallurgica 46 (4): 299–303
Miller M K et al (2005) Stability of Ferritic MA/ODS Alloys at High Temperatures. Intermetallics 13 (3–4): 387–392
Miller M K et al (2006) Characterization of Precipitates in MA/ODS Ferritic Alloys. Journal of Nuclear Materials 351 (1–3): 261–268
Miracle D B (1993) The Physical and Mechanical Properties of NiAl. Acta Metallurgica et Materialia 41 (39): 649–684
Mitra R (2006) Mechanical Behaviour and Oxidation Resistance of Structural Silicides. International Materials Reviews 51 (1): 13–64
Mitra R (2018) Intermetallic Matrix Composites. Woodhead Publishing, Duxford, UK
Morris D, Gunther S (1997) Room and High Temperature Mechanical Behavior of a Fe_3Al-based Alloy with α-α" Microstructure. Acta Materialia 45 (2): 811–822
Mueller A J et al (2000) Evaluation of Oxide Dispersion Strengthened (ODS) Molybdenum and Molybdenum-Rhenium Alloys. International Journal of Refractory Metals and Hard Materials 18 (4–5): 205–211
Muktinutalapati N R (2011) Materials for Gas Turbines – An Overview. In: Benini E (ed) Advances in Gas Turbine Technology, IntecOpen, p 293–314
Mutry K L, Charit I (2008) Structural Materials for Gen-IV Nuclear Reactors: Challenges and Opportunities. Journal of Nuclear Materials 383 (1–2): 189–195

Noebe R D et al (2009) High Work Output Ni-Ti-Pt High Temperature Shape Memory Alloys and associated Processing Methods.

Ölander A (1932) An Electrochemical Investigation of Solid Cadmium-Gold Alloys. Journal of the American Chemical Society 54 (10): 3819–3833

Parthasarathy T A et al (2002) Oxidation Mechanisms in Mo-reinforced Mo_5SiB_2(T2)– Mo_3Si Alloys. Acta Materialia 50 (7): 1857–1868

Patra A et al (2015) Tungsten Based ODS Alloys-A Comprehensive Survey. Lap Lambert Academic Publishing

Perepezko J H (2009) The Hotter the Engine, the Better. Science 356 (5956): 1068–1069

Perepezko J H et al (2014) Structural Intermetallics, Alloy Design, Processing, and Applications. Advanced Materials and Processes 172 (9): 22–26

Petrovic J J, Vasudevan A K (1999) Key Development in High Temperature Structural Silicides. Materials Science and Engineering A 261: 1–5

Pettifor D G (1988a) Structure Maps for Pseudobinary and Ternary Phases. Materials Science and Technology 4 (8): 675–691

Pettifor D G (1988b) Theoretical Predictions of Structure and Related Properties of Intermetallics. Materials Science and Technology 8 (4): 345–349

Pickens J R (1991) High-Strength Aluminum P/M Alloys. In: ASM International (ed) ASM Handbook Volume 2 – Properties and Selection: Nonferrous Alloys and Special-Purpose Materials. ASM International

Pitt D M et al (2001) SAMPSON Smart Inlet SMA Powered Adaptive Lip Design and Static Test. In: 19th AIAA Applied Aerodynamics Conference, Fluid Dynamics and Co-located Conferences, Anaheim, CA, USA

Pitt D M et al (2002) SAMPSON Smart Inlet Design Overview and Wind Tunnel Test: II. In: Proceedings Volume 4698, Smart Structures and Materials 2002: Industrial and Commercial Applications of Smart Structures Technologies. SPIE's 9th Annual International Symposium on Smart Structures and Materials, San Diego, CA, United States, 2002

Pitt D M et al (2002) SAMPSON Smart Inlet Design Overview and Wind Tunnel Test: I. In: Proceedings Volume 4698, Smart Structures and Materials 2002: Industrial and Commercial Applications of Smart Structures Technologies. SPIE's 9th Annual International Symposium on Smart Structures and Materials, San Diego, CA, United States, 2002

Polvani R S et al (1976) High Temperature Creep in a Semi-Coherent NiAl-Ni_2AITi Alloy. Metallurgical Transactions A 7: 33–40

Portebois L et al (2014) Effect of Boron Addition on the Oxidation Resistance of Silicide Protective Coatings: A Focus on Boron Location in as-coated and oxidized coated Niobium Alloys. Surface and Coatings Technology 253: 292–299

Potgieter J H, Maledi N B (2014) High Temperature Corrosion Resistance of Pt-Based Superalloys in 0.2% SO_2-N_2 Gas. The Open Materials Science Journal 8: 18–26

Prahlad H, Chopra I (2001) Design of a Variable Twist Tilt-rotor Blade Using Shape Memory Alloy (SMA) Actuators. In: Proceedings Volume 4327, Smart Structures and Materials 2001: Smart Structures and Integrated Systems. SPIE's 8th Annual International Symposium on Smart Structures and Materials. Newport Beach, CA, United States, p 46–59

Ray R et al (1989) Carbide-dispersion-strengthened B2 NiAl. Materials Science and Engineering A 119: 103–111

Reese R T, Vick C P (1983) Soviet Nuclear Powered Satellites. British Interplanetary Society Journal 36: 457–462

Reuss S, Vehoff H (1990) Temperature Dependence of the Fracture Toughness of Single Phase and Two Phase Intermetallics. Scripta Metallurigica et Materialia 24 (6): 1021–1026

Reviere R D et al (1992) Processing Microstructure and Low-Temperature Properties of Directionally Solidified NiAl/NiAlNb Alloys. Materials Letters 14 (2–3) 149–155

Richerson D W (2004) Ceramic Components for Gas Turbine Engines: Why Has It Taken So Long? Ceramics Engineering and Science Proceedings 25 (3): 3–32

Ruggieri R T et al (2008). Development of a 1/4-Scale NiTinol Actuator for Reconfigurable Structures. In: Proceedings Volume 6930, Industrial and Commercial Applications of Smart Structures Technologies 2008. SPIE Smart Structures and Materials + Nondestructive Evaluation and Health Monitoring, San Diego, CA, United States, 2008

Saiyed N et al (2000) Acoustics and Thrust of Separate-Flow Exhaust Nozzles with Mixing Devices for High-Bypass-Ratio Engines. Available via DIALOG. https://ntrs.nasa.gov/archive/nasa/casi.ntrs.nasa.gov/20000083968.pdf. Accessed 24 Feb 2019

Sakidja R et al (2005) Synthesis of Oxidation Resistant Silicide Coatings on Mo-Si-B Alloys. Scripta Materialia 53 (6) 723–728

Sankar M et al (2015) Microstructure, Oxidation Resistance and Tensile Properties of Silicide Coated Nb-Alloy C-103. Materials Science and Engineering A 645 (1): 339–346

Saunders S R et al (1997) Behaviour of Fecralloy and Iron Aluminides Alloys in Coal Gasification Atmospheres Containing HCl. Materials Science Forum 251–254: 583–590

Scheppe F et al (2002) Nickel Aluminides: A Step Toward Industrial Applications. Materials Science and Engineering A 329–331: 596–601

Schmidt F F, Ogden H R (1963) The Engineering Properties of Tantalum and Tantalum Alloys. Available via DIALOG. https://apps.dtic.mil/dtic/tr/fulltext/u2/426344.pdf. Accessed 8 Nov 2018

Schoijet M, Girifalco L A (1967) Theory of Diffusion in Ordered Alloys of the β-Brass Type. Solid State Communications 29 (3): 481–495

Schoijet M, Girifalco L A (1968) Diffusion in order-disorder Alloys. The Face Centered Cubic AB3 Alloy. Journal of Physics and Chemistry of Solids 29 (6): 911–922

Sikka V K et al (2000) Advances in Processing of Ni_3Al-based Intermetallics and Applications. Intermetallics 8 (9–11): 1329–1337

Singh K, Chopra I (2002) Design of an Improved Shape Memory Alloy Actuator for Rotor Blade Tracking. In: Proceedings Volume 4701, Smart Structures and Materials 2002: Smart Structures and Integrated Systems. SPIE's 9th Annual International Symposium on Smart Structures and Materials, San Diego, CA, United States, 2002

Sofla A Y N et al (2010) Shape Morphing of Aircraft Wing: Status and Challenges. Materials & Design 31 (4): 1284–1292

Soleimani-Dorcheh A et al (2014) On Ultra-High Temperature Oxidation of Cr-Cr_3Si Alloys: Effect of Germanium. Materials and Corrosion 65 (12): 1143–1150

Stoloff N S (2000) Emerging Applications of Intermetallics. Intermetallics 8 (9–11): 1313–1320

Stone H W (1996) Mars Pathfinder Microrover A Small, Low-Cost, Low-Power Spacecraft. In: Proceedings of the 1996 AIAA Forum on Advanced Developments in Space Robotics.

Strelec J K et al (2003) Design and Implementation of a Shape Memory Alloy Actuated Reconfigurable Wing. Journal of Intelligent Material Systems and Structures 14 (4–5): 257–273

Strutt P R et al (1976) Creep Behavior of the Heusler Type Structure Alloy $NiAl_2Ti$. Metallurgical Transactions A 7: 23–31

Sturm D et al (2007) The Influence of Silicon on the Strength and Fracture Toughness of Molybdenum. Materials Science and Engineering A 463 (1–2): 107–114

Testa C et al (2005) Feasibility Study on Rotorcraft Blade Morphing in Hovering. Proceedings of SPIE – The International Society for Optical Engineering 5764: 171–182

Trinkle D R, Woodward C (2005) The Chemistry of Deformation: How Solutes Soften Pure Metals. Science 310 (5754): 1665–1667

United Nations General Assembly (1992) Principles Relevant to the Use of Nuclear Power Sources in Outer Space – Resolution 47/68. Available via DIALOG. https://www.un.org/ga/documents/gares47/list47.htm. Accessed 27 Sept 2016

Van Humbeeck J (2012) Shape Memory Alloys with High Transformation Temperatures. Materials Research Bulletin 47 (10): 2966–2968

Vilasi M et al (1993) Crystal Structure of Tri-Niobium Tri-Iron Chromium Hexasilicide Nb=3Fe=3Cr=lSi6: an Intergrowth of Zr4Co4Ge7 and Nb2Cr4Si5 Blocks. Journal of Alloys and Compounds 194 (1): 127–132

Vilasi M et al (1998a) New Silicides for New Niobium Protective Coatings. Journal of Alloys and Compounds 264 (1–2): 244–251
Vilasi M et al (1998b) Phase Equilibria in the Nb-Fe-Cr-Si System. Journal of Alloys and Compounds 269 (1–2): 187–192
Vorberg S et al (2004) Pt-Al-Cr-Ni Superalloys: Heat Treatment and Microstructure. JOM 56 (9): 40–43
Vorberg S et al (2005) A TEM Investigation of the γ-γ' Phase Boundary in Pt-based Superalloys. JOM 57 (3): 49–51
Wadsworth J et al (1988) Recent Advances in Aerospace Refractory-Metal Alloys. International Materials Reviews 33 (1): 131–150
Wang C C, Akbar S A (1993) Diffusion in Ordered Alloys and Intermetallic Compounds. Acta Metallurgica et Materialia 41 (10): 2807–2813
Wang L et al (2016) Microstructure and Mechanical Properties of NiAl-based Hypereutectic Alloy obtained by Liquid Metal Cooling and Zone Melted Liquid Metal Cooling Directional Solidification Techniques. Journal of Materials Research 31 (5): 646–654
Wang L et al (2017) Microstructure Evolution and Room Temperature Fracture Toughness of as-cast and Directionally Solidified Novel NiAl-Cr(Fe) Alloy. Intermetallics 84: 11–19
Ward-Close C M et al (1996) Intermetallic-Matrix Composites – A Review. Intermetallics 4 (3): 217–229
Wenderoth M et al (2005) On the Development and Investigation of Quaternary Pt-Based Superalloys with Ni Additions. Metallurgical and Materials Transactions A 36 (3): 567–575
Whittenberger J D (1999) 1300 K Creep Behavior of [001] oriented Ni-49Al-1Hf (at%) Single Crystals. Materials Science and Engineering A 286 (1): 165–183
Whittenberger J D et al (1990a) 1300 K Compressive Properties of Several Dispersion Strengthened NiAl Materials. Journal of Materials Science 25: 2771–2776
Whittenberger J D et al (1990b) Preliminary Investigation of a NiAl Composite prepared by Cryomilling. Journal of Materials Research 5 (2): 271–277
Whittenberger J D et al (1992a) Compressive Strength of Directionally Solidified NiAl-NiAlNb Intermetallics at 1200 and 1300 K. Scripta Metallurgica et Materialia 26 (6): 987–992
Whittenberger J D et al (1992b) Influence of Grain Size on the Creep Behavior of HfC-dispersed NiAl. Materials Science and Engineering A 151: 137–146
Whittenberger J D et al (1999) Microstructure and 1000-1400 K Mechanical Properties of Cryomilled NiAl-0.7Zr. Journal of Materials Research 14 (6): 2418–2429
Whittenberger J D et al (2000) Elevated Temperature Compressive Strength Properties of Oxide Dispersion Strengthened NiAl after Cryomilling and Roasting in Nitrogen. Materials Science and Engineering A 291 (1–2): 173–185
Wiedemann C et al (2005). Size Distribution of NaK Droplets released during RORSAT Reactor Core Ejection. Advances in Space Research 35 (7): 1290–1295
Witkin D B, Lavernia E J (2006) Synthesis and Mechanical Behavior of Nanostructured Materials via Cryomilling. Progress in Materials Science 51 (1): 1–60
Wojcik C C (1991) High Temperature Niobium Alloys. In: Stephens J J, Ahmad I (eds) High Temperature Niobium Alloys. The Minerals, Metals and Materials Society, Warrendale, PA, USA, p 1–12
Yamabe-Mitarai Y et al (2004) Platinum-Group-Metals-Based Intermetallics as High-Temperature Structural Materials. JOM 56 (9): 34–39
Yamabe-Mitarai Y, Murakami H (2014) Mechanical Properties at 2223 K and Oxidation Behavior of Ir Alloys. Intermetallics 48: 86–92
Yang J M et al (1997) Microstructure and Mechanical Behavior of in-situ Directional Solidified NiAl/Cr(Mo) Eutectic Composite. Acta Materialia 45 (1): 295–305
Yang R (1992) Equilibria and Microstructural Evolution in the β/β'/γ' Region of the Ni-Al-Ti System: Modeling and Experiment. Acta Metallurgica et Materialia 40 (7): 1553–1562
Yang R et al (1991) A Microstructural Study of a Ni_2AlTi-Ni(Al,Ti)-Ni_3(Al,Ti) Three-Phase Alloy. Journal of Materials Research 6 (2): 343–354

Yang R et al (1992) Three-Phase β/β'/γ' Ni-Al-Ti-(Cr,Fe) Alloys for High Temperature Use. Materials Science and Engineering A 152 (1–2): 227–236
Yu J L et al (2017) Mechanical Properties and Fracture Behavior of an Nb-Silicide in situ Composite. Intermetallics 90: 135–139
Yu K O et al (1993) Investment Casting of NiAl Single-Crystal Alloys. JOM 45 (5): 49–51
Yvon P, Carré F (2009) Structural Materials Challenges for Advanced Reactor Systems. Journal of Nuclear Materials 385 (2): 217–222
Zhao J-C, Westbrook J H (2003) Ultrahigh-Temperature Materials for Jet Engines. MRS Bulletin 28 (9): 622–630

Further Reading

Lagoudas D C (2007) Shape Memory Alloys – Modelling and Engineering Applications. Springer
Lecce L, Concilio A (2015) Shape Memory Alloy Engineering: for Aerospace, Structural and Biomedical Applications. Butterworth-Heinemann
Leo D J (2007) Engineering Analysis of Smart Material Systems. John Wiley & Sons Inc.
Liu C T et al (1992) Ordered Intermetallics – Physical Metallurgy and Mechanical Behaviour. Springer Netherlands
Otsuka K, Ren X (2005) Physical Metallurgy of Ti-Ni-Based Shape Memory Alloys. Progress in Materials Science 50 (5): 511–678
Sauthoff G (1995) Intermetallics. VCH, Weinheim
Schwartz M (2002) Encyclopedia of Smart Materials – 2 Volume Set. John Wiley & Sons
Stoloff N S, Sikka V K (1996) Physical Metallurgy and Processing of Intermetallic Compounds. Chapman & Hall, New York

Appendices

Appendix 1: Wrought Aluminum Alloys and Relevant Compositions

International designation	Si	Fe	Cu	Mn	Mg	Cr	Zn	Ti	Others
2001	0.20	0.20	5.2–6.0	0.15–0.50	0.20–0.45	0.10	0.10	0.20	Ni = 0.05 Pb = 0.003 Zr = 0.05
2011	0.40	0.70	5.0–6.0	–	–	–	0.30	–	Bi = 0.20–0.60 Pb = 0.20–0.60
2014	0.50–1.20	0.70	3.9–5.0	0.40–1.20	0.20–0.80	0.10	0.25	0.15	–
2017	0.20–0.80	0.70	3.5–4.5	0.40–1.00	0.40–0.80	0.10	0.25	0.15	–
2020	–	–	4.5	0.50	–	–	–	–	Li = 1.2
2021	0.20	0.30	5.8–6.8	0.20–0.40	0.02	–	0.10	0.02–0.10	Sn = 0.03–0.08 V = 0.05–0.15 Zr = 0.10–0.25 Cd = 0.05–0.20
2023	0.10	0.15	3.6–4.5	0.30	1.0–1.6	0.10	–	0.05	Sc = 0.01–0.06 Zr = 0.05–0.15
2024	0.50	0.50	3.8–4.9	0.30–0.90	1.2–1.8	0.10	0.25	0.15	–
2024A	0.15	0.20	3.7–4.5	0.15–0.80	1.2–1.5	0.10	0.25	0.15	–
2027	0.12	0.15	3.9–4.9	0.50–1.2	1.0–1.5	–	0.20	0.08	Zr = 0.05–0.15
2036	0.50	0.50	2.2–3.0	0.10–0.40	0.30–0.60	0.10	0.25	0.15	–
2050	0.08	0.10	3.2–3.9	0.20–0.50	0.20–0.60	0.05	0.25	0.10	Ag = 0.20–0.70 Ga = 0.05 Li=0.70–1.30 V = 0.05 Zr = 0.06–0.14
2055	0.07	0.10	3.2–4.2	0.10–0.50	0.20–0.60	–	0.30–0.70	0.10	Ag = 0.20–0.70 Li = 1.0–1.3 Zr = 0.05–0.15
2060	0.07	0.07	3.4–4.5	0.10–0.50	0.6–1.1	–	0.30–0.50	0.10	Ag = 0.05–0.50 Li = 0.60–0.90 Zr = 0.05–0.15

(continued)

S. Gialanella, A. Malandruccolo, *Aerospace Alloys*, Topics in Mining, Metallurgy and Materials Engineering, https://doi.org/10.1007/978-3-030-24440-8

International designation	Si	Fe	Cu	Mn	Mg	Cr	Zn	Ti	Others
2065	0.10	0.10	3.8–4.7	0.15–0.50	0.25–0.80	–	0.30	0.10	Ag = 0.15–0.50 Li = 0.8–1.5 Zr = 0.05–0.15
2076	0.12	0.15	2.9–3.8	0.10–0.50	0.05–0.40	–	0.10–0.50	0.10	Li = 1.0–1.4 Zr = 0.05–0.15
2090	0.10	0.12	2.4–3.0	0.05	0.25	0.05	0.10	0.15	Li = 1.9–2.6 Zr = 0.08–0.15
2091	0.20	0.30	1.8–2.5	0.10	1.1–1.9	0.10	0.25	0.10	Li = 1.7–2.3 Zr = 0.08–0.15
2094	0.12	0.15	4.4–5.2	0.25	0.25–0.80	–	0.25	0.10	Li = 0.70–1.4 Zr = 0.04–0.18
2097	0.12	0.15	2.5–3.1	0.10–0.60	0.35	–	0.35	0.15	Li = 1.2–1.8 Zr = 0.08–0.16
2098	0.12	0.15	3.2–3.8	0.35	0.25–0.80	–	0.35	0.10	Ag = 0.25–0.60 Li = 0.80–1.3 Zr = 0.04–0.18
2099	0.05	0.07	2.4–3.0	0.10–0.50	0.10–0.50	–	0.40–1.0	0.10	Be = 0.0001 Li = 1.6–2.0 Zr = 0.05–0.12
2124	0.20	0.30	3.8–4.9	0.30–0.90	1.2–1.8	0.10	0.25	0.15	–
2195	–	–	4.0	–	0.40	–	–	–	Ag = 0.40 Li = 1.0 Zr = 0.11
2196	–	–	2.9	–	0.50	–	–	–	Ag = 0.40 Li = 1.75 Zr = 0.11
2198	–	–	3.2	0.50	0.50	–	0.35	–	Ag = 0.40 Li = 1.0 Zr = 0.11
2199	0.05	0.07	2.6	0.30	0.30	–	0.55	–	Li = 1.8 Zr = 0.09
2218	0.90	1.00	3.5–4.5	0.20	1.2–1.8	0.10	0.25	–	Ni = 1.7–2.3
2219	0.20	0.30	5.8–6.8	0.20–0.40	0.02	–	0.10	0.02–0.10	V = 0.05–0.15 Zr = 0.10–0.25
2296	–	–	2.45	0.28	0.60	–	0.25	–	Ag = 0.43 Li = 1.6 Zr = 0.11
2297	–	–	2.8	0.30	0.25	–	0.50	–	Li = 1.4 Zr = 0.11
2397	–	–	2.8	0.30	0.25	–	0.10	–	Li = 1.4 Zr = 0.11
2519	0.25	0.30	5.3–6.4	0.10–0.50	0.05–0.40	–	0.10	0.02–0.10	V = 0.05–0.15 Zr = 0.10–0.25
2618	0.10–0.25	0.9–1.3	1.9–2.7	–	1.3–1.8	–	0.10	0.04–0.10	Ni = 0.9–1.2
3002	0.08	0.10	0.15	0.05–0.25	0.05–0.20	–	0.05	0.03	V = 0.05
3021	0.50	0.70	0.20–0.60	0.05–0.80	0.10	0.10	0.10	0.10	–
3030	0.15	0.35	0.10	0.10–0.70	0.05	0.05	0.05–0.50	0.05–0.35	–
3102	0.40	0.70	0.10	0.05–0.40		–	0.30	0.10	–
3130	0.15	0.20	0.05	0.10–0.40	0.05	–	0.05–0.30	0.05	–
4004	9.0–10.5	0.80	0.25	0.10	1.0–2.0	–	0.20	–	–

(continued)

International designation	Si	Fe	Cu	Mn	Mg	Cr	Zn	Ti	Others
4007	1.0–1.7	0.40–1.0	0.20	0.8–1.5	0.20	0.05–0.25	0.10	0.10	Ni = 0.15–0.70 Co = 0.05
4043	4.5–6.0	0.80	0.30	0.05	0.05	–	0.10	0.20	–
4047	11.0–13.0	0.80	0.30	0.15	0.10	–	0.20	–	–
4104	9.0–10.5	0.80	0.25	0.10	1.0–2.0	–	0.20	–	Bi = 0.02–0.20
4115	1.8–2.2	0.70	0.10–0.50	0.6–1.2	0.10–0.50	–	0.20	–	–
4143	4.7–6.0	0.80	0.30	0.05	0.15–0.30	–	0.10	0.20	–
4147	11.0–13.0	0.80	0.25	0.10	0.10–0.50	–	0.20	–	–
4343	6.8–8.2	0.8	0.25	0.10	–	–	–	–	Ni = 0.20
4643	3.6–4.6	0.80	0.10	0.05	0.10–0.30	–	0.10	0.15	–
5005	0.30	0.70	0.20	0.20	0.50–1.1	0.10	0.25	–	–
5017	0.40	0.70	0.18–0.28	0.6–0.8	1.9–2.2	–	–	0.09	–
5022	0.25	0.40	0.20–0.50	0.20	3.5–4.9	0.10	0.25	0.10	–
5024	0.25	0.40	0.20	0.20	3.9–5.1	0.10	0.25	0.20	Sc = 0.10–0.40 Zr = 0.05–0.20
5025	0.25	0.25	0.10	0.20	4.5–6.0	0.20	0.25	0.05–0.20	Be = 0.0009 Sc = 0.05–0.55 Zr = 0.10–0.25
5050	0.40	0.70	0.20	0.10	1.1–1.8	0.10	0.25	–	–
5051	0.40	0.70	0.25	0.20	1.7–2.2	0.10	0.25	0.10	–
5205	0.15	0.70	0.03–0.10	0.10	0.6–1.0	0.10	0.05	–	–
5251	0.40	0.50	0.15	0.10–0.50	1.7–2.4	0.15	0.15	0.15	–
5854	0.40	0.40	0.10	0.50	2.8–3.6	0.30	0.20	0.15	Li = 0.0009 Pb = 0.01 Ca = 0.0009 Na = 0.0009 Mn+Cr = 0.10–0.60
6056	0.70–1.3	0.50	0.50–1.1	0.40–1.0	0.6–1.2	0.25	0.10–0.70		Zr+Ti = 0.20
6063	0.20–0.60	0.35	0.10	0.10	0.45–0.90	0.10	0.10	0.10	–
6156Cl						–			
7010	0.12	0.15	1.5–2.0	0.10	2.1–2.6	0.05	5.7–6.7	0.06	Ni = 0.05 Zr = 0.10–0.16
7040	0.10	0.13	1.5–2.3	0.04	1.7–2.4	0.04	5.7–6.7	0.06	Zr = 0.05–0.12
7050	0.12	0.15	2.0–2.6	0.10	1.9–2.6	0.04	5.7–6.7	0.06	Zr = 0.08–0.15
7056	0.10	0.12	1.2–1.9	0.20	1.5–2.3	–	8.5–9.7	0.08	Zr = 0.05–0.15
7075	0.40	0.50	1.2–2.0	0.30	2.1–2.9	0.18–0.28	5.1–6.1	0.20	Zr+Ti = 0.25 when required for extruded or forged products
7150	0.12	0.15	1.9–2.5	0.10	2.0–2.7	0.04	5.9–6.9	0.06	Zr = 0.08–0.15
7349	0.12	0.15	1.4–2.1	0.20	1.8–2.7	0.10–0.22	7.5–8.7	–	Zr+Ti = 0.25
7449	0.12	0.15	1.4–2.1	0.20	1.8–2.7	–	7.5–8.7	–	Zr+Ti = 0.25
7475	0.10	0.12	1.2–1.9	0.06	1.9–2.6	0.18–0.25	5.2–6.2	0.06	–
8090	0.20	0.30	1.0–1.6	0.10	0.6–1.3	0.10	0.25	0.10	Li = 2.2–2.7 Zr = 0.04–0.16

(continued)

International designation	Si	Fe	Cu	Mn	Mg	Cr	Zn	Ti	Others
8091	0.30	0.50	1.6–2.2	0.10	0.50–1.2	0.10	0.25	0.10	Li = 2.4–2.8
									Zr = 0.08–0.16
8093	0.10	0.10	1.0–1.6	0.10	0.9–1.6	0.10	0.25	0.10	Li = 1.9–2.6
									Zr = 0.04–0.14

Chemical composition of wrought aluminum alloys of the 2XXX, 3XXX, 4XXX, 5XXX, 6XXX, 7XXX, 8XXX groups. Compositions are in wt%. Aluminum content is balance (Lumely 2011; Polmear et al. 2017; Prasad et al. 2014; The Aluminum Association 2015)

Appendix 2: Alloying Elements in Cast and Wrought Aluminum Alloys

Element	Symbol	Effects
Chromium	Cr	Generally added to Al-Mg, Al-Mg-Si, and Al-Mg-Zn alloys. Cr additions prevent recrystallization during heat treatment. Above 0.35%, it tends to form coarse intermetallics with other impurities or with Ti and Mn. The presence of Cr in wrought products leads to the formation of fine dispersoid phases that inhibit nucleation and grain growth. This latter aspect is interesting during heat treatments and hot working of both Al-Mg-Si and Al-Mg-Zn alloys. A critical issue related to the presence of chromium is the increase in quenching sensitivity of heat-treatable alloys in case the hardening phases precipitate on pre-existing Cr-containing particles
Copper	Cu	Copper is one of the most important alloying elements in Al. Its content ranges between 2% and 10%. Both cast and wrought alloys can be subjected to an increase in mechanical properties after solution treatment and subsequent aging. This is possible thanks to the formation of Cu-containing precipitates (see Sect. 3.2.3.3). Strengthening effect is maximum for wt% ranging between 4% and 6%, depending on the alloy and on the influence of the other alloying elements that may have synergistic hardening effects. In 7XXX series, copper enhances the aging rate by increasing the degree of supersaturation. In general, Cu increases the alloy resistance to stress corrosion but decreases the resistance to general corrosion of Al-Zn-Mg alloys
Iron	Fe	In Al-Cu-Mg cast alloys, iron increases strength and hardness at elevated temperatures. The iron-containing alloys are also characterized by improved load-bearing capabilities and dimensional stability. An excess of Fe leads to the formation of the several intermetallic phases, among which Cu_2FeAl_7. This decreases the amount of copper available to precipitate formation, with potentially deleterious effects on the effectiveness of the heat treatments that may result less effective in forming the strengthening phases. In Al-4Cu-0.5 Mg wrought alloy, an Fe content lower than 0.5% lowers the tensile properties after heat treatment. This phenomenon occurs, unless a sufficient concentration of Si is present for the formation of the FeSi intermetallic. The strength properties remain unaltered when a sufficient amount of Si combines

(continued)

Element	Symbol	Effects
		with Fe to form FeSi, although this phase tends to lower the fracture toughness of the alloy
Lithium	Li	Lithium reduces the density of the alloy and increased the yield strength. Lithium forms ordered δ' (Al_3Li) precipitates that are responsible for the increase in mechanical properties. Lithium content may vary depending on the alloy (see Table 3.12). Further details about the role in aluminum alloys are given in Sect. 3.2.4
Magnesium	Mg	Magnesium can be added to Al-Cu alloys, resulting in an increase in strength induced by solution treatment and quenching. Mg shows a great influence on the aging characteristics of the alloy. Minor additions (0.5%) can lead to significant variations in properties after aging treatments. The influence of Mg on corrosion resistance of Al-alloys depends on the alloy itself and on the thermal treatment. Strengthening effect of Mg on wrought alloys can be maximized by means of cold-working processes before aging. This effect is significant in artificially aged products and not that much with naturally aged and cold-worked alloys. When present in combination with silicon, magnesium can form the Mg_2Si intermetallic that in the β'-metastable condition is an important strengthening element (see Sect. 3.2.3.3)
Manganese	Mn	Manganese addition leads to an increase in tensile strength, as well as in yield strength. It prevents recrystallization during heat treatment. Commercial alloys usually do not contain more than 1% Mn since its presence can lead to a loss in ductility
Nickel	Ni	Nickel is used to improve high-temperature strength of both cast and wrought Al-Cu-Mg alloys. Although 0.5% Ni reduces the tensile properties of heat-treated wrought Al-4Cu-0.5 Mg-alloys at room temperature
Silicon	Si	Silicon combines with magnesium leading to the formation of the β' (Mg_2Si) strengthening phase (see Sect. 3.2.3.3). In 6XXX series alloys, Mg to Si ratio is approximately kept around 1.73:1 in order to promote the formation of the Mg_2Si intermetallic. The maximum solubility of Mg_2Si in the Al matrix is 1.85%, and it decreases with temperature. Mg_2Si precipitation upon age hardening occurs by formation of GPZs and very fine precipitates. An excess in Si leads to segregation at the grain boundaries, causing grain boundary fracture in recrystallized structures. In Al-Cu-Mg alloys, Si additions reduce cracking tendency of the material
Silver	Ag	Silver is added to increase the strength after heat treatment and aging
Ti	Titanium	Titanium additions are used for grain refinement of castings and ingots. Presence of boron enhances the effect of titanium
Zinc	Zn	Zinc forms intermetallics with Mg ($MgZn_2$). An increase in Zn content, as well as the combined increase in both Mg and Zn contents, decreases the corrosion resistance of the alloy. In 7XXX alloy series, Zn and Mg have the control of the aging process
Zirconium	Zr	Zirconium concentrations in the 0.1–0.3% range are used to inhibit recrystallization and recovery thanks to the formation of a fine precipitation of intermetallics within the alloy. Zr additions are used in an increasing number of alloys, in particular those belonging to the Al-Zn-Mg family, for increasing recrystallization temperature, thus allowing a

(continued)

Element	Symbol	Effects
		fine grain size to be retained even at higher treatment temperatures. The microstructure of wrought products can be controlled by means of Zr additions. These Zr-containing alloys are less quench sensitive, as compared to those having Cr additions. As-cast grain size can be reduced by means of Zr additions. Indeed, some superplastic alloys employ high Zr contents in order to retain a relatively fine microstructure during high-temperature forming processes

Summary of the effects of the most important alloying elements in aluminum alloys. Concentrations in wt% (Polmear et al. 2017; Prasad and Wanhill 2017)

Appendix 3: Alloying Elements in Magnesium Alloys

Element	Symbol	Effects
Aluminum	Al	Aluminum is a common alloying element in Mg-alloys, and it is used in both wrought and cast alloys. Its main function consists in the formation of the $Mg_{17}Al_{12}$ strengthening phase. The maximum solubility of Al in Mg is 12.7% at 473 °C. When the Al content exceeds the 6%, the alloy can be heat-treated. However, increasing Al content leads the alloy to be more prone to stress corrosion cracking. Al is also able to improve casting properties since it widens the solidification range of the alloy in which it is contained
Beryllium	Be	The maximum solubility of beryllium in Mg is 0.001%. It has beneficial effects because it reduces the oxidation of liquid Mg. However, its use is not recommended in cast alloys because it induces grain coarsening
Calcium	Ca	Calcium is used as grain refiner. It enhances the creep as well as the corrosion resistance. Ca can improve the deformability of alloys subjected to rolling processes (e.g., sheet production), although its content must be kept below 0.3% in order to avoid the alloy to become more prone to crack formation after welding processes. Cast alloys that contain Ca show a reduced oxidation of the liquid surface of the melt and are less prone to oxidation during heat treatment
Cerium[a]	Ce	Cerium has positive effects on the elongation, even in small amounts (0.2%), thus improving the alloy deformability
Cobalt	Co	Cobalt has detrimental effects on the corrosion resistance. Together with Cu, Fe, and Ni, it is considered as a harmful impurity
Copper	Cu	Copper has detrimental effects on the corrosion resistance. A significant drop of this property occurs when Cu content exceeds 0.05%. However, its presence can lead to an improvement of room-temperature and high-temperature strength
Gadolinium[a]	Gd	Gadolinium improves both the corrosion resistance, the high-temperature strength, and the creep resistance
Iron	Fe	Iron is regarded as a detrimental impurity since its presence, like that of copper, reduces the corrosion resistance of the alloy. In order to obtain

(continued)

Element	Symbol	Effects
		the maximum corrosion resistance, 0.005% is the highest Fe content for Mg-alloys
Lithium	Li	The maximum solubility of Li in Mg is 5.5% at 588 °C. Lithium additions reduce the density of the alloy. Furthermore, Li reduces the strength of the alloy while enhancing its ductility
Manganese	Mn	Manganese leads to an improvement of the alloy weldability. It has very limited effects on the ultimate and yield strength of the alloy and can be used for grain-refining purposes. When added to Mg-Sc alloys, it improves the creep resistance thanks to the formation of Mn_2Sc and other Mn_xSc_y compounds
Neodymium[a]	Nd	Neodymium has a high solubility limit in magnesium. The formation of stable precipitates, both within the grain and at the grain boundaries, determines an improvement in the strength of the alloy
Nickel	Ni	Similar to Fe and Cu, nickel is considered as a detrimental impurity for what concerns the corrosion resistance. In order to obtain the highest corrosion resistance, the maximum nickel content is 0.005%. However, the formation of Mg_2Ni intermetallic phase is able to enhance the room-temperature strength
Silicon	Si	The main effect of silicon is to increase the fluidity of liquid magnesium; therefore, cast alloys usually contain this element. When added together with REs, it forms stable silicides that improve high-temperature creep and stress resistance properties. In association with iron leads to a reduction in the corrosion resistance properties.
Silver	Ag	Silver improves the mechanical properties of Mg-alloys that have been subjected to age hardening
Strontium	Sr	Strontium is usually added together with other higher concentration alloying elements. Sr does not have significant influence on yield and ultimate strength of the alloy. However, its presence has been reported to improve the creep performances of the alloy
Thorium	Th	Since it is a radioactive element, the use of thorium is limited to a few alloys. The usual Th content in Mg-alloys ranges between 2% and 3%. Its addition leads to an improvement in creep resistance properties up 370 °C. Th improves high-temperature strength as well. The weldability of Zn-containing alloys is improved by thorium additions
Yttrium[a]	Y	Yttrium has high solubility in Mg (~12.6%). High-strength alloys such as WE54 and WE43 contain Y in the 4–5% range. Yttrium is added together with RE elements to improve the high-temperature strength, creep performances, and corrosion resistance
Zinc	Zn	Zinc introduces both solid solution and age hardening in Mg-alloys. It is used in combination with Al in order to increase the strength of the alloy without reducing its ductility. Moreover, Zn has positive effects on the high-strength precipitation hardenable alloys containing REs, thorium and zirconium. Its presence enhances the corrosion resistance, avoiding the detrimental effects of Fe and Ni impurities
Zirconium	Zr	The maximum solubility of zirconium in Mg is 0.50% at 654 °C, while the maximum solubility at room temperature is 0.28%. It is a strong grain refiner, especially in alloys containing Ag, REs, Th and Zn. The addition of Zr to alloys containing REs and Zn leads to an increase in

(continued)

Element	Symbol	Effects
		both strength and ductility. The grain-refining action of zirconium fades away in alloys containing Al, C, Fe, and Si and gaseous inclusions of H_2, N_2, and O_2, since it reacts with these elements. The development of Mg-Zr alloys for aerospace applications resulted in alloys with enhanced strength, in particular at elevated temperatures. A good example of Zr-containing alloy is WE43, which long-term temperature stability is maintained up to 250 °C

Effect of the most important alloying elements in magnesium alloys, including impurities. Concentrations in wt% (Prasad and Wanhill 2017; Zhang and Zhao 2012)

[a]REs-containing Mg-alloys have been extensively investigated in recent years. For REs atoms, it is not easy to diffuse in the alloy, and this has positive effects on the recrystallization temperature. Furthermore, they form small and stable precipitates. The addition of REs usually leads to high aging and solid solution effects, thus increasing high-temperature stress resistance, creep resistance, and corrosion resistance. In some cases, the addition of specific RE elements leads to an improvement of the ductility as well. This is the case of Ce, Gd, Nd, and also Y, not a true RE element. Finally, the addition of REs leads to a narrowing of the solidification range of the alloy. This latter feature is beneficial for the minimization of the micro-shrinkage porosities in castings and for the reduction of the probability to form cracks during welding processes

Appendix 4: Alloying Elements in Steels

Element	Symbol	γ-stabilization	Carbides formation	Main alloying effects
Aluminum	Al	No	No	Al is a strong deoxidizer and can be used for this purpose during alloy refinement, either alone or in association with Mn and Si. Al can form non-soluble inclusions (e.g., Al_2O_3) in molten steel, both during primary melting and refining process. It is a very effective element when used as grain refiner. Its presence promotes precipitation hardening in stainless steels. It is used in nitriding steels leading to the formation of very hard aluminum nitrides (1000 HV). Al is used for increasing corrosion resistance in low-carbon steels and for increasing oxidation resistance in heat-resistant steels
Antimony	Sb	No	–	It is considered as an impurity and it is not intentionally added. Moreover, it is difficult to remove during melting and refining processes. Its presence lowers the surface quality of continuously cast billets. It causes temper embrittlement in alloy steels. Sb has a strong tendency to

(continued)

Element	Symbol	γ-stabilization	Carbides formation	Main alloying effects
				segregate, especially at grain boundaries and during cooling processes
Beryllium	Be	–	–	Beryllium improves the corrosion resistance of steels
Boron	B	No	–	Boron is usually added in very limited amounts, ranging from 0.0005% to 0.0035%. B tends to segregate at grain boundaries, thus improving hardenability of steels. Boron is a strong nitride former, and this feature interferes with the hardening effect of elemental B. The addition of a specific amount of Ti to the steel protects B from nitride formation because Ti combines preferentially with N, thus avoiding boron nitride formation
Calcium	Ca	–	–	Ca is a strong deoxidizer. Concerning steelmaking processes, the *SiCa treatment* can be used in order to modify inclusion morphology and type. In particular, the process helps removing alumina from the molten bath thanks to the combination of Ca and S. This process leads to $CaO{\cdot}xAl_2O_3$ inclusions surrounded by liquid CaS. Thanks to this phenomenon, Al_2O_3 is removed from the liquid steel and contributes to the formation of the liquid slag. SiCa treatment can reduce sulfur and oxygen content in the alloy. In *killed steels*, CaS is present as spherical particles. Ca, Al, and Si tend to form low-melting compounds that results in better machinability of the steel. Some problems arise when CaS-based inclusions become too large, promoting the formation of fatigue cracks
Carbon	C	Yes	Yes	Carbon is, with iron of course, the base element in steels. It is inexpensive and can be easily added in various forms during steelmaking. During solidification, carbon shows a moderate tendency to macrosegregation, being more prone to segregating at microstructural defects, such as grain boundaries and dislocations. Carbon is an interstitial, both in *fcc* and in *bcc* iron, although with different solubility limits, and has a strengthening effect, enhancing tensile strength, yield strength, and hardness. This is possible

(continued)

Element	Symbol	γ-stabilization	Carbides formation	Main alloying effects
				through carbide dispersion strengthening and solid solution strengthening. Strengthening effect is proportional to the C content up to a threshold value (around 1% for carbon steel). Notch toughness and weldability of steels decrease at higher carbon concentrations
Chromium	Cr	No	Yes	Chromium is one of the most important alloying elements in steels, it is a carbide former, and for increasing Cr/C ratios, Cr_2C, and, the most common, Cr_7C_3 and $Cr_{23}C_6$ would form. Cr carbides increase hardness and wear resistance. Chromium carbides precipitation at the grain boundary (i.e., sensitization) lowers the corrosion resistance of stainless steels, in particular in the heat-altered zone (HAZ) after welding processes. For this reason, titanium is introduced as preferential carbide-forming element. Chromium implies an increase in mechanical properties, in hardenability, in oxidation and in general corrosion resistance. Stainless steels usually contain more than 13% Cr, and this leads to the formation of a thin, stable, and adherent Cr_2O_3-based layer on their surface causing an increase in the corrosion resistance in many different environments
Cobalt	Co	Yes	No	The use of cobalt as alloying element has been recently limited by its toxicity. As a consequence of this, Co-free alloys have been developed. By adding cobalt to carbon steel, the material undergoes a reduction in its hardenability. When combined with Cr, it increases the hardenability of Cr-Mo steels. Its use is generally limited to the production of tool steels, maraging steels, creep-resistant steels, and high-speed steels and for steels used in high-temperature applications. Co is a grain boundary stabilizer at high temperature and contributes to maintaining high-temperature strength and temper as well. This latter feature is the reason for the increase in useful lifetime of tool steel items. Cobalt promotes precipitation hardening and acts as a stabilizer agent for martensite increasing the M_s temperature and reducing the concentration of retained austenite in alloy steels

(continued)

Element	Symbol	γ-stabilization	Carbides formation	Main alloying effects
Copper	Cu	Yes	No	Copper is generally regarded as an impurity in steels, and its content is kept below 0.04%. It cannot be oxidized and removed during refining of steels, and the best approach for reducing Cu content in steel is through dilution. A particular class of low-alloy steels contains an appreciable amount of Cu for increasing atmospheric corrosion resistance. This class of steels is named weathering steels, and thanks to the presence of Cu, P, Ni, and Cr, they are able to form a protective layer on the surface. This family of steels is also known as COR-TEN. PH stainless steels use Cu for increasing the strength thanks to the precipitation of Cu-rich intermetallics during aging treatment. Copper is also added to austenitic stainless steels, and when exceeding 1% it causes an improvement in the corrosion resistance against HCl and H_2SO_4 and against stress corrosion phenomena (see Sect. 8.1.6)
Hydrogen	H	–	–	Hydrogen is an interstitial atom, it is soluble in liquid steel, and its presence can be very detrimental to the material. Modern steelmaking processes imply degassing stages for removing this element from liquid steel. The presence of hydrogen can lead to hydrogen embrittlement (see Sect. 8.1.7)
Lead	Pb	–	–	Lead is insoluble in iron, and it is used for increasing machinability of the material. When added to steel, it tends to form a particle dispersion throughout the material. This is particularly useful while machining, since it results both in a good lubrication and increased chip-breaking ability. However, leaded steel market has experienced a contraction in the last decades because of the environmental hazard and health risk associated with lead. Therefore, modern steelmaking processes tend to prefer the use of sulfur for the production of steels with improved machinability
Manganese	Mn	Yes	Yes	Manganese is both a γ-stabilizer element and a moderate carbide-forming element; it dissolves in Fe_3C and forms

(continued)

Element	Symbol	γ-stabilization	Carbides formation	Main alloying effects
				alloyed cementite as well. Other typical Mn carbides are $Mn_{23}C_6$ and $Mn_{15}C_4$. Manganese is present, as substitutional, in almost all steel grades in a quantity $\geq$0.30%. In steelmaking processes, Mn is used both as deoxidizer agent (often in association with Al, Si, or both) and for S removal, through the formation of MnS, thus avoiding the deleterious formation of FeS. Manganese enhances the steel hardenability for concentrations exceeding 0.8%. Steels with Mn concentrations larger than 2% are used for special purposes, such as railway track connectors and rock crushers. These steels belong to a special class named Hadfield manganese steels, also known as austenitic manganese steels
Molybdenum	Mo	No	Yes	Mo is an α-stabilizer element. It is a strong carbide former element, Mo_2C and Mo_6C being the main phases found in steels. Mo carbides are important for steel grades used in wear-resistant applications (e.g., tool steels). Molybdenum is added for minimizing temper embrittlement, but it cannot avoid it completely. It is an effective substitutional solution hardener and can induce secondary hardening during tempering of quenched steels and in austenitic alloys at high temperatures. It acts also as grain refiner when added to fine-grained steels. Mo improves corrosion resistance; it is used both in ferritic and austenitic stainless steels, acting as a passivation-enhancing element in stainless steels, thus promoting a fast formation of the passivating Cr_2O_3 layer, also increasing its stability. Mo is also present in duplex stainless steels. Super-austenitic stainless steels have a Mo content $\geq$ 6%; the high Mo content reduces the susceptibility to pitting corrosion but can lead to the formation of brittle intermetallic compound (chi-phase), which is responsible for the reduction in deformability of the alloy. Mo is also one of the alloying elements used in Maraging steels (see Sect. 5.4.1), in view of its positive effects on hardening, plasticity and ductility in the aged condition

(continued)

Element	Symbol	γ-stabilization	Carbides formation	Main alloying effects
Nickel	Ni	Yes	No	Nickel is a strong γ-stabilizer and a non-carbide-forming element, and thereby it is a necessary alloying element in austenitic stainless steels. Due to its position in the Ellingham diagram (see Sect. 8.2.1), it does not undergo oxidation during steelmaking processes, meaning that the amount of Ni is constant over time during the process. The substitutional solid solution of nickel in iron has a limited effect on increasing the alloy yield strength. However, this element improves the hardenability of steel, in particular when combined with Cr and Mo. In this condition the material undergoes also an increase in fatigue resistance and in impact toughness. It reduces the ductile to brittle transition temperature. For this reason, nickel is used as alloying element in steels for cryogenic applications. If combined with Al, it forms with aging treatments the intermetallic Ni_3Al precipitates, on which precipitation hardening of PH steels is based
Niobium	Nb	No	Yes	Niobium is a strong carbide-, nitride-, and carbonitride-forming element. Niobium (named columbium in the US literature) introduces a grain refinement effect in steels. It is used in HSLA (high-strength low-alloy) steels in order to enhance the strength by maintaining good toughness through precipitation strengthening and the formation of Nb-carbonitrides, and it is also able to prevent carbide dissolution and reprecipitation in undesirable locations. For this reason, it is added to some ferritic stainless steels (e.g., AISI 436 and 437), austenitic stainless steels (e.g., AISI 347, 348, and 384), and also to 17-4 PH stainless steels
Nitrogen	N	Yes	No	Nitrogen is an important γ-stabilizer element, which forms, together with carbon and metals, carbonitrides. Nitrogen may also partially replace nickel as austenite stabilizer in austenitic steels, in many of which it is used and results in solution or precipitation strengthening.

(continued)

Element	Symbol	γ-stabilization	Carbides formation	Main alloying effects
				This is the case of AISI types 201, 202, 205, 304N, 304LN, 316N, and 316LN. Nitrogen is also used in duplex stainless steels for increasing pitting corrosion resistance and strength
Oxygen	O	–	–	Oxygen is considered as an impurity in steels. Usually, the oxygen content has to be lower than 60 ppm. The solubility of oxygen in iron is very low, and this leads to undesired oxides formation or porosities
Phosphorus	P	No	No	Phosphorus is generally regarded as an impurity in steels, and its content is usually below 0.02%. Upon increasing the P content, the material experiences a decrease in toughness and ductility. It has a strong tendency for segregation at the grain boundary, and this leads to deleterious effects like temper embrittlement. However, for specific steel grades, a high P content can have beneficial effects. Low-alloy structural steels with high P content show better atmospheric corrosion resistance. The addition of P can cause precipitation hardening and yield strength improvement in some austenitic Cr-Ni steels. Re-phosphorized steels are a special class of alloys with improved strength and enhanced machinability
Rare earth metals		–	–	Rare earth metals are a group of 17 elements belonging to the III B group. They can have strong deoxidizing and desulfurizing effects. These elements have also a tendency to segregate at the grain boundaries. Rare earth metals have effects on shape modification and control of inclusions; in particular, La and Ce are added for the sulfide shape control. Rare earth metals can also increase the hardenability of steels
Silicon	Si	No	No	During steel refinement it is used as deoxidant, either alone or in combination with Al and Mn or both. It does not form carbides in steel, but its presence may lead to the formation of SiO_2 inclusions. Silicon retards the formation of Fe_3C during tempering and increases yield strength, hardenability, and wear resistance in heat-treated steels. It is also

(continued)

Element	Symbol	γ-stabilization	Carbides formation	Main alloying effects
				added to a particular family of steels called *transformer steels*. They have silicon content around the 3% and are characterized by low core losses and high magnetic permeability. Silicon dissolves into martensite, and it is able to retard martensite decomposition up to 300 °C
Sulfur	S	No	–	Like P and O, sulfur is generally considered as an impurity, with concentrations below 0.02%. The tendency of sulfur for segregating at the grain boundary is very high, and this causes the reduction of the high-temperature ductility in alloy steels. This element is insoluble in Fe and can form deleterious low melting compounds, such as FeS. However, similar to P, sulfur is intentionally added to a special class of steels that have a S content between 0.08% and 0.35% resulting in an improved machinability of the alloy
Tantalum	Ta	–	Yes	Like niobium, tantalum is a strong carbide former. It has a very high affinity to carbon; in some austenitic stainless steels (i.e., AISI type 347 and AISI 348), it is used as carbide stabilizing agent
Tin	Sn	–	–	Tin is considered as a detrimental impurity, in view of its effects on steel properties. One of the deleterious consequences of its presence is the temper embrittlement and the quality of the surface of continuously cast steels
Titanium	Ti	–	Yes	Titanium is considered as a very strong deoxidizer due to its high affinity to oxygen. During melting and refining, the addition of Ti has to be carried out carefully in order to avoid its oxidation and subsequent migration of oxides into the slag. Titanium is also an effective former of nitrides and carbides, and its effects are similar to those of V and Nb. However, Ti carbides and nitrides are more stable if compared with Nb- and V-based ones. Ti nitrides act as grain boundary stabilizers. Ti is also used as carbide stabilizing agent in austenitic stainless steels, ferritic and PH steels. Titanium improves intergranular

(continued)

Element	Symbol	γ-stabilization	Carbides formation	Main alloying effects
				corrosion (i.e., sensitization) in austenitic stainless steels. This element is also used in Fe-Ni-Cr alloys (e.g., alloy A286) for the formation of γ' precipitates. In maraging steels, Ti forms the intermetallic hardening H-phase (see Table 5.6). Furthermore, the addition of Ti prevents boron nitrides formation, so that boron can retain its elemental form
Tungsten	W	No	Yes	Tungsten is an α-stabilizer and carbide former element. Carbides (WC and W_2C) are very hard and of same stoichiometry and structure of Mo carbides. Tungsten induces secondary hardening during tempering treatment of quenched steels, limits grain growth at high temperature, and together with Mo, is the most important alloying element in tool steels
Vanadium	V	–	Yes	Vanadium forms extremely hard carbides and nitrides. V carbides have higher hardness than Cr, Mo, and Fe carbides. Vanadium is used for grain refinement, resulting in strength and toughness increase. Vanadium can also dissolve into austenite thus increasing its hardenability. Vanadium addition introduces large secondary hardening effect after tempering, with a general improvement of the high-temperature hardness of the material
Zirconium	Zr	No	Yes	Zirconium in steel limits the austenite phase field. It is a carbide- and nitride-forming element. Its effects on steel are similar to those of titanium, although it is very expansive and, therefore, not so common in steels

There are at least two possible criteria for the classification of alloying elements in steels. The first is based on the stabilization of the austenite (γ) phase. The second one is based on their tendency to form carbides. The carbide formation requires that concentration of the carbon would locally overcome a threshold value, i.e., its solubility limit in the relevant alloy; otherwise the element would remain in solution. According to the aforementioned criteria, the information concerning the most important alloying elements, and also impurities, in steels are given in this appendix. Concentrations in wt% (Campbell 2008, Kutz 2002, Totten 2007)

Appendix 5: Alloying Elements in Superalloys

Element	Symbol	Co-based superalloys	Fe-based superalloys	Ni-based superalloys
Aluminum	Al	Better oxidation resistance	Precipitation of the γ' phase	Moderate solid solution hardening effect
		Formation of the β-CoAl phase	Delayed formation of the hexagonal η Ni_3Ti phase	Precipitation of the γ' phase
		–	–	Better oxidation resistance
Boron	B	Better creep strength and ductility	Better creep strength and ductility	Moderate solid solution hardening effect
		Refinement of grain boundary	Retards formation of grain boundary η Ni_3Ti	Inhibition of carbides coarsening
		Workability improvement	–	Better grain boundary strength
		–	–	Better creep strength and ductility
Carbon	C	Carbides formation	Carbides formation	Carbides formation
		Decrease in ductility	Fcc matrix stabilization	Moderate solid solution hardening effect
		Refinement of grain boundary	–	–
Cobalt	Co	Matrix element	–	Increase in γ solidus temperature
		–	–	Moderate increase in the γ' volume fraction in some alloys
		–	–	Increase in γ solvus temperature
		–	–	Better resistance to hot corrosion under salt deposit
Chromium	Cr	Precipitation of $M_{23}C_6$ and M_7C_3 carbides	Hot corrosion and oxidation resistance improvement	Hot corrosion and oxidation resistance improvement
		Hot corrosion and oxidation resistance improvement	Solid solution hardening	Precipitation of $M_{23}C_6$ and M_7C_3 carbides
		Promotes TCP phase formation	–	Moderate solid solution hardening effect
		–	–	Moderate increase in the γ' volume fraction

(continued)

Element	Symbol	Co-based superalloys	Fe-based superalloys	Ni-based superalloys
		–	–	Promotes TCP phase formation
Hafnium	Hf	–	–	Creep strength and ductility improvement
		–	–	Improvement in grain boundary strength
		–	–	Formation of the HfC carbide
		–	–	Promotes eutectic γ/γ' formation
Iron	Fe	Workability improvement	Matrix element	Decrease in the oxidation resistance
		Fcc matrix stabilization	–	Promotes TCP phase formation
		–	–	Workability improvement
Molybdenum	Mo	Solid solution hardening	Solid solution hardening	High solid solution hardening effect
		Formation of the Co_3Mo intermetallic	Precipitation of the M_6C carbide	Moderate increase in the γ' volume fraction
		Promotes TCP phase formation	–	Increase in alloy density
Niobium	Nb	Precipitation of MC and M_6C carbides	Precipitation of the γ'' phase	High solid solution hardening effect
		Formation of the Co_2Nb intermetallic	Precipitation of the NbC carbide	Large increase in the γ' volume fraction
		Reduction in the surface stability	Precipitation of the δ-Ni_3Nb phase	Formation of the NbC carbide
		–	–	Precipitation of the γ'' phase
		–	–	Promotes γ' phase formation
		–	–	Precipitation of the δ-Ni_3Nb phase
Nickel	Ni	Fcc matrix stabilization	Fcc matrix stabilization	Matrix element
		Decrease in hot corrosion resistance	Inhibition of TCP phase formation	–
Rhenium	Re	–	–	Moderate solid solution hardening effect
		–	–	Retards γ' coarsening
		–	–	Increase in the γ/γ' lattice mismatch

(continued)

Element	Symbol	Co-based superalloys	Fe-based superalloys	Ni-based superalloys
Tantalum	Ta	Precipitation of MC and M_6C carbides	Precipitation of the γ'' phase	High solid solution hardening effect
		Formation of the Co_2Ta intermetallic	Precipitation of the TaC carbide	Precipitation of the TaC carbide
		Reduction in the surface stability	–	Large increase in the γ' volume fraction
		–	–	Oxidation resistance improvement
Titanium	Ti	Precipitation of the TiC carbide	Precipitation of the γ' phase	Moderate solid solution hardening effect
		Formation of the Co_3Ti intermetallic	Precipitation of the TiC carbide	Precipitation of the γ' phase
		Formation of the Ni_3Ti intermetallic (with sufficient amount of Ni)	–	Precipitation of the TiC carbide
		Reduction in the surface stability	–	–
Tungsten	W	Effective solid solution hardening	Effective solid solution hardening	Effective solid solution hardening effect
		Formation of the Co_3W intermetallic	Precipitation of the M_6C carbide	Moderate increase in the γ' volume fraction
		Promotes TCP phase formation	–	Formation of the M_6C carbide
		–	–	Increase in alloy density
		–	–	Promotes TCP phase formation
Zirconium	Zr	Formation of MC carbide	Creep strength and ductility improvement	Moderate solid solution hardening effect
		Creep strength and ductility improvement	Retards formation of grain boundary η Ni_3Ti	Inhibition of carbide coarsening
		Reduction in the surface stability	–	Better grain boundary strength
		Refinement of grain boundary	–	Better creep strength and ductility

Role of alloying elements in superalloys. Concentrations in wt% (Geddes et al. 2010)

Appendix 6: Precipitates and Secondary Phases in Superalloys

Phase	Symbol	Crystal structure	Lattice parameters [nm]	Formula	Other information
Delta	δ	Orthorhombic (ordered Cu_3Ti)	$a_0 = 0.3624–0.511$ $b_0 = 0.3624–0.4251$ $c_0 = 0.7406–0.4556$	Ni_3Nb	δ phase is an intergranular precipitate that has detrimental consequences on the creep response of the alloy. Observed in Incoloy 718. It has different morphologies depending on the formation mechanism. An acicular shape is observed when formed in the 815–980 °C range. It forms by cellular reaction at low aging temperatures. Intragranular precipitation occurs at high aging temperatures
Eta	η	hcp ($D0_{24}$)	$a_0 = 0.5093$ $c_0 = 0.8276$	Ni_3Ti	The η phase has been observed in all the three families of superalloys when high Ti/Al ratios are present. Its formation occurs as intergranular precipitates or intragranular acicular platelets in a Widmanstätten pattern

(continued)

Phase	Symbol	Crystal structure	Lattice parameters [nm]	Formula	Other information
Gamma prime	γ'	fcc (ordered $L1_2$)	0.3561 for pure Ni_3Al to 0.3568 for $Ni_3(Al_{0.5}Ti_{0.5})$	Ni_3Al $Ni_3(Al, Ti)$	Main strengthening phase in Ni- and Fe-Ni-based superalloys. Its shape varies from spherical (e.g., in Nimonic 80 and Waspaloy) to cuboidal in more recent Ni base superalloys. Variations in the Al/Ti ratio and in the Mo content influence the morphology of the phase. In Ni-Fe-based alloys is metastable. When the mismatch with the matrix is high, extended exposure above 700 °C leads to conversion in the deleterious η (Ni_3Ti) or δ (Ni_3Nb) phases
Gamma double prime	γ''	bct (ordered $D0_{22}$)	$a_0 = 0.3624$ $c_0 = 0.7406$	Ni_3Nb	γ'' is the main strengthening phase in Incoloy 718. Usually, precipitates have the form of coherent disc-shaped particles that form on the {100} planes with an average diameter of 600 Å and a thickness ranging between 50 and 90 Å. In case of overaging of the alloy, γ'' phase converts into the deleterious δ phase

(continued)

Phase	Symbol	Crystal structure	Lattice parameters [nm]	Formula	Other information
Laves	–	Hexagonal	$a_0 = 0.475–0.495$	Co_2Ti	They are classified as TCP phases typical of Co and Fe base superalloys. Observed in Co-based superalloys with Ti content greater than 5 wt%
				Fe_2Mo	
			$c_0 = 0.770–0.815$	Fe_2Nb	
				Fe_2Ta	
				Fe_2Ti	
MC	–	Cubic	$a_0 = 0.430–0.470$	HfC	Usually primary carbide that forms directly from the melt. MC carbides appear as globular, irregularly shaped particles
				NbC	
				TaC	
				ThC	
				TiC	
				ZrC	
M_7C_3	–	Hexagonal	$a_0 = 1.398$	Cr_7C_3	M_7C_3 carbide has generally a blocky intergranular shape. It has been observed in some Co-based superalloys and in other alloys such as Nimonic 80A after exposure at temperatures exceeding 1000 °C
			$c_0 = 0.4523$		
M_6C	–	fcc	$a_0 = 1.085–1.175$	Co_3Mo_3C	It is a randomly distributed carbide. The metal is commonly molybdenum. It can be originated from the melt or from the decomposition of MC carbide at high temperatures
				Fe_3Mo_3C	
				Fe_3Nb_3C	
				Fe_3W_3C	
				Fe_4W_2C	
				Nb_3Co_3C	
				Ta_3Co_3C	
M_3B_2	–	Tetragonal	$a_0 = 0.560–0.620$ $c_0 = 0.300–0.330$	Fe_3B_2	Borides appear very similar to carbides; they have been observed in both Fe-Ni- and Ni-based superalloys when B content is 0.03%, or greater
				Mo_2FeB_2	
				Nb_3B_2	
				Ta_3B_2	
				V_3B_2	

(continued)

Phase	Symbol	Crystal structure	Lattice parameters [nm]	Formula	Other information
M_2SC	–	hcp	$a_0 = 0.3234–3.386$ $c_0 = 11.233–11.834$	(Nb, Ti, Zr)2_SC	This phase is observed in alloys with high levels of S or low C/S ratios
$M_{23}C_6$	–	fcc	$a_0 = 1.050–1.070$ according to composition	$Cr_{23}C_6$ (Co, Fe, Mo, W)$_{23}C_6$	The $M_{23}C_6$ carbide usually forms at grain boundaries in the form of either globules, lamellae, or platelets. The most common composition implies Cr as metal (particularly in Co-based superalloys); however, Co, Fe, Mo, and W can substitute Cr. In Co superalloys it can originate from the decomposition of M_6C carbides at high temperature
MN	–	Cubic	$a_0 = 0.4240$	NbN TiN ZrN	MN nitrides can be observed in alloys containing Nb, Ti, or Zr. They are characterized by square to rectangular shapes and usually are used by carbides as nucleation sites. Nitrides are not soluble below the melting temperature
Mu	μ	Rhombohedral	$a_0 = 0.475$ $c_0 = 2.577$	(Fe, Co)$_7$(Mo, W)$_6$ Co_2W_7	It generally appears as coarse, irregular Widmanstätten platelets. It forms at high temperatures in alloys with high content of Mo or W

(continued)

Phase	Symbol	Crystal structure	Lattice parameters [nm]	Formula	Other information
Sigma	σ	Tetragonal	$a_0 = 0.880–0.910$ $c_0 = 0.450–0.480$	$(Fe, Ni)_x(Cr, Mo)_y$	It is classified as a TCP phase typical of Fe-Ni- and Co-based superalloys. It is less common in Ni base superalloys. σ phase has detrimental effects on both mechanical properties and corrosion resistance

An overview on the main precipitate and secondary phases that can be found in all the three superalloy families (ASM International 1991; Geddes et al. 2010)